Marine Science

Peter Castro, Ph.D.

California State Polytechnic University, Pomona

Michael E. Huber, Ph.D.

Jacobs Australia

Original Artwork By

William C. Ober, M.D.
Washington & Lee University

Claire E. Ober, B.A., R.N.

Mc
Graw
Hill
Education

Dedication

To all future marine biologists
—Peter Castro—

To Erin and Mason
—Michael Huber—

About the cover photo: A tiger shark *(Galeocerdo cuvier)* swims off the coast of the Bahamas. Sharks have many adaptations that allow them to be successful predators in the ocean: an excellent sense of smell, a lateral line that senses vibrations in the water, and sharp, replaceable teeth. Despite their reputation, however, sharks are quite vulnerable to human impacts such as overfishing and habitat destruction.

cover: Image Source/Getty Images

MHEonline.com

Send all inquiries to:
McGraw-Hill Education
8787 Orion Place
Columbus, OH 43240

ISBN: 978-0-02-142265-4
MHID: 0-02-142265-6

Printed in the United States of America.

1 2 3 4 5 6 7 8 9 DOW 20 19 18 17 16 15

About the Authors

Peter Castro, Ph.D.

Peter Castro realized that he had to become a marine biologist during a high school field trip to the coral reefs in his native Puerto Rico. He obtained a B.S. in biology from the University of Puerto Rico, Mayagüez, but left the warm Caribbean for warm Hawai´i to obtain a Ph.D. in marine zoology from the University of Hawai´i, Manoa. His first experience with cold water was a year of post-doctoral research at Hopkins Marine Station of Stanford University in Monterey Bay, California. He is currently Professor Emeritus at California State Polytechnic University, Pomona. He also holds a B.A. in history and art history from his home institution, something that took him 18 years to accomplish as a part-time student. He is fluent in five languages and taught marine biology (in English and Spanish) as a Fulbright Scholar at Odessa State University in the former Soviet Union. His research specialty is the biology of crustaceans symbiotic with reef corals and other invertebrates, research that has taken him anywhere where the water is warm enough to dive. He has also been doing research for almost the last two decades on the systematics of deep-water crabs, mostly, of all places, in Paris, France. His research also has taken him in the last couple of years to Singapore, Germany, and Brazil. Dr. Castro has so far published 58 peer-reviewed papers on his research. He is currently editor-in-chief of the *Journal of Crustacean Biology*, associate editor for *Zootaxa*, and editor for the volume on brachyuran crabs for the *Treatise on Zoology*.

Michael Huber, Ph.D.

Michael became fascinated by aquatic organisms when he caught his first trout on an Alaskan lake at age two. His interest in marine biology grew, and he went on to obtain B.S. degrees in zoology and oceanography from the University of Washington in Seattle. He received his doctorate from Scripps Institution of Oceanography for research on a group of symbiotic coral crabs. He remained at Scripps as a research biologist, working on such diverse research topics as the genetics and cell biology of unicellular algae and bioluminescence in midwater organisms. In 1988 he moved to the Biology Department at the University of Papua New Guinea, where he had the opportunity to work on some of the world's most spectacular coral reefs and was Head of the University's Motupore Island Research Station. He also became increasingly involved in marine environmental science. This interest continued to grow when he left Papua New Guinea in 1994 to become the Scientific Director of James Cook University's Orpheus Island Research Station on Australia's Great Barrier Reef. In 1998 he became a full-time environmental advisor, providing scientific information and advice on marine environmental issues and the development of conservation programs to international agencies, governments, and private industry. Dr. Huber is a Member Emeritus of GESAMP, an international scientific body that advises the United Nations system on marine environmental issues.

Mike lives in Brisbane, Australia, with his wife and two children. His hobbies are fishing, diving, swimming, jazz and rock music, reading, and gardening.

Contents

Contents

Features

Preface

Interest in the world ocean started for many of us while we were high school students. Our fascination probably started after watching a television documentary, visiting an Internet site on whales or El Niño, or perhaps a visit to a marine theme park. For some it was a visit to the shore, the keeping of a marine aquarium, or, if lucky, while first snorkeling or scuba diving in a coral reef. Concern with our impact on the marine environment, particularly the conservation of marine life, is another motivation for taking a high school course in marine science.

Marine Science was especially written for the marine science courses that have become so popular in high schools from coast to coast. It has been developed taking advantage of our experiences in writing *Marine Biology*, a text used by students around the world and now in its tenth English edition as well as editions in six other languages. *Marine Science* is nevertheless not a watered-down version of *Marine Biology*, having been written and developed with high school students in mind. *Marine Science* is a rigorous but highly readable introduction to the oceans. Although it emphasizes marine life and its relationship with the ocean environment, the textbook has extensive material on the physical features of the world ocean. We have developed *Marine Science* guided by the theme of how marine life interacts with its non-living environment and how the oceans and their inhabitants are being affected by human impact. We follow a global perspective to emphasize that the oceans are interconnected and that life on Earth is greatly affected by the world ocean.

Up-To-Date Content

In addition to its coverage on the living organisms and communities of the oceans, *Marine Science* has three chapters dedicated to the physical and chemical factors that make the oceans unique. As with *Marine Biology*, the data presented in this book is vetted and the most up-to-date information is presented to students. Examples of recently researched and updated information include the following:

- Chapter 3 coverage on the physical and chemical properties of sea water, including the importance of sound transmission for animal communication, and the impacts of ocean acidification

- Chapter 4 coverage on waves and tides

- Chapter 5 coverage on atmospheric circulation, ocean circulation, and how the two are linked

- Expanded information on marine fungi

- New section on chlorophyll and accessory pigments in marine primary producers

- Updated, to follow more current views, the position of chaetognaths as protostomes and echiurans as annelids

- Updated species counts for all invertebrate and vertebrate groups

- Updated status of great whales and dolphins

- Added information on the biology (physiology, ecology) of mangroves and seagrasses

- Added information on bioturbation, fjord communities, and coral caves

- Added recent information on a tropic cascade involving sea otters and the recovery of seagrass meadows

- Expanded section on the human impact on seagrasses

- Added new information on aquaculture and the use of methane hydrates as a source of energy

- Expanded and updated information on marine protected areas

- Added new information on endangered marine species, including updating of our unique table on Red List and CITES lists for marine species

- Expanded data on plastic pollutants and artificial reefs

- Coverage of cutting edge marine science in unique, *Marine Science* feature strands: Marine Science in Action, Humans and the Ocean, Habitat Spotlight, and Nature of Science.

Organization

Marine Science is organized into four units. Unit 1 (Chapters 1 through 5) introduces students to marine science and the physical and chemical factors that shape ocean environments. Chapter 1 describes the history of marine science. It also explains the fundamentals

of the scientific method, emphasizing that science is a process, an ongoing human endeavor. We think it is critical that students understand how and why science works, that science has limitations, and that there is still much to be learned. Chapters 2 is a basic introduction to marine geology and the physical features of the sea floor. Chapters 3 through 5 are an introduction to the major chemical and physical factors shaping the ocean environments. These chapters give students an overview of seawater chemistry, wave formation, what controls the tides, and how oceanic and atmospheric circulation are linked. *Marine Science* includes more information on these subjects, and places greater stress on their importance to understanding marine ecosystems, than other texts. While these topics are introduced in Unit 1, they are reinforced throughout the book. Wave refraction, for example, is described in conjunction with intertidal communities (Chapter 13) and estuarine circulation is discussed as part of the ecology of estuaries (Chapter 14). This approach emphasizes the importance of the physical and chemical environment to the organisms of the sea throughout the book.

Unit 2 (Chapters 6 through 11) surveys the diversity of marine life from the perspective of organismal biology. As in Unit 1, we provide introductory information that is reviewed and expanded upon in later chapters. Chapter 6, "Fundamentals of Biology," briefly reviews some essential biological concepts. In covering basic biology we have tried to balance the needs of a spectrum of students ranging from those with no prior instruction to those who have taken a number of biology courses. Depending on the level of their students, teachers may choose to cover Chapter 6 in class, assign it as review reading, or omit it and rely on the in-text glossary entries in later chapters to remind students of the definitions of key terms. In discussing the various taxa we emphasize functional morphology, ecological and physiological adaptations, and economic importance or other significance to humanity. Classification and phylogeny are not stressed, though we do present cladograms illustrating widely accepted phylogenetic schemes for invertebrates and vertebrates. As in the rest of the book we have selected organisms from around the world for photographs, line drawings, and color paintings, but organisms from the coasts of North America are emphasized. Organisms are referred to by their most widely accepted common

names. One or two common or important genera are noted in parentheses the first time a group is mentioned in a chapter, but we have not attempted to provide comprehensive lists of genera.

Unit 3 of the book (Chapters 12 through 18) presents an ecological tour of the major environments of the world ocean, commencing with an introduction to some fundamental principles of marine ecology in Chapter 12. The remaining six chapters of Unit 3 proceed from nearshore to offshore and from shallow to deep water, describing the physical characteristics of each environment and the adaptations and interactions of the organisms that live there. This admittedly arbitrary sequence follows the teaching sequence of the greatest number of our reviewers, but the chapters are designed so that they can be covered in any sequence according to teachers' preferences and needs. Most chapters include generalized food webs with standardized color coding to indicate the nature of the trophic relationships.

Finally, Unit 4 looks at the many ways in which humans interact with the world ocean: our use of and impact on the marine environment and the influence of the ocean on the human experience. The section presents an up-to-date, comprehensive view of issues and concerns shared by many students. The chapter on resource utilization (Chapter 19) looks not only at traditional uses, such as fisheries, aquaculture, and oil and gas extraction, but also at more modern aspects, such as the emerging technologies to generate energy from the sea, the pharmacological use of marine natural products, and the application of genetic engineering and other technologies in aquaculture. Chapter 20 discusses human-induced degradation of the marine environment, balanced by an examination of marine conservation and habitat restoration. In closing the book on this note, we hope to stimulate students to reflect on the past and future significance of the world ocean to all our lives.

Acknowledgments

The illustrations were expertly done by Bill Ober and Claire Ober, who contributed their talent and knowledge to develop the artwork that was exclusively designed for *Marine Science*.

The World of
Marine Science

Marine Science, adapted from Castro's *Marine Biology*, broadens its coverage with the addition of three new oceanography chapters that help students better understand the oceans' physical aspects. This expanded content coupled with a new stunning design, learning system, updated features, and data analysis labs, make *Marine Science* the ideal fit for a marine science course.

The *Marine Science* Learning System

The Castro text incorporates a variety of study aids to help students better understand the ocean and all its complexities.

The **unit opener** sets up chapter content that will be covered. It lists the chapters included in the unit to give the students a preview of what they will study.

There are four **Unit Projects**, one for each unit, that build upon each other and allow students to develop their research and analysis skills in the study of a global challenge related to the material they are covering in the text. Teachers' can use these as an independent study for individual students or groups of students or as a classroom activity that can be worked on throughout the school year.

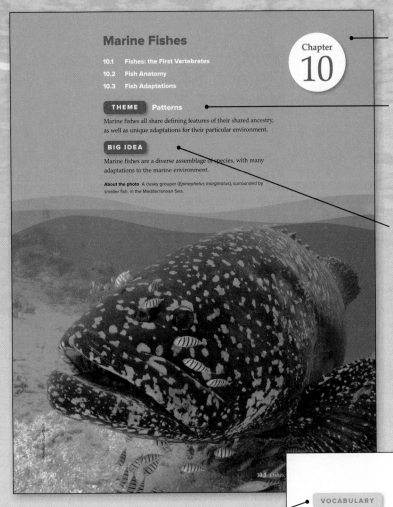

Marine Fishes

10.1 Fishes: the First Vertebrates
10.2 Fish Anatomy
10.3 Fish Adaptations

THEME Patterns

Marine fishes all share defining features of their shared ancestry, as well as unique adaptations for their particular environment.

BIG IDEA

Marine fishes are a diverse assemblage of species, with many adaptations to the marine environment.

About the photo A dusky grouper (*Epinephelus marginatus*), surrounded by smaller fish, in the Mediterranean Sea.

10.1 *Fishes*

The **chapter opener** previews what will be covered in the chapter.

Each chapter opens with a **Theme**. These themes highlight how the concepts covered in *Marine Science* are applicable across all fields of science and engineering. These unifying concepts include themes such as patterns, cause and effect, energy and matter, and structure and function.

The **Big Idea** is the main take away for the chapter. This Big Idea is what students should glean through their study of the chapter content.

Vocabulary indicates the important terms covered in the section.

Each section opens with a **Main Idea** that summarizes the important points of the section.

The **Key Questions** are used to guide student reading and ensure that they understand the main idea of the chapter.

The **Dive In!** introduces students to the chapter content with an interesting and accessible example.

VOCABULARY

vertebrates
vertebrae
cartilage
placoid scales
caudal fin
dorsal fins
pectoral fins
demersal
cycloid
ctenoid
operculum
fin rays
swim bladder

DIVE IN!

What characterizes vertebrates? Most chordates are vertebrates. While vertebrates are traditionally defined by possessing vertebrae, or a spine, this is not always the case. Hagfishes, which are a type of jawless fish, do not have true vertebrae. They do, however, share several morphological and molecular characteristics with vertebrates. Because of their genetic similarities to lampreys (another jawless fish that DO have vertebrae), most biologists have placed hagfishes in the subphylum Vertebrata.

10.1 Fishes: the First Vertebrates

Main Idea
Scientists categorize fishes into three main groups based on body structure.

Key Questions
1. Which characteristics of fishes separate them from the rest of phylum Chordata?
2. What are the three traditional groups of fishes?

Vertebrates (subphylum Vertebrata) share four fundamental characteristics of the phylum Chordata with protochordates, the invertebrate chordates like lancelets and sea squirts. Vertebrates differ from these other chordates in having a backbone, also called the vertebral column or spine, which is a dorsal row of hollow skeletal elements, usually bone, called **vertebrae**. The vertebrae enclose and protect the nerve cord, also called the spinal cord, which ends in a brain that is protected by a skull made of cartilage or bone. Vertebrates also are characterized by a bilaterally symmetrical body and the presence of an internal skeleton, or endoskeleton.

Fishes are the oldest and structurally the simplest of all living vertebrates. They also are the most abundant vertebrates in terms of both species and individuals. About 32,000 species of fishes are presently known to science. They make up about half of all species of vertebrates on Earth. Most species of fishes, about 58%, are marine. Many new species are being discovered every year, but a good number are on the brink of extinction.

There is some disagreement as to how to classify the major groups of fishes. Three groups are traditionally recognized. One of the schemes proposed to show the relationships of these three groups with the rest of the vertebrates is given in Figure 10.1.

Jawless Fishes

The most primitive fishes living today are the jawless fishes (Agnatha). Because they lack jaws, they feed by suction with the aid of a round, muscular mouth and rows of teeth. The body is cylindrical and elongated like that of eels or snakes (Fig. 10.2). They lack the paired fins and scales of most fishes, but, like fishes and other vertebrates, the brain is protected by a skull. Some biologists do not even consider jawless fishes to be vertebrates because they lack true vertebrae.

Hagfishes Hagfishes, or slime eels (*Myxine*, *Eptatretus*), are jawless fishes that feed mostly on dead or dying fishes (Fig. 10.2, p. 268). They are known as the slime eel because of the abundant mucus produced by glands on the skin. The mucus is used by hagfishes to protect themselves while feeding with their

Figure 10.30 An egg of the swell shark (*Cephaloscyllium ventriosum*) containing a one-month-old embryo. The large, white structure is the yolk sac. The complete egg case of another shark is shown in the inset.

Figure captions provide additional information and insight into content being addressed.

female, which gives birth to live young. The young are larger than in oviparous species. Such fishes are known as **ovoviviparous**. Most ovoviviparous fishes are cartilaginous. A total of 300 embryos, still in egg cases, were found ready to emerge inside the reproductive tract of a female whale shark. Some rockfishes are among the few marine bony fishes that can be classified as ovoviviparous. Most bony fishes are oviparous and spawn their eggs for external fertilization.

In some ovoviviparous sharks, the embryos rely on other sources of nutrition once they have consumed the yolk. In the sandtiger shark (*Carcharias taurus*; see Fig. 10.3) only two pups, which are large (up to 1 m) and active, are born. Each survived in one of the two branches of its mother's reproductive tract by eating its brothers and sisters. When that source of food is gone, they consume unfertilized eggs produced by their mother's ovaries.

Viviparity Some sharks and rays have embryos that actually absorb nutrients from the walls of the mother's reproductive tract. These sharks are said to be **viviparous**. The surfperches, which are bony fishes, are also viviparous. Their young have large fins that absorb nutrients from the walls of the mother's uterus. Some sharks go further and have evolved to possess what some call a placenta, tissue that provides nourishment to the embryo. It is not a true placenta, however, because it develops from the walls of the mother's uterus rather than from the embryo as in the case of the true placenta of practically all mammals, including humans. All viviparous fishes are therefore said to show aplacental viviparity.

There is now evidence of yet another method of reproduction among sharks. A female bonnethead shark (*Sphyrna tiburo*), a hammerhead shark, gave birth to a pup after three years in an aquarium together with two females of the same species but no males. Though it was possible that the female had stored sperm in her body as the result of sex with a male shark prior to capture, the DNA of the pup matched the DNA of one of the three females with no evidence of DNA from a male. Though very rare among vertebrates, **parthenogenesis**, the development of an egg into a normal embryo without fertilization by sperm, has also been reported in a few bony fishes.

Chapter **vocabulary words** are bolded and highlighted to draw student attention. The definitions of these terms can be found in the glossary at the back of the book.

10

Throughout the text there are four feature strands. These include **Nature of Science, Marine Science in Action, Habitat Spotlight,** and **Humans and the Ocean**. Each explores a unique topic in marine science exploring habitats, human and ocean interaction, and other topics of interest. Each of these features has a corresponding activity available in Connect. This activity allows students to expand upon what they've studied in each feature.

The Endangered Sea Turtles

Humans are by far the most formidable and destructive enemies of sea turtles. Humans exploit turtles for a variety of reasons, from food to jewelry to cosmetics. Turtles are also suffering the effects of climate change, habitat destruction, and overfishing.

Direct collection The green turtle is especially esteemed for its meat, and its cartilage is used to make turtle soup. Sea turtles can live for months without food or water, and in the days before refrigeration sailors kept them alive aboard ship as a source of fresh meat, storing them on their backs for months. Coming ashore by the thousands, females were an easy catch. Thousands of sea turtles are illegally killed every year for meat in coastal regions of Latin America, Southeast Asia, and India. Sea turtle eggs, particularly those of leatherbacks, are said to be an aphrodisiac. There is still a flourishing, though technically illegal, market for sea turtle eggs in many coastal regions. Eggs are eaten or fed to pigs or cattle.

The polished shell of the hawksbill is the source of tortoiseshell used to make jewelry, combs, and other articles. Baby sea turtles are stuffed and sold as souvenirs. Sea turtle leather, which is soft and durable, is much prized for shoes, boots, handbags, and wallets. Sea turtle oil is also used in cosmetics.

Development and fishing Much more harmful is the fact that many nesting areas have been turned into resorts or public beaches. Females searching for nesting beaches avoid lights because dark areas along the horizon indicate land, and lights along a beach look like a starry horizon. Artificial lighting also disorients baby turtles after hatching, so they do not head to sea and therefore die, often eaten by seabirds and other predators.

Sea turtles also become entangled in gill and drift fishing nets, including abandoned ones, and drown because they cannot swim to the surface to breathe. Many choke to death after swallowing plastic bags they think are jellyfishes. Shrimp nets are estimated to kill thousands of sea turtles a year. They have been especially deadly to the Kemp's ridley turtle in the Gulf of Mexico, once very common but now so rare that only a few hundred breeding females remain. It is considered the most endangered sea turtle.

All seven species of sea turtles are classified as threatened worldwide because their numbers are low. The number of leatherbacks has decreased by 95% since the early 1980s, mostly victims of fishing nets, and only a few thousand nesting leatherbacks remain in the Pacific Ocean. The six species of sea turtles found in United States waters are protected under the Endangered Species Act of 1973. Four species are officially classified as endangered and are in danger of extinction: the leatherback, hawksbill, Kemp's ridley, and the green sea turtle (in Florida).

Green turtle (*Chelonia mydas*).

Conservation The numbers of green turtles, once very common, have been increasing in some areas, undoubtedly the result of protection. Shrimp nets in the United States are required to have turtle exclusion devices, or TEDs, to allow sea turtles to escape once caught in the nets. But sea turtles are not protected at all in many countries and, in those where they are, enforcement is difficult because it is impossible to protect all coasts and nesting grounds from fishers and egg hunters. Stricter worldwide enforcement of conservation practices, the establishment of reserves and national parks to protect nesting beaches, the control of pollution and fishing practices, the restocking of former nesting areas, and the regulation of trade in sea turtle products might help save them.

The **custom artwork** in *Marine Science* is a result of collaboration between Peter Castro, Michael Huber, and Bill and Claire Ober. Together the authors and illustrators developed unique and specialized art, graphs, and maps that illustrate the key concepts of the text.

In Context references are used to provide additional information about terms or concepts with which students need to be familiar.

Each section concludes with a set of **Section Review** questions. These questions can be used to review section content and ensure student understanding.

Figure 10.15 (a) Cross section through the skin of a fish, showing the lateral line sensory system. The nerves collect nerve impulses from the sensory organs and transmit the information to the brain. (b) In addition to the lateral line, cartilaginous fishes have ampullae of Lorenzini, a network of jelly-filled canals capable of detecting electrical fields in the water.

IN CONTEXT

Hormones Molecules that act as chemical messengers within the body.
• 6.1, The Ingredients of Life

the "male." Hermaphroditism also is found among several deep water fishes, an adaptation to the depths of the ocean where it can be difficult to find members of the opposite sex.

A variation of hermaphroditism among fishes is sex reversal or **sequential hermaphroditism**, in which individuals begin life as males but change into females (protandry) or females change into males (protogyny). These changes are controlled by sex hormones but triggered by social cues such as the absence of a dominant male. Sex reversal occurs in several groups of marine fishes, but it is most prevalent among sea basses and groupers, parrotfishes, and wrasses. Some rather complicated reproductive strategies have been discovered among these fishes.

In at least some species of anemone-fishes, all individuals begin as males. Each sea anemone is inhabited by a single, large female that mates only with a large, dominant male.

All the other fishes that live on the anemone are small, non-breeding males. If the female disappears or is experimentally removed, her mate changes into a female and the largest of the non-breeding males becomes the new dominant male. The new female can start spawning as soon as 26 days after her sex change. Males of some wrasses form harems of many females. If the male disappears, the largest, dominant female immediately begins to act as a male and within a relatively short period of time changes color and transforms into one that is capable of producing sperm.

Section Review

1. **Explain** How do marine fishes stay hydrated in such a salty external environment?

2. **Compare** What adaptations are common across groups of marine fishes?

Fishes

Chapter 10 Review

REVIEW QUESTIONS

Multiple Choice

1. Which is not a characteristic of jawed fishes?
 a. endothermy
 b. gills
 c. skin covered by scales
 d. two-chambered heart

2. Cartilagenous fishes and bony fishes are different in that only
 a. bony fishes have paired fins.
 b. bony fishes have a keen sense of smell.
 c. bony fishes have an operculum.
 d. cartilaginous fishes have a complete skeleton.

3. Bony fishes are divided into which two groups?
 a. hagfishes and lampreys
 b. sharks and ray-finned fishes
 c. ray-finned fishes and lobe-finned fishes
 d. jawless fishes and cartilaginous fishes

Short Answer

4. What is the evolutionary significance of lobe-finned fishes?

5. Compare and contrast cartilaginous fishes and bony fishes.

6. Describe an adaptation of a marine fish for feeding on small prey like tiny crustaceans.

Critical Thinking

7. Hagfishes and lampreys are the only living representatives of a very ancient group. Why do you suppose there are still some of these jawless fishes around?

8. A deep-water shark, new to science, is collected for the first time. The specimen is studied in detail, but its stomach is empty. How can you get a rough idea of its feeding habits? The specimen is female, and its reproductive tract is found to contain 20 eggs. Can you tell the type of development characteristics of this species?

DATA ANALYSIS LAB

How can we learn more about basking sharks?
Marine scientists know very little about the lives of basking sharks, including where they live, mate, and give birth. In an effort to learn more, scientists tagged 18 sharks off the coast of Massachusetts.

Data and Observations
The illustration shows the previously known ranges of basking sharks together with new information retrieved from the tags.

Atlantic Ocean

Key:
Previously known range of basking sharks
Migration path of tagged sharks

Think Critically

1. What did marine scientists learn from the data collected by the tags?

2. After analyzing the data from the tags in this study, the marine scientists hypothesized that the pregnant females and juveniles live in deep, tropical waters for months at a time. Describe a method in which this hypothesis could be tested.

**Data obtained from: Hamady, L. L., 2011. Clues in shark vertebrae reveal where they've been. Oceanus, vol. 49, no. 2, Winter, pp. 10–13.*

Chapter 10 *Review* **301**

The **Review Questions** are comprised of multiple choice, short answer, and critical thinking questions used to review the most important concepts addressed in the chapter.

Each chapter review includes a **Data Analysis Lab.** These labs use real-world current data and are directly related to the content covered in the chapter. Students must use the data provided to answer critical thinking questions. These labs are an excellent opportunity for students to develop their data analysis skills.

Digital Advantage

McGraw-Hill Connect® is a digital teaching and learning environment that saves time while improving student performance.

- Teachers have access to a variety of resources, including question banks, robust reporting, and a complete lab manual, lab manual teacher's guide, and teacher's manual.
- Auto-graded tests for each chapter are correlated to the Ocean Literacy Standards.
- The **Inquiry Activities** provided in Connect allow students to expand upon what they've studied in the **Nature of Science, Marine Science in Action, Habitat Spotlight,** and **Humans and the Ocean** features.
- Digital images and PowerPoint slides are available to help teachers build dynamic presentations.
- Tegrity Digital Lecture Capture enables teachers to record lectures and make them available to students anytime, anywhere, on just about any device.
- An adaptive **SmartBook®** ensures that students get the most out of each minute spent studying.

SmartBook® is the first and only adaptive reading experience designed to improve the way students read and learn. Powered by the intelligent diagnostic and adaptive LearnSmart® engine, SmartBook uses assessments that consider both confidence level and responses to content questions to identify what a student knows and doesn't know. As the student reads, the material constantly adapts to ensure the student is focused on the content he or she needs the most to close any knowledge gaps. The result is that students learn faster, study more efficiently, and retain more knowledge, allowing teachers to focus valuable class time on higher-level concepts.

Supplementary Resources

Marine Science Laboratory Manual

Our NEW *Marine Science Laboratory Manual* offers 34 new labs ideal for any marine science classroom. These labs can be performed anywhere in the country and do not require access to the ocean. Each lab is designed to be completed in one lab period. The Marine Science Teacher's Manual identifies the best time to use each lab in the course of your chapter instruction. A **Lab Manual Teacher's Guide** is located in Connect; it provides guidance for teachers on lab procedure as well as answers to the lab questions.

- **Material lists** are provided to help students and teachers prepare the lab and save time.
- Each lab begins with a **Problem** that sets up the purpose of the lab.
- **Objectives** help to focus students on what should be learned from the lab.
- A detailed **Procedure** is laid out for students to aid in their discovery.
- Questions and charts help students to demonstrate what they've learned from the lab.

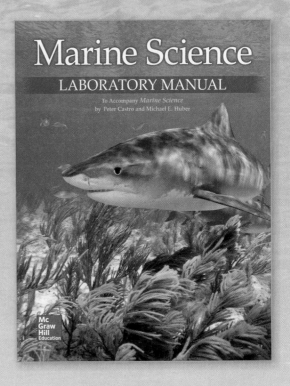

Marine Science **Teacher's Manual**

Marine Science provides a robust teacher's manual. This teacher's manual, available through Connect, provides a detailed pacing guide for each chapter, Chapter Summaries, additional resources, answers to the Section and Chapter Review Questions, and many supplementary activities.

- The pacing guide indicates how many days you should designate for instruction for each section of the book.
- The **Main Idea** and **Key Questions** from the student edition are reviewed.
- Section content is correlated to the **Ocean Literacy Standards**.
- A suggested sequence for teaching Laboratory Manual and feature-based inquiry activities is provided.
- In addition, the teacher's manual also provides **Vocabulary** and **Key Question Activities** as well as **Chapter Projects**.

Unit

1

The Ocean Environment

A colony of gentoo penguins (*Pygoscelis papua*) prepares to dive into ocean. Gentoo penguins nest and give birth on land, but are well-adapted to swim and feed in the icy Antarctic waters. The oceans influence both marine and terrestrial life, as well as the very climate of Earth.

UNIT PROJECT

Analyze a Challenge

Think about challenges related to marine science which are currently facing society. In coastal and inland communities alike, there might be challenges such as pollution, invasive species, or waste management. Communities also face public health issues, such as outbreaks of diseases, and ethical concerns surrounding science and technology. Communities turn to fellow citizens, scientists, engineers, and other professionals to solve these problems. Many times these problems are not clearly defined or easy to solve. They involve many variables and points of view. Your task is to analyze a global challenge related to marine science,and to specify qualitative and quantitative criteria and constraints for solutions that account for societal needs and wants.

Chapter 1

Principles of Marine Science

THEME **Cause and Effect: Mechanism and Explanation**

Marine scientists use the scientific method to investigate and reach conclusions about the properties of the oceans and the organisms that live in them.

BIG IDEA

A deep curiosity and quest for knowledge has driven ocean exploration for many centuries; scientists use the observations they have made to propose hypotheses about ocean processes and marine organisms.

About the photo Scientists collect eggs during coral spawning off the coast of Curaçao, in the southern Caribbean.

Nature/UIG/Getty Images

1.1 An Introduction to Marine Science

Marine science is a broad field that covers the link between oceanography and marine biology. Marine scientists study everything from the physics of ocean currents to the communication behavior in marine mammals. The ocean is vast, home to countless strange and wonderful creatures. It is often the beauty, mystery, and variety of life in the sea that attract students to a course in marine science. Even professional marine scientists feel a sense of adventure and wonder in their studies.

There are also many practical reasons to study marine science. Life on Earth probably originated in the sea, so the study of marine organisms teaches us much about all life on Earth, not just marine life. Many medical advances, have been underpinned by research on marine organisms, such as studies of the animal immune system in sea anemones and sea star larvae, the fertilization of sea urchin eggs, squid nerve cells, and how barnacles adhere to rocks.

The oceans also represent a vast source of human wealth. It provides food, medicines, and raw materials, offers recreation to millions, and supports tourism all over the world. The oceans, however, are not without their dangers. The awesome powers of the seas have taken down many ships, and coastal communities can be flattened by cyclonic storms or seismic waves. Some organisms harm humans directly by causing disease or attacking people. Marine organisms can erode piers, sea walls, and other structures in the ocean, foul ship bottoms, and clog pipes.

At a much more fundamental level, marine systems help determine the very nature of our planet. The ocean plays a large role in determining the climate, with the deep-ocean currents distributing heat across the globe. Marine organisms produce around half of the oxygen we breathe and help regulate Earth's climate. Our shorelines are shaped and protected by marine life, at least in part, and some marine organisms even help create new land. In economic terms, it has been estimated that the ocean's living systems are worth more than $20 *trillion* a year.

To make full and wise use of the sea, to solve problems marine organisms create, and to predict the effects of human activities on the ocean, we must learn all we can about marine systems. In addition, the oceans provide valuable clues to Earth's past, the history of life, and even our own bodies. This is the challenge, the adventure, of marine science.

Since marine science is such a broad field, many marine researchers specialize in a sub-field, such as marine biology or oceanography. Nearly all the disciplines of biology are represented in marine biology. There are marine biologists who study the basic chemistry of living things, for example. Others are interested in whole organisms: how they behave, where they live and why, and so on. Other marine biologists adopt a global perspective and look at the way entire oceans function as systems. Marine biology is thus both part of a broader science and itself made up of many different disciplines, approaches, and viewpoints.

Marine scientists might also focus on oceanography, the scientific study of the oceans. Like marine biology, oceanography has many branches. Geological oceanographers, or marine geologists, study the sea floor. Chemical oceanographers study ocean chemistry, and physical oceanographers study waves, tides, currents, and other physical aspects of the sea. Marine biology is most closely related to biological oceanography, so closely, in fact, that the two are difficult to separate. Sometimes they are distinguished on the basis that marine biologists tend to study organisms living relatively close to shore, whereas biological oceanographers focus on life in the open ocean, far from land. Another common distinction is that marine biologists tend to study marine life from the perspective of the organisms (for example, studying what an organism eats), while biological oceanographers tend to take the perspective of the ocean (for example, studying how food energy cycles through the system). In practice there are so many exceptions to these distinctions that many marine scientists consider marine biology and biological oceanography to be the same.

A marine scientist's interests may also overlap broadly with those of terrestrial biologists, chemists, or geologists. Many of the basic ecosystem processes are similar whether they occur on land or in the sea. Nevertheless, marine science does have a flavor all its own, partly because of its history.

Figure 1.1 Pacific islanders have navigated the Pacific for millennia in canoes like this Hawaiian double-hulled canoe (*wa'a kalulua*).

The History of Marine Science

People have been living by the sea since the dawn of humanity, and seafood is thought to have been crucial to early human survival and migration. The earliest known stone blades, from 165,000 years ago, were discovered in a seaside cave in South Africa, along with piles of shells from Stone Age clambakes and the earliest traces of ochre pigment, thought to be used for symbolic body painting and decoration. Ancient bone or shell harpoons and fishhooks have also been found, as well as the earliest known jewelry in the form of shell beads from as long as 110,000 years ago. As they used its resources, people steadily gained a store of practical knowledge about the sea.

Figure 1.2 Romans often used marine animals in the magnificent mosaics that decorated the floors and walls of their villas. This mosaic depicts in great detail several types of fishes, an octopus capturing a lobster, a squid, and other Mediterranean marine life.

Navigation and exploration Knowledge of the ocean and its organisms expanded as people gained skills in seamanship and navigation. Ancient Pacific Islanders had detailed knowledge of marine life, which their descendants still retain (Fig. 1.1). They were consummate mariners, using clues such as wind, wave, and current patterns to navigate over vast distances. The Phoenicians were the first accomplished Western navigators. By 2000 B.C. they were sailing around the Mediterranean Sea, Red Sea, eastern Atlantic Ocean, Black Sea, and Indian Ocean.

The ancient Greeks and Romans had considerable knowledge of nearshore organisms in the Mediterranean region (Fig. 1.2). The Greek philosopher Aristotle is considered by many to be the first marine biologist. He described many forms of marine life and recognized, among other things, that gills are the breathing apparatus of fish.

During the centuries known as the Dark Ages, scientific inquiry, including the study of the oceans, came to a grinding halt in most of Europe. Much of the knowledge of the ancient Greeks was lost or distorted. Not all exploration of the ocean stopped, however. During the ninth and tenth centuries the Vikings continued to explore the northern Atlantic. Arab traders were also active during the Middle Ages, voyaging to eastern Africa, southeast Asia, and India. In the Far East and the Pacific, people also continued to explore and learn about the sea.

During the Renaissance, spurred in part by the rediscovery of ancient knowledge preserved by the Arabs, Europeans again began to investigate the world around them, and several undertook voyages of exploration. Christopher Columbus sent reports back to Europe about the "New World" in 1492. In 1519 Ferdinand Magellan embarked on the first expedition to sail around the globe. Many other epic voyages contributed to our knowledge of the oceans. Fairly accurate maps, especially of places outside Europe, began to appear for the first time.

Early naturalists Before long, explorers became curious about what lived in the ocean they sailed. An English sea captain, James Cook, was one of the first to make scientific observations along the way and to include a full-time naturalist among his crew. In a series of three great voyages, beginning in 1768, he explored all the oceans. He was the first European to see the Antarctic ice fields and to land in Hawai'i, New Zealand, Tahiti, and a host of other Pacific islands. Cook was the first to use a chronometer, an accurate timepiece that enabled him to determine his longitude precisely, and therefore prepare reliable charts. From the Arctic to the Antarctic, from Alaska to Australia, Cook extended and reshaped the European conception of the world. He brought back specimens of plants and animals and tales of strange new lands. Though Cook was generally respectful and appreciative of indigenous cultures, he was killed in 1779 in a fight with native Hawaiians at Kealakekua Bay, Hawai'i.

By the nineteenth century, it had become common for vessels to take a naturalist along to study the organisms encountered. Perhaps the most famous of these shipboard naturalists was Charles Darwin, another Englishman. Beginning in 1831, Darwin sailed around the world on HMS *Beagle* for five years, horribly seasick most of the time. The *Beagle*'s primary mission was to map coastlines, but Darwin made detailed observations of all aspects of the natural world. This set off a train of thought that led him, years later, to propose the theory of evolution by natural selection. Though best known for the theory of evolution, Darwin made many other contributions to marine science. He explained, for example, the formation of the distinctive rings of coral reef called atolls. He used nets to capture the tiny, drifting organisms known as plankton, which marine biologists continue to do today (Fig. 1.3). Darwin's many interests also included barnacles, attached crustaceans. Specialists still refer to his treatise on them.

In the United States the most important early exploratory voyage was probably the United States Exploring Expedition of 1838–1842, often called the "Wilkes Expedition" after its leader, Lt. Charles Wilkes of the U.S. Navy. The expedition included only 11 naturalists and artists, derisively called "clam diggers" by the rest of the crew, and some historians conclude that it was more about projecting American influence than scientific discovery. Wilkes was by all accounts a vain and cruel man who promoted himself to Captain as soon as he left port, and upon his return was court martialed for flogging his crew to excess. Only two of the expedition's six ships made it home. Nevertheless, the Wilkes Expedition's achievements were impressive. The expedition charted 2,400 km of the coast of Antarctica, confirming it as a continent, as well as the coast of the Pacific Northwest of North America. It explored some 280 islands in the South Pacific, collecting information about peoples and cultures as well as flora and fauna. The 10,000 biological specimens included some 2,000 previously unknown species (Fig. 1.4). The expedition, the first international survey sponsored by the U.S. government, also laid a foundation for government funding of scientific research.

Figure 1.3 These marine scientists are hauling in a net known as a "bongo net" used to capture minute marine plankton. One is signaling instructions to the winch operator.

The *Challenger* Expedition By the middle of the nineteenth century, a few lucky scientists were able to undertake voyages specifically to study the oceans, instead of having to tag along on ships doing other jobs. One was Edward Forbes, who in the 1840s and 1850s carried out extensive trawling of the sea floor, mostly around his native Britain but also in the Aegean Sea and other places. Forbes died prematurely in 1854, at the age of 39, but was the most influential marine biologist of his day. He discovered many previously unknown organisms and recognized that sea-floor life varies at different depths. Perhaps his most important contribution, however, was to inspire new interest in the life of the sea floor.

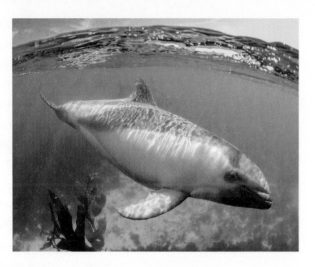

Figure 1.4 Peale's dolphin (*Lagenorhynchus australis*), named after the Wilkes Expedition naturalist who first described it, is one of 2,000 marine and terrestrial species discovered by the expedition.

Forbes's contemporaries and successors, especially from Britain, Germany, Scandinavia, and France, carried on his studies of sea-floor life. Their ships were poorly equipped and the voyages short, but their studies produced many interesting results. They were so successful, in fact, that British scientists managed to convince their government to fund the first major oceanographic expedition, under the scientific leadership of Charles Wyville Thompson. The British navy supplied a light warship to be fitted out for the purpose. The ship was named HMS *Challenger*.

Challenger underwent extensive renovations in preparation for the voyage. Laboratories and quarters for the scientific crew were added, and gear for collecting samples in deep water was installed. Though primitive by modern standards, the scientific equipment on board was the best of its day. Finally, in December 1872, *Challenger* set off.

Over the next three and a half years, *Challenger* and her crew sailed around the world, gathering information and collecting water, sediment, and biological samples (Fig. 1.5, p. 10). The sheer volume of data gathered was enormous—it took 19 years to publish the results, which fill 50 thick volumes. *Challenger* brought back more information about the ocean than had been recorded in all previous human history.

It was not just the duration of the voyage or the amount of information collected that set the *Challenger* expedition apart from earlier efforts. The expedition set new standards for ocean research. *Challenger*'s scientists collected data in a more systematic way than previous expeditions, and kept meticulous records. For the first time, scientists began to get a coherent picture of what the ocean was like. They also learned about the enormous variety of marine life, for *Challenger* brought back thousands of previously unknown species. Thus, the *Challenger* expedition laid the foundations of modern marine science.

Other expeditions continued the work begun by *Challenger,* and major oceanographic cruises continue to this day. In many ways, though, the voyage of the *Challenger* remains one of the most important in the history of oceanography.

The Growth of Marine Labs Even before the *Challenger* set off, marine scientists were excited about the organisms brought back by ocean expeditions. Unfortunately, the vessels had quarters for only a few scientists. Most scientists only got to see the dead, preserved specimens that the ships brought back to port. Such specimens revealed much about marine life around the world, but scientists wanted to know how the organisms actually lived: how they functioned and what they did. Living specimens were essential for this, but ships usually stayed in one place for only a short time, making long-term observations and experiments impossible.

As an alternative to ships, scientists began to work at the seashore. Among the first were two Frenchmen, Henri Milne Edwards and Victor Andouin, who around 1826 began making regular visits to the shore to study marine life. Other scientists soon followed suit. These excursions offered the opportunity to study live organisms, but there were no permanent facilities and only a limited amount of equipment could be taken along. Eventually, scientists set up permanent laboratories where they could keep organisms alive and work over long periods. The first such laboratory was the Stazione Zoologica, founded in Naples, Italy, in 1872—the same year the *Challenger* embarked. The laboratory of the Marine Biological Society of the United Kingdom was founded at Plymouth, England, in 1879.

The first major American marine laboratory was the Marine Biological Laboratory at Woods Hole, Massachusetts. It is hard to pinpoint the exact date when this laboratory was established. The first marine laboratory at Woods Hole was started by the United States Fish Commission in 1871, but it did not flourish. Several other short-lived laboratories subsequently appeared in the area. Harvard biologist Louis Agassiz, who also studied many of the specimens collected by the Wilkes Expedition, established a laboratory on nearby Cape

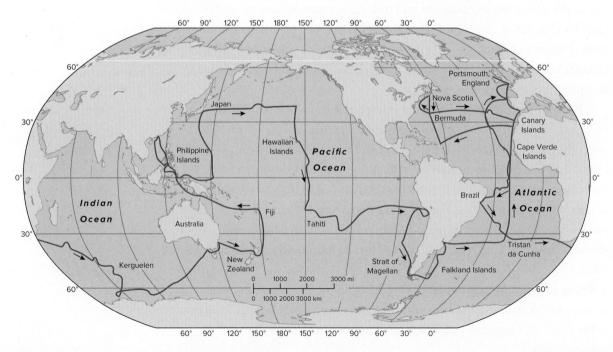

Figure 1.5 The route of the *Challenger* expedition, which from 1872 to 1876 conducted the first systematic survey of the world ocean.

Ann in 1873. In 1888 this laboratory moved to Woods Hole and officially opened its doors as the Marine Biological Laboratory. It is still one of the world's most prestigious marine labs.

After these early beginnings, other marine laboratories were established. Among the earliest in the United States were the Hopkins Marine Station in Pacific Grove, California (Fig. 1.6), Scripps Institution of Oceanography in La Jolla, California, and the Friday Harbor Marine Laboratory in Friday Harbor, Washington. In the ensuing years, more laboratories appeared all over the world, and new ones are being established even today.

The onset of World War II had a major effect on the development of marine science. A new technology, sonar, or *sound navigation and ranging*, was developed in response to the growing importance of submarine warfare. Sonar is based on the detection of underwater echoes—a way of detecting sound in the sea (Fig. 1.7). The ocean, long thought of as a silent realm, was suddenly found to be full of sound, much of it made by animals. During wartime, learning about these animals was no longer the casual pursuit of a few interested marine biologists but a matter of national security. As a result of this urgency, several marine laboratories, such as Scripps and the Woods Hole Oceanographic Institution (established in 1929), underwent rapid growth. When the war ended, these labs not only remained vital research centers, but continued to grow.

The years immediately after World War II saw the refinement of the first practical **scuba,** or *self-contained underwater breathing apparatus.*

Figure 1.6 An early marine biology class at Stanford University's Hopkins Marine Station. The station, established in 1892, is the third oldest in the United States.

Signal

Echo

Sea floor

Figure 1.7 A ship uses sonar by "pinging," or emitting a loud pulse of sound, and timing how long it takes the echo to return from the sea floor. The water depth can be determined from the return time.

Observing the Ocean

The ocean is a vast, interconnected network, and conditions in one place can be affected by events occurring far away. Storms, earthquakes, and the interactions and movements of marine organisms often occur suddenly, making them hard to capture unless you happen to be in just the right place at the right time. Traditional techniques for monitoring and studying the ocean will always have an important place in marine science, but they can't provide the continuous coverage of the ocean that is needed to really understand how the ocean functions.

New technologies are allowing scientists to observe the oceans in ways that would have seemed like science fiction not long ago. For example, the Argo system consists of 3,000 or more automated floats scattered throughout the ocean. Each float continuously measures parameters such as water temperature, salinity, and oxygen and then transmits that data via satellite. Argo floats provide huge amounts of new information, especially from areas that are difficult to study from ships.

Monitoring networks Oceanographers are wiring the sea floor, providing power and communications for an amazing array of instruments that can measure currents, test water chemistry, detect sea-floor tremors, and track biological activity. The Long-term Ecosystem Observatory (LEO) was the first such system in the United States. LEO, now known as the Coastal Ocean Observation Laboratory (COOL), was installed in 1996 and has grown to include underwater gliders, shore-based radar, and moored instruments. COOL is being integrated with similar systems into a single network covering the United States from Maine to Florida.

In the Pacific, several undersea monitoring networks record data in areas of interest. The Victoria Experimental

Some components of ocean observing systems being developed around the world

Marine biologists fit a southern elephant seal (*Mirounga leoninia*) with electronic instruments.

Network Under the Sea (VENUS), the Monterey Accelerated Research System (MARS) and the North East Pacific Time-integrated Undersea Networked Experiments (NEPTUNE) measure undersea data from British Columbia to California. They are part of an even larger program, the Ocean Observatories Initiative. Similar networks are being developed in Europe, Japan, the Gulf of Mexico, and the Arctic Ocean. Exciting new devices are on the way, such as genomic sensors to identify planktonic DNA and docking stations for free-ranging autonomous underwater vehicles.

Benefits Ocean observatories save lives and money. In 2006 COOL helped forecast the track of Tropical Storm Ernesto for emergency response authorities and the public. Observing networks are providing early warning of tsunamis and earthquakes, tracking the effects of climate change, monitoring fish populations, and making shipping more efficient. These networks have also brought unexpected benefits, in part because they are used by scientists from many disciplines. For example, marine geologists use the NEPTUNE network to monitor for earthquakes, but the endangered fin whale sings at a frequency that interferes with these measurements, so some geologists developed software tools to identify and filter out fin whale songs. Marine biologists then used this software to track the whales. In another example, forensic scientists used VENUS's underwater video to monitor the decomposition of pig carcasses to help determine the time of death of human bodies recovered from the sea.

Living monitors Not all ocean observing systems are inanimate—large marine animals, such as sea lions and whales, are being recruited to help. To get a firsthand look at what these animals do beneath the surface without interfering with their natural behavior, scientists developed "crittercam." Crittercams are small, compact cameras directly attached to marine animals. Crittercams have provided some of the most amazing underwater discoveries -humpback whales using curtains of bubbles to herd herring, social diving behavior in chinstrap penguins, and the movements of endangered sea turtles in Mexico. Marine animals are also used as "autonomous underwater samplers," in which animals are equipped with transmitters that measure temperature, depth, and salinity. This allows oceanographers to record diving behavior as well as collect data from places that might not be accessed any other way.

Like all cutting-edge endeavors, developing ocean observing systems is challenging work. High-tech electronics can be fickle even in the laboratory, much less the depths of the sea. A NEPTUNE scientist says, "We're learning a lot, which is another way of saying that things are breaking." But most of NEPTUNE's instruments are transmitting photos, audio, and video—in vast amounts, and these systems are only getting more reliable and advanced every day.

Figure 1.8 Scuba is an important tool in the work of many marine scientists. This scientist is using an apparatus called a respirometer to measure the production and consumption of oxygen by organisms on a coral reef.

Figure 1.9 The R/V *Thomas G. Thompson*, operated by the University of Washington, was the first of a new generation of research vessels.

The basic technology was developed in occupied France by the engineer Émile Gagnan to allow automobiles to run on compressed natural gas. After the war, Gagnan and fellow Frenchman Jacques Cousteau modified the apparatus, using it to breathe compressed air under water. Cousteau went on to devote his life to scuba diving and the oceans.

Using scuba, marine biologists could, for the first time, descend below the surface to observe marine organisms in their natural environment (Fig. 1.8). They could now work comfortably in the ocean, collecting specimens and performing experiments, though they were still limited to relatively shallow water, generally less than 50 m.

Figure 1.10 *Alvin*, a deep-sea submarine operated by the Woods Hole Oceanographic Institution, is one of the most famous vessels in the history of marine science.

Marine Biology Today

Oceanographic ships and shore-based laboratories are as important as ever to marine science. Today many universities and other institutions operate research vessels (Fig. 1.9). Modern ships are equipped with the latest equipment for navigation, sampling, and studying the creatures that are collected. Many, like *Challenger*, were originally built for other uses, but growing numbers of vessels are purpose-built for scientific research at sea.

In addition to ships as we normally think of them, some remarkable craft are used to study the marine world. High-tech submarines can descend to the deepest parts of the ocean, revealing a world that was once inaccessible (Fig. 1.10). Various odd-looking vessels ply the oceans, providing specialized facilities for marine scientists (Fig. 1.11).

Marine laboratories, too, have come a long way since the early days. Today, labs dot coastlines around the world and are used by an international community of scientists. Some are equipped with the most up-to-date facilities available. Others are simple field stations, providing a base for scientists to work in remote areas. There are even undersea habitats where scientists can live for weeks at a time, literally immersed in their work (Fig. 1.12, p. 16). Marine laboratories are important centers not only of research, but also of education. Many offer hands-on undergraduate courses in which students can study marine science firsthand, and most provide facilities where graduate students begin their careers in marine science.

Figure 1.11 R/V *FLIP*, short for *f*loating *i*nstrument *p*latform, operated by Scripps Institution of Oceanography, provides a stable platform for research at sea. (*a*) Most of the hull consists of a hollow tube that floats while the vessel is towed into position. (*b*) When the hull is flooded and sinks, *FLIP* swings into a vertical position.

Figure 1.12 A diver swims outside *Aquarius*, the world's only underwater marine science laboratory. *Aquarius* is located in the Florida Keys Marine Sanctuary at a depth of about 20 m. The living quarters are in the large cylinder at the upper left.

New technology offers exciting opportunities for studying the oceans. It goes almost without saying that computers and electronics have had profound impact. Satellites peer down at the ocean and, because they are so far away, can view huge areas of the ocean all at once to capture the big picture (Fig. 1.13, p. 17). Much of our knowledge of large-scale features like ocean currents has been provided by this **remote sensing** technology. Satellites only see the surface of the ocean, however, and a lot of the action is a long way down. Submarines are one way to penetrate the depths, but scientists are making increasing use of underwater robots, including remotely operated vehicles (ROVs), which are controlled from the surface, and autonomous underwater vehicles (AUVs), which operate independently of direct human control. Marine scientists are also developing an array of instruments that sit on the bottom, float in place, drift with the currents, or are even attached to animals (see "Observing the Ocean," above). Space technology has a role to play here as well; many oceanographic instruments relay their data through satellites.

Figure 1.14 High-tech meets low-tech: The robotic arm of the ROV *Ventana* captures a pom-pom anemone *(Liponema brevicornis)* in an ordinary kitchen colander. © MBARI 1998

0.1
0.2
0.4
0.6
0.8
1
10

Phytoplankton
Pigment
Concentration
(mg/m³)

NASA/GSFC

Figure 1.13 A satellite image showing the abundance of photosynthetic organisms in the ocean, as indicated by the amount of pigment in the water. This photo was taken by the Coastal Zone Color Scanner (CZCS), which was mounted on the *Nimbus-7* satellite. It is actually a composite of information gathered over nearly an eight-year period. Advances in computer and space technology made this image possible.

Marine scientists today use every available tool in their study of the sea, even some decidedly low-tech ones (Fig. 1.14). Information about the ocean pours in at an ever-increasing pace. Much is yet to be learned, however, and the oceans remain a realm of great mystery and excitement.

. .

Section Review

1. Identify Which groups of people were the main contributors to pre-Renaissance knowledge of oceans and marine biology?

2. Discuss What contributions to marine science did the voyage of the HMS *Challenger* from 1872–1876 yield?

3. Describe What modern technology is used by marine scientists to gather data about the oceans?

1.2 The Scientific Method

VOCABULARY

scientific method

induction

deduction

hypothesis

testable

alternative hypothesis

experiment

variables

controlled

scientific theory

Main Idea

Like all types of scientists, marine scientists use the scientific method to study the properties of the oceans and the organisms that live in them.

. .

Key Questions

1. Why are observations such an important part of the scientific method?

2. Why might a hypothesis or scientific theory be modified or rejected?

DIVE IN!

How important are observations? In the 1800s, a U.S. Naval officer figured out that there was an ice-free passage through the Arctic Ocean. Whales in the Atlantic had harpoons from the Pacific imbedded in them. He knew that whales needed to come to the surface to breathe. He concluded that there must be an ice-free passage somewhere in the Arctic Ocean. Larger ships are now using this route, the Northwest Passage, as a result of the rapid melting of the Arctic Ice

Marine science is an adventure, to be sure, but it is also a science. All scientists share a certain way of looking at the world. Students of marine science need to be familiar with this approach and how it affects our understanding of the natural world, including the ocean.

We live in an age of science. Advertisers constantly boast of "scientific" improvements to their products. Newspapers regularly report new breakthroughs, and many television stations have special science reporters. Governments and private companies spend billions of dollars every year on scientific research and education. Why has science come to occupy a position of such prestige in our society? The answer, quite simply, is that it works! Science is among the most successful of human endeavors. Modern society could not exist without the knowledge and technology produced by science. Everyone's lives have been enriched by scientific advances in medicine, agriculture, communication, transportation, art, and countless other fields.

Much of the practical success of science results from the way it is done. Scientists do not see the world as a place where things just happen, for no reason. They assert instead that the universe can be explained by physical laws. Scientists don't go about discovering these laws haphazardly; they proceed according to time-tested procedures. The set of procedures that scientists use to learn about the world is called the **scientific method**.

Scientists sometimes disagree over the fine points of the scientific method. As a result, they may apply the method in slightly different ways. In spite of these minor differences, most scientists do agree on the basic principles of the scientific method, which should be seen as a flexible framework guiding the study of nature and not a rigid set of rules.

Observation: The Currency of Science

The goal of science is to discover facts about the natural world and principles that explain these facts. At the heart of the scientific method is the conviction that we can learn about the world only through our senses or with tools that

extend our senses. Microscopes, for example, extend our vision to help us see what is otherwise too small to see. Thus, scientific knowledge is fundamentally derived from the observation of nature. Scientific conclusions are based on observations, and not on preexisting ideas of how the world is or should be.

One of the advantages of relying on observations is that they can be verified by others. A person's thoughts, feelings, and beliefs are internal. No one really knows what goes on in the mind of another. On the other hand, the world studied by scientists is external to any single person. Different people can look at the same object. Sensory perception may be imperfect, and scientists, like anyone else, are not always impartial, but the object is there for all to see. Thus, there is a way to check and validate any one person's observations.

Observation is critical to all phases of the scientific method. To begin with, it allows us to describe the natural world. The only way to learn what organisms live in a particular part of the ocean, how many of them there are, how fast and how large they grow, when and how they reproduce, what they eat, how they behave, and so on, is to observe that part of the ocean and the organisms living there. Exploration and description are vitally important parts of marine science, constantly revealing new information. Previously unknown species, for example, are discovered almost every week. New technologies, such as underwater cameras that reveal the behavior of whales or genetic techniques that have uncovered vast numbers of previously unknown marine microbes, regularly improve our ability to observe the sea and make new discoveries. Each finding leads to new observations. The discovery of completely unknown and unexpected ecosystems at deep-sea hot springs, for example, led biologists to look for—and find—similar ecosystems in other parts of the ocean.

As they observe more and more about the world, scientists inevitably seek to explain their observations—*why is that species of seaweed found only in a certain depth range?*—and to make predictions—*will the fishing be good next year?* The desire to explain and predict in turn guides yet more observation.

Two Ways of Thinking

To describe, explain, and make predictions about the natural world, scientists use two basic forms of thinking. In **induction**, one uses observations to arrive at general principles. Reasoning from general principles to specific conclusions is called **deduction**. Scientists once argued about which way of thinking is acceptable, but now generally agree that both induction and deduction are indispensable.

Induction When using induction, a scientist starts with a series of individual observations. Ideally, he or she has no goal or preconceptions about the outcome and is completely objective. The scientist then uses these observations to reach a general conclusion. For example, suppose a marine scientist examined a sailfish (Fig. 1.15, p. 20), a shark (Fig. 1.16, p. 20), and a tuna (Fig. 1.17, p. 20) and found that they all had gills. Because sailfishes, sharks, and tunas are all fishes, he might draw the general conclusion *All fishes have gills.* This is an example of induction.

Figure 1.15 The Atlantic sailfish (*Istiophorus albicans*) attacking a school of sardines. The long projection on the snout is called the bill.

The scientist must be careful in using induction. The step from isolated observations to general conclusion critically depends on the number and quality of the observations and on recognizing their limitations. If the biologist stopped after examining the sailfish, which happens to have a bill, he might use induction to make the false conclusion *All fishes have bills.* Even after examining all three fishes, he might have concluded *All marine animals have gills* instead of just *All fishes have gills.* This is where deduction comes into play.

Deduction In deduction, scientists start with a general statement about nature and predict what the specific consequences would be if that statement is true. The general statement might be based on hunch or intuition, but it is usually based on observations. Suppose our marine biologist used induction to make the general statement *All marine animals have gills.* He might then reason that if all marine animals have gills and whales are marine animals, then whales must have gills. The biologist has used a general statement about all marine animals to make a statement about a particular kind of marine animal.

Testing Ideas

Scientists are never content to simply make statements about the world and let it go at that. Instead, they are obsessed with testing the statements to see if they are, in fact, true. Both induction and deduction lead the scientist to make statements that *might* be true. A statement that might be true is called a **hypothesis**. A crucial feature of the scientific method is that all hypotheses are tested, usually again and again. This insistence on testing is one of the great strengths of the scientific method. Incorrect hypotheses are usually quickly weeded out and discarded.

Figure 1.16 A tiger shark (*Galeocerdo cuvier*). The five vertical gill slits can be seen just in front of the pectoral fins.

Figure 1.17 A school of southern bluefin tuna (*Thunnus maccoyii*).

©Francois Gohier/VWPics/Alamy

Figure 1.18 A sperm whale (*Physeter catodon*) exhales bubbles of air from its lungs through the blowhole on top of its head.

Constructing the Hypothesis Scientific hypotheses must be stated in a way that allows them to be critically tested. What this means is that it must be possible, at least potentially, to *disprove* the hypothesis if it is false. Sometimes this is simple. For example, the hypothesis that whales have gills is easy to test. All the scientist has to do is examine a whale to see if it has gills. By doing so, he would find that whales have lungs, not gills (Fig. 1.18). He would have proven the hypothesis *Whales have gills* false. He would also disprove the more general hypothesis *All marine animals have gills*. The steps our marine scientist used to construct and test these hypotheses are illustrated in Figure 1.19, p. 22. This line of reasoning is not entirely imaginary. Aristotle used similar logic in the fourth century B.C., observing not only that whales breathe with lungs and not gills, but that unlike most fishes they give birth to live young instead of laying eggs. Unfortunately, Aristotle's recognition that whales and other marine mammals are not fishes was lost to Western science for more than two millennia.

Hypotheses are often much more complicated than whether or not an animal has gills. Marine organisms are affected by weather other factors. Marine scientists increasingly express their understanding of how such factors interact by constructing models, in the form of mathematical and computer models that predict what will happen in a given set of circumstances. Although complex, such models are really just hypotheses that can be tested by comparing the predictions with what is actually observed.

People sometimes make the mistake of proposing hypotheses that cannot be fairly tested. Someone who believes in mermaids might say, *There are mermaids in the ocean.* The problem with this hypothesis is that it could never be *proven* to be false. An army of marine scientists could spend their entire careers looking for a mermaid without finding one, but the true believer could always say, *The mermaids are there, you just didn't find them.* No matter how hard they looked, the scientists could never prove that there are no mermaids. Therefore, the statement *There are mermaids in the ocean* is not a valid scientific hypothesis because it is not **testable**.

IN CONTEXT

The term "fish" is used for a single individual or for more than one individual of the same species.

The term "fishes" is used to refer to more than one species of fish.

Figure 1.19 An example of the use of the scientific method. Two hypotheses were derived from the same observations. When tested by further observations, one hypothesis (left) is accepted and one (right) is rejected.

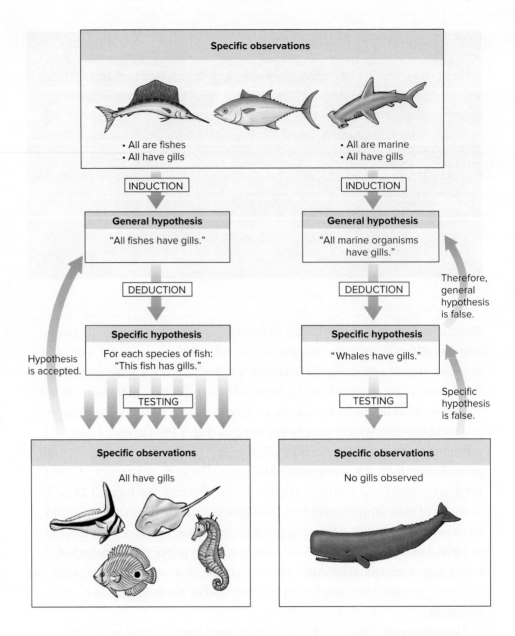

Specific observations

- All are fishes
- All have gills

- All are marine
- All have gills

INDUCTION

INDUCTION

General hypothesis

"All fishes have gills."

General hypothesis

"All marine organisms have gills."

DEDUCTION

DEDUCTION

Therefore, general hypothesis is false.

Specific hypothesis

For each species of fish: "This fish has gills."

Specific hypothesis

"Whales have gills."

Hypothesis is accepted.

Specific hypothesis is false.

TESTING

TESTING

Specific observations

All have gills

Specific observations

No gills observed

The Nature of Scientific Proof It must be possible to disprove a hypothesis, at least in principle, before the hypothesis is a scientific one. But how can a hypothesis be proved true? This question has always troubled scientists, and the answer may trouble you too. In general, no scientific hypothesis can be absolutely proved true. For example, consider the hypothesis that all fishes have gills. It is easy to see that this hypothesis can be proved false by finding a fish without gills. But even though every fish so far examined has gills, this does not prove that *all* fishes have gills.

Somewhere out there may lurk a fish without them. Just as it cannot be proved that there are no mermaids, it can never be proved that all fish have gills.

In science, then, there are no absolute truths. Knowing this, scientists could throw up their hands and look for another line of work. Fortunately, most scientists have learned to accept and deal with the lack of absolute certainty that is inherent in science by making the best of the available evidence.

All scientific hypotheses are examined and tested, poked and prodded, to see if they agree with actual observations of the world. When a hypothesis withstands all these tests, it is conditionally accepted as "true" in the sense that it is consistent with the available evidence. Scientists speak of *accepting* hypotheses, not *proving* them. They accept the hypothesis that all fish have gills because every attempt to reject it has failed. At least for now, the hypothesis fits the observations. The good scientist, however, never quite forgets that any hypothesis, even a personal favorite, could suddenly be thrown out the window by new information. No hypothesis is exempt from testing, or immune to being discarded if it conflicts with the evidence. The bottom line in science is observation of the world, *not* preconceived human ideas or beliefs.

Testing the Hypothesis Because hypotheses generally can't be proved to be true, scientists, somewhat surprisingly, spend their time trying to disprove, not prove, hypotheses. More confidence can be placed in a hypothesis that has stood up to hard testing than in an untested one. Thus, the role of the scientist is to be a skeptic.

Often scientists are trying to decide among two or more **alternative hypotheses**. After looking at the sailfish, shark, and tuna, our imaginary marine scientist advanced two possible hypotheses: that all fish have gills and that all marine animals have gills. Both hypotheses were consistent with his observations to that point. After examining a whale, he rejected the second hypothesis and, in doing so, strengthened the first one. He arrived at the best hypothesis by a process of elimination.

Real marine scientists rarely have it as easy as our imaginary one, who was able to construct and then test his hypotheses about gills with just a few simple observations. Hypothesis testing usually requires carefully planned, painstaking observations. Occasionally a new observation or set of observations leads to the complete rejection of an accepted hypothesis, which has been termed a "scientific revolution." Such discoveries make headlines, but most of the time the scientific process is a gradual one in which hypotheses are continually refined and modified, and new alternatives proposed, as more information becomes available.

Hypotheses can often be tested by making the right kind of observations of the natural world, at the right time and place.
Improved observing systems and continuing ocean exploration are constantly revealing more of the ocean's secrets.

Sometimes, however, the conditions needed to test a hypothesis do not occur naturally and scientists must manipulate nature, that is, perform an **experiment**, to make the necessary observations.

Suppose another marine scientist decides she wants to find out how water temperature affects the growth of mussels. She might be able to find two places, one warm and one cold, and measure how fast mussels grow at each place. The temperature at any given place changes all the time, however, and she would probably have difficulty finding two locations where one is always warmer than the other. Even if she does, the difference in temperature would not stay the same and there would probably be many other differences between

Carbonate Experiments on the Reef

Experiments performed out in the ocean are as important to marine science as those done in the lab. One of many examples is a research program being conducted on remote Pacific island coral reefs. Human activities are increasing the amount of carbon dioxide (CO_2) in the atmosphere, a proportion of which diffuses into the ocean. This may stimulate the growth of photosynthetic organisms such as seaweeds, which use CO_2 to produce glucose and grow. The increased CO_2 concentrations also make the ocean slightly more acidic, which interferes with the formation of calcium carbonate skeletons and shells in organisms that have them. This includes reef corals and many coral reef algae, known as calcareous algae. Increasing CO_2 levels are likely to affect coral reefs, and in particular may encourage "fleshy," noncalcareous algae to take over from corals and calcareous algae.

The limitations of lab studies Marine scientists have a lot of information about how some reef corals and algae respond to changes in CO_2 and acidity from laboratory experiments. There are, however, hundreds of different corals and algae on reefs, many of which are difficult to grow in the laboratory. Conditions on reefs are highly variable between locations and over time. In the oceans, reef organisms and water chemistry aren't static, with growth and death of reef-builders having a direct impact on the surrounding water chemistry. This complexity can't be replicated in the lab, so biologists need to do experiments in the field to determine the effects of increasing CO_2 on natural reefs.

Field experiments In one series of experiments, biologists placed "benthic tents," clear plastic pyramids, over different combinations of corals and algae growing on a reef. The tents let natural light through, but retained the volume of water around the organisms. The scientists installed electronic sensors inside the tents, and pumped out water samples to measure the different organisms acidity, oxygen levels, and calcium carbonate chemistry as they grew.

The scientists were also studying how changes in acidity affect the growth of unconfined corals and algae, by using "minireefs." The minireefs are just cinderblocks on which small pieces of coral and algae are attached. The biologists placed the minireefs in different parts of the reef where the acidity regime is known to vary naturally, and installed a sensor to measure the acidity. They measured the organisms' growth rates, and compared them to the acidity regime experienced. Early results indicated that calcareous algae grow best if the acidity levels are fairly constant, and are inhibited by wide fluctuations in acidity.

The best of both worlds Even in the field, the laboratory has its uses. Light is one of the main factors controlling growth of corals and algae, and on a natural reef the light level constantly changes with the time of day and cloud cover. To measure how light affects organisms in field experiments, biologists take specimens into a shipboard laboratory, where they measure their growth under precisely controlled light levels. This is typical of marine science, in which lab and field experiments complement each other to answer complex questions about life in the sea.

the two places. The mussels might be different, for example. They might be eating different foods or different amounts of food. There might be pollution or an outbreak of disease at one of the sites. In any natural situation, there would be countless factors other than temperature that could explain differences in mussel growth. Factors that might affect observations are called **variables**.

Faced with all these variables, the scientist decides to perform an experiment. She collects mussels from one place and divides them at random into two groups. Now she knows that the mussels in the two groups are pretty much the same. She places the two groups into tanks where she can control

the water temperature and grows one group in warm water, the other in cold. She feeds all the mussels the same amounts of the same food at the same time, protects the mussels from pollution and disease, supplies both holding tanks with seawater from the same source, and keeps all the other living conditions exactly the same for both groups. Because all these variables are the same for both groups, the scientist knows that they cannot be responsible for any differences observed in mussel growth. The only difference between the two groups is temperature.

To prevent a variable from affecting the experiment, the scientist has two options. One is to artificially keep the variable from changing—for example, by giving all the mussels exactly the same food. The other is to make sure that any changes that do occur are identical for both groups. By supplying both tanks with seawater from the same source, for example, our scientist ensures that any changes in the quality of the water affect both groups of mussels equally. Variables that are prevented from affecting an experiment are said to be **controlled**, and the experiment is called a controlled experiment (Fig. 1.20). Since the biologist has controlled the effects of other variables while growing the mussels at different temperatures, she can be more confident that any observed difference in growth rate between the two groups is due to temperature.

Similarly, the scientist could study how food supply affects mussel growth by keeping the mussels at the same temperature but giving them different amounts of food. Experiments thus allow the effects of different variables to be separated. The way that variables interact can also be studied. The mussels could be maintained in different combinations of temperature and food supply, for example, to see whether the temperature at which they grow fastest depends on how much food they get. Experiments aren't just for the laboratory—many important experiments are done out in the real world (see "Carbonate Experiments on the Reef," above).

The Scientific Theory Many people think of a theory as a shaky proposition, and most of us have heard people ridicule some idea because it was "only a theory."

Figure 1.20 (*a*) Many different variables might produce differences between groups of mussels at different locations. (*b*) Controlling the variables in an experiment allows the effects of a single factor to be tested.

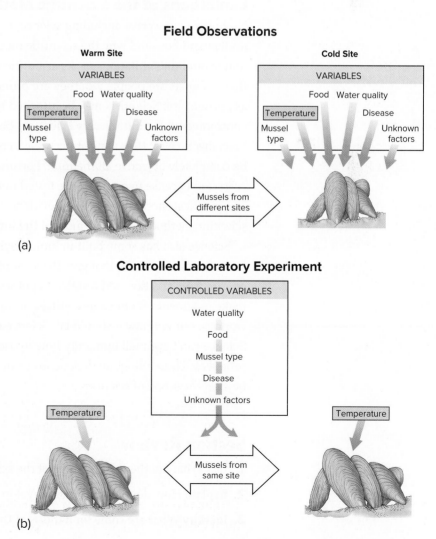

Field Observations

Warm Site

VARIABLES

Food Water quality

Temperature Disease

Mussel type Unknown factors

Cold Site

VARIABLES

Food Water quality

Temperature Disease

Mussel type Unknown factors

Mussels from different sites

(a)

Controlled Laboratory Experiment

CONTROLLED VARIABLES

Water quality

Food

Mussel type

Disease

Unknown factors

Temperature

Temperature

Mussels from same site

(b)

The public usually reserves such scorn for controversial or unpopular theories. The theory of gravity, for instance, is rarely criticized for being "only a theory" even though physicists have no accepted explanation of why gravity exists. People often say "theory" in everyday conversation to mean speculation or just one of several possible explanations for something, but scientists don't use the term scientific theory to refer to a controversial or provisional hypothesis. A hypothesis, or set of hypotheses, is not considered **scientific theory** until alternatives have been ruled out and the hypothesis has passed every possible test. A scientific theory is supported by overwhelming evidence and represents a comprehensive explanation of our observations of how the world works. It is an established scientific principle that guides the search for new knowledge by leading to new, testable hypotheses.

It must be remembered, however, that a theory is still a hypothesis, albeit a well-tested one. As with other hypotheses, theories cannot be absolutely proved and are accepted as true only as long as they are supported by evidence. Good scientists accept theories for the time being because the best available evidence supports them. They also recognize that any theory could be overturned by new evidence.

Limitations of the Scientific Method

No human enterprise, including science, is perfect. Just as it is important to understand how and why the scientific method works, it is important to understand the limitations of the scientific method. For one thing, remember that scientists are people too; they are prone to the same human shortcomings as anyone else. Scientists may be attached to favorite theories even when confronted with contradictory evidence—being wrong can be hard to accept. Like anyone else, they may let personal biases affect their thinking. No one can be completely objective all the time. Fortunately, factual errors are usually corrected because hypotheses are tested not just by one person but by many. The practical success of science is evidence that the self-checking nature of the scientific method does work most of the time.

Science also has some built-in limitations. Ironically, these limitations arise from the same features that give the scientific method its power: the insistence on direct observation and testable hypotheses. This means that science cannot make judgments about values, ethics, or morality. Science can reveal how the world is, but not how it should be. Science cannot decide what is beautiful. Science can't even tell humanity how to use the knowledge and technology it produces. These things all depend on values, feelings, and beliefs, which are beyond the scope of science.

· ·

Section Review

1. **List** What are the general steps of the scientific method?

2. **Explain** How does a hypothesis develop into a scientific theory?

3. **Identify** What are some limitations of the scientific method and science?

Ocean Modeling

Oceanographers have deployed an incredible number of instruments to collect data on ocean currents, species distribution, ocean-atmosphere flux, and more. These analytical instruments, however, aren't haphazardly placed. Scientists start with a certain question in mind. For example, a scientist might have a hypothesis concerning how many blue whales live in the North Pacific or need to make a prediction regarding the intensity and direction of a hurricane. Instruments deployed in select locations to help collect data that might support or disprove their hypothesis.

Since the ocean is such a large place, it would be nearly impossible to locate and count all the blue whales at the same time, especially since they are on the move! Likewise, scientists cannot just look at one instrument to determine where a hurricane is heading or how much damage it might do. This is where scientists turn to scientific modeling. Scientific models incorporate variables that are important to the study into a mathematical equation which can help to explain a phenomenon that cannot be directly measured. Scientific models are used in many areas of marine science, like to estimate changes in a population, describe biogeochemical processes, or predict future events like the effect of rising global temperatures on ocean currents.

This image, captured by the Tropical Rainfall Measuring Mission (TRMM) satellite, shows the height of rain columns within Hurricane Irene. The higher the rain column, the more energy a storm can generate. The red color indicates a higher rain column.

Variables Scientific models, especially ones that look at ecosystem processes, are complex. Think about all the variables that might be needed to estimate the number of blue whales. Such variables might include: the number of

whale calls heard in a certain location over a certain period of time, the temperature or depth of water blue whales are known to live, the amount of krill (the whale's food) in the area, how fast a blue whale moves, the life cycle of a whale, or the number of offspring a blue whale is likely to have. Hurricanes direction and intensity are also affected by many parameters including wind speed and direction, humidity, pressure, and oceanic and atmospheric temperatures. Many models must be compiled and run on supercomputers, which can store and analyze thousands of data points at once.

Refinement A well-made model may give a close approximation to the actual number of blue whales living in a certain area or where a Category 4 hurricane may hit, but it is only an approximation. Based on the data collected, a model might suggest that there are no blue whales living in a certain part of the North Pacific or that the Category 4 hurricane may not make landfall. New data is always being acquired- there might be reports of a blue whale sighting or a radio transmitter signal of a blue whale recorded by a sonobuoy in that area or the wind may make an unexpected turn and push the hurricane inland. Then, the scientists need to evaluate why their model was not accurate. Scientific models are always a work in progress and can be updated and amended as new discoveries are made.

Scientists are constantly developing new hurricane models. By incorporating historical data into their models, scientists can predict the approach of storms earlier, giving residents plenty of time to know if they need to board up their homes or evacuate. Scientists are continuously adding data from new technology to increase the accuracy of their models as well. Ocean winds are measured from satellites from space and rom an instrument flown from planes right above the ocean surface, like the stepped-frequency microwave radiometer which monitors the microwaves naturally radiated from ocean foam. Recently, scientists have also started to provide storm surge predictions, modeling how high the ocean levels will rise based on the geography of the coastline. Predicting storm surges may lead to increased awareness of where evacuation are necessary, where to build new communities, and the importance of natural coastal buffers, such as estuaries and wetlands.

Hal Pierce, SSAI/NASA-GSFC

Chapter **1** Review

REVIEW QUESTIONS

Multiple Choice

1. All of the following are examples of remote sensing except
 a. AUVs.
 b. ROVs.
 c. satellites.
 d. scuba.

2. Factors that may affect variables are called
 a. experiments.
 b. hypotheses.
 c. theories.
 d. variables.

3. Which is true about a valid scientific hypothesis?
 a. It is the same as a scientific theory.
 b. It is a testable statement.
 c. It can be scientifically proved to be true.
 d. It is never rejected.

Short Answer

4. What role do marine laboratories play in the field of marine science?

5. What tools do marine scientists use to make observations about oceans and marine organisms?

6. Compare and contract induction and deduction.

Critical Thinking

7. Most of the major advances in marine biology have come in the last 200 years. What do you think are the reasons for this?

8. Recall that the statement "There are mermaids in the ocean" is not a valid scientific hypothesis. Can the same be said of the statement "There are *no* mermaids in the ocean?" Why?

9. Imagine you are a marine scientist and you notice that a certain type of crab tends to be larger in a local bay than in the waters outside the bay. What hypotheses might account for this difference? How would you go about testing these hypotheses?

DATA ANALYSIS LAB

How can AUVs help scientists learn more about hurricanes? Autonomous Underwater Vehicles (AUVs) can collect vast amounts of data under extreme conditions. Scientists can use AUVs to predict the path of a hurricane with great accuracy. Predicting the intensity of a hurricane has proven to be more difficult. For example, scientists overestimated the intensity of Hurricane Irene, which made landfall on August 28, 2011 in New Jersey.

Data and Observations

The graph shows water temperature readings recorded by an AUV located off the coast of New Jersey as Hurricane Irene passed over the water.

Changes in Sea Water Teperature, New Jersey Coast

Think Critically

1. Explain what happened to water temperature as the leading edge of Hurricane Irene passed over the AUV on August 28.

2. Scientists hypothesize that storm waves lead to mixing of ocean layers, cooling the sea surface temperature by up to 8°C. Think about how hurricanes are fueled. How might this change in temperature affect the intensity of the storm?

3. If this hypothesis were supported with further data, how would this help scientists predict the intensity of hurricanes?

*Data obtained from: Schofield, O., S. Glenn, and M. Moline, 2013. The robot ocean network. **American Scientist**, vol. 101, no. 6, Nov./ Dec., pp. 434–441.*

Plate Tectonics and the Structure of Ocean Basins

Chapter

2

2.1 **Structure and Formation of Earth**

2.2 **How Plate Tectonics Formed the Oceans**

2.3 **Modern Ocean Basins**

THEME **Stability and Change**

The ocean floor is always changing, even if it is at a very slow rate, due to the movement of Earth's tectonic plates.

BIG IDEA

Through exploration and data collection and analysis, scientists have been able to map with great accuracy the seafloor and its features, including continental margins, abyssal plains, mid-oceans ridges, trenches, volcanoes, and hydrothermal vents.

About the photo This type of hydrothermal vent, a "white smoker," releases extremely hot, mineral-rich water into the South Pacific.

Image courtesy of Submarine Ring of Fire 2004 Exploration, NOAA Vents Program

2.1 Structure and Formation of Earth

habitats

Big Bang

density

core

mantle

crust

basalt

granite

DIVE IN!

How far can we drill into Earth's surface? The deepest hole drilled to date is 12 km, still only within Earth's crust. In order to reach Earth's mantle, scientists are now trying to drill on the ocean floor, rather than on land. Ocean crust is thinner than continental crust, so a drill can go further. Areas off the coast of Hawai'i, California, and Mexico are being considered. Scientists' goals are to collect temperature, pressure, and mineral data to learn more about the processes that occur within Earth's interior.

Main Idea

Earth formed about 4.5 billion years ago and materials settled within the planet according to their density.

Key Questions

1. What are the three main layers of Earth?
2. What is the main difference between oceanic crust and continental crust?

The oceans are not just places where the land happens to be underwater. The sea floor is geologically distinct from the continents. It is locked in a perpetual cycle of birth and destruction that shapes the oceans and controls much of the geology and geological history of the continents. Geological processes beneath the sea affect not only the oceans, but dry land as well.

The Water Planet

Most of the processes that mold ocean basins occur slowly, over hundreds of millions of years. On this timescale, where a human lifetime is but the blink of an eye, solid rocks flow like liquid, entire continents move across the face of Earth, and mountains grow from flat plains. To understand the sea floor, we must learn to adopt the unfamiliar perspective of geological time.

At first glance geology might not seem to have much to do with marine science, but geological processes profoundly influence marine **habitats**, the natural environments where marine organisms live. Geological processes sculpt the shoreline, determine the water depth, control whether the bottom is muddy, sandy, or rocky, create new islands and undersea mountains for organisms to colonize, and determine the nature of marine habitats in countless other ways. Indeed, much of the history of life in the oceans has been determined by geological events.

Our planet is very much a water planet, unique in having large amounts of liquid water—the oceans—on its surface. The oceans not only cover 71% of the globe, but also regulate its climate and atmosphere.

The Structure of Earth

Earth and the rest of the solar system are thought to have originated about 4.5 billion years ago from a cloud or clouds of dust. This dust is thought to be the debris remaining from a great cosmic explosion called the **Big Bang**, which astrophysicists estimate occurred about 13.7 billion years ago. The dust particles collided with each other, clumping into larger particles. These larger particles collided in turn, joining into pebble-sized rocks that collided to form larger rocks, and so on. The process eventually built up Earth and other planets.

So much heat was generated as the early Earth formed that the planet was probably molten. This allowed materials to settle within the planet according to their **density**. Density is the mass of a given volume of a substance. Obviously, a pound of styrofoam weighs more than an ounce of lead, but most people think of lead as "heavier" than styrofoam. This is because lead weighs more than styrofoam if *equal volumes* of the two are compared. In other words, lead is denser than styrofoam. The density of a substance is calculated by dividing its mass by its volume. If two substances are mixed, the denser material tends to sink and the less dense to float.

While the young Earth was molten, the densest material flowed toward the center of the planet, while lighter materials floated toward the surface. The light surface material cooled into a thin crust. Eventually, the atmosphere and oceans began to form. If Earth had settled into orbit too far away from or too close to the sun, or our atmosphere had formed differently, the planet would be either so hot that all the water would have evaporated into space or so cold that all water on Earth would be perpetually frozen. Fortunately for us, both our planet's orbit and its atmosphere are such that liquid water, and therefore life as we know it, can exist.

Internal Structure The internal structure of Earth reflects the planet's early beginnings. As materials sank or floated according to their density, they formed layers like those of an onion (Fig. 2.1). The innermost layer, the **core**, is mostly iron. The pressure in the core is more than a million times that at Earth's surface, and the temperature is estimated to be over 4,000 °C. The core is made up of a solid inner core and a liquid outer core. It is thought that swirling motions of the liquid material in the iron-rich outer core produce Earth's magnetic field.

The layer outside Earth's core is the **mantle**. Though most of the mantle is thought to be solid, it is very hot—near the melting point of the rocks. Because of this, much of the mantle slowly flows like a liquid, swirling and mixing over hundreds of millions of years.

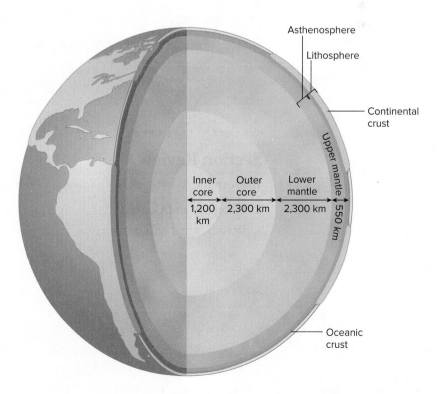

Figure 2.1 The interior of Earth is divided into the core, mantle, and crust. The core is subdivided into the inner core and the outer core. The mantle is also subdivided into upper and lower layers. The upper mantle has a largely fluid portion known as the asthenosphere. The upper portion of the mantle, along with the crust, forms the lithosphere.

Asthenosphere

Lithosphere

Continental crust

Upper mantle

Inner core	Outer core	Lower mantle	
1,200 km	2,300 km	2,300 km	550 km

Oceanic crust

The **crust** is the outermost, and therefore best-known, layer of Earth. Compared with the deeper layers, it is extremely thin, like a rigid skin floating on top of the mantle. The composition and characteristics of the crust differ greatly between the oceans and the continents.

Continental and Oceanic Crusts The geological distinction between ocean and continents results from physical and chemical differences in the rock that makes up the crust (Table 2.1). Oceanic crust, which makes up the sea floor, consists of a type of mineral called **basalt** that has a dark color. Most continental rocks are of a general type called **granite**, which has a different chemical composition than basalt and is lighter in color. Oceanic crust is denser than continental crust, though both are less dense than the underlying mantle. Ocean crust is also much thinner than continental crust, so it doesn't retain as much of the heat coming up from the mantle. The continents can be thought of as thick blocks of relatively warm, light crust floating on the mantle, much as icebergs float in water. Oceanic crust floats on the mantle, too, but because it is denser it doesn't float as high. That is why the continents lie high and dry above sea level and oceanic crust lies below sea level, covered with water.

Oceanic and continental crusts also differ in age. The oldest oceanic rocks are less than 200 million years old, quite young by geological standards. Continental rocks, on the other hand, can be as much as 3.8 billion years old.

Table 2.1 **Comparison of Continental and Oceanic Crusts**

Oceanic Crust (Basalt)	Continental Crust (Granite)
Density about 3.0 g/cm³	Density about 2.7 g/cm³
Only about 5 km thick	20 to 50 km thick
Geologically young	Can be very old
Dark in color	Light in color
Rich in iron and magnesium	Rich in sodium, potassium, calcium, and aluminum

· ·

Section Review

1. **Describe** What are the characteristics of each layer of Earth?

2. **Compare and Contrast** What are the similarities and differences between oceanic crust and continental crust?

2.2 How Plate Tectonics Formed the Oceans

Main Idea

The movement of Earth's plates creates many seafloor features including mid-ocean ridges, trenches, volcanoes, volcanic islands, and island arcs.

. .

Key Questions

1. How does seafloor spreading explain the features of mid-ocean ridges?
2. Describe key changes to Earth's surface and when they occurred.

VOCABULARY

continental drift

plate tectonics

mid-ocean ridge

faults

trenches

sediment

magnetic anomalies

rifts

sea-floor spreading

tectonic plates

subduction

shear boundary

lithogenous sediment

weathering

biogenous sediment

calcerous ooze

siliceous ooze

microfossils

carbon dating

For centuries people viewed the world as static and unchanging. Evidence of geological change is all around, however, from catastrophic earthquakes and volcanic eruptions to the slow erosion of river valleys. People eventually understood that the face of the planet does indeed change. Today scientists recognize Earth as a world of constant transformation where even the continents move.

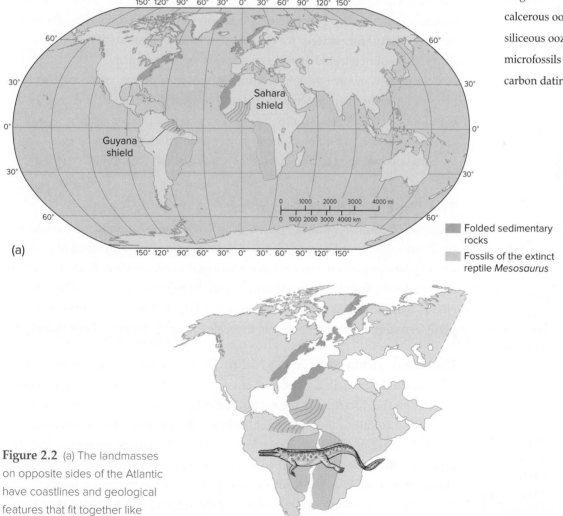

Figure 2.2 (a) The landmasses on opposite sides of the Atlantic have coastlines and geological features that fit together like pieces of a puzzle (b)..

Figure 2.3 The major features of the sea floor.

Early Evidence of Continental Drift

As early as 1620 the English philosopher, writer, and statesman Sir Francis Bacon noted that the coastlines on opposite sides of the Atlantic fit together like pieces of a puzzle (Fig. 2.2). It was later suggested that the western hemisphere might once have been joined to Europe and Africa. Coal deposits and other geological formations, for example, match up on opposite sides of the Atlantic. Fossils collected on opposing coasts are also similar.

On the basis of such evidence, Alfred Wegener, a German geophysicist, proposed the first detailed hypothesis of **continental drift** in 1912. Wegener hypothesized that all the continents had once been joined in a single "supercontinent," which he named Pangaea. He thought Pangaea began breaking up into the continents we know today about 180 million years ago.

The Theory of Plate Tectonics

Wegener's hypothesis was not widely accepted because he could not explain *how* the continents moved. Later proposals of continental drift also failed to provide a workable mechanism, but evidence continued to accumulate. In the late 1950s and the 1960s, scientists were able to put all the evidence together. They concluded that the continents *did* drift, as part of larger plates in the process of **plate tectonics**, a phenomenon that involves the entire surface of our planet.

Discovery of the Mid-Ocean Ridge In the years after World War II, sonar allowed the first detailed surveys of large areas of the sea floor. These surveys resulted in the discovery of the **mid-ocean ridge** system, a continuous chain of submarine volcanic mountains that encircles the globe like the seams on a baseball (Fig. 2.3). The mid-ocean ridge system is the largest geological feature on Earth. At regular intervals the mid-ocean ridge is displaced to one side or the other by cracks, or **faults**, in Earth's crust known as transform faults. Occasionally the submarine mountains of the ridge rise so high that they break the surface to form islands, such as Iceland and the Azores (Fig. 2.4).

The mid-ocean ridge in the Atlantic, called the Mid-Atlantic Ridge, runs right down the center of the Atlantic Ocean, parallel to the opposing coastlines. The ridge forms an inverted Y in the Indian Ocean and runs up the eastern side of the Pacific (Fig. 2.3). The main section of ridge in the Eastern Pacific is called the East Pacific Rise.

Surveys of the sea floor also revealed the existence of a system of deep depressions in the sea floor called **trenches** (Fig. 2.3). Trenches are especially common in the Pacific.

Significance of the Mid-Ocean Ridge When the mid-ocean ridge system and trenches were discovered, geologists realized that geological activity is concentrated around them. Earthquakes are associated with ridges, and volcanoes with trenches (Fig. 2.5).

The nature of the sea floor is also related to the mid-ocean ridge. Sea-floor rock at the ridge is very young, and gets progressively older moving away from the ridge. Right at the ridge crest there is little, if any, **sediment,** loose material like sand and mud that settles to the bottom, but sediment gets thicker and thicker at greater distance from the ridge. The deepest sediment, that resting directly on the sea-floor rock, gets older away from the ridge (Fig 2.6).

One of the most important findings came from studying the magnetism of rocks on the sea floor. Earth's magnetic field periodically reverses direction. When a magnetic compass points north as it does today, the magnetic field is arbitrarily called "normal." During a "reversal," the field is opposite that of today and the compass would point south. In the recent past, geologically speaking, reversals have occurred about every 300,000 years, though the interval has sometimes been much longer or shorter, and there have been times in Earth's history when no reversals occurred over tens of millions of years. Reversals are thought to be caused by changes in the motion of material in the iron-rich outer core.

Figure 2.4 A portion of the Mid-Atlantic Ridge above the sea surface in Iceland.

Figure 2.5 The world distribution of earthquakes and volcanoes. Compare these with the locations of mid-ocean ridges and trenches shown in Figure 2.3.

- Shallow-focus earthquakes
- ▲ Volcanoes
- • Deep-focus earthquakes

©Oddur Sigurðsson

Life Below the Sea Floor

You might expect that the deepest-living organisms to inhabit the depths of the ocean's trenches. In fact, life extends deeper—much deeper—than that. Biologists have discovered that microbes live not only on, but *in* the sea floor, inhabiting deep layers of sediment and tiny fissures in the rocks of the crust. Rich communities of microbes have been found more than 3 km below the sea floor and probably extend to at least 10 km. The microbes grow through chemosynthesis, using chemical energy bound up in the rocks. Similar organisms occur in large numbers on newly formed crust that has just cooled from magma erupting at mid-ocean ridges.

The deep biosphere To study this deep biosphere, scientists use deep-ocean drilling ships to drill deep into Earth's crust. The drilling, of course, disturbs the microbial communities, but using remotely operated vehicles (ROVs), the scientists install sea-floor observatories over the drill hole that can last for up to ten years. The observatories allow the microbial communities to return to normal, and are then used to make measurements, collect samples, and perform experiments. This is a relatively new field of research because the tools required, such as the ROVs, sea-floor observatories, and genetic biotechnologies, have only recently become available.

Until recently microbes were the only life known from deep below the sea floor. All of those microbes represent a food source, so perhaps it is not surprising that worms were recently discovered 3.6 km deep in a fissure. As you might expect, the worms appear to feed on deep-Earth bacteria.

The importance of microbes Microbes on and below the sea floor not only change life on the sea floor, they also alter the crust itself. Microbes use minerals from the crust, breaking down some and forming new ones. Geologists think that about two-thirds of the 4,300 minerals in Earth's crust would never have formed if not for marine microbes. While this is remarkable, photosynthesis by microbes has had an even greater impact than chemosynthesis. Photosynthesis produces the oxygen in Earth's atmosphere, and for billions of years microbes dominated oxygen production on Earth. Oxygen has been critical to the formation of the minerals in Earth's crust. Most contain oxygen and are believed not to exist elsewhere in the solar system where oxygen is not available. Granite, which is the dominant mineral in continental crust, is thought to be the result of the oxygen and energy produced by marine microbes and it is unique to Earth. The production of granite led to the formation and stabilization of the continents. Thus, we may be indebted to primordial marine microbes for the very rocks under our feet.

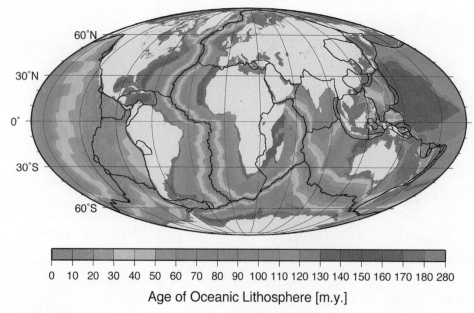

Figure 2.6 Examination of the age of sea-floor rocks reveals stripes that parallel the mid-ocean ridges (shown as black lines). The youngest crust (red) is found nearest the ridge, and the oldest (blue) the farthest away.

Age of Oceanic Lithosphere [m.y.]

Many rocks contain tiny magnetic particles. When the rock is molten these particles are free to move, and they act like tiny compasses, pointing north or south depending on whether Earth's field is normal or reversed (Fig. 2.7). When the rocks cool, the particles are frozen in place and keep their orientation even when the magnetic field changes. It is therefore possible to tell the orientation of Earth's magnetic field at the time the rocks cooled.

Sea-floor rocks have patterns of magnetic bands, or "stripes," that run parallel to the mid-ocean ridge. The bands represent zones in which the rocks alternate between normal and reversed magnetization. The bands are symmetric around the ridge, so that the pattern on one side of the ridge is a mirror image of the pattern on the other side. These bands, or **magnetic anomalies** (see Fig. 2.7), could only have formed if the normally magnetized rock cooled from molten material at different times than the reverse-magnetized bands. The sea floor, then, was not formed all at once but in strips that parallel the mid-ocean ridge.

Formation of the Sea Floor The discovery of magnetic anomalies, together with the other evidence, finally led to an understanding of plate tectonics. The jump from the various observations of the sea floor and mid-ocean ridges to the theory of plate tectonics is a good example of the use of induction in science.

Huge slabs of oceanic crust are separating at the mid-ocean ridges, creating cracks in the crust called **rifts**. When a rift occurs it releases some of the pressure on the underlying mantle, like removing the cap from a bottle of soft drink. The reduced pressure allows hot mantle material to melt and rise up through the rift. The ascending mantle material, or magma, pushes up the oceanic crust around the rift to form the mid-ocean ridge (see Fig. 2.7). When

IN CONTEXT

Induction: The development of a generalized conclusion from a series of individual observations.
• 1.2, The Scientific Method

KEY:
■ Normal
░ Reversed

Mid-ocean ridge

Time

Time

Magma

Figure 2.7 Cross section of the sea floor at a mid-ocean ridge showing the mechanism of sea-floor spreading. As the sea floor moves away from the rift, molten material rises from the mantle and cools to form new sea floor. When the rocks cool, they "freeze" whatever magnetic orientation, normal or reversed, is present at the time. The entire floor of the ocean was created at the mid-ocean ridge in this manner.

this molten material reaches Earth's surface, it cools and solidifies to form new oceanic crust. The process repeats itself as the sea floor continues to move away from the mid-ocean ridge. The entire process by which the sea floor moves away from the mid-ocean ridges to create new sea floor is called **sea-floor spreading** (see Fig. 2.7), and the ridges are also called spreading centers.

Sea-floor spreading explains many features of the mid-ocean ridge. Right at the ridge crest the crust is new and has not had time to accumulate sediment. As the crust moves away from the ridge it ages and sediment builds up. This explains why the sediment gets thicker and the rocks get older moving away from the ridge. Sea-floor spreading also explains the pattern of magnetic stripes. As new sea floor is created, it "freezes" the magnetic field prevailing at the time and preserves that magnetization as it moves away from the ridge. Eventually Earth's magnetic field reverses, starting a new stripe.

Sea-Floor Spreading and Plate Tectonics

Sea-floor spreading is only part of the story of plate tectonics. Earth's surface is covered by a fairly rigid layer composed of the crust and the uppermost part of the mantle. This layer, averaging about 100 km thick, is called the lithosphere, meaning "rock sphere." The lithosphere is broken into a number of plates called lithospheric plates or **tectonic plates** (Fig. 2.8). A plate can contain continental crust, oceanic crust, or both. The lithosphere floats on a denser, more plastic layer of the upper mantle called the asthenosphere. While the crust, mantle, and core are distinguished by their chemical composition, the distinction between the lithosphere and the asthenosphere is based on how easily the rock flows.

Figure 2.8 The division of Earth's surface into lithospheric plates. Some areas are not fully understood, and there are some small plates not shown here.

Plate Separations Mid-ocean ridges form the edges of many plates. Here, the lithospheric plates move apart and new sea floor—that is, new oceanic lithosphere—is created. If the plate includes a block of continental crust, the continent is carried along with the plate as it moves away from the ridge.

This is the mechanism of continental drift. The plates spread at around 2 to 18 cm per year. For comparison, human fingernails grow at about 6 cm per year. The spreading rate varies from place to place and over time.

As new lithosphere is created, old lithosphere is destroyed somewhere else. Otherwise, Earth would have to expand constantly to make room for the new lithosphere. Lithosphere is destroyed at trenches, another important type of boundary between plates. A trench is formed when two plates collide and one of the plates dips below the other and sinks back down into the mantle (Figs. 2.9 and 2.10). This downward movement of the plate into the mantle is called **subduction**, and trenches are also called subduction zones.

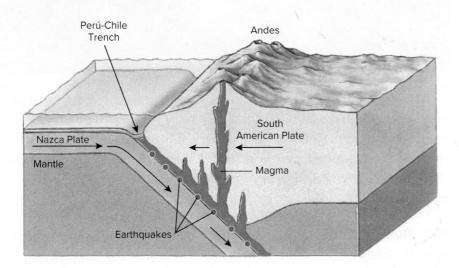

Figure 2.9 Some trenches are formed by the collision of continental and oceanic plates. In this example the Nazca Plate dips beneath the South American Plate. The lighter materials from the sinking plate rise back to the surface as they melt, creating the volcanic Andes.

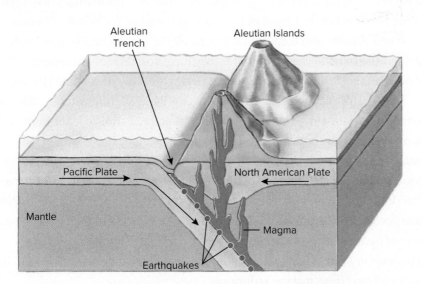

Figure 2.10 Trenches can also be caused by the collision of two oceanic plates. The oceanic portion of the North American Plate has collided with the Pacific Plate. In this case, the Pacific Plate dipped below the North American Plate, but either of the two plates might have done so. The volcanoes associated with the trench have produced the Aleutian Island arc.

Figure 2.11 Mount
Veniaminof, an active volcano
on the Alaska Peninsula. The
peninsula is geologically part
of the Aleutian Island chain
that has formed behind the
Aleutian Trench (see Fig. 2.10).

As the plate sinks, it weakens under the heat and pressure of the mantle and begins to break up, causing earthquakes. Eventually the plate gets so hot it melts. Some of the molten material rises back to the surface to form volcanoes. The rest continues to sink into the mantle. Some of this material may eventually be recycled, rising again at another mid-ocean ridge several hundred million years later.

Plate Collisions The collision that produces a trench can be either between an oceanic plate and a continent or between two oceanic plates. When an oceanic plate collides with a continent, it is always the oceanic plate that descends into the mantle. This is because the continental block is less dense than the oceanic plate and floats on top. That is why very old rocks are found only on the continents: Because oceanic crust is eventually destroyed at trenches, it never gets old by geological standards. Continental crust, on the other hand, is not destroyed in trenches and can last for billions of years.

When oceanic and continental plates collide, continental volcanoes develop behind the trench. These volcanoes may form coastal mountain ranges. The Andes on the Pacific coast of South America are a good example (see Fig. 2.9).

The oceanic plate lifts the edge of the continent slightly when it slides down under the continent. As the slab of cold lithosphere sinks deeper in the mantle, however, it tends to drag the interior of the continent down with it. The central part of the continent subsides, sometimes so much that seawater floods in to form shallow seas. After tens of millions of years the continent breaks free of the sinking plate and floats back up to its normal elevation, and the shallow sea disappears. This is thought to have occurred in North America when a huge slab of oceanic lithosphere called the Farallon Plate slid under the west coast. That explains why Denver, Colorado, the "mile-high city," is surrounded by rocks formed from marine sediments.

When two oceanic plates collide, one of the plates dips beneath the other to form the trench. Again the trench is associated with earthquakes and volcanoes (Fig. 2.11). The volcanoes may rise from the sea floor to create chains of volcanic islands. As viewed on a map, trenches are curved because of Earth's

40 Chapter 2 *Plate Tectonics and the Structure of Ocean Basins*

M.E. Yount/USGS

spherical shape. The volcanic island chains associated with the trenches follow the trenches' curvature and are called island arcs. Examples include the Aleutian and Mariana islands (see Fig. 2.10).

Occasionally two continental plates collide. Because of the relatively low density of continental crust, both plates float and neither is subducted. Therefore, no trench is formed. Instead, the two continental blocks push against each other with such tremendous force that the continents become "welded" together. The force is eventually too much for the rocks, which buckle and fold like an accordion. The huge folds form mountain ranges. The Himalayas, for example, were formed when India collided with the rest of Asia.

Shear Boundaries There is a third type of plate boundary in addition to trenches and mid-ocean ridges. Sometimes two plates move in such a way that they slide past each other, neither creating nor destroying lithosphere. This type of plate boundary is called a **shear boundary** (Fig. 2.12). In the fault where the two plates move past each other there is immense friction between the plates. This friction prevents the plates from sliding smoothly. Instead they lock,

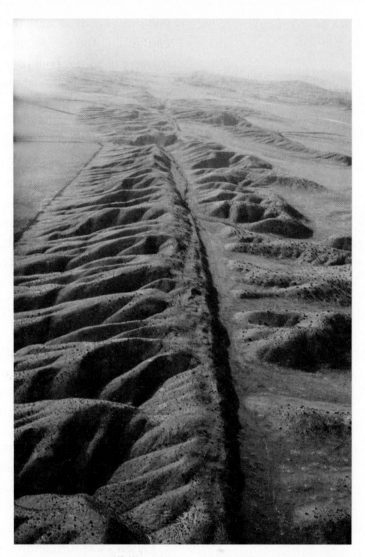

Stocktrek/Stockbyte/Getty Images

Figure 2.12 The San Andreas fault in central California is a shear boundary.

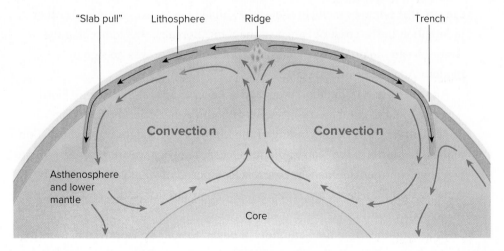

Figure 2.13 The movement of lithospheric plates was long thought to be driven by large-scale convection currents in the asthenosphere and lower mantle. Today the prevailing hypothesis is that the plates move mainly because of slab pull, in which old, cold, and dense lithosphere sinks into the mantle and pulls the rest of the plate behind it.

and stress builds up until the plates break free and slip all at once, causing an earthquake. The San Andreas Fault in California is the largest and most famous example of a shear boundary (see Fig. 2.12).

Geologists once thought the most likely explanation for what makes the plates move was convection, in which heat from Earth's core causes the mantle to swirl like thick soup heating in a saucepan. It was thought that the convection currents carry the overlying plates along like icebergs in a current (Fig. 2.13, p. 41). Since the 1990s, however, more and more geologists have come to the view that mantle convection is probably not the main cause of plate motion. The most widely accepted hypothesis today is that oceanic lithosphere cools as it ages, becoming denser. Eventually it sinks into the mantle, forming a trench and pulling the rest of the plate along behind it. This "slab pull" causes the plates to separate at the mid-ocean ridge, allowing fresh magma to well up from the mantle. Convection may still have a role in plate motion, and the upwelling mantle material may also help push the plates apart.

The Geological History of Earth

Earth's surface has undergone dramatic alterations. The continents have been carried long distances by the moving sea floor, the ocean basins have changed in size and shape, and new oceans have been born.

Continental Drift and the Changing Oceans About 200 million years ago all the continents were joined in the supercontinent Pangaea, just as Wegener proposed. Antarctica was in approximately the same place it is today, but all the other continents were in different positions (Fig. 2.14a). India was attached to Antarctica and Africa rather than to Eurasia as it is now.

Pangaea was surrounded by a single, vast ocean called Panthalassa. Panthalassa, which covered all the rest of the planet, was the ancestor of the modern Pacific Ocean. A relatively shallow sea, the Tethys Sea, separated Eurasia from Africa. The Tethys Sea, the precursor of the present-day Mediterranean, was home to many of the world's shallow-water organisms. Another indentation in the coast of Pangaea, the Sinus Borealis, was to become the Arctic Ocean. Before Pangaea began to break up, there was no sign of the modern Atlantic or Indian oceans.

Approximately 180 million years ago, a new rift appeared between North America and the combined continents of South America and Africa (Fig. 2.14b). This rift was the beginning of the Mid-Atlantic Ridge, and its formation marked the birth of the North Atlantic Ocean. Pangaea was now separated into two large continents. One was Laurasia, composed of what is now North America and Eurasia. South America, Africa, Antarctica, India, and Australia made up the southern continent of Gondwana.

At around the same time another rift began to split up Gondwana, marking the beginning of the Indian Ocean. South America and Africa began to move to the northeast, and India, separated from the other continents, began to move north.

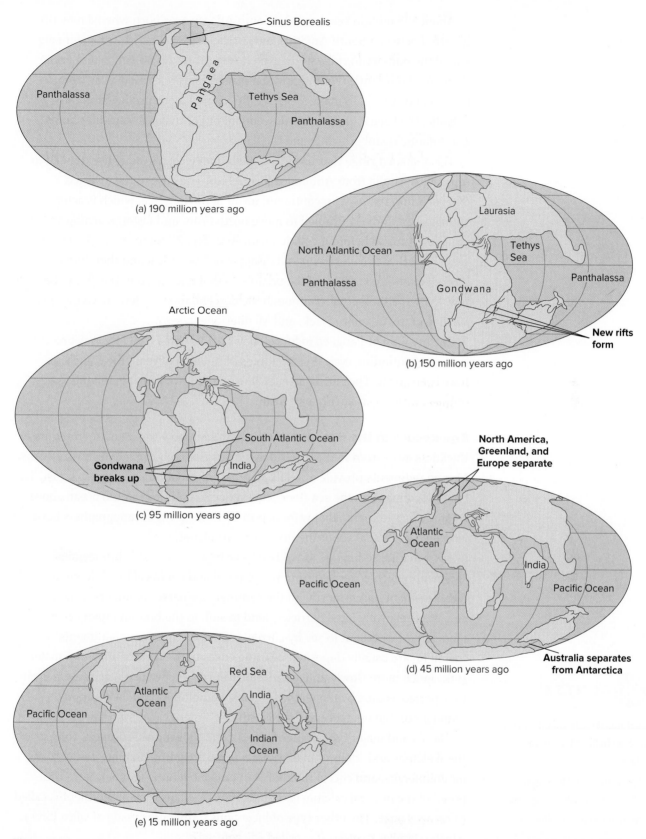

Figure 2.14 (a) The supercontinent Pangaea as it probably looked 190 million years ago. (b) The world 150 million years ago, some 30 million years after Pangaea began to break up. The northern Atlantic Ocean was born when one rift separated Laurasia from Gondwana. A second rift that appeared between Africa and Antarctica formed the young Indian Ocean. (c) The world 95 million years ago after the southern Atlantic Ocean was born as a new rift between South America and Africa. (d) The world 45 million years ago, 15 million years after the rift in the Indian Ocean extended between Antarctica and Australia, separating the two continents. (e) The world about 15 million years ago.

About 135 million years ago the South Atlantic was born when a new rift occurred between South America and Africa (Fig. 2.14c). This rift eventually joined the mid-ocean ridge in the North Atlantic to form a single mid-ocean ridge. As the Atlantic Ocean grew, the Americas were carried away from Eurasia and Africa. To make room for the new sea floor produced in the Atlantic, the Pacific Ocean—the descendant of Panthalassa—steadily shrank. The Atlantic is still growing, and the Pacific shrinking.

The Y-shaped ridge that produced the Indian Ocean gradually extended to separate Australia from Antarctica (Fig. 2.14d). The base of the inverted Y extended into the African continent, forming the Red Sea, which is actually a young ocean. India continued to move north faster and faster, reaching speeds of up to 20 cm/yr before it collided with Asia (Fig. 2.14e) to create the Himalayas. The Tethys Sea closed up and disappeared as Africa, and then India, moved up into Asia. Most of the organisms that once lived in the Tethys Sea are now known only as fossils, although the descendants of a few surviving species still live in the Caspian, Black, and Mediterranean seas.

The breakup of Pangaea to form today's continents is only the most recent turn of a continuing cycle. For hundreds of millions of years the continents have been adrift. They alternately collide and re-form into larger landmasses, only to break up and drift apart again.

The Record in the Sediments We have seen how the increase in sediment thickness away from mid-ocean ridges gave a clue to how plate tectonics works. Marine sediments provide a wealth of other information about Earth's past. The type of sediment on the sea floor often reflects conditions in the ocean above. By studying sediments that were deposited in the past, oceanographers have learned a great deal about the history of our planet.

Most marine sediments are of two basic types. The first is **lithogenous sediment**, which is derived from the physical and chemical breakdown, or **weathering**, of rocks, mostly on the continents. Coarse sediments, which consist of relatively large particles, tend to sink to the bottom rapidly rather than being carried out to sea by currents. Coarse lithogenous sediments, therefore, are usually deposited near the edges of the continents. Finer material sinks much more slowly and is carried far out to sea by currents. Some is even transported as dust by the wind. The most common kind of lithogenous sediment on the open ocean floor is a fine sediment called red clay.

The second major type of marine sediment, **biogenous sediment**, consists of the skeletons and shells of marine organisms, including diatoms, radiolarians, foraminiferans, and coccolithophorids. Some biogenous sediments are composed of the mineral calcium carbonate ($CaCO_3$). This type of sediment is called **calcareous ooze**. The other type of biogenous sediment is made of silica (SiO_2), which is similar to glass. It is called **siliceous ooze**.

Though large fossils like whale bones and shark teeth can be found, most of the organisms that produce biogenous sediments are microscopic or nearly so. Such sediment particles are sometimes called **microfossils**, since each particle represents the preserved remains of a dead organism (Fig. 2.15). Microfossils

IN CONTEXT

Diatoms Single-celled algae with a shell, or test, made of silica.

Coccolithophorids Single-celled algae covered with plates made of calcium carbonate.

Foraminiferans (Forams) Protozoans, often microscopic, with a calcium carbonate shell.

Radiolarians Single-celled protozoans with a test made of silica.

tell scientists what organisms lived in the ocean in the past. Because some of these organisms are known to prefer cold or warm water, microfossils also give clues to ancient ocean temperatures. Ocean temperatures are determined by Earth's climate and ocean currents.

The climate of Earth in the past can also be determined by the chemical composition of the microfossils. Various methods including **carbon dating**, a procedure in which the ratios of different atomic forms, or isotopes, of carbon are measured, can be used to determine the age of microfossils. It is also possible to tell the temperature of the water in which the organisms lived by measuring the ratios of magnesium (Mg) to calcium (Ca) or of different isotopes of oxygen in the microfossils. Thus, microfossils have preserved a detailed record of Earth's past climate. Though the record in the sediments is not always easy to read, it is supplemented by other information. The ratio of the elements strontium (Sr) and calcium in ancient coral skeletons, for example, also records past ocean temperatures. Ice cores from polar areas like Greenland and Antarctica also preserve a record of past temperatures, as well as samples of the ancient atmosphere in the form of tiny bubbles trapped in the ice. These and other studies are providing an increasingly detailed picture of Earth's past climate.

Climate and Changes in Sea Level The climate of Earth has fluctuated rhythmically through much of its history, alternating between warm, or interglacial, periods and cold periods, or ice ages (Fig. 2.16). Earth is presently in an interglacial period. During ice ages, great glaciers build up on the continents.

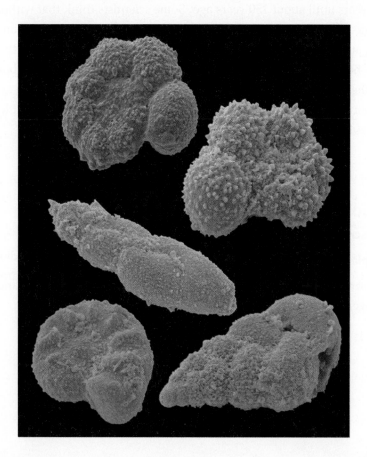

Figure 2.15 The fossilized tests of several species of foraminiferans.

Andrew Syred/Science Source

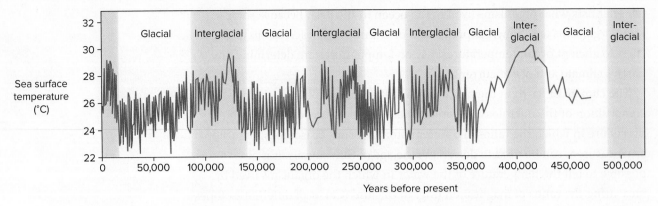

Figure 2.16 A history of Earth's climate over the past half-million years as determined from foraminiferan microfossils. The red line shows the average sea surface temperature in the equatorial Pacific Ocean as determined from the magnesium-to-calcium ratio in the microfossils. The white and blue bands indicate major glacial and interglacial periods, respectively, recorded in oxygen isotope ratios. Note that the shifts between ice ages and interglacial periods involve relatively small changes in average temperature.

Because large amounts of water are trapped as ice instead of flowing to the sea in rivers, there is less water in the ocean. Thus, sea level falls during ice ages.

The Pleistocene Epoch, which began 2.6 million years ago, was the last major period of glaciation. During the Pleistocene a series of ice ages was interspersed by brief warm periods of melting. The peak of the last ice age occurred about 18,000 years ago. At that time, vast ice sheets as thick as 3 km covered much of North America and Europe. Sea level was about 130 m lower than it is today.

Sea level continues to rise, though the rate of melting slowed during the past 3,000 years until about 150 years ago. Some scientists think that without the influence of humans Earth would be entering another ice age. Human impact on the atmosphere, however, has intensified the greenhouse effect, the retention of heat in the lower atmosphere as a result of an increase in carbon dioxide and other gases. Global temperatures and the rate of glacial melting are now increasing rapidly, and sea level is projected to continue to rise for at least the next century.

· ·

Section Review

1. Describe How does seafloor spreading occur?

2. Explain Why are subduction zones important?

3. Describe What is the difference between lithogenous sediment and biogenous sediment?

2.3 Modern Ocean Basins

Main Idea

The sea floor is divided into two main regions—the continental margin and the deep-sea floor.

· ·

Key Questions

1. What are the three main regions of continental margins?

2. What are the main features of the deep-sea floor?

VOCABULARY

world ocean

continental margins

continental shelf

continental slope

continental rise

submarine canyons

shelf break

active margin

passive margin

abyssal plain

abyssal hills

seamounts

guyots

central rift valley

hydrothermal vents

sulfides

black smokers

The structure of the ocean floor is dominated by the workings of plate tectonics. Because this is a global process, the major features of the sea floor are quite similar from place to place around the world.

The Geography of the Ocean Basins

The oceans are not distributed equally with respect to the Equator. About two-thirds of Earth's land area is found in the northern hemisphere, which is only 61% ocean. About 80% of the southern hemisphere is ocean.

The oceans are traditionally classified into four large basins (Fig. 2.17). The Pacific is the deepest and largest ocean, almost as large as all the others combined (Table 2.2). The Atlantic Ocean is a little larger than the Indian Ocean, but the two are similar in average depth. The Arctic is the smallest and shallowest ocean. A number of shallow seas, such as the Mediterranean Sea, Gulf of Mexico, and South China Sea, are connected or marginal to the main ocean basins.

Though we usually treat the oceans as four separate entities, they are actually interconnected. This connection is most obvious when the world is viewed from the South Pole (Fig. 2.18). From this view it is clear that the Pacific, Atlantic, and Indian oceans are large branches of one vast system. The connections among the major basins allow seawater, materials, and some organisms to move from one ocean to another. Because the oceans are actually one great interconnected system, oceanographers often speak of a single **world ocean**. They also refer to the continuous body of water that surrounds Antarctica as the Southern Ocean.

Continental Margins The sea floor is divided into two main regions: the continental margins, which represent the submerged edges of the continents, and the deep-sea floor itself.

The **continental margins** are the boundaries between continental and oceanic crust. Most sediment from the continents settles to the bottom soon after entering the sea and accumulates on the continental margins, where sediment deposits may be up to 10 km thick. Continental margins generally consist of a shallow, gently sloping **continental shelf**, a steeper **continental slope**, and another gently sloping region, the **continental rise**, at the base of the continental slope (Fig. 2.19).

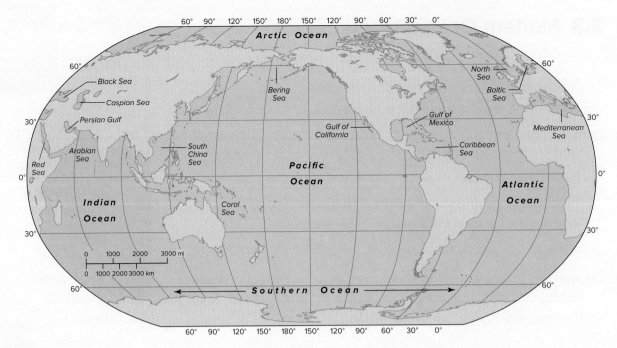

Figure 2.17 The major ocean basins and some of the marginal seas.

The continental shelf The shallowest part of the continental margin is the continental shelf. Though they make up only about 8% of the ocean's surface area, continental shelves are the biologically richest part of the ocean, with the most life and the best fishing. The shelf is composed of continental crust and is really just part of the continent that presently happens to be under water. During past times of low sea level, in fact, most of the continental shelves were exposed. At these times, rivers and glaciers flowed across the continental shelves and eroded deep canyons. When sea level rose, these canyons were submerged and gave rise to much larger **submarine canyons**.

Table 2.2 **Average Depths and Total Areas of the Four Major Ocean Basins**

Ocean	AREA		AVERAGE DEPTH		
	Millions of km²	Millions of mi²	Meters	Feet	Deepest Place
Pacific	166.2	64.2	4,188	13,741	Marianas Trench, 10,994 m
Atlantic	86.5	33.4	3,736	12,258	Puerto Rico Trench, 8,605 m
Indian	73.4	28.3	3,872	12,704	Java Trench, 7,725 m
Arctic	9.5	3.7	1,330	4,364	Molloy Deep, 5,608 m

The continental shelf extends outward at a gentle slope that in most places is too gradual to see with the naked eye. The shelf varies in width from less than 1 km on the Pacific coast of South America to more than 750 km on the Arctic coast of Siberia. The continental shelf ends at the **shelf break**, where the slope

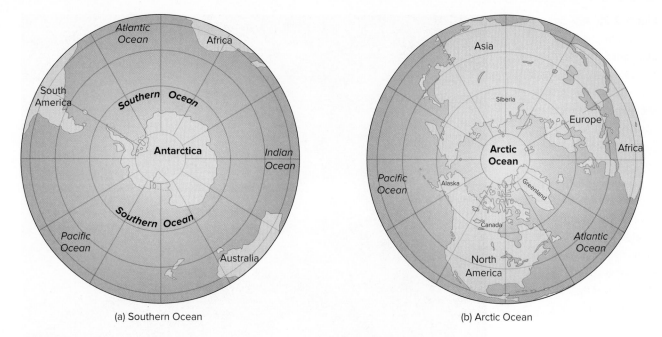

(a) Southern Ocean (b) Arctic Ocean

Figure 2.18 (*a*) South polar view of the world. The major ocean basins can be seen as extensions of one interconnected world ocean. The ocean that surrounds Antarctica is often called the Southern Ocean. (*b*) North polar view of the world showing the partially enclosed Arctic Ocean.

abruptly gets steeper (Fig. 2.20). The shelf break usually occurs at depths of 120 to 200 m but can be as deep as 400 m.

The precise edge is difficult to define, but is of more than academic interest because, under international law, countries have the right to control the resources on their adjacent continental shelves. Recently there has been intense interest in detailed mapping of the continental shelves in the Arctic Ocean, because shrinking Arctic sea ice, a result of global warming, is set to give countries, including Russia, the United States, and Canada, unprecedented access to oil and other seabed resources—if they can claim them.

The continental slope The continental slope is the closest thing to the exact edge of the continent. It begins at the shelf break and descends down to the deep-sea floor. Submarine canyons beginning on the continental shelf cut across the continental slope to its base at a depth of 3,000 to 5,000 m (Fig. 2.21). These canyons channel sediments from the continental shelf to the deep sea.

The continental rise Sediment moving down a submarine canyon accumulates at the canyon's base in a deposit called a deep-sea fan, similar to a river delta. Adjacent deep-sea

Figure 2.19 An idealized continental margin consists of a continental shelf, continental slope, and a continental rise. Seaward of the continental rise lies the deep-sea floor, or abyssal plain. These basic features vary from place to place.

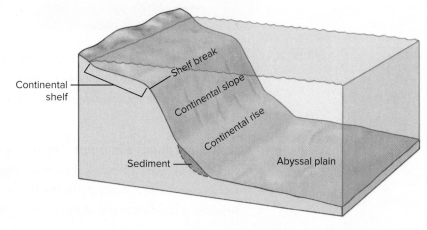

Figure 2.20 A section of the continental shelf (upper right) off Atlantic City, New Jersey, approximately 30 km wide. The white arrows indicate the shelf break. The red arrow shows the head of a submarine canyon called Tom's Canyon. The linear marks on the shelf are iceberg scours made during the last ice age. This image was created with a sophisticated form of sonar known as multibeam sonar. The steepness of the continental slope is exaggerated.

Figure 2.21 Multibeam sonar image of the continental margin off California. Numerous submarine canyons cut across the continental shelf and down the continental slope to the deep-sea floor. The largest is Monterey Canyon, which extends nearly to the shore at Monterey Bay (arrow).

(t)©Center for Coastal and Ocean Mapping, University of New Hampshire, courtesy of Larry Mayer. (b)Reprinted by permission of Dr. William Haxby, Columbia University, The Earth Institute

fans may merge to form the continental rise. The rise consists of a thick layer of sediment piled up on the sea floor. Sediment may also be carried along the base of the slope by currents, extending the continental rise away from the deep-sea fans.

Active and passive margins The nature of the continental margin, and therefore of coastal biological habitats, depends to a large extent on the plate tectonic processes occurring in the region. The continent of South America provides a good example of the relationship between the continental margin

Andes

Pacific
Ocean

South
American
Plate

Atlantic
Ocean

Nazca Plate

Active margin

Pacific
Ocean

Little or
no shelf

Steep
slope (∼6°)

Trench, no rise

Passive margin

Wide shelf

Gentle slope
(∼2°)

Atlantic
Ocean

Continental rise

Figure 2.22 The opposite sides of South America have very different continental margins. The leading edge, or west coast, is colliding with the Nazca Plate. It has a narrow shelf and steep slope and has a trench rather than a continental rise. The trailing edge, the Atlantic coast, has a wide shelf, a relatively gentle continental slope, and a well-developed continental rise. The steepness of all the slopes is exaggerated for illustrative purposes.

and plate tectonics (Fig. 2.22). The South American Plate (see Fig. 2.8) consists of both the continent itself and the part of the Atlantic sea floor created by the Mid-Atlantic Ridge. South America is carried westward with the plate as new sea floor is created in the Atlantic. The west coast of South America is colliding with the Nazca Plate, leading to the formation of a trench (see Figs. 2.3 and 2.9). Trenches are zones of intense geological activity, including earthquakes and volcanoes, so this type of continental margin is called an **active margin**. The west coast of North America also has a type of active margin, but it is more complex than that of South America.

As the colliding plate descends into the trench, some of the sediment gets scraped off, folded, and "plastered" onto the continental margin. The edge of the continent is lifted by the oceanic plate passing below, and the coast is built up by volcanoes. These processes give active margins steep, rocky shorelines, narrow continental shelves, and steep continental slopes. Because the sediments at the base of the continental slope are either carried down into the trench or scraped onto the continent, active margins usually lack a well-developed continental rise.

South America's east coast, on the other hand, is not a boundary between plates and is therefore relatively inactive geologically. The continental margin here can be thought of as the trailing edge left when South America separated from Africa. This type of margin is called a **passive margin** (Fig. 2.23).

Figure 2.23 Aerial view of the coast along southern Cape Cod, Massachusetts, showing the ocean, barrier islands, sand dunes, and a lagoon with underwater sandbars.

©Steve Dunwell/Photographer's Choice/Getty Images

2.3 *Modern Ocean Basins* **51**

The Hawaiian Islands, Hot Spots, and the Great Mantle Plume Debate

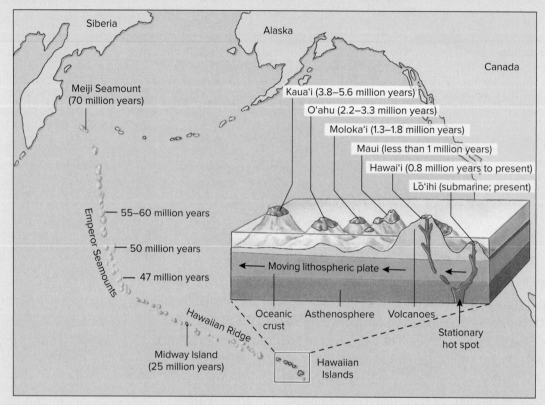

The Emperor Seamounts and Hawaiian Island chain. Some geologists think that the chain was formed not by a hot spot as shown here but by a crack in the lithosphere.

The Hawaiian Islands are part of a chain of volcanoes called the Hawaiian Ridge, which continues in a string of seamounts called the Emperor Seamount chain. The oldest and northernmost island in the chain, the Meiji Seamount, is 70 million years old. The islands get progressively younger as you travel south east along the island chain, all the way to Loʻihi, which lies submerged off of the island of Hawaii. As volcanic

islands age they erode, eventually forming steep cliffs and soil suitable for vegetation. Each of the islands in the Hawaiian Ridge has distinct geological and vegetative features.

Island formation This pattern is widely attributed to a hot spot, where a plume of magma rises from deep in the mantle to erupt in volcanic activity. As the Pacific Plate moves over the stationary hot spot, new eruptions of magma break out producing the line of volcanoes. The bend between the Hawaiian Ridge and the Emperor Seamounts is thought to have occurred when the Pacific Plate changed direction.

Passive margins typically have flat coastal plains, wide shelves, and relatively gradual continental slopes. Because there are no tectonic processes to remove it, sediment accumulates at the base of the continental slope. Passive margins therefore usually have a thick continental rise.

Deep-Ocean Basins Most of the deep-sea floor lies at a depth of 3,000 to 5,000 m, averaging about 4,000 m. The deep-sea floor, or **abyssal plain**, rises at a very gentle slope of less than 1 degree toward the mid-ocean ridge. Though relatively flat, it often has submarine channels, low **abyssal hills**, plateaus, rises, and other features. The abyssal plain is dotted with volcanic islands and

The ancient Hawaiians had a remarkably similar explanation for the formation of their islands. They recognized that the islands were formed by volcanoes, and that the eroded, forested Kaua'i was the oldest of the islands, and the still-erupting Hawai'i the youngest. Traditions explain that Pele, the Goddess of Fire, originally lived on Kaua'i but was repeatedly attacked by the Goddess of the Sea and progressively moved to younger islands until reaching her present home, the Kilauea Volcano on Hawai'i.

Pele's house: volcanic eruption at Kilauea, Hawai'i.

Hot spots and superplumes Geologists have identified around 50 other hot spots across the world, most under oceanic plates. There are also a few hot spots beneath continents. The most famous is responsible for the geysers and bubbling pools of Yellowstone. Geologists hypothesize that even larger masses of hot magma called superplumes rise from deep in the mantle, lifting up great areas of lithosphere. The plateau region in southern Africa is thought to have been formed by a superplume.

Mantle plumes and hot spots are accepted by most geologists, but there is some dissent. Some argue that the pressure deep in the mantle is too great to allow mantle plumes to rise, and claim there is not enough evidence of the unusual heat deep in the mantle that should underlie hot spots. If the bends in the various Pacific island chains all formed when the Pacific Plate changed direction as claimed by the hotspot hypothesis, they should all be the same age. However, the island chains differ in age by up to 10 million years. This and other evidence indicates that hot spots may move around, contrary to the long-held hypothesis that they remain stationary under the moving plates.

Some geologists claim that the Hawaiian and similar island chains were formed by cracks in the lithosphere that allow hot magma to rise and gradually extend, like a crack in a windshield. Recent seismic experiments using sensors both on the Hawaiian Islands and on the surrounding deep-sea floor, however, provide strong support for the mantle plume hypothesis.

Island endemics The formation of the Hawaiian Islands by a hot spot explains an unusual feature of the Hawaiian Islands. About a quarter of the shallow-water fish species and one-fifth of shallow-water molluscs there are endemic, meaning they are found nowhere else. This is one of the highest rates of marine endemism anywhere, but even the oldest island, Kaua'i, at only 5 million years old, is far too young for so many unique species to have evolved there. This shows that these species probably originated on much older islands in the chain, which is evidence that the Hawaiian Seamount was formed by a hotspot.

submarine volcanoes called **seamounts**. Distinctive flattopped seamounts called **guyots** (pronounced "*gee*-ohs") are common in parts of the Pacific. Guyots and many other seamounts were once islands, but are now several hundred meters beneath the sea surface because the lithosphere has sunk into the mantle under the weight of the island, and also because sea level has risen. The abyssal plain and seamounts are home to a tremendous variety of marine life.

At trenches, where the plate descends into the mantle, the sea floor slopes steeply downward. Trenches are the deepest parts of the world ocean. The Mariana Trench in the western Pacific is the deepest place of all, at 10,994 m deep.

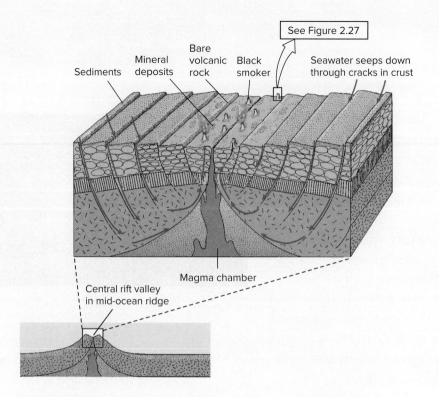

Figure 2.24 A cross section of a mid-ocean ridge showing how seawater seeps down through cracks in the crust, is heated, and reemerges in underwater hot springs.

Figure 2.25 Colorful red and white tubeworms grow around a black smoker on the Juan de Fuca Ridge, offshore of Oregon and Washington. Black smokers are common in hydrothermal vent areas.

The Mid-Ocean Ridge and Hydrothermal Vents The mid-ocean ridge itself is an environment that is unique in the ocean. As noted previously, the ridge is formed when material rising from the mantle pushes up the oceanic crust. Right at the center of the ridge, however, the plates are pulling apart. This leaves a great gap or depression known as the **central rift valley**. The floor

and sides of the valley are riddled with crevices and fractures. Seawater seeps down through these cracks until it gets heated to very high temperatures by the hot mantle material (Fig. 2.24). The heated water then forces its way back up through the crust and emerges in **hydrothermal vents**, or deep-sea hot springs.

The water coming from many hydrothermal vents is warm, perhaps 10 to 20 °C, much warmer than the surrounding water. At some vents, however, the water is blisteringly hot, up to 350 °C. The water is so hot that when scientists first tried to measure its temperature the thermometer they were using started to melt! To take accurate readings, they had to return with a specially designed thermometer.

As the hot water seeps through cracks in Earth's crust, it dissolves a variety of minerals, mainly **sulfides**. When the mineral-laden hot water emerges at the vent, it mixes with the surrounding cold water and rapidly cools. This causes many of the minerals to solidify, forming mineral deposits around the vents. **Black smokers** (Figs. 2.25 and 2.26) are chimney-like structures that progressively build up around a vent as the minerals solidify. The "smoke" is actually a dense cloud of mineral particles. Vents can also produce "white smokers," "snowblowers," or even "blue smokers" depending on the minerals released and other factors.

When hydrothermal activity was first discovered it was thought to be confined to mid-ocean ridges. Hydrothermal vents, complete with black smokers and other mineral deposits, have since been found behind trenches. They result from the same volcanic activity that creates island arcs. Relatively cool (40 to 75 °C) vents have also been discovered near, but not at, mid-ocean ridges. These vents produce chimneys of carbonate rather than sulfide minerals and are caused by chemical reactions between seawater and newly formed oceanic crust rather than by volcanic activity. One such chimney rises 60 m above the sea floor, making it the tallest hydrothermal vent known.

Deep-sea hot springs are of great interest not only to geologists, but also to biologists. The discovery of unexpectedly rich marine life around hydrothermal vents was one of the most exciting finds in the history of marine biology. These organisms are discussed in Chapter 18.

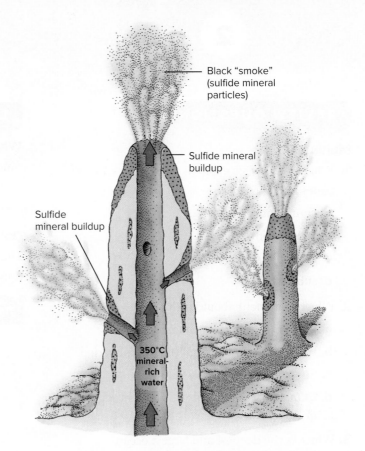

Black "smoke" (sulfide mineral particles)

Sulfide mineral buildup

Sulfide mineral buildup

350°C mineral-rich water

Figure 2.26 Cross section of a black smoker. Minerals are deposited as a precipitate when hot, mineral-laden water emerging from the rift zone meets the cold ocean water. Over time these mineral deposits build up the chimney of the black smoker.

· ·

Section Review

1. Describe What are the characteristics of the continental shelf, the continental slope, and the continental rise?

2. Explain How do hydrothermal vents form?

Chapter ② Review

Multiple Choice

1. Which hypothesis explains what makes Earth's plates move?

 a. continental drift

 b. convection

 c. slab pull

 d. subduction

2. Which statement about Earth is true?

 a. Earth is presently in an ice age.

 b. Earth is presently in an interglacial period.

 c. Sea levels on Earth are falling.

 d. Climate has remained the same through much of Earth's history.

3. Which is the deepest and largest ocean?

 a. Atlantic

 b. Arctic

 c. Indian

 d. Pacific

Short Answer

4. Why are continents higher than oceanic crust?

5. Compare and contrast a collision of an oceanic plate and an oceanic plate with a collision of an oceanic plate and a continental plate.

6. What happens when two continental plates collide?

7. What are the differences between active and passive margins?

Critical Thinking

8. Why are most oceanic trenches found in the Pacific Ocean?

9. Scientists who study ancient marine life forms usually obtain fossils not from the sea floor, but from areas that were once undersea and have been uplifted onto the continents. Why?

10. What are some of the major pieces of evidence for the theory of plate tectonics? How does the theory explain these observations?

What influenced the formation of minerals on Earth?

Since Earth formed, it has passed through distinct stages, each affecting the formation, concentration, and distribution of its 4,400 known minerals.

Data and Observations

The table shows major events in the history of mineral formation on Earth.

Time Period	Events
4.4 – 2.4 bya	Earth's first 1,500 minerals formed as a result of repeated melting and cooling of Earth's crust (due to subduction and uplift of Earth's plates) and chemical weathering in early oceans and the atmosphere. Minerals included quartz, feldspar, mica, jadeite, and uranium.
2.2 – 2.0 bya	The appearance of photosynthetic organisms, which released oxygen into the atmosphere, caused iron in Earth's basalt crust to oxidize. More than 2,500 new minerals formed, including hematite, rhodonite, turquoise, and calcium carbonate.
2.0 – 1.0 bya	During a time of little change biologically, few new minerals formed.
700 – 500 mya	Earth cycled through periods of freezing and warming. Although few new minerals were formed, the concentration of existing minerals changed greatly, including an increase of fine-grained clay minerals at Earth's surface and carbonate minerals in warm, shallow oceans.
500 mya – present	Life on land began to thrive, including the appearance of land plants and fungi. The appearance and success of these two groups led to increased breakdown of rock through chemical and physical weathering, and clay minerals continued to increase at Earth's surface.

Think Critically

1. Which processes affected the formation and concentration of minerals on Earth?

2. Which evidence supports the statement that, on Earth, "Most minerals owe their existence to life."?

Data obtained from: Hazen, R., 2010. Evolution of minerals. Scientific American, vol. 302, no. 3, March, pp. 58–65.

Ocean Chemistry and Structure

Chapter

3

THEME Patterns

By collecting samples and analyzing data, scientists determine the chemical and physical properties of seawater.

BIG IDEA

The chemical and physical properties of seawater affect in many ways conditions in the ocean, on land, and in the atmosphere.

About the photo Icebergs float in Jökulsárlón lagoon, Iceland. In this photo you can see all three phases of water.

3.1 The Unique Nature of Pure Water

VOCABULARY

temperature

heat

calories

evaporation

density

specific heat

heat capacity

evaporative cooling

cohesion

surface tension

adhesion

viscosity

salts

ions

dissociate

Main Idea

The hydrogen bonds between water molecules make water different from any other substance on Earth.

Key Questions

1. What is the only substance on Earth that naturally occurs in all three states?
2. Why is water often referred to a universal solvent?

Most of us never give water a second thought unless we are hot or thirsty. Water quenches our thirst because it makes up most of our bodies. Marine organisms, too, are mostly water—80% or more by weight in most cases, and in jellyfishes, or sea jellies, over 95%. Water not only fills the ocean, it makes life possible. Seawater is primarily salt and water—more than 99.5% of the ocean is composed of water and sodium chloride. Other salts, trace elements, organic molecules, dissolved gases, and nutrients make up the rest of seawater. This chapter explores the chemical composition and structure of seawater and discusses the chemical alterations that the ocean is experiencing due to global climate change.

Molecular Structure

All matter is made of atoms. Only 118 or so known substances (chemists argue about the exact number) are composed of a single kind of atom; these are called elements. In all other substances, two or more atoms are chemically combined into larger particles called molecules. Water molecules are made up of one oxygen atom, which is relatively large, and two small hydrogen atoms. The oxygen and hydrogen atoms have weak, opposite charges that create electrical attractions, or hydrogen bonds, between adjacent water molecules (Fig. 3.1). Hydrogen bonds are much weaker than the bonds that hold the oxygen and hydrogen atoms in the water molecule together, but they make water different from any other substance on Earth.

DIVE IN!

Why do coastal areas have less extreme weather? The fact that water can absorb or lose a lot of heat before its temperature changes means that coastal areas experience more moderate temperatures than inland areas. The oceans help keep the temperature ranges in coastal cites lower than cities further inland.

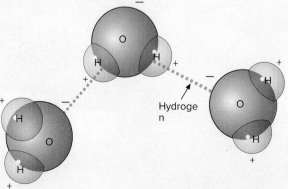

Figure 3.1 The different ends of water molecules have opposite electrical charges. The oxygen (O) end has a weak negative charge, the hydrogen (H) end a slight positive charge. Opposite charges attract each other like the opposite poles of a magnet.

Temperature and Heat

The particles (atoms and molecules) of matter are constantly in motion. **Temperature** reflects the *average* kinetic energy of the particles—the higher the average kinetic energy is, the faster the particles move and the higher the temperature will be. Temperature is measured in degrees, most commonly by thermometers, and can be expressed on Celsius (C), Kelvin (K), or Fahrenheit (F) scales.

Heat, on the other hand, is the *total* kinetic energy of all the particles in a substance or object. It is measured in **calories**, which equals the amount of heat necessary to raise the temperature of 1 gram of water from 14.5 °C to 15.5 °C (+1 °C). To understand the difference between temperature and heat, we can compare a hot mug of coffee to the Atlantic Ocean off the coast of Maine. The particles in the coffee are moving faster on average than the particles in the ocean. Therefore, the mug of coffee has a higher temperature. Because the ocean is so large, however, there are more particles in motion compared to the coffee, so there is greater total kinetic energy; thus, the ocean has more heat.

Heat can be transferred from a substance of high temperature to a substance with a lower temperature. The transfer of heat causes the average kinetic energy to change within the two substances, resulting in a temperature change, a change of state, or both. As an example, we can use the reference to the hot mug of coffee. If you brought the mug of hot coffee outside on an ice-cold day, the mug and the coffee would transfer energy to the cold air. The loss of energy would cause the molecules in the mug and the coffee to slow down, and the energy gained by the air would cause the air particles to speed up. The mug and the coffee would experience a decrease in temperature, and the temperature of the air would increase (slightly, because that energy would be dispersed across the air particles). If it is left outside long enough, the coffee could transfer enough energy to the air to change its state and freeze solid.

The Three States of Water Matter can exist in three different states, or phases—solid, liquid, or gas. Water is the only substance that naturally occurs in all three states on Earth. In liquid water, hydrogen bonds hold most of the molecules together in small groups (Fig. 3.2).

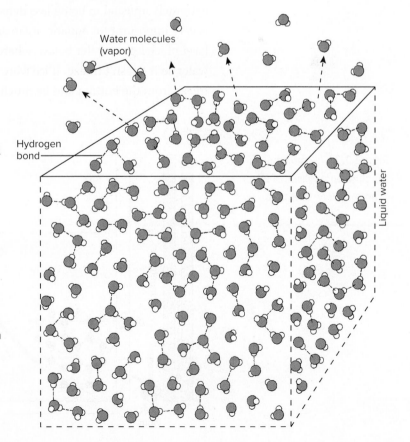

Figure 3.2 The molecules in liquid water form groups of various sizes held together by hydrogen bonds. The molecules are moving too rapidly to be held permanently in place, so the groups constantly break up and re-form. Evaporation occurs when molecules break free of hydrogen bonds and enter the gaseous state.

The molecules are in constant motion, however, and because the hydrogen bonds are weak, the groups constantly break apart and re-form. This phenomenon happens so quickly that each hydrogen bond lasts a few trillionths of a second. When heat is added to a substance, the kinetic energy will increase and the molecules will move faster. If a molecule moves fast enough, it will break free of all the hydrogen bonds and escape from the liquid phase into the gaseous or vapor phase. This is the process of **evaporation**.

In water vapor, the molecules are not held together by hydrogen bonds. They are separate and much farther apart than in the liquid (see Fig. 3.2). As temperature rises, so does the evaporation rate, because the molecules move faster and more of them escape the hydrogen bonds. When the water gets hot enough, nearly all the bonds are broken and many molecules enter the vapor state at once; in other words, the water boils.

When liquid water cools, the molecules not only move slower (their kinetic energy decreases), they pack closer together and take up less space; so, the volume of water decreases. Because the volume decreases without changing the mass, the water gets denser. **Density** is the degree of compactness of a substance. As seawater gets colder, therefore, it gets denser. As we shall see, cold seawater tends to sink in the ocean. Freshwater also gets denser as it gets colder, but only down to a temperature of about 4 °C. Below 4 °C, fresh water gets less dense as it cools.

Water freezes when the molecules move so slowly that the hydrogen bonds take over, locking the molecules into a fixed, three-dimensional pattern known as a crystal. In ice crystals, the molecules are farther apart than in liquid water, so water expands as it freezes. Because the same mass of water occupies more volume as ice than as liquid water, ice is less dense and floats. Water is extremely unusual in being less dense as a solid than as a liquid, a property that is very important for aquatic organisms, both freshwater and marine. A floating layer of ice leaves water below, where organisms can live, and it insulates the water so it doesn't freeze. If ice were denser than liquid water, the ocean would freeze from the bottom and be much less hospitable to life.

Figure 3.3 The molecular structure and phases of water change with temperature.

Specific Heat The molecular structure of water changes with temperature. In ice, hydrogen bonds hold the vibrating molecules in a hexagonal pattern. As heat is added, the ice warms up and the molecules vibrate more rapidly until they break free of the crystal structure. When this happens, the ice melts. While the ice is melting, added heat is absorbed by breaking hydrogen bonds, not by increasing the temperature. When the ice is completely melted, additional heat causes the temperature to rise again. Some molecules gain enough speed to break free of all the bonds and evaporate. At 100°C, nearly all of the hydrogen bonds are broken and the water boils.

Figure 3.4 In the heat of the sun, this seaweed (*Hormosira banksii*) in New Zealand is beginning to shrivel. Shore organisms that are frequently exposed to the air are subjected to much more extreme temperatures than submerged organisms.

In ice, vibrating water molecules are held in place in the ice crystal by hydrogen bonds. The bonds must be broken before the molecules can begin to move around, which is what happens when ice melts (Fig. 3.3). Because of this, ice melts at a much higher temperature than similar substances that do not form hydrogen bonds. If it were not for the hydrogen bonds, ice would melt at about – 90 °C instead of 0 °C!

Not only does ice melt at a comparatively high temperature, it absorbs a lot of heat when it melts. The amount of energy required to raise a mass of substance one degree Celsius is the **specific heat** of that substance. Because of the energy-absorbing hydrogen bonds, water has a high specific heat. The amount of heat required to melt a substance is called its latent heat of melting. Water has a higher latent heat of melting than any other commonly occurring substance. The reverse is also true: A great deal of heat must be removed from liquid water to freeze it. Thus, it takes a long period of very cold weather before a body of water will freeze.

In melting ice, added heat breaks more hydrogen bonds rather than increasing the speed of molecular motion, so the temperature of the ice-water mixture remains at a constant 0 °C until all the ice melts. That is why ice keeps drinks cold: added heat goes into melting the ice, not raising the temperature.

When all the ice has melted, added heat again makes the molecules move faster, so it raises the water temperature. Some of the heat energy, however, still goes into breaking hydrogen bonds rather than increasing the speed of molecular motion; therefore, it takes a lot of heat to raise the temperature. The amount of heat needed to raise a substance's temperature by a given amount, or its **heat capacity**, reflects how much heat the substance can store. Water has one of the highest heat capacities of any naturally occurring substance. Water's ability to absorb a lot of heat with a relatively small increase in temperature is why it is used as a coolant, in car engines for example. More importantly, for marine organisms, water's high heat capacity means that most of them are not subjected to the rapid and sometimes drastic temperature changes that occur on land (Fig. 3.4).

(a)

(b)

Figure 3.5 Water molecules at the surface form strong cohesive bonds between each other, creating a skin-like barrier (a). Some organisms, such as marine water striders, are able to use this property to walk on water (b).

Water also absorbs a lot of heat when it evaporates; that is, it has a high latent heat of evaporation. Again, this is due to hydrogen bonds: only the faster-moving molecules, those with the most energy, can break the bonds to enter the gaseous phase. Because the fastest molecules leave the liquid phase, those left behind have a lower average velocity and, therefore, a lower temperature. This is known as **evaporative cooling**, which is why evaporating perspiration cools our skin.

Cohesion, Surface Tension, and Viscosity

As mentioned earlier, hydrogen bonds are constantly forming and breaking between water molecules in the liquid state. These bonds cause water molecules to stick together. Because water contains a large number of hydrogen bonds at any given moment, water has more **cohesion** than other liquids. Cohesion is the sticking together of particles of the same substance.

The cohesive property of water explains water's **surface tension**, which is the measure of the elastic tendency of liquids. The cohesion between the water molecules at the surface of a liquid is greater than the **adhesion** (sticking together between particles of different substances) between water and air. These cohesive bonds cause the water molecules to arrange into an ordered membrane at the water surface, forming a layer similar to a thin skin (Fig. 3.5a). This property is what makes it possible for organisms such as water striders to walk on water (Fig. 3.5b). Surface tension is also a key component in wave formation, allowing the wind to maintain contact with the surface of the water and transfer more energy into the body of water (wave formation will be discussed in-depth in Chapter 4).

Water's ability to pour, or its resistance to flow, is a measure of its **viscosity**. The higher the viscosity, the more resistant to flow a liquid will be. Water has a lower viscosity than liquids such as honey and maple syrup because it has less

<image name="copyright" >© Lanna Cheng, Scripps Institution of Oceanography</image>

internal friction. Its hydrogen bonds, however, make it more viscous than many other liquids. Viscosity is affected by temperature, where colder liquids have a higher viscosity than warmer ones. This is true for honey, which is easier to pour when it is warm, and for the ocean, where the surface water in the tropics is less viscous than the water at the poles. Marine organisms, especially plankton, have evolved specific adaptations to stay afloat in the less viscous tropical waters. These include oil globules, spines, and extra appendages.

Water as a Solvent

Water can dissolve more things than any other natural substance and is often called the universal solvent. Water is especially good at dissolving **salts**. Salts are made up of particles that have opposite electrical charges. Such electrically charged particles, which can be single atoms or groups of atoms, are known as **ions**. For example, ordinary table salt, or sodium chloride (NaCl), consists of a positively charged sodium ion (Na^+) combined with a negatively charged chloride ion (Cl^-). The oppositely charged ions are joined by the same electrical attraction that creates hydrogen bonds between water molecules. Ions have much stronger electrical charges than the opposite ends of the water molecule, however, so these ionic bonds are much stronger than hydrogen bonds. If no water is present, then the ions bind strongly together to form salt crystals.

When a salt crystal is placed in water, the strongly charged ions attract the water molecule (with their weak charges) like iron filings to a magnet. A layer of water molecules surrounds each ion, insulating it from the surrounding ions (Fig. 3.6). This greatly weakens the ionic bonds that hold the salt crystals together. The ions pull apart, or **dissociate**, and the salt dissolves.

Sodium ion
Chloride ion

⊕ Sodium ion
⊖ Chloride ion
🔴 Water molecule

Figure 3.6 Ions from a table salt (sodium chloride) crystal dissociating in water.

. .

Section Review

1. **Explain** the difference between temperature and heat.

2. **Identify** three ways in which hydrogen bonds give water its unique chemical properties.

3. **Determine** Which is denser, ocean water at the Equator or ocean water at the North Pole? Explain your reasoning.

3.2 Physical and Chemical Properties of Seawater

VOCABULARY

Deep Sound Channel

solutes

salinity

rule of constant proportions

precipitation

profile

anthropogenic

nutrients

gas exchange

buffering

chemical equilibrium

DIVE IN!

Can you float in salt water without a raft? In the Dead Sea, you can. The water is so dense due to the high salinity that people can easily float on the surface of the water. There are many photos of people floating casually in the water while reading a magazine or newspaper. The salt, however, will cause any small cuts to hurt a lot!

Figure 3.7 Not all ions in seawater enter the ocean at the same place. Positive ions like sodium and magnesium come mostly from the weathering of rocks and are carried to the sea by rivers. Negative ions like chloride and sulfide enter the ocean at hydrothermal vents and from volcanos.

Main Idea

The chemical and physical properties of seawater include its salinity, pressure, density, sound propagation, light absorption, and dissolved gas contents.

. .

Key Questions

1. What are sources of dissolved substances in seawater?
2. How does salinity affect the density of seawater?

The characteristics of seawater are due to the nature of pure water and to the materials dissolved in it. Some substances dissolved in seawater are produced by the chemical weathering of rocks on land and are carried to the sea by rivers (Fig. 3.7). Other materials come from Earth's interior. Most of these are released into the ocean at hydrothermal vents and fissures along the rift valley of the mid-ocean ridges. Some materials are released into the atmosphere from volcanoes and enter the ocean via rain and snow.

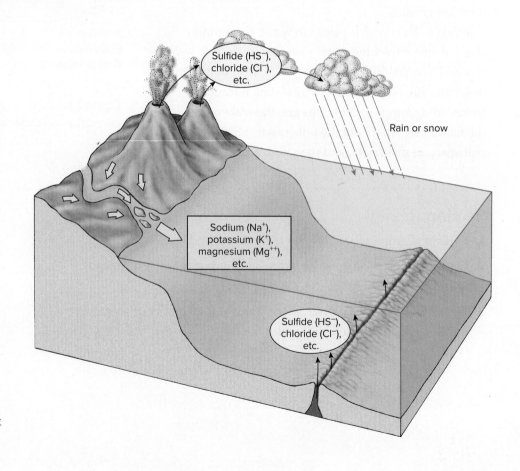

Physical Properties

The physical properties of seawater shape life in the oceans. Fundamental properties such as light level, pressure, and transmission of light and sound are drastically different in the oceans than they are on land.

Transparency One of the most biologically important properties of seawater is that it is relatively transparent, so sunlight can penetrate into the ocean. This is vital because photosynthetic organisms need light to grow. If seawater weren't transparent, then there would be little photosynthesis in the sea, and then only right at the surface.

Sunlight contains all the colors of the visible spectrum of light, which is seen in the rainbow, but not all the colors penetrate seawater equally well. Clear ocean water is most transparent to blue light. Other colors are absorbed more than blue, so as depth increases, more colors are absorbed, and soon only blue light remains (Fig. 3.8). Things that appear red on the surface look gray or black at depth because there is no red light to reflect off them and be seen (Fig. 3.9). At greater depths—about 1,000 m in the clearest water—even the blue light gets absorbed, and only darkness remains.

The transparency of water is strongly affected by suspended or dissolved material. Obviously, muddy water is not as clear as clean water. Large quantities of plankton also reduce water transparency. Water near coasts often contains a lot of material brought in by rivers that give coastal waters a greenish tint and makes them less transparent than the deep-blue waters of the open ocean. Scientists have a number of sophisticated electronic instruments to make precise measurements of water clarity and light penetration. For surface waters, however, a surprisingly simple and elegant instrument called a Secchi disk is sufficient (Fig. 3.10).

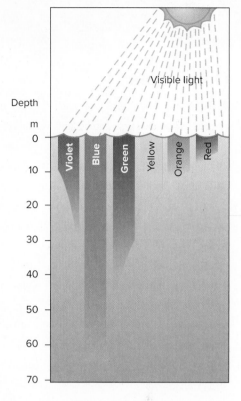

Figure 3.8 Different colors of light penetrate to different depths in the ocean.

(a)

(b)

Figure 3.10 Secchi disks are commonly used for measuring water clarity in surface waters. The disk is slowly lowered to where it just ceases to be visible. The greater this depth, the clearer the water.

Figure 3.9 At a depth of 30 m, only blue light remains, and under natural lighting, this sea star (*Thromidia catalai*) appears light blue, with the tips of its arms almost black (a). This is because wavelengths from the red, orange, and yellow spectrum of visible light cannot penetrate this deep. When lit with an electronic flash, the sea star's true colors are revealed (b).

Ocean Optics

What color is the ocean? When asked this question, most people will quickly respond: "Blue!" More observant individuals might respond: "Which ocean? The Atlantic Ocean off of the coast of Maine often looks dark greenish-blue, but parts of the Pacific can look almost translucent turquoise." Ocean color can be indicative of many things, from underlying sediment composition to an out-of-control algal bloom.

The color of the ocean is a result of how beams of sunlight interact with the surface of the water. Water molecules absorb mainly the red part of the light spectrum, whereas wavelengths in the blue regions are not absorbed as well. Instead, blue wavelengths are scattered away from the ocean's surface, giving the ocean its overall blue appearance.

Why does ocean color vary? As you know, the ocean is not made up of pure water. When light hits the ocean, it also interacts with sediments, dissolved organic material, and living organisms. The variation in color is the result of all that other stuff in the water. In particular, chlorophyll-containing microbes, such as phytoplankton, can drastically change the ocean's color. As photosynthesis occurs, phytoplankton absorb light with red and blue wavelengths, while reflecting green light. Part of the ocean that is very green might be full of photosynthetic phytoplankton. Likewise, a milky blue patch in the ocean could be the result of a large population of coccolithophores, phytoplankton that have bright, white calcium carbonate exoskeletons – excellent at reflecting light. If you look closely at the photo above, you can pick out areas where different species of phytoplankton are blooming - there are dark green patches near the coastline and a vast milky blue patch of coccolithophores as you move offshore. You can also see the sediments carried by the Mississippi River.

The optical properties of phytoplankton in the oceans aid scientists trying to study them. Phytoplankton blooms can often stretch from several hundred to thousands of kilometers in size, but may last for only a few days. To monitor a phytoplankton bloom thousands of kilometers in size by boat would be almost impossible. The bloom might disperse long before researchers on a boat have enough time to take

The sediment carried by the Mississippi River is visible as brown plumes. The river water is also rich in nutrients, leading to blooms of phytoplankton near the delta.

and analyze enough samples to determine the size of the bloom or even identify its existence! Luckily, the use of remote satellite technology and the optical properties of the ocean allow scientists to monitor the color of the ocean.

Color scanning technology In 1978 NASA sent the first ocean color scanner technology into space. Using sensitive detectors, this instrument scanned the wavelength of visible light reflected back from the ocean, converted it to an electronic signal, and sent it back down to earth. This instrument allowed scientists to indirectly map chlorophyll and primary productivity in a manner that previously would have taken decades. Remote color sensing technology has continued to advance and aid in our understanding of the ocean. Scientists can now track phytoplankton populations as well as upwelling events and identify potentially harmful algal blooms all across the world using space-based color data from the ocean. Knowing the color of the ocean allows scientists to make better predictions concerning the overall health of the ocean, as well as how severe weather events and changes climate may affect marine ecosystems.

NASA Image by Robert Simmon, based on Landsat data provided by the UMD Global Land Cover Facility

Pressure Another factor that changes dramatically with depth in the ocean is pressure. Organisms on land are under 1 atmosphere (atm) of pressure at sea level—the weight of all the air above them. Marine organisms, however, are under the weight of water as well as the atmosphere. Because water is much heavier (denser) than air, marine organisms are under much more pressure than organisms on land. The pressure increases dramatically with depth because the amount of water above increases (Fig. 3.11). With each 10 m of increased depth, another atm of pressure is added.

As pressure increases, gases are compressed. Gas-filled structures inside organisms, such as air bladders, floats, and lungs, shrink or collapse. This limits the depth range of many marine organisms. It also means that submarines and scientific instruments must be specially engineered to withstand pressure. This greatly increases the difficulty, expense, and sometimes the danger of studying the sea. The reverse effect also causes problems: organisms that contain gas-filled structures are often injured when they are brought up from the deep (Fig. 3.12).

Water depth	Total pressure		
	atm	bar	psi
0	1	1.0	14.7
10 m	2	2.0	29.4
20 m	3	3.0	44.1
30 m	4	4.1	58.8
40 m	5	5.1	73.5
50 m	6	6.1	88.2
60 m	7	7.1	102.9
70 m	8	8.1	117.6
80 m	9	9.1	132.3

Figure 3.11 The pressure at any place depends on the weight pressing down from above. As the pressure increases, flexible gas-filled structures such as the red balloon are compressed. Pressure can measured in atmospheres (atm), bars, pounds per square inch (psi), and other units.

Figure 3.12 Many fishes, like this longfin grouper (*Epinphalus quoyanus*), have an internal air sac or swim bladder. If the fish is brought to the surface too quickly, the bladder blows up like a balloon because of the decreased pressure.

Sound Propagation As mentioned in the previous section, water is denser than air. The higher density is primarily due to the compactness of the water molecules, which is much greater than air molecules. This compactness causes sound to travel 4.5 times faster in water than in the air (340 m/s in air and approximately 1520 m/s in the shallow ocean). Sound is the oscillation between compressions and rarefactions of the water, which causes a change in pressure. A receiver, such as an ear or a hydrophone, detects these changes in pressure. The speed of sound varies throughout the ocean as a function of salinity, temperature, and pressure—higher salinity, temperature, and pressure causes sound to travel faster.

The depth at which sound travels slowest varies with latitude, from a few hundred meters deep in tropical and temperate latitudes to the surface around the poles (Fig. 3.13). This region is known as the **Deep Sound Channel**, and sound waves, especially low-frequency waves, can travel uninterrupted for thousands of kilometers. This region is important for whales that are communicating long distances and for the military when it is operating submarines.

For many marine species, hearing is an important sensory system, just like in humans. Animals like fishes, marine mammals, and shrimps use sound for communication, navigation, reproduction, and predator and prey detection. Studies have shown that as human activity in our oceans increases, it is becoming noisier underwater. How this will affect marine life is yet to be determined, but responses such as behavioral changes, hearing loss, increases in stress, and death have been observed.

Chemical Properties

The chemical properties of seawater are part of what distinguish the ocean from freshwater habitats. The ocean is defined by its salinity, but the chemical properties are far more complex than just a bit of salt.

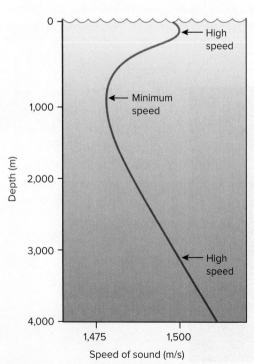

Figure 3.13 Sound speed changes with water depth.

Salt Composition Seawater contains at least a little of almost everything, but most of the **solutes**, or dissolved materials, are made up of a surprisingly small group of ions. In fact, only six ions compose more than 99% of the material dissolved in seawater (Table 3.1). Sodium and chloride account for about 85%, which is why seawater tastes like table salt.

When seawater evaporates, the ions in it are left behind and combine to form various salts. **Salinity** is defined as the total amount of salt dissolved in seawater. Salinity is usually expressed as the number of grams of salt in 1000 grams of seawater. If 35 grams of salt are dissolved into 1000 grams of seawater, for example, the seawater has a salinity of 35 parts per thousand, or 35‰.

Today salinity is typically measured using electronic instruments. Because of their electrical charges, ions are good electrical conductors. The conductivity of seawater, therefore, reflects the salinity. Oceanographers express salinity determined from conductivity measurements in practical salinity units (psu) rather

than parts per thousand, but the two units are numerically the same: 35 psu is equal to 35‰.

The salinity of water greatly affects the organisms that live in it. Most marine organisms will die in freshwater. Even slight changes in salinity harm some. Those that live in river mouths, or other places where the salinity is prone to fluctuations, have evolved specific mechanisms to cope with salinity changes.

The chemist William Dittmar analyzed seawater samples from the *Challenger* expedition and found that the relative percentage of the major ions in seawater remained constant even though the total amount of salt varied slightly from place to place. Chloride ions, for example, nearly always make up 55.03% of the amount of salt that is present. This principle is called the **rule of constant proportions**.

The proportions of different ions vary slightly, especially near hydrothermal vents or in coastal areas where rivers bring in large amounts of positively charged ions (see Fig. 3.7). In some places, intense biological activity also affects ion ratios. Still, in most of the oceans, the proportions remain remarkably constant.

This indicates that for the most part the oceans are chemically well mixed and that ocean salinity varies almost entirely as a result of the addition or removal of pure water rather than the addition or removal of salts. If salinity varied by adding or removing any particular salt, then the relative amounts of the ions in the seawater would change. If magnesium chloride ($MgCl_2$) were added, for example, then the proportions of magnesium and chloride would go up. Thus, although marine organisms are exposed to changes in total salinity, they rarely have to deal with changes in the ratios of the various ions. This makes it easier for them to control their internal salt and water balance.

Water is removed from the ocean primarily by evaporation and to a lesser extent by freezing. When seawater freezes, the ions are excluded from the ice that is formed. They are left behind in the unfrozen water, increasing its salinity, and the ice is almost pure water. That is why icebergs are not salty. Water is added by **precipitation**—rain and snow—and, to a lesser extent on short timescales, by the melting of glaciers and polar ice.

Table 3.1 The Composition of Seawater of 35‰ Salinity

Ion	Concentration ‰	Percentage of Total Salinity
Chloride (Cl⁻)	19.345	55.03
Sodium (Na⁺)	10.752	30.59
Sulfate (SO₄⁻)	2.701	7.68
Magnesium (Mg²⁺)	1.295	3.68
Calcium (Ca²⁺)	0.416	1.18
Potassium (K⁺)	0.390	1.11
Bicarbonate (HCO₃⁻)	0.145	0.41
Bromide (Br⁻)	0.066	0.19
Borate (H₂BO₃⁻)	0.027	0.08
Strontium (Sr²⁺)	0.013	0.04
Fluoride (F⁻)	0.001	0.003
Other dissolved material	<0.001	<0.001

Figure 3.14 Niskin bottles are one of several types of specially designed bottles for collecting seawater.

The average salinity of the ocean is about 35‰. The open ocean varies relatively little, between about 33‰ and 37‰, depending mostly on the balance between evaporation and precipitation. Partially enclosed seas might have more extreme salinities. The Red Sea, for instance, is in a hot, dry region where evaporation predominates over precipitation, so it is very salty—about 40‰. Near coasts or in enclosed basins, runoff from rivers can have a strong effect. River runoff gives the Baltic Sea a typical salinity at the surface of only about 7‰, for example.

Salinity, Temperature, and Density We have already seen that temperature greatly affects water's density. Salinity also influences seawater density: the saltier the water is, the denser it is. The density of seawater, therefore, depends on its temperature and its salinity.

Temperature in the open ocean varies between about -2 and +30 °C. Temperatures below 0 °C are possible because salt water freezes at a colder temperature than pure water. This is one reason why the ocean is less prone to freezing than lakes and rivers.

Temperature in the ocean varies considerably more than salinity; so as a practical matter, density is usually controlled more by temperature than salinity. Exceptions occur, however, and the temperature and salinity of ocean water still need to be measured to determined density.

Temperature and salinity can be measured by lowering specially designed bottles and thermometers on a wire to the desired depth (Fig. 3.14). Then a weight called a messenger is released to slide down the wire, triggering the bottles to snap shut and trap a water sample that can be analyzed for salinity. The thermometers are also triggered to "freeze" the temperature reading so it doesn't change as the thermometer is retrieved.

A series of bottles can be attached to the cable to measure temperature and salinity, and therefore density, at several depths at once (Fig. 3.15). A graph that shows the temperature at different depths in the ocean is called a temperature **profile**. A temperature profile obtained from a given location can be thought of as showing the temperature in a vertical shaft of water, or water column, extending down from the surface. Profiles can also be plotted for salinity, density, or any other characteristic. In a profile, the vertical axis appears to be upside down, as show in Figure 3.15.

Using bottles and thermometers is time consuming and expensive. Today oceanographers usually use electronic sensors to quickly and accurately record salinity, temperature, and depth throughout the water column, rather than just at certain depths. Conductivity-temperature-depth meters (CTDs) are widely used in conjunction with other instruments (Fig. 3.16, p. 72). Expendable bathythermographs (XBTs), which measure temperature but not salinity profiles, are, as the name implies, disposable.

Though temperature, salinity, and various other things can be measured electronically, many other measurements still require a water sample, so Niskin bottles and similar devices remain in use. These days, oceanographers usually mount all the open bottles on a single rack (see Fig. 3.16) rather than stringing them along the wire. To collect samples from different depths, each bottle is

closed at a different time while the rack is lowered. The rack is easier to use than a series of individual bottles.

Even with electronic instruments, measurements from ships can be made in only one place at a time. To study a large area, the ship has to move from place to place. In the meantime, conditions might change because of currents or the weather. One alternative is to use a number of ships at once. This is expensive, however, and is not done often. Today, oceanographers increasingly make their measurements with automated instruments.

Another way to get the big picture is to use satellites. Satellites can measure conditions only near the surface, but they give instantaneous coverage of a large area. Furthermore, a series of measurements can be made in a short time. This makes it possible to follow rapid changes in surface conditions that result from currents and weather.

Trace Elements, Nutrients, and Organics The "other dissolved material" listed in Table 3.1 includes substances such as trace elements. Trace elements, usually metals and metalloids, are present in tiny amounts in the ocean. They can enter the ocean through a variety of sources, including river runoff, groundwater seeps, the atmosphere, hydrothermal activity, and **anthropogenic** (human-derived) inputs. Most trace elements are removed from the ocean when they are incorporated into sinking particles and subsequent buried on the seafloor.

Many of these elements are also **nutrients**, dissolved materials that are critically important to marine life. Nutrients are raw materials other than

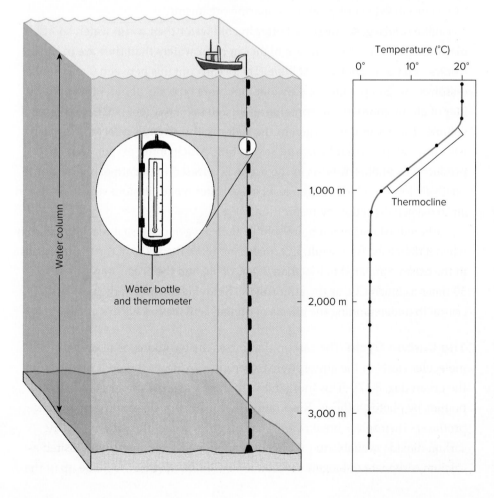

Figure 3.15 (a) Temperature profiles like this one can be created by dropping sample bottles and thermometers to measure several depths at once. (b) The results are plotted on a profile, or graph showing how temperature or another property varies with depth. Temperature is measured only where the bottles are located (dots) and the temperature between the bottles (red line) must be inferred.

carbon, such as vitamins and minerals, which are necessary for organisms to function. These nutrients allow organisms to convert energy in the form of glucose into other compounds. Compounds containing nitrogen (N), phosphorus (P), and iron (Fe) are particularly important, and their availability determines the biological productivity of much of the oceans.

Organics are materials containing carbon, and they too make up a small percentage of oceanic compounds. Most organics are produced by marine organisms through metabolic processes and excreted as a solid or in dissolved form. Solid organics include fecal pellets and molted tests, whereas many dissolved organics are released into the ocean via death and cell lysis (breaking of the membrane). When the organics come in contact with seawater, bacteria and other microbes quickly break them down.

Dissolved Gases Gases as well as solid materials are dissolved in seawater. For organisms, the most important gases in the ocean are oxygen (O_2), carbon dioxide (CO_2), and nitrogen (N_2). All three are found in the atmosphere and dissolve in seawater at the sea surface. Sometimes the reverse occurs, and the sea surface releases gases to the atmosphere. This process, known as **gas exchange**, occurs between the ocean and the atmosphere. Gases can also be produced *in situ* (locally within the ocean). Oxygen is produced largely via photosynthesis of marine plants and phytoplankton and CO_2 is produced via respiration of marine organisms.

Unlike solids, gases dissolve better in cold water than warm water, so dissolved gas concentrations are higher in polar waters than they are in the tropics. Oxygen is not very soluble. The ocean contains between 0 and 8 ml of dissolved oxygen per liter of seawater, but more typically about 4 to 6 ml/L. A liter of air, by comparison, contains about 210 ml of oxygen, or 21% of the total volume. The amount of oxygen in the water is also strongly affected by organisms through photosynthesis and respiration. Much of the oxygen that is produced by photosynthesis in the ocean is released to the atmosphere. The relatively low dissolved oxygen level in seawater makes it more susceptible than air to oxygen depletion by respiration.

Carbon dioxide is much more soluble than oxygen because it reacts chemically when it dissolves. As a result, CO_2 makes up more than 80% of the dissolved gas in the ocean compared to less than 0.04% of air, and the ocean stores more than 50 times as much CO_2 as the atmosphere. Studying the ocean, therefore, is critical to understanding the effects of human activities on Earth's climate.

The Carbon Cycle The carbon that forms the backbone of all organic molecules starts in the atmosphere as carbon dioxide, and then it dissolves in the ocean (Fig. 3.17). This inorganic carbon is converted into organic compounds by photosynthesis. Respiration by consumers, decomposers, and the producers themselves breaks down the organic compounds and makes the carbon dioxide available to producers again. Some carbon is also deposited as calcium carbonate in biogenous sediments (sediments that are made up of the

Figure 3.16 Water sampling bottles (red arrow) are now usually mounted in a "rosette" to a single frame. A variety of electronic instruments (yellow arrows) that measure physical factors can also be mounted on the frame, which is then lowered through the water column.

skeletons and shells of marine organisms) and coral reefs. Sediments contain 25 times as much carbon as the ocean water.

Under certain conditions, some of this calcium carbonate dissolves back into the water. Dissolved CO_2 molecules from the atmosphere and calcium carbonate combine with seawater (H_2O) in a process called **buffering**. This buffering forms carbonic acid (H_2CO_3), a weak acid that is found in carbonated drinks. Because these excess carbonic acid molecules throw off the **chemical equilibrium** of the water molecules, some of the carbonic acid dissociates into a bicarbonate ion (HCO_2^-) and a hydrogen ion (H^+). Some of these bicarbonate ions dissociate into carbonate (CO_3^{-2}) and another hydrogen ion when the bicarbonate concentration becomes too high and is off balance. Ocean buffering can be summarized by the following equation:

$$CO_2 + H_2O \leftrightarrows H_2CO_3 \leftrightarrows HCO_3^- + H^+ \leftrightarrows CO_3^{-2} + 2\ H+$$

The equation is in chemical equilibrium when all the compounds are in balance with each other. Currently, the ocean's acid balance is slightly on the alkaline side at around pH 8.0 – 8.3 (pure water is a neutral pH 7.0). Humans have nevertheless greatly affected the global carbon cycle by burning fossil fuels and increasing the amount of CO_2 in the atmosphere.

Figure 3.17 The carbon cycle connects the oceans, land, and the atmosphere.

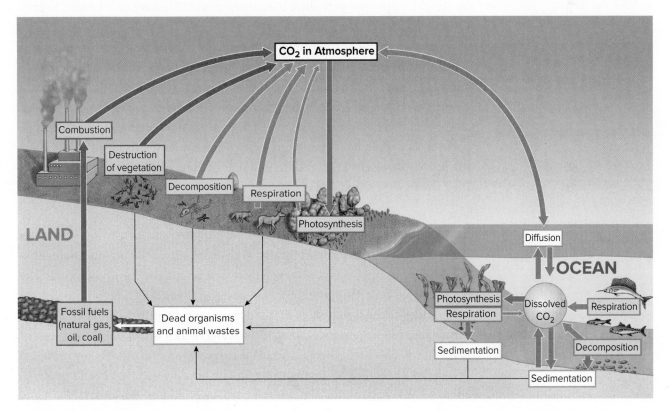

Section Review

1. Recall How deep into the ocean can light penetrate in clear water?

2. Identify What are the two most common ions found in seawater?

3. List three ways in which gases can enter the ocean.

3.3 The Effects of Increasing CO_2 on Ocean Chemistry

VOCABULARY

ocean acidification

calcified

DIVE IN!

What is the level of atmospheric CO_2 that most scientists consider "acceptable"? Some scientists feel that the levels of atmospheric carbon dioxide need to be maintained at 350 ppm in order for conditions on Earth to remain similar to the conditions to which life on Earth is adapted and in which human societies developed. This number is based on scientists studying past climate data and climate change. CO_2 levels have been higher than 350 ppm since 1988. CO_2 levels reached 400 ppm in 2013. At our current global rate of fossil fuel use, levels of CO_2 will reach 450 ppm by 2034.

Main Idea
Increased CO_2 in seawater affects both abiotic and biotic factors in the marine environment.

Key Questions
1. What has happened to carbon dioxide levels in the atmosphere since the 1850s? Why has this change occurred?
2. How does an increase in carbon dioxide in the atmosphere affect the chemistry of ocean water?

Since about 1750 and particularly since the expansion of the Industrial Revolution in the mid-nineteenth century, human activity has caused atmospheric CO_2 levels to increase drastically (Fig. 3.18). CO_2 concentrations have increased by about 1.5 ppm per year since measurements on Mauna Loa, Hawai'i, began in 1958 (Fig. 3.19). The sawtooth pattern in Figure 3.19 reflects the seasonal effect of trees in the northern hemisphere: New growth and the emergence of leaves in spring draws down CO_2, but the death and decay of the leaves in autumn releases CO_2 back into the atmosphere. The CO_2 concentration now exceeds 400 ppm, which is 40% higher than any other time prior to 1750 in at least the last 800,000 years. The rate of increase is more important because the parts of the carbon cycle that remove CO_2 from the atmosphere—dissolution in the ocean, photosynthesis, and deposition in sediments—cannot keep up with the increase.

The Chemistry of Ocean Acidification
Rising CO_2 levels will profoundly affect the chemistry of the ocean. About 30% of the CO_2 we emit into the atmosphere dissolves into the ocean each year. The more CO_2 we emit, the more the ocean absorbs. Because CO_2 is continually being added to the water, more H^+ ions are entering solution. These extra hydrogen ions increase the acidity of the water, which is measured as lower pH. You will recall that the ocean's pH is currently around 8.0. The increase in dissolved CO_2 from fossil-fuel burning has already lowered the ocean's pH by about 0.1 pH units, and it could drop by 0.3 – 0.5 units by the year 2100 if CO_2 emissions continue to grow as they are now. These small changes in pH are more important than they might sound because the pH scale is logarithmic. A pH decline of 1 unit represents a 10-fold increase in the concentration of hydrogen ions, and a pH drop of 0.1 means a 30% increase. The decreasing ocean pH is widely called **ocean acidification**, but the ocean is not expected to become acidic, only somewhat less alkaline. That does mean there isn't a problem, however, for marine organisms.

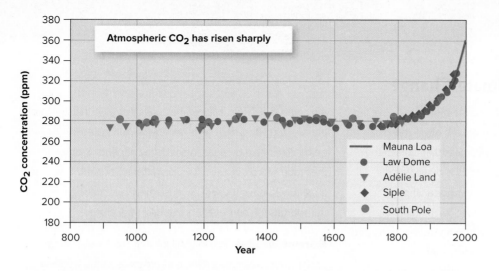

From the equation on p. 73, you can see that an increase in CO_2 results in an increase of CO_3^{-2} and H^+. This creates an imbalance between the bicarbonate and carbonate ions ($HCO_3^- \leftrightarrows CO_3^{-2} + H^+$). Some of the carbonate ions recombine with the hydrogen ions to rebalance the system. To replace the CO_3^{-2} that was lost in this reaction, calcium carbonate ($CaCO_3$) in the ocean dissolves into calcium (Ca^+) and carbonate (CO_3^{-2}) to bring the equation back into chemical equilibrium. Thus, lower pH makes calcium carbonate dissolve easier.

The Effects of Ocean Acidification on Organisms

The organisms that are most at risk from ocean acidification are those with shells, skeletons, or other structures made of calcium carbonate ($CaCO_3$). When the solubility balance in the ocean is altered, organisms have more trouble making their shells and other calcareous structures, requiring more time and energy in an acidic environment. Combined, calcium carbonate dissolution and inhibition of structure formation can negatively impact species spanning the entire food chain.

Figure 3.19 Atmospheric CO_2 readings since 1958 at Hawai'is Mauna Loa.

Coral Reefs and Climate Change

Coral reefs a major hot spot of marine biodiversity, rivaling rainforests in productivity. In fact, reefs are often called the "rainforests of the ocean." They cover less than 0.1% of the ocean's bottom, yet they host at least 25% of the world's marine species. They occur in shallow tropical and subtropical waters that are oligotrophic (low in nutrients). Coral reefs protect shorelines by absorbing wave energy and they are the nursery ground for many fish species.

Corals, and their associated coralline algae, are known as ecosystem engineers, because their very presence changes an environment. Corals have calcium carbonate ($CaCO_3$) skeletons, and coralline algae incorporate $CaCO_3$ into their cell walls. These skeletons provide structure and habitat for the myriad fishes and invertebrates that live in coral reefs. An increase in atmospheric CO_2 causes more CO_2 to diffuse into the ocean, shifting the equilibrium of carbonate species in the water. With lower concentrations of $CaCO_3$, corals and coralline algae will have difficulty continuing to grow.

Coral reefs are also threatened by the increased sea surface temperature and by extreme weather events that are predicted to increase in frequency and intensity. The National Oceanic and Atmospheric Administration (NOAA) has named the impacts from climate change as the greatest threat to the world's coral reefs

Temperature When water temperatures rise above their normal tolerance range, the corals become stressed. Stressed corals expel their zooxanthellae, the symbiotic algae that live within the coral and provide up to 90% of the coral's energy needs. Without zooxanthellae, the corals "bleach," or turn white, which makes them more susceptible to disease. It is difficult for corals to recover from a bleaching event. If the stress that caused the bleaching is prolonged or if there is not another food source available to the corals while they are bleached, they will not survive.

To make matters worse, the high temperatures that cause bleaching are usually associated with high solar irradiance, which favors the growth of macroalgae over corals. More sunlight and higher water temperatures provide ideal conditions for quick algal growth. Fertilizer pollution and overfishing (decreasing herbivores on the reef) can cause macroalgae to grow over the top of the corals. Once algae take over there is little chance for

the corals to survive. Without the corals that build the structure of the reef, the ecosystem shifts from being dominated by corals to being dominated by algae. Animals that depend on corals for their food source, like sea stars, parrotfishes, butterfly fishes, and angelfishes, leave the reef because they have nothing to eat and herbivorous fish and invertebrates take their place.

Extreme weather You might think that coral reefs, being under water, would be protected from the extreme weather events that affect the surface, but this is not the case. Large storms with high wave energy, like hurricanes, stir up sediments that can bury corals, or block out the light that the zooxanthellae need to photosynthesize. Large waves can break off pieces of coral from the reef, which can be very hard for them to recover from because of the slow growth of most coral species.

Some corals are able to adapt to the effects of climate change by keeping large energy reserves to survive bleaching events. Other species are more resistant to large wave action because they don't have delicate branches. Even so, researchers agree that climate change is happening faster than the corals are able to adapt. Keeping corals healthy by reducing other stressors, such as overfishing and increased nutrient pollution in coastal areas, will be the best way to help them cope with the effects of climate change.

Rising sea surface temperatures and other environmental stressors are leading to a predicted increase in bleaching events.

The ecological consequences of ocean acidification are profound, though not as well understood as the chemistry behind the process. Corals and certain groups of plankton, such as pteropods (Fig. 3.20), are most vulnerable because of the type of calcium carbonate they make. Plankton communities are likely to be radically altered, especially in the Southern Ocean where scientists use pteropods as an indicator of ecosystem health. As acidification becomes more severe, it will affect other important plankton, such as foraminifera. Ocean acidification might enhance the growth of plankton that remove CO_2 from the atmosphere or foster the formation of cooling clouds and thus help mitigate climate change. It is also possible that acidification might reduce the ocean's CO_2 uptake and accelerate climate change by increasing its release to the atmosphere. If it ends up negatively impacting plankton, all marine ecosystems will be affected, because plankton form the basis of the food web in most of these ecosystems.

Figure 3.20 Pteropods are important food sources for larger marine animals. They are particularly vulnerable to ocean acidification because of their thin calcium carbonate shells.

In addition to plankton, acidification will affect mussels, clams, snails, sea urchins, coralline algae, barnacles, and a host of other organisms that have **calcified** structures. Researchers have already observed pH-related changes in intertidal communities on the coast of Washington State—mussels, barnacles, and coralline algae are declining as pH levels fall in favor of non-calcified competitors.

Not all the effects of ocean acidification are related to calcification. Greater CO_2 concentrations can cause greater mortality in marine larvae. Ocean acidification can suppress the immune systems of many animals, reduce metabolic rates, and alter its internal chemical balance. Reduced pH can also change behavior. Under acidified conditions, clownfishes become repulsed by the smell of anemones, their natural habitat. By avoiding anemones, clownfishes allow themselves to be more vulnerable to predation. Ocean acidification has also been shown to make squids more sluggish, probably making them less effective predators and more vulnerable to their own predators.

Scientists recently made the unexpected finding that acidification makes underwater sound go farther, potentially affecting whales and other animals that use sound to communicate or that are disturbed by ship noise or other artificial sounds. Studies of the effects of acidification are relatively new, and no doubt other effects will be discovered. The oceans are acidifying, and although we don't fully understand the effects, we do know that the consequences are potentially huge.

· ·

Section Review

1. **Explain** Why is the rate of atmospheric CO^2 increase such a cause for concern?

2. **Define** What is ocean acidification?

3. **Describe** How will ocean acidification affect marine organisms with calcified structures?

Chapter ③ Review

REVIEW QUESTIONS

Multiple Choice

1. The area of the ocean in which sound waves, especially low frequency waves, can travel uninterrupted for thousands of kilometers is called the

 a. Deep Scattering Layer.

 b. Deep Sound Channel.

 c. depth profile.

 d. thermocline.

2. Inorganic carbon that dissolves in the ocean from the atmosphere is converted into organic compounds by

 a. buffering

 b. cellular respiration.

 c. photosynthesis.

 d. sedimentation.

3. How do rising atmospheric levels of CO_2 affect ocean chemistry?

 a. The pH of ocean water decreases.

 b. The pH of ocean water increases.

 c. There is a decrease of CO_3^{-2} and H^+ in ocean water.

 d. It becomes more difficult for calcium carbonate to dissolve in ocean water.

Short Answer

4. Why is the high heat capacity of water important to marine organisms?

5. What is salinity? How is it measured and expressed?

6. Explain the process of buffering that occurs in oceans.

Critical Thinking

7. In unusually cold winters the northern Black Sea sometimes freezes, while the nearby Adriatic Sea usually does not, even though it is just as cold. Freshwater runoff gives the surface of the Black Sea a low salinity of about 18‰. What would you guess about the salinity of the Adriatic?

8. Will ocean acidification affect organisms with silica skeletons?

DATA ANALYSIS LAB

How does an increase CO_2 in the atmosphere affect variables in seawater? The Mauna Loa Observatory is located on the Mauna Loa Volcano on the island of Hawaii. Scientists at the Observatory have been collecting data on climate change, including atmospheric levels of carbon dioxide, methane, and ozone, since the 1950s. The Observatory is considered the baseline station for atmospheric data.

Data and Observations

The graph shows three sets of data—the concentration of CO_2 in the atmosphere and in the ocean, as well as ocean pH levels near the Mauna Loa recording station in Hawaii.

Think Critically

1. Using a line of best-fit, describe the relationship between the concentration of CO_2 in the atmosphere and the concentration of CO_2 in seawater at this recording station?

2. Using a line of best-fit, describe the relationship between the concentration of CO_2 in seawater and the pH of seawater at this recording station?

3. Correlation does not imply causation. Explain the experiment you could perform to show how each avariable actually affects the other.

*Data obtained from: Arguez, A., et al., 2008. State of the Climate 2007. Bulletin of the American Meteorological Society, vol. 89, no. 7, July, pp. 1–181.

Waves and Tides

Chapter
4

THEME Patterns

By making observations and collecting and analyzing data, scientists can find patterns that help them better understand the nature of waves and tides and predict their patterns.

BIG IDEA

Waves and tides are influenced by many factors and, in turn, affect marine organisms as well as humans.

About the photo A wave breaks as it approaches the shore off Cape Kiwanda, Oregon.

Design Pics/Natural Selection Craig Tuttle

4.1 Introduction to Wave Energy and Motion

Main Idea

Waves carry energy across the sea surface but do not transport water.

Key Questions

1. What are the three most common generating forces of waves?
2. What are the two restoring forces that cause the water surface to return to its undisturbed state?

When you think of the ocean, chances are that waves are one of the first things that come to your mind. Anyone who has swum in, sailed on, or simply walked beside the sea is familiar with waves and tides. Waves happen in every kind of body of water, whether you think of waves crashing on the shore, a surfer riding the perfect wave, the waves following a passing ship, or the ripples in a puddle when the wind picks up. In this chapter, we will describe the origin and parts of a wave and their characteristics, and we will describe the world's largest waves, the tides.

Wave Formation

Waves are started by disturbances called **generating forces**. Although there are many generating forces for waves, we will discuss the three most common ones—wind, earthquakes, and landslides.

How Waves Start When wind blows across the surface of a body of water, it creates friction between the air and the water. The drag along the water causes small **capillary waves** to form. Capillary waves are the smallest of the wind driven waves. They are also called ripples (Fig. 4.1). Patches of these waves are seen across the surface of the water on a windy day, and they disappear when the wind dies down. As the wind continues and capillary waves grow bigger, the surface of the water becomes rougher. When the smooth surface of the water is disturbed, more energy is transferred from the air to the water, making it easier for the wind to grip the water, forming even larger waves. If the wind continues to blow, it pushes the peaks of the waves up and stretches out the troughs. Larger waves are formed as the wind continues to increase, and they move away from their source slightly faster than the wind that formed them.

After the waves have moved away from the wind or storm that formed them, they settle into **swells**. Swells are evenly spaced waves with smoothly rounded crests and troughs. Swells start out as progressive wind waves, but as they organize into evenly spaced waves, the shorter waves from the storm dissipate, so only long waves carrying a large amount of energy are left. They are a subset of progressive wind waves even after they settle into swells. Swells carry energy very long distances across ocean basins. Groups of swell waves generated by

DIVE IN!

What are rogue waves? A rogue wave is usually defined as a wave that is greater than twice the size of surrounding waves and can reach heights of up to 33 m. One area of the ocean that is known for producing rogue waves is off the southeast coast of South Africa, where large swells meet the fast-moving coastal current. Rogue waves can cause severe damage to ships as well as injuring people aboard.

Figure 4.1 Capillary waves, the smallest wind-driven waves, form on the surface of the water.

Figure 4.2 Swell waves are generated by the wind and can travel long distances across ocean basins.

large storms in the Pacific Ocean between 40°S and 50°S have been traced traveling north through the Pacific to the shores of Alaska. Swell waves very closely follow the ideal pattern of waves, and when you think of gentle rolling waves in the open ocean, you are imagining swell waves. Swell waves are still categorized as progressive wind waves, but they are different from the progressive wind waves that are formed in storms; they are organized into a steady pattern of evenly spaced crests and troughs (Fig. 4.2).

A second generating force for waves is earthquakes. When the seafloor is disturbed, the water above it moves up or down with the seafloor. This causes a large amount of water to be displaced, and it can cause waves to move away from the source of the earthquake. These waves, called **tsunamis**, can travel very long distances and cause great damage to coastal areas. Tsunamis are not produced with every earthquake, and they can be very hard to track.

The third most common wave-generating force is a disturbance created by landslides or ice breaking off of glaciers. This process is similar to throwing a stone into a lake. When a large piece of land or ice enters the ocean, it moves the water out of the way. As the land or ice settles or sinks, the water that was displaced moves back to its original position. This movement creates a series of waves that move away from the location where the material originally entered the water.

Many factors influence the formation of waves. The first is wind speed. Higher wind speeds create larger waves. **Fetch**, or the distance of open water that the wind can blow across without changing distance, also influences wave formation. Areas that have a large fetch will produce more waves than areas with a small fetch; similarly, the wider the fetch, the more waves that can form (Fig. 4.3).

More waves are formed when wind continues for a long time. Water depth also influences wave formation. Waves are more easily produced in shallow areas than in deep water.

Wave height is dependent on three factors: wind speed, fetch, and wind duration. The largest waves form when all three factors are maximized. Figure 4.4 shows the average wave height of the world's oceans. The most favorable conditions for wave formation occur between 40°S and 50°S, where there are large areas of open ocean with few landmasses to limit the fetch of the wind. Sailors call this area the "roaring forties" and "furious fifties" for the large and notorious storms that produce large waves. Predominantly westerly winds blow almost continuously in this area, which combined with a large fetch, produce very high waves. Large waves can be produced anywhere, but this area is infamous for the largest waves in the world.

Wave Dissipation Waves do not carry energy forever. The energy is dissipated, and the water particles stop moving because of restoring forces. A **restoring force** is a force that causes the water surface to go back to its undisturbed state. There are only two restoring forces—surface tension and gravity. Surface tension is responsible for restoring small capillary waves. Surface tension isn't strong enough to restore waves any bigger than small ripples of waves. Gravity is a powerful force that is responsible for restoring large waves.

Wave Energy and Motion

A wave has two different types of energy—potential and kinetic. The potential energy comes from the height of a wave. Potential energy is the energy that an object has because of its position above the surface of Earth. The higher the wave is, the more potential energy the water within that wave has. The kinetic energy of a wave comes from the motion of the water particles. Kinetic energy is the energy of objects in motion, so the water particles in a wave exert kinetic energy when they move.

Orbital Motion Waves carry energy across the sea surface but they do not actually transport water. How can this be? It is important to remember that a

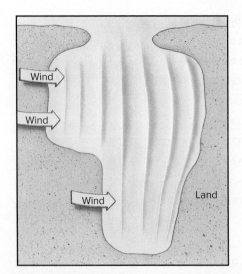

Figure 4.3 Wind speed is equal across this bay, but the fetch is longer on the wider portion, so the waves are bigger there.

Figure 4.4 Average wave heights across the globe. Areas in red have the highest average wave height, while areas in purple have the lowest.

Courtesy NASA/JPL-Caltech.

wave is *energy* moving through the water, not just the physical movement of the water you actually see. Wave energy travels in a particular direction, the direction of propagation. As the wave energy moves through the water, it moves water particles up and down in an orbital motion. When a water particle moves with a wave, the wave energy causes the particle first to move up, then forward with the wave energy. The energy of the wave then continues forward, but the particle of water moves down under its original position and back behind where it originally was. When the next wave arrives in that spot, the particle moves up and forward again, completing its orbit. Each particle of water in the wave only moves in circles, and only the energy of the wave continues to move forward (Fig. 4.5).

Direction of propagation

Figure 4.5 Water particles move in a circular pattern, the do not move in the direction of the wave. Anything floating on or in the water rear the surface moves in circles as well.

Wave Anatomy The highest point of a wave is called its <mark>crest</mark>, and the lowest point is its <mark>trough</mark>. Wave crests or troughs can be close together or far apart. The height of a wave is the distance from the crest to the trough, but scientists more often use the <mark>amplitude</mark> of a wave, which is one half of the wave height. The <mark>wavelength</mark> is the distance from crest to crest or trough to trough. The equilibrium surface is the undisturbed ocean surface, or sea level (Fig. 4.6).

Wave speed is determined by wavelength and wave <mark>period</mark>, which is the time it takes for two consecutive crests or troughs to pass the same point. The <mark>frequency</mark> is the reciprocal of the wave period. It is equal to the wavelength divided the wave period. In deep water, wave speed is dependent on the wavelength of the wave, not on the depth of the water. In shallow water, wave speed depends on water depth, not on the wavelength. After a wave is formed, its speed can change, but its period remains the same. The period of a wave is determined by the generating forces of the wave; so even as the wave moves, its period cannot change.

Figure 4.6 An idealized series of waves, with labeled anatomy.

The surface of the ocean is much more complex than waves moving in a single direction with the wind that formed them. It is often a confused jumble of many different waves going in different directions because they were formed by winds of different speeds blowing in different directions for varying lengths of time. The intricate surface of the ocean results from the interactions of all the waves.

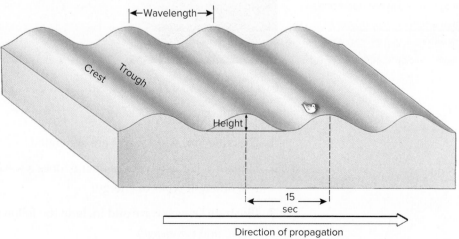

Wavelength

Crest Trough

Height

15 sec

Direction of propagation

Wave reinforcement (waves in phase)

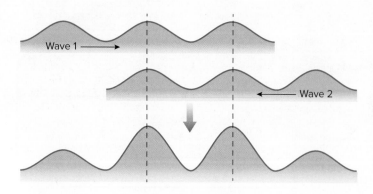

Wave 1

Wave 2

Wave cancellation (waves out of phase)

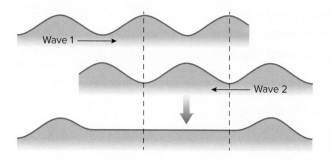

Wave 1

Wave 2

Figure 4.7 Wave cancellation results when waves meet out of phase. When waves meet in phase, however, wave reinforcement occurs.

Figure 4.8 Rogue waves — large waves in the open ocean that seemingly come out of nowhere — can be caused by wave reinforcement.

Wave Interaction **Wave cancellation** occurs when the crest of one wave meets the trough of another, and the sea surface ends up being intermediate between the two, effectively canceling out the waves. If the crests of two waves collide, however, they add together to produce a higher wave which is called **wave reinforcement** (Fig. 4.7). Wave reinforcement can cause very large waves to form seemingly out of nowhere. These rogue waves can be as high as a 10 story building and can cause great damage. Oceanographers once dismissed rogue waves as rare or even nonexistent and nothing more than a sailors' myth, but they have now confirmed that rogue waves do occur. They might account for a significant fraction of the 100 or so large ships that are lost, or disappear, at sea each year (Fig. 4.8).

Section Review

1. **Describe** how wind waves are generated.

2. **Explain** What two types of energy does a wave have? How can a wave have two different types of energy?

3. **Illustrate** Draw a wave and include the following labels: *crest, trough, wave height,* and *wavelength.*

4.2 Types of Waves

Main Idea
There are several types of waves, each with its own characteristics and different ways of traveling through the ocean.

. .

Key Questions
1. List different types of waves.
2. What does it mean when a wave is refracted?

VOCABULARY

deep-water waves

shallow-water waves

progressive wind waves

refracted

longshore transport

rip currents

tsunamis

intermediate waves

internal waves

standing waves

node

antinodes

There are two broad categories of waves: **deep-water** and **shallow-water waves**. The two categories of waves have different characteristics and travel differently through the ocean. Whether a wave is a deep-water wave or a shallow-water wave depends on the depth of the water and the wavelength. A wave can transition from being a deep-water wave to a shallow-water wave and vice versa depending on the depth of the water. Waves that are in between a deep-water wave and a shallow-water wave are called **intermediate** waves, having characteristics of both types (Fig. 4.9).

Deep-Water Waves

A deep-water wave is a wave that travels in water that is deeper than one half of its wavelength. The speed of deep-water waves depends on the wavelength of the wave, not on the depth of the water. Like all ocean waves, deep-water waves do not move water; they carry and move energy from one place to another. There are two main types of deep-water waves. Most waves that you see in the ocean or in large lakes are deep-water waves known as **progressive wind waves**. These waves are generated by wind, restored by gravity, and progress in a particular direction. We have already discussed one special type of progressive wind wave—swell waves. Deep-water waves form when swell

DIVE IN!

What is the tsunami warning system? The tsunami warning systems monitor ocean basins for earthquakes and the tsunamis that can be generated as a result. The warning systems help people take measures to get out of harm's way before a tsunami reaches shore. Tsunami warning systems in the United States are similar to tornado warnings. A broadcast by the Emergency Alert System is broadcast on television and radio, and sirens may go off as well. Designated evacuation routes exist in many coastal communities.

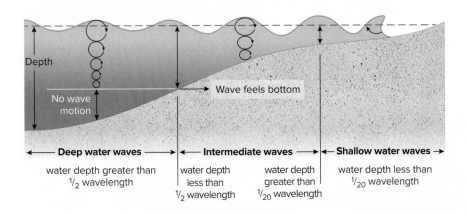

Figure 4.9 Differences between deep- water, intermediate, and shallow-water waves.

Figure 4.10 As waves approach the shore, the bottom of the wave slows down before the top, causing the wave to break and create surf.

waves originate and travel in water that is deeper than one half of their wavelength. As swell waves approach the shore, they transition from deep-water to shallow-water waves. Wind-generated waves can be deep or shallow, depending on the depth of the water where the wave travels and the wavelength of the wave.

Shallow-Water Waves

Shallow water waves travel in water that is less than 1/20th of their wavelength. Their speed depends on the depth of the water, not on the wavelength of the wave. As a deep-water wave approaches the coast, its bottom starts to "feel" the bottom of the ocean (Fig. 4.10). The interaction of the wave and the seafloor changes the pattern of the orbits made by the water particles in the wave from circles to flattened circles, or ellipses. The friction between the wave and the seafloor causes the wave to slow down even more. As the waves behind it catch up, they get closer together, giving them a shorter wavelength. The waves pile up, becoming higher and steeper. A wave breaks, or spills over, when the bottom of the wave is slowed down by the seafloor but the top of the wave does not slow down, creating what is known as surf (Fig. 4.11). The energy that was originally put into the wave by the wind is expended on the shoreline as the wave breaks. If you are on a beach on a day when the waves are large, then you can feel the energy of a wave as it crashes onto the beach. Large surf waves seem to shake the ground and are so loud that they can make conversation difficult.

Figure 4.11 Waves change shape as they move from the open ocean to the coast. As the water gets shallower, the orbital motion of water particles shifts to become more elliptical. The drag of the wave across the bottom is what causes waves to break and form surf.

Figure 4.12 Waves refracting as they approach a beach

Figure 4.13 Various structures are built on beaches to prevent erosion due to longshore transport.

Wave Refraction Waves are **refracted**, or bent, as they move from deep water to shallow water. When waves approach the shore, they usually do so at an angle. When the bottom of the waves slow down as the water gets shallower, the part of the wave that remains in deeper water continues to move at the original speed. Because of this, the waves "bend" and the waves along the surf are parallel to the shore (Fig. 4.12). Waves can also bend and wrap around an island depending on the seafloor surrounding the island.

Longshore transport As discussed above, waves usually don't travel perpendicularly towards the shore. As a wave moves towards the shore, the particles of water are transported toward the shore and down the beach in the

Water transported in surf zone

Wave crests

Rip current

Wave direction

Figure 4.14 The formation of a rip current

direction of the waves. This is called **longshore transport**. Longshore transport is responsible for moving sand and sediments down a beach. When the wave feels the bottom, the energy in the wave picks up sand and sediment on the seafloor. The energy of the wave moves the sand with it, so the sand ends up moving down the beach (Fig. 4.13). This also causes water to build up along the beach. The areas where it can return to the ocean are sometimes narrow. This creates **rip currents**, or areas of rapid seaward-flowing water (Fig. 4.14). These can be dangerous to swimmers. Beaches that are prone to rip currents often have signs warning swimmers, and most water-safety courses teach ocean goers to swim parallel to the beach if they get caught in a rip current.

Tsunamis Sudden movements of Earth's crust, such as an earthquake or an underwater volcano eruption, can cause **tsunamis**. Tsunamis, or seismic sea waves, can travel very long distances and cause great damage to coastal areas. If a large portion of Earth's crust is moved because of seismic activity, then the water column above it might be displaced. Gravity causes the water to be pulled back toward its equilibrium state, which creates a wave with an extremely long wavelength (200 km long or more). Because of their extremely long wavelength, tsunamis are actually shallow-water waves. They can travel very long distances across entire ocean basins at over 600 km per hour while carrying great amounts of energy. Because they carry so much energy, they have the potential to cause great damage when they reach the shore (Fig. 4.15).

Tsunamis can be several meters high at their origin, but that height is distributed over the entire length of the wave. If you add the other swells and disturbances in the ocean on top of that, then a tsunami can be hard to track across open water, which is another reason they can be so dangerous. When tsunamis reach shallow water, they break just like any other shallow-water wave, but they can create very large surges of water that can flood the coast for up to 5 to 10 minutes before flowing back toward the sea. Because tsunamis are hard to follow across ocean basins and because not all earthquakes produce

Figure 4.15 Tsunamis carry a great deal of energy, and can cause significant damage when they hit coastal communities.

Tech. Sgt. DeNoris Mickle/USAF

(a)

Sea surface

Low density water

Pycnocline →

High density water

Internal wave propagation direction

(b)

Figure 4.16 Waters of different densities form internal waves below the sea surface (a). Internal waves are sometimes visible from the sea surface. In the Strait of Gibraltar, stratified water flowing between the saltier Mediterranean Sea (bottom right) and the relatively fresher North Atlantic Ocean(upper left) forms a large internal wave 100 m below the surface. The swell-like oscillations are visible from space (b).

tsunamis, oceanographers and geologists do not have a completely accurate tsunami warning system. Computer models help predict when a tsunami will arrive in an area, and pressure sensors across the ocean can help track a tsunami across the water. But even with these modern technologies, tsunamis can cause great damage to coastal areas.

Other Types of Waves

Waves traveling in depth between half and 1/20th of their wavelength are intermediate waves and have characteristics of deep- and shallow-water waves. If the wave is closer in depth to half of its wavelength, then it acts more like a deep-water wave. If it is closer in depth to 1/20th of its wavelength, then it acts more like a shallow- water wave. Intermediate waves show characteristics between deep and shallow-water waves.

Not all waves form at the boundary between air and water. Waves can form at any boundary between water of two different densities. Waves forming under the surface are called **internal waves**. Internal waves regularly form where water is stratified in layers of different densities (pycnoclines)(Fig. 4.16a). Where the waters meet, they oscillate and form a wave below the sea surface (Fig. 4.16b). Internal waves vary greatly in wavelength, and can have periods from seconds to hours. Internal waves are important in ocean mixing, especially in areas that have a large amount of freshwater runoff. Internal waves that form between the less dense fresh water and the more dense seawater help mix the two together. Internal waves can also play an important part in transporting eggs of fishes and other animals that reproduce in the open ocean to the coastal areas where the young fish grow into adults.

Deep-water, shallow-water, and internal waves are progressive waves, or waves that progress in a particular direction. **Standing waves**, in contrast, are

Waves That Kill

Surfers and beachgoers who've been hit by a wave know how forceful they can be. Most waves are pretty harmless, but coastal storm waves cause damage every year, and sometimes deaths. Occasionally, seismic sea waves, known as tsunamis, unleash all of the awesome power of the sea.

Also called tidal waves, tsunamis are unrelated to tides and are produced by seismic disturbances like earthquakes, landslides, and volcanoes. Tsunamis are much longer and faster than ordinary waves. They can have wavelengths of 240 km and travel at speeds greater than 700 km/h. In the open ocean, tsunamis are typically less than 1 m high and are hardly noticeable. They grow as they reach shore, but most tsunamis do little damage.

Fatal tsunamis Once in a while, though, tsunamis are devastating. In 1883 the volcanic island of Krakatoa in the Indian Ocean exploded, unleashing tsunami waves around half the globe that killed over 35,000 people.Fatal tsunamis occur every year or two on average, mostly in the Pacific because of seismic activity around the Pacific Rim. A 1946 tsunami in Hawai'i killed 159 and another in 1960, the result of Chilean earthquake, killed 61. The U.S. west coast suffered deadly tsunamis caulst by both the 1960 Chile earthquake and the great Alaskan earthquake of 1964. A 1998 tsunami in Papua New Guinea left 2,200 known dead. Many other places have been hit. Between 1992 and 2011, tsunamis killed more than a quarter of a million people.

The deadliest tsunami struck in the Indian Ocean on December 26, 2004, triggered by a massive earthquake off the Indonesian island of Sumatra. Within minutes of the earthquake, waves at least 10 m high (estimpates range as high as 30 m) smashed onto the Aceh, on the Sumatran coast. In less than eight hours tsunami waves radiated around the Indian Ocean, killing an estimated 230,000 people in 12 different countries. Survivors suffered the loss of their homes, businesses, farms, hospitals, schools, and water and electricity supplies.

In the two years following the Indian Ocean tsunami, fatal tsunamis struck Haiti, Chile, Indonesia, and Japan. In Japan, in addition to killing nearly 16,000 people, the massive waves damaged a nuclear power plant in Kagoshima, releasing radioactive material into the atmosphere and ocean. This made it the worst nuclear disaster since Chernobyl in 1986.

Worldwide death toll from tsunamis between the years 1992 and 2011.

Epicenter 26 Dec 04 • Epicenter 28 Mar 05

Waves from the Indian Ocean tsunami of December, 26 2004 and the resultant toll of dead and missing. As with many other tsunamis, there was actually a series of waves rather than a single one. Most places were hit by three waves, with the second being the most destructive.

Surviving tsunamis All people must do to survive a tsunami is evacuate to higher ground, but they need to know the tsunami is coming. A tsunami warning system was established in the Pacific for Hawai'i and the west coast of North America in the 1960s, and ha probably saved hundreds of lives. Another system was established in the Indian Ocean following the 2004 tsunami, too late for its victims but hopefully a future life-saver. Unfortunately, many coasts are not covered by warning systems, and even where they exist word may not get out to rural areas or people who aren't tuned in to media. In Hawai'i, media and other electronic systems are supplemented by old-fashioned sirens, but this is the exception and not the rule. An even greater problem is that there may not be sufficient time to issue warnings. Fortunately, educating the public about the relationships between earthquakes and tsunamis can be far more effective than tsunami warning systems. People on the coast need to understand that an earthquake might bring a tsunami. Many in earthquake-prone Japan knew to evacuate after the Sendai earthquake, or there would have been many more dead. In both the 2004 Indian Ocean tsunami and the one in Chile in 2010, schoolgirls were credited with saving hundreds of lives by raising the alarm. But in many places tsunami education still lags. The ocean breifly recedes as a tsunami approaches, and people sometimes move into harm's way to look at the exposed marine live. Where small earthquakes are common people can become complacent; even when they know to move to high ground they may not respond appropriately.

Coastal resiliancy Despite all of our scintific advances, we can't stop tsunamis. We can, however, make wiser choices of where to build on the coast and use better engineering. Since the 2011 tsunami in Japan, higher seawalls are bing built around many nuclear plants. Coastal areas can also increase resiliency by protecting natural habitats such as reefs, mangrove forests, and salt marshes that, in turn, protect the coast. As tsunami waves move over natural barriers like coral reefs, their energy is diminished through friction, mitigating the tsunami wave intensity that ultimately makes landfall.

Humans will always have to live with the threat of tsunamis as long as they chose to live along coastal environments, but with advanced warning systems, strong educational awareness programs, and natural habitat protection they don't have to die with them.

The March 11, 2011 tsunami in Sendai, Japan washed away buildings and complete sections of the city in its path

STR/AFP/Getty Images

(a)

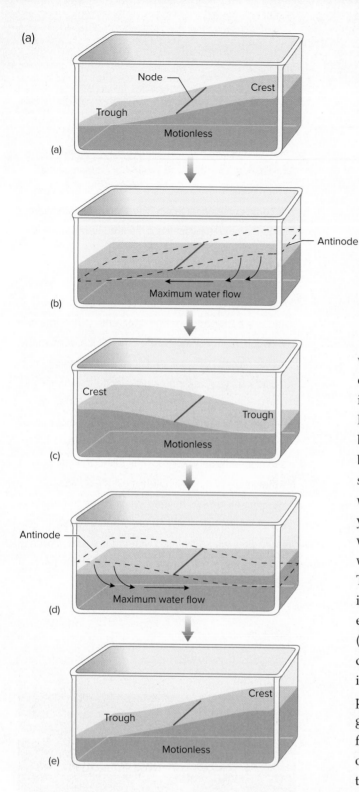

(a)

Node
Crest
Trough
Motionless

(b)

Antinode
Maximum water flow

(c)

Crest
Trough
Motionless

(d)

Antinode
Maximum water flow

(e)

Trough
Crest
Motionless

Figure 4.17 Instead of progressing in one direction, standing waves oscillate around a single point called a node (a). Standing waves that form regularly in bays and inlets are popular tourist attractions and surfing spots (b).

(b)

waves that do not progress because they form in enclosed bodies of water such as bays and estuaries. Instead, they reflect back upon themselves. Imagine you have a rectangular basin with a flat bottom and sides. If you were to partially fill the basin with water and lift one side up, then rapidly set it back down, you would see that the water would flow back and forth within the container. If you did this repeatedly, you would see one area where the water stayed roughly the same height, with water oscillating up and down on either side. The point around which a standing wave oscillates is called the **node**. The high and low points at each end of a standing wave are called the **antinodes** (Fig. 4.17a). Standing waves form in enclosed coastal basins for the same reason they would form in your imaginary rectangular basin – instead of progressing forward, the wave hits a barrier (the geology of the basin) and reflects the energy back, forming a stable node around which the antinodes oscillate. The node of the standing wave is sometimes at the opening to the bay, so there is minimal water movement at the opening and larger movement at the end on the shore where the antinode is located (Fig. 4.17b)

National Oceanic & Atmospheric Administration (NOAA)

Section Review

1. **Contrast** a deep-water and a shallow-water wave.

2. **Explain** how a tsunami forms.

3. **Describe** What are internal waves and where do they form?

4.3 The Tides

Main Idea
Tides result from the gravitational pull of the moon and the sun and from the rotation of the Earth, moon, and sun.

. .

Key Questions
1. What are tides?
2. What are the three main tidal patterns?

The sea surface has been rising and falling in the rhythmic pattern known as tides for billions of years. The tides are a dominant influence on nearshore sea life. They alternately expose and submerge organisms on the shore, drive the circulation of bays and estuaries, trigger spawning, and influence the lives of marine organisms in countless other ways (Fig. 4.18).

Tides do not affect the middle of the ocean basins. They have a pronounced effect on nearshore areas. In fact, there is an entire ecosystem called the intertidal zone that is determined by the tides and would not exist if it weren't for the rising and falling of the tides each day (Fig. 4.19).

DIVE IN!

When do we observe the highest high tides? They occur when the moon is either new or full and closest to Earth during its revolution. Every 28 days, during its elliptical orbit around Earth, the Moon reaches its nearest proximity to Earth. This is referred to as a perigee. Three to four times a year a new moon or a full moon will occur when the moon is in perigee. This produces perigean spring tides, a time when the highest high tides occur. In some low-lying coastal areas, perigean spring tides can cause minor flooding.

©Jeff Foott/Discovery Channel Images/Getty Images

Figure 4.18 Spawning in these grunion (*Leuresthes tenuis*) is precisely timed to coincide with the highest tides, when grunion can reach the uppermost parts of the beach. Similarly, the hatching of the eggs also corresponds to high tide so that the young can swim away.

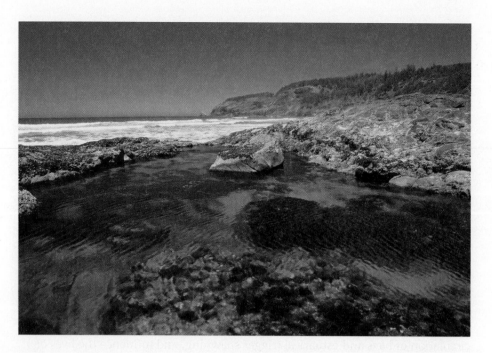

Figure 4.19 The rise and fall of the tides forms the ecosystem known as the intertidal zone, which has unique inhabitants and characteristics.

What Causes Tides?

Tides are essentially very large waves that travel around Earth. The tides are caused by the gravitational pull of the Moon and the Sun and by the rotation of Earth, the Moon, and the Sun. The Moon and Earth are held together by gravitational attraction. The Moon's gravity is strongest on the side of the planet that is closest to the Moon. The Moon's gravity pulls the water in the ocean toward the Moon, so if Earth were covered entirely by water, then the ocean would bulge toward the Moon (Fig. 4.20). On the opposite side of Earth, the Moon's gravitational pull is weaker, so the water does not move toward the Moon. In fact, it bulges away from the Moon because of centrifugal force, the same force that keeps water in a bucket if you swing it around very fast (see Fig. 4.20). Because of gravity, you would expect the water to flow from the bucket to the ground, but centrifugal force pushes the water back into the bucket and keeps it from spilling. This centrifugal force on Earth arises because, strictly speaking, the Moon does not revolve around Earth. Instead, Earth and the Moon rotate together around their combined center of mass, which lies at a point inside Earth that is offset slightly from the actual center of the Earth (due to the mass of the Moon). This is called the Earth-Moon system. The offset causes the Earth-Moon system to wobble slightly, like an unbalanced tire, and it creates a centrifugal force that pushes water away from the Moon. Thus, on a water-covered Earth, the water would form two bulges on opposite sides of the planet, one bulge toward the Moon where the Moon's gravity dominates and the other bulge away from the Moon where centrifugal force predominates. The water would be relatively deep under the bulges and shallow away from the bulges because the pull of gravity is stronger than the centrifugal force (see Fig. 4.20).

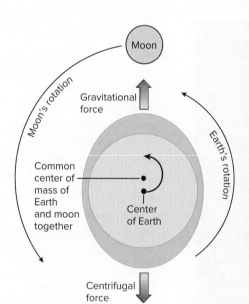

Figure 4.20 The Earth-Moon system. The Moon's gravity causes water to bulge toward it on the side of Earth closest to the Moon. On the opposite side, farthest from the Moon, centrifugal force causes water to bulge in the opposite direction.

In addition to the rotation of Earth and Moon, Earth is spinning like a top on its own axis. As it does this, any given point on the planet's surface will alternatively lie under a bulge and then away from it. High tide occurs when the point is under a bulge. Because Earth takes 24 hours to complete a rotation, the point will have two high tides and two low tides every day. The Moon actually advances a little in its own orbit in the course of 24 hours. It takes the point on Earth an extra 50 minutes to catch up and align directly with the Moon again. A full tidal cycle, therefore, takes 24 hours and 50 minutes, causing the high and low tides in an area to vary daily (Fig 4.21).

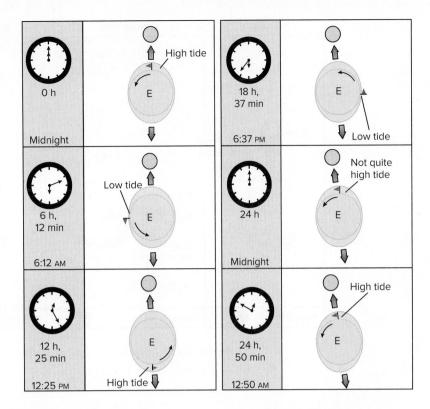

Figure 4.21 Because the Moon moves while Earth is rotating, a full tidal cycle takes longer than one 24 hr day.

The Sun produces tidal bulges in the same way as the Moon. Although the Sun is much larger than the Moon, it is 400 times farther away, so the effect of the Sun on the tides is only about half as strong as the Moon's. When the Sun and the Moon are in line with each other, which happens at the full and new moons, their effects add together. At these times, the tidal range, or the difference in water level between successive high and low tides, is large. Such tides are called **spring tides** because they seem to surge up like a spring of water. The name is unrelated to the season because spring tides occur throughout the year. When the Sun and the Moon are at right angles, their effects partially cancel each other. During these **neap tides**, the tidal range is small. Neap tides occur when the Moon is in the first and third quarters. Because the Moon is the dominant cause of the tides, a full tidal cycle is the same as a lunar cycle, which is 29½ days. When the Moon is in the first quarter, the Sun and the Moon are at right angles to each other, which causes a neap tide. One week later during the second quarter, the Moon is full and the Sun, the Moon, and Earth are in alignment, so the combined gravitational pull of the Sun and the Moon will cause larger spring tides. The third quarter occurs one week later when the Sun and the Moon are back at right angles to each other and neap tides occur. In the fourth quarter during the new moon, another spring tide is produced (Fig. 4.22).

Tide levels also change with the seasons because the distance between Earth and the Sun changes with the seasons. During winter in the northern hemisphere, Earth is closer to the Sun than it is during the summer, so the Sun has a stronger influence on tides. No matter what the season, the Moon always has the strongest influence on the tides.

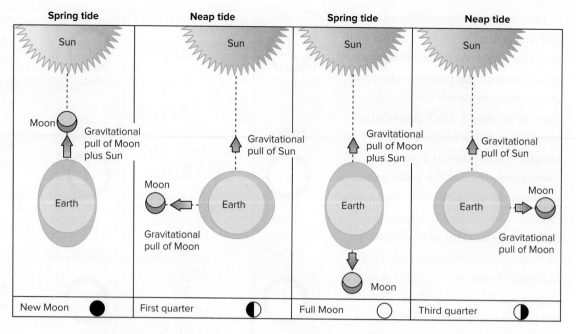

Figure 4.22 Spring and neap tides result from the interaction between the gravitational pulls of the Moon and the Sun.

Tidal Patterns

Because of the continents and the shape of the seafloor, tides in the real world behave differently than they would on a water-covered Earth. There are three main tidal patterns: **semidiurnal**, **mixed semidiurnal**, and **diurnal** (Fig. 4.23).

Most places have semidiurnal tides as predicted, meaning that they have two high and two low tides of approximately the same height each day. The east coast of North America and most of Europe and Africa have semidiurnal tides. Mixed semidiurnal tides occur when an area has two high tides and two low tides of different heights each day. Most of the west coast of the United States and Canada have mixed semidiurnal tides. Diurnal tides occur when there is only one high and one low tide each day. Diurnal tides are uncommon. They occur on the cost of Antarctica and in parts of the Gulf of Mexico, Caribbean, and the Pacific. The distinctions between the types of tides are not absolute; in most places, the tidal pattern varies shift between the different types. Southern

Figure 4.23 Tidal patterns can be semidiurnal (a), mixed semidiurnal (b), or diurnal (c). In many places the tidal pattern fluctuates.

California, for example, has a predominantly mixed semidiurnal tide, but on some days the tides are almost diurnal (Fig. 4.24).

Tides vary from place to place and time to time because of the effects of land and the seafloor. On a water-covered Earth with a perfectly flat seafloor, the tides would move in one smooth, symmetrical wave. In reality, however, the continents prevent that single wave from sweeping around the globe, essentially breaking it into several separate waves within individual ocean basins. Islands, ridges, basins, canyons, reefs, and other features further affect the waves. The tide at a given place thus depends on its location relative to these features. The tides are also influenced by variation in the orbits of the Sun and the Moon relative to the Equator, and the planets have small effects. Even the weather can affect the tides. Strong winds, for example, can pile water upon the shore, causing higher tides than would otherwise occur. In addition to these factors, tides are affected by ocean currents and the Coriolis effect. To predict the tides at a given place, oceanographers must combine observations of actual tides at that place with theoretical equations. They are then able to produce remarkably accurate tide tables that predict the time and height of high and low tides. Tide tables are available for most coastal areas.

Tide Levels The highest level that a tide rises in a day is called high water. Consequently, the lowest point that a tide drops is called low water. In a mixed semidiurnal tide, it is necessary to distinguish between the two high and two low tides; that is, between the higher and lower high water and the higher and lower low water. The tidal range is the difference between the highest high tide and the lowest low tide. The tidal range is not a constant number because it changes depending the high and low tides of the day.

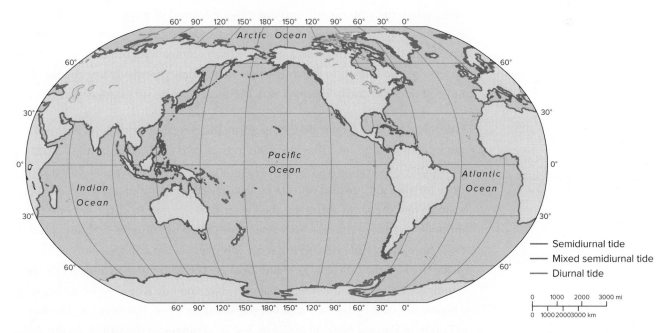

Figure 4.24 The worldwide distribution of semidiurnal, mixed semidiurnal, and diurnal tides.

Figure 4.25 A generalized tidal cycle

Average tides are calculated for a given place using many years of data. This helps oceanographers calculate tide levels. Average low tide is especially important in coastal navigation, where oceanographers use the average low tide to determine the actual depth of the water and when it is safe for boats to navigate through shallow water. Tide levels are given in distance from the mean sea level, which is why it is possible to have a minus tide. A minus tide occurs when a low tide is lower than the average low water level (Fig. 4.25).

If you are at the beach during a rising tide, called a **flood tide**, then you might notice that the beach is shrinking because the water is moving toward the shore. Similarly, if you notice that the beach growing, then you are probably at the beach during the **ebb tide**, when the water is rushing back toward the sea. The period of time between the flood and ebb tides is called slack water, when the water is still and not moving toward or away from the shore. Fishermen pay special attention to the flood and ebb tides because some fish travel with the tides. The waves in an area can change with the tides as well, so surfers are also known to watch the tides. The tidal currents during the flood and ebb tides can move very fast and, therefore, can be very dangerous.

Tidal Range The largest tidal ranges, the difference in water level between successive high and low tides, are seen in narrow basins. Tides in these long, narrow bays and estuaries produce a standing wave, or a wave that doesn't move forward and instead reflects on itself, oscillating within an enclosed basin. When the node of the standing wave is at the mouth of the bay and the antinode is at the shore, the tidal level is amplified because the standing wave oscillates around the node; so the wave is highest at the antinode or the shore. Inside the bay, the tide level is higher because the water level is affected by the standing wave and by the tide; outside the bay, the tide would be lower because it is only affected by the tide and not the standing wave. For example, the Bay of Fundy in Canada has one of the largest tidal ranges in the world. At the mouth of the long, narrow bay, the tidal range is only about 2 m, but at the shore the tidal range is as high as almost 11 m (Fig. 4.26). Accordingly, short,

(a)

(b)

(tl)SimplyCreativePhotography/Getty Images, (tr)Design Pics/David Chapman/Getty Images, (b)Janet S. Robbins

wide bays or estuaries can reduce the tidal range because no standing waves form; near the edges of the bay, the tide will be apparent, but as you move along the shore (closer to the middle of the bay) the tide will not be as high as it is at the edges. Estuaries that have a large freshwater input might have decreased tides because of the difference in density between the incoming tide and the freshwater inputs. The saltier water coming in with the tide is denser than the input of fresh water from rivers, so the denser water travels along the bottom and is hidden by the freshwater—the tidal range might not be visible on the shore.

Tidal Bores Tides can produce waves that can travel very long distances. These **tidal bores** can be up to 8 m high and can travel as fast as 64 km/hr. This makes them potentially dangerous phenomena that can cause great damage. Tidal bores affect shipping routes through estuaries and near river mouths. They are relatively rare and don't occur in every river and estuary. Tidal bores most commonly occur in shallow estuaries and river mouths that have very large tidal amplitudes. When the flood tide approaches the estuary and shallow water, it feels the bottom and starts to slow down, but because of the large amount of water and force behind it, when the tidal bore breaks, it continues to move toward the shore and up the river. Bores can be dangerous because of their force and speed. An open area of beach can be covered by a tidal bore in minutes. The world's largest tidal bore occurs in the Qiantang River in China. The dramatic tidal range of the Bay of Fundy also creates a tidal bore in the upper potion of the bay (Fig. 4.27)

Figure 4.26 The Bay of Fundy, in Canada, has one of the most extreme tidal ranges in the world. The difference between low tide (a) and high tide (b) is almost 11 m.

Figure 4.27 A warning sign posted at a beach in the Bay of Fundy, warning visitors about the dangerous tidal bore.

Between the Tides

The intertidal zone is the area that is below water at high tide and above water at low tide. It is often called the littoral zone and can contain many different types of habitats. One of those habitats is tide pools, which are formed when the tide recedes and water is left in depressions in the rocks. They can be big or small, depending on the substrate and type of tides in the area.

Life in the intertidal Life in tide pools can be very challenging to the organisms that live inside of them. Tide pools can go through many different extreme conditions over the course of one day. Seawater is regularly supplied to the tide pools by high tides and waves, but fresh water from rain or rivers can greatly decrease the salinity in the pools. Alternately, high temperatures can cause evaporation, markedly increasing salinity. During neap tides or when there is little wave action, tide pools may dry up completely, so the organisms living in them must either move to another pool or have a way of tolerating desiccation. The waves that bring much needed nutrients and water to the tide pools can also wash animals away, so they must have a way to attach to the substrate during times of strong wave action.

Tide pool living isn't all about tolerating unfavorable conditions. Seaweeds grow easily because of high light levels and regular supply of nutrients from waves and high tides. Many animals, like snails and sea urchins, feed on this abundant food source. Tide pools often are in rock, so there is a varied substrate with many nooks and crannies for

animals to hide in to avoid predators. Many sea birds and other shore dwelling animals feed in tide pools, and some species of fish even use tide pools as nurseries.

Zonation Life in the intertidal zone is a race for light, space, and food. Many organisms are adapted to live in one of four zones: the splash zone, or the high, middle, or low intertidal zone. The splash zone is highest up on the shore, and receives the least input from waves or tides. It may only be submerged during spring tides. Lichens and snails are common in the splash zone. The high intertidal zone, below the splash zone, is exposed to the air most of the time but is submerged during very high tides. Barnacles are often found in the high intertidal zone. The middle intertidal zone is usually submerged, except for extreme low tides when it is exposed. Common organisms in the middle intertidal zone include sea urchins and sea stars. The low intertidal zone is the lowest on the shore and is submerged most of the time. It has the largest number of organisms and an abundance of seaweeds.

Marine scientists know a lot about the intertidal zone because it can be so easy to study during low tides. At high tide, however, high wave action can make experiments or research almost impossible. Many seminal experiments in marine ecology have been conducted in the intertidal zone. The habitats and organisms of the intertidal zone will be further explored in Chapter 13.

Organisms in the intertidal occur as specific zones based on their ability to withstand desiccation and wave action. Here, different types of seaweed form bands.

Mandy Lindeberg, NOAA/NMFS/AKFSC

Tide Measurements and Tables

Because so many factors determine when tides occur, we cannot simply predict the tides by looking at the Moon's location in the sky. Tide charts can be slightly off for a particular area because of the shape of the coast. Coastlines are very diverse, so the tide table for a beach that is just a few miles away might be slightly different than the observed tides. As previously stated, the average tides in an area are used to calculate the different tide levels. These average tides together with theoretical tidal equations help oceanographers predict tide levels throughout the world.

There are 196 tidal stations in the United States, where tides are measured and kept in a large database. This number sounds like a lot of stations, but because they are not on every beach on every coast, local features can change the tides along the coast (Fig. 4.28). Underwater bathymetry, such as reefs and underwater canyons, and the shape of the coastline greatly affect how tides affect local areas.

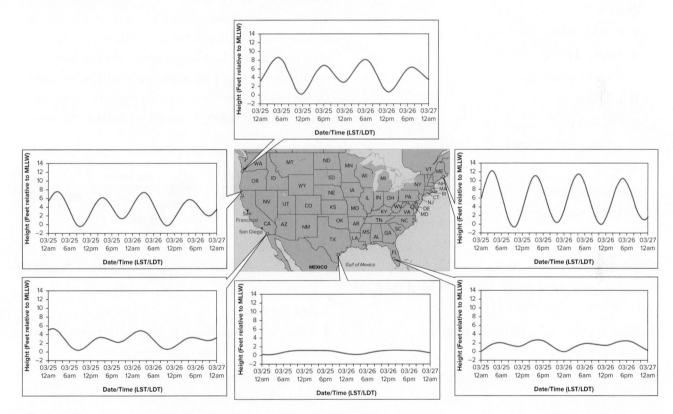

Figure 4.28 Six of the 196 tidal stations on the coasts of the United States. These stations monitor actual tides along with other physical and chemical features of the seawater. You can see the variation in tidal patterns, which are based on the shape of the continental shelf and local shoreline.

. .

Section Review

1. **Explain** What causes tides?

2. **Compare and contrast** spring tides and neap tides.

3. **Illustrate** Draw a diagram of a mixed semidiurnal tide and label the high high water, low high water, high low water, and low low water points.

Chapter 4 Review

REVIEW QUESTIONS

Multiple Choice

1. Which area of the ocean has the most favorable conditions for wave formation to occur?

 a. between 0° and 10°N

 b. between 0° and 10°S

 c. between 40° and 50°N

 d. between 40° and 50°S

2. Which is responsible for moving sand down the beach?

 a. circular orbit

 b. elliptical orbit

 c. longshore transport

 d. rip currents

3. A tide with one low and one high water per day is which type of tidal pattern?

 a. diurnal

 b. mixed diurnal

 c. mixed semidiurnal

 d. semidiurnal

Short Answer

4. Explain how wind speed, fetch, and wind duration influence the formation of waves.

5. What is the role of centrifugal force in forming tides?

6. Why do tides vary from place to place and time to time?

Critical Thinking

7. Most tsunamis occur in the Pacific Ocean. How would you explain this?

8. If you owned a seaside home and a bad storm brought heavy winds and high surf to your coastline, would you prefer it to be during a new moon or a quarter moon? Why?

DATA ANALYSIS LAB

Why does comparing real-time data against computer models matter? On December 26, 2004 one of the most devastating tsunamis in history swept across the Indian Ocean basin. More than 200,000 people were killed and billions of dollars in damages were incurred. With this particular tsunami, scientists were able to collect the most real-time data than ever before, giving scientists new insight into the behavior of tsunamis.

Data and Observations

The graph shows the relative height of sea level in the Indian Ocean on December 26, 2004, about 2 hours after the earthquake that generated the tsunami occurred. The blue lines show sea height as recorded by the Jason-1 satellite, the green portions show sea height as predicted by a computer model at the time.

Think Critically

1. How do the data predicted by the model compare to the real-time data from the satellite?

2. How can the accuracy of computer models of tsunamis help both scientists and the public?

3. How might the continued comparison of real-time data to computer models help improve the accuracy of future models?

*Data obtained from: Geist, E. L., V. V. Titov, and C. E. Synolakis, 2006. Tsunami: Wave of change. Scientific American, vol. 294, no. 1, January, pp. 56–63

Ocean and Atmospheric Circulation

THEME Patterns

After analyzing data, scientists can explain certain patterns that occur in both atmospheric and oceanic circulation.

BIG IDEA

The ocean and the atmosphere interact leading to, among other things, heat being transported from on area to another around Earth.

About the photo A severe storm approaches off the coast of Volusia County, Florida.

5.1 Atmospheric Circulation

Main Idea
The uneven heating of Earth's surface is the driving force behind both wind and ocean currents.

Key Questions
1. What are the results of the relationship between the Sun, the atmosphere, and the ocean?
2. What are the most important greenhouse gases?

The ocean never rests. Throughout its depths, currents move and mix ocean waters, transporting heat, nutrients, pollutants, and organisms. Ocean circulation profoundly affects marine organisms and their habitats, as well as Earth's climate and, therefore, all terrestrial habitats. The atmosphere and the ocean are closely linked, and their dynamic interactions create the weather and climate that we experience over the surface of Earth. Winds, storms, droughts, and clouds are by-products of the relationship between the Sun, the atmosphere, and the ocean.

To understand physical oceanic processes, you must understand how they interact with other global processes.

Atmospheric Structure
This planet's atmosphere consists of 90 km of well-mixed gases above Earth's land and water. The gas molecules are closest together near the surface, and they spread out with altitude. In other words, the air gets thinner. This is evident even on tall mountains: climbers on peaks such as Mt. Everest require oxygen tanks near the summit.

The atmosphere is broken down into four distinct layers separated by three distinct boundaries (Fig. 5.1). The layer closest to the surface, the **troposphere**, contains 90% of the atmosphere's gas by mass. The troposphere decreases in temperature with an increase in altitude, until it reaches −60 °C at the **tropopause**, approximately 12 km high. This is because the air at lower altitudes experiences more heat radiated from Earth's surface than the air higher up. The temperature levels out at the tropopause, which is found 12–20 km high. From there, the temperature climbs with an increase in altitude through the **stratosphere** (20–50 km). The stratosphere contains **ozone**, which is a form of oxygen that is more reactive than the oxygen found near the surface. Each molecule of ozone consists of three oxygen atoms (O_3), whereas normal oxygen contains two (O_2). The chemical structure of ozone makes it capable of absorbing ultraviolet light from the Sun. The ozone molecules that are closer to the Sun (at higher altitude) absorb more ultraviolet light, which is why you see a temperature increase with altitude. Without this layer of ozone, organisms on Earth would not be protected from the Sun's harmful ultraviolet radiation. Temperature

DIVE IN!

What are the doldrums? The doldrums are areas in the Atlantic and Pacific Oceans where very little, if any, horizontal wind is generated. The doldrums are about 5 degrees north and south of the equator. Air heated by the Sun rises vertically in this area. For ships that depend solely on wind as a source of energy to move, the doldrums are a difficult area to get through. Early explorers that sailed across the Atlantic Ocean reported spending weeks in the doldrums with no wind to propel their ships.

increases until the **stratopause**, where the temperature levels off again, approximately 50 km up. Because so little ozone is found higher than this region, the temperature again drops with altitude. This area of temperature decline is the **mesosphere** (55–90 km). The highest section of the atmosphere, the **thermosphere**, starts above the mesosphere, and extends to 500 km above Earth's surface. As in the stratosphere, the temperature increases with altitude, but unlike the stratosphere, this increase is due to the small amount of regular oxygen (O_2) that efficiently absorbs light energy from the Sun.

Earth's weather takes place in the lowest layer, the troposphere. Here, the air and oceans interact to create wind, hurricanes, blizzards, tornadoes, waves, and currents. For the rest of this section, we will be focusing on the events that take place in this region of the atmosphere.

Heating the Atmosphere

Our climate is ultimately controlled by light energy from the Sun. Some reflects back into space, but most is absorbed by Earth's atmosphere and surface (including the ocean) and converted from light to heat.

Uneven Heating The Sun's light energy is not equally distributed across Earth's surface. Radiation is greatest at the equator, and it declines toward the poles. This is because the angle of the Sun's rays changes with Earth's latitude (Fig. 5.2, p. 106). At the equator, rays perpendicularly contact Earth's surface if Earth is not tilted (the angle between the ray and Earth is 90°). Earth, however, tilts on an axis as it rotates around the Sun. As Earth tilts, the angle between the Sun's rays and Earth's surface changes. When Earth is most tilted toward the Sun, the Sun's rays strike Earth directly (90°, perpendicular) at 23.5°N, and when Earth is most tilted away from the Sun, the Sun's rays strike Earth directly at 23.5°S. The longitudinal lines at 23.5°N and 23.5°S are called the Tropic of Cancer and the Tropic of Capricorn, respectively. The region known as the tropics falls between these two latitudes. The angle between sunlight and Earth's surface continues to decline until it reaches the poles, where rays are almost parallel to the surface.

The Sun's rays are most concentrated over Earth's surface when they are perpendicular to Earth's surface (at the equator when Earth is not tilted). Rays that strike Earth farther from the equator spread out more over Earth's surface, and are thus not as concentrated. Additionally, sunlight must travel through less of Earth's atmosphere at the equator compared to higher latitudes.

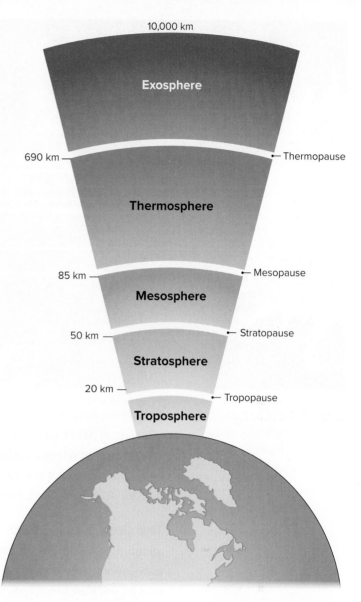

Figure 5.1 The layers of Earth's atmosphere

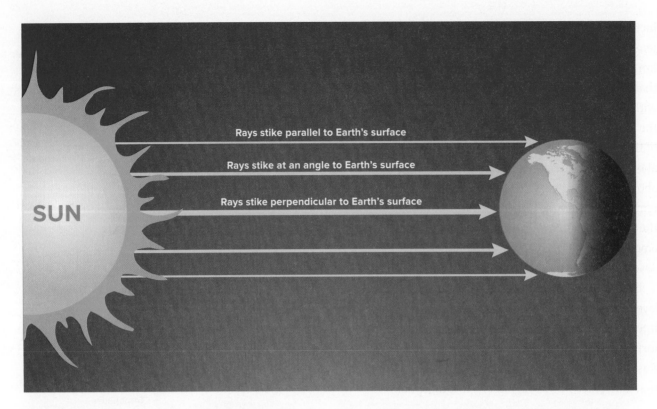

Rays stike parallel to Earth's surface

Rays stike at an angle to Earth's surface

Rays stike perpendicular to Earth's surface

SUN

Figure 5.2 Due to Earth's axis and shape, solar rays hit the surface at different angles at different latitudes.

The atmosphere dissipates and reflects some of this sunlight, reducing the intensity of light that reaches the crust and oceans. The combination of rays spreading over a larger surface area and traveling through more atmosphere with latitude is what causes the tropics to be warmer than the poles. This uneven heating of Earth is the driving force behind wind and currents.

The Greenhouse Effect The atmosphere is composed of many gases, including some known as **greenhouse gases**. Like the glass of a greenhouse, trace amounts of greenhouse gases in the atmosphere are transparent to light energy, but they block heat transmission back to space. The heat is retained and warms Earth (Fig. 5.3). Without this greenhouse effect, Earth's surface would average about −19°C, some 33°C colder than the current average, and it would be inhospitable most of Earth's organisms — including the human population.

The most important greenhouse gases are ordinary water vapor, carbon dioxide (CO_2), methane (CH_4), and nitrous oxide (N_2O). Water vapor is not a very potent greenhouse gas, but so much of it is in the atmosphere that it accounts for most of the greenhouse effect. But atmospheric water is a product, not a cause, of the greenhouse effect. If other greenhouse gases did not warm the planet in the first place, then all the surface water would be ice. Because other greenhouse gases raise Earth's temperature, the planet has liquid water. Some water evaporates into the atmosphere and warms the planet even more than the greenhouse gases that started the process. Water vapor, however, drops out of the atmosphere as precipitation within days; if other gases did not maintain the warming, then the planet would freeze. As an analogy, a guitar amplifier produces the volume, but it is the electric guitar that generates the sound. Similarly, water vapor amplifies the greenhouse effect, but other greenhouse gases drive it. The most important of these is CO_2.

In the natural carbon cycle, atmospheric CO_2 levels are controlled primarily by air-sea exchange and biological activity—except on timescales of hundreds to millions of years, where massive geological events can dramatically alter CO_2 levels. There are natural records of past CO_2 levels for scientists to analyze. Tiny bubbles of air trapped in polar ice, for example, record what the atmosphere was like when the ice formed. The temperature at the time can be determined from isotope ratios in the ice itself. **Isotopes** are different forms of an element; the number of protons is the same, but the number of neutrons is different. Because different isotope ratios are influenced by specific environmental factors, they can be used to directly compare ancient atmospheric composition and air temperatures. Antarctic ice cores show that over the last 400,000 years and incorporating multiple ice ages, atmospheric CO_2 fluctuated between only about 180 and 300 parts per million (ppm) (Fig. 5.4, p. 108). Other greenhouse gases have also fluctuated within a narrow band. These fluctuations are not enough to produce ice ages, which are mostly driven by small variations in Earth's orbit around the Sun. Volcanic activity has a minor impact, mainly by injecting **aerosols**, microscopic particles, into the upper atmosphere, where they increase the reflection of sunlight and slightly cool Earth. Thus, for hundreds of thousands of years, greenhouse gas concentrations have fluctuated in concert with other natural cycles.

Vertical and Global Circulation

Much like the waters of the oceans, the air in the atmosphere is constantly in motion. The uneven heating of Earth's surface, as well as its rotational pattern, are responsible for the wind and weather patterns we experience on land and at sea.

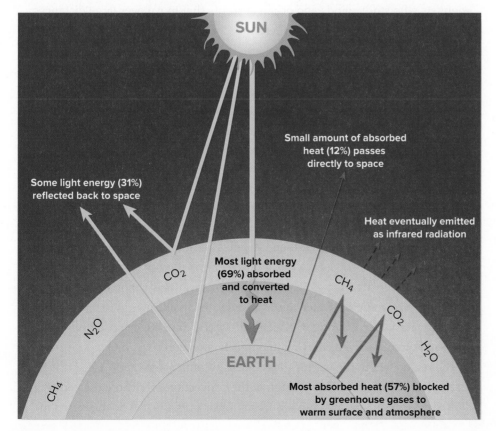

Figure 5.3 A schematic diagram of the greenhouse effect, including the major greenhouse gases

Figure 5.4 The amount of CO_2 in the atmosphere, as measured from Antarctic ice cores, is much higher currently than at any point in the last 400,000 years.

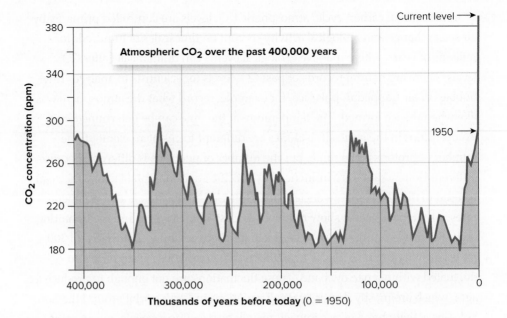

Coriolis Effect Because Earth is round and it rotates, anything that moves over its surface tends to turn a little rather than move in a straight line. One way to understand this curving and deflecting, called the **Coriolis effect**, is to try a simple experiment. Put a piece of paper on a record turntable or some other rotating surface. Try to draw a straight line. The line will come out curved because the paper is spinning, moving under the pen. Figure 5.5 shows another way to understand the Coriolis effect.

The Coriolis effect is too slight to notice when you're walking or driving in a car. It is very important and noticeable, however, for airplane pilots and for scientists who study winds and ocean currents that move over large distances (Fig. 5.5). In the northern hemisphere the Coriolis effect always deflects things to the right. In the southern hemisphere, things are deflected to the left. The magnitude of the Coriolis effect increases as you move toward the poles. This is because the circumference of Earth decreases with an increase in latitude. Therefore, the effect (veering to the right) on an airplane traveling from 80°N to 90°N will be greater than the effect on an airplane traveling from 0°N to 10°N . The patterns of winds and surface currents cannot be understood without accounting for the Coriolis effect.

Vertical Air Circulation **Wind** is the movement of molecules in the air. The winds in our atmosphere are driven by heat energy from the Sun. As you just learned, the equator is warmer than the poles because most of the Sun's energy is absorbed near the equator. As solar energy heats the equator, the air there becomes less dense and rises. This heat also creates water vapor, by evaporating oceans and fresh surface water. Air from adjacent areas flows in to replace the rising equatorial air, creating wind. This process creates a **convection cell** of two layers of horizontal winds moving in opposite directions (Fig. 5.6).

The region where the warmer, less dense air and water vapor rises creates an area of low air pressure. The portion of the convection cell where the cooler, denser air sinks forms a high-pressure area. Air pressure can be measured in

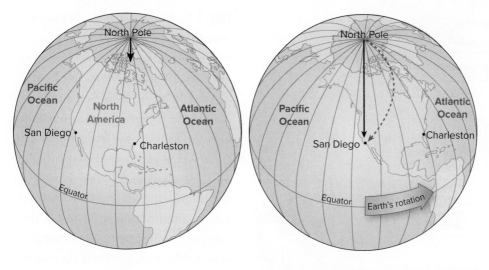

Figure 5.5 The Coriolis effect is very important for long-distance navigation, such as airplane travel. (a) A plane leaves the North Pole heading for Charleston, SC. (b) If the plane does not account for the Coriolis effect, the rotation of the Earth would cause it to land in San Diego, CA instead. From the ground, the plane would appear to follow a curved path (red). From space, an observer would see the plane's path as straight.

millibars, millimeters, or inches of mercury (torrs), all using a barometer. Low-pressure areas, which are areas of less-than-average air pressure, are usually associated with wet, cloudy conditions; high-pressure areas (those of higher-than-average air pressure) tend to have drier air and clear skies.

If Earth did not rotate, the Sun would cause the warm, wet air in the tropics to rise and move toward the poles (see Fig. 5.6). The water vapor would cool, form clouds, and eventually precipitate down toward Earth in the form of rain. The air would move toward the poles where it would cool, become denser, and sink back to Earth. The continuous cycling of this air from the equator toward the poles and back would form two cells, one in the northern hemisphere, and one in the southern hemisphere.

Global Air Circulation The rotation of Earth causes the winds to not move straight from the equator to the poles. This is because the Coriolis effect bends the winds. This bending stops the rising air at the equator from reaching the poles; instead, it forms three distinct convection cells over each hemisphere

Figure 5.6 Air near the equator absorbs increased solar radiation and rises. Air from higher latitudes moves in to replace the rising air, causing wind.

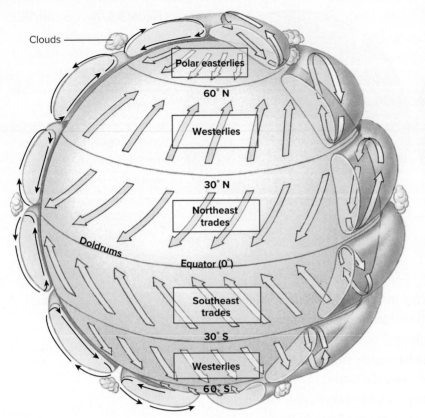

Clouds

Polar easterlies

60° N

Westerlies

30° N

Northeast trades

Doldrums

Equator (0°)

Southeast trades

30° S

Westerlies

60° S

Figure 5.7 The major wind patterns on Earth are created by the rising of sun-warmed air and the sinking of cold air. The wind fields are pictured here as they would look on an imaginary water-covered Earth. In actuality they are modified by the influence of the continents.

(Fig. 5.7). In the two cells that form over the tropics (one each in the northern and southern hemispheres), air rises at the equator and sinks at around 30°N and 30°S. Some of this sinking air is pushed back toward the equator, forming the **trade winds**. The trade winds approach the equator at an angle of about 45°. They blow from the northeast in the northern hemisphere and from the southeast in the southern hemisphere. Over the open ocean, where they are not affected by land, the trade winds are the steadiest winds on Earth. These winds also pick up a lot of water vapor from the warm ocean water, resulting in sustained winds that are warm and humid.

Some of the air that sinks at 30°N and 30°S, instead of forming the trade winds, moves from the west toward higher latitudes. These winds, because they are coming from the west, are known as the **westerlies**, and they lie at the middle latitudes. The westerlies move in the opposite direction to the trade winds and can be found between 30°N and 60°N, and 30°S and 60°S. These winds also pick up moisture as they move over oceans toward the poles. Around 60°N and 60°S, this moist air begins to rise again, forming the highest-latitude convection cells, between 60°N and 90°N and 60°S and 90°S. The air sinking at the poles (90 °N and 90 °S) travels back over Earth's surface from the northeast (northern hemisphere) and southeast (southern hemisphere), forming the high-latitude **polar easterlies**. The polar easterlies are the most variable winds of all (see Fig. 5.7).

Regions where moist air of low density rises (0°, 60°N, and 60°S) are areas of low pressure that are characterized by clouds, rain, and high humidity. Latitudes where air sinks (30°N, 30°S, 90°N, and 90°S) have higher pressure, drier air, and clearer skies.

· ·

Section Review

1. **Illustrate** Draw a diagram that shows the layer of Earth's atmosphere and how temperature changes as altitude increases. Include one fact about each layer of the atmosphere in your diagram.

2. **Explain** How do greenhouse gases warm the Earth?

3. **Identify** and describe the major wind patterns on Earth.

5.2 Surface Circulation

> ## Main Idea
> Surface currents are produced by wind and influenced by the Coriolis force.
>
> .
>
> ## Key Questions
> **1.** What drives surface currents?
> **2.** What are gyres?

VOCABULARY

Ekman spiral

Ekman layer

Ekman transport

equatorial currents

gyres

upwelling

monsoons

equatorial upwelling

The wind drives the strongest ocean currents, which occur in the surface layer. Wind and surface currents are ultimately driven by heat energy from the Sun. Surface currents and the wind direction are also influenced by the Coriolis effect. Friction on the ocean surface from the major wind fields of the atmosphere causes the sea surface to move, creating currents. All the major surface currents of the open ocean, in fact, are driven by the wind and ultimately by heat energy from the Sun.

Ekman Transport

When pushed by the wind, the upper-most layer of surface water begins to move. Instead of going in the same direction as the wind, the surface current moves off at an angle of 45° because of the Coriolis effect (Fig. 5.8).

This top layer pushes on the layer of water below, and again the Coriolis effect comes into play—the second layer does not move in the same direction as the top one; it moves slightly to the right and slightly slower. This process passes downward through the water column—each layer is pushed by the layer above it and pushes the one below it, in a pattern called the **Ekman spiral** after the Swedish oceanographer who discovered it. The effect of the wind decreases with depth, so that progressively deeper layers move more and more slowly. Eventually, at a depth of a few hundred meters at most, the wind is not felt at all. The upper part of the water column that is affected by the wind is called the **Ekman layer**. Though each micro-layer moves in a different direction, taken as a whole the Ekman layer moves at 90° from the wind direction. This process is called **Ekman transport**.

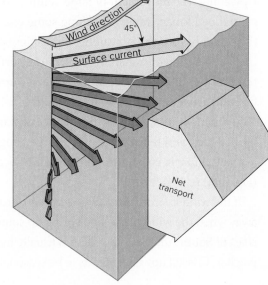

Figure 5.8 The water in an Ekman layer is transported at right angles to the wind direction.

What is the fastest current? The Gulf Stream has an average speed of 6.4 km/h. At the surface, where the current is the fastest, its speed can reach 9 km/h. At the northern end of the current, where it widens, its surface speed slows to 1.6 km/h. The Gulf Stream runs north along the eastern coast of the United States and was observed by early explorers and vessels carrying goods to trade. After working with boat captains who traveled throughout the Atlantic Ocean, Benjamin Franklin was the first to publish a map of the Gulf Stream in 1769.

Cold currents
Warm currents
Winds

0 1000 2000 3000 mi
0 1000 2000 3000 km

Figure 5.9 The major surface currents of the oceans. In the main ocean basins the currents combine to form large circular systems called gyres.

Global surface currents

A comparison of wind and surface-current patterns demonstrates the consequences of the Coriolis effect. While the trade winds move toward the equator, for example, the **equatorial currents** that these winds produce move parallel to the equator (Fig. 5.9).

Under the influence of the Coriolis effect, the wind-driven surface currents combine into huge, circular systems called **gyres**. These gyres span ocean basins, flowing clockwise in the northern hemisphere and counterclockwise in the southern hemisphere.

The currents shown in Figure 5.9 are average patterns over large distances and a long-time span. At a given place on a given day, the current is often different—currents shift with the seasons and the weather. On the continental shelf, currents are strongly affected by the bottom, the shape of the coastline, and the tides.

Upwelling

Upwelling, the movement of large volumes of bottom water to the surface, is caused by Ekman transport. Along some coasts, mainly on the eastern sides of the ocean basins, the prevailing winds blow parallel to the coast in a direction that causes Ekman transport to carry the surface layer offshore. At the edge of the continental shelf, deeper water moves upward to the surface to replace it, typically from a depth of 50 to 300 m (Fig. 5.10, p. 114).

On some coasts, such as the Pacific coast of South America, upwelling is fairly steady and takes place over a large geographic area. On other coasts, such as the California coast, upwelling tends to occur as localized, short-lived events (Fig. 5.11, p. 114).

If the wind is strong for a few days, then strong upwelling occurs, only to fade away when the winds cease and spring up somewhere else. Even on the Pacific coast of South America, upwelling is usually most intense in small, localized patches. Coastal upwelling tends to be seasonal, occurring mainly during the time

Tall Ships and Surface Currents

For purely practical reasons, winds and surface currents were among the first oceanic phenomena to be observed and documented. For centuries ships were at the mercy of the wind, and the names sailors gave to various regions reflected their knowledge of global wind patterns. Many of these names are still used today. The trade winds were named by the traders who relied on them during their voyages. The equatorial region where winds are light and variable because of the rising air masses is called the doldrums. The area around 30°N or 30°S, where the trade winds and westerlies are moving apart, often had unpredictable and varying winds. Sailors would often end up stranded and short of drinking water, and were forced to throw their dying horses overboard. To this day these are known as the horse latitudes.

15th century navigators Sailors also knew about surface currents. A clever navigator could shorten a passage by

Outward voyage ——————▶
Return voyage – – – – ▶
Gyre currents ——————▶

The route of early Portuguese ships on trading voyages to the west coast of Africa. The southbound route is shown by a solid line, the northward journey by a dotted line. The prevailing currents are indicated by red arrows.

weeks or months by riding favorable currents and avoiding unfavorable ones. In the 15th century, Portuguese sailors under the guidance of Prince Henry the Navigator made careful observations of currents along the west coast of Africa. Soon they were using their knowledge of these currents on their trading voyages. On the southbound journey, the ships sailed close to shore while in the Northern Hemisphere, riding the Canary Current. When they crossed the Equator they swung to the west to avoid the northward flowing Benguela Current. On the voyage home, they took the opposite path, completing a figure eight.

Early mariners knew about other currents as well. Christopher Columbus noted the existence of the North Atlantic Equatorial Current on his third voyage to the New World. While searching for the Fountain of Youth, the Spanish explorer Juan Ponce de León described the Florida Current. In the Pacific, fishermen recorded their knowledge of the Peru and Kuroshio Currents.

The Gulf Stream Even Benjamin Franklin has a place in this story. While serving as deputy colonial postmaster he noticed that mail ships routinely made the trip to Europe two weeks faster than they returned. Upon questioning seafarers he learned of the mighty Gulf Stream and asked his cousin, a sea captain, to sketch it on a nautical chart. Franklin developed the sketch into the first published chart of the Gulf Stream, which is remarkably accurate. He understood that the Gulf Stream is a vast "river" of warm water flowing through the cold waters of the surrounding Atlantic Ocean and instructed ships to stay in the warm water of the Gulf Stream when sailing east but to avoid it on the westward journey. Later, on voyages as emissary to England, Franklin regularly measured sea temperature and published his findings in a report about the Gulf Stream.

Figure 5.10 When prevailing winds blow in the right direction along the coast, Ekman transport carries surface waters offshore. Deeper, nutrient-rich water rises to replace it.

of year when winds are strong and blow in the right direction along the coast. Such seasonality is most pronounced on the coast of Somalia, in eastern Africa (Fig. 5.12). This area is dominated by **monsoons**, strong winds that blow northward in summer and southward in winter. The summer monsoon causes very intense coastal upwelling, which ceases when the winds change to the winter monsoon.

Figure 5.11 This satellite image shows summer upwelling along the California coast. The red area is warm surface water off Southern California, and the arrow identifies San Francisco Bay. The purple to the north is very cold deep water that has upwelled along the coast. The black area to the left of the image is obscured by clouds, and the black area on the right is land.

Mark Abbott/NOAA

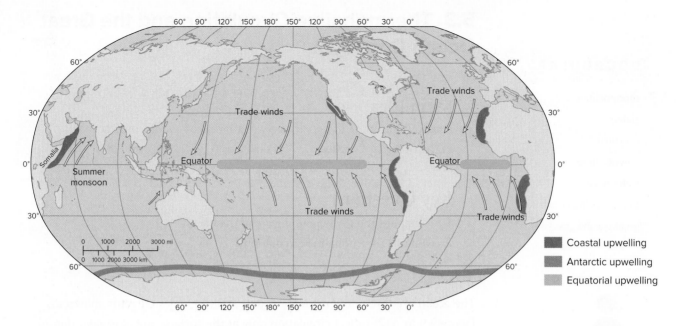

Figure 5.12 Prevailing winds and major upwelling areas of the world

Coastal winds are not the only source of upwelling. The Coriolis effect also produces **equatorial upwelling**, especially in the Pacific. The direction of the Coriolis effect changes from right to left at the equator. Thus, the north equatorial currents transport surface water to the right (north), whereas the south equatorial currents transport water to the left (south)(Fig. 5.13).

As a result, in equatorial regions the sea surface is being pulled apart, or diverging; deep water moves up to fill the void, typically from a depth of a few hundred meters. Equatorial upwelling is not as intense as coastal upwelling, but can occur over much larger areas.

Upwelling also occurs in a ring around Antarctica as a result of complex wind and current patterns between the continent and the Antarctic Circumpolar Current (see Figure 5.9). Although coastal and equatorial upwelling typically come from depths of a few hundred meters at most, Antarctic upwelling brings water from as deep as 4,000 m to the surface.

Figure 5.13 The Coriolis effect influences the equatorial currents differently on opposite sides of the equator. Surface water therefore moves away from the equator.

Section Review

1. Describe Ekman transport.

2. Illustrate how coastal upwelling occurs.

3. Explain What causes equatorial upwelling?

5.3 Thermohaline Circulation and the Great Ocean Conveyor

VOCABULARY

thermoclines

stable

overturn

downwelling

water mass

thermohaline circulation

great ocean conveyor

DIVE IN!

What are water masses? Water masses form when a large volume of surface water sinks without mixing very much with the surrounding water. Once the water mass sinks, its temperature and salinity do not change. Oceanographers can use salinity and temperature to trace the water mass as it moves through the ocean. Antarctic Bottom Water is the deepest water in the ocean, sinking all the way to the ocean floor, then moving throughout the Atlantic Ocean basin. North Atlantic Deep Water (NADW) has a salinity of about 34.6 psu and a temperature between 3 °C and 2 °C. It does not sink as deep as AABW, but still circulates throughout the Atlantic basin at a depth of 3,000 m.

Main Idea

Due to differences in density between layers of water, some currents in the ocean are vertical.

· ·

Key Questions

1. What are the three main layers of the ocean?

2. What is thermohaline circulation?

The ocean is a three-dimensional habitat, and the major currents shown in Figure 5.9 (p. 112) reflect circulation only at the surface, not at greater depths. Over much of the ocean, surface waters are isolated from deeper waters by differences in density, which is why oceanographers painstakingly measure the two factors—temperature and salinity—that determine seawater density.

Three-Layer Ocean

Because the densest water sinks, the ocean is usually layered, or stratified; the densest water is on the bottom, and the least dense water is at the surface. This can be seen in typical profiles of salinity, temperature, and density (Fig. 5.14). Deep water is normally cold and dense, whereas the surface water is relatively warm and less dense.

Surface layer Although there are actually many thin layers of water, each having a slightly different density, it is not a bad approximation to view the ocean as having three principal layers (see Fig. 5.14a). The surface layer is usually 100 to 200 m deep. The surface layer is mixed by wind, waves, and currents, so it is also known as the mixed layer. The surface layer is not always well mixed, however. Sometimes, usually in the spring and summer in temperate and polar latitudes, the very uppermost part of the surface layer gets heated by the Sun. The warm water floats in a shallow lens on top, and there is a sharp transition to the cooler water below. Salinity varies near the surface because of precipitation, evaporation, and river runoff. Divers often notice these sudden changes in temperature over small depth intervals, called **thermoclines**. When the weather becomes colder, wind and waves again mix the water column and these temporary, shallow thermoclines break down.

Intermediate layer The intermediate layer lies below the surface layer, typically to 1,000 to 1,500 m deep. The permanent thermocline, a transition zone between warm surface water and cold water below, defines the intermediate layer. The main thermocline should not be confused with the much shallower, seasonal thermoclines in the surface layer. The main thermocline breaks down

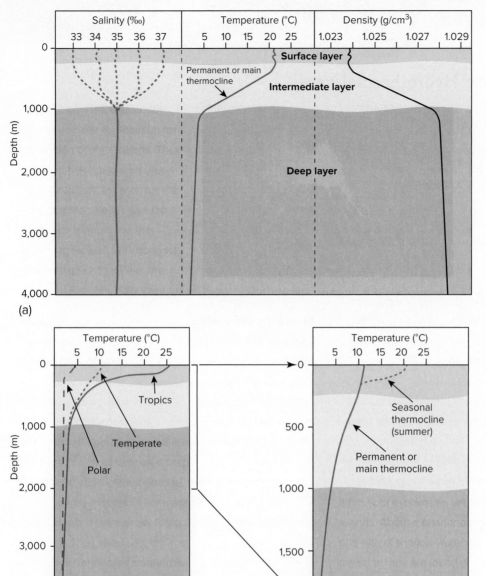

Figure 5.14 (a) Typical profiles of salinity, temperature, and density in the open ocean. (b) The surface temperature varies with latitude, with the highest surface termparature occuring in the tropics. Temperatures in deep water are more stable. (c) In temparate and polar waters, seasonal thermoclines may develop in the surface layer.

only occasionally and then only in a few places. The main thermocline is a feature of the open ocean. The waters over the continental shelf are not deep enough to have a main thermocline; in shelf waters, the mixed surface layer often extends all the way to the bottom.

Bottom layer Below about 1,500 m lie the deep and bottom layers. Technically, deep water and bottom water are different, but they are similar in being uniformly cold, typically less than 4°C.

Stability and Overturn

Most of the time surface water, being warmer and less dense, floats on top of the denser water below, unless wind or wave energy stirs up the water column. Such a water column is said to be **stable**. How stable the water column is depends on the density difference between the layers. If the surface layer is

Larval Transport Near Hydrothermal Vents

Because currents are so important to marine life, marine scientists often integrate physical oceanography into their studies. This is true not only at the scale of entire ocean basins, but also at much smaller scales. Recent studies of currents in the vicinity of hydrothermal vents provide an example.

Hydrothermal vent communities Hydrothermal vents support rich communities of worms, bivalves, shrimps, and other animals. Each vent is like an oasis, often separated from similar communities by distances of hundreds or thousands of kilometers. The vent communities can be short-lived, as the vents that support them can die out or erupt, wiping out the community under a wave of lava. One of the mysteries about vent communities is how new vents are colonized, or re-colonized after an eruption, since the animals are not capable of traversing the cold, foodless void between vents. Vent organisms produce young that have a larval form, which swim only weakly and are carried from place to place in the currents. Another question is whether some larvae from a vent community settle at the same vent to sustain the populations there, or if they are all carried away by the currents so that the community depends on importing larvae from other communities.

Are larvae the answer? After a vent in the Pacific off the coast of Mexico was wiped out by a violent eruption, biologists found that new larvae apparently came in from more than 300 km away. This was a surprise, because they had thought that the larvae spend only 30 to 40 days in the water column before settling to the bottom, and in that time the prevailing currents could transport them 100 km at most.

So, the biologists teamed with physical oceanographers to further study current patterns near the vents. The first step was to put meters along the vents and in the water column above to measure the speed and direction of the currents. it isnt feasible to deploy enough current meters over a large enough area to measure the broader circulation pattern. Instead, the current measurements were used to develop a computer model of the circulation. The model predictions were tested by releasing a chemical tracer that can be detected in very low amounts, and taking water samples at various distances around the vent to determine where the currents carried the tracer.

Surprising discoveries The results brought more surprises. The working hypothesis had been that larvae probably rode plumes of warm water rising from the vents several hundred meters up into the water column, and were carried horizontally from there by currents. Instead, the results indicate that the fastest currents flow along the bottom at the edges of the rift zone where the vents are located. The study also showed that the current swirls around the vent in eddies, initially flowing away from the vent and then coming back to it. Thus, it seems likely that some larvae from the vent community would return to maintain the population, at least at this study site.

Even the relatively fast currents measured near the bottom did not flow fast enough to carry larvae 300 km in their 30- to 40-day larval span. Are there faster currents? Do the larvae spend longer in the water column than thought? Are there other transport mechanisms that the scientists missed? Also, the current eddies observed may be due to the current swirling around some nearby seamounts. Do other vent sites have similar circulation? For marine scientists to answer these and other questions, there's only one thing to do: head back to the deep sea!

only slightly less dense than the layer below, then less energy is required to mix the two layers and the water column will have low stability. A highly stable water column on the other hand results from a large density difference between deep and shallow water, and it requires more energy to mix.

Sometimes water columns become unstable, meaning that the surface water becomes denser than the water below. The surface water sinks and mixes with deeper water. This process is known as **overturn** (Fig. 5.15). Overturn follows a

regular seasonal pattern in temperate and polar regions. The summer sun warms the surface layer, making it much less dense than the water below. In autumn, the surface water begins to cool, and storm winds mix colder deep water to the surface. If the surface water gets cold enough during the winter, it becomes denser than deeper water and sinks, which is called **downwelling**. Because surface water with the same temperature and density is descending through the water column, the temperature and density profiles are vertical straight lines, and the oceanographers identify conditions of overturn by looking for such straight-line profiles. Overturn usually occurs in temperate and polar regions during the winter when the surface water cools.

When overturn occurs over only a small area or when the surface water is only slightly denser than the water below it, the downwelling water simply mixes with deeper water, so the mixed layer extends deeper in the water column. This mixing is important for the productivity of temperate and polar waters. When downwelling is intense, however, a large volume of water might sink without mixing much with the surrounding water. The processes that change salinity in the open ocean—precipitation, evaporation, and freezing—occur only at the surface. Temperature also changes primarily at the surface, through evaporative cooling, solar heating, or the exchange of heat with the atmosphere. After surface water has sunk, therefore, its salinity and temperature don't change. From then on, the volume of water, or **water mass**, has a fingerprint, a characteristic combination of temperature and salinity. Oceanographers can use this to follow the movement, or circulation, of water masses over great distances. Because this form of circulation is driven by changes in density, which in turn is determined by temperature and salinity, this form of circulation is known as **thermohaline circulation**.

Figure 5.15 The progression of overturn during summer (a), autumn (b), and winter (c).

The Great Ocean Conveyor

After water masses leave the surface, they sink to a depth determined by their density. Water of intermediate density descends only part way, to a depth where it is denser than the water above it but less dense than the water below it. To sink all the way to the bottom of the ocean, surface water must become very dense; that is, cold and relatively salty. Most commonly, this is due to the formation of sea ice, which forms from freshwater, leaving behind the salts in the still-liquid ocean to form a brine. This occurs at only a few locations, and only intermittently. The main places where surface overturn reaches the bottom are in the Atlantic Ocean, south of Greenland, and just north of Antarctica (Fig. 5.16, p. 120). The water masses that originate in these locations sink and spread along the seafloor, forming the North Atlantic Deep Water and the Antarctic Bottom Water.

After sinking, the water spreads through the Atlantic and into the other ocean basins. The water eventually rises to the surface and flows back to the Atlantic, where the cycle begins again (Fig. 5.17, p. 120). This global thermohaline

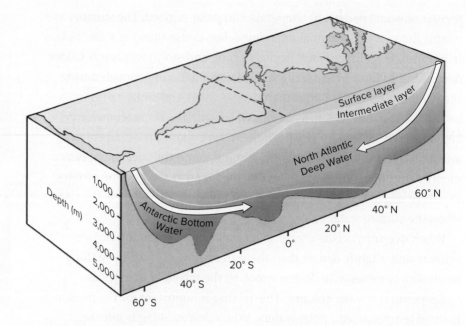

Figure 5.16 The deepest water in the deep and bottom layers of the Atlantic Ocean consists largely of two water masses: the North Atlantic Deep Water and the Antarctic Bottom Water.

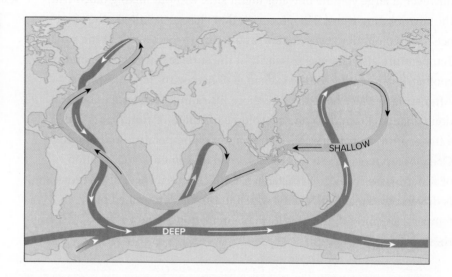

Figure 5.17 The deep ocean circulation is part of a global pattern known as the great ocean conveyor. This circulation pattern plays an important role in global weather and climate.

circulation, called the **great ocean conveyor**, mixes the oceans on a timescale of 1000 years. The great ocean conveyor is also critical in regulating Earth's climate, and fluctuations in the conveyor have produced rapid climate changes, even ice ages, in the past. The conveyor also brings dissolved oxygen to the deep sea. This effect is enhanced because oxygen dissolves best in cold water. The great ocean conveyor doesn't flow steadily and uniformly like an actual conveyor belt. Instead, it constantly varies in strength and exact position in a series of superimposed cycles, or oscillations. The best known of these is the El Niño-Southern Oscillation (ENSO or El Niño), but many other such oscillations, operating on different timescales and in different parts of the world, are now known.

· ·

Section Review

1. Describe the three layers of ocean water.

2. Describe the great ocean conveyor.

5.4 Interactions between Oceans and the Atmosphere

Main Idea

The interaction of the ocean and the atmosphere influences Earth's climate and weather.

. .

Key Questions

1. What is the El Niño-Southern Oscillation?
2. What is a tropical cyclone?

VOCABULARY

Gulf Stream

oscillations

El Niño-Southern Oscillation (ENSO)

cyclonic storms

storm surges

The ocean and atmosphere interact over long and short timescales to influence Earth's climate and weather. Until recently, climatic changes have been caused by natural perturbations such as solar activity, volcanic eruptions, meteor impacts, changes to Earth's wobble, and the shifting of continental plates, among other things. This section will explore how the ocean and atmosphere work together to influence Earth's climate and weather.

Oceans Act as Open Systems

Recall that water is particularly good at storing and transporting heat because of its high heat capacity. The warm currents on the western sides of the gyres carry vast amounts of solar heat from the equator to higher latitudes. Cold currents flow in the opposite direction on the eastern sides. Ocean currents thus act like a giant thermostat, warming the poles, cooling the tropics, and regulating the climate of the planet.

The role of surface currents in transporting heat is reflected in the sea-surface temperature (Fig. 5.18, p. 122). Surface temperature is higher on the western sides of the oceans, where currents carry warm water away from the equator, than on the eastern sides, where cold currents flow toward the equator. Because of this, tropical organisms such as corals extend into higher latitudes on the western sides. Cold-loving organisms such as kelps, on the other hand, occur closest to the equator on the eastern sides. This is also why you see many Californian surfers wearing wetsuits, even in the summer. The water temperatures off the Southern California coast are more similar to those in Connecticut in the summer months, even though Connecticut is almost 1300 km north of Southern California.

Currents transport heat to other parts of the ocean, and they have great influence on Earth's climate as well. The reverse is also true—local weather and climate can impact the ocean. The world's ocean and atmosphere are deeply connected, and it is impossible to understand one system without having knowledge of the other.

DIVE IN!

How are cyclones named? Cyclones in the Atlantic Basin are named using a series of six lists of chosen names that are recycled every six years. The lists are maintained by the World Meteorological Organization. The names on the list do not change unless a storm is so severe that it results in a high number of deaths and financial costs and it would be upsetting or unsuitable to use the name again. The name is deleted and another name is chosen in its place. Recently "retired" names include Katrina, Sandy, and Ingrid. There are similar lists for other regions of the world, including the Eastern North Pacific Basin and the Central North Pacific Basin.

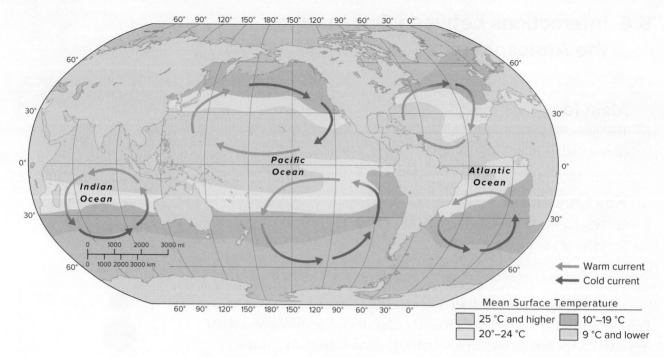

Figure 5.18 The average sea-surface temperatures of the oceans are strongly affected by the ocean gyres.

Mean Surface Temperature

25 °C and higher 10°–19 °C
20°–24 °C 9 °C and lower

⟵ Warm current
⟵ Cold current

IN CONTEXT

Heat capacity is the amount of heat needed to raise a substance's temperature by a given amount.

• 3.1, The Unique Nature of Pure Water

Regional weather has the ability to alter oceanic conditions. For example, winds influence currents and waves. If the wind is traveling in the same direction as the ocean, the magnitude of the current speed and wave height will be amplified. If the winds are coming from an opposite direction, then water movement will be dampened, causing currents to travel slower and waves to be smaller. The temperature of the atmosphere also influences the temperature of the ocean, and vice versa. This is true seasonally—the air and water heat up in the summer and cool down in the winter—and over shorter time spans. If an area experiences a drop or an increase in temperature, then the ocean nearby will show a similar change. Because of the ocean's high heat capacity, however, this change will be a smaller magnitude than the change in air temperature. Cloud coverage can also affect sea surface temperature. Clouds block the Sun and in turn decrease oceanic temperature. Clouds can also produce rain, which will affect the ocean by decreasing surface salinity in localized areas.

The ocean affects the atmosphere in similar ways. Heat is transferred between the two systems, so warm waters can create warmer climates. It is evident when you compare Eastern Canada to Western Europe. In the North Atlantic, the **Gulf Stream** current travels from the tropics up the coast of the United States, until it turns east around Cape Hatteras, North Carolina. The Gulf Stream brings warm water to Europe, creating a climate that is much milder than expected for that particular latitude. This is why you find palm trees in Ireland but pine trees in Newfoundland on the opposite side of the Atlantic. The ocean's high heat capacity also creates milder climates close to the water, compared to inland areas where Earth's crust does a poor job of

storing heat. These areas are known as maritime climates. Lastly, water evaporating from the ocean will enter the atmosphere, which tends to create humidity and clouds and to increase rainfall.

These are examples of how the atmosphere and the ocean interact to create the climate and weather patterns around the world. Specific climate and weather events that are associated with the ocean—oscillations and tropical cyclones—are discussed next.

Oscillations

Oscillations are cyclical changes in ocean circulation that affect climate. The Pacific Decadal Oscillation has a cycle of approximately 10 years, and it influences the climate in North America and east Asia. The North Pacific Gyre Oscillation, operates on a 10- to 15-year cycle. In the Atlantic, the North Atlantic Oscillation seems to influence weather profoundly in Europe, although new research suggests that it might not be a real climate oscillation at all. Several other ocean cycles that affect global weather patterns have also been discovered, but the most influential and famous oscillation is the El Niño-Southern Oscillation.

In recent years, **El Niño-Southern Oscillation** (ENSO) has received wide-spread attention from marine scientists and, because of its impact on humans, from economists, politicians, and the general public. The term El Niño originally referred to a change in the surface currents along the coasts of Perú and Chile. For much of the year, the winds along this coast blow from south to north, producing strong upwelling. The upwelling brings nutrients to the surface, making these waters one of the world's richest fishing grounds. Every year, usually in December, upwelling decreases, and the water gets warmer. Locals have been familiar with this event for centuries because it signals the end of the peak fishing season. Because this change in currents comes around Christmas, they called it El Niño, or The Child. Every few years, however, the change is much more pronounced than usual. The surface water gets much warmer, and upwelling ceases completely. Primary production drops to almost nothing, and the fishes that normally teem in these waters disappear. Fisheries along the Perú-Chile coast are devastated, and seabirds die in huge numbers.

La Niña, the other extreme, brings weather conditions that are roughly the opposite of El Niño's. Places that experience drought during El Niño, for example, can expect wet weather during La Niña years.

The "Southern Oscillation" part of ENSO has also been known for a long time. Like El Niño, the Southern Oscillation was first noted because of its impact on humans. People in India depend on the summer monsoon winds to bring rain for their crops. Sometimes the monsoons fail, producing famine and hardship. Beginning in 1904, a British administrator named Gilbert Walker set out to predict when the monsoons would fail by analyzing weather records from around the world. He never achieved his goal, but he discovered a major atmospheric phenomenon that he named the Southern Oscillation.

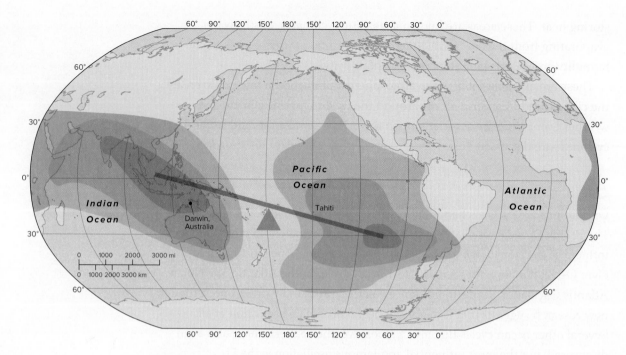

Figure 5.19 In the Southern Oscillation, atmospheric pressure seesaws between two large systems half a world apart. The shaded areas represent the linked zones of atmospheric pressure. The darker the blue, the more closely the atmospheric pressure is linked to pressure on the other side of the oscillation.

The Southern Oscillation refers to a long-distance linkage in atmospheric (or barometric) pressure. When the pressure is unusually high over the Pacific Ocean, it tends to be unusually low over the Indian Ocean, and vice versa (Fig. 5.19). Over a period of months or years, the air pressure swings back and forth like a giant seesaw that extends halfway around the world. The changes in atmospheric pressure bring dramatic changes in wind and rainfall, including the failure of the summer monsoon.

Modeling ENSO In the late 1950s, it became apparent that El Niño and the Southern Oscillation were two sides of the same coin—instead of being isolated phenomena in a particular region, they are part of a complex interaction of ocean and atmosphere that links the entire planet. El Niño corresponds closely to dramatic changes in the Southern Oscillation. Scientists use the term "ENSO" to signify that El Niño and the Southern Oscillation are two aspects of the same global phenomenon.

When the strongest El Niño in at least a century occurred during 1982 and 1983, it became clear that El Niño has profound global effects on weather, wildlife, and society. The 1982–83 El Niño brought extreme weather to much of the planet—storms and floods in some places and in others drought, forest fires, and heat waves. These extreme conditions focused scientific and public attention on understanding and dealing with El Niño.

In the ensuing years, scientists carefully analyzed information from the 1982–1983 El Niño, as well as milder El Niño events in 1986–87 and 1991–95. Based on what they learned, they developed a Pacific-wide system of oceanographic and atmospheric monitoring equipment in the hope of getting early

warning of new ENSO events. The system was soon put to the test. In late 1996, there were early signs, and by early 1997, the prediction was official—another El Niño was on its way.

The prediction turned out to be accurate, with a vengeance. The 1997–98 El Niño was one of the strongest known El Niño of all time, even stronger than the 1982–83 El Niño. Its severity caught even the scientists who predicted it by surprise, and the devastating effects were felt over much of the globe.

How El Niño form El Niño begins to form when the atmospheric seesaw of the Southern Oscillation brings lower than usual atmospheric pressure in the southeastern Pacific and higher than usual pressure to the Eastern Indian Ocean (see Fig. 5.19). This weakens the trade winds and, as a consequence, coastal upwelling along South America also weakens or ceases altogether.

Not only does the interrupted upwelling bring a failure in fisheries, the absence of cold water coming up from depth means that seawater near the coast is warmer than usual. In non-El Niño years the trade winds push a mass of very warm surface water, the Pacific Warm Pool, to the far western Pacific (Fig. 5.20). When the trade winds slack off during an El Niño, the Warm Pool moves eastward, warming the surface near the coast of South America even more. The high surface temperatures result in increased evaporation and more moisture in the atmosphere. Thus, El Niño brings high rainfall to the coasts of South America and Southern California.

Impacts of El Niños El Niño affects different parts of the globe in different ways (Fig. 5.21). In the eastern Pacific, El Niño years bring torrential rains which can cause widespread floods and mudslides. Hurricane activity increases

Satellite data processed by Dr. Xiao-Hai Yan at University of Delaware

Figure 5.20 This satellite image shows the temperature of the ocean surface. The coldest water is shown in blue, the warmest in red. The Pacific Warm Pool is identified with an arrow. During El Niño years, the Pacific Warm Pool moves east to the central Pacific.

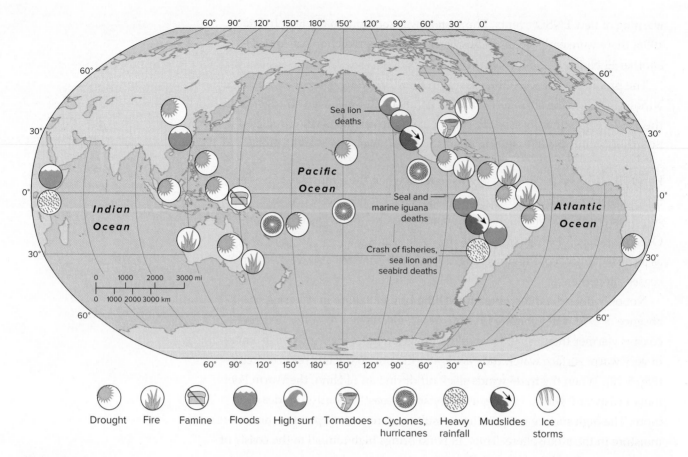

Figure 5.21 A sampling of some of the worldwide events linked to a strong El Niño.

in the Pacific. Because of the weaker trade winds, however, Atlantic circulation patterns change and hurricanes are less intense in the Atlantic and Caribbean. The reduced storm activity, however, can bring drought and wildfires to eastern South America, including the Amazon basin, and the input of energy to the atmosphere from the warm surface waters along the coast can increase tornadoes in areas where tornadoes are uncommon.

On the other side of the Pacific and into the Indian Ocean, El Niño events produce a more or less reversed pattern. The eastward movement of the Pacific Warm Pool means cooler sea temperatures, a dryer atmosphere, and lower rainfall for Indonesia, New Guinea, and Australia, which can suffer drought and wildfires during an El Niño. El Niño also produces more far-reaching effects, including torrential rain and floods in East Africa and drought in southern Africa.

Disease outbreaks from water contamination sometimes followed storms and floods. Warmer temperatures lead to increases in mosquito-borne diseases such as malaria, encephalitis, and dengue fever. In North America, outbreaks of bubonic plague and an often fatal respiratory disease known as hantavirus are associated with a severe El Niño. This is because the growth in vegetation after the high rainfall provides more food for rats and mice, which spread the diseases.

As previously mentioned, the 1997–98 El Niño was particularly intense. Worldwide, several thousand people were killed in floods and storms, and many secondary effects were recorded. Crop and livestock losses were enormous, bringing hardship to farmers, especially in poorer countries, and in some places famine was experienced. The financial losses from agricultural failure amounted

to hundreds of millions of dollars, and markets were affected widely as prices rose because of reduced supply. The fisheries crash in Perú and Chile had similar effects—much of the catch is processed into animal feed, and farmers as far away as Ireland faced higher feed prices. Fires, often lit by farmers using the opportunity to clear land, raged out of control in drought-stricken Indonesia and the Amazon basin. Thousands of square miles of rainforest were destroyed. The fires in Indonesia were so bad that much of Southeast Asia endured months of haze and air pollution and an accompanying increase in respiratory ailments. The thick smoke even caused several plane crashes. Mass coral bleaching occurred on many reefs around the world due to high water temperatures. Even more unusual problems occurred. After its floods were over, for example, Kenya suffered a plague of a kind of beetle that, when it is crushed, releases a poison that causes burning, itching, and blindness.

Learning from ENSO Damage estimates from the record-breaking El Niño of 1997–98 range upward of $20 billion, but it could have been much worse. Because scientists could predict El Niño, the world had advance warning and time to prepare. Many countries intensified their disaster preparedness efforts and accumulating food reserves. Industry and agriculture adjusted their planning. Cattle ranchers in Australia, for example, reduced their herds, cutting the losses from drought. In southern Africa, farmers planted more drought-resistant crops. Countries around the world continue to monitor El Niño development and plan accordingly. Every El Niño is different, however, and in a given place the exact effects are hard to predict. In 1997–98, for example, parts of East Africa expected drought based on past El Niño events. Instead, they had terrible floods. Nonetheless, when an El Niño can be correctly predicted, there are huge benefits for society.

The effects of El Niño are not necessarily all bad, or at least not bad for everyone. Cyclone activity was down in Hong Kong and the Philippines, as were hurricanes in the Caribbean. Farmers lucky enough to escape El Niño's wrath enjoyed higher prices for their crops. In Chile and the southwestern United States, heavy rains filled dams and replenished water supplies. Heavy snowfall delighted skiers and resort owners. Deserts bloomed when the rains brought spectacular displays of wildflowers. Indeed, El Niño's periodic soakings might be a vital part of desert ecology. El Niño brought a relatively mild winter to much of the northern United States and Europe, saving millions of dollars on heating bills. While the normal fisheries in Perú and Chile crashed, warm-water fishes such as dolphin and tunas appeared in record numbers. Salmon fishers in Oregon even caught striped marlin, normally confined to the subtropics.

The effects of ENSO are not harmful in and of themselves as much as they are because of the extreme changes in the normal patterns that they create. Heavy rain where it is rare might cause floods, and the lack of heavy rain in places that depend on it might causes drought and famine. Indeed, scientists now realize that El Niño is not really an abnormal occurrence, but simply one extreme of a regular weather cycle.

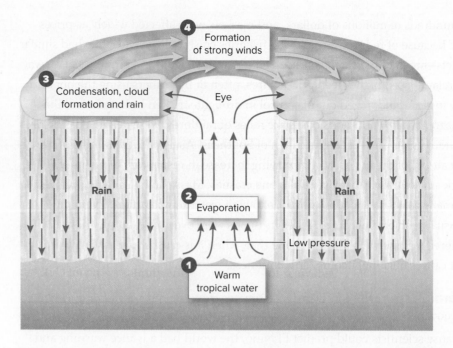

Figure 5.22 Cyclonic storms form over warm surface waters and are driven by energy from evaporation and condensation.

Cyclonic Storms

Cyclonic storms are rotating storms that are generated over the warm waters of the tropics and that occur over much shorter time spans than ENSO. Depending on the region, tropical cyclones might be called "hurricanes," "cyclones," or "typhoons," but all are characterized by heavy rain, strong winds, and an eye in the middle of the storm.

Movement of Cyclonic Storms In tropical areas of low pressure, considerable evaporation from the sea surface takes place, introducing large amounts of warm, humid air into the atmosphere. This warm air rises, creating an eye of very low pressure; then it cools at altitude and condenses into clouds and rain (Fig 5.22). The condensation releases great quantities of energy, which is converted into strong winds. For a storm to be considered a hurricane in the North Atlantic and Northeastern Pacific, it must have sustained winds of at least 119 km/hr. Sustained winds in the most intense storm on record, Typhoon Tip, which hit the Philippines in 1979, reached 310 km/hr. As long as a tropical cyclone is traveling over warm water, it will continue to generate energy via evaporation and condensation. When a cyclone passes over land or cooler water, it loses its energy source. The cyclone will expend its energy and dissipate.

The sustained wind speeds of cyclonic storms are used to classify them into five different categories. This classification scheme is known as the Saffir-Simpson hurricane scale (SSHS). The categories, wind speeds, and predicted damages can be found in Table 5.1, on p. 130. Note that these are *sustained* wind speeds - storms may have even stronger gusts of wind.

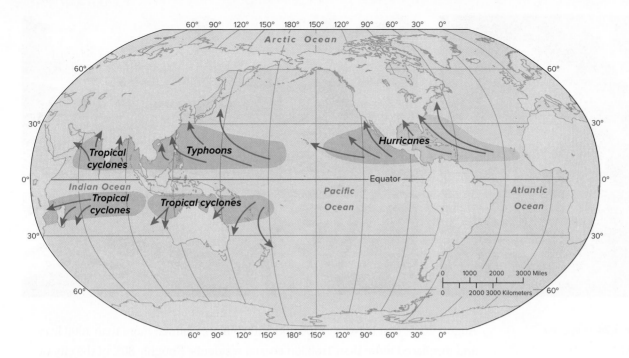

Figure 5.23 Areas around the globe where cyclonic storms form

The rotation of tropical cyclones is due to the Coriolis effect, which is why they form in the tropics, but not between 5°N and 5°S, where the Coriolis effect is weakest (Fig. 5.23). Cyclones rotate counterclockwise in the northern hemisphere, and clockwise in the southern hemisphere. They can span up to 4000 km in diameter and cause extensive damage from coastal areas to hundreds of kilometers inland.

Impacts of Cyclonic Storms Tropical cyclones cause destruction by heavy rains and strong winds and by secondary effects such as intense storm surges and tornadoes. **Storm surges** form from a combination of (1) the difference in atmospheric pressure between the perimeter of the storm and the eye and (2) from spiraling winds. As the eye of the storm approaches land, pressure and water build up and flood coastal areas. This flooding can even more extensive if it coincides with high tide. Storm surges are usually the most destructive aspect of hurricanes, in some events reaching more than 5 m high.

Almost 2 million people have died worldwide as a result of tropical cyclones since 1800, and 90% of those deaths are due to storm surge. The deadliest cyclone on record, the Bhola Cyclone, killed between 300,000 and 500,000 people in Bangladesh in November 1970. It is estimated that the storm directly impacted over 3.6 million people. A 10 m high storm surge rushed into the Ganges Delta when the cyclone made landfall, and wind gusts as high as 222 km/h were reported. Fishing vessels, homes, and agricultural lands were all destroyed in the cyclone, leaving 75% of the surviving population dependent on government assistance.

Hurricane Katrina, which first made landfall in Miami, Florida, before traveling across the Gulf of Mexico to reach New Orleans, Louisiana, has been the world's costliest storm to date. The storm, which reached the United States at the end of

Figure 5.24 Hurricane Katrina, which made landfall in the United States in August 2005 (a), caused significant damage to communities along the Gulf of Mexico (b).

August 2005, cost between $100 and $200 billion, claimed more than 1000 lives, and misplaced more than 1million coastal residents. Roughly 80% of the city of New Orleans flooded when the barrier to Lake Pontchartrain collapsed under the rising level of the lake due to rain and storm surge (Fig. 5.24).

Tropical cyclones are predicted to become more intense and more common as the climate continues to change, according to the 2014 Intergovernmental Panel on Climate Change (IPCC) Report. Some coastal communities have begun preparing for these changes, from engineering storm barriers in the Netherlands to developing plans to relocate populations of people in the South Pacific's Carteret Islands in Papua New Guinea.

Table 5.1: **The Saffir-Simpson Hurricane Scale**

Category	Sustained Wind Speeds	Comments
1	119-153 km/hr	Dangerous winds, damage to roofs and gutters, can topple trees and power lines
2	154-177 km/hr	Very dangerous winds, roofs and siding damage expected, near total power-loss expected
3	178-208 km/hr	Devastating damage expected, damage to exterior structures like decks or gables, loss of electricity and water service
4	209-251 km/hr	Catastrophic damage expected, loss of roofs and exterior walls, extended power outages, areas may be uninhabitable post-storm
5	≥252 km/hr	Catastrophic damage expected, high percentage of homes destroyed, most areas in the path of the storm will be uninhabitable post-storm

. .

Section Review

1. **Explain** How does heat transfer between the atmosphere and the ocean affect regional climates?

2. **Identify** What changes to the atmospheric pressure over the Pacific result in an El Niño event?

3. **Describe** What is the most destructive aspect of most tropical cyclones?

5.5 Ocean Circulation and Humans

DIVE IN!

How do positive feedback cycles contribute to climate change? One example is when ice shelves and sea ice melt. Ice is a reflective surface, reflecting about 90% of the radiation hitting Earth from the Sun. As sea ice melts, more water is exposed. Open water absorbs heat, leading to further ice melt. Another example is the melting of permafrost. As permafrost melts, it releases methane, a greenhouse gas. The more methane gas released, the more intense the greenhouse effect, which, in turn, leads to more melting permafrost. Under normal conditions, permafrost is considered to be a carbon sink. If Earth's temperature continues to increase and enough permafrost melts, permafrost will become a source of carbon rather than a sink.

Increasing evidence shows that natural occurrences can alter climate and that humans have the ability to change climate as well. Over the past 250 years we have been adding CO_2 to the atmosphere at a rate that will continue to affect the atmosphere and the ocean.

Anthropogenic Impacts on Climate

Much of the increase in atmospheric CO_2 is a direct result of human activities. In recent decades about 75% of the increase is a result of burning coal, oil, and other fossil fuels, and most of the rest has come from intensified land use, particularly the burning of forests and other biomass. The main human sources of methane are fossil fuels, the digestive systems of cattle and sheep, rice paddies, and landfills. Human emissions of greenhouse gases are the only thing that can explain the rapid global climate change that has occurred in recent decades (Fig. 5.25).

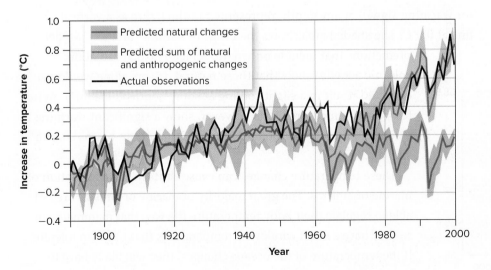

Figure 5.25 The green line shows the results of models that consider only solar and volcanic activity on global temperatures, not an increase in greenhouse gases. The red line shows the results of models that do take into account greenhouse gas effects, and these results follow the measured temperature observations (black line).

Since the start of the Industrial Revolution around 1750, 555 GtC (gigatons of carbon) have been released via fossil fuel combustion. Other prominent sources include cement production (375 GtC) and deforestation and changes in land use (180 GtC). The oceans have absorbed 30% of the CO_2, which is contributing to ocean acidification. Earth has incorporated another 30%, through carbon encorporated into plants during photosynthesis which is eventually buried. The remaining 40% of CO_2 has accumulated in the atmosphere.

These increasing concentrations of CO_2 and other greenhouse gases in the atmosphere can have profound impacts on climate and the ocean by altering ocean circulation, rainfall patterns, storm duration and frequency, temperatures, sea level, and the extension of ice caps. Shifts driven by climate change in some of these processes have already occurred, and they have the potential to affect humans and have profound effects on marine life.

The increase in greenhouse gases is causing Earth to warm. This is evident even when natural variability such as ENSO is considered. Since 1880, the global temperature has risen by 0.85°C, and scientists predict that the temperature in 2100 will be 0.3°C to 4.8°C higher than in the years 1986–2005 (Fig. 5.26). This may not seem like a lot, but even a 3°C temperature rise will make Earth its warmest in 3 million years, far warmer than the 10,000-year experience of human civilization. The year 2014 was the warmest since record keeping started in 1880. The combined global land and ocean average temperature for the year was 0.69°C above the twentieth-century average of 13.9°C.

Sea Surface Temperature The increase in atmospheric temperature is also causing the temperature of the ocean to increase. Sea surface (the upper 75 m of the water column) temperatures have been increasing since 1971 by approximately 0.11°C per decade. Figure 5.27 shows predicted versus actual warming experienced by the North Atlantic. If the North Atlantic hadn't warmed or cooled since 1960 there would be no temperature change (or **warming signal**) at any depth, represented in the figure as a vertical line at 0.0°C. The shaded area shows the range of natural variation. Model predictions that include only natural factors (solar radiation, volcanic activity) show that these natural drivers alone would bee expected result in a slight cooling at some depths (blue line). Actual temperature measurements however, show a significant warming to a depth of more than 700m (red line). Other ocean basins show a similar trend.

These temperature changes can cause shifts in the distribution of marine organisms. The great majority of marine organisms are not able to regulate their body temperature the way that mammals can, so they have a small window of temperatures that they can tolerate. If the temperature of the oceans changes, they will likely have to move to new locations to survive, or become extinct. Even organisms that can control their body temperature, such as dolphins and seals, might have to relocate to follow their food source. This will not only

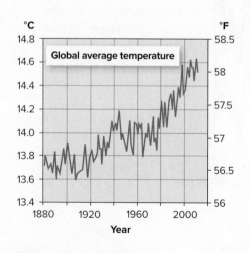

Figure 5.26 Global average temperature, 1886–2011

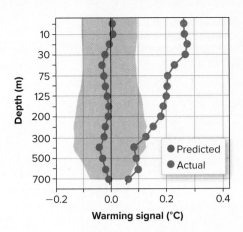

Figure 5.27 Models that incorporate only natural factors predict that the North Atlantic should have experienced a neutral to negative warming signal over the last ~50 years (blue line). Actual temperature measurements, however, show a positive warming signal from the surface to a depth of 700m (red line).

change the ecosystem, but will affect the hundreds of millions of humans around the world that rely on local fisheries for food and income.

Sea Level Rise The warming oceans are causing sea levels to rise. This rise is due to the expansion of surface waters as they warm and to the melting of glaciers and ice caps. Approximately 50% of the sea level rise so far is due to **thermal expansion**, and approximately 25% is due to glacier loss, with the rest due to changes in features such as **permafrost** (a layer of frozen polar soil) and ice caps. This has been happening since the end of the last ice age, but the rate is accelerating. Sea level rise has averaged 1.5 mm/yr since 1870, but the average rise over the past 20 years has been 3.2 mm/yr. In total, global sea level has risen 19 cm since 1901. Particularly worrying is the recent discovery that the great ice sheets covering Greenland and Antarctica are losing ice to the oceans despite the fact that warming has also produced snowfall increases in some places. Scientists didn't expect the ice sheets to start melting until next century. Antarctic sea-ice shelves are shrinking, or even collapsing and drifting away as icebergs, as large parts of the Larsen B Ice Shelf did on the Antarctic Peninsula in 2002. This doesn't immediately affect sea level because the ice is already floating. But, the ice shelves act as dams to slow the flow of glaciers on land, and when the ice shelves are gone, the glaciers move to the ocean faster. This slowing down of ice formation in the Arctic and Antarctic will reduce the differences in density between the downwelling water that drives the great ocean conveyor and the other water masses. Eventually, this could slow down thermohaline circulation in the world's ocean, which would ultimately change climate throughout the globe. The distribution of heat would be altered, which would exacerbate the predicted changes to climate and completely alter marine species distributions as well.

The sea ice that covers the Arctic Ocean has been shrinking for at least 30 years, with the summer sea ice minimum decreasing each decade by approximately 12%. It is getting thinner, and the area that remains covered by ice in summer is steadily getting smaller. In 2007, summer ice cover in the Arctic was the lowest since measurements began in 1979, and the summers of 2011 and 2012 each broke the preceding record (Fig. 5.28). Temperatures in the Arctic are increasing at twice the rate of the rest of the planet, which is having a

Figure 5.28 Sea ice extent in the Arctic, from 1978–2014

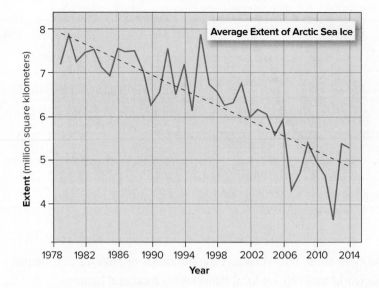

strong influence on the Arctic ice. Like the Antarctic ice shelves, melting Arctic sea ice doesn't directly raise sea level because the ice is already floating, but the melting accelerates global warming because it decreases Earth's **albedo**. Albedo is the reflectivity of Earth's surface, or the amount of solar energy reflected back into space. The bright white ice reflects solar energy back to space, whereas dark, open water absorbs the heat. Sea ice currently reflects 90% of the radiation hitting Earth from the Sun, playing a major role in maintaining the temperature of the globe.

Sea level rise will increasingly threaten coastal areas, which are home to more than 40% of the world's population and most of its largest cities, including New York, Los Angeles, London, Shanghai, Cairo, and Rio de Janeiro. Recent studies predict that northeastern North America will initially be most affected by sea level rise. This won't be a case of cities washed away overnight in a Hollywood-style inundation. Instead, coastal erosion will steadily increase, requiring even more expensive seawalls and other protective measures or the gradual abandonment of buildings, refineries, dockyards, and other shoreline infrastructure. The latest studies predict that sea level will rise between 0.26 m and 0.98 m by 2100. Saltwater incursion will cause increasing problems with storm-water drainage, water supplies, and sewage systems. Coastal areas will be more vulnerable to storms, floods, and other disasters—not just because of the rising seas but also because storms and floods are expected to become more extreme, and possibly more frequent and widespread. Increasing coastal vulnerability combined with more extreme weather could well mean more catastrophes such as Hurricane Katrina and Superstorm Sandy.

Climate Change

When news outlets and journalists report on anthropogenic changes to our climate, they often use the term "global warming." As you have learned, however, this is an incomplete and sometimes inaccurate summation of the changes occurring on Earth. Scientists favor the umbrella term "global climate

change," which encompasses not just temperature change but the full suite of predicted climate impacts.

Impact on Humans More extreme storms will be in part due to a predicted speeding up of the water cycle, another product of increased greenhouse gases. The higher temperatures will promote evaporation, and rainfall will increase globally. The increase will not, however, be evenly distributed. Climate change is expected to intensify existing patterns, so places that are wet now will experience more frequent and more severe floods and drought and forest fires will become worse in arid regions.

The area that is affected by monsoons will likely expand, and although the wind intensity might decrease, the rainfall from each storm is likely to intensify. ENSO events are anticipated to increase in intensity as well. People who depend on melting snow and glaciers for their summer water supplies (for example, in the western United States, western Europe, and western China) will suffer increasing late-summer water shortages as the snow season shortens, melting takes place earlier in the year, and glaciers shrink and disappear (Fig. 5.29). This is and will continue to be evident in oceans, where areas of high evaporation will become more saline and areas that experience large amounts of rainfall will become fresher. Globally, an estimated 1 to 4 billion people will face increasing water shortages.

Although increased atmospheric CO_2 might have a slight fertilizing effect on some crops, overall agricultural yields are expected to decline, mostly in developing countries. Declining wheat yields in a number of countries might already reflect the early impacts of climate change. Human health will also suffer, and the World Health Organization estimates that climate change is already killing 150,000 people per year, and another 1°C will double that. As to the cost, an authoritative estimate is that the climate change price tag for "business as usual" will be at least 5% of the entire global economy each year. When broader implications come into play, the bill could rise to 20%. These might include the risk of feedbacks such as the slowdown of the great conveyor. This would mean that everybody on the planet gets 20% less goods and services on average than they would without anthropogenic climate

Figure 5.29 The terminus of Bear Glacier, Kenai Fjords National Park, Alaska, in 2002 (a) and 2007 (b). The glacier retreated more than 2 km in the 5 years between photos.

change—a 20% climate tax that would be paid by everyone and that can't be reduced by legislation.

The predicted economic impact of climate change has been likened to the impact of the world wars and the Great Depression of the twentieth century. As of now, the only way to avoid the worst impacts of climate change would be to reduce our emission of greenhouse gases.

Impacts on Marine Ecosystems It is impossible to predict all the ways that climate change will impact marine organisms, both positively and negatively. Some species, such as kelps and sea jellies, are predicted to fare well with the increase in oceanic CO_2. This is because of the potential fertilizing effect of CO_2 in the case of kelp, which use CO_2 in photosynthesis. For sea jellies, current climate change predictions call for a reduction in calcareous competitors due to ocean acidification. Certain invasive species might be able to expand into new habitats and thrive as well. On the other hand, as discussed in Section 3.3, organisms with calcified structures are predicted to suffer, including bivalve mollusks, corals, and certain types of plankton.

In addition to physiological effects, climate change will also impact marine organisms on a habitat-wide scale. Biologists are unsure how capable habitats such as mangroves and salt marshes are able to keep up with sea level rise, and these habitats support a wide range of fishes and invertebrates. It is unlikely that light-dependent habitats, such as coral reefs and seagrasses, will be able to grow fast enough to outpace the increase in water depth, essentially drowning from lack of sunlight.

Despite these dire predictions, there is hope. Scientists, conservationists, and governments in many countries are working together to restore and protect a variety of resources. Action being taken includes writing climate change management plans, creating stronger fishing regulations to protect specific populations, and growing coral in nurseries to replace those negatively affected by climate change. Though it is and will continue to be a lot of work, activities such as these are critical to maintaining healthy oceans.

Section Review

1. **Explain** What are possible effects of sea level rising due to increased temperatures on Earth?

2. **Explain** How could increase sea surface temperature affect the great ocean conveyor? Why is this important?

3. **Summarize** What are some possible ways that climate change could impact marine organisms, both positively and negatively?

Chapter 5 Review

REVIEW QUESTIONS

Multiple Choice

1. The tendency of objects moving large distance on Earth's surface to bend to the right in the Northern Hemisphere and to the left in the Southern Hemisphere is due to
 - **a.** downwelling.
 - **b.** Ekman transport.
 - **c.** the Coriolis effect.
 - **d.** the greenhouse effect.

2. Which is a transition zone between warm surface water and colder water below?
 - **a.** El Nino
 - **b.** great ocean conveyor
 - **c.** thermocline
 - **d.** water mass

3. Which is not a step in the formation of a cyclone?
 - **a.** considerable evaporation takes place
 - **b.** large amounts of warm, humid air are introduced into the atmosphere
 - **c.** an 'eye' of low pressure is created
 - **d.** condensation releases energy, which is converted into weak winds

Short Answer

4. Why does coastal upwelling commonly result in productive fisheries?

5. Why is the great ocean conveyor important?

6. What are some negative effects of an El Nino?

7. What are the two main factors contributing to sea level rise?

Critical Thinking

8. It is estimated that stabilizing CO_2 emissions will cost 1-3% of the entire global economy every year through the year 2050. Present an argument that even the maximum investment of 3% per year is worth it to avoid the 5-20% "climate tax" if we do nothing to stabilize CO_2 emissions.

DATA ANALYSIS LAB

Tracking the Debris Path after a Tsunami On March 11, 2011, a tsunami hit the coast of Japan. The wave washed millions of tons of debris into the ocean, much of which ended up in the Kurishio Current. This allowed scientists a unique opportunity to predict and track the path of the debris as it was carried by ocean surface currents.

Data and Observations

The figures show the predicted path of the debris two years after the tsunami (2013) and three years after the tsunami (2014). The red dot shows where the debris originally entered the water.

Think Critically

1. Describe how the debris moved after entering the water in March 2011.

2. Predict in which direction the debris will move after March 2014. Explain your reasoning.

3. Where do you think the debris that does not break down will ultimately end up?

Data obtained from: Eriksen, M., 2012. Tracking tsunami flotsam. Natural History, vol. 120. No. 7, July/Aug., pp. 18–23.

The Organisms of the Sea

The oceans are home to diverse assemblages of organisms adapted to the challenges of life in the sea. Fishes like these Creole wrasses (*Clepticus parra*) travel in groups called schools, which may provide individuals with protection from predators and increased swimming efficiency. Though the corals below them may look like plants, they are in fact a colony of invertebrate organisms closely related to jellyfish. Unit 2 covers the diversity of life in the sea – from the minute prokaryotes to the large and complex mammals.

UNIT PROJECT

Design a Solution

Global challenges are often too large to address as a single problem. That isn't to imply that these problems are too big to solve. Rather, these types of problems often become much more solvable when looked at as a collection of individual challenges that are much smaller in scope. Your task is to design a solution to a complex real-world problem facing the marine environment by breaking it down into smaller, more manageable problems.

Fundamentals of Biology

THEME Patterns

There are basic features shared by all living things, including the molecules and cells we are made of, reproduction, and adaptations to the environments in which we live.

BIG IDEA

Marine organisms have numerous adaptations to survive in the variety of oceanic habitats.

About the photo Stromatolites, like these in Shark Bay, Australia, are ancient colonial organisms with cells that resemble the very first cells on Earth.

Jason Edwards/Getty Images

6.1 The Ingredients of Life

Main Idea

Organisms are made up of four main groups of organic molecules that are responsible for body structure and performing basic life functions.

. .

Key Questions

1. What are the four main organic molecules that make life possible?
2. What are the reactants and products of photosynthesis and cellular respiration?

VOCABULARY

metabolism

homeostasis

organic compounds

carbohydrates

proteins

amino acids

enzymes

lipids

nucleic acids

genome

photosynthesis

chlorophyll

carbon fixation

autotrophs

heterotrophs

respiration

anaerobic

aerobic

primary production

Now that we have discussed some of the major features of the marine environment, we can turn our attention to life in the sea. Perhaps the most basic question of all is, What *is* life? Most people have a pretty good feel for what the word *living* means, but biologists have never been able to agree on a precise definition. The best we can do is describe the common properties of living things.

Organisms are not random collections of material, but have a precise chemical and physical organization. They all use energy, have the ability to do work, maintain themselves and grow. They do this through a vast number of chemical reactions that are collectively called **metabolism**. Living things also use energy to maintain stable internal conditions different from their surroundings, which is known as **homeostasis**. They can sense and react to their external environment. Finally, all life forms reproduce to perpetuate their kind, and use nucleic acids to record their individual characteristics and pass them on to their offspring.

The basic features of living things are shared by all organisms, not just those that live in the sea. Here we will pay particular attention to marine organisms.

The Building Blocks

Life requires an intricate series of interactions among an immense variety of chemicals. The most important of these is one of the simplest: water. As the universal solvent, water provides the medium in which all the other molecules dissolve and interact. Water is the base of a complex "chemical soup" inside all organisms where the chemical reactions of metabolism take place.

Besides water, most of the chemicals that make life possible are **organic compounds**, molecules that contain atoms of carbon, hydrogen, and usually oxygen. Organic compounds are high-energy molecules; that is, it takes energy to make them (Fig. 6.1). The ability to use energy to make, or synthesize, organic compounds is an important characteristic of living things. Organic compounds break down when left to themselves. When this happens they release the energy that went into their construction. Organisms have developed the ability to control this breakdown, often in order to harness the energy that is released.

How fast do kelp grow? The giant kelp is a brown alga that grows in "forests" along the temperate Pacific coast. Like other algae, the giant kelp uses energy from sunlight to carry out photosynthesis. Under favorable conditions, giant kelp can grow up to 60 cm/day, but on average grows about 30 cm/day. As a result, giant kelp forests are the basis of robust ecosystems that support a variety of marine organisms, from invertebrates, to fishes, to mammals. Giant kelp is harvested as a source of algin, a compound used to make foods and other products. Because of its fast growth rate, it is possible that giant kelp could be considered a renewable resource and potential fuel source.

Carbohydrates Most organic molecules belong to one of four main groups. **Carbohydrates** are composed mainly of only the basic elements of other organic molecules: carbon, hydrogen, and oxygen. The simplest carbohydrates are glucose and a few other simple sugars. More complex carbohydrates are formed by the combination of simple sugars into chains. Ordinary table sugar, for example, consists of two simple sugar molecules joined together. Starches and other complex carbohydrates consist of much longer chains, which may contain components other than simple sugars.

Organisms use carbohydrates in many ways. Simple sugars play a vital role in the most basic metabolic processes, as we will see. Organisms often use complex carbohydrates like starches to store energy reserves.

Some carbohydrates are structural molecules, which provide support and protection. For example, some animals use chitin, a modified carbohydrate, as a skeletal material. Plants and many algae produce a structural carbohydrate called cellulose, the main ingredient of wood and plant fibers. Other carbohydrates have roles in reproduction, metabolism, and interactions among cells—in fact, in nearly every biological function—that scientists are only beginning to understand.

Proteins **Proteins** are another major group of organic molecules. Like complex carbohydrates, proteins are composed of chains of smaller subunits. In proteins the subunits are **amino acids**, which contain nitrogen as well as the carbon, hydrogen, and oxygen typical of all organic compounds.

Proteins have an extraordinary variety of functions. Muscles are made largely of protein. **Enzymes** are proteins that speed up, or catalyze, specific chemical reactions. Without enzymes, most metabolic reactions would proceed very slowly or not at all. There are also many structural proteins, including those in the skin, hair, and skeletons of some marine animals. Some hormones, chemicals that act as messengers to help different parts of the body work together, are proteins. Other proteins carry oxygen in the blood and muscles. Proteins also act as poisons, send chemical signals, produce light, act as antifreeze in the blood of some Antarctic fishes, and have countless other functions.

Figure 6.1 Chemical reactions can be divided into those that require energy and those that release it. Organisms need energy to make complex molecules from simple ones, much as energy is required to push a rock uphill. The breakdown of energy-rich complex molecules is a "downhill" process that releases energy.

Figure 6.2 The thick layers of blubber on this whale are composed almost entirely of lipid. The blubber allows whales to go without eating for long periods, provides buoyancy, and protects them from cold. Most industrial whaling has ceased but so-called traditional whaling, such as that undertaken by these Alaskan Inuit, continues legally in some places.

Lipids Fats, oils, and waxes are examples of **lipids**, another group of organic chemicals. Lipids are often used for energy storage (Fig. 6.2). Another useful property of lipids is that they repel water: Many marine mammals and birds use a coating of oil to keep their fur or feathers dry. Some marine organisms that are exposed to the air at low tide have a coating of wax that helps keep them from drying out. Lipids also provide buoyancy and insulation from the cold. Some hormones are lipids rather than proteins.

Nucleic Acids **Nucleic acids** store and transmit the basic genetic information of life. Nucleic acid molecules are chains of subunits called nucleotides, which consist of a simple sugar joined to molecules, called nitrogen bases, which contain phosphorus and nitrogen. One type of nucleic acid is DNA (deoxyribo-nucleic acid). In most organisms, DNA molecules specify the organism's "recipe"—all the instructions for its construction and maintenance. An organism's complete genetic information is called its **genome**. The organism inherits this genetic information, in the form of DNA, from its parents and passes it on to its offspring. The DNA chains are often tens or hundreds of millions of nucleotides long, but the nucleotides contain only four different types of nitrogen bases: adenine, cytosine, thymine, and guanine. The sequence of the different nucleotides on the chain forms a code, like a four-letter alphabet, that contains the genetic information. Four letters doesn't sound like much, but Morse code can record messages using only two: dots and dashes. DNA's "words" are genes, sequences of DNA that specify the order of amino acids in a single protein.

IN CONTEXT

Water is called the **universal solvent** because it can dissolve more different substances than any other liquid.

Molecules Combinations of two or more atoms; most substances are composed of molecules rather than individual atoms.

- 3.1, The Unique Nature of Pure Water

Another group of nucleic acids is RNA (ribonucleic acid). Like DNA, RNA consists of chains of only four types of nucleotides. Three are the same as in DNA, but the fourth nucleotide in RNA is uracil instead of thymine. Some RNA molecules act as intermediaries in assisting in the conversion of the information stored in DNA into proteins, and in some microbes RNA stores the information itself. For decades these were thought to be the only roles for RNA, but we now know that RNA influences many if not most biochemical processes. Small RNAs regulate the expression of genes stored in DNA into proteins, trigger immune responses and defend against viral infections, regulate embryonic development, and catalyze reactions much as enzymes do. Discovery of the critical and widespread roles of RNA ranks among the most exciting discoveries in biology in decades, and has triggered a wave of new research, including many possible applications in medicine.

Because organisms' nucleic acids define so much about them, determining their nucleotide sequences has become one of the hottest fields of biological research. Complete genome sequences are known for only a small but increasing number of organisms, most notably humans. Partial nucleic acid sequences are available for many more organisms and contain a wealth of information. These sequences can be used as genetic "bar codes" to identify species, including ones that have never been collected, and shed light on details of their metabolism, lifestyle, and even relationships to other organisms.

The Fuel of Life

The molecules that make up living things interact in many complex chemical systems. The most basic of these have to do with the capture, storage, and use of energy or, more simply put, with the production and use of food. These systems store and transfer energy using ATP (adenosine triphosphate) as a common "energy currency." ATP is a high-energy molecule based on adenosine, one of the nucleotides in nucleic acids. Energy is stored as chemical energy by converting a related but lower-energy molecule, ADP (adenosine diphosphate), into ATP. When the ATP is broken back down into ADP the energy is transferred to other molecules to be used in metabolism. In an average day a person cycles through 57 kg of ATP. Whales must use tons!

Microorganisms use a variety of metabolic pathways to process energy. Most other organisms, however, largely rely on only two: photosynthesis and respiration.

Figure 6.3 The bull kelp (*Nereocystis luetkeana*) captures the sunlight streaming down through the water to perform photosynthesis. The kelp uses the light energy to grow. Other organisms eat the kelp, making use of the energy that originally came from the sun.

Photosynthesis: Making the Fuel Most organisms ultimately get their energy from the sun (Fig. 6.3). In **photosynthesis**, algae, plants, and some other organisms capture the sun's energy and use it to make glucose, a simple sugar, some of which is converted into other organic compounds. Most other organisms use these organic molecules as a source of energy. Organic material contains a tremendous amount of energy. As food it fuels our bodies and those of most other creatures. In such forms as oil, gas, and coal, it heats our homes, runs our factories, and powers our cars.

Nick Norman/National Geographic/Getty Images

Photosynthesis begins when an organism absorbs solar energy, sunlight, with chemicals called photosynthetic pigments (Fig. 6.4). The most common photosynthetic pigment is **chlorophyll**, which gives plants their characteristic green color. Most algae have additional pigments that often mask the green chlorophyll. Because of these pigments, algae may be brown, red, or even black instead of green.

In a long series of enzyme-controlled reactions, solar energy captured by chlorophyll and other pigments is converted into chemical energy, in the form of ATP, and then used to make glucose, with carbon dioxide (CO_2) and water (H_2O) as the raw materials. Carbon dioxide is one of very few carbon-containing molecules not considered to be an organic compound. Photosynthesis, then, converts carbon from an inorganic to an organic form. This is called fixing the carbon, or **carbon fixation**. The solar energy absorbed by chlorophyll is stored as chemical energy in the form of glucose. The glucose is then used to make other organic compounds (see "Primary Production," below). ATP may be the "energy currency" of the cell, but fixed carbon—organic matter—is the "currency" in which energy flows through food webs and ecosystems.

Photosynthesis produces oxygen gas (O_2), as a by-product. All the oxygen gas on Earth, in the atmosphere we breathe and in the ocean, was produced by photosynthetic organisms, which constantly replenish Earth's oxygen supply. We rely on photosynthesis not only for food, but for the air we breathe.

Organisms that are capable of photosynthesis can obtain all the energy they need from sunlight and don't need to eat. They are called **autotrophs** ("self-feeders"). Plants are the most familiar autotrophs on land. In the ocean, however, bacteria and algae are the most important autotrophs.

Many organisms cannot produce their own food and must obtain energy from organic matter produced by autotrophs. These organisms, including all animals, are called **heterotrophs**.

Respiration: Burning the Fuel

Both autotrophs and heterotrophs perform **respiration** to use the solar energy stored in organic chemicals by photosynthesis. The chemical reactions are different, but the net result of oxygen-dependent, or **aerobic**, respiration is essentially the reverse of photosynthesis (Fig. 6.5, p. 146). Sugars are broken down using oxygen, and carbon dioxide and water are given off. This process is sometimes called cellular respiration to distinguish it from the physical act of breathing. Since

Figure 6.4 In photosynthesis, carbon dioxide and water are used to make glucose. The energy for this process comes from sunlight. Oxygen is given off as a by-product.

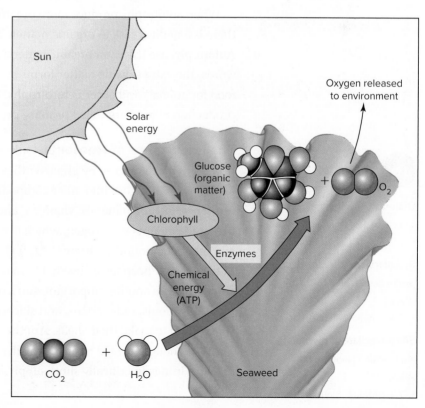

Sun

Solar energy

Oxygen released to environment

Glucose (organic matter)

O_2

Chlorophyll

Enzymes

Chemical energy (ATP)

CO_2

H_2O

Seaweed

High energy

Glucose + O₂

Solar
energy

Autotrophs
only

Photosynthesis

Respiration

Chemical
energy to
organism

Autotrophs
and
heterotrophs

H_2O + CO_2

Low energy

Figure 6.5 Respiration is basically the opposite of photosynthesis. Respiration breaks down glucose, using oxygen, and produces carbon dioxide and water. The energy that originally came from the sun is then made available to the organism.

IN CONTEXT

Diatoms Unicellular algae with a shell, or test, made of silica.
• 7.3, Single Celled Eukaryotes

Radiolarians Unicellular protozoans with a test made of silica.
• 7.3, Single Celled Eukaryotes

Silicoflagellates Unicellular algae with a skeleton made of silica.
• 7.3, Single Celled Eukaryotes

breathing acts to provide oxygen for cellular respiration, the two are related.

Cellular respiration is similar to burning wood or oil in that organic material is broken down, oxygen is consumed, and energy is released. That's why dieters talk about "burning calories." In respiration, however, the energy in organic material is not released in flames. Instead, it is stored in ATP.

If oxygen isn't available, many organisms can use a different form of respiration called **anaerobic** respiration. Anaerobic respiration does not require oxygen, but it isn't as efficient as aerobic respiration and provides less energy to the organism. Many animals temporarily switch to anaerobic respiration, however, when extreme exertion depletes the oxygen in their muscles and blood. Some organisms that live in oxygen-depleted environments, like muddy sediments or the guts of fishes, use anaerobic respiration all the time. **Aerobic** respiration is much more widespread than anaerobic respiration, however, and in this text the general term "respiration" refers to aerobic, or oxygen-dependent, respiration.

Primary Production The sugars produced by photosynthesis supply both raw material and energy, via respiration, for the manufacture of other organic compounds. Through elaborate chemical processes, some of the glucose formed during photosynthesis is converted into other types of organic molecules—carbohydrates, proteins, lipids, and nucleic acids. The energy for these conversions, in the form of ATP, comes from metabolizing much of the rest of the glucose. Thus, most of the glucose produced by photosynthesis is either converted to other organic compounds or respired to fuel this conversion (Fig. 6.6).

When autotrophs produce more organic matter than they use in respiration, there is a net increase in organic matter. This is called **primary production**. Autotrophs use the extra organic material to grow and reproduce. In other words, the extra organic matter forms more living material, which means more food for animals and other heterotrophs. Organisms that perform this primary production of food are called primary producers, or sometimes just producers.

The Importance of Nutrients Only carbon dioxide, water, and sunlight are needed to make glucose by photosynthesis, but additional materials are needed to convert the glucose into other compounds. These raw materials, called **nutrients**, include minerals, vitamins, and other substances. Primary producers use large amounts of nitrogen, which is needed to make proteins, nucleic acids, and other compounds. Nitrate (NO_3^{-1}) is the most important form of nitrogen in the ocean. Phosphorus, needed to make ATP and nucleic acids among other compounds, is another important nutrient. Its main source is phosphate (PO_4^{-3}). Diatoms, radiolarians, and silicoflagellates require large amounts of silica (SiO_2) to make their shells. Another important nutrient is iron in various forms. Although needed in much smaller amounts than nitrogen, phosphorus, or silica, iron is in critically short supply in large parts of the ocean.

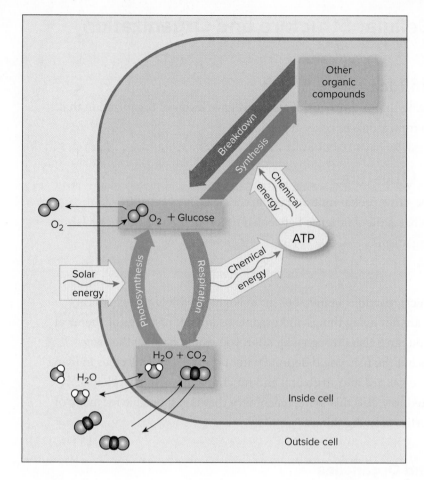

Figure 6.6 In photosynthesis, autotrophs capture energy from the sun and store it as chemical energy in simple sugars like glucose. Some of the sugars are broken down in respiration, and the energy is stored in ATP. The thin black arrows represent molecules moving in and out of the cell.

Section Review

1. **Construct** Make a table to record information about the four main organic molecules that make life possible. The table column headings should be "Molecule," "Description," and "How Used by Living Things."

2. **Explain** Why is the statement "Cellular respiration is basically the opposite of photosynthesis" true?

6.2 Cellular Structure and Organization

VOCABULARY

cell membrane

cytoplasm

organelles

cytoskeleton

prokaryotes

cell wall.

ribosomes

eukaryotes

nucleus

endoplasmic reticulum

Golgi apparatus

mitochondria

chloroplasts

tissues

organs

organ systems

population

species

communities

ecosystem

Main Idea

All living things are made of cells and have levels of organization that range from atoms to ecosystems.

. .

Key Questions

1. Describe prokaryotic and eukaryotic cells.
2. What is the difference between a unicellular and a multicellular organism?

The chemical reactions of metabolism make life possible, but chemical reactions are not living things and organisms are not just bags of chemical soup. If you ran a fish through a blender, you would have all the same molecules but the fish would definitely not be alive. The molecules in living things are *organized* into structural and functional units that work in a coordinated way. It is this living machinery that is responsible for many of the properties of living things.

Cells and Organelles

The basic structural unit of life is the cell. All organisms are made of one or more cells. Cells contain all the molecules needed for life packaged in a living wrapper called the **cell membrane**, or plasma membrane. This membrane isolates the gelatinous contents of the cell, or **cytoplasm**, from the outside world. The cell membrane allows some substances to pass in or out but prevents others from doing so.

There are a number of structures within the cytoplasm. Among the most important are membranes similar to the outer cell membrane. These interior membranes act as sites for such chemical processes as photosynthesis and respiration. Membranes may also divide the cell into compartments that enclose still more complex structures specialized for particular tasks. These membrane-bound structures are known as **organelles**.

Cells also have a complex internal framework made of protein fibers. This framework, called the **cytoskeleton**, supports the cell, allows it to move and change shape, and helps it divide.

Structurally Simple Cells: Prokaryotes
Prokaryotic cells are primitive in the sense that they are the most ancient of the two major types of cells. They are also the structurally simplest and smallest cells. Prokaryotic cells are distinguished by the absence of most kinds of organelles (Fig. 6.7), though some prokaryotes do have certain organelles or organelle-like structures and recent research has revealed a more complex cytoskeleton than was previously known. Organisms with prokaryotic cells are called **prokaryotes**. Bacteria are the best known group

of prokaryotes. Archaea are the second prokaryotic group. Most prokaryotes are microscopic, and most are tiny even by the standards of microscopic organisms, but there a few giant bacteria that can be seen with the naked eye.

Outside the cell membrane of prokaryotes lies a supportive **cell wall**. Inside, the cell has a single, ring-like molecule of DNA. Attached to the cell membrane and scattered through the cytoplasm are small structures made of protein and RNA called **ribosomes** that make the cell's proteins. In photosynthetic prokaryotes, folds of the plasma membrane contain photosynthetic pigments. Some prokaryotes have one or more protein filaments known as flagella (singular, flagellum) that extend outside the cell membrane. These cells move by rotating their flagella like tiny propellers.

Structurally Complex Cells: Eukaryotes Eukaryotic cells, the second major cell type, are more organized and complex than prokaryotic cells (Fig. 6.8, p. 150). Various membrane-bound organelles found only in **eukaryotes** do specialized jobs within the cell. The **nucleus** contains chromosomes that carry most of the cell's DNA. The nucleus thus holds the cell's genetic information and directs most of its activities. The nucleus can be thought of as the cell's headquarters.

The "factories" of eukaryotic cells are two structures, mostly folded membranes, called the **endoplasmic reticulum** and the **Golgi apparatus**. These organelles make, package, and transport many of the organic molecules that the cell needs. Ribosomes sit in some of the endoplasmic reticulum. Organic molecules, water, and other materials are often stored in **vacuoles**.

Respiration in eukaryotes takes place in **mitochondria**. Mitochondria are the cell's power plants, breaking down organic molecules to provide energy. Plant and algal cells have two important structures that animals don't. First, they have **chloroplasts**, chlorophyll-containing organelles where photosynthesis takes place. Second, except for some single-celled algae, they have a cell wall.

Eukaryotic cells often have flagella, which are called cilia when they are short and numerous. Eukaryotic flagella and cilia differ from prokaryotic flagella in being enclosed in a membrane, having a more complex structure, and beating back and forth instead of rotating. Some eukaryotes use their flagella or cilia to swim, as prokaryotes do, but many eukaryotes use flagella and cilia to push water and particles through or over their bodies instead of pushing their bodies through the water. Many marine animals, for example, use cilia to push food particles to the mouth. Cilia in your lungs carry away dust and other irritants.

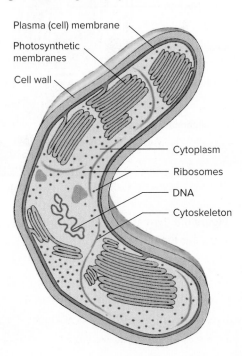

Plasma (cell) membrane
Photosynthetic membranes
Cell wall
Cytoplasm
Ribosomes
DNA
Cytoskeleton

Figure 6.7 Prokaryotic cells have a plasma membrane lying inside a cell wall. The plasma membrane is where respiration and photosynthesis occur. Prokaryotic cells also contain scattered ribosomes and a ring-like molecule of DNA.

Golgi apparatus

Nucleus

Plasma (cell) membrane

Mitochondrion

Endoplasmic reticulum

Ribosome

Cytoplasm

Cytoskeleton

(a)

Endoplasmic reticulum

Nucleus

Mitochondrion

Ribosome

Golgi apparatus

Cell wall

Plasma (cell) membrane

Chloroplast

Vacuole

Cytoskeleton

(b)

Figure 6.8 Typical cells of (*a*) animals and (*b*) algae and plants, and some of the structures they contain.

From Snack to Servant: How Complex Cells Arose

Prokaryotes were the first life on Earth. There are fossil bacteria thought to be 3.8 billion years old, nearly as old as the oceans themselves. It was another billion years or more before eukaryotic cells appeared. Strangely enough, this advance in cell design, shared by all multicellular life, may have started with a simple "meal."

Biologists believe that mitochondria, the organelles responsible for cell respiration, were once free-living bacteria. They are similar in size to many bacteria. Respiration in both bacteria and mitochondria takes place in their membranes, and mitochondria have folded internal membranes similar to those found in many bacteria. Although most of a eukaryotic cell's DNA is contained in the nucleus, mitochondria have a small amount of their own DNA. This DNA is in the form of a single, circular molecule, like the chromosomes of bacteria. Additionally, mitochondria within the replicate by dividing, much like bacteria reproduce by cell division.

Endosymbiosis Because of these similarities, biologists believe that mitochondria were originally bacteria that came to live inside other cells. A relatively large cell that had already evolved some chemical features of today's eukaryotes may have engulfed a smaller one but been unable to digest it, or a disease-causing bacterium may have invaded the larger cell but not killed it. Both phenomena have been observed in the laboratory in living cells. Either way, the small cells eventually took up permanent residence in the larger cells, using them as hosts. Scientists called this endosymbiotic theory. Symbiosis is an ecological relationship where two organisms depend on one another and both benefit. Since this symbiotic relationship occurs within a cell, it is known as endosymbiosis.

For a small cell living within a larger cell to be symbiosis, both cells must have benefited from the arrangement. The host cells could not use oxygen in respiration. The smaller bacteria could, and provided this ability to their hosts. Over time the host cells transferred most of the symbiotic bacteria's genes into their own nuclei, and thus came to control the symbionts. The symbionts and host cells became more and more dependent on each other. Eventually the symbionts became mitochondria.

Photosynthetic eukaryotes Chloroplasts, the organelles that perform photosynthesis, are also thought to have arisen from symbiotic bacteria. Chloroplasts probably arose when a eukaryotic cell containing mitochondria engulfed a photosynthetic bacterium that wasn't digested and became symbiotic. The exact nature of this symbiont is hotly debated, but it probably looked something like a symbiotic bacterium called Prochloron that lives in some sea squirts and other invertebrates. The eukaryotic host cells gained the symbiont's photosynthetic ability and were the first algae. The algae eventually evolved different types of chloroplasts, which were later exchanged when one kind of alga engulfed another.

The hypothesis that mitochondria and chloroplasts arose from bacteria, once highly controversial, is now almost universally accepted by biologists. Some biologists even classify mitochondria and chloroplasts as symbiotic bacteria rather than organelles. All multicellular organisms, including humans, carry in their very cells the legacy of these bacterial partnerships, perhaps the result of an ancient snack.

Prochloron. This bacterium was once thought to be closely related to the symbiotic, photosynthetic bacterium that gave rise to chloroplasts, but genetic studies indicate that a slightly different group of bacteria was involved.

Figure 6.9 Sponges are the only multicellular animals that are at the cellular level of organization. They have specialized cells, though these cells don't form tissues.

Levels of Organization

A cell is a self-contained unit that can carry out all the functions necessary for life. Indeed, many organisms get by just fine with only a single cell. One-celled organisms are called unicellular. Multicellular organisms have more than one cell. The human body, for instance, contains something like 100 trillion cells. In multicellular organisms the cells interact and communicate, and cells become specialized for particular jobs. Nearly all prokaryotes are unicellular, but some bacteria group together, interact, and specialize for specific functions in ways that border on multicellularity. Many eukaryotes are unicellular, but the majority are multicellular.

In multicellular organisms, specialized cells that perform the same task may be organized into more complex structures. The extent of specialization and organization is referred to as the level of organization (Table 6.1).

At the cellular level of organization, each cell is essentially an independent, self-sufficient unit. It can perform all the functions needed to sustain itself and reproduce, and there is little or no cooperation among cells.

Only a few multicellular organisms remain at the cellular level not having true **tissues** or organs (Fig. 6.9). In most, groups of cells act together to do a particular job. These specialized, coordinated groups of cells are called tissues. Cells that are specialized to contract and do work, for example, are bound together to form muscle tissue. Nervous tissue, specialized to collect, process, and transmit information, is another example.

In most animals, organization doesn't stop at the tissue level. Their tissues are further organized into **organs** that carry out specific functions. The heart is an example of an organ. There are several different kinds of tissues that make up the heart. Muscle tissue, for example, allows the heart to contract and pump blood, and nervous tissue controls the muscle.

Table 6.1 Levels of Organization of Biological Systems

Level	Description	Examples	
Ecosystem	A community or communities in a large area, together with their physical environment	Nearshore ecosystem	One or more organisms
Community	All the populations in a particular habitat	Rocky shore community	
Population	A group of organisms of the same species that occur together	All the mussels on a stretch of rocky shore	
Individual	A single organism	One mussel	
Organ system	A group of organs that work in cooperation	Digestive system	Within individual organisms
Organ	Tissues organized into structures	Stomach	
Tissue	Groups of cells bound together and specialized for the same function	Muscle tissue	
Cell	Independent cells, the fundamental unit of living things	Muscle cell, single-celled organisms	
Organelle	A complex structure within cells, bound by a membrane	Nucleus, mitochondrion	Within cells
Molecule	Combinations of atoms that are bound together	Water, proteins	
Atom	The fundamental unit of all matter	Carbon, phosphorus	

Increasing complexity →

Different organs act together in **organ systems**. The digestive system, for instance, includes many organs: mouth, stomach, intestine, and so on. Animals usually have a number of organ systems, including nervous, digestive, circulatory, and reproductive systems.

Complex organization also exists at levels above the individual organism. A **population** is a group of organisms of the same kind, or **species**, that lives in one place. Mussels on a stretch of rocky shore, for example, form a population (Fig. 6.10).

Populations of different species that occur in the same place form **communities**. A community is more than just a collection of organisms that happen to use the same habitat. The characteristics of communities are determined by the way the organisms *interact*—which organisms eat each other, which compete, and which depend on each other. Rocky shores, for example, are home to many organisms in addition to mussels: seaweeds, crabs, barnacles, snails, and sea stars, to name a few (Fig. 6.11). It is the interactions among all these organisms that give the rocky shore community its own unique structure.

A community, or often several communities, together with the physical, or non-living, environment, makes up an **ecosystem**. The rocky shore community, for instance, along with other nearby communities, is part of a larger ecosystem that includes the tides, currents, nutrients dissolved in the water, and other physical and chemical aspects of the area.

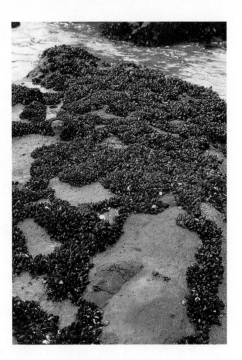

Figure 6.10 Blue mussels (*Mytilus galloprovincialis*) growing on a rocky shore in New Zealand are part of a population.

Figure 6.11 The community that lives on this rocky shore is made up of populations of mussels, barnacles, seaweeds, and many other species.

Section Review

1. **Compare and Contrast** Make a Venn diagram to compare and contrast the characteristics of prokaryotic and eukaryotic cells.

2. **Construct** Make a diagram that shows the levels of organization within an individual organism in increasing complexity.

6.3 Perpetuating Life

Reproduced with permission from G.M. Hallengraeff and M. Gustaf, /lt/ Plankton: A Microscopic World./xit/ 1998, CSIRO/Robert Brown and Associates

VOCABULARY

heredity

daughter cell

mitosis

asexual reproduction

fission

sexual reproduction

gametes

germ tissue

meiosis

eggs

sperm

gonads

fertilization

zygote

recombination

embrio

larval

broadcast spawning

hermaphrodites

reproductive strategy

Main Idea
Successful reproduction is critical to the survival of a species.

Key Questions
1. What are the basic steps involved in asexual and sexual reproduction?
2. What is reproductive strategy? Give one example of a reproductive strategy.

One of the most basic characteristics of living things is the ability to reproduce—to generate offspring similar to themselves. Any life form that failed to replace itself with new individuals of its own kind would soon vanish from the planet. It is only by reproducing that a species ensures its own survival.

Organisms must do two things when they reproduce. First, they must produce new individuals to perpetuate the species. Second, they must pass on to this new generation the characteristics of the species in the form of genetic information. The transfer of genetic information from one generation to the next is called **heredity**.

Organisms achieve the first objective, the production of offspring, in many different ways. But there are relatively few ways through which they pass on genetic information. This similarity of hereditary mechanisms is evidence that all living things are fundamentally related.

Modes of Reproduction

Individual cells reproduce mainly by dividing to form new **daughter cells** (Fig. 6.12). In prokaryotes, where the DNA that carries the genetic information is not enclosed in a nucleus, the cell divides into two new ones by a simple process called cell fission. Before dividing, the cell copies, or replicates, its DNA. Each daughter cell gets one of the copies, as well as the rest of the cellular machinery it needs to survive.

In eukaryotes the DNA lies on several different chromosomes. Before a cell divides, each chromosome is replicated. The most common form of eukaryotic cell division, **mitosis**, is a complex process that takes place in the nucleus. Mitosis ensures that each daughter cell gets a copy of every chromosome. As in bacterial cell division, the end result is two daughter cells that are exact duplicates of the original, with the same genetic information.

Figure 6.12 Two cells of a dinoflagellate (*Dinophysis tripos*) form from the division of a single cell. The two cells, joined here, will soon separate.

Asexual Reproduction Cell division is the primary way that single-celled organisms reproduce. Because a single individual can reproduce itself without the involvement of a partner, this form of reproduction is called **asexual reproduction**. All forms of asexual reproduction are similar in that the offspring inherit all the genetic characteristics of the parent. They are, in fact, exact copies, or clones.

Many multicellular organisms also reproduce asexually. Some sea anemones, for example, split in half to create two smaller anemones. This process is known as **fission**. Another common form of asexual reproduction is budding. Instead of dividing into two new individuals, the parent develops small growths, or buds, that break away and become separate individuals (Fig. 6.13a). Many plants reproduce asexually by sending out various kinds of runners that take root and then sever their connection to the parent (Fig. 6.13b). Because it is so common in plants, asexual reproduction is sometimes called vegetative reproduction, even when it occurs in animals.

Sexual Reproduction Most multicellular and some unicellular organisms reproduce part or all of the time by **sexual reproduction**, in which new offspring arise from the union of two separate cells called **gametes**. Usually each of the gametes comes from a different parent.

DIVE IN!

How are corals adapted for reproduction? Corals are sessile animals, and males and females are not always in close proximity. The answer: Sexual reproduction in a coral reef is a highly coordinated spawning event. Biologists think that long-term cues, like changes in temperature and length of day, guide the timing of spawning. Biologists have also found that corals have photosensitive chemicals that respond to moonlight. The result of all of these cues is one large, synchronized spawning event. Trillions of eggs and sperm from different coral species are released within minutes along a coral reef. This phenomenon has been described as the largest spawning event on the planet.

(a)

(b)

Figure 6.13 Budding is a common form of reproduction in marine animals, including this mushroom coral (*Fungia fungites*) (a). This seagrass (*Halodule pinifolia*) reproduces asexually by sending out rhizomes, or "runners," that take root and form new plants (b).

Organisms that reproduce sexually have a special kind of tissue called **germ tissue**. Whereas all the other cells in the body divide only by mitosis, germ cells are capable of a second type of cell division, **meiosis**.

In most cells of eukaryotic organisms the chromosomes occur in pairs, with each chromosome of a pair storing similar genetic information. Such cells are diploid, designated $2n$. Meiosis produces daughter cells that have copies of only half the parents' chromosomes—one of each pair. Cells with half the normal number of chromosomes are haploid, designated n or $1n$. The haploid daughter cells produced in meiosis are the gametes. In some seaweeds and microscopic organisms, all the gametes are the same. Most organisms, however, have two types of gametes: female gametes called **eggs** and male gametes called **sperm**. In animals the germ tissue is usually contained in gonads, the organs that produce the gametes. Ovaries are the female **gonads**, which produce eggs. In animals the male gonads, or testes, produce sperm. Seaweeds also have gonads that produce sperm, while in flowering plants sperm develop within pollen, which is produced by organs in the flower called anthers.

Eggs are large cells, usually an organism's largest. They contain all the cellular machinery—organelles and cytoplasm—characteristic of regular cells. The eggs of many species also contain large amounts of energy-rich yolk. Sperm, by contrast, are usually an organism's smallest cells, with almost no cytoplasm and few of the typical organelles. Sperm are little more than tiny packages of chromosomes equipped with mitochondria-powered flagella that enable the sperm to swim (Fig. 6.14).

Sperm are attracted to eggs from the same species. When egg and sperm come into contact, they fuse; in other words, **fertilization** occurs. The genetic material contained in the two gametes combines. Since both haploid gametes had $1n$ chromosomes—or half the normal number—the fertilized egg, called a **zygote**, is back to the normal diploid, or $2n$, number of chromosomes. It has

Figure 6.14 (*a*) The egg of the variegated sea urchin (*Lytechinus variegatus*) is many times larger than sperm from the same species. (*b*) The head of a sperm cell is made up almost entirely of the cell nucleus and contains the genetic material. The midpiece contains mitochondria that power the tail, or flagellum, with which the sperm swims.

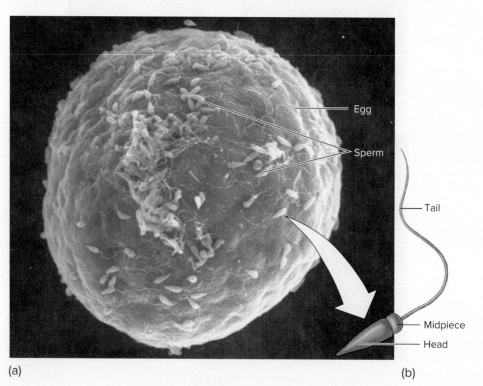

Egg

Sperm

Tail

Midpiece

Head

(a)

(b)

DNA from *both* parents, however, and is therefore different from either. This **recombination** of genetic information to produce offspring that differ slightly from their parents is the most important feature of sexual reproduction.

The fertilized egg begins to divide by ordinary mitosis. Now called an **embryo**, it is nourished by the yolk. Through an extraordinarily complex process called embryological development, the embryo eventually becomes a new adult of the species. Along the way, most marine organisms pass through immature, or **larval**, stages that often look completely different from the adults.

Reproductive Strategies

The goal of reproduction for any organism is to pass its hereditary characteristics on to a new generation. There is a nearly endless variety of ways to achieve this (Fig. 6.15). Some species release millions of eggs and sperm into the water, where fertilization occurs, and have no further interaction with their offspring. This is called **broadcast spawning**. Others issue only a few offspring and invest a lot of time and energy in caring for them. Some species have many different larval stages, whereas others develop directly from egg to adult. Some reproduce asexually, some reproduce sexually, and some do both. In sexually reproducing species, males and females may always be separate individuals, or there may be **hermaphrodites**, individuals that have both male and female reproductive organs either at the same time or at different times of life.

The particular combination of methods used by a given species is called its **reproductive strategy**. The strategy used by a species depends on its size, on where and how it lives, on what kind of organism it is—in fact, on just about everything about it and its environment.

(a) (b) (c) (d)

Figure 6.15 Marine organisms have many different reproductive strategies, a few of which are shown here. (a) Corals (*Acropora millepora*) are broadcast spawners. (b) Male yellowhead jawfish (*Opistognathus aurifrons*) aerating and incubating eggs, an example of parental investment in protecting their young. (c) New Zealand fur seals (*Arctocephalus forsteri*) take care of their young for much longer than most marine animals. (d) Marine plants like seagrasses (*Enhalus acaroides*) flower to reproduce.

Section Review

1. **Describe** What are the differences between asexual and sexual reproduction?

2. **Identify** Which factors influence the type of reproductive strategy used by a species?

6.4 The Diversity of Life in the Sea

IN CONTEXT

Scientific Theory A *hypothesis* that is accepted as "true" because it has passed test after test and is supported by a large body of evidence.
• 1.2, The Scientific Method

Main Idea
Marine organisms are classified into the same groups to which all living things belong—all major groups of living things are found in the oceans.

. .

Key Questions
1. What is the biological species concept?
2. What is an advantage of binomial nomenclature?

There are so many different living things in the sea that it almost boggles the mind. From microscopic bacteria to gigantic whales, marine organisms come in all shapes, sizes, and colors. The ways in which they live are just as varied. Making sense of all this diversity might seem a hopeless task. Fortunately, there is a unifying concept that helps make the bewildering diversity of life comprehensible. This concept is the **theory of evolution**. Remember that scientists do not use the term "theory" lightly. Evolution, the gradual alteration of a species' genetic makeup, is supported by a vast body of evidence. The *ways* in which evolution occurs, on the other hand, never cease to fascinate biologists.

Natural Selection and Adaptation

Evolution occurs because individual organisms have genetic differences in their ability to find food and avoid being eaten, in their success at producing off-spring, in their metabolism, and in countless other attributes. The best-adapted individuals, those most successful at meeting the challenges of the environment, produce more offspring on average than those that are not so well adapted. This process was called **natural selection** by Charles Darwin, the nineteenth-century English naturalist who, along with another Englishman, Alfred Wallace, first proposed the modern theory of evolution.

The best-adapted individuals not only produce more offspring, they also pass their favorable characteristics on to those offspring. The favorable traits therefore become more common, and over generations the population as a whole gradually becomes more similar to the best-adapted individuals. Thus, the result of natural selection is that the population continually adapts to its environment—that is, it evolves.

Every population is constantly adapting to its environment. The world is an ever-changing place, however, and organisms are always being faced with new challenges. Populations either adapt to the changes in the environment or become extinct, making way for others. As a result, evolution is an endless process.

Classifying Living Things

The adaptation of various populations to different environments has produced a fantastic variety of life forms. It is difficult to study and discuss the millions of life forms on Earth unless we first classify them, or sort them into groups. One of the goals of biological classification is to give a universally accepted name to different types of organisms so that scientists from all over the world can call a species by the same name.

Biological Species Concept We have loosely defined a species as a "type" of organism, but what exactly does that mean? After all, fishes are all the same general type of animal, but a sailfish is obviously not the same species as a tuna. No definition is perfect, but here we will define species as populations of organisms that have common characteristics and can successfully breed with each other. This definition is often referred to as the **biological species concept**.

Successful breeding means that the offspring produced are fertile and can themselves propagate the species. Dogs, to use a familiar example, are all the same species because all breeds of dog—no matter how different they may look—can interbreed to produce fertile offspring. Organisms that cannot breed with each other are not members of the same species no matter how similar they look. When two populations are unable to interbreed successfully, they are said to be **reproductively isolated**.

Biological Nomenclature Organisms are identified biologically by two names, the name of their **genus** and a species name or, more correctly, **species epithet**. A genus is a group of very similar species. Dogs, for example, have the scientific name *Canis familiaris*. They are closely related to other species in the genus *Canis*, like wolves, *Canis lupus*, and coyotes, *Canis latrans*. Similarly, the genus *Balaenoptera* contains several related species of whales, including the blue whale, Balaenoptera musculus; the fin whale, *Balaenoptera physalus*; and the minke whale, *Balaenoptera acutorostrata*. This two-name system is called **binomial nomenclature**. It was first introduced in the eighteenth century by the Swedish biologist Carolus Linnaeus. At that time Latin was the language of scholars, and Latin is still used for scientific names, along with Greek. By convention, binomial names are always underlined or italicized. The first letter of the generic name, but not the species epithet, is capitalized. When the name of the genus has been previously mentioned it may be abbreviated. Thus, the blue whale could be identified as *B. musculus*. The species epithet is never used by itself, but only in conjunction with the genus name or initial.

Students often ask why biologists have to use complicated, hard-to-pronounce Latin or Greek names rather than ordinary common names. The trouble with common names is that they are not very precise. The same common name may be applied to different species, and the same species may have different common names.

The name "spiny lobster," for example, is applied to many different species that are not even all in the same genus. Australians call spiny lobsters "crayfish." Americans reserve the name "crayfish" for freshwater relatives of lobsters

DIVE IN!

Viruses in the ocean? Yes, more than you might think. A liter of ocean surface water contains about 100 billion viruses, most of which have yet to be identified or classified. Some of the largest viruses on Earth are found in the ocean. One of these "giant" viruses infects a marine protist, *Cafeteria roenbergensis*. The *Cafeteria* virus has a drastically greater number of base pairs of DNA, compared to a typical virus, and even has its own genes. Viruses do not typically contain genes, they are known for taking over the cellular machinery of the organism they infect and using it to replicate. The discovery of "giant" viruses has caused scientists to take a closer look whether viruses are living or not.

When Fishes Stepped on Land

Organisms evolve in tiny steps, with each change only incrementally different from before, so a key challenge in evolutionary biology is to understand how complex and radically different new structures could have evolved. Fins are great for swimming, and legs are great for walking, but how did animals get from one to the other while still being able to catch food, avoid predators, and find mates every step of the way? Recent breakthroughs have helped us understand how fishes moved onto land to become tetrapods (animals with two pairs of limbs).

The "missing link" In 2006, paleontologists discovered fossils of a fish (*Tiktaalik roseae*) that appears very close to the so-called missing link between fishes and land vertebrates. *Tiktaalik* lived about 375 million years ago, 10–15 million years before the earliest known fossil tetrapods. The arrangement of bones in *Tiktaalik's* fins is almost perfectly intermediate between the limbs of early tetrapods and the fins of lobe-finned fishes. In particular, *Tiktaalik's* fin bones foreshadow the elbow, wrist, and fingers of tetrapods. *Tiktaalik* was probably unable to walk, but it could at least use its fins to do push-ups. *Tiktaalik* also had other features that are refined in tetrapods, such as the beginnings of a neck. Like its fins, *Tiktaalik's* skull and associated vertebrae have features in between those of fishes and tetrapods. Scientists doubt that *Tiktaalik* is the direct ancestor of tetrapods, but it is probably a very close cousin. Every summer *Tiktaalik's*

discoverers return to the Canadian Arctic, where they found its fossils, in hope of new discoveries about the emergence of tetrapods, which 300 million years later would return to the sea as whales (see "The Whales That Walked to Sea," p. 191).

The genetics of change Meanwhile, biologists are learning the genetic basis for these changes. The key is that evolution does not have to completely reinvent the wheel—or the limb. The genome is flexible enough that the same basic system, with only minor tweaks, can control the development of a range of different structures. The earliest fishes did not have paired fins on the side, only median fins, single fins along the midline like the dorsal fin of sharks. In today's sharks the dorsal fin and paired fins develop from different cells in different parts of the embryo, but biologists recently showed that the same genetic programs control the embryonic development of both. These genes are also involved in the development of median fins in lampreys, primitive fishes that branched off before paired fins emerged, and the development of wings in chick embryos. Thus, the same genetic mechanism controls the development of median fins, paired fins, limbs, and wings. Paired fins and limbs did not evolve from scratch, but emerged through the application of the molecular system that was already producing median fins. The researchers are testing this hypothesis by studying lancelets, the nearest invertebrate relatives of vertebrates.

Lobe-finned Fish
380 million years ago

Tiktaalik

Early Tetrapod
365 million years ago

Figure 6.16 *Panulirus interruptus*, the Pacific spiny lobster, is sometimes called a rock lobster or a crayfish depending on where you live.

(Fig. 6.16). Things get even more confusing when people speak different languages. In Spanish, spiny lobsters are called *langostas*, which can also refer to grasshoppers. Another example is the name "dolphin," which is applied not only to the lovable relatives of whales but also to a delicious game fish. Dolphin fish are called *dorado* in Latin America and *mahimahi* in Hawai'i. At the local crab shack this confusion is only a minor annoyance. To biologists, however, it is essential to precisely identify whatever species they are discussing. The use of scientific names that are accepted worldwide avoids confusion.

Confusion over common names can have practical implications as well. After a dive on a coral reef near Australia, one of the authors of this book mentioned to his dive buddy that he had seen a gray reef shark. Gray reef sharks, although not so dangerous as to force you out of the water, can be aggressive and merit some attention and respect. The diving partner, who was familiar with the area, replied that the shark was not a gray reef, but a "graceful whaler." "Looks just like a gray reef shark to me," said the author but, reassured, he continued diving, blissfully ignoring the "graceful whalers" that swam by. Only later did he discover that "graceful whaler" was just a local name for the gray reef shark, *Carcharhinus amblyrhynchos*!

Phylogenetics: Reconstructing Evolution

The goals of biological classification include not only assigning agreed-upon names to organisms but also grouping them according to their relatedness. Most people instinctively understand that a seal is related to a sea lion, for example, or that oysters and clams are related. They might be hard-pressed, though, to explain exactly what they mean by "related." To a biologist, saying

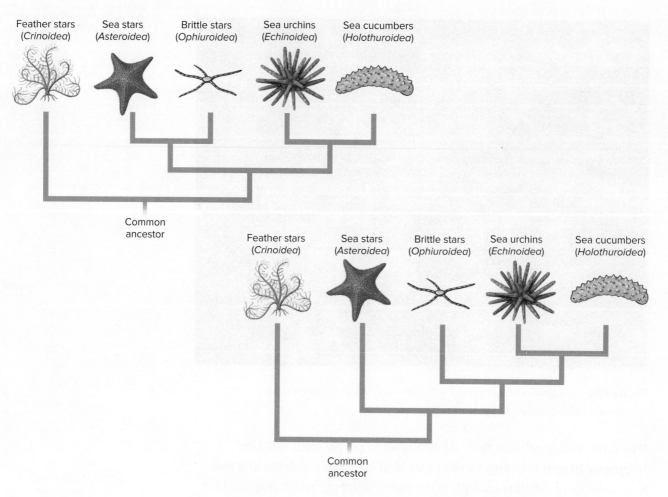

Figure 6.17 Alternative proposals for the phylogeny of the major groups of echinoderms. Conflicting hypotheses like these are common in phylogenetics.

that two groups of organisms are "related" means that they share a common evolutionary history, or **phylogeny**. In particular, it means that both groups evolved from a common ancestral species. Closely related groups evolved from common ancestors relatively recently, while the common ancestors of distantly related groups occurred further in the past. The study of such evolutionary relationships is called **phylogenetics**.

It is almost always difficult to determine the phylogenetic relationships among organisms. Very few groups have a good record of fossils that trace their evolutionary history. Usually the fossil record is incomplete or nonexistent. Biologists have to try to piece together an organism's evolutionary history from other evidence. Body structure, reproduction, embryological and larval development, and behavior all provide clues. Biologists rely increasingly on molecular studies and, in particular, on DNA and RNA sequences. Because evolution is genetic change by definition, the study of the nucleic acids that store organisms' genetic information reveals a great deal about evolution. Unfortunately, different methods of studying phylogeny frequently yield different results, and experts in a particular group often disagree about its phylogeny (Fig. 6.17). Classification schemes change and new arguments arise as fresh information becomes available.

Table 6.2 Biological Classification of Humans (*Homo sapiens*) and the Bottlenose Dolphin (*Tursiops truncatus*)

	Taxonomic Level	Example
Human		
	Domain	Eukarya
	Kingdom	Animalia
	Phylum	Chordata
	Class	Mammalia
	Order	Primates
	Family	Hominidae
	Genus	*Homo*
	Species	*Homo sapiens*
Dolphin		
	Domain	Eukarya
	Kingdom	Animalia
	Phylum	Chordata
	Class	Mammalia
	Order	Cetacea
	Family	Delphinidae
	Genus	*Tursiops*
	Species	*Tursiops truncatus*

Note: The term "division" is used in place of "phylum" in the classification of plants.

As noted previously, biological classification seeks to group organisms according to their relatedness. Species that are very closely related, for example, are grouped into the same genus. Genera with similar phylogenies can be grouped together into a larger group, called a family. This process can be continued to sort organisms into progressively larger groups. All the members of a group, or taxon (plural, taxa), have certain common characteristics and are thought to share a common ancestry. Taxa are systematically arranged in a hierarchy that extends from the most general classification, the domain, down to the species (Table 6.2).

The Tree of Life In centuries past, all known organisms were classified as either plants (kingdom Plantae) or animals (kingdom Animalia). As biologists learned more about the living world, however, they discovered many organisms that didn't fit, and the classification scheme was changed to recognize additional kingdoms. A system of five kingdoms became widely accepted. In addition to the Animalia and Plantae, this traditional system included fungi (kingdom Fungi), bacteria (kingdom Monera), and protists (kingdom Protista). Fungi are multicellular, heterotrophic organisms that are neither plants nor animals. Protists are an extremely varied group of organisms, some plant-like and some animal-like, that includes both unicellular eukaryotes and multicellular seaweeds. Biologists now recognize that protists are a collection of taxa with different evolutionary histories rather than a true kingdom, although for convenience they are still often grouped together under the name "protists."

In recent years, studies of the RNA and cellular chemistry of prokaryotes have led biologists to conclude that there are two main groups of prokaryotes that are as different from each other as they are from all eukaryotic organisms. To recognize this, some classification systems place these two groups of prokaryotes into two new kingdoms, the Bacteria and the Archaea. Other systems recognize a new taxon, the domain, that is even more general than the kingdom; in this system the Bacteria and Archaea are separate domains and each includes a number of different kingdoms. In this text we use the three-domain system of classification (Fig. 6.18), but not all biologists agree with this system.

Viruses are not included in this system because biologists are not in agreement that they are alive or where to fit them in the tree of life. Viruses consist of little more than a handful of genes, encoded in either RNA or DNA, enclosed in a capsule of protein. Viruses are best known for causing disease in humans. Viruses also infect marine organisms, from prokaryotes to marine mammals, and even other viruses, and in doing so have been a driving force in evolution and play a major role in marine food webs.

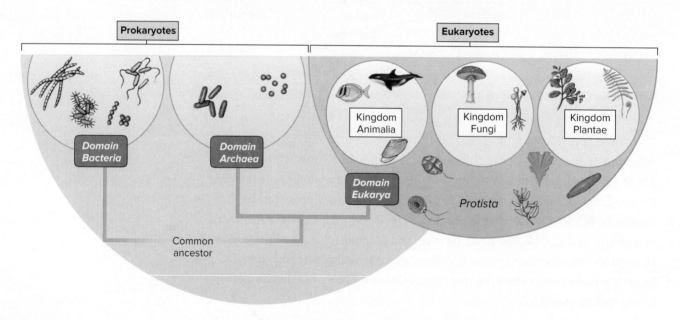

Figure 6.18 The major groups of living things.

Section Review

1. **Identify** What are the three domains and the different groups or kingdoms within each domain?

2. **Describe** Which tools and/or characteristics do scientists use to determine phylogenetic relationships among organisms?

6.5 Challenges of Life in the Sea

Main Idea
Marine organisms are uniquely adapted to survive in the conditions unique to the marine environment.

. .

Key Questions
1. For which particular conditions do marine organisms need to adapt to for survival?
2. Why is a cell membrane particularly important to marine organisms?

VOCABULARY

planktonic

benthic

nekton

diffusion

selectively permeable

osmosis

active transport

osmoconformers

osmoregulate

urea

ectotherms

endotherms

homeotherms

surface-to-volume ratio

Every habitat, the natural environment where an organism lives, has its own unique set of characteristics and presents special challenges to the organisms that live there. Marine organisms must cope with problems unlike those of land dwellers. Even within the marine environment there are different habitats, each of which poses special difficulties. **Planktonic** organisms—those that drift in the water—face very different conditions than **benthic** organisms—those that live on the bottom—or **nekton**—organisms that are strong swimmers. Organisms have evolved innumerable ways to adapt to the conditions of their habitats.

Many adaptations of marine organisms have to do with maintaining homeostasis. The living machinery inside most organisms is sensitive and can operate properly only within a narrow range of conditions. Living things have therefore evolved ways to keep their internal environments within this range regardless of the external conditions.

Salinity
Many enzymes and other organic molecules are sensitive to ion concentrations. Marine organisms are thus immersed in a medium—seawater—with the potential to profoundly alter their metabolism. To fully understand the problems posed by salinity, we must know something about how dissolved ions and molecules behave.

Diffusion and Osmosis In solution, ions and molecules move around just like water molecules do. If the molecules are concentrated in one part of the solution, this random movement spreads them out until they are evenly distributed (Fig. 6.19). The overall result is that the molecules move from areas of high concentration to areas of low concentration. This process is called **diffusion**.

Whenever the internal composition of a cell differs from that on the outside, substances tend to move in or out of the cell by diffusion. If the surrounding seawater contains more sodium, say, than the cell's cytoplasm, sodium tends to diffuse into the cell. This is not good if the organism is sensitive to sodium. Similarly, substances that are concentrated inside cells tend to diffuse out.

DIVE IN!

What is the advantage of staying warm? Bluefin tuna are known for migrating thousand of kilometers from the cold waters of the North Atlantic or Pacific oceans, were they feed, to tropical waters to breed. Tuna also move from warmer surface waters to colder water as they dive for food. Despite swimming through water temperatures that range from 30°C to 3°C, tuna are able to maintain their body temperatures due to a complex blood flow system. Scientists hypothesize that tuna have a larger habitat because they can handle significant temperature changes without their metabolic rate being affected. As a result of being endothermic, tuna can dive deep and swim fast to hunt for food as well as avoid predators.

IN CONTEXT

Ions *Atoms* or groups of atoms that have a positive or negative charge. The most common ions in seawater are chloride and sodium.

Solute Ions, organic molecules, or anything else that is dissolved in a solution.

Salinity The total amount of *salts* dissolved in water. Salts are combinations of ions..

Many of the cell's precious molecules—like ATP, amino acids, and nutrients—are much more concentrated in the cell than in seawater. Therefore, they tend to leak out of the cell.

One answer to this problem is to have a barrier that stops molecules from diffusing in and out. The cell membrane is just such a barrier. It blocks the passage of the common ions in seawater as well as many organic molecules. The membrane cannot be a complete barrier, however, because the cell needs to exchange many materials—like oxygen and carbon dioxide—with its sur-roundings. The cell membrane is **selectively permeable**; that is, it allows some substances to enter and leave the cell but blocks others. Water and other small molecules pass readily through the selectively permeable cell membrane, for example, whereas more complex molecules like ATP and many proteins do not.

The selective permeability of the cell membrane solves the problem of diffusion of ions and organic molecules, but it creates a new problem. Like any other molecule, water diffuses from areas where it is concentrated to areas of lower concentration. If the total concentration of *solutes* inside the cell is higher than that outside, the concentration of *water* will be lower. Because water molecules are free to cross the cell membrane, they will stream into the cell, causing it to swell (Fig. 6.20). On the other hand, the cell will tend to lose water and shrivel if the total salt concentration outside the cell is higher than inside. The diffusion of water across the cell membrane is known as **osmosis**.

Diffusion and osmosis always move materials from high concentration to low. The selective permeability of the cell membrane can control this move-ment, but not reverse it, and cells often need to move materials in the opposite direction. They may, for example, need to rid themselves of excess sodium even though the seawater outside is more concentrated than the inside of the cell, or to take up sugars and amino acids from the surrounding water even though they are already more concentrated inside the cell. In **active transport**, proteins in the cell membrane pump materials in the opposite direction to which they would move by diffusion. This process requires energy, in the form of ATP.

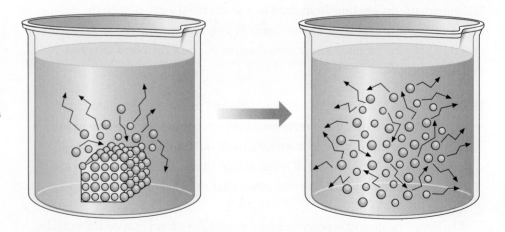

Figure 6.19 When a crystal of salt dissolves in water, the ions are concentrated in one part of the solution but eventually spread out because of their random movement. This process, called diffusion, results in the flow of ions from areas of high concentration to areas of low concentration.

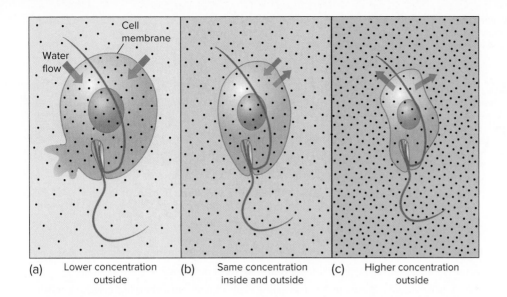

| (a) | Lower concentration outside | (b) | Same concentration inside and outside | (c) | Higher concentration outside |

Figure 6.20 Osmosis. (a) If there is a higher concentration of dissolved material inside a cell than outside, then water will move into the cell. (b) If the internal and external concentrations are the same water moves in and out of the cell equally. (c) When the outside water has a higher concentration of dissolved material, the cell loses water.

Active transport is so important that it represents over one-third of cells' total energy expenditure.

Regulation of Salt and Water Balance Marine organisms have adapted to the problems of maintaining the proper balance of water and salts in various ways. Some don't actively maintain salt and water balance at all; their internal concentrations change as the salinity of the water changes. Such organisms are called **osmoconformers**. Because they don't maintain salt and water balance, many osmoconformers can only live in a narrow salinity range. Others tolerate relatively large changes in the salinity of the surrounding seawater (and therefore their tissues), but even they experience osmotic problems outside their tolerance range. In areas like the open ocean, where salinity doesn't fluctuate much, osmoconformers have few difficulties.

Other marine organisms **osmoregulate**, or control their internal concentrations to avoid osmotic problems. Osmoregulators still have limits, but are generally able to tolerate salinity changes better than osmoconformers. One way osmoregulators adapt to different salinities is by adjusting the concentration of solutes in their body fluids so that the overall concentration of their fluids remains relatively constant. It doesn't matter if there are the same dissolved chemicals inside and out as long as the total *amount* of dissolved material is the same. Some organisms osmoregulate by changing the amount of just one particular chemical to match changes in the salinity outside. Sharks, for example, increase or decrease the amount of a chemical called **urea** in their blood. A single-celled marine alga called *Dunaliella* uses a different chemical, glycerol, to do the same thing. *Dunaliella* may be the champion osmoregulator. It can maintain its water balance in water ranging from nearly fresh to more than nine times saltier than normal seawater.

By contrast, many osmoregulators maintain blood concentrations that are different from that of the surrounding seawater. Most marine fishes have body fluids considerably more dilute than seawater (Fig. 6.21a, p. 168) and tend to lose water by osmosis. They replace the lost water by drinking seawater. They also conserve water by producing only a very small amount of urine. Some of

Water loss by
osmosis through
gills and skin
35⁰/₀₀

14⁰/₀₀

Drinks
seawater

Salts pass
through gut

Small volume
of relatively
salty urine

Salts excreted
by gills

(a) Marine fish

Red
blood
cells

Salts

Water gain by
osmosis through
gills and skin
0⁰/₀₀

14⁰/₀₀

Does
not
drink

Salt absorbed
by gills

Large volume
of dilute urine

No salts
enter gut

(b) Freshwater fish

Figure 6.21 Regulation of salt and water balance. (a) In most marine fishes, water flows by osmosis from the body fluids to the seawater. To avoid dehydration, the fish drinks seawater. It excretes as little urine as possible. Salts that are absorbed are excreted, mostly through the gills but also in the urine. (b) Freshwater fishes tend to gain water by osmosis. They refrain from drinking and produce large amounts of dilute urine. Salts are absorbed by the gills to replace those lost in the urine.

Figure 6.22 Sea turtles have glands near the eyes that excrete excess ions, producing salty "tears."

the excess salts that are taken in with the seawater they drink are never absorbed and pass right through the gut. Marine fishes do absorb some of the salts they drink and must actively rid themselves of, or excrete, these excess salts. Some salts are excreted in the urine, but because so little urine is produced this is not enough. Most of the excess salts are excreted through the gills.

Freshwater fishes have the reverse problem: Their blood has a higher concentration of salt than the surrounding water, and they take in water by osmosis. Their adaptations are opposite those of marine fishes (Fig. 6.21b).

Marine birds, reptiles, and some marine plants, also have special cells or glands to rid themselves of excess salt (Fig. 6.22). Marine plants and most algae have the added advantage of a fairly rigid cell wall that helps them resist the swelling caused by osmotic water gain.

Temperature

Organisms are greatly affected by temperature. Metabolic reactions proceed faster at higher temperatures and slow down dramatically as it gets colder. A general rule is that most reactions occur about twice as fast with a 10 °C rise in temperature. At extreme temperatures, however, most enzymes cease to function properly.

Most marine organisms are adapted to live in particular temperature ranges. Many polar species, for instance, have enzymes that work best at low temperatures, and they cannot tolerate warm water. The reverse is true of many tropical species. As a result, temperature plays a major role in determining where different organisms are found in the ocean (Fig. 6.23).

Organisms are often categorized according to how their metabolism affects their body temperature. All organisms generate metabolic heat, but in most organisms this heat is quickly lost to the environment and does not raise body temperature. These organisms are known as **ectotherms** and are commonly called "cold-blooded." In **endotherms**, on the other hand, metabolic heat is retained and raises the body temperature above that of the surroundings. Endotherms are often called "warm-blooded." Endotherms include mammals, birds, and some large fishes, including some tunas and sharks.

Another way to categorize organisms is by whether they can keep their body temperature more or less constant regardless of the temperature of the environment. **Homeotherms** can maintain a stable internal temperature. Mammals and, to a lesser extent, birds are able to regulate their internal temperature so that it stays much the same even when the external temperature varies. Being endotherms, they retain heat, but they also control their metabolism, producing more heat as needed by burning up fats and other energy-rich molecules. This allows them to remain highly active regardless of water temperature, but it also requires a lot of energy, so homeotherms need to eat more frequently. To reduce the "heat bill"—the energy cost of maintaining their body temperature:

Table 6.3 Classification of Animals by Their Response to Changes in External Temperature

	Poikilotherms: Temperature varies with external temperature	**Homeotherms:** Regulate body temperature so it doesn't vary as much as external temperature
Ectotherms: Most metabolic heat is rapidly lost to the environment; body temperature matches external temperature	Invertebrates, most fishes, marine reptiles	None
Endotherms: Retain significant metabolic heat; internal temperature stays warmer than external temperature	Some large sharks, tunas, and billfishes	Mammals and birds

Figure 6.23 Marine organisms are often restricted to specific regions that correspond to average water temperatures. The major divisions are the polar, temperate, subtropical, and tropical regions. This map refers only to surface waters; deep water is uniformly cold.

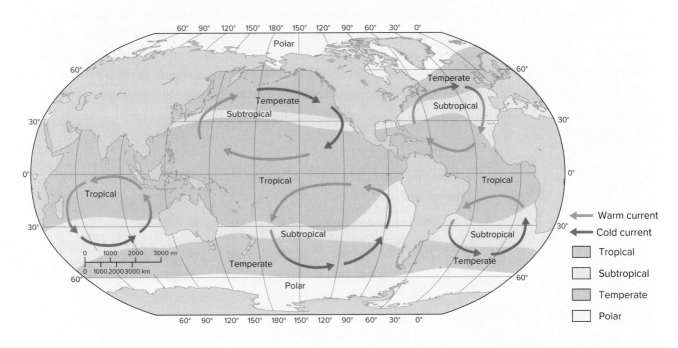

Volume	Surface area	S/V ratio
1 cm³	6 cm²	6:1
8 cm³	48 cm²	6:1
8 cm³	24 cm²	3:1

Figure 6.24 A cube that is 1 cm on a side (top) has a volume of 1 cm³ and, because it has six sides, a surface area of 6 cm². Eight such cubes (middle) will have both eight times the volume and eight times the surface area, so the surface-to-volume (S/V) ratio will stay the same: 6:1. If the eight cubes are combined into a single large one (bottom), the S/V ratio is reduced by half, to 3:1.

Mammals and birds are insulated with hair, blubber, or feathers. The classification of organisms according to how they respond to changes in the temperature of their environment is summarized in Table 6.3.

In contrast, the body temperature of **poikilotherms** changes along with the temperature of the surroundings. Most ectotherms are also poikilotherms. As the temperature of the water rises and falls so does their body temperature, and consequently their metabolic rate. This can make ectothermic animals sluggish if the water is colder than they are used to. Endothermic tunas and sharks are also poikilotherms. Although they retain metabolic heat generated by their large muscles and are warmer than the surrounding water, their internal temperature still goes up and down as water temperature changes.

Surface-to-Volume Ratio

Adaptations to salinity and temperature are needed because salts and heat can flow in and out of organisms—if organisms were not affected by their environment, they would not have to adapt. Organisms also exchange nutrients, waste products, and gases with the environment. Such materials often move in and out across the surface of marine organisms, so the amount of surface area is very important. More precisely, it is the amount of surface area relative to the total volume of an organism—in other words, the **surface-to-volume ratio**, or S/V ratio—that determines how rapidly heat and materials flow in and out.

One thing that determines the S/V ratio is the size of the organism. As organisms grow larger, their volume increases faster than their surface area. Small organisms have a larger S/V ratio than big ones (Fig. 6.24). Small organisms, especially single-celled ones, can therefore rely on simple diffusion across their surfaces for the exchange of materials. Larger organisms must develop supplementary mechanisms, like respiratory and excretory systems.

Section Review

1. **Describe** What is the difference between an osmoconformer and an osmoregulator?

2. **Describe** How do ectotherms differ from endotherms?

3. **Explain** Why is it important that organisms maintain a good S/V ratio?

Chapter 6 Review

Multiple Choice

1. When autotrophs produce more organic matter than they use in respiration, there is a net increase in organic matter. This is called
 a. anaerobic respiration
 b. diffusion
 c. primary production
 d. osmoregulation

3. Which organisms regulate their internal temperature so that it stays constant?
 a. ectotherms
 b. endotherms
 c. homeotherms
 d. poikilotherms

4. The transfer of genetic information from one generation to the next is
 a. cell division
 b. heredity
 c. meiosis
 d. mitosis

Short Answer

5. What are organelles? Name three examples of organelles and their functions.

6. Explain how evolution occurs through natural selection.

7. Describe how sharks regulate salt and water balance and body temperature.

Critical Thinking

8. During the day, algae carry out both photosynthesis and respiration, but at night, when there is no light, they can only perform respiration. Small, isolated tide pools on rocky shores are often inhabited by thick growths of seaweeds, which are algae. Would you expect the amount of oxygen in the water to differ between night and day? How?

DATA ANALYSIS LAB

How did multicellular animals evolve? Biologists have many questions about the origins of multicellular animals. Scientists believe that the common ancestor of multicellular animals resembled single-celled organisms called choanoflagellates. How these microscopic filter feeders may have evolved into complex, multicellular animals is very much a topic of debate among scientists.

Data and Observations

Dr. Claus Nielsen developed a 6-step process (shown below) through which multicellular animals may have evolved from single-celled organisms.

Individual cells stick together more closely due to specialized molecules. Substances, including food, move between cells leading to faster growth.

A primitive gut forms, allowing the larvae to feed and extending the larval stage. Some larvae may reproduce, skipping the adult sponge stage altogether.

A nervous system evolves. With increased chemical communication among cells, predators can find prey and prey can escape predators.

A middle layer of tissue, the mesoderm, evolves. This allows the development of specialized systems for movement, respiration, and waste processing.

Bilateral symmetry develops. Sense organs become concentrated at the head end of the body.

A one-way gut develops, with the mouth and anus separated. This allows for more efficient digestion.

Think Critically

1. Why would changes in food intake and digestion be important in the evolution of complex, multicellular animals?

2. How does the development of a nervous system and sense organs relate to increasing complexity of multicellular animals?

Data obtained from: Holmes, B., 2009. Spongebob the ancestor: The evolution of animals. New Scientist, vol. 202, no. 2706, May, pp. 38–41.

Chapter 7

The Microbial World

THEME Energy and Matter: Flows, cycles, and conservation

The organisms presented in this chapter—viruses, bacteria, archaea, algae, protozoans, and fungi—all play important roles in the cycling of matter and energy in marine ecosystems.

BIG IDEA

Viruses, protists, and fungi all play important roles in marine food webs because they act as primary producers, decomposers, or carry out other processes that make nutrients available to primary producers.

About the photo: Microscopic diatoms, like these *Thalassionema nitzchioides* from Long Island Sound, Connecticut, are abundant in the open ocean.

20 μm

Dr. Yaqin 'Judy' Li, Milford Laboratory/NEFSC/NMFS/NOAA

7.1 Viruses

Main Idea
Viruses are prevalent in marine environments and play a significant role in recycling dissolved organic matter within a food web.

. .

Key Questions
1. What is the general structure of viruses?
2. Which organisms in the marine environment do viruses infect?

VOCABULARY

microorganisms

viruses

capsid

retroviruses

lysogenic

bacteriophages

dissolved organic matter (DOM)

This chapter, the first in our survey of life in the ocean, is devoted to marine **microorganisms**, which are easily the most abundant forms of marine life. Microorganisms live everywhere in the ocean, from the deepest trenches to the highest tide pools, and even below the sea floor. The various groups of marine microorganisms discussed in this chapter have little in common and span all three biological domains. They do share having a small size, much smaller than most other forms of life, hence the name of "microorganisms" or "microbes."

Marine microorganisms have played critical roles in the evolution of life on our planet. Indeed, without microorganisms there would be no life on Earth. Not only did life as we know it today evolve from microorganisms, but without them the planet could not support other forms of life.

Viruses, remarkably, span the non-living and the living. Unlike living organisms, a virus is a particle not made up of a cell. Viruses consist of only a short chain of genetic material containing relatively few genes and protected by an outer protein coat, or **capsid**. They are parasites that reproduce and develop only when infecting a living cell. Nearly all viruses are minute, roughly between 20 and 200 nanometers (nm) (see Appendix A for relative size). They can be seen only with the most powerful microscopes. A recently discovered giant marine virus (Fig. 7.1, p. 174) is an exception, being 440 nanometers in diameter and thus visible through an ordinary light microscope!

Types of Viruses
Several types of viruses have been described. **Retroviruses** store their genetic information in the form of single-stranded RNA, in contrast to the double-stranded DNA of most viruses and all forms of life. Retroviruses are responsible for deadly human diseases, such as AIDS and some types of leukemia. Some viruses reproduce by attaching into a living cell and injecting their nucleic acid into the cell. The viral nucleic acid then commands the cell to produce, or replicate, copies of the virus, which are ultimately released as the infected cell bursts. The new generation of viruses then infects other cells or is released from the organism to infect new hosts. Another type, the **lysogenic** viruses, reproduce by inducing their nucleic acid to become part of the complete genetic information contained in the DNA, or genome, of the host cell. The infecting

DIVE IN!

What is infectious salmon anemia? It is a highly contagious viral disease that affects salmon in aquaculture operations and could affect wild salmon. The average mortality rate if a farm is infected is 30%, with mortality rates soaring to 90% if the infection is not controlled. Over the years outbreaks have occurred in major aquaculture operations across the globe. Since the farmed fish are kept in large pens in open ocean waters, scientists are concerned that the disease could transfer from farmed salmon to wild salmon. There is no treatment for the virus and scientists continue to collect data and monitor the disease.

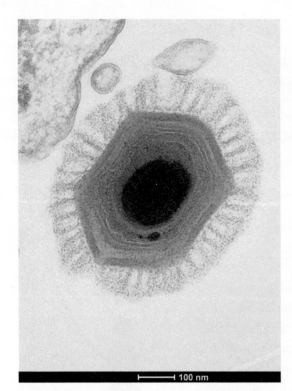

Figure 7.1 *Megavirus*, one of the largest viruses ever recorded, has a diameter of 440 nanometers and bigger than some bacteria, was discovered off the coast of Chile in 2010.

100 nm

virus will then, sometimes long afterwards, directs the production not of host proteins but of new viruses. Many more viruses are then released as the infected cell is destroyed. Of particular importance in the ocean are the highly diverse **bacteriophages**, viruses that specifically target—and destroy—bacteria.

Marine Viruses

Until recently, marine viruses were practically ignored by marine scientists. They are nearly invisible, so it was not until the development of new techniques that viruses were shown to be extremely common in marine environments and to play an important role in the complex cycles of life in the ocean. Though viruses are not referred to as living "organisms" in the strict sense of the word, viruses constitute the most abundant and diverse life-like particles in the ocean. Virus numbers in water samples can be determined by staining them with chemicals and observing them under a microscope, something that can be done on a research vessel. Their numbers in seawater samples can be staggering. Viruses have been found to be extremely abundant in deep-sea sediments.

The abundance of viruses in the water is directly related to the abundance of microbial life, particularly bacteria, which they infect and destroy in order to reproduce. Viruses also infect other marine microbes that form part of the phytoplankton, the very important primary producers that are part of the drifting plankton. The bursting of the cells of bacteria and phytoplankton as the result of viral infections also releases into the water large amounts of organic molecules and cell debris, which make up **dissolved organic matter (DOM)**. DOM is readily taken in by bacteria and other microbes. These microbes are eaten by small zooplankton, which are then eaten by larger zooplankton, so that the energy contained in DOM is not lost but ultimately is available to all kinds of animals, from small fishes to whales. The bursting of cells as a result of viral infections also releases some essential nutrients, which can then be made available to primary producers.

Viruses are also responsible for diseases affecting marine life other than microbes. Some viruses affect molluscs and crustaceans of commercial importance, while others cause serious diseases in fishes, sea turtles, and marine mammals. Humans are also vulnerable. Oysters and mussels that filter hepatitis viruses from sewage-contaminated water commonly infect humans eating them raw or improperly cooked.

Tiny Cells, Big Surprises

It is surprising how little biologists know about the lives of marine microbes. Our knowledge is rapidly growing as we apply new techniques and approaches to studying the private lives of marine microorganisms that call the ocean home.

Sampling marine microbes Information about marine microbes was traditionally gathered by collecting organisms from water samples and growing them in the lab. Many organisms, however, do not grow in the lab so their presence in samples went undetected. The vast majority of bacteria and marine microbes are poorly understood. Some groups of microbes remained undiscovered for a long time. Most archaea, for instance, weren't discovered until the 1970s.

Viruses, prokaryotes, and microscopic eukaryotes seem to be everywhere but are very difficult to collect and characterize. Now, biologists collect fragments of nucleic acid from the water and sequence them. The presence of organisms can be detected by identifying characteristic DNA or RNA sequences, a research approach called metagenomics. Even a partial sequence can yield information about the unique characteristics of these microbes.

New discoveries By using RNA sequencing technology, scientists discovered that archaea are far more prevalent in the oceans than previously thought. In 2001 two groups of archaea were found to be extremely abundant in water samples collected from below the photic zone—that is, where light does not penetrate (see Chapter 18). The samples showed that archaea were rare at the surface but were just as common as bacteria below 1,000 m depth. DNA sequences in samples from the Sargasso Sea have indicated the presence of at least 1,800, and there may be as many as 50,000, new species of microbes.

Similar techniques have been used to identify previously unknown eukaryotic microorganisms. These eukaryotes are

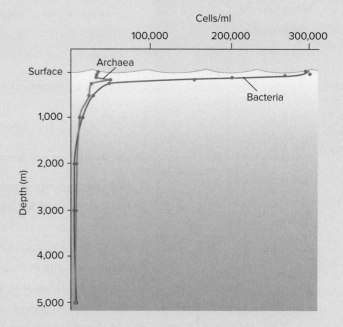

Depth distribution of bacteria and archaea in the subtropical North Pacific.

smaller than 3 μm in size, making them difficult to collect and study, unlike other unicellular but larger eukaryotes in the plankton, which can be collected using plankton nets. In 2001 two new groups related to dinoflagellates were discovered near Antarctica. Similar but unique groups of eukaryotic microorganisms (including yet another new group related to dinoflagellates) were reported at the same time by another team working at depths of 75 m in the tropical Pacific. The fact that some of these microorganisms, collected in relatively shallow water where light still penetrates, are photosynthetic, adds a new dimension to the discovery. They represent a previously unknown source of primary production in typically unproductive tropical waters. Other groups of minute algae have been discovered in recent years, including three groups that are related to dinoflagellates. Other groups certainly remain to be discovered.

Section Review

1. **Explain** Why is the abundance of viruses in water directly related to the abundance of microbial life in the water?

2. **Describe** What is the role of viruses in marine food webs?

7.2 Prokaryotes

VOCABULARY

decay bacteria

decomposers

calcareous

endolithic

red tides

epiphytes

endophytes

photosynthetic

chemosynthetic

anoxic

nitrogen fixation

DIVE IN!

Did cyanobacteria shape the history of Earth? The oldest known fossils, dating back to 3.5 billion years ago, belong to cyanobacteria. They were one of the first organisms to carry out photosynthesis. Ancient cyanobacteria were so productive that they changed the concentration of gases in Earth's atmosphere. Scientists agree that without increase in oxygen from cyanobacterial photosynthesis, other organisms that rely on celluar respiration would not have been able to evolve.

Main Idea
Depending on the species, bacteria and archaea are important primary producers, decomposers, and nitrogen fixers in marine environments.

· ·

Key Questions
1. What are the main characteristics of bacteria and archaea?
2. What is the difference between autotrophs and heterotrophs?

Prokaryotes are the smallest and structurally simplest true living organisms and the oldest forms of life on Earth. Nonetheless, they carry out nearly all the chemical processes performed by supposedly more complex organisms, as well as many that are unique to prokaryotes (Table 7.1). Most of these chemical processes actually evolved first among prokaryotes.

In contrast to viruses, prokaryotes are made up of cells. Prokaryotic cells are enclosed by a protective cell wall. The cell, or plasma, membrane lies immediately inside the cell wall. Prokaryotes not only lack a nucleus and most other membrane-bound organelles but also differ from eukaryotes in the circular shape of the DNA molecules that encode their genetic information, the size of their ribosomes, and a number of other ways.

The Prokaryotic Domains

Despite their similarities, the two prokaryotic domains, Bacteria and Archaea, also have important differences, including the chemistry of their cell walls, cell membranes, and the cellular machinery that manufactures proteins. In fact, genetic analyses indicate that bacteria and archaea are as different from each other as they are from humans.

Bacteria Bacteria (domain Bacteria) appear to have branched out very early on the tree of life and are genetically distant from both archaea and eukaryotes. Because they are structurally simple, bacteria (singular, bacterium) are classified as prokaryotes. Nevertheless, they have evolved a great range of metabolic abilities. They are abundant in all parts of the ocean.

Bacterial cells have many shapes, including spheres, spirals, rods, and rings (Fig. 7.2, p. 178), depending on the species. The unique chemistry of bacterial cell walls (see Table 7.2) makes them rigid and strong. A stiff or slimy covering often surrounds the cell wall as additional protection or a means of attaching to surfaces. The cells are much smaller than those of single-celled eukaryotes (see Appendix A for relative size). About 250,000 average bacterial cells would fit on the period at the end of this sentence. There are exceptions, however, such as a sediment bacterium discovered off southwest Africa, with cells as wide as 0.75 mm, large enough to be seen with the naked eye (Fig. 7.3, p. 178). Another

Table 7.1 Examples of Metabolic Reactions in Marine Prokaryotes

Name of Reaction	Energy Source	Other Raw Materials	By-product(s)	Production of Organic Carbon	Major Groups	Comments and Examples
Photosynthesis	Light	Carbon dioxide, water	Oxygen	Yes	Cyanobacteria	Light captured by chlorophyll, essentially the same process performed by algae and plants
	Light	Hydrogen sulfide	Sulfur or sulfate	Yes	Purple and green bacteria	Light captured by bacteriochlorophyll
Chemosynthesis						Energy supplied by a variety of inorganic molecules, of which only a few examples are included here
Sulfur oxidation	Hydrogen sulfide	Oxygen	Sulfate	Yes	Sulfur-oxidizing bacteria	*Bacteria in wetland sediments, mudflats, deep-sea hydrothermal vents*
Aerobic ammonium	Ammonium	Oxygen	Nitrite	Yes	Nitrifying bacteria and archaea	*Bacteria in wetland sediments, mudflats, oxidation deep-sea hydrothermal vents; open-water archaea*
Nitrite oxidation	Nitrite	Oxygen	Nitrate	Yes	Nitrifying bacteria	*Bacteria in wetland sediments, mudflats, deep-sea hydrothermal vents*
Iron oxidation	Reduced iron	Oxygen	Oxidized iron	Yes	Iron-oxidizing bacteria	*Bacteria in sediments, deep-sea hydrothermal vents*
Aerobic respiration	Organic matter	Oxygen	Carbon dioxide, water	No	Heterotrophic, aerobic bacteria and archaea; decay bacteria and archaea	Organic matter broken down using oxygen to release energy
Anaerobic respiration						Organic matter broken down in the absence of oxygen to release energy; less efficient than aerobic respiration
Sulfate reduction	Organic matter	Sulfate	Hydrogen sulfide	No	Sulfate-reducing bacteria and archaea	
Nitrate reduction, denitrification	Organic matter	Nitrate, nitrite	Nitrogen gas, ammonia	No	Denitrifying bacteria	
Methanogenesis	Organic matter	Hydrogen, carbon dioxide	Methane	No	Methanogens	*Archaea in many environments*
Nitrogen fixation	ATP	Nitrogen gas	Ammonium	No	Nitrogen-fixing bacteria and archaea	Ammonium and other compounds produced from ammonium can be used as nutrients by primary producers.
Anaerobic ammonium oxidation	Ammonium	Nitrite	Nitrogen gas	No	Some bacteria and archaea	Ammonium converted to nitrogen in the absence of oxygen to release energy
Light-mediated ATP synthesis	Light		ATP	No	Archaeum in highly saline pools	Light captured by bacteriorhodopsin

Figure 7.2 A high-magnification photograph taken with an electron microscope showing *Cyclobacterium marinus*, a ring-forming marine bacterium.

Figure 7.3 Three cells of *Thiomargarita namibiensis*, the largest known bacterium. Cells can be 0.75 mm wide.

Prokaryotes Organisms with cells that do not have a nucleus or most other organelles.
- 6.2, Cellular Structure and Organization; Figure 6.7

Eukaryotes Organisms with cells that contain a nucleus and other organelles that are enclosed by membranes.
- 6.2, Cellular Structure and Organization; Figure 6.8

giant bacterium (0.57 mm long) is found inside the intestines of coral reef fishes. In large numbers, marine bacteria are sometimes visible as whitish hairs on rotting seaweed or iridescent or pink patches on the surface of stagnant pools in mudflats and salt marshes.

Bacteria can grow to extremely high numbers in favorable environments such as detritus, particles of dead organic matter. **Decay bacteria** (or **decomposers**, those that carry out decomposition)break down waste products and dead organic matter and release nutrients into the environment. They are vital to life on Earth because they ensure the recycling of essential nutrients, as in the recycling of dissolved organic matter in oceanic food webs. Most organic matter is sooner or later broken down by decomposition, though in very deep, cold water the process is slower than elsewhere. Decay bacteria play another crucial role because they constitute a major part of the organic matter that feeds countless bottom-dwelling animals. Even many organic particles in the water column are composed mostly of bacteria. Some marine bacteria are also involved in degrading oil and other toxic pollutants that find their way into the environment. The same process of decomposition is unfortunately involved in the spoilage of valuable fish and shellfish (molluscs and crustaceans) catches. Other types of marine bacteria cause diseases in marine animals and humans.

Bacteria are found everywhere in the marine environment, on almost all surfaces and in the water column. In fact, the most abundant forms of life on the planet seem to be bacteria, found in surprisingly high numbers in open water. Their role, though probably highly significant due to their vast numbers, remains unknown. Bacteria have also been found in seemingly harsh environments, such as in sediment more than 3 km below the sea floor.

Cyanobacteria, once known as blue-green algae, are a group of photosynthetic bacteria (see "Autotrophs," below). In addition to having the pigment chlorophyll *a*, as do eukaryotic photosynthetic organisms, most contain a bluish pigment called phycocyanin. Most marine cyanobacteria also have a

reddish pigment, phycoerythrin. The visible color of the organism depends on the relative amounts of the two pigments. When phycocyanin predominates, the bacteria appear blue-green; when phycoerythrin predominates, they appear red.

Cyanobacteria were among the first photosynthetic organisms on Earth. They are thought to have had an important role in the accumulation of oxygen (O_2) in our atmosphere. Fossil stromatolites, massive **calcareous** (made of $CaCO_3$) mounds formed by cyanobacteria, are known to date back some 3 billion years. Stromatolites are still being formed in tropical seas (Fig. 7.4).

Cyanobacteria are widely distributed in the marine environment. *Prochlorococcus*, a minute cyanobacterium, is thought to be the most abundant photosynthetic organism in the ocean. It is particularly common in the nutrient-poor tropical and subtropical waters. Many species of cyanobacteria can tolerate wide ranges of salinity and temperature, while others are found in unexpected places, like inside the hair of polar bears! Some cyanobacteria, called **endolithic**, burrow into calcareous rocks and coral skeletons. Others form thick, dark crusts along the wave-splashed zone of rocky coasts. Some exploit oxygen-poor sediments, including polluted sites. Cyanobacteria that are members of the plankton can rapidly multiply and change the color of the water. Some so-called **red tides** are caused by cyanobacteria that contain a red pigment. A few species can cause skin rashes on swimmers and divers.

Some cyanobacteria live on the surface of seaweeds and seagrasses. Photosynthetic organisms that live on algae or plants are called **epiphytes**. Some biologists, however, use this term to refer to *all* organisms living on algae or plants. Other cyanobacteria live inside algae and are called **endophytes**.

©Charlie Arneson

Figure 7.4 Stromatolites, calcareous mounds deposited by cyanobacteria, are frequently found as fossils. These, however, are living stromatolites growing in shallow water in the Exuma Cays, Bahama Islands.

Symbiotic Bacteria—the Essential Guests

Some bacteria have evolved to live in symbiotic associations with other marine organisms. Some symbiotic bacteria are parasites that may cause a disease. Others benefit their hosts. In many cases the symbiotic bacteria are even sheltered in special tissues or organs that evolved in the host, an example of coevolution, in which two species evolve in response to each other.

All eukaryotic organisms, including humans, shelter bacteria without which they could not live. The chloroplasts and mitochondria of eukaryotic cells evolved from symbiotic bacteria, as you learned in Chapter 6. These bacteria have become an integral part of all complex cells.

The role of symbionts There are many cases of symbiotic bacteria among marine organisms. For example, they are involved in digesting the wood ingested by shipworms (*Teredo*) (which happen to be bivalve molluscs, not worms). Like all wood-eating animals, shipworms lack cellulase, the enzyme needed to digest cellulose, the main component of wood. Wood is a surprisingly common habitat in many marine environments, so everything from driftwood to boat bottoms is exploited by shipworms, thanks to their symbiotic bacteria.

Symbiotic bacteria are also responsible for the light, or bioluminescence, produced by some fishes, squids, octopuses, and other animals. The bacteria are usually sheltered in special light-producing organs, called photophores. These animals use light to communicate with other members of their species, lure prey, blend with the light that filters from the surface, and perform other functions. Flashlight fishes (*Anomalops*), which live in shallow tropical waters, lodge their symbiotic bacteria in an organ beneath each eye. A shutter mechanism controls the emission of light, so that the fish can "blink" at night. Groups of fish blink in synchrony to attract prey.

Symbionts and human health Marine symbiotic bacteria can also affect human health. Pufferfishes store a toxin that is deadly to any predators (including humans) that eat them. The fish, or *fugu*, is a delicacy in Japan. Licensed chefs must prepare the fish; otherwise, the toxin (**tetrodotoxin**) can cause death. Tetrodotoxin is a neurotoxin (that is, it affects the

nervous system), and is one of the most powerful poisons known. There is no antidote. The toxin is stored in the liver and gonads of the fish, so the internal organs must be expertly removed. Mistakes do happen and numerous deaths (including the suicides of disgraced cooks) take place every year in Japan. Cooks may still be guilty, but not the puffers: the toxin is not produced by puffers but apparently by symbiotic bacteria. The fish are actually resistant to tetrodotoxin as a result of a genetic mutation that changes the site where the toxin binds to nerve cells.

Being a neurotoxin, tetrodotoxin may block pain signals in humans. A derivative being tested as a pain reliever for cancer and other diseases would have the advantage of not being addictive, as are morphine and similar drugs.

The production of powerful toxins that are used by other organisms is just one example of the amazing abilities of bacteria—invisible to the eye but powerful giants when it comes to their role in the environment.

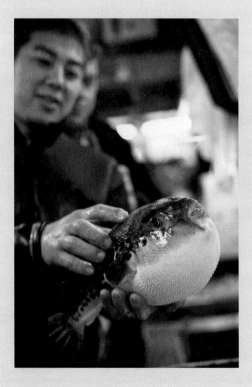

A Japanese pufferfish, or *fugu* (*Takifugu niphobles*).

Andia/Photoshot

Archaea Archaea (domain Archaea) are among the simplest, most primitive forms of life. Some look very similar to the oldest fossils ever found, cells estimated to be at least 3.8 billion years old. Archaea are thought to have had an important role in the early evolution of life. Like bacteria, their cells are small and can be spherical (Fig. 7.5), spiral, or rod-shaped. In fact, until recently archaea (singular, archaeum) were thought to be bacteria. The origins of Archaea remain unclear. Despite being prokaryotic, however, there is evidence that archaea are more closely related to eukaryotes than to bacteria.

Some groups of archaea were discovered only recently, first in extreme environments on land, such as hot sulfur springs, saline lakes, and highly acidic or basic environments. Archaea were thus named "extremophiles," meaning "lovers of extremes." Archaea were subsequently found in extreme marine environments, such as in very deep water, where they survive at pressures of 300 to 800 atmospheres. Some archaea live at the high temperatures of hydrothermal vents (Fig. 7.5). Some of these cannot grow in temperatures less than 70° to 80 °C, and the strain shown in Figure 7.5 can live at 121 °C, the highest of any known organism. Other archaea depend on extremely salty environments, such as coastal salt pans and deep ocean basins.

Because the first groups of microorganisms to be classified as archaea were extremophiles, it was thought that *all* archaea were extremophiles; indeed, the two terms became almost synonymous. New techniques based on detecting nucleic acid sequences, however, have shown that archaea are common in many marine environments, not only in the water column but in sediments as well. Some archaea are also known to live symbiotically in sponges, sea cucumbers, and fishes. Thus, the hypothesis that archaea are restricted to extreme environments was proven false.

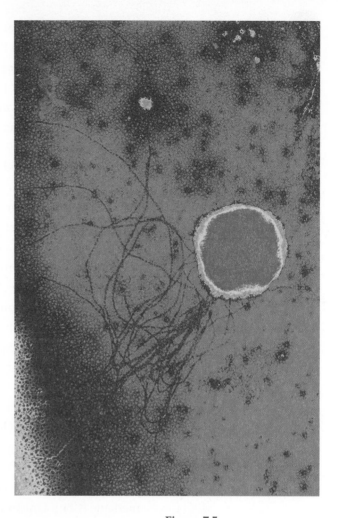

Figure 7.5
This colored transmission electron micrograph shows hyperthermophilic (extreme-heat-loving) archaea known as strain 121. This strain was isolated from a hydrothermal vent in the Pacific Ocean.

Prokaryote Metabolism

Prokaryotes show an amazing variety of chemical reactions that are involved in the transfer of energy and in the manufacture of chemical compounds. Some actually obtain energy in more than one way. Metabolism is all the chemical reactions that take place in an organism.

Autotrophs Autotrophic bacteria and archaea make their own organic compounds and thus are primary producers. Energy sources vary. Some autotrophic bacteria, such as the cyanobacteria previously mentioned, are photosynthetic or photoautotrophic. They contain chlorophyll or other photo-synthetic pigments (see Table 7.2) to trap light energy to manufacture organic

IN CONTEXT

Hydrothermal Vents Undersea hot springs associated with mid-ocean ridges and other geologically active environments.
• 2.3, Modern Ocean Basins

Hypothesis A statement about the world that might be true.
• 1.2, The Scientific Method

Autotrophs Organisms that can use energy (usually solar energy) to make organic matter.

Heterotrophs Organisms that cannot make their own food and must use the organic matter produced by autotrophs.
• 6.1, The Ingredients of Life

compounds from carbon dioxide (CO_2); cyanobacteria release oxygen in the process (see Table 7.1). Photosynthesis takes place on folded membranes within the bacterial cell rather than in chloroplasts, as in algae and plants. Photosynthetic bacteria account for much of the primary production in many open-ocean areas. Even bacteria living at depths where no surface light penetrates have been suggested as using the faint glow of hydrothermal vents to perform photosynthesis.

Though the end products can be similar, the biochemistry of bacterial photosynthesis is different from that of algae and plants. It also varies considerably among different groups of bacteria. Some photosynthetic bacteria, for example, have a kind of chlorophyll unique to prokaryotes (bacteriochlorphyll) and produce sulfur (S) instead of oxygen (see Table 7.1). Some bacteria could use a pigment called proteorhodopsin to capture light energy and store it in ATP. They do not have chlorophyll or use the energy to manufacture organic compounds from carbon dioxide and thus are not photosynthetic. An archaeum (*Halobacterium*) similarly lacks chlorophyll but contains a pigment, bacteriorhodopsin, that converts light energy directly into ATP. Marine microbes can therefore use light as an energy source in many ways, not just in photosynthesis.

Other bacterial autotrophs, called **chemosynthetic**, or chemoautotrophic, derive energy not from light but from chemical compounds, including hydrogen sulfide (H_2S) and other sulfur, nitrogen, and iron compounds (see Table 7.1). Chemosynthesis is also involved in altering some of the minerals found in sea-floor rocks. Some chemosynthetic archaea, the methanogens, produce methane (CH_4). One example is *Methanopyrus*, an archaeum that has been isolated from sediments near hydrothermal vents.

Heterotrophs Most marine bacteria, like animals, are heterotrophs that obtain energy from organic matter by respiration (see Table 7.1). Many heterotrophic bacteria and archaea are decomposers. Respiration in aerobic bacteria and archaea uses oxygen, as in most other groups of organisms, whereas in anaerobic species respiration does not require oxygen. Anaerobic bacteria do not grow when oxygen is present and thrive when it is absent, as in sediments that are devoid of oxygen, or **anoxic**. This is the case in bacteria that use sulfate (SO_4^{-2}) instead of oxygen to perform respiration, producing in the process the hydrogen sulfide (H_2S) that is to blame for the rotten-egg smell of some sediments.

Nitrogen Fixation Many planktonic cyanobacteria (such as *Trichodesmium*) carry out **nitrogen fixation**, converting gaseous nitrogen (N_2), which is not available as a nitrogen source to primary producers, into ammonium (NH_4^{+1}). Most of the ammonium is then transformed into nitrates (NO^{-1}) and other nitrogen compounds (see Table 7.1) that can be finally used by primary producers as a nitrogen source for the manufacture of essential proteins and other nitrogen-containing compounds. Other bacteria and some archaea are also known to be nitrogen fixers. Nitrogen fixation requires energy, which is provided by ATP generated from photosynthesis or other reactions.

A Fourth Domain of the Tree of Life?

Most biologists classify all organisms into three domains, a system that relies heavily on the chemistry of cells. Such differences led to the recognition that archaea are different from bacteria and that each of these two groups belongs in a separate domain, even though both look similar under the microscope. A three-domain system of classification thus far works fine. Recent studies that use increasingly sophisticated techniques have nevertheless hinted that the three-domain system is too simple and that a fourth domain might exist hidden somewhere in the ocean.

Why a fourth domain? It all started when microbiologist J. Craig Venter transformed his yacht (the *Sorcerer II*) into a research vessel and, from 2003-2007, lead a team of scientists to collect seawater from around the world. The water samples were analyzed by a team from Venter's institute and the University of California Davis using metagenomics, a technique where the genomes of unknown organisms are recovered directly from the environment, not from laboratory cultures as is traditionally done. The genomes in Venter's samples were cut into pieces, the nucleic acid chains sequenced, and the sequenced pieces put back together and matched with the help of computers.

The analysis of the water samples showed sequences that were different from any studied before. These included sequences related to two superfamilies of "housekeeping genes," those genes involved in the maintenance of the basic cell functions. Housekeeping genes are nearly universal among living organisms, and are even carried by some viruses, which do not consist of cells and are not strictly living. The first superfamily, *recA*, includes genes that are involved in controlling the repair of DNA chains and in the exchange of pieces of genetic information between DNA chains in the chromosomes. The second superfamily, *rpoB*, is involved in passing the information contained in DNA into RNA, the first step in the manufacture of proteins.

Any potential evolutionary relationships between the sequences found in the seawater samples and known sequences of the same superfamilies found in the other domains were used by computers to draw phylogenetic trees (similar to those shown in Figure 6.18). Some branches of these trees did not fit with the data on the known sequences, thus suggesting that they did not belong in any of the three known domains but instead belong in an unknown group of microorganisms. These branches might therefore represent an unknown branch, probably a new domain, in the tree of life.

Other explanations It is possible, however, that the sequences belong to an unknown group of marine viruses, which are not included in any of the recognized domains because they are not classified as living organisms. The discovery of these sequences has triggered all sorts of questions, the aim of current and future research. Do the unusual gene sequences belong to a microorganism with an unknown cellular organization and function or do they instead belong to a bizarre virus? The questions are fascinating and challenging. Do they live in the water column or are they associated with more "complex" organisms? The unusual gene sequences contain complete genes. Can we predict from the known sequences the biology and role of the unknown organisms (or viruses)? What did they evolve from or evolve into? Future confirmation of an unexpected fourth domain of the tree of life could lead to significant new discoveries and hypotheses as did the discovery of Archaea. It could prove to be another example of the open, ever expanding, and self-correcting nature of science.

Section Review

1. **Explain** What is the importance of bacteria in marine ecosystems?
2. **Identify** In which marine environments are cyanobacteria found?
3. **Identify** In which marine environments are archaea found?

7.3 Single Celled Eukaryotes

DIVE IN!

How harmful are harmful algal blooms (HABs) to humans? Shellfish can accumulate toxins from HABs, leading to shellfish poisoning when consumed by humans. Amnesic Shellfish Poisoning (ASP) is caused by a toxin produced by diatoms in the genus *Pseudonitzschia*. ASP symptoms include nausea and vomiting. Severe cases result in dizziness, seizures, and coma. Paralytic Shellfish Poisoning (PSP) is caused by toxins produced by dinoflagellates. PSP symptoms include tingling, numbness, and lack of muscle control. In severe cases, muscle paralysis can result in people being unable to breathe.

Main Idea

Single-celled eukaryotes and fungi play important roles in marine ecosystems as primary producers and decomposers, respectively.

. .

Key Questions

1. Compare and contrast unicellular algae and protozoans.
2. Why are fungi classified into their own kingdom?

As you learned at the beginning of this chapter, microorganisms can be found in all three domains of life. We have seen examples of Bacteria and Archaea, and now we will discuss the microorganisms found in Domain Eukarya.

Unicellular Algae

Algae (singular, alga) are a very diverse group of simple, mostly aquatic (that is, marine and freshwater), mostly photosynthetic organisms. Being eukaryotic, their cells contain a nucleus and other organelles enclosed by membranes. Photosynthesis takes place in chloroplasts—green, brown, or red organelles with layers of internal membranes that contain photosynthetic pigments. The color of algae is a result of the pigments and their concentration. In contrast to the land plants with which we are familiar, algae lack true leaves, stems, roots, and flowers. They have relatively simple reproductive structures, and their nonreproductive cells are also mostly simple and unspecialized.

Biologists used to refer to algae as plants. Many of the unicellular algae, however, show animal-like characteristics. Some swim by moving their flagella, and distinguishing these free-swimming algae from some of the structurally simpler animals can be difficult at first glance. Some species carry out photosynthesis as plants do, and very similar species move and eat food particles as animals do. Some species do both and are claimed by both botanists (plant biologists) and zoologists (animal biologists). These unicellular organisms are collectively called protists for convenience, under the kingdom Protista even if the various groups of protists have different evolutionary histories. Multicellular seaweeds are also considered to be protists, mostly because they lack the specialized tissues of plants. Seaweeds, together with other multicellular primary producers, the marine plants, are discussed in Chapter 8.

Diatoms Diatoms (division, or phylum, Bacillariophyta) are unicellular, though many species aggregate into chains or star-like groups. Diatom cells are enclosed by cell walls made largely of silica (SiO_2), a glass-like material. This glassy shell, or **frustule**, consists of two tightly fitting halves often resembling a flat, round, or elongated box (Fig. 7.6). The frustule typically has intricate perforations and ornaments such as spines or ribs, making diatoms strikingly

Figure 7.6 Diagrammatic representation of a diatom cells.

beautiful when seen under a microscope. The frustule allows light to pass through, so that the conspicuous golden brown chloroplasts can capture light energy for photosynthesis. The minute perforations allow dissolved gases and nutrients to enter and exit. The sinking of open-ocean diatoms below the well-lit surface layer is often slowed by oil droplets in their cells and spines on the frustules.

The characteristic color of diatoms is due to yellow and brown **carotenoid** pigments, in addition to two types of chlorophyll, *a* and *c* (see Table 7.2). Diatoms are efficient photosynthetic factories, producing much-needed food (the food being the diatoms themselves), as well as oxygen, for other forms of life. They are a potential source of biofuels as alternatives to fossil fuels. Diatoms are very important open-water primary producers in temperate and polar regions. In fact, the huge number of diatom cells in the ocean account for a hefty share of the organic carbon and oxygen produced on planet Earth.

Around half the estimated 12,000 living species of diatoms are marine. Most are planktonic, but many produce a stalk-like structure for attachment to rocks and other surfaces in shallow water, where they contribute significantly to primary production. The brownish scum sometimes seen on mudflats or glass aquaria often consists of millions of diatom cells. Some are able to glide slowly on surfaces. A few are colorless, having no chlorophyll, and live as heterotrophs on the surfaces of seaweeds. Some planktonic diatoms produce a potent toxin, domoic acid, which accumulates in shellfish (clams, oysters, mussels) and plankton-feeding fishes that eat the diatoms. The toxin causes a serious, sometimes fatal, illness of the nervous system in humans and marine mammals that eat contaminated shellfish or fish.

Diatoms reproduce mostly by cell division, a type of asexual reproduction. The overlapping halves of the frustule separate, and each secretes a new, smaller half. The smaller, lower frustule of the parent cell becomes the larger frustule of a daughter cell, which then secretes a new, smaller frustule (Fig. 7.7). Full size is regained by the development of specialized stages known as **auxospores**. Auxospores then re-establish the normal cell size that display the frustule characteristic of the species.

Types of reproduction:

Asexual (or *vegetative*) The production of new individuals by simple division (or other means) without involving *gametes*, so that the offspring are genetically identical to the parents.

Sexual The production of new individuals by the formation of *gametes* (sperm and eggs), so that the offspring are genetically different from the parents.

• 6.3, Perpetuating Life

Asexual reproduction

Smaller frustules

Auxospore

Sexual reproduction

Fertilization

Sperm cell

Egg

Figure 7.7 The usual method of reproduction in diatoms is by cell division. Most, but not all, frustules get smaller after each successive division. Resistant cells called auxospores are produced two ways: directly from the expansion of a smaller frustule or by sexual reproduction when an egg is fertilized by a sperm cell, both of which are liberated from separate frustules.

Favorable environmental conditions, such as abundant nutrients and light, trigger periods of rapid reproduction called **blooms**. Such blooms may even occur under Arctic and Antarctic ice. This is a general phenomenon that also occurs in other algae. During blooms most diatoms get progressively smaller, not only because of asexual reproduction but also as a result of depletion of silicate from the water by the growing population.

The glassy frustules of dead diatoms eventually settle to the bottom of the sea floor, where they can form thick deposits of siliceous material that cover large portions of the ocean floor. Such biogenous sediments are known as diatomaceous ooze, a type of siliceous ooze. Huge fossil deposits of these sediments can be found inland in various parts of the world. The siliceous material, or **diatomaceous earth**, is mined and used in products such as filters for swimming pools, for clarifying beer, as temperature and sound insulators, and as mild abrasives in toothpaste.

Dinoflagellates The dinoflagellates (division, or phylum, Dinophyta) make up another large group of planktonic, unicellular organisms. Their most outstanding characteristic is the possession of two flagella, one wrapped around a groove along the middle of the cell and one trailing free (Fig. 7.8). These flagella direct movement in any direction. Most dinoflagellates have a cell wall armored with plates made of cellulose, the characteristic component of the cell walls of seaweeds and land plants. The plates have spines, pores, or other ornaments.

Though most dinoflagellates photosynthesize, many can also ingest food particles. A few have a light-sensitive pigment spot that acts as a crude "eye". It has been suggested that during their evolution dinoflagellates gained the

Biogenous Sediments
Sediments made of the skeletons and shells of marine organisms; *siliceous ooze* is made of silica, *calcareous ooze* of calcium carbonate.

- 2.2, How Plate Tectonics formed the Oceans

(a)

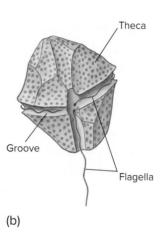

Theca

Groove

Flagella

(b)

Figure 7.8 Dinoflagellates like *Gonyaulax polyedra* have a cell wall, or theca, which consists of cellulose plates. The theca is marked by grooves for the flagella; only one groove is visible here (arrow).

ability to function as primary producers by capturing and using chloroplasts from other algae. Almost all known dinoflagellates, around 1,200 living species, are marine. Dinoflagellates are important planktonic primary producers, especially in warm water.

Dinoflagellates reproduce almost exclusively by simple cell division. They sometimes form blooms that color the water red, reddish-brown, yellow, or other unusual shades. Some of these dinoflagellates release toxic substances, and seafood collected during red tide periods can be poisonous, as in the case of ciguatera, a tropical fish poisoning. Other dinoflagellates are noted for the production of light, or bioluminescence. Though bioluminescence has also been observed in some bacteria and many types of animals, dinoflagellates are generally responsible for the diffuse bioluminescence sometimes seen on the sea surface. This effect is, of course, seen only at night. It is especially bright if the water is disturbed by a boat or when a wave crashes on the shore.

A group of round, golden brown dinoflagellates called zooxanthellae live in close association with a variety of animals. Animals harboring zooxanthellae range from sponges and sea anemones to giant clams. It is perhaps in reef-building corals where zooxanthellae are most significant. They fix carbon dioxide by photosynthesis, release organic matter used by the coral, and help in the formation of the coral skeleton.

A few highly modified dinoflagellates are parasites of seaweeds and some marine animals. Like zooxanthellae, these specialized forms reveal their true nature by having life cycles that include free-swimming stages resembling typical dinoflagellates.

IN CONTEXT

Cellulose A complex carbohydrate characteristic of plants and other primary producers.

• 6.1, The Ingredients of Life

The Bay of Fire

Imagine entering a quiet tropical bay at night. There are no lights on shore. On a moonless night, the spectacle is unforgettable. As the propeller of your boat churns the black water, an intense blue-green light is left behind, like an eerie trail of cool fire. Outside the bay you see a few sparks of light, but inside the bay things are very different. Long streaks of light shoot like fireworks beneath the boat. They are created by fleeing fish. A swim in the bay is even more spectacular. Your dive into the water is accompanied by a flood of light, and sparks scatter out as you wave your arms.

The bioluminescent bay Such light effects result from an unusual concentration of bioluminescent dinoflagellates, a natural and regular phenomenon in some locations such as *Bahía Fosforescente*, or Phosphorescent Bay, in southwestern Puerto Rico. The bay is small, about 90 acres, fan-shaped, and it does not exceed 4.5 m in depth.

Bioluminescent Bay is a more apt name because the light is produced by living organisms. The bay's primary source of bioluminescence is *Pyrodinium bahamense*, a unicellular photosynthetic dinoflagellate about 40 µm in diameter. The dinoflagellates are very concentrated within the bay because of its shallow depth and limited connection to the open ocean. The bay is connected to the open sea by a narrow, shallow channel. As water containing the dinoflagellates flows into the bay, evaporation from the shallow basin increases the salinity, and therefore the density, of the surface water , causing it to sink. *Pyrodinium* cells stay near the surface so they are not carried out of the bay as the denser water flows outward along the bottom of the shallow entrance. The tidal range is only about a foot at the most, so water exchange with the outside is limited. The bay thus acts as a trap that keeps the dinoflagellates from leaving.

Why *Pyrodinium*? Of all the planktonic organisms that enter the bay, why is *Pyrodinium* favored? A key factor seems to be the thick mangrove trees bordering the bay. Mangrove leaves fall into the water, where intense bacterial decomposition increases the organic nutrient levels in the water. Some nutrients essential to the growth of *Pyrodinium*, perhaps vitamins, may be released by bacteria or other microorganisms.

Bahía Fosforescente is protected to keep the critical natural balance intact, but bioluminescence has noticeably decreased during recent years. A few other bays, termed "bio bays," in Puerto Rico fortunately show even more spectacular bioluminescence, like Laguna Grande on the northeastern tip of the island and Puerto Mosquito on the island of Vieques.

Bahía Fosforescente, Puerto Rico.

One such dinoflagellate is *Pfiesteria*, sometimes called the "phantom dinoflagellate" because it spends most of its life as a harmless resting stage, or **cyst**, in the sediment. It has been suggested that coastal pollution in the form of excessive nutrients can trigger blooms of *Pfiesteria*. *Pfiesteria* and *Pfiesteria*-like microorganisms are responsible for deadly open sores on fishes. *Pfiesteria* causes the sores by producing a recently discovered toxin, not by actively feeding on the fishes' tissues as first thought. These parasites are also known to be harmful to crabs, oysters, and clams and have been implicated in sores and in symptoms such as temporary memory loss in humans.

Other Unicellular Algae Three additional groups can be abundant in some areas and thus very significant primary producers in the plankton. Silicoflagellates (division, or phylum, Chrysophyta) are characterized by a star-shaped internal skeleton made of silica (Fig. 7.9) and two flagella of different lengths. Fossil silicoflagellates are common in marine sediments and can be used to date the sediments. Typical coccolithophorids (division, or phylum, Haptophyta) are flagellated, spherical cells covered with button-like, ornamented structures called coccoliths made of calcium carbonate (Fig. 7.10). Coccoliths can be found as fossils in chalk sediments that were uplifted from the ocean floor such as the white cliffs of Dover in England. Cryptophytes (division, or phylum, Cryptophyta) have two flagella and lack a skeleton. Like other eukaryotes, cryptophytes have a chloroplast derived from a prokaryote, but the chloroplast is contained within a reduced eukaryotic cell that lives symbiotically inside the cryptophyte cell. Members of these three groups of algae are so small that hundreds could fit into a large diatom or dinoflagellate cell.

Figure 7.9 Skeleton of *Dictyocha speculum*, a silicoflagellate.

Figure 7.10 Coccolithophorids, such as *Umbilicosphaera sibogae* from Australia, are tiny, single-celled phytoplankton that can be significant primay producers in the open ocean. The plates that cover the cell are made of calcium carbonate.

Protozoans: The Animal-Like Protists

Protozoans are structurally simple and very diverse eukaryotic organisms that are traditionally considered to be animal-like. Most biologists agree that protozoans (meaning "first animals") actually comprise several groups of unrelated origins. Having a single cell is about the only thing that protozoans—which show an enormous diversity in structure, function, and lifestyle—have in common. Though a few form colonies, most protozoans are single-celled, unlike true animals, and are visible only under a microscope. Like animals, protozoans are heterotrophs and ingest food. Some, however, also contain chlorophyll and photosynthesize.

There is considerable disagreement as to how to classify the estimated 50,000 species of protozoans, though they are generally considered to be protists along with unicellular algae and seaweeds.

The minute size and apparent simplicity of protozoans disguise a complex nature. As in all unicellular organisms, each cell might be described as a "super cell" because it can perform many of the same functions carried out by the multitude of cells in structurally more complex organisms.

Protozoans inhabit water everywhere, living not only in marine and freshwater environments but even inside other organisms. Many kinds of marine protozoans can be readily collected from sediments rich in organic debris, the surface of seaweeds, the guts of animals, and plankton samples.

Foraminiferans Foraminiferans (phylum Foraminifera), often called forams, are marine protozoans that usually have a shell, or **test**, made of calcium carbonate ($CaCO_3$). The test is usually microscopic and can have several chambers (Fig. 7.11a), which increase in size as the foram grows. Foraminiferans have long, thin, retractable **pseudopodia**—extensions of the cytoplasm. The pseudopodia protrude through pores in the test and form a network used to trap diatoms and other organisms suspended in the water. Food is then moved into the interior of the cell as if on a conveyor belt.

Most foraminiferans are benthic, and thus live on the bottom, either free or attached. Attached forams develop into conspicuous growths as much as 5 cm in diameter (Fig. 7.11b). Each growth consists of a single cell that forms a test, though some can be covered with sand grains or other materials instead. The tests of bottom-living foraminiferans can be important contributors of calcareous material on coral reefs and sandy beaches. Relatively few species are planktonic, but these can be very abundant. Their tests are smaller and thinner than those that live on the bottom and can have delicate spines that aid in flotation. The tests of planktonic foraminiferans eventually sink to the bottom in such high numbers that large stretches of the ocean floor are covered by foraminiferan ooze, a type of calcareous ooze. Many limestone beds around the world are products of foraminiferan sediments that were uplifted from the ocean floor.

Most species of forams are known only as microfossils. The distribution of these microfossils in sediments is important to geologists. Tests of warm-water species are slightly larger and more porous than those from colder water. Past water temperatures and levels of glacial ice at the poles can be determined from the distribution of certain marker species as well as chemical analyses of foram tests. Their distribution is a good indicator of the right type of environment for the formation of oil, as well as the age of sediments.

Figure 7.11 Foraminiferans. (a) Foraminiferans typically have calcareous tests that consist of a spiral arrangement of chambers. The thin, long strands are thin spines with pseudopodia running along them, which are used to capture food. (b) *Homotrema rubrum* is a foraminiferan that forms bright red calcareous growths. It is so common in Bermuda that its skeletons are responsible for the island's famous pink beaches.

(a)　　　　　　　(b)

(l)Courtesy of Dr. John R. Dolan, Laboratoire d'Oceanologique de Villefranche-sur-Mer/NOAA Fisheries Collection, (r)©Charlie Arneson; Observatoire Oceanologique de Villefranche

Radiolarians Radiolarians (phylum Radiolaria) are planktonic marine protozoans that secrete elaborate and delicate skeletons made of glass (silica) and other materials. The typical skeleton is spherical with radiating spines (Fig. 7.12), though the structure varies. Thin pseudopodia capture food, as in foraminiferans.

Most radiolarians are microscopic, but some form sausage-shaped colonies that reach 3 m (9 ft) in length, making them true giants among protozoans. Radiolarians inhabit open water throughout the oceans. When abundant, the remains of their skeleton settle to the bottom and form a siliceous ooze known as radiolarian ooze. This ooze is more extensive in deep water because the radiolarian skeleton is more resistant to dissolving under pressure than those of forams.

Ciliates Ciliates (phylum Ciliophora) are protozoans that have many hair-like **cilia**, which are used in locomotion and feeding. The most familiar ciliates are freshwater forms, such as *Paramecium*. Many marine ciliates are found creeping over seaweeds and in bottom sediments. Some live on the gills of clams, in the intestines of sea urchins, on the skin of fishes, and in other unusual places. Other ciliates live attached to surfaces, even forming branched colonies of tiny individuals. Tintinnids are common ciliates that drift in the water and build vase-like cases, or **loricas**, loosely fitting tests that drift in the water (Fig. 7.13). The cases can be transparent or made of bits of particles. Planktonic ciliates are important in the so-called microbial loop of open-water food webs, which facilitates the flow of energy from dissolved organic matter to the larger plankton.

Fungi

Fungi (kingdom Fungi) are eukaryotic and mostly multicellular, though some, such as molds and yeasts, are unicellular. Multicellular fungi typically form long filaments called **hyphae**. Fungi are heterotrophs that lack chloroplasts and chlorophyll and cannot perform photosynthesis. As in plants, however, fungi have cell walls. The cell wall of fungi is made of chitin, a highly resistant carbohydrate found in the skeleton of many animals, whereas it is made of cellulose in plants.

There are at least 1,500 known species of marine fungi, mostly microscopic, and mostly living in or on other organisms. Fungi absorb nutrients from their environment by secreting digestive enzymes, and many, as in bacteria, decompose detritus. They can decompose the cellulose in the cell walls of seaweeds and plants. Cellulose, which is abundant in driftwood and the dropped leaves of seagrasses and salt-marsh plants, is not digested by most bacteria. Fungi are the most important decomposers of dropped mangrove leaves and thus contribute to the recycling of nutrients in mangrove forests. A similar process occurs with dropped seagrass leaves, drift seaweed, and in the driftwood that accumulates in deep water along tropical rain forests. Marine fungi are also important in the breakdown of detritus and other organic matter that accumulates as the result of plankton blooms and pollution. Some marine fungi are parasites of seagrasses or borers in mollusc shells. Others are

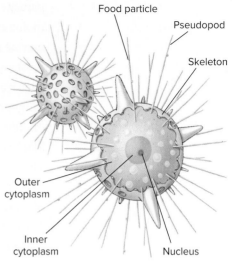

Figure 7.12 A typical radiolarian cell consists of a dense central portion surrounded by a less dense zone that is involved in the capture of food particles and in buoyancy.

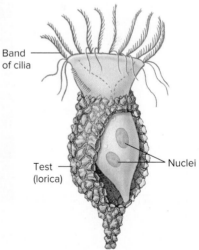

Figure 7.13 *Tintinnopsis* is a marine tintinnid that forms a vase-like lorica made of sand grains. Specialized cilia on one end are used in feeding.

parasites that cause diseases of economically important seaweeds, sponges, molluscs, lobsters, and fishes, as well as corals ad marine mammals. Some marine fungi are being investigated as a source of antibiotics for use in medicine.

Many fungi live in symbiotic associations, mostly with green algae or cyanobacteria, to form unique entities, the **lichens**. In lichens the long hyphae of the fungi provide support, whereas the algae or cyanobacteria provide food from photosynthesis. Marine lichens can be typically found as thick, dark brown or black patches in the wave-splashed zone of exposed rocky shores (Fig. 7.14). They are tolerant to exposure to air for long periods of time.

By comparison with the multitude of lichens on land, there are very few types of marine lichens. Their role on the ecology of rocky shores is largely unknown. The cyanobacteria of some marine lichens are known to be nitrogen-fixers. Some lichens are able to loosen bits of the rocks where they live by growing into the rock. A few types live on the shells of rocky shore barnacles and molluscs.

Molecular techniques similar to those used in identifying viruses and marine microbes are being applied to the sampling of fungi in places such as deep-sea sediments and anoxic environments. These are fungi that are often difficult to culture in the lab. The number of species of marine fungi will undoubtedly increase with time.

Figure 7.14 Encrusting marine lichens, cyanobacteria, and some microscopic algae are often visible as a black band on wave-splashed rocky shores, as in this exposed rocky point south of Monterey Bay, California.

. .

Section Review

1. **Compare and Contrast** What are similarities and differences between diatoms and dinoflagellates?

2. **Describe** What are the main characteristics of foraminiferans, radiolarians, and ciliates?

3. **Identify** What is the role of algae, protozoans, and fungi in marine ecosystems?

Table 7.2 Most Important Characteristics of Marine Microbes

Groups	Distinguishing Features	Photosynthetic Pigments	Major Food Reserves	Major Cell-Wall or Cell-Membrane Components	Ecology
Viruses	Non-cellular, parasites of all organisms	None	None	Capsid of protein, sometimes lipids	Infect all marine organisms, release dissolved organic matter (DOM) from destroyed cells
Bacteria	Prokaryotic, unicellular, some form colonies	Chlorophyll *a*, phycobilins, carotenoids in cyanobacteria; bacteriochlorophyll in other photosynthetic bacteria	Variety of types	Cell wall of peptidoglycan containing muramic acid, chains of amino sugars, and amino acids in cyanobacter	Essential role in nutrient cycles; autotrophs and heterotrophs, nitrogen fixers; components of detritus, some cause diseases, symbiotic in many organisms, some produce blooms
Archaea	Prokaryotic, unicellular	Bacteriorhodopsin in light-capturing archaea	Variety of types	Cell wall made up of variety of compounds but no muramic acid	Autotrophs and heterotrophs, nitrogen fixers
Diatoms	Eukaryotic, unicellular, some form chains, mostly planktonic	Chlorophyll *a*, *c*, carotinoids	Chrysolaminarin, oil	Cell membrane secretes frustule of silica, pectin	Important primary producers, frustules are components of siliceous sediments
Dinoflagellates	Eukaryotic, unicellular, two flagella, mostly planktonic, some bottom-dwelling	Chlorophyll *a*, *c*, carotenoids	Starch, oil	Cell membrane secretes cellulose plates (thecae)	Primary producers, symbiotic with other organisms as zooxanthellae, red tides, many are bioluminescent, some cause diseases
Silicoflagellates	Eukaryotic, unicellular, two flagella, planktonic	Chlorophyll *a*, *c*, fucoxanthin	Starch	Cell membrane secretes silica skeleton	Primary producers, skeletons are components of siliceous sediments, some produce blooms
Coccolithophorids	Eukaryotic, unicellular, one or two flagella, planktonic	Chlorophyll *a*, *c*, carotenoids	Starch	Cell membrane secretes calcium carbonate plates	Important primary producers, plates are components of calcareous sediments, some produce blooms
Cryptophytes	Eukaryotic, unicellular, usually two flagella, planktonic	Chlorophyll *a*, *c*, carotenoids, phycopilins	Variety of types	Cell membrane made up mostly of protein	Important primary producers
Foraminiferans	Eukaryotic, unicellular, bottom-dwelling and planktonic	None	Variety of types	Cell membrane secretes calcium carbonate shell (test)	Heterotrophs, shells are components of calcareous sediments
Radiolarians	Eukaryotic, unicellular, some colonial, mostly planktonic	None	Variety of types	Cell membrane secretes silica shell	Heterotrophs, shells are components of siliceous sediments
Ciliates	Eukaryotic, unicellular, planktonic and bottom-dwelling	None	Variety of types	Cell membrane made up mostly of lipids	Heterotrophs, some associated with marine animals
Fungi	Eukaryotic, unicellular or multicellular, mostly bottom-dwelling	None	Variety of types	Cell wall of cellulose, chitin, and other compounds	Decomposers, many cause diseases, symbiotic with algae or cyanobacteria in lichens

Chapter 7 Review

REVIEW QUESTIONS

Multiple Choice

1. Which group makes up the most abundant and diverse life-like particles in the ocean?

 a. archaea

 b. bacteria

 c. protists

 d. viruses

2. Which are multicellular eukaryotes that are heterotrophs that form hyphae?

 a. diatoms

 b. dinoflagellates

 c. foraminiferans

 d. fungi

Short Answer

3. What are different types of viruses?

4. Why is the nitrogen fixation carried out by cyanobacteria, bacteria, and some archaea in marine environments important?

5. Give three examples of symbiotic relationships that algae, protozoans, and fungi have with other organisms.

Critical Thinking

6. Scientists use the particular structure of nucleic acids and other chemical differences to separate the archaea from the bacteria. Can you think of other characteristics that could be used to distinguish not only between these two domains but also between them and protists?

7. What might be the adaptive significance of bioluminescence in a dinoflagellate? Are there any disadvantages of bioluminescence in a planktonic organism?

DATA ANALYSIS LAB

How does climate change affect the distribution of plankton? A group of scientists analyzed sediment core samples taken off the coast of California to look for changes in the population sizes of foraminiferans.

Data and Observations

The changes in population sizes of three warm-water foram species are shown in the graph. The core samples extend back 300 and 1,400 years (red and black lines, respectively; *G bulloides* was not counted in the earliest samples).

Think Critically

1. What is the trend in the population sizes for these three species since 1900?

2. From 1901 to 2013, sea surface temperatures rose at an average of 0.13°F per decade. How do the population changes of these warm-water species coincide with increased sea surface temperatures due to global warming?

3. Since the year 2000, sea surface temperatures have increased by about 0.5°F. Predict what the graph would look like if it were extended to 2015.

Data obtained from: Field, D., T. Baumgartner, C. Charles, V. Ferreira-Bartina, and M. Ohman, 2006. Planktonic Foraminifera of the California Current Reflect 20th-Century Warming. Science, vol. 311, no. 5757, May, pp. 63–66.

Multicellular Primary Producers: Seaweeds and Plants

8.1 Multicellular Algae: The Seaweeds

8.2 Flowering Plants

THEME Structure and Function

Seaweeds and marine flowering plants are adapted to survive in many shallow-water marine environments.

BIG IDEA

Seaweeds and marine flowering plants are important in marine environments, providing food and habitats for a range of organisms.

About the photo Several species of kelp, including giant kelp (*Macrocystis pyrifera*), make up this kelp forest off Santa Catalina Island, CA.

Island Effects/E+/Getty Images

8.1 Multicellular Algae: The Seaweeds

VOCABULARY

macroalgae

seaweeds

thallus

blades

pneumatocysts

stipe

holdfast

encrusting

spores

zoospores

sporophyte

gametophyte

alternation of generations

carpospores

algin

carrageenan

agar

IN CONTEXT

Primary Producers Organisms that manufacture organic matter from carbon dioxide (CO_2), usually by photosynthesis.

Autotrophs Organisms that can use energy (usually solar energy) to make organic matter.

• 6.1, The Ingredients of Life

Main Idea

Seaweeds are multicellular protists that act as primary producers in marine environments worldwide.

Key Questions

1. What are the three main groups of seaweeds? Why are they classified this way?
2. What is the general structure of seaweeds?

As inhabitants of land, our perception of the world of photosynthetic organisms is based mostly on plants such as trees, ferns, and mosses. Many fascinating photosynthetic organisms populate the oceans, but for the most part they are very different from the land plants that surround us. Nearly all marine photosynthetic organisms, in fact, are not considered true plants at all and therefore are not members of the kingdom Plantae. Non-plant photosynthetic organisms include photosynthetic bacteria and unicellular algae (both described in Chapter 7) and the seaweeds covered here. Some biologists, however, consider some or all seaweeds to be plants. The true marine plants, seagrasses, salt-marsh grasses, and mangroves are also covered in this chapter.

Seaweeds and marine true plants share the fact that they are all primary producers capable of using light energy to perform photosynthesis. In other words, these organisms are autotrophs.

The most familiar types of marine algae are those popularly known as seaweeds. "Algae" (singular "alga") and "seaweeds" are interchangeably used, except that, to complicate our lives, unicellular "algae" are not seaweeds but all seaweeds are algae. Just remember that "red algae" and "red seaweeds" mean the same thing! "Seaweed" is a rather unfortunate name. For one thing, the word "weeds" does not do justice to these conspicuous and often elegant inhabitants of rocky shores and other marine environments. Some biologists opt for the more formal name of macroalgae. On the other hand, the term "seaweeds" is useful in distinguishing them from the unicellular algae surveyed in Chapter 7 and the true marine plants described later in this chapter. By definition, seaweeds are all multicellular; unicellular green and red algae are therefore not considered seaweeds. All seaweeds and unicellular green and read algae are nevertheless considered protists (kingdom Protista), whereas seagrasses, salt-marsh grasses, and mangroves are true plants (kingdom Plantae).

Like unicellular algae, seaweeds are eukaryotic. The structures of seaweeds, however, are far more complex than those of unicellular algae. Reproduction is also more elaborate. Though more complex than unicellular algae, seaweeds nevertheless lack the highly specialized structures and reproductive mechanisms of the mostly terrestrial true plants, most especially the flowering plants.

The range of variation observed among seaweeds is spectacular. Those we see on rocky shores at low tide are usually small and sturdy as an adaptation to withstand waves. Some small, delicate ones live on other seaweeds. Kelps found offshore in cold waters are true giants that form dense underwater forests.

The multicellular condition of seaweeds allows many adaptations not available to unicellular forms. The ability of seaweeds to grow tall and rise off the bottom provides new opportunities as well as challenges, particularly that of wave action.

Seaweeds play an important role in many coastal environments. They transform solar energy into chemical energy in the form of organic matter and make it available to a long list of hungry creatures, which can include humans. There are always exceptions, and there are a few seaweeds that are not primary producers but parasites of other seaweeds. Other organisms live on or even within the tissues of seaweeds, and more importantly, seaweeds also take carbon dioxide from their environment and release oxygen to be used by organisms both in the ocean and on land. It has been estimated that seaweeds, together with the unicellular algae in the water, produce 90% of all the oxygen in the atmosphere.

General Structure

Seaweeds show a wide range of growth forms and complexity of structures. Nevertheless, several unifying features are worth mentioning. Seaweeds lack the true leaves, stems, and roots of plants. The complete body is known as the **thallus** (plural, thalli) whether it is a filament, a thin leafy sheet, or a giant kelp.

The leaf-like, flattened portions of the thallus of many seaweeds are known as **blades** (Fig. 8.1, p. 198). They have a large surface area and are the main photosynthetic regions, though all portions of the thallus are able to photosynthesize in light as long as they have chlorophyll. Blades are technically not true leaves because they have no veins or specialized transport tissue like those found in land plants. Another difference is that, in contrast to true leaves, the upper and lower surfaces of blades are identical to each other. Gas-filled bladders known as **pneumatocysts** (Fig. 8.1) sometimes keep the blades close to the sea surface, thereby maximizing their exposure to the sunlight. The mixture of gases in the pneumatocysts of some seaweeds includes carbon monoxide, a gas that is toxic to humans.

Some seaweeds have a distinct, stem-like structure to provide support, the **stipe**, from which blades originate. It is long and tough in the large kelps. A structure that looks like roots, the **holdfast**, attaches the thallus to the bottom. Holdfasts are particularly well developed in the kelps (Fig. 8.1). They are not involved in any significant absorption of water and nutrients and do not penetrate through the sand or mud, as true roots do.

Most seaweeds cannot anchor in soft sediments and are therefore restricted to hard bottoms. Water and nutrients, which bathe the entire thallus, are picked up directly across the surface without the need of roots. Also, in contrast to the leaves and stems of plants, the stipe and holdfast usually lack tissues specialized for water and nutrient transport.

DIVE IN!

Do seaweeds have medicinal properties? Some compounds in red algae have been found to have anti-viral properties, particularly against HIV and herpes simplex viruses. Certain compounds found in a variety of seaweeds have antibiotic properties. One area of research is investigating the use of one compound to treat a bacterial lung infection that people who have cystic fibrosis are susceptible to. Studies are also being conducted as to the effectiveness of certain macroalgal compounds in the treatment of lung cancer and colon cancer.

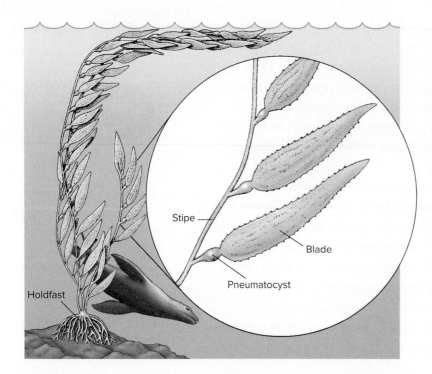

Figure 8.1 The giant kelp (*Macrocystis*).

Photosynthetic Pigments

Chlorophyll is the essential photosynthetic pigment in practically all photosynthetic organisms: single-cell algae and other microorganisms, seaweeds, and true-plants. Because of its chemical structure, chlorophyll is remarkable for its ability to "capture" sun energy. There are actually three main types of chlorophyll: *a*, *b*, and *c*. Chlorophyll *a* is the primary photosynthetic pigment found in all organisms that carry out photosynthesis, an exception being some bacteria and archaea.

Accessory photosynthetic pigments, such as carotenoids, are chemical compounds that also capture sun energy but can only pass it on to chlorophyll *a*. Chlorophyll *a* and accessory pigments absorb particular wave lengths of the sunlight spectrum so that by working together they broaden the range of wave lengths, or absorption spectrum, used in photosynthesis. Chlorophyll *a*, which is green, is more efficient in absorbing the violet-blue and red portions of the light spectrum, not green, which is reflected, explaining why chlorophyll appears green. Carotenoids, which can be red, orange, or yellow, absorb those parts of the spectrum not used by chlorophyll and conversely reflect red to yellow back to us. Chlorophyll *b*, which is only found in some groups of seaweeds (see Table 8.1), is an accessory pigment that absorbs somewhat similar wave lengths as does chlorophyll *a*. Chlorophyll *c*, also an accessory pigment, is characteristic of some single-cell microorganisms (protists) such as diatoms and dinoflagellates as well as brown seaweeds (see Table 8.1). Still other types of chlorophylls are found among bacteria and archaea.

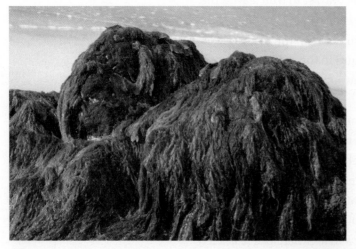

Figure 8.2 *Ulva*, or sea lettuce, is a green alga that is common on rocks where fresh water reaches the sea.

Types of Seaweeds

The classification of seaweeds takes into consideration not only structure but also the types of photosynthetic pigments and the food products they store (see Table 8.1). There are three types of seaweeds: the green, brown, and red algae. It is not always easy to recognize the groups by their color in nature because the relative proportion of chlorophyll and other pigments (see Table 8.1) can vary depending on factors such as water depth and age. Some pigments protect seaweeds from the sun's ultraviolet radiation. Chemical analysis, however, reveals the characteristic pigments of each group.

Green Algae Most green algae (division, or phylum, Chlorophyta) live in freshwater and terrestrial environments. Only around 10% of the estimated 7,000 species are marine; many of these marine species are unicellular and only about 1,800 are multicellular. This, however, does not mean that multicellular green algae are uncommon in the sea. Certain species dominate in environments with wide variations in salinity, such as bays and estuaries and isolated tide pools on rocky coasts.

Figure 8.3 The dead man's fingers (*Codium fragile*) form branched clumps on rocky shores. Its chloroplasts remain alive, and still photosynthesize, in sea slugs that feed on the seaweed.

Most multicellular green algae have a simple thallus compared to the other two groups of seaweeds. Their pigments and food reserve are the same as those in plants (see Table 8.1), so it is thought that land plants evolved from green algae. Chlorophyll in both green algae and plants is not normally masked by any accessory pigments, and green algae typically have a bright green thallus.

Filamentous green algae can be common on rocks in shallow water and other seaweeds, as well as in rocky shore tide pools. The filaments of these species can be branched or unbranched. Species of *Enteromorpha* have a thin thallus in the form of a hollow tube. They sometimes flourish in polluted areas. Sea lettuce (*Ulva*; Fig. 8.2) forms paper-thin sheets whose shape varies depending on environmental factors. Different species of *Ulva* are widespread, from polar to tropical waters. *Valonia* forms large spheres or clusters of spheres in the tropics and subtropics.

Several other green algae consist of thin filaments or tubes (siphons) formed by a single giant cell with many nuclei. Such is the case in *Caulerpa*, which is restricted to the tropics and subtropics. Its many species show a great variety of shapes. Dead man's fingers (*Codium*; Fig. 8.3) is a green alga that extends from tropical to temperate waters, including both coasts of North America. It consists of multinucleated filaments woven into a spongy, often branching thallus. The thallus of *Halimeda*, a calcareous green alga, consists of numerous segments with deposits of calcium carbonate. The accumulation of its dead, calcified segments plays an important role in the formation of coral reefs.

Brown Algae The color of the brown algae (division, or phylum, Phaeophyta), which actually varies from olive green to dark brown, is due to a preponderance of yellow-brown accessory pigments, particularly fucoxanthin, over chlorophyll. Almost all the approximately 1,800 known species are marine. Brown algae are often the dominant primary producers on temperate and polar rocky coasts and include the largest and most complex seaweeds.

The simplest brown algae have a finely filamentous thallus, as in the widely distributed *Ectocarpus*. The thallus is flat and branched in *Dictyota* and fan-shaped and lightly calcified in *Padina* (Fig. 8.4, p. 200). Both are tropical and subtropical. The thallus of most species of *Desmarestia* is branched in many ways. *Desmarestia* is found in cold waters. It ranges from the Antarctic, where it is one of the dominant species, to temperate shores elsewhere.

© Bill Ober

Figure 8.4 *Padina*, a brown alga, consists of clusters of flat blades that are rolled into circles. This species is from the Hawaiian Islands.

Some of the most conspicuous of all brown algae are those exposed at low tides at the middle and upper levels of rocky shores. Their thick, leathery thalli can withstand exposure to air. Many species have gas-filled floats. Known locally as rockweeds, or wracks, *Fucus* is found on the Atlantic and Pacific coasts of North America and other temperate shores (Fig. 8.5). The knotted seaweed (*Ascophyllum*; Fig. 8.6) is found along temperate Atlantic coasts. In warm waters, including the Gulfs of Mexico and California, these temperate species are replaced by sargasso weed. Sargasso weed has spherical air bladders that keep the small, leaf-like blades afloat at the sea surface. Most species grow on rocks, but at least two float offshore in huge masses. The Sargasso Sea, an area in the Atlantic north of the West Indies, takes its name from sargasso weed. The seaweed drifts in other regions of the world as well. It is particularly common in the Gulf of Mexico.

The kelps are the most complex and largest of all brown algae. Most kelps are found below the low tide level in temperate and subpolar latitudes. In these environments they can occur in great abundance, providing food and shelter for many other organisms.

Some kelps consist of a single large blade, up to 3 m in length, as in the many species of *Laminaria*. Their blades are harvested for food in several parts of the world. Several blades can grow from a single holdfast. In some species of *Laminaria* the blade is split or branched. In *Agarum* and *Alaria*, a conspicuous rib runs along the middle of the single blade. The blade of *Alaria* can be as long as 25 m. *Postelsia*, commonly known as the sea palm because of its appearance (Fig. 8.7), grows on intertidal rocks exposed to heavy waves. It occurs in thick clusters from central California to British Columbia. Two branched forms, the feather-boa kelp (*Egregia*) and the southern sea palm (*Eisenia*), are also common on Pacific rocky shores.

In the Pacific the largest kelps are found in deeper water just below the lowest tide level. The bull kelp, *Nereocystis*, consists of a whip-like stipe up to 30 m long with a large, spherical pneumatocyst at the upper end. Another large

Figure 8.5 The spiral rockweed (*Fucus spiralis*) is common on rocky shores on the Atlantic coasts of temperate North America and Europe. Its thallus lacks the air bladders that characterize a similar species, the bladder rockweed (*Fucus vesiculosus*).

Figure 8.6 The knotted rockweed (*Ascophyllum nodosum*) occurs on the North American and European coasts of the North Atlantic.

kelp is *Pelagophycus*. It is similar to the bull kelp, but it has impressive antler-like branches.

Macrocystis (see Fig. 8.1), the giant kelp, is the largest of the kelps and the world's largest bottom-dwelling organism. Its massive holdfast is attached to hard bottoms from the intertidal to depths of 25 m or more. Several long stipes grow from the holdfast, from which elongated blades develop (Fig. 8.8a, p. 203). At the base of each blade a gas-filled pneumatocyst eventually develops, which helps keep the blades close to the surface. Individuals as long as 100 m have been recorded. It has been estimated that such kelps can grow as much as 14 cm per day in optimal conditions. Many individuals, each with many fast-growing and intertwined stipes, form dense and very productive kelp beds, or forests (Fig. 8.8b), in the colder waters of the North and South Pacific. Kelp beds are harvested by chopping off the tops for the extraction of several natural products. Kelp beds are among the richest, most productive environments in the marine realm. They show very high primary production, the rate of production of organic matter. The ecological significance and other aspects of the biology of kelp beds will be discussed in Chapter 15.

Red Algae There are more species of marine red algae (division, or phylum, Rhodophyta) than of marine green and brown algae combined. Among other features, they have red accessory pigments called phycobilins (phycocyanin and phycoerythrin), which mask chlorophyll (see Table 8.1). Most species actually are red, though some can have different colors depending on their daily exposure to light. The group is essentially marine; only a few of the approximately 6,000 species live in fresh water or soil. Red algae inhabit most shallow-water marine environments. Some are harvested for food and for the extraction of various products (see "Economic Importance," below).

Figure 8.7 Stands of the sea palm (*Postelsia palmaeformis*) are common on rocky shores exposed to heavy wave action on the Pacific coast of North America.

Seaweeds for Gourmets

Seaweeds (raw, cooked, or dried) are used as food in many cultures. Seaweeds are good sources of some vitamins (particularly B12), minerals, fiber, and antioxidants. Seaweeds can add variety and taste to bland foods and may be used to wrap such foods as sushi. Mariculture of edible seaweed is a growing business in many parts of the world. Seaweeds are harvested by hand and sold at health food stores or on the Internet. Connoisseurs use seaweed in salads, soups, omelettes, casseroles, and sandwiches. Cookbooks and websites containing seaweed recipes are available to whet the appetites of the most demanding gourmets.

Eating seaweed across the globe Many seaweed species are featured in cuisine across the world. Green and red seaweeds (or "limu" in Hawai'ian) are beloved by the Hawaiians, especially in tuna poke. In the British Isles, purple laver, a red alga, is enjoyed as a main ingredient in laverbread. After washing and boiling the seaweed, it is formed into flat cakes, rolled in oatmeal, and fried. It is also eaten as a hot vegetable or fried with bacon. Red algae can also be found in desserts in Canada, New England, and northern Europe. Kelp, which are brown algae, are eaten by people from Alaska to Chile.

Seaweed cultivation In many parts of the world, seaweed is not just a cuisine, but an industry. Across Asia, several species are carefully cultivated, supporting multimillion-dollar operations. In Japan seaweed culture is a very old tradition and it is used extensively in their cuisine. Species of *Laminaria* and *Alaria* are dried and shredded, then prepared in various ways. *Laminaria japonica* is heavily cultivated in Japan and when processed is known as *kombu*. The seaweeds are even used to make tea and candy. *Undaria*, or *wakame*, is other popular kelp that is best when consumed fresh or cooked for a very short time. Another species of *Porphyra* is cultivated to make thin sheets of *nori*, widely used in soups and to wrap *sushi*. Are you hungry for feather-boa burgers, French-fried sea palm, and *wakame* shakes?

Seaweed Cake*

1 1/2 cups salad oil

2 cups sugar

3 eggs

2 cups grated or chopped seaweed: bull kelp (*Nereocystis*), ogo (*Gracilaria coronopifolia*), *Eucheuma* (Hawaiian species), or local equivalent

2 cups grated carrots

1 cup crushed, drained pineapple (or 1 cup grated coconut, preferably fresh)

2 1/2 cups flour

1 teaspoon baking soda

1 teaspoon salt

1 teaspoon cinnamon

1 cup chopped walnuts (optional)

Harvesting kelp (*Laminaria japonica*) in Japan.

Mix together sugar and salad oil. Add eggs individually, beating well after each egg is added. Add seaweed, carrots, and pineapple (or coconut). Sift together flour, baking soda, salt, and cinnamon and add to mixture; mix well. Add chopped walnuts if desired. Bake in oblong or bread-loaf pan at 160 °C (350 °F) for 45 to 50 minutes. Cake may be covered with buttercream frosting. Enjoy.

*Adapted from I. A. Abbot, *Limu, An Ethnobotanical Study of Some Hawaiian Seaweeds*, 4th edition, National Tropical Botanical Garden, Lawa'i, Kauai'i, Hawai'i, 1996.

© Isabella Abbot, University of Hawaii at Manoa

(a) (b)

Figure 8.8 (a) The growing end of a stipe of the giant kelp (*Macrocystis pyrifera*). (b) A kelp forest in California.

The structure of the thallus of red algae does not show the wide variation in complexity and size seen in brown algae. Some reds have become greatly simplified, at least in their structure, by becoming parasites of other seaweeds. A few have lost all trace of chlorophyll and have become heterotrophs, depending entirely on their host for nutrition. Most red algae are filamentous, but the thickness, width, and arrangement of the filaments vary a great deal. Dense clumps are more common on the upper levels of rocky shores that are exposed at low tide; longer and flatter branches predominate in areas less exposed to air and in deeper water. These variations are observed among the many species of *Gelidium* and *Gracilaria* that are found worldwide. *Endocladia* forms wiry clumps on rocky shores from Alaska to Southern California.

Some species of *Gigartina* have large blades as long as 2 m. These are among the most massive of the red algae. Numerous species of *Porphyra* are common on rocky shores above the lowest tide marks from polar to tropical coasts (Fig. 8.9). The most common growth form is a thallus with thin, large blades. *Rhodymenia* is common in the North Atlantic. Its blades can reach 1 m in length. Irish moss (*Chondrus*) is another North Atlantic red alga. It can tolerate wide ranges of temperature, salinity, and light, and its shape varies greatly in response to these physical factors.

The coralline algae are red algae that deposit calcium carbonate within their cell walls. They are important in several marine environments. The calcified thallus takes a variety of shapes: thin disks growing over other seaweeds, branches with many joints, smooth or rough **encrusting** growths on rocks (Fig. 8.10, p. 204). The color of live coralline algae typically varies from light to intense reddish-pink; dead ones are white. Warm-water coralline algae are actively involved in the formation and development of coral reefs. Other species thrive in temperate and polar waters, often attaining a large size.

Figure 8.9 Many species of *Porphyra*, a red alga, inhabit temperate, polar, and tropical rocky shores around the world. Some are of economic importance.

Figure 8.10 *Corallina*, a coralline alga.

© Charlie Arneson

IN CONTEXT

Mitosis The resulting cells are identical to the original cell, having their chromosomes in pairs (diploid cells, or 2*n*).

Meiosis The resulting cells are haploid (*n*, or 1*n*), as in gametes (spores in seaweeds), because they contain half the number of the parent's chromosomes.

• 6.3, Perpetuating Life

Life History

Reproduction is a complex affair in seaweeds. Asexual, or vegetative, reproduction is common. It can be more important than sexual reproduction in most species. Fragments of the thallus can often grow into new individuals, as occurs in the floating masses of *Sargassum* of Sargasso Sea fame. Some seaweeds produce **spores**, which are cells specialized for dispersing to new locations or persisting through unfavorable conditions. Some spores are protected by resistant cell walls; others have flagella for movement and are known as **zoospores**.

The production of gametes is a key event in sexual reproduction. Gametes from two different individuals fuse, so that the new generation contains genetic information from both parents. Genetic variation is thus ensured generation after generation. Gametes produced by all members of a seaweed species can be similar in appearance or can consist of larger, non-motile eggs and smaller sperm that can swim by means of flagella. Male gametes in the red algae lack flagella and are non-motile. They can be released in strands of slime. Male and female gametes can be formed in the same thallus, but the chances are good that fusing gametes will be from separate thalli.

Cells of seaweeds (and of us all—clam, fish, or human) divide and produce identical cells by mitosis. Seaweeds can also produce haploid spores or gametes by meiosis. The existence of diploid and haploid cells is fundamental in understanding the often complex life histories of seaweeds. Their life histories can be divided into four basic types.

The first type (Fig. 8.11a) is the most common among all three groups of seaweeds, and it involves two types of thalli. The first is a diploid (2*n*) **sporophyte** generation that through meiosis produces not gametes but haploid (*n*, or 1*n*) spores. Except in the red algae, these spores are typically motile. They divide and develop into the second kind of thallus, a haploid (*n*) **gametophyte** generation. The gametophyte is the one that produces haploid gametes. In

some species there are separate male (sperm-producing) and female (egg-producing) thalli; in others, both types of gametes are produced by every thallus. The gametes are released and, on fertilization, produce a diploid (2*n*) zygote that develops into the diploid sporophyte. A life history with two generations, a sporophyte and a gametophyte, is an example of the phenomenon of **alternation of generations**. In some algae, such as sea lettuce (*Ulva*) and the brown *Dictyota*, the sporophyte and gametophyte are structurally identical, or isomorphic. On the other hand, in kelps (*Laminaria*, *Macrocystis*, and others) the large plant we see is the sporophyte, whereas the gametophyte is minute and barely visible, thus being polymorphic because there are at least two different-looking generations. A similar type of alternation of generations takes place in flowering plants, when a sporophyte dominates over a minute gametophyte.

The second type of life history, unique to the red algae, is more complex, involving alternation of three generations (Fig. 8.11b). It is similar to the life history illustrated in Figure 8.11a, but a third generation, a diploid carposporophyte, results from the fusion of gametes. **Carpospores**, diploid spores produced by the carposporophyte, develop into sporophytes.

Figure 8.11 Generalized patterns of sexual reproduction in the life histories of seaweeds and marine flowering plants. These patterns are not absolute, and many modifications have been described. In some groups certain reproductive stages have different names.

NO ALTERNATION OF GENERATIONS

Example: rockweeds

(a)

Example: some green algae

(b)

ALTERNATION OF GENERATIONS

Examples:
• sea lettuce (sporophyte same size as gametophyte),
• kelps (sporophyte much larger than gametophyte)

(c)

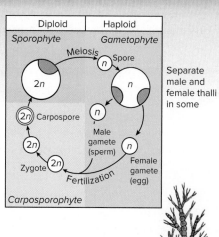

Carposporophyte

Example: many red algae

(d)

Example: all flowering plants

(e)

Marine Algae as Biofuels

Land plants such as corn and soybeans are currently used as sources of biofuels, alternatives to the increasingly scarce and polluting fossil fuels. Biofuels help offset the effects of global climate change because the plants used to produce biofuels take up carbon dioxide (CO_2) from the atmosphere. Although CO_2 is released when biofuels are burned, the uptake of CO_2 during growth is greater than or equal to atmospheric deposition avoiding a long-term increase in CO_2, unlike fossil fuels. Biofuels are also cleaner than fossil fuels and they are renewable resources that can be quickly and naturally replenished.

Biofuels from the sea Marine unicellular algae are some of the most promising sources of biofules. Fast-growing microalgae are among the most efficient primary producers, typically more efficient than land plants. Instead of transforming solar energy into organic matter that is fermented into ethanol, the biofuel most commonly obtained from land plants, many marine algae transform the glucose produced by photosynthesis into lipids, which are stored as energy-rich oils that can be used as biofuels. Like many primary producers, marine algae increase growth under high-CO_2 conditions. CO_2, a greenhouse gas that makes up roughly 10% of emissions from fossil-fuel power plants, can be separated from other cases by the use of chemicals, "captured" and buried on land or under the sea floor, an expensive process. Pumping this CO_2 into ponds or tanks used to grow marine algae increases algal production, reduces atmospheric pollution, and is less costly than other methods of long-term carbon storage, known as carbon sequestration or capture.

In addition, algae can be grown in arid regions along the coast, avoiding the use of valuable agricultural land for food crops. The use of seawater rather than fresh water, already a scant resource in many parts of the world, is another advantage of marine algae over land plants or freshwater algae. Some microalgae are heterotrophs, and can be selectively bred to grow in the dark, enabling around the clock production. Furthermore, in contrast to land-based biofuels, marine biofuels do not increase food prices, as has happened with corn and soybeans. Marine algae can be fertilized with nutrients from wastewater, whereas the use of reactive-nitrogen compounds, found in wastewater, on land plants is a major cause of eutrophication in marine environments.

Are algal biofuels sustainable? There are nevertheless some drawbacks in the production of marine biofuels. The algae must be grown in quantities large enough to make the process economically feasible. The right algae must be grown under very particular conditions. They are very sensitive to toxic substances accumulating in the water, water pollutants, or even predation by contaminating grazers. Extractions of lipid must be efficient enough in order to not get other chemical components in the process. As a relatively new technology, marine algae biofuel production cost is far higher than the cost of conventional fossil fuel production. Solutions to these problems involve the expertise of marine biologists, engineers, and other professionals.

Numerous pilot facilities around the world are at work aiming to produce marine biofuels in a sustainable way. One such project is OMEGA (Offshore Marine Encosures for Growing Algae), which was developed by NASA's space program in the United States and envisioned for long space flights. OMEGA uses freshwater algae that are grown in treated wastewater contained in flexible plastic tubes that float offshore. The fast-growing unicellular algae produce oil that could be used as a source of biofuels, fertilizer, and animal food. Sources include oils from marine unicellular green algae (such as some species of *Chlorella* and *Dunaliella*) and marine diatoms (species of *Navicula*). Biofuels from genetically engineered algae are already added to jet fuel used by several airlines. Fast-growing seaweeds such as kelps and some red algae are being considered for biofuel production of ethanol. Floating "seaweed farms" have been envisioned in Japan, Europe, and elsewhere, which together with other marine biofuels will help power the road vehicles, planes, or perhaps even space craft of the future.

The third type of life history (Fig. 8.11c, p. 205) is perhaps the easiest to understand because it is similar to that of animals, including humans. There is no alternation of generations. Thus, there is only one thallus, and it is diploid. The thallus produces haploid gametes by meiosis. After fertilization the resulting zygote develops into a new diploid thallus. This type of life history is observed in some brown algae and in some green algae.

In the fourth type of life history, which occurs in some green algae (Fig. 8.11d, p. 205), the dominant thallus is haploid and produces haploid gametes. On fertilization the gametes form a diploid zygote. It is in the zygote where meiosis takes place, resulting in haploid spores. Each of these spores develops into a haploid individual, the only kind of thallus in the cycle.

There are many known variations on these basic life history schemes, and probably others waiting to be discovered. Other aspects of the life history of seaweeds are also interesting. For example, the development of gametes or spores can be influenced by the amounts of nutrients in the water, by temperature, or by day length. High levels of nitrogen nutrients in the water cause the development of asexual spores in sea lettuce, but low levels stimulate the development of gametes instead. The release of gametes and spores can be triggered by the splashing of water in an incoming tide (and therefore by the cycles of the Moon) or by chemical messengers received from cells of the opposite sex. In some seaweeds the release of male and female gametes is timed to take place at about the same time.

Economic Importance

People have used seaweeds since time immemorial. Samples from the oldest human settlement in the western hemisphere, in southern Chile, show that the first Americans probably used seaweeds as food and medicine. People around the world still harvest seaweeds to be used in many ways. The most obvious use is as a food source. People from different cultures have discovered that many seaweeds are edible, especially some of the red and brown algae. They are consumed in a variety of ways. The farming, or aquaculture, of seaweed is big business in China, Japan, Korea, and other nations.

Seaweeds produce several types of gelatinous chemicals, called phycocolloids, that are used in food processing and in the manufacture of different products. These phycocolloids are valuable because of their ability to form viscous suspensions or gels even at low concentrations.

One important phycocolloid, **algin** (which comprises alginic acid and its salts, the alginates), is used extensively as a stabilizer and emulsifier in the manufacture of dairy products such as ice cream, cheese, and toppings, which need to be smooth and not likely to separate. Algin is also used in the baking industry to prevent frostings and pies from becoming dry. As a thickener and emulsifier it is used in the pharmaceutical and chemical industries and in the manufacture of various products, from shampoo and shaving cream to plastics and pesticides. Algin also has uses in the making of rubber products, paper, paints, and cosmetics. One of its biggest applications is in the textile industry, where algin thickens the printing paste to provide sharper prints. A major

Figure 8.12 The *Kelstar*, a kelp harvester based in San Diego, California. It is 55 m long and has a capacity of 600 tons.

source of algin for commercial uses is the giant kelp (*Macrocystis*). The west coast of temperate North America, particularly California, is home to extensive giant kelp forests, making this an important algin-producing area. The forests are leased from the state of California, and large barges equipped with rotating blades cut and collect the stipes and fronds to a depth of 1 to 2 m below the surface (Fig. 8.12). The stipes quickly grow back toward the surface. Another source of algin is *Laminaria*, which is harvested in the North Atlantic.

A second phycocolloid, **carrageenan**, is obtained from red algae such as Irish moss (*Chondrus*) in the North Atlantic and *Eucheuma* in the tropics. Several species of *Eucheuma*, for instance, are farmed in the Philippines. Carrageenan is especially valued as an emulsifier. It is used to give body to dairy products and an amazing variety of processed foods, including instant puddings.

Another phycocolloid extracted for its ability to form jellies is **agar**. Agar is used to protect ham, fish, and meats during canning, in low-calorie foods (because it is not digestible by humans), and as a thickener. It is also used in laxatives and other pharmaceuticals and in cosmetics. Biologists use agar as a medium in which to grow bacteria and molds. It is also widely used in research involving the analysis of proteins and DNA. Agar is obtained commercially from several red algae, especially from species of *Gelidium, Gelidiella,* and *Pterocladiella.*

Seaweed can also be used for fertilizer, food additives in animal feeds, and wound dressings in hospitals. Coralline algae are sometimes used in Europe to reduce the acidity of soils, and some red seaweeds are marketed as nutritional supplements. Seaweeds are also being investigated as a source of biofuels.

· ·

Section Review

1. **Construct** Make a table like the one shown below to describe the main characteristics of green, algae, brown algae, and red algae found in marine environments.

	Pigments	Morphology	Examples
Green Algae			
Brown Algae			
Red Algae			

2. **Identify** Name two ecological roles of seaweeds.

3. **Describe** How are seaweeds economically important?

8.2 Flowering Plants

Main Idea
Although most flowering plants thrive on land, there are some that are adapted to marine environments.

. .

Key Questions
1. What are the characteristics of flowering plants?
2. What are the main groups of flowering plants found in marine environments?

VOCABULARY

angiosperms

pollen grain

embryo sac

halophytes

DIVE IN!

How do seagrasses get their common names? It depends what eats it, what it looks like, and where it grows. Green sea turtles swim through and graze on the thick mats of turtle grass (*Thalassia*) found throughout the warm waters of the Caribbean. Manatee grass (*Syringodium*) is found in the Gulf of Mexico and throughout Florida's estuaries, where manatees spend the winter feeding on blades of manatee grass. Paddle grass (*Halophila*), found in the Gulf of Mexico and the Caribbean, has blades that resemble small ping-pong paddles. Surf grass (*Phyllospadix*) is found in the surf zone along the rocky shores of North America from Alaska to Mexico.

The 250,000 species of flowering plants, or **angiosperms** (variously classified as division, or phylum, Anthophyta or Magnoliophyta), are the dominant plants on land, but a few do live in the ocean. Like other land plants (ferns, conifers, and other non-flowering plants), and in sharp contrast to algae (which are protists), flowering plants have true leaves, stems, and roots, all provided with specialized tissues to transport water, nutrients, and the food manufactured by photosynthesis. As such, they are grouped in the kingdom Plantae.

Few of these "higher" plants are successful in the oceans. Of all flowering plants, only the seagrasses are truly marine. They often live submerged by seawater, rarely exposed at low tide. Salt-marsh grasses and mangroves inhabit estuaries and shores protected from wave action. They are not completely at home in the ocean, and usually only their roots are covered by water at high tide. There are also many flowering plants adapted to colonize coastal areas exposed to salt-laden winds and occasional sea spray, though they do not tolerate immersion in seawater. Such plants can be found on sand dunes or living along the edges of salt marshes.

Reproduction in flowering plants involves alternation of generations between a dominant sporophyte and minute gametophytes. The gametophytes develop in the sporophyte's flowers and are thus dependent on the sporophyte. The male gametophyte, which produces the haploid male gametes, is contained within the **pollen grain**. The female gametocyte is the **embryo sac**, which produces the haploid egg cells. Both gametocytes develop either within a single flower or in separate male and female flowers.

Seagrasses

Seagrasses superficially resemble grass but actually are not grasses at all. The closest relatives of certain seagrasses seem to be members of the lily family, so we know that seagrasses evolved from land plants.

Seagrasses have adapted to life in the marine environment. They have horizontal stems called rhizomes that commonly grow beneath the sediment (Fig. 8.13, p. 210). Roots and erect shoots grow from the stems.

Seagrass Reproduction The sexual reproduction of seagrasses involves tiny flowers (Fig. 8.13b). The pollen grains, which contain the sperm, are produced by male flowers that typically develop in plants that are separate from plants producing female flowers. Some seagrasses produce long, thread-like pollen grains instead of the minute and round pollen of land flowering plants. Pollen, which is often produced in sticky strands, is carried to female flowers for fertilization, or pollination, of the egg. Water currents are involved in the dispersal of pollen, but marine animals, such as small crustaceans, could also be involved. It has been suggested that seagrass flowers are inconspicuous because insects are not needed for pollination, as in most land plants. Tiny seeds, which in some species develop inside small fruits within the flower or in floating fruits, result from successful fertilization. Seeds and Fruits are dispersed by water currents and perhaps in the feces of the sea turtles, fishes, ducks, and other animals that browse on the plants. Seagrasses readily reproduce asexually by extending rhizomes under the substrate (Figs. 8.13). Some seagrass meadows actually consist of a single gigantic, long-living, genetically identical clone.

Diversity of Seagrasses About 70 species of seagrasses are known. Most are tropical and subtropical, but several species are common in colder waters. Most seagrasses are restricted to muddy and sandy areas in shallow water, but some, like surf grass (see below), live on rocky shores. Different species of seagrasses vary in their maximum depth, but all are limited by the penetration of light through the water.

Most species of seagrasses are found in the tropical Pacific and Indian oceans. Typical seagrass meadows in this region consist of several species of seagrasses living together. Among the most common are turtle grass (*Thalassia*), paddle grass (*Halophila*; Fig. 8.13c), strapweed (*Amphibolis*), and ribbonweed (*Posidonia*). Turtle grass (Fig. 8.13d) is the most common seagrass in the tropical and subtropical Atlantic. Extensive meadows are found to depths of about 10 m, deeper in clear water. Manatee grass (*Syringodium*; Fig. 8.13a) is often found together with turtle grass.

Eelgrass (*Zostera*) is the most widely distributed of the nearly 60 species of seagrasses known. It is found in many temperate and tropical regions of the

Figure 8.13 Some common seagrasses.

(a) Manatee grass (*Syringodium*)

(b) Eelgrass (*Zostera*)

(c) Paddle grass (*Halophila*)

(d) Turtle grass (*Thalassia*)

(e) Surf grass (*Phyllospadix scouleri*)

© Peter Castro

world, where it inhabits shallow, well-protected coastal waters such as bays and estuaries. It has distinctively flat, ribbon-like leaves (see Fig. 8.13b). It is common in oxygen-poor sediments. Like tropical seagrasses, thick *Zostera* meadows, or beds are highly productive and provide shelter and food to a variety of animals, some of considerable economic importance. Several species of *Posidonia*, which are found in the Mediterranean and along the southern coast of Australia, also play an important role in shallow coastal waters. The Mediterranean *Posidonia* meadows, sometimes known as Neptune grass, are threatened by pollution and fishing trawls.

Surf grass (*Phyllospadix*; Fig. 8.13e) is an unusual seagrass because it is an inhabitant of rocky coasts exposed to wave action, as its common name implies. Some species can become exposed at low tides. It is found on the Pacific coast of North America.

Seagrasses form extensive meadows that have a most important ecological role in shallow-water environments. They provide shelter to a variety of organisms including some of considerable economic importance, their roots and leaves decrease water turbulence that helps stabilize soft bottoms, and are highly productive, providing a rich source of detritus that directly or indirectly feeds many animals. The ecological importance of seagrass meadows and the threats to their existence because of human intrusion are discussed in detail in Chapter 15.

Salt-Marsh Plants

Cordgrasses (*Spartina*) are true members of the grass family (Fig. 8.14). At least 14 species are known. They are not really marine species but, rather, land plants tolerant of salt. Unlike seagrasses, which are true marine species, cordgrasses do not tolerate total submergence by seawater.

They live in salt marshes and other soft-bottom coastal areas throughout temperate regions worldwide (Fig. 8.15). Cordgrass salt marshes show a high primary production and provide habitat and breeding grounds for many

Figure 8.14 Cordgrass (*Spartina*) is a flowering plant that dominates the seaward margin of salt marshes in temperate and subarctic regions around the world.

Figure 8.15 The world distribution of salt marshes and mangrove forests. Mangroves tend to replace salt marshes in tropical regions. The two overlap in areas of the Gulf of Mexico, southern Australia, and New Zealand. Mangroves were introduced by humans to the Hawaiian Islands.

James Lynch, USGS

Figure 8.16 The red mangrove (*Rhizophora mangle*). Notice the long roots extending into the mud, exposed here at low tide. Other species of mangroves can be found farther inland.

species important to fisheries. They also offer protection against erosion, the result of the network of horizontal stems that extends under the sediment, and provide natural water purification systems. Flowers, and to a much lesser extent leaves, provide food to insects and other grazers. Most organic matter is made available to other organisms in the form of detritus.

Cordgrasses inhabit the zone above mudflats that becomes submerged by seawater only at high tide, so their leaves are always partly exposed to air. Salt glands in the leaves excrete excess salt. Other salt-tolerant plants, or **halophytes**, such as pickleweed (*Salicornia*), salt grasses (*Distichlis*), and rushes (*Juncus*) can be found at higher levels on the marsh.

The ecological significance of salt-marsh plants as a result of their high productivity, a source of detritus, shelter for many marine and terrestrial species, and protection against erosion is highly significant. Salt marshes are nevertheless rapidly disappearing from their natural environment victims of coastal development. These and other issues related to the ecology of salt-marshes is discussed in Chapter 14.

Mangroves

Mangroves are trees and shrubs adapted to live along tropical and subtropical shores around the world (Fig. 8.15). They are essentially land plants that can tolerate salt to varying degrees. Luxuriant and very productive mangrove forests, or mangals, flourish along muddy or sandy shores protected from wave action (Fig. 8.16).

Mangroves include around 70 mostly unrelated species of flowering plants. They are adapted in various ways to survive in a salty environment where water loss from leaves is high and sediments are soft and poor in oxygen. Adaptations become more crucial in mangroves living right on the shore, such as species of the red mangrove (*Rhizophora*), which are found throughout the tropics and subtropics (Figs. 8.16 and 8.17). The extreme northern and southern limits of the red mangrove are those areas in which killing frosts begin (see Fig. 8.15). Salt marshes replace red mangrove forests in areas exposed to frosts. Other important species of mangroves include the black (*Avicennia*) and white (*Laguncularia*) mangroves. The largest number of species of these and other mangroves are found along the shores of the tropical Indian and Pacific oceans, but a good number of species are abundant in the Atlantic and Gulf of Mexico.

(a) (b)

Figure 8.17 A seedling of the red mangrove (*Rhizophora mangle*) (a) as it appears in the tree and (b) one that has taken root in the soft sediment.

(t)© Peter Castro, (bl & br) ©Charlie Arneson

The leaves of the red mangrove are thick, an adaptation to reduce water loss. As in several other mangroves, seeds germinate while still attached to the parent tree (Fig. 8.17a). They develop into elongated, pencil-shaped seedlings as long as 30 cm before falling from the parent. Successful seedlings stick in the soft, muddy sediment like a knife thrown into a lawn, or float in the water to be carried by currents to new locations (Fig. 8.17b).

Many species of marine and terrestrial organisms live among the roots or branches of mangroves. Like seagrasses and salt-marsh plants, mangroves show high primary production, but relatively few animals graze on the hard-to-digest leaves, with most organic matter being consumed as detritus. The most important species of mangroves, their distribution, the ecological significance of mangrove forests, and human impact are discussed in detail in Chapter 14.

Table 8.1 Most Important Characteristics of Seaweeds and Marine Plants

Group	Distinguishing Features	Photosynthetic Pigments	Major Food Reserves	Major Cell-Wall Components	Ecology
Green algae	Eukaryotic protists, unicellular and multicellular; mostly bottom-dwelling, some planktonic; alternation of generations in many	Chlorophyll *a*, *b*, carotenoids	Starch	Cellulose, calcium carbonate in calcareous algae	Primary producers; calcareous algae are important sources of calcareous deposits in coral reefs
Brown algae	Eukaryotic protists, multicellular; bottom-dwelling ; alternation of generations present in some	Chlorophyll *a*, *c*, carotenoids (fucoxanthin and others	Laminarin, oil	Cellulose, alginates	Primary producers; dominant components of kelp forests
Red algae	Eukaryotic protists, multicellular; bottom-dwelling; alternation of three generations	Chlorophyll *a*, phycobilins (phycocyanin, phycoerythrin), carotenoids	Starch	Agar, carageenan, cellulose, calcium carbonate in coralline algae	Primary producers; coralline algae are important sources of calcareous deposits in coral reefs
Flowering plants	Eukaryotic plants, multicellular; bottom-dwelling ; alternation of dominant sporophyte and minute male and female gametophytes	Chlorophyll *a*, *b*, carotenoids	Starch	Cellulose	Dominant primary producers in seagrass meadows, salt marshes, and mangrove forests; nursery grounds for many species; help stabilize soft bottoms, protect coast from turbulence; food for many species in the form of detritus

Section Review

1. **Describe** How are seagrasses, cordgrasses, and mangroves adapted to living in marine environments?

2. **Explain** What is the ecological importance of flowering plants in the marine environment?

REVIEW QUESTIONS

Multiple Choice

1. The phrase *alternation of generations* is used to describe organisms that have

 a. a life history with one generation, isomorphic

 b. a life history with one generation, a sporophyte

 c. a life history with two generations, a sporophyte and a gametophyte

 d. a life history with two generations, isomorphic and polymorphic

2. Which term is used to describe salt-tolerant plants?

 a. angiosperm

 b. eukaryote

 c. halophytes

 d. sporophytes

3. Which statement about seaweeds and flowering plants that live in marine environments is false?

 a. They are all primary producers.

 b. They are all eukaryotes.

 c. They are all unicellular.

 d. They all contain chlorophyll *a*.

Short Answer

4. What are the basic characteristics of seaweeds?

5. Describe reproduction in seaweeds.

Critical Thinking

6. The actual color may vary among some species of red and brown algae. A particular species is collected and is described as "dark redbrown." How could we tell if this alga is a member of the Rhodophyta or of the Phaeophyta?

7. Only very few flowering plants have invaded the oceans, but those that have are very successful. What are some possible reasons for the small number of marine flowering plants? How do those that have taken the step manage to thrive in some environments?

DATA ANALYSIS LAB

Do invertebrates forage on pollen from *Thalassia testudinum*? The male flowers of *Thalassia* release pollen shortly after sunset. Biologists analyzed video data of two male *Thalassia* flowers: one that was releasing pollen and one that was not, on the same night in a lagoon near Puerto Morelos, Mexico.

Data and Observations

The graphs show the number of visits per minute by small invertebrates to a male flower. The "S" indicates the time of sunset and the "P" indicates when the flower begins releasing pollen.

(a)

(b)

Think Critically

1. What are the differences in data between the flower releasing pollen and the one that is not? What does this indicate?

2. Describe an experiment you could perform to test the hypothesis that the invertebrates were consuming the pollen.

Data obtained from: van Tussenbroek, B. I., L. V. Monroy-Velazquez, and V. Solis-Weiss, 2012. Meso-fauna foraging on seagrass pollen may serve in marine zoophilous pollination. Marine Ecology Progress Series, vol. 469, pp. 1–6.

Marine Animals Without a Backbone

THEME Patterns

The study of the major characterisitics of the invertebrate phyla help marine biologists discover the relationships between these groups.

BIG IDEA

Studying marine invertebrates allows biologists to identify adaptations that help these animals to inhabit different marine environments.

About the photo Octopuses, like this *Octopus vulgaris*, are members of the phylum Mollusca.

9.1 Sponges, Cnidarians, and Comb Jellies

VOCABULARY

vertebrates
invertebrates
sessile
collar cells
osculum
suspension feeders
filter feeders
deposit feeders
spicules
spongin
metamorphosis
radial symmetry
nematocysts
polyp
medusa
bell
coral reefs
extracellular
nerve net
ciliary combs

Main Idea

Sponges, cnidarians, and comb jellies are among the simplest animals.

Key Questions

1. What makes sponges unique among animals?
2. What is radial symmetry?
3. What are nematocysts?

Most species of multicellular organisms inhabiting our planet are animals (kingdom Animalia). In contrast to photosynthetic organisms like algae and plants, we animals cannot manufacture our own food and must therefore obtain it from others. The need to eat has resulted in the evolution of myriad ways of obtaining and processing food, as well as equally diverse ways of avoiding being eaten.

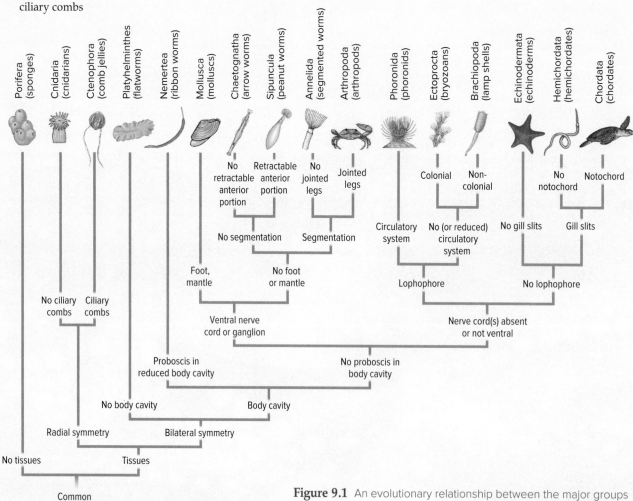

Figure 9.1 An evolutionary relationship between the major groups of marine invertebrates.

Our survey of the marine animals follows the traditional classification into two major groups: the **vertebrates**, which have a backbone (a row of bones called vertebrae), and the **invertebrates**, animals without a backbone.

At least 97% of all species of animals are invertebrates, the subject of this chapter. Different interpretations have been proposed to explain the relationships between the major groups of invertebrates. The scheme given here (Fig. 9.1) will serve as a guide in our survey.

Sponges

Sponges are best described as aggregations of specialized cells. Sponges have a cellular level of organization, meaning that the cells are largely independent of each other and do not form true tissues and organs (see Table 9.1). They are among the structurally simplest multicellular animals (Fig. 9.1).

Nearly all sponges are marine. All are **sessile**, living permanently attached to the bottom or some other hard surface. They show an amazing variety of shapes, sizes, and colors but share a relatively simple body plan. Numerous tiny pores, or ostia, on the surface allow water to enter and circulate through a series of canals where plankton and organic particles are filtered out and eaten (Fig. 9.2a). This network of canals and a relatively flexible skeletal framework give most sponges a characteristic spongy texture. Because of this unique body plan, sponges are classified as the phylum Porifera, or "pore bearers."

Sponges could perhaps be similar to the first multicellular animals, which were probably simple colonies in which some cells became specialized for such functions as feeding and protection. Sponge cells are very plastic and easily change from one type to another. If experimentally separated, the cells can even regroup and form a new sponge (Fig. 9.3, p. 218).

Morphology The architecture of sponges is best understood by examining the simplest kind of sponge (Fig. 9.3a). The outer surface is covered with flat

DIVE IN!

Carnivorous sponges? Almost all species of sponges are filter feeders, consuming plankton and organic particles filtered from the water. Eleven species of sponges, however, are known to be carnivorous. These sponges, found on the seafloor deep in the Pacific, have hairs made up of microscopic hooks that capture small crustaceans such as amphipods. When an animal is captured, the sponge begins to engulf it and digest it.

IN CONTEXT

Tissues Specialized, coordinated groups of cells.

Organs Structures consisting of several types of tissues grouped together to carry out particular functions.

• 6.3, Perpetuating Life

Figure 9.2 Sponges consist of complex aggregations of cells that carry out specific functions. Collar cells trap food particles in both (a) simple and (b) complex sponges.

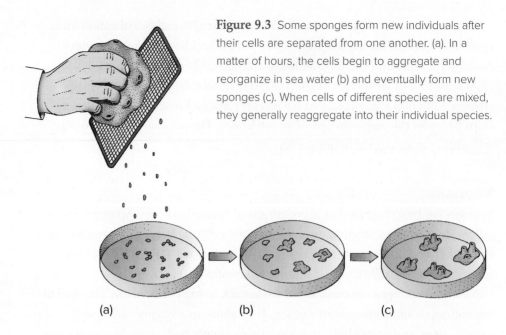

Figure 9.3 Some sponges form new individuals after their cells are separated from one another. (a). In a matter of hours, the cells begin to aggregate and reorganize in sea water (b) and eventually form new sponges (c). When cells of different species are mixed, they generally reaggregate into their individual species.

(a)　　　　　(b)　　　　　(c)

Plankton Organisms that drift with the currents.

Detritus Particles of dead organic matter.

Figure 9.4 Feeding on particulate organic matter can be classified as suspension or deposit feeding. The difference between these two types of feeding is not always well defined. Fanworms, for instance, are tube-dwelling polychaetes that switch back and forth between suspension and deposit feeding, depending on the strength of the water current.

cells called pinacocytes and occasional tube-like pore cells, or porocytes, through which a microscopic canal allows water to enter. Water is pumped into a larger feeding chamber lined with **collar cells**, or choanocytes. Each choanocyte has a flagellum that creates currents and a thin collar that traps food particles, which are then ingested by the body of the cell. Water then leaves through the **osculum**, a large opening on top of the sponge.

Sponges are an example of **suspension feeders**, animals that eat food particles suspended in the water. Because sponges actively filter the food particles, they are a type of suspension feeder known as **filter feeders** (Fig. 9.4). In contrast, other animals are **deposit feeders** that eat detritus settling on the bottom.

Most marine sponges show a more complex arrangement in which the collar cells are restricted to chambers connected to the outer pores by a network of canals (Fig.9.2b). Water exits not through a single osculum but through several oscula, each of which serves as the exit for many canals. This increased complexity is associated with increased size, which demands higher water flow through the sponge and therefore a larger surface area of collar cells.

As sponges get larger, they need structural support. Most have **spicules**, transparent siliceous or calcareous supporting structures of different shapes

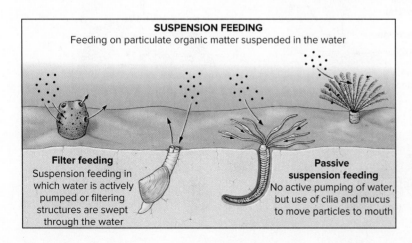

SUSPENSION FEEDING
Feeding on particulate organic matter suspended in the water

Filter feeding
Suspension feeding in which water is actively pumped or filtering structures are swept through the water

Passive suspension feeding
No active pumping of water, but use of cilia and mucus to move particles to mouth

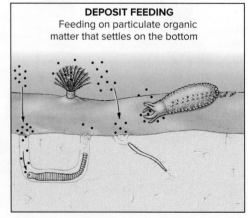

DEPOSIT FEEDING
Feeding on particulate organic matter that settles on the bottom

and sizes (Fig. 9.2a). Many also have a skeleton of tough, elastic fibers made of a protein called **spongin**. Spongin can be the only means of support, or it can be found together with spicules. When present, spongin and spicules are mostly between the outer and inner layers of cells. Wandering cells, or amebocytes, secrete the spicules and sponging, and some of these wandering cells also transport and store food particles. Some can even transform themselves into other types of cells, quickly repairing any damage to the sponge.

Reproduction Many sponges reproduce asexually when branches or buds break off and grow into separate sponges identical to the original one. Sponges also reproduce sexually by producing gametes. Unlike most animals, sponge gametes are not produced by gonads. Instead, specialized collar cells or amebocytes develop into -gametes (Fig. 9.5). The gametes are like those of other animals: large, nutrient-rich eggs and smaller sperm cells with a flagellum. Most sponges are hermaphrodites, animals in which individuals have both male and female gonads. Some species, however, have separate males and females. Sponges typically release sperm into the water. The release of gametes into the water is called broadcast spawning. The eggs, however, are usually retained inside the body and fertilization takes place internally after sperm enter the sponge.

The early stages of development take place inside the sponge. A tiny, flagellated sphere of cells is eventually released into the water (Fig. 9.5). This planktonic larva, called the parenchymula larva in most sponges, is carried by currents until it settles on the bottom and develops into a minute sponge. Most marine invertebrates have characteristic larvae that eventually change into juveniles that resemble adults. This drastic change from the larva to a juvenile is called **metamorphosis** (Fig. 9.5).

IN CONTEXT

Siliceous Made of silica (SiO_2).

Calcareous Made of calcium carbonate ($CaCo_3$).
- 2.2, The Origin and Structure of the Ocean Basins

Gametes Reproductive cells with half the genetic material of each parent, usually produced by organs called gonads: sperm (male gametes produced by the testes) and eggs (female gametes produced by the ovaries).
- 6.4, Perpetuating Life

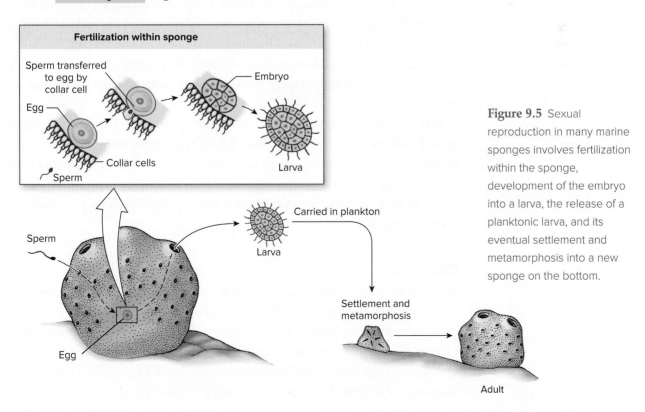

Figure 9.5 Sexual reproduction in many marine sponges involves fertilization within the sponge, development of the embryo into a larva, the release of a planktonic larva, and its eventual settlement and metamorphosis into a new sponge on the bottom.

(a) (b) (c)

Figure 9.6 (a) A tubular sponge from the Caribbean. (b) An encrusting sponge from Hawai'i. (c) A coralline sponge in Puerto Rico.

Diversity There are approximately 15,000 species of sponges, but only 8,550 are thus far known, almost all marine. Sponges live from the poles to the tropics, but the largest number of species inhabits shallow tropical waters. Sponges grow into branching, tubular (Fig. 9.6a), round, or volcano-like masses that sometimes reach a huge size. Encrusting sponges form thin, sometimes brightly colored growths on rocks or dead coral (Fig. 9.6b).

Glass sponges, such as the Venus flower basket sponge (*Euplectella*), live anchored in deep-water sediments and have a lace-like skeleton of fused siliceous spicules. Boring sponges (*Cliona*) bore thin channels through calcium carbonate, such as oyster shells and corals. In the sclerosponges, or coralline sponges (*Ceratoporella*), a calcium carbonate skeleton forms beneath the body of the sponge, which also contains siliceous spicules and sponging (Fig. 9.6c). Sclerosponges were first known as fossils, but living specimens were discovered in underwater caves and on steep coral reef slopes after the advent of scuba diving.

Some marine sponges are of commercial importance. Bath sponges (*Spongia*) are still harvested in a few locations in the Gulf of Mexico and the eastern Mediterranean in what was a once-flourishing industry. Bath sponges, not to be confused with synthetic sponges, consist of the spongin fibers remaining after cells and debris are washed away. Some marine sponges produce potentially useful chemicals.

Cnidarians

The next level of organizational complexity among animals after the sponges involves quite a big step: the evolution of multicellular animals with tissues that perform specific functions. This development makes it possible for organisms to swim, respond to external stimuli, and engulf prey, among other things. Cnidarians, sometimes called coelenterates (phylum Cnidaria), include the sea anemones, jellyfishes (or sea jellies), corals, and their relatives.

Besides having a tissue level of organization, cnidarians display **radial symmetry**, where similar parts of the body are arranged and repeated around a central axis (Figs. 9.7). If a radially symmetrical animal were cut like a pizza, all the resulting slices would be similar. Animals with radial symmetry look the same from all sides and have no head, front, or back. They do, however, have an oral surface, where the mouth is, and an aboral surface on the opposite side (Fig. 9.7).

Cnidarians have a centrally located mouth surrounded by tentacles, slender, finger-like extensions used to capture and handle food. The mouth opens into a gut, where food is digested. The cnidarian gut is a blind cavity with only one opening, the mouth. Cnidarians capture small prey by discharging their **nematocysts** (or cnidae), unique stinging structures found within cnidocytes, specialized cells in the tentacles (see Fig. 9.8).

Cnidarians occur in two basic forms (Fig. 9.7): a **polyp**, a sac-like attached stage with the mouth and tentacles oriented upward, and a bell-like **medusa** (commonly known as jellyfish, or sea jelly), which is like an upside-down polyp adapted for swimming. The life history of some cnidarians includes both polyp and medusa stages. Others spend their entire lives as either polyp or medusa.

The characteristic larva of most cnidarians is the planula, a cylindrical, ciliated stage consisting of two layers of cells. After a time in the plankton, the planula settles on the bottom and metamorphoses into a polyp or develops into a medusa.

Figure 9.7 The flower-like appearance of many cnidarians is a consequence of their radial symmetry. In both the medusa and polyp, tentacles are arranged and repeated around a central axis that runs through the mouth.

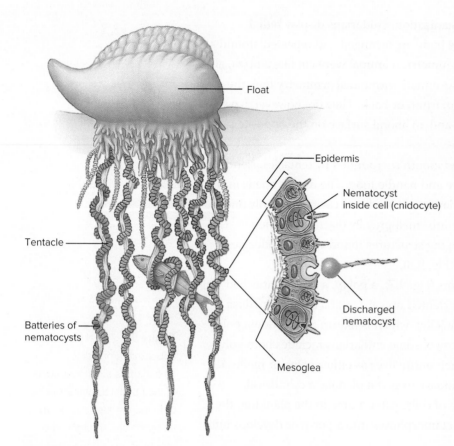

Float

Epidermis

Nematocyst inside cell (cnidocyte)

Tentacle

Discharged nematocyst

Batteries of nematocysts

Mesoglea

Two layers of cells form the body wall of cnidarians. One of these, the epidermis (see Figs. 9.7 and 9.8), is external, whereas the other, the gastrodermis, lines the gut. There is also a narrow, gelatinous middle layer, or mesoglea, that usually doesn't contain cells. In medusae this layer is expanded to form a gelatinous, domed bell, hence their common name of jellyfish, or sea jelly.

Types of Cnidarians The presence of tissues allows cnidarians to perform more complex functions than sponges can. Their basic body plan, though structurally simple, has been very successful. Some 10,000 species are known, almost all of which are marine.

Figure 9.8 A diagrammatic representation of the Portuguese man-of-war (*Physalia physalis*). The long tentacles, here contracted, are armed with nematocysts notorious for their ability to sting swimmers.

Hydrozoans The hydrozoans (class Hydrozoa) have a wide range of forms and life histories. Many consist of feathery or bushy colonies of tiny polyps. They attach to pilings, shells, seaweeds, and other surfaces (Fig. 9.9). The polyps can be specialized for feeding, defense, or reproduction.

Reproductive polyps produce minute, transparent medusae (Fig. 9.10). These medusae, usually planktonic, release gametes into the water. The fertilized eggs develop into free-swimming planula larvae. Each planula settles on the bottom and develops into a polyp. This first polyp divides repeatedly and develops into a colony of many interconnected polyps. There are, however, several variations of this life cycle. For instance, some hydrozoans lack a polyp stage, and instead their planula develops into a medusa. A few lack a medusa stage, and instead the polyp produces gametes directly. Siphonophores are hydrozoans that form drifting colonies of polyps. Some polyps in a siphonophore colony are specialized as floats, which can be gas-filled, as in the Portuguese man-of-war (*Physalia physalis;* see Fig. 9.8), or contain droplets of oil. Other polyps in the same animal form long tentacles used to capture prey. Toxins from the nematocysts can produce painful reactions in swimmers or divers (see "The Case of the Killer Cnidarians").

Figure 9.9 Colonial hydrozoans include this feather-like colony of the hydroid *Macrorhynchia phillipina*. It is a common species in warm and some temperate waters around the world.

© Bill Ober

Male medusa
Testis
Female medusa
Ovary
Sperm
Egg
Fertilization
Embryo
Planula larva
Settlement and metamorphosis
Young colony
Mature colony

Feeding polyp
Medusae buds
Tentacles
Mouth
Reproductive polyp

Life Cycle of a Hydrozoan

Eudendrium, worldwide

Figure 9.10 The life cycles of hydrozoans follow different patterns. A common one involves a sessile, asexually reproducing colony of polyps that releases planktonic, sexually reproducing medusae.

Figure 9.11 Scyphozoan medusae are larger and more complex than hydrozoan medusae. This example is the sea nettle (*Chrysaora quinquecirrha*), which is found from Cape Cod to the Gulf of Mexico.

Scyphozoans The larger jellyfishes, or sea jellies, common in all oceans are quite different from the often tiny hydrozoan medusae. These large medusae (Fig. 9.11) are the dominant stage of the life cycle of scyphozoans (class Scyphozoa). Scyphozoan polyps are very small and release juvenile medusae. A few species lack a polyp stage altogether. The rounded body, or **bell**, of some scyphozoan medusae can reach a diameter of 3 m across in some deep-water species. Scyphozoans swim with rhythmic contractions of the bell, but their swimming ability is limited and they are easily carried by currents. Some can sting swimmers. Jellyfishes in general have become more common and widespread worldwide in recent years, an indication of the declining health of the oceans or perhaps a natural fluctuation in numbers.

Cubomedusae The cubomedusae, or box jellyfishes (class Cubozoa), are cnidarians that have a small, mostly transparent, cube-shaped medusa with at least four tentacles. The life cycle includes a minute polyp as in scyphozoans. Box jelly-fishes, particularly the sea wasp, are among the most dangerous marine animals known, giving extremely painful and sometimes fatal stings (see "The Case of the Killer Cnidarians").

The Case of the Killer Cnidarians

The stings of most cnidarians are harmless to humans, but there are exceptions. Delicate and innocent-looking, some cnidarians are among the most dangerous marine animals. The sinister side of these creatures is due to the potent toxins released by their nematocysts.

Portuguese man-of-war The Portuguese man-of-war (*Physalia*), a siphonophore, is found in warm waters around the world. Though its blue, sail-like float can be seen fairly easily, its long tentacles are nearly invisible. Armed with thick batteries of nematocysts, these tentacles may reach 50 m in length, wrapping around a prey item or unlucky human appendage. *Physalia* may congregate by the thousands, sometimes forcing the closure of beaches. Pieces of tentacle that wash ashore can be as nasty as the whole animal.

Physalia stings are very painful, like being repeatedly burned with a hot charcoal, and the pain may last for hours. Red lines appear wherever tentacles have touched the skin, and welts usually follow. The pain has even been reported to cause unconsciousness. Sometimes after a sting, there can be nausea and difficulty in breathing. The tentacles may damage the cornea if they touch the eye. Allergic reactions to the toxin may cause shock and even death, and swimmers may drown because of pain or shock.

What if you get stung? If you are stung, the best thing to do is to stay calm. Carefully wash the area with seawater, but don't rub the area or wash with fresh water because this stimulates firing of the nematocysts. Alcohol will inactivate the nematocysts. The toxin is a protein, and some recommend using meat tenderizer, which contains papain, a protein-digesting enzyme derived from the papaya fruit. The meat tenderizer is of little help in relieving the pain, however, because the poison is injected into the skin while the meat tenderizer remains on the surface. Soaking the stung skin in hot water also helps denature (break down) the protein and relieve the pain. Severe reactions should be treated in a hospital.

Cubomedusae An order of medusae, the cubomedusae, releases even more powerful toxins. The sea wasp, *Chironex fleckeri*, found in the Indian Ocean, has been responsible for many known deaths. Its stings cause immediate, extreme pain. Death due to heart failure may follow within minutes, especially in children. Skin that touches the tentacles swells up, and the purple or dark brown lines that are left are slow to heal. Fortunately, an antivenin has been developed. Otherwise, the recommended first aid is to dowse the sting with vinegar.

There are other tropical cubomedusae that give severe stings, particularly in Australia and the West Indies. The irukandji (including *Carukia barnesi* and several other species) of northern Australia have caused several deaths in recent years. Unlike the sea wasp, irukandji normally live offshore but currents occasionally sweep them into shallow water.

Cubomedusae are more common along the shore during summer. Their transparent bells are difficult to see in the water. Most are small, but in the sea wasp the bell may reach 25 cm in diameter and the tentacles may stretch to 4.5 m. Irukandji have a bell only 2.5 cm in diameter and four tentacles of about the same length, so they are almost impossible to spot in the water.

The sting of *Chiropsalmus*, a cubomedusa.

David Doubilet/National Geographic/Getty Images

Anthozoans Most cnidarian species, more than 6,000, are anthozoans (class Anthozoa), solitary or colonial polyps that lack a medusa stage. The anthozoan polyp is more complex than hydrozoan or scyphozoan polyps. The gut, for instance, contains several thin partitions, or septa, which provide additional surface area for the digestion of large prey. Septa also provide support, which allows the polyp to be larger than the polyps of other cnidarians. Sea anemones are common and colorful anthozoans that often have large polyps. Colonial anthozoans occur in an almost infinite variety of shapes. Corals include various groups of mostly colonial anthozoans. Many of these have calcium carbonate skeletons and, though such corals occur in cold waters, in tropical waters they often form **coral reefs**. Gorgonians, such as sea fans (Fig. 9.12), are colonial anthozoans that secrete a tough, branching skeleton made in part of protein. Precious corals are gorgonians with fused red or pink calcareous spicules in addition to the protein skeleton. Black corals, which are neither gorgonians nor stony corals, secrete a hard, black, protein skeleton. Precious and black corals are carved into jewelry, but are being brought to near extinction in the process. Some anthozoans form fleshy colonies with large polyps and no hard skeletons. Examples of these are the soft corals, sea pens, and sea pansies.

Figure 9.12 Sea fans are gorgonians with branches that grow in only one plane and have many cross-connections.

Biology of Cnidarians Practically all cnidarians are carnivores, animals that prey on other animals. Many capture and digest prey much bigger than that of filter feeders such as sponges. Cnidarians use their nematocysts primarily to capture prey. Each nematocyst consists of a fluid-filled capsule containing a thread that can be quickly ejected (see Fig. 9.8, p. 222). The thread can be sticky or armed with spines, or it can be a long tube that wraps around parts of the prey. Some nematocysts contain toxins.

Digestion After ingestion, food passes into the gut, where it is digested. The initial phase of digestion is said to be **extracellular** because it takes place outside cells. Intracellular digestion within cells lining the gut completes the breakdown of food.

Behavior Though cnidarians lack a brain or true nerves, they do have specialized nerve cells, or neurons. These cells interconnect to form a **nerve net** that transmits impulses in all directions. This simple nervous system can produce some relatively sophisticated behaviors. Some anemones can tell whether other members of the same species are also members of the same clone, a group of genetically identical individuals, by using special nematocysts. They are known to attack and even kill anemones from other clones using special nematocysts. Some scyphozoan medusae have primitive eyes, but the eyes of cubomedusae are known to form images. Medusae also have statocysts, small, calcareous bodies in fluid-filled chambers surrounded by sensitive hairs. Statocysts give medusae a sense of balance.

Comb Jellies

The comb jellies, or ctenophores (phylum Ctenophora), are an exclusively marine group of about 180 species. Their radially symmetrical and gelatinous body resembles that of a medusa, but a closer look reveals some unique traits (Fig. 9.13). Ctenophores swim with eight rows of **ciliary combs**, long cilia fused at the base, like combs, that beat in waves. The continuous beating of the ciliary combs refracts light, creating a prism-like, multicolor effect. Body length varies from a few millimeters in the sea gooseberry (*Pleurobrachia*) to 2 m in the elongated Venus's girdle (*Cestum*).

Comb jellies are common in both warm and cold waters. They are carnivores with a voracious appetite. Swarms of comb jellies consume large numbers of fish larvae and other plankton. Many capture their prey using two long tentacles armed with sticky cells named colloblasts. A few species have nematocysts, possibly obtained by eating jellyfishes or siphonophores.

Figure 9.13 This comb jelly (*Mnemiopsis leidyi*) displays the rows of ciliary combs characteristic of the group. Four rows are visible here, the middle ones appearing as multicolored bands.

. .

Section Review

1. **Describe** Make a table like the one shown below to describe the basic characteristics of sponges.

	Description
Locomotion	
Level of Organization	
Feeding	
Structural Support	
Reproduction	

2. **Explain** What is the importance of neurons and a nerve net in cnidarians?

3. **Identify** What are the similarities and differences between cnidarians and ctenophores?

9.2 Marine Worms and Some Small Invertebrate Phyla

VOCABULARY

bilateral symmetry

anterior

posterior

dorsal

ventral

mesoderm

proboscis

hydrostatic skeleton

segmentation

coelom

parapodia

setae

morphology

closed circulatory system

gills

capillaries

trochophore

swarming

Main Idea

Some of the worm phyla show the first occurances of features common in more complex groups, including organs and organ systems, bilateral symmetry, a coelom, and a closed circulatory system.

. .

Key Questions

1. What is bilateral symmetry? What are some of the advantages of bilateral symmetry?
2. What is a closed circulatory system?

Radial symmetry works fairly well in animals that attach to surfaces or drift in currents, but animals that crawl or swim in one direction have different needs (Fig 9.14a). Most animals show **bilateral symmetry**, the arrangement of body parts in such a way that there is only one way to cut the body and get two identical halves (Fig. 9.14b). Bilaterally symmetrical animals, including humans, have a front, or **anterior**, end and a rear, or **posterior**, end. At the anterior end is a head with a brain, or at least an accumulation of nerve cells, and sensory organs such as eyes. Bilaterally symmetrical animals similarly have a back, or **dorsal** surface, that is different from the belly, or **ventral** surface. Bilateral symmetry allows animals to be more active in the pursuit of prey and to develop more sophisticated behaviors than those of radially symmetrical animals.

Flatworms

The structurally simplest bilaterally symmetrical body plan is seen in the flatworms (phylum Platyhelminthes), so called because they are dorsoventrally flattened—that is, they have flat backs and bellies. Flatworms also are the simplest animals in which tissues are organized into real organs and organ systems.

Morphology The presence of a central nervous system in which information is stored and processed is of special significance. In flatworms it typically consists of a simple brain, which is just an aggregation of nerve cells in the head. There are also several nerve cords running from the brain through the length of the worm. The nervous system coordinates the movements of a well-developed muscular system. The gut is similar to those of cnidarians and ctenophores in having only one opening to the outside, the mouth. The space between the outer and inner

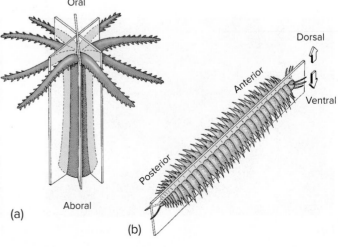

Figure 9.14 The radial symmetry of a soft coral's polyp (a) in contrast to the bilateral symmetry of a worm (b). Bilateral symmetry implies the development of an anterior end with a head, a brain, eyes, and all the other features demanded by more complex behaviors.

(l)©Leslie Newman & Andrew Flowers/Science Source, (r)©Dr. Nico K. Michiels, Institute of Evolution and Ecology, University Tuebingen, Germany

DIVE IN!

Which are some of the weirdest worms that live in the ocean? Species of the polychaete worm *Osedax* are referred to as "bone eaters" because they feed on the decaying bones of animals that sink to the bottom of the ocean. In most species of *Osedax*, females are very large and males are parasitic dwarfs. In *O. priapus*, which was discovered in 2014, the males are the same size as the females. They are free-living and feed on their own, unlike their counterparts in other species. Since the dwarf males in other species live attached to the female, they do not need to move to mate. Biologists questioned how male *O. priapus* mated with female worms. They determined that the males, which anchor in bone as do the females, stretches out its body, up to ten times its normal length, in order to reach a nearby female to mate.

tissue layers, however, is no longer thin or gelatinous as in cnidarians and ctenophores but is filled with tissue. In developing embryos this middle layer of tissue, the **mesoderm**, gives rise to muscles, the reproductive system, and other organs—not only in flatworms but also in structurally more complex animals. The gut and other internal organs are surrounded by tissue, since there is no body cavity, a feature of more specializedanimals.

Diversity There are some 30,000 known species of flatworms. The most commonly seen marine flatworms are the turbellarians, a group that consists mostly of free-living carnivores. Most are small, but some are obvious because of their striking color patterns (Fig. 9.15). Some turbellarians live inside or on the surface of oysters, crabs, and other invertebrates.

Flukes, or trematodes, are the largest group of flatworms, with roughly 18,000 species, many of which are marine. All flukes are parasites, which live in close association with other animals and feed on their tissues, blood, or intestinal contents. Like most parasites, flukes have complex life histories with amazing reproductive abilities, a key to their success. Adult flukes always live in a vertebrate. The larvae inhabit invertebrates like snails or clams or vertebrates like fishes. The larva must then be eaten by the vertebrate destined to harbor the adult. Flukes are common in fishes, seabirds, and whales.

Tapeworms, or cestodes, are parasitic flatworms that, with a few exceptions, have a long body consisting of repeated units. These unique worms live inside the intestine of most species of vertebrates, including marine ones. The head of the worm attaches to the walls of the gut by suckers, hooks, or other structures. Tapeworms lack a gut or mouth. They absorb nutrients from their host's intestinal contents directly across the body wall. Their larvae are found in invertebrates or vertebrates. Tapeworms can reach a prodigious length. The record appears to be a species found in whales known to reach 40 m, which makes it the longest invertebrate!

Figure 9.15 (a) A turbellarian flatworm (*Pseudobiceros gratus*) from Australia's Great Barrier Reef. (b) Turbellarian flatworms (*Pseudoceros bifurcus*) joust in a mating ritual known as penis fencing. The worms are hermaphrodites, and each tries to penetrate the skin of the other with its needle-like penis to inject its sperm. The first to succeed acts as the male and avoids the energy costs of healing the wound and developing eggs.

(a)

(b)

How to Discover a New Phylum

It is not very difficult to discover a new marine invertebrate species. Discovering a new phylum is a different story.

The founding species of following phyla escaped description until relatively recently. All three are exclusively marine. Each has an interesting story of how it was discovered.

Phylum Gnathostomulida The first species of what eventually became the new phylum Gnathostomulida was not officially described until 1956. Gnathostomulids are a group of about 100 species of minute worms living among sediment particles around the world. They are similar to the flatworms but possess unique features, including a set of toothed jaws to scrape bacteria, diatoms, and other organisms from sand grains.

Phylum Loricifera In 1961 Robert Higgins predicted the existence of a group that lived in the spaces between clean, coarse sediment particles in deep water. He actually found a specimen in 1974 but, unfortunately, did not realize it was something new.

One year later, Reinhardt Kristensen collected a specimen, but it was destroyed while being prepared for examination. In 1982, he was working with a large sample off the coast of France. To save time he washed the sample with fresh water instead of following the standard method. It happened to loosen the grip of the animals on the sediment, and Kristensen got a complete series of larval and adult specimens.

The microscopic animals Kristensen found have a body encased by six plates. The head, which can be retracted, bears a set of spines and a mouth at the end of a cone. Kristensen got together with Higgins, and they concluded that the specimens Higgins examined in 1974 and those subsequently found by Kristensen were members of a new phylum. Thus, the new phylum Loricifera (meaning "armor bearer") was officially born in 1983 when Kristensen published a paper in a German scientific journal. The first species was named *Nanaloricus mysticus* (mystic, or enigmatic, dwarf armor), and the larva was baptized the Higgins larva in honor of Higgins—a nice consolation prize, indeed! Some 20 additional species have so far been described. About 100 new species have been collected but not yet officially described.

Phylum Cycliophora This new phylum comes from another unexpected location: the hairs around a Norway lobster's mouth. Though first observed in the 1960s, the phylum Cycliophora was not described until 1995. The first species discovered, *Symbion pandora*, has a tiny, bottle-shaped sac with a disk-like mouth. Cilia around the mouth sweep food particles that come off the lobster's mouth at mealtime.

This strange lifestyle coincides with a bizarre life cycle. The animal alternates sexual and asexual generations. Dwarf males, which live attached to females, exist only to produce sperm. Females also reproduce asexually, but they produce only females that develop inside box-like structures inside their mothers. The new generation of females is set free from the boxes as it bursts out from the mothers (like the Greek myth of Pandora's box, hence the name of the first cycliophoran). A second species of *Symbion* was described from the American lobster, and a third species was discovered on the European lobster.

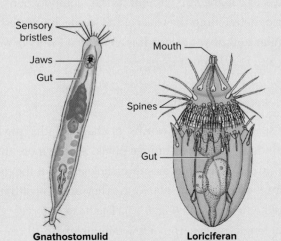

Gnathostomulid

Sensory bristles
Jaws
Gut

Mouth
Spines
Gut

Loriciferan

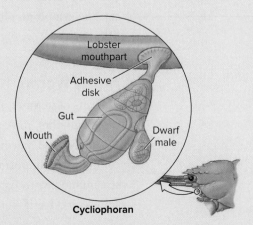

Lobster mouthpart
Adhesive disk
Gut
Mouth
Dwarf male

Cycliophoran

Captured
polychaete worm

Proboscis

Eyes

Figure 9.16 Ribbon worms use a proboscis to entangle prey. The proboscis secretes toxins and can be armed at the tip with a spine. Once the prey is captured, the proboscis is pulled back and it's dinnertime.

Ribbon Worms

Though they look like long flatworms, ribbon, or nemertean, worms (phylum Nemertea) show several features that indicate a more complex degree of organization. Their digestive tract is complete, with a gut that includes a mouth and an anus to get rid of undigested material. They also have a reduced body cavity and a circulatory system, by which blood transports nutrients and oxygen to tissues. The most distinctive feature of ribbon worms, however, is their **proboscis** (Fig. 9.16), a long, fleshy tube used to entangle prey. It is everted from a cavity above the mouth like the finger of a glove. All ribbon worms are predators that feed on worms and crustaceans.

There are about 1,300 species of ribbon worms, most of which are marine. Found in all oceans but most common in shallow temperate waters, some are nocturnal and not easily seen; others are brightly colored and found under rocks at low tide. Ribbon worms are incredibly elastic, and the proboscis can extend a meter or more beyond the body. One species reaches 30 m long.

Nematodes

Nematodes (phylum Nematoda), commonly known as roundworms, are hardly ever seen, but their numbers in sediments, particularly those rich in organic matter, can be staggering. Many species are parasitic, and most groups of marine organisms have nematode parasites. Nematodes are perfectly adapted to live in sediments or in other organisms. They are mostly small, with slender and cylindrical bodies that are typically pointed at both ends. Nematodes that inhabit sediments feed mostly on bacteria and organic matter. The gut, which ends in an anus, lies within a well-developed body cavity filled with fluid that transports nutrients. A layer of muscles in the tough but flexible body wall pushes and squeezes against the fluid, which acts as a **hydrostatic skeleton** that provides support and aids in locomotion.

The actual number of species of nematodes is debatable. Estimates vary between 15,000 and 80,000 species, but there could be half a million undiscovered. The adults of *Anisakis* and a few related nematodes inhabit the intestine of seals and dolphins. Their larvae, however, are found in the flesh of many types of fishes and can infect humans who eat raw or poorly cooked fish. Often these larvae are vomited or coughed up without further complications. Sometimes, however, larvae penetrate into the walls of the stomach or intestine, causing symptoms similar to those of ulcers. It is a risk that lovers of raw-fish dishes such as *sashimi* and *ceviche* must take.

Arrow Worms

In terms of number of species, the arrow worms, or chaetognaths (phylum Chaetognatha), rank among the smallest animal phyla. A little more than 100 species, all marine, are known. They are nevertheless one of the most common and important members of the plankton. They are almost transparent, streamlined, with fish-like fins and tail (Fig. 9.17). The head has eyes, grasping spines, and teeth. Total length ranges from a few millimeters to 10 cm.

Arrow worms are voracious carnivores with efficient sensory structures to detect their prey. They prey on small crustaceans, the eggs and larvae of fishes and other animals, other arrow worms, and practically anything else that is small. They spend most of their time motionless in the water but will swim in rapid, darting movements to grab their prey.

Segmented Worms

A large group of perhaps as many as 20,000 species, the segmented worms, or annelids (phylum Annelida), includes earthworms and many marine worms. Their body plan includes innovations that are found in some of the more structurally complex groups of animals. The body consists of a series of similar compartments, or segments, a condition known as **segmentation** (Fig. 9.18a, p. 232). Segmentation can be clearly seen in the rings of the familiar earthworm. The gut goes through all the segments and lies in a cavity known as a **coelom** (Fig. 9.18c). The coelom is entirely surrounded by tissue that develops from mesoderm, in contrast to the simpler body cavity of nematodes. It is present in all of the remaining structurally complex phyla. The coelom of segmented worms is filled with fluid and divided by partitions that correspond to the external segments. The importance of the coelom and segments is that they act as a hydrostatic skeleton and can be contracted in sequence by means of muscles in the body wall. When longitudinal muscles contract they shorten the segments, lengthening them when the muscles relax, whereas contraction and relaxation of circular muscles reduces and increases segment diameter, respectively (Fig. 9.18c). These movements, plus the flexibility given by segmentation, make annelids efficient crawlers and burrowers.

Polychaetes Almost all marine annelids are polychaetes (class Polychaeta), which are common and important in many environments (Fig. 9.18d). Each of the body segments of most polychaetes has a pair of flattened extensions, or **parapodia**, which are provided with stiff and sometimes sharp bristles, or **setae** (Fig. 9.18b).

Morphology Like all annelids, polychaetes have a circulatory system that transports nutrients, oxygen, and carbon dioxide. Circulating blood always remains within distinct blood vessels (Fig. 9.18c), making it a **closed circulatory system**. Muscular contraction of vessels helps blood circulation. Wastes from the coelom are removed by an excretory organ pair in each segment (Fig. 9.18c).

In small animals, oxygen—essential in the release of energy through respiration—can easily move from the water across the body wall to all the tissues. In the larger and relatively more active polychaetes, however, obtaining enough oxygen from the water is a potential problem. Polychaetes have solved this problem by evolving **gills** on the parapodia or elsewhere. The gills are thin-walled extensions of the body wall that have many blood vessels called **capillaries**, which allow for the easy absorption of oxygen. Capillaries are also found on the walls of the parapodia (Fig. 9.18c). This absorption of oxygen, along with the elimination of carbon dioxide, is known as gas, or respiratory, exchange.

Eyes
Grasping spines
Mouth
Brain
Intestine
Lateral fin
Ovary
Anus
Testis
Tail fin

Figure 9.17 Arrow worms, or chaetognaths, are efficient planktonic predators. *Sagitta elegans* ("elegant arrow"), illustrated here, is widely distributed. Notice that individuals have both testes and ovaries. A copepod, a favorite prey, clearly shows in the intestine of the arrow worm shown above. © *Cheryl Hopcroft, Institute of Marine Science, Fairbanks, AK*

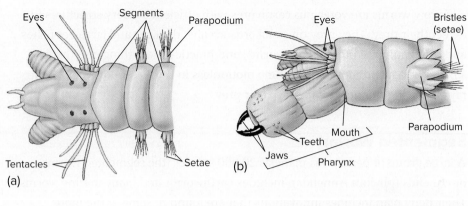

Figure 9.18 This sandworm (*Nereis*) illustrates the meaning of the name polychaetes—"many setae, or bristles." (a) Dorsal view of the head, with the pharynx retracted, showing the sensory tentacles and eyes. (b) Side view of the head, showing the large pharynx in an extended position. (c) Section across a segment. (d) Dorsal view of the worm.

(a) Eyes, Segments, Parapodium, Tentacles, Setae

(b) Eyes, Bristles (setae), Parapodium, Pharynx, Mouth, Teeth, Jaws

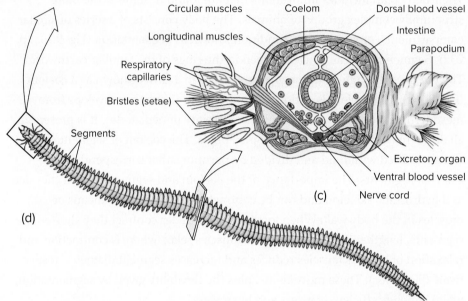

(c) Circular muscles, Coelom, Dorsal blood vessel, Intestine, Parapodium, Longitudinal muscles, Respiratory capillaries, Bristles (setae), Excretory organ, Ventral blood vessel, Nerve cord

(d) Segments

The life history of many polychaetes involves a planktonic larval stage known as the **trochophore**, which has a band of cilia around the body. The trochophore is of considerable interest because it is also one of the larval stages in some other groups of invertebrates. The types of larvae, among other characters, are used by biologists to deduce evolutionary relationships among different groups of invertebrates.

Diversity The approximately 10,000 species of polychaetes are almost entirely marine. Length varies a great deal but is typically 5 to 10 cm. Many polychaetes crawl on the bottom, hiding under rocks or coral (Fig. 9.19a). These crawling worms, such as most sandworms (*Nereis*), are mostly carnivores. They feature heads provided with several pairs of eyes and other sense organs (see Fig. 9.18) used to search for small invertebrates. A proboscis, often armed with jaws, is used to capture prey. The parapodia are well developed and are used in locomotion.

Other polychaetes burrow in mud or sand. Many, like bloodworms (*Glycera*), capture small prey. Others, like lugworms (*Arenicola*), are deposit feeders (see Fig. 9.4).

Many polychaetes live in temporary or permanent tubes, either singly or in aggregations (Fig. 9.19b). The tubes are be made of mucus, protein, bits of seaweed, cemented mud particles, sand grains, or tiny fragments of shells.

(a)

(b)

Figure 9.19 Polychaetes are common inhabitants of most marine bottoms. (a) The anterior end of a free-living polychaete (*Hermodice carunculata*), a fireworm that feeds on corals. The bright red structures are gills. (b) *Sabella melanostigma*, a feather-duster worm, inhabits leathery tubes.

Figure 9.20 Diagrammatic representation of a beard worm, or pogonophoran. (a) Most secrete and live in tubes buried in the soft sediment. (b) Only the upper end of the tube protrudes, with the tentacle or tentacles extending from it. (c) Worm removed from its tube.

Tube-dwelling polychaetes usually have reduced parapodia. Some, such as *Terebella* and related forms, are suspension feeders. Their tentacles have cilia and mucus that catch organic particles in the water and move them to the mouth (see Fig. 9.4). Fanworms, or feather-duster worms (*Sabella*), use feathery tentacles covered with cilia to capture, sort, and transport particles(Fig. 9.19b). Serpulids (*Serpula*) and spirorbids (*Spirorbis*), also suspension feeders, extend feather-like tentacles from calcium carbonate tubes they build on rocks and other surfaces.

Polychaetes are also successful at other lifestyles. Species of *Tomopteris* are planktonic throughout life. Their parapodia are flat and expanded to help in swimming. In the tropical Pacific the bodies of the Palolo worm (*Eunice*) periodically break off, and the posterior half swims up to the surface to spawn. This behavior, known as **swarming**, is timed in some areas with the phases of the moon, reaching its peak just after full moon. This bit of information is useful because in some places people gather the worms for food.

Some polychaetes live on the external surface of such invertebrates as sea stars and sea urchins. Several species live in the burrows of other invertebrates or inhabit shells occupied by hermit crabs.

Beard worms, also known as pogonophorans, and vestimentiferans are highly specialized polychaetes. They lack a mouth and gut. Except for sponges and tapeworms, this phenomenon is uncommon in animals. A tuft of one to many thousands of long tentacles characterize beard worms (Fig. 9.20), the tentacles being responsible for the group's common name. The tentacles appear to be involved in absorbing -nutrients dissolved in the water. Some beard worms have symbiotic bacteria that use the nutrients to manufacture food, which in turn is used by the worms.

(a) Retracted worm

(b)

(c)

IN CONTEXT

Symbiosis The living together in close association of two different species.

• 12.1, The Organization of Communities

About 130 species of beard worms are known. They are mostly restricted to deep water, which helps explain why they remained unknown until 1900. Beard worms and vestimenti-ferans were once grouped as separate phyla. The total length of beard worms ranges from 10 cm to 2 m.

Nearly 20 species of vestimentiferans are found at hydrothermal vents and deep-sea whale remains. These tubeworms depend on symbiotic bacteria to obtain nutrients.

Leeches Bloodsucking leeches (class Hirudinea) live mostly in fresh water, but marine species can be found attached to marine fishes and invertebrates. Leeches are highly specialized annelids distinguished by a sucker at each end and no parapodia.

Echiurans

Long considered a separate phylum, echiurans are now considered annelids (class Echiura). They look like soft, externally unsegmented sausages buried in the mud or in coral. They are similar to peanut worms (see below) in shape and size except for having a non-retractable, spoon-like or forked proboscis. All of the approximately 230 species are marine. Echiurans are deposit feeders that use the proboscis to gather organic matter. The "fat innkeeper" (*Urechis caupo*) of the Pacific coast of North America lives in U-shaped tubes in mud.

Peanut Worms

Often called peanut worms, the sipunculans (phylum Sipuncula) have soft, unsegmented bodies with a coelom. Some biologists consider them to be annelids. Peanut worms burrow in muddy bottoms, rocks, and corals or hide in empty shells. All are marine, living mostly in shallow water. The long anterior portion contains a mouth and a set of small lobes or branching tentacles (Fig. 9.21). These can be pulled into the remaining portion of the body, and the worm then becomes a compact bundle that looks like a large peanut. Peanut worms are 1 to 35 cm long. Approximately 150 species are known, all deposit feeders.

Figure 9.21 *Antillesoma antillarum*, a peanut worm (or sipunculan), burrows in coral and rocks in the Caribbean. The anterior portion, which is extended in this specimen, can be retracted.

©Charlie Arneson

Lophophorates

Three groups of marine invertebrates have a unique feeding structure, the **lophophore**. It consists of a set of ciliated tentacles arranged in a horseshoe-shaped, circular, or coiled fashion. Lophophorates are suspension feeders, using their cilia to create feeding currents. They share other traits: lack of segmentation, bilateral symmetry, a coelom, and a U-shaped gut.

Bryozoans Bryozoans ("moss animals") form colonies on seaweed, rocks, and other surfaces. Approximately 6,000 species, almost all marine, are grouped in the phylum Ectoprocta. Bryozoan colonies consist of minute interconnected individuals, or **zooids**, that secrete skeletons of a variety of shapes. The colonies can be encrusting or take an upright form that looks like tufts of crusty lace. A close look reveals the rectangular, round, or vase-like compartments occupied by zooids. The lophophore is retractable. The U-shaped gut ends in an anus outside the edge of the lophophore.

Phoronids Phoronids (phylum Phoronida) can be easily confused with polychaetes. They are worm-like and build tubes made in part of sand grains. A horseshoe-shaped or circular lophophore and a U-shaped gut confirms they are not polychaetes. All of the as many as 32 known species are marine, burrowing in sand or attaching their tubes to rocks and other hard surfaces in shallow water. Though one California species can reach 25 cm, most phoronids are only a few centimeters long.

Brachiopods There are close to 400 living species of lamp shells, or brachiopods (phylum Brachiopoda). But as many as 30,000 species are known only as fossils. Lamp shells have a shell with two valves, like the unrelated clams. The two valves are dorsal and ventral to the body, in contrast to the lateral (right and left) valves of clams. Furthermore, lamp shells have a conspicuous lophophore, consisting of at least two coiled and ciliated arms, that occupies most of the space between the valves. Most brachiopods are found attached to rocks or burrowing in soft sediments.

. .

Section Review

1. **Identify** What evolutionary advances are seen in the phyla of bilaterally symmetrical worms?

2. **Describe** Why are trochophores of interest to biologists?

9.3 Molluscs: The Successful Soft Body

Main Idea

Molluscs make up the largest group of marine animals; they have unique features including a muscular foot and a radula.

Key Questions

1. What are some different classes of molluscs and which animals are examples of each class?

2. What is an open circulatory system?

Snails, clams, octopuses, and other familiar forms are members of the phylum Mollusca. Molluscs have been very successful: there are more species of molluscs in the ocean than of any other animal group. They exhibit an immense diversity of structures and habits, thrive on practically every conceivable type of diet, and occupy all marine environments, from the wave-splashed zone of rocky shores to hydrothermal vents in the deep sea. There are nearly 100,000 known living species of molluscs on both sea and land but there may be as many as 200,000 species surpassed only by the arthropods (and perhaps nematodes) as the largest phylum of animals.

Most molluscs have a soft body enclosed in a calcium carbonate shell (Fig. 9.22). The body is covered by a **mantle**, a thin layer of tissue that secretes the shell. The unsegmented body is typically bilaterally symmetrical. There is a

Littorina obtusata, North Atlantic Ocean

© Bill Ober

Figure 9.22 The general body plan of a gastropod, indicating the most important internal structures. In many species the head and foot can be retracted into the shell, leaving a tough operculum blocking the shell opening.

ventral, muscular **foot**, usually used in locomotion. Most molluscs have a head that normally includes eyes and other sensory organs. A feature unique to molluscs is the **radula**, a ribbon of small teeth that is used to feed, usually by rasping food from surfaces (Fig. 9.22). The radula is -modified in carnivorous molluscs (Figs. 9.23 and 9.24c, p. 238). The radula is made largely of **chitin**, a highly resistant carbohydrate also found in other invertebrates. Gas exchange is through paired gills. The body cavity, a coelom, is much reduced, being restricted to small cavities around the heart and a few other organs.

Types of Molluscs

All molluscs have the same basic body plan, but it is often greatly modified. The shell, for example, is internal in squids and absent in octopuses and a few other groups. Portions of the body are coiled and asymmetrical in snails, and in some molluscs the radula is modified or even absent.

Gastropods The gastropods (class Gastropoda), are the largest, most common, and most varied group of molluscs. Snails are the most familiar gastropods, but the group includes other forms such as limpets, abalones, and nudibranchs (Fig. 9.24). There may be as many as 80,000 to 100,000 species, mostly marine. A typical gastropod can best be described as a coiled mass of vital organs enclosed by a dorsal shell (see Fig. 9.22). The shell rests on a ventral creeping foot (the term "gastropod" means "stomach footed") and is usually coiled.

Most gastropods use their radula to scrape algae from rocks, as in periwinkles, limpets (*Megathura*; Fig. 9.24a), and abalones (*Haliotis*; Fig. 9.24b). Some, like mud snails (*Hydrobia*), are deposit feeders on soft -bottoms. Whelks (*Nucella, Buccinum*), oyster drills (*Murex, Urosalpinx*), and cone shells (*Conus*; Fig. 9.24c), are carnivores. They prey on clams, oysters, worms, and even small fishes. The violet snail *Janthina* has a thin shell and produces a bubble raft out of mucus to float on the surface, looking for siphonophores, its prey. Sea hares (*Aplysia*), whose shells are small, thin, and buried in tissue, graze on seaweeds.

Nudibranchs, or sea slugs, are gastropods that have lost the shell altogether. Colorful branches of the gut or exposed gills make nudibranchs among the most beautiful of all marine -animals (Fig. 9.24d). They prey on sponges, hydroids, and other -invertebrates. As a defensive mechanism, nudibranchs often produce noxious chemicals or retain undischarged nematocysts taken undigested from their prey.

Bivalves Bivalves (class Bivalvia) include clams, mussels, oysters, and similar molluscs. Approximately 10,000 living species are known. In

How can a tiny octopus be so deadly? The greater blue-ringed octopus, *Hapalochlaena lunulata*, is found in shallow waters off the coasts of Australia, Japan, and some Pacific islands. Its body is only about 5 cm long and it is considered to be one of the deadliest marine animals on Earth. It houses symbiotic bacteria which produce venom called tetrodoxin that acts on the nervous system of animals and causes paralysis. If a human is bitten, the paralysis can lead to respiratory arrest. A victem could survive if artificial respiration can be performed until the venom wears off. If not, the bite can be fatal. There is no antidote for the venom.

Figure 9.23 The Cooper's nutmeg snail (*Cancellaria cooperi*) seeks out electric rays, which rest partially buried in sand. It then extends its long proboscis, makes a tiny cut in the ray's skin with the radula at the end of the proboscis, and then sucks the ray's blood. A high-magnification photo (inset) shows the teeth of the radula.

Figure 9.24 Gastropods come in all shapes, colors, and habits. (a) The giant keyhole limpet (*Megathura crenulata*), from the Pacific coast of North America. (b) The red abalone (*Haliotis rufescens*). (c) Cone shells, such as *Conus geographus*, are carnivorous snails that bury themselves in sand, waiting for prey such as small fishes. (d) A flashy nudibranch (*Hermissenda crassicornis*).

(a) (b)

(c) (d)

bivalves the body is laterally compressed (that is, flattened sideways) and enclosed in a shell with two parts, or **valves** (Fig. 9.25). The **umbo**, the upper hump near the hinge of each shell, is the oldest part of the shell. Growth from the umbo proceeds in the form of concentric growth lines. There is no head to speak of, and no radula. The gills, expanded and folded, are used not only to obtain oxygen but also to filter and sort small food particles from the water. The inner surface of the shell is lined by the mantle, so that the whole body lies in the mantle cavity, a large space between the two halves of the mantle. Strong muscles, the adductor muscles, are used to close the valves.

Clams (*Macoma, Mercenaria*) use their shovel-shaped foot to burrow in sand or mud. When the clam is buried, water is drawn in and out of the mantle cavity through **siphons** formed by the fusion of the edge of the mantle (Fig. 9.25a and d). This allows clams to feed and obtain oxygen while buried in sediment.

Not all bivalves are burrowers. Mussels (*Mytilus*), for instance, secrete strong **byssal threads** that attach them to rocks and other surfaces. Oysters (*Ostrea, Crassostrea*) cement their left shell to a hard surface, often the shell of another oyster (Fig. 9.26a). Aphrodisiac or not, they have been swallowed by food lovers for thousands of years. Pearl oysters (*Pinctada*) are the source of most commercially valuable pearls. Pearls are formed when the oyster secretes shiny layers of calcium carbonate to coat irritating particles or parasites lodged between the mantle and the iridescent inner surface of the shell, which is called mother-of-pearl. Cultured pearls are obtained by carefully inserting a tiny bit of shell or plastic into the mantle. Some

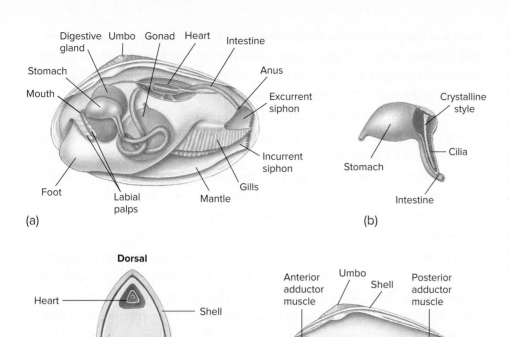

Figure 9.25 A laterally compressed body is the most distinctive feature of bivalves, illustrated here by a clam. (a) The generalized internal organs and structures of a bivalve. (b) Food is digested in the stomach with the help of the crystalline style. (c) The overall body shape of a bivalve. (d) The path of the particles from the incurrent siphon to the mouth is indicated by arrows.

scallops live unattached and can swim for short distances by rapidly ejecting water from the mantle cavity and clapping the valves (Fig. 9.26b). The largest bivalve is the giant clam (*Tridacna*), which grows to more than 1 m in length.

Many bivalves bore in coral, rock, or wood. The shipworm bores in mangrove roots, driftwood, and wooden structures such as boats and pilings. They use their small valves to excavate the wood, which is eaten. Symbiotic bacteria in the shipworm's gut digest the wood. The rasping valves lie at the inner end of a tunnel lined with calcium carbonate, and a tiny siphon protrudes from the entrance at the other end. Shipworms are an example of **biofouling**, the undesirable accumulation of organisms that settle on the bottoms of boats, pilings, and other submerged structures.

Cephalopods The cephalopods (class Cephalopoda), include the octopuses, squids, cuttlefishes, and other fascinating creatures. Cephalopods are predators that adapt the molluscan body plan to an active way of life. Nearly all are agile swimmers with a complex nervous system and a reduction or loss of the shell. All known 800 living species are marine. The foot of cephalopods (the name means "head-footed") is modified into arms and tentacles, usually equipped with suckers that are used to capture prey. The large eyes, usually set on the sides of the head, are remarkably like ours. The body, rounded in octopuses and elongated in squids, is protected by a

(a)

(b)

Figure 9.26 (a) Oysters such as *Crassostrea virginica* are harvested commercially around the world. (b) Some bivalves, like scallops and the file clam (*Lima scabra*), live free on the bottom and can swim by clapping their shells. Others bury themselves in sand or mud.

thick and muscular mantle. The mantle forms a mantle cavity behind the head that encloses two or four gills. Water enters through the free edge of the mantle and leaves through the siphon, or funnel, a muscular tube formed by what remains of the foot, which projects under the head. Cephalopods swim by forcing water out of the mantle cavity through the siphon. The flexible siphon can be moved around, allowing the animal to move in practically any direction, an example of jet propulsion in nature.

Octopuses (*Octopus*) have eight long arms and lack a shell (Fig. 9.27). They are common bottom dwellers. Including arms, the size varies from 5 cm in the dwarf octopus (*Octopus joubini*) to a record of 9 m in the Pacific giant octopus (Fig. 9.28a).

Octopuses are efficient hunters, with crabs, lobsters, and shrimps among their favorite dishes. They bite their prey with a pair of beak-like jaws. The radula can help rasp away flesh. Some may also secrete a paralyzing substance, and some have a highly toxic bite. Most, however, are harmless. They use crevices in rocks, and even discarded bottles and cans, as homes. Their shelters are identified by the presence of rocks, which they move around, and by the remains of crabs. Like most other cephalopods, they can distract potential predators by emitting a cloud of dark fluid produced by the **ink sac** (see Fig. 9.27).

Squids are better adapted for swimming than are octopuses (Figs. 9.28b and 9.29, p. 242). The body is covered by the mantle, which has two triangular fins (Fig. 9.29). The arrangement of the internal organs is similar to that of the octopus, except that the squid's body is elongated. Squids can remain motionless in one place or move backward or forward just by changing the direction of the siphon. Eight arms and two tentacles, all with suckers, circle the mouth. The tentacles are long and retractable and have suckers only at the broadened tips. They can be swiftly shot out to catch prey. The shell is reduced to a stiff **pen** embedded in the upper surface of the mantle. Adult size varies from tiny individuals of a few centimeters in length to the largest living invertebrate, the deep-sea colossal squid (*Mesonychoteuthis hamiltoni*). Very few specimens have been caught, but the largest, caught in 2007, had a total length of about 10 m. The giant squids (*Architeuthis*) are also found in deep water. One species supposedly grows to 20 m in length but weighs less than the colossal squid. Although not as large, the Humboldt, or jumbo, squid (*Dosidicus gigas*) can reach 1.8 m. Once found from Baja California, Mexico, to southern South America, it has been gradually extending its range as far north as Alaska. A voracious predator, the invading squid could threat northern Pacific fisheries.

Cuttlefishes (*Sepia*) resemble squids in having eight arms and two tentacles, but the body is flattened and has a fin running along the sides. Cuttlefishes, which are not fish at all, have a calcified internal shell that aids in buoyancy. This shell is the "cuttlebone" sold as a source of calcium for cage birds.

The chambered nautilus (*Nautilus*) has a coiled external shell containing a series of gas-filled chambers that serve as a buoyancy organ. The shell can be up to 25 cm in diameter. The body, which occupies the outer, largest chamber, has 60 to 90 short, suckerless tentacles used to capture prey.

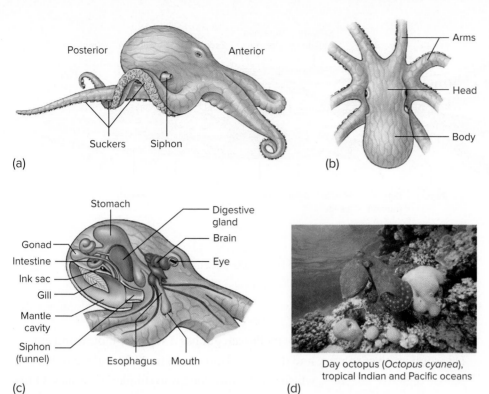

Posterior Anterior

Suckers Siphon

(a)

Arms

Head

Body

(b)

Stomach

Digestive gland

Gonad

Brain

Intestine

Eye

Ink sac

Gill

Mantle cavity

Siphon (funnel)

Esophagus Mouth

(c)

Day octopus (*Octopus cyanea*), tropical Indian and Pacific oceans

(d)

Figure 9.27 External (a, b, and d) and internal (c) structure of the octopus.

Other Molluscs Close to 600 species of chitons (class Polyplacophora) are known, all marine. They can be readily identified by the eight overlapping shell plates that cover their slightly arched dorsal surface (Fig. 9.30, p. 243). Their internal organs are not coiled as in snails.

Most chitons live on shallow, hard bottoms where they use a rasping radula to feed on algae. Many of them return to a homesite after feeding. One species, however, captures small crustaceans and other invertebrates with a flap-like extension of the mantle that surrounds the mouth.

The monoplacophorans (class Monoplacophora) are represented by only a handful of species of limpet-like molluscs. They were known only as fossils until the discovery of live individuals in 1952. They have now been collected, mostly from deep water, in scattered locations around the world. Monoplacophorans are peculiar because their gills and other organs are repeated along the body, which is reminiscent of the segmentation of annelids.

The 500 to 600 species of tusk shells, or scaphopods (class Scaphopoda), have an elongated shell, open at the top and tapered like an elephant tusk. They live in sandy or muddy bottoms. The narrow top end of the shell protrudes from the bottom, while the foot projects from the wide end. Many species have thin tentacles with adhesive tips. They are used to capture, young bivalves and other small organisms from the sediment. Tusk shells are most common in deep water, but empty shells sometimes wash ashore.

Biology of Molluscs

As you have learned, molluscs are a diverse group of organisms. They occupy a variety of habitats and ecological niches. They have wide ranging feeding and behavioral patterns to thrive across the marine environment.

(a)

(b)

Figure 9.28 (*a*) The Pacific giant octopus (*Enteroctopus dofleini*). (*b*) Mating market squid (*Doryteuthis opalescens*) from the Pacific coast of North America.

(t)Richard Carey/iStock/Getty Images, (c)Mauricio Handler/National Geographic/Getty Images, (b)Mark Conlin/Oxford Scientific/Getty Images

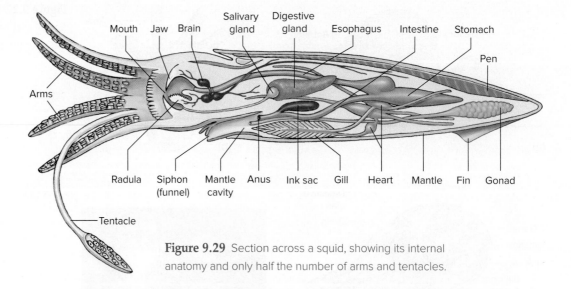

Mouth Jaw Brain Salivary gland Digestive gland Esophagus Intestine Stomach Pen

Arms

Radula Siphon (funnel) Mantle cavity Anus Ink sac Gill Heart Mantle Fin Gonad

Tentacle

Figure 9.29 Section across a squid, showing its internal anatomy and only half the number of arms and tentacles.

Feeding and Digestion The molluscan gut has a separate mouth and anus. Digestion involves salivary and digestive glands (see Figs. 9.22 and 9.29) that release digestive enzymes, which break down food into simpler molecules. Other aspects of the digestive system differ among groups and according to diet.

Grazers such as chitons, limpets, and many snails have a rasping radula that removes minute algae from surfaces or cuts through large seaweeds. Their relatively simple digestive system can efficiently process large amounts of hard-to-digest plant material. Digestion is partly extracellular in the gut cavity and partly intracellular in the digestive glands. Some shell-less gastropods that feed on seaweeds keep the seaweeds' chloroplasts intact. The chloroplasts are kept in the digestive gland, where they can photosynthesize and provide nourishment for the gastropod.

Carnivorous snails have a radula modified to drill, cut, or even capture prey. The radula and mouth are contained in a proboscis that can be protruded to strike the prey. Jaws can even be present. In these snails digestion is extracellular and takes place in the stomach.

Bivalves ingest food particles that are filtered and sorted out by the cilia on the gills. The radula is absent, and food enters the mouth trapped in long strings of mucus. An enzyme-secreting rod in the stomach, the **crystalline style** (see Fig. 9.25b), continually rotates the food to help in its digestion. The stomach contents eventually pass into a large digestive gland for intracellular digestion. Giant clams not only filter food but also obtain nutrients from zooxanthellae living in tiny branches of the gut that extend into the clam's expanded mantle. Specialized cells focus light to zooxanthellae that live below the mantle. This extra nourishment perhaps allows the clams to attain their giant size.

All cephalopods are carnivores that have to digest large prey. The stomach is sometimes connected to a sac in which digestion is rapidly and efficiently completed. It is entirely extracellular.

Molluscs have a circulatory system that transports nutrients and oxygen. A dorsal, muscular heart pumps blood to all tissues. Most molluscs have an

IN CONTEXT

Enzymes Substances that speed up specific chemical reactions.

• 6.1, The Ingredients of Life

Chloroplasts Cell organelles in plants and other primary producers in which the process of photosynthesis takes place.

• 6.2, Cellular Structure and Organization

open circulatory system in which blood flows out of vessels into open blood spaces. Cephalopods, on the other hand, have a closed circulatory system in which the blood always remains in vessels and can be more effectively directed to oxygen-demanding organs such as the brain.

Figure 9.30 Chitons use their strong foot and the flexibility provided by the eight articulated shells to fit tightly to the irregular surface of rocky shores.

Nervous System and Behavior The nervous system of molluscs varies in complexity. Gastropods and bivalves do not have a single brain but, rather, several sets of **ganglia**, or "local brains," clusters of nerve cells located in different parts of the body (see Fig. 9.22, p. 236).

Cephalopods have the most complex nervous system not just of molluscs but of all invertebrates. Some of the separate local brains of other molluscs are fused into a single, large brain (see Figs. 9.27c and 9.29, p. 241 and 242) that coordinates and stores information received from the environment. Different functions and behaviors of cephalopods are controlled by particular regions of the brain, as in humans (see "The Complex Brain of the Octopus"). Giant nerve fibers rapidly conduct impulses, allowing cephalopods to captureprey or escape at amazing speeds. The strikingly complex eyes of cephalopods reflect the development of their nervous system. Octopuses and cuttlefishes have a remarkable capacity for learning. Most cephalopods display color changes correlated with particular behaviors and moods, from intricate sexual displays to camouflage. Some cuttlefishes flash two large, black spots resembling eyes, perhaps to fool potential predators. Octopuses are known to use tools and even show play behavior. Some species change color and behavior to mimic, or imitate, poisonous fishes and sea snakes!

Reproduction and Life History Most molluscs have separate sexes, but some species are hermaphrodites, animals in which individuals have both male and female gonads. In bivalves, chitons, tusk shells, and some gastropods, sperm and eggs are released into the water and fertilization is external. Fertilization is internal in cephalopods and most gas tropods. When cephalopods mate, the male uses a modified arm to transfer **spermatophores**, elongated packets of sperm, to the female (Fig. 9.28b). Males of gastropods that copulate have a long, flexible penis.

Some molluscs have a trochophore larva like polychaetes, a characteristic often used as evidence for close affinities between molluscs, the segmented worms, and other groups. In gastropods and bivalves the trochophore usually develops into a **veliger**, a planktonic larva that has a tiny shell. In many gastropods, however, part or all of development takes place within strings or capsules of eggs. Cephalopods lack larvae, and the young develop from large, yolk-filled eggs.

Female octopuses protect their eggs sometimes for years in deep-water species, which are often attached to crevices or holes in rocks, until they hatch. The female usually dies afterward because she eats little or nothing while guarding the eggs.

The Complex Brain of the Octopus

The remarkable complexity of the nervous system and behavior of octopuses (Class: Cephalopoda) and their kin is well known. The key to this complexity is a centralized and complex brain unique among invertebrates.

Brain structure Comparisons between the octopus brain with that of vertebrates, particularly humans, are intriguing and almost inevitable. Like the vertebrate brain, the octopus brain consists of two major halves connected by nerve fibers. The ganglia, clusters of neurons (nerve cells), form lobes known to control particular functions. Like mammalian brains, some of the lobes are folded to increase their surface area and therefore their potential functions. Another similarity is that some of the lobes have miniaturized neurons, which allow concentrating more of them per volume. There are between 300 to 500 million neiruons in the brane of some octupuses, comparable to the number in a cat's brain. The number of neurons in some parts of the brain actually increases as the octopus grows in size. There are even additional neurons concentrated outside the brain, in the arms and eye region, suggesting limited brain control for the use of the arms and perhaps visual memory outside the brain.

Brain evolution The cephalopod and vertebrate brains evolved independently from each other. Their development in different but parallel paths, however, has stimulated the search for the origin of complex behavior and intelligence among animals.,and that includes us. The mapping of nerve impulses in the octopus brain, for instance, has demonstrated that the lobe where memory and learning takes place is organized very much like the region of the vertebrate brain devoted to these functions. Because the cephalopod neurons are simpler than those in vertebrates (lacking, for example, the outer sheath that speeds up impulses in vertebrates), these results suggest that what is important is how the brain is organized, not the relative complexity of its components.

Octopus Connectome Project A team of neurobiologists from several nations has started the Octopus Connectome Project, which is mapping out the neurons in the octopus brain to find out how they are connected. The study of the octopus connectome, a detailed map of the neural connections in the brain (a field known as "connectomics") hopes to unravel the detailed structure of individual neurons in the octopus brain and relate such information to how the brain works. Our understanding of neuronal function and nerve impulses began with studies of the nervous system of another cephalopod, the squid. For years the squid giant nerve fibers served as the model for studying the nerve impulse. It may just be that the relatively simpler brain of the octopus can enable us to better understand neuronal connections and how information processing in these connectomes leads to higher-level nervous functions such as learning and problem solving.

Vertical lobe
Frontal lobe
Optic lobe
Nerves to mouth region
Esophagus
Nerves to arms

Section Review

1. **Describe** What are the basic characteristics of molluscs?

2. **Summarize** What adaptations do cephalopods have to live an active way of life?

9.4 Arthropods: The Armored Achievers

Main Idea
The success of arthropods in adapting to all types of environments is due in part to a protective exoskeleton and jointed appendages.

. .

Key Questions
1. What are some of the different groups of arthropods and which animals are examples of each group?
2. What are the main characteristics of arthropods?

VOCABULARY

exoskeleton

molt

carapace

pereopods

maxillipeds

cephalothorax

abdomen

antennae

compound eyes

pleopods

nauplius

Arthropods (phylum Arthropoda) make up the largest phylum of animals, with more than a million known species and several million remaining undiscovered. Three out of four are arthropods. The largest group of arthropods by far, however, are the insects, which, though the dominant arthropods on land, are rare in the sea. The overwhelming majority of marine arthropods are crustaceans (subphylum Crustacea), a group that includes barnacles, shrimps, lobsters, crabs, and a huge variety of less familiar animals.

The arthropod body is segmented and bilaterally symmetrical. In addition to a flexible, segmented body arthropods have jointed appendages, such as legs and mouthparts, that are moved by sets of attached muscles. Another characteristic of arthropods is a tough, non-living external skeleton, or **exoskeleton**, composed of chitin and secreted by the underlying layer of tissue. The exoskeleton and jointed appendages provide protection, support, flexibility, and increased surface area for muscle attachment.

To grow, arthropods must **molt**, or shed their exoskeleton (Fig. 9.31, p. 246). A new shell develops under the old one prior to molting, then hardens after the old skeleton is discarded and the animal takes in water to expand itself. Most arthropods are small because the rigid exoskeleton imposes limitations on size. We will never see arthropods as big as giant squids or whales, but the legs of a giant spider crab (*Macrocheira*) can reach 3 m long.

The Small Crustaceans Small crustaceans are everywhere: in the plankton, on the bottom, among sediments, on and in other animals, and among seaweeds.

Copepods are extremely abundant and important in the plankton. They use their mouthparts to filter out or capture food. Some species are so common that they are among the most abundant animals on Earth. Many planktonic species keep from sinking by using their enlarged first pair of antennae (Fig. 9.32, p. 246) to swim. Many species are parasitic, some being so simplified that they look like little more than small bags of tissue.

DIVE IN!

How big were ancient marine arthropods? A recent discovery of a fossilized arthropod that lived almost 500 million years ago has scientists estimating that the marine organism was up to 2 m long—the largest organism on Earth at the time. *Aegirocassis benmoulae*, the giant segmented animal, appeared to be an arthropod, and was a filter feeder, capturing plankton and other small organisms. The discovery of the fossilized remains may help scientists learn more about how the legs of arthropods evolved.

Figure 9.31 This is not really a live Galápagos shore crab (*Grapsus grapsus*) but its exoskeleton, or molt. The old exoskeleton covered the entire external surface of the crab, even its mouthparts and eyes. After molting, which is regulated by hormones, the soft, helpless crab must find shelter for a few days until its new, larger skeleton hardens.

Crustaceans

There are approximately 70,000 known species of crustaceans, but there could be as many as 150,000 undescribed species. Most are marine. The crustacean body plan is repeated in myriad forms, from the familiar to the not-so-familiar.

Barnacles are filter feeders that usually live attached to surfaces, including living surfaces like whales and crabs. Many are very particular about the type of surface on which they live. Some are among the most important types of fouling organisms.

Common barnacles look almost like molluscs because their bodies are enclosed by heavy, calcareous plates (Fig. 9.33). The plates on the upper surface open to allow the feathery filtering appendages (cirri), which are actually the legs, to sweep the water. Some barnacles have become highly modified parasites and lack plates. All barnacles, however, have typical crustacean larvae, which swim and attach to surfaces before metamorphosing into adults.

Figure 9.32 The structurally simplest crustaceans, like this planktonic copepod, tend to be small and have appendages that are less specialized. One exception here is the first antenna, which is specialized for swimming. All appendages are paired, but the ventral view shows only those on one side.

(a)

(b)

Figure 9.33 (a) Barnacles conceal a crustacean body beneath thick plates. They lie on their backs and use their legs to filter feed. (b) The thatched barnacle (*Tetraclita squamosa*) from the Pacific coast of North America.

Amphipods are small crustaceans with a curved body that is flattened sideways (Fig. 9.34). Most amphipods are under 2 cm in length, but some deep-water species are much larger. The head and tail typically curve downward, and the appendages are specialized according to function. Beach hoppers, common in shore debris, are strong jumpers that spring about by briskly stretching their curved bodies. Other amphipods crawl among seaweeds. Burrowing in the skin of whales and living as part of the plankton are some other lifestyles of this large, mostly marine group of over 8,000 species.

Isopods are found in many of the same environments as amphipods. They are about the same size as amphipods, but isopods are easily identifiable because the main part of the body has legs that are similar to each other and the body is dorsoventrally flattened and thus has a flat back (Fig. 9.35). Pill bugs are common land isopods that are similar to many marine species. Fish lice (no relation to lice of birds and land mammals, which are insects) and other isopods are parasites of fishes and other crustaceans.

Krill, or euphausiids, are planktonic, shrimp-like crustaceans up to 6 cm long. The head is fused with some of the body segments to form a distinctive **carapace** that covers the anterior half of the body like armor. Most krill are filter feeders that feed on diatoms and other plankton. Krill are extremely common in polar waters, aggregating in gigantic schools of billions of individuals. They are an almost exclusive food source for many Antarctic whales, penguins, and fishes. Other species of krill live in the deep sea.

Figure 9.34 Most amphipods, like this beach hopper (*Orchestoidea*), can be recognized by a body that is curved and flattened sideways.

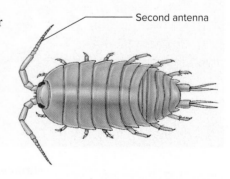

Figure 9.35 The sea louse, or sea roach (*Ligia oceanica*), is neither a louse nor a roach but an isopod. It feeds mainly on decaying seaweeds carried in to shore by the waves.

Figure 9.36 The American lobster (*Homarus americanus*) illustrates the basic body plan of decapod crustaceans. One significant omission from this drawing is the feathery gills, which lie in a chamber on each side of the cephalothorax.

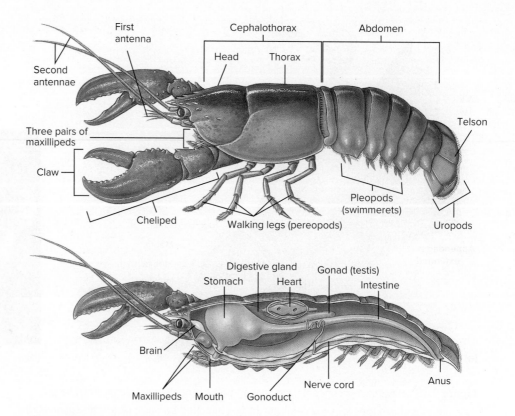

Shrimps, Lobsters, and Crabs With around 15,000 living species, the decapods (meaning "10 legs") are the largest group of crustaceans. They include the shrimps, lobsters, and crabs. Decapods are also the largest crustaceans in size. Many are prized as food and are of great commercial importance.

Decapods feature five pairs of legs, or **pereopods**, the first of which is heavier and usually has claws used for feeding and defense (Fig. 9.36). Decapods also have three pairs of **maxillipeds**, which are closer to the mouth, turned forward, and specialized to sort out food and push it toward the mouth. Maxillipeds are used as filtering devices in decapods that eat small food particles. The carapace is well developed and encloses the part of the body known as the **cephalothorax**. The rest of the body is called the **abdomen**.

Shrimps and lobsters tend to have laterally compressed bodies with distinct and elongated abdomens, the "tails" we like to eat so much. Shrimps are typically scavengers, specialists in feeding on bits of detritus on the bottom. Many colorful shrimps, particularly in the tropics, live on the surface of other invertebrates (Fig. 9.37) or remove parasites from the skin of fish. Other shrimps live in deep water. Ghost and mud shrimps burrow in muddy bottoms. Lobsters, such as the American, or Maine, lobster (*Homarus*) and the clawless spiny lobster (*Panulirus*) are mostly nocturnal and hide during the day in rock or coral crevices (Fig. 9.36). They are scavengers and predators that crush molluscs and sea urchins. Hermit crabs, which are not true crabs, are also scavengers.

Figure 9.37 This shrimp (*Dasycaris zanzibarica*) is camouflaged to live on the branching arms of crinoids in the tropical Indian and Pacific oceans.

© C.H.J.M. Fransen, The Netherlands

248 **Chapter 9** *Marine Animals Without a Backbone*

Figure 9.38 Hermit crab (*Dardanus*) with sea anemones living on its shell. The hermit crab is protected by the nematocysts on the sea anemone's tentacles, and the sea anemones get to eat food particles that are let loose when the hermit crab feeds

They hide their long, soft abdomens in empty gastropod shells. Some hermit crabs cover their shells with sea anemones (Fig. 9.38) or sponges for added protection and camouflage. One type, however, doesn't hide its abdomen (Fig. 9.39).

In the true crabs, or brachyurans, the abdomen is small and tucked under the compact and typically broad cephalothorax. The abdomen is visible as a flat, V-shaped plate in males; in females it is expanded and U-shaped for carrying eggs (Fig. 9.40, p. 250). Crabs can be highly mobile, and most can easily move sideways when in a hurry. They make up the largest and most diverse group of decapods, more than 7,000 species strong. Most are scavengers or predators, but some have specialized diets such as seaweeds, organic matter in mud, or even coral mucus. Many crabs live along rocky shores or sandy beaches, exposed to air much of the time. Land crabs, which are found in the tropics, spend most of their lives on land, returning to the sea only to release their eggs.

(a)

(b)

Figure 9.39 (a) The coconut crab (*Birgus latro*) is a large, land-dwelling hermit crab that does not use a shell as an adult. (b) Juveniles are marine, starting as planktonic larvae before settling to the sea floor. They use shells for a home as they crawl out of the sea to begin life on land.

Figure 9.40 The abdomen is V-shaped in male crabs (top) and larger and U-shaped in females (bottom). This is the European, or shore, crab (*Carcinus maenas*), an introduced pest on the Atlantic and Pacific coasts of North America and other regions around the world.

Biology of Crustaceans

Crustaceans are specialized for life in water. Most possess gills and their chitinous skeleton is usually hardened by calcium carbonate. The appendages are specialized for swimming, crawling, attaching to other animals, mating, and feeding. Crustaceans possess two pairs of **antennae** (see Fig. 9.32) in contrast to insects, which have only one pair. Antennae, are usually involved in sensing the surroundings.

Feeding and Digestion Filter feeding is very common in copepods and many of the small planktonic crustaceans. Stiff, hair-like bristles on some appendages are used to catch food particles in the water. Particles are carried to the bristles by currents induced by the beating of other appendages. Still other specialized appendages move food from the bristles to the mouth. Appendages are adapted for piercing and sucking in parasitic copepods and isopods.

Food passes to a stomach that typically has chitinous teeth or ridges for grinding and bristles for sifting. The stomach, two-chambered in decapods, is connected to digestive glands that secrete digestive enzymes and absorb nutrients. Digestion is essentially extracellular. The intestine ends in an anus.

As in molluscs, absorbed nutrients are distributed by an open circulatory system. Gas exchange is typically carried out by gills attached to some of the appendages. In most decapods the gills lie in a chamber under the carapace, where they are constantly bathed by water. In land crabs, however, the gill chamber, though moistened, is filled with air and acts almost as our lungs do.

Nervous System and Behavior The nervous system of the structurally simplest crustaceans is ladder-like, but it is more centralized in decapods. Crustaceans have a small, relatively simple brain (see Fig. 9.36) but well-developed sensory organs. Most have **compound eyes**, which consist of a bundle of up to 14,000 light-sensitive units grouped in a mosaic. The compound eyes of decapods are at the end of movable stalks and can be used as periscopes. Crustaceans have a keen sense of "smell" being very sensitive to chemicals in the water. Many crustaceans have a pair of statocysts for balance.

Crustaceans are among the most behaviorally complex invertebrates. They use a variety of signals to communicate with each other. Many of these signals involve special body postures or movements of the legs and antennae, which are often -distinctively marked or colored to make the signals more conspicuous. This type of communication has been shown to be very important in settling disputes between neighbors and in courtship. Courtship behavior can be especially elaborate.

Reproduction and Life History The sexes are separate in most crustaceans. Gametes are rarely shed into the water. Instead, males use specialized appendages to transfer sperm directly to the female. Even hermaphroditic species transfer sperm between individuals. Barnacles, for instance, have a penis that

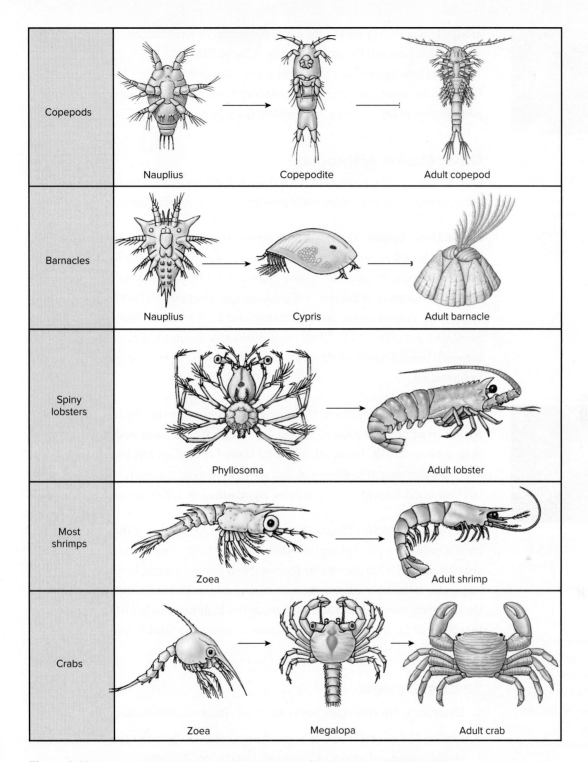

Copepods	Nauplius	Copepodite	Adult copepod
Barnacles	Nauplius	Cypris	Adult barnacle
Spiny lobsters	Phyllosoma		Adult lobster
Most shrimps	Zoea		Adult shrimp
Crabs	Zoea	Megalopa	Adult crab

Figure 9.41 The eggs of marine crustaceans hatch into planktonic larvae that undergo consecutive molts. Each molt adds a new pair of appendages, with the ones already present becoming more specialized. The last larval stage eventually metamorphoses into a juvenile. The larvae and adults shown here are not drawn to scale. In most cases, each arrow represents several molts.

can stretch to reach other barnacles in the neighborhood. Mating in decapods usually takes place immediately after the female molts, while the exoskeleton is still soft. Females of many species can store sperm for long periods and use it to fertilize separate batches of eggs. In amphipods and isopods eggs are brooded in a chamber formed by lateral extensions of the body. In decapods and other

Figure 9.42 *Limulus polyphemus* is a horseshoe crab that grows as long as 60 cm including the tail. It feeds on clams and other small, soft-bottom invertebrates.

Figure 9.43 Sea spiders, or pycnogonids, are often found on sponges, sea anemones, jellyfishes, and bryozoans. Most shallow-water species are small, with body length rarely over 1 cm. Some deep-sea sea spiders are bigger, as are species in Arctic and Antarctic waters, like the one shown here.

groups females carry their eggs using **pleopods**, or swimmerets, specialized appendages beneath the body (see Fig. 9.36, p. 248).

Most crustaceans have planktonic larvae that look nothing like the adult. Probably the most characteristic crustacean larva is the **nauplius**, but the type and number of larval stages vary widely from group to group (Fig. 9.41).

Other Marine Arthropods

Crustaceans are the most abundant group of arthropods in the oceans, but other groups play important ecological roles in the marine environment.

Horseshoe Crabs The horseshoe crabs (class Merostomata) are the only surviving members of a group that is widely represented in the fossil record. The five living species of horseshoe crabs are not true crabs but "living fossils," not unlike forms that became extinct long ago. Horseshoe crabs live on soft bottoms in shallow water on the Atlantic and Gulf coasts of North America and Southeast Asia (Fig. 9.42). Their most distinctive feature is a horseshoe-shaped carapace that encloses a body with five pairs of legs. They emerge on beaches to reproduce.

Sea Spiders Sea spiders (class Pycnogonida) only superficially resemble true spiders. Four or more pairs of jointed legs stretch from a small body (Fig. 9.43). A large proboscis with the mouth at the tip is used to feed on soft invertebrates such as sea anemones and hydrozoans. Sea spiders are most common in cold waters, but they occur throughout the oceans. Approximately 1,000 species are known.

Insects Insects (class Insecta) are distinguished from other arthropods by having only three pairs of legs as adults. They are the most diverse group of animals on Earth but are rare in the sea. Most marine insects live at the water's edge, where they scavenge among seaweeds, barnacles, and rocks. Many inhabit the decaying seaweed that accumulates at the high tide mark. One marine insect that is found far from shore is the marine water strider (*Halobates*).

· ·

Section Review

1. **Describe** What distinguishes crustaceans from other arthropods?

2. **Review** Make a table like the one shown below to describe the characteristics of marine arthropods besides crustaceans.

	Description	Habitat
Horseshoe crabs		
Sea spiders		
Insects		

9.5 Echinoderms

Main Idea
Echinoderms have unique features, including five-way symmetry and a water vascular system

. .

Key Questions
1. What is a water vascular system?
2. What are the classes of echinoderms? Name one distinguishing feature of each class.

VOCABULARY

lophophore

zooids

pentamerous

endoskeleton

water vascular system

tube feet

ampullae

madreporite

pedicellariae

evisceration

coelomic fluid

respiratory trees

regeneration

Up to this point, many of the marine invertebrates we've discussed look similar to what we see on land, and many of them have terrestrial members. The echinoderms are a wholly aquatic group, and have unique feeding and body structures not seen on land.

Echinoderms: Five-Way Symmetry

Sea stars, sea urchins, sea cucumbers, and several other forms make up the echinoderms (phylum Echinodermata). Echinoderms are radially symmetrical, like cnidarians and comb jellies. The radial symmetry of echinoderms, however, is only a secondary development. Their planktonic larvae are bilaterally symmetrical, and only the adults develop radial symmetry. Unlike cnidarians and comb jellies, echinoderms have **pentamerous** radial symmetry—symmetry based on five parts (Fig. 9.44). As might be expected in a radially symmetrical

Figure 9.44 (a) Aboral and (b) oral surfaces of a sea star (*Asterias vulgaris*) common on the Atlantic and Gulf coasts of North America. (c) Internal structure, with an arm cut across to show the relationship between tube feet, the internal sacs, and the canals that make up the water vascular system.

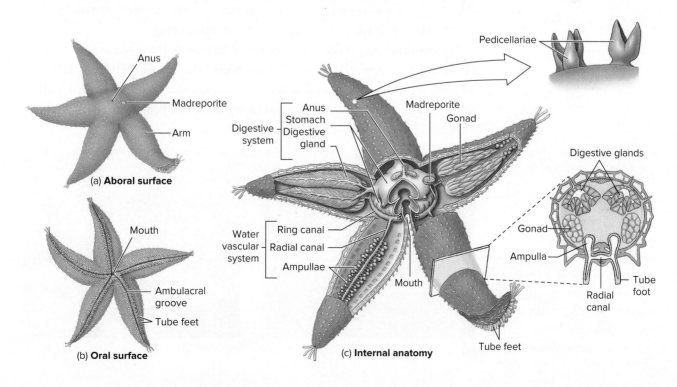

(a) **Aboral surface**
Anus
Madreporite
Arm

(b) **Oral surface**
Mouth
Ambulacral groove
Tube feet

(c) **Internal anatomy**
Pedicellariae
Anus
Stomach
Digestive gland
Digestive system
Madreporite
Gonad
Water vascular system
Ring canal
Radial canal
Ampullae
Mouth
Tube feet
Digestive glands
Gonad
Ampulla
Radial canal
Tube foot

DIVE IN!

How many arms do sea stars really have? The traditional definition of a sea star is that they have pentamerous radial symmetry, or symmetry based on five parts. Many sea stars have five arms, however, some have an even number of arms, and some can have up to 50 arms! The sunflower sea star, *Pycnopodia helianthoides*, is a large sea star found in the Pacific northwest that has between 16 and 24 arms. The nine-armed sea star, *Luidia senegalensis*, has nine arms and lives in calm, shallow waters along the Atlantic coast ranging from Florida to Brazil.

animal, echinoderms lack a head. They have no anterior or posterior end or dorsal or ventral side. It is useful instead to refer to one surface of echinoderms as oral, because that is where the mouth is located, and the opposite surface as aboral (Fig. 9.44a and b).

Echinoderms typically have a complete digestive tract, a well-developed coelom, and an **endoskeleton**, a skeleton that is internal just like ours. It is secreted within the tissues, rather than externally, like the exoskeleton of arthropods. Though sometimes it looks external, as in the spines of sea urchins, the endoskeleton is covered by a thin layer of ciliated tissue. Spines and pointed bumps give many echinoderms a spiny appearance, hence the name Echinodermata, meaning "spiny-skinned."

Unique to echinoderms is the **water vascular system**, a network of water-filled canals (Fig. 9.44c). **Tube feet** are muscular extensions of these canals. They are extended when filled with water, sometimes by the action of muscular sacs, the **ampullae**, that extend inside the body opposite the tube feet. Tube feet often end in a sucker and are used for attachment, locomotion, and the reception of chemical and mechanical stimuli. In sea stars and sea urchins the system connects to the outside through the **madreporite**, a porous plate on the aboral surface.

Echinoderms are radially symmetrical as adults. They are characterized by an endoskeleton and a unique water vascular system.

Types of Echinoderms The echinoderms are a large group of a little more than 7,000 known species, many more known only as fossils, all marine.

Sea Stars Sea stars (class Asteroidea), sometimes called starfishes, clearly display the distinctive echinoderm body plan (Fig. 9.44). Most species have five arms that radiate from a central disk, though some have more than five (Fig. 9.45a), sometimes close to 50. Hundreds of tube feet protrude from the oral surface along radiating channels on each arm called ambulacral grooves (Fig. 9.44b). Sea stars can move in any direction, though usually slowly, by reaching out their tube feet and pulling themselves along.

The endoskeleton of sea stars consists of interconnected calcium carbonate plates that form a relatively flexible framework. This allows their arms to be somewhat flexible. The aboral surface of many sea stars is often covered with spines modified into minute, pincer-like organs called **pedicellariae** (Fig. 9.44c). They help keep the surface clean.

Figure 9.45 (*a*) The sunflower sea star (*Pynopodia heliothonides*). (*b*) *Linckia guildingi*, from the Caribbean, shows a remarkable ability for regeneration; here the larger arm regenerated the central disk and four small arms to form a complete individual.

(a)

(b)

(l)© Brendan Hunter/Getty Images, (r)© Charlie Arneson

Most of the over 1,800 known species of sea stars are predators of bivalves, snails, barnacles, and other attached or slow-moving animals. Typical examples include *Asterias* (Fig. 9.44), common on rocky shores from the North Atlantic to the Gulf of Mexico, and *Pisaster*, its counterpart on the North Pacific.

Brittle Stars Brittle stars (class Ophiuroidea) also have a star-shaped body architecture. The five arms, however, are long, very flexible, and sharply demarcated from the central disk (Fig. 9.46). The swift, snake-like movements of the arms are used in locomotion, sometimes to capture food. The tube feet, which lack suckers, are used in feeding. Unlike sea stars, the gonads and other internal organs of brittle stars are contained within the central disk and do not extend into the arms.

Most brittle stars eat detritus and small animals they pick up from the bottom of the water. Particles are collected by the tube feet and passed from foot to foot to the mouth. They lack an anus. There are more species of brittle stars (nearly 2,100) than of any other group of echinoderms. Brittle stars are widely distributed but not always visible, often hiding under rocks and corals or covering themselves with mud or sand.

Sea Urchins In sea urchins (class Echinoidea) the endoskeleton forms a round, rigid, shell-like test with movable spines and pedicellariae (Figs. 9.47, p. 256). Locomotion is achieved by the movable spines, jointed to sockets in the test, and the sucker-tipped tube feet. The flat and radiating body plan of sea stars can be transformed into a sea urchin by dropping the arms to form a sphere. The five rows of ambulacral grooves with their tube feet now extend along the outer surface of the sphere (Fig. 9.48, p. 256). The mouth is on the bottom and the anus on top. The plates that make up the test can be seen in a sea urchin cleaned of spines and tissue. Bands of pores along the ambulacral grooves correspond to the bands of tube feet.

Sea urchins graze on attached or drifting seaweeds and seagrasses. In the process they also ingest detritus and encrusting animals such as sponges and bryozoans. The mouth, directed downward, has an intricate system of jaws and muscles called Aristotle's lantern (Fig. 9.48), which is used to bite off algae and other bits of food from the bottom. Five pairs of protactor muscles pull the lantern down, exposing the teeth; five pairs of retractor muscles move the lantern and teeth in. Additional muscles are responsible for grasping and other movements of the teeth.

Sea urchins are a common sight on rocky shores throughout the world. Examples are species of *Arbacia* from the Atlantic Ocean and Gulf of Mexico and *Strongylocentrotus* from most North Atlantic and Pacific coasts. Sea urchins in the tropics show an even richer variety of shapes and sizes, particularly on coral reefs.

Not all of the approximately 1,100 species in the class Echinoidea have round tests with prominent spines. Heart urchins and sand dollars (Fig. 9.47b) are adapted to live in soft bottoms by having flattened bodies and short spines.

Figure 9.46 *Ophiothrix oerstedii* is a brittle star typically found on sponges and other habitats in the Caribbean.

Figure 9.47 (a) The green sea urchin (*Strongylocentrotus droebachiensis*) is found on rocky shores and kelp forests on the Atlantic, Arctic, and Pacific coasts of North America. (b) *Mellita sexiesperforata*, the six-hole sand dollar from sandy bottoms along the southern United States and the Caribbean.

(a)

(b)

They are deposit feeders that use their tube feet and sometimes strands of mucus to pick up organic particles.

Sea Cucumbers In yet another modification of the echinoderm body plan, sea cucumbers (class Holothuroidea) are superficially worm-like (Fig. 9.49). They do not have spines and appear to lack radial symmetry. The basic body plan of sea cucumbers is elongated along the oral-aboral axis, as if they were pulled from the mouth and anus and stretched. The animal lies on one side, where the five rows of tube feet are concentrated. The oral and aboral surfaces are at the ends. The endoskeleton consists of microscopic, calcareous spicules scattered

Figure 9.48 Internal structure of a typical sea urchin, with a simplified diagram of the Aristotle's lantern.

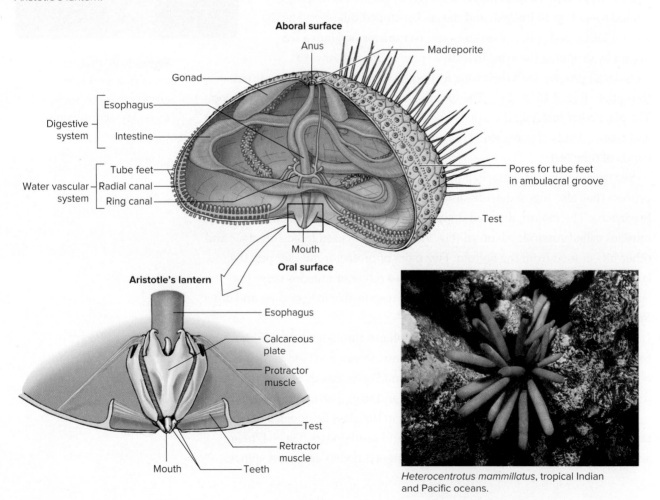

Heterocentrotus mammillatus, tropical Indian and Pacific oceans.

throughout the warty, often tough skin. Like sea urchins, most species have five rows of tube feet extending from mouth to anus. Sea cucumbers in general are among the largest echinoderms, reaching 2 m in length in some species.

Many sea cucumbers are deposit feeders. The tube feet around the mouth are modified into branched tentacles that are used to pick up organic matter from the bottom or scoop sediment into the mouth (see Fig. 9.49). Some sea cucumbers burrow or hide and extend only their tentacles to obtain food directly from the water.

Many sea cucumbers have evolved novel defensive mechanisms that compensate for the lack of a test and spines. When disturbed, some species discharge sticky, sometimes toxic filaments through the anus to discourage potential predators. Others resort to a startling response, the sudden expulsion of the gut and other internal organs through the mouth or anus, a response known as **evisceration**. It is assumed that evisceration distracts the offender while the sea cucumber, which will eventually grow back the lost organs, escapes.

Crinoids Crinoids (class Crinoidea) are suspension feeders that use outstretched, feathery arms to obtain food from the water. Crinoids are represented by a little more than 600 living species of feather stars (Fig. 9.50) and sea lilies. More than 5,000 species are only known as fossils. Sea lilies are restricted to deep water. They attach permanently to the bottom, but some are known to crawl on hard bottoms. Feather stars perch and crawl in both shallow and deep waters, especially in the tropical Indian and Pacific oceans. They are also capable of graceful swimming.

The body plan of crinoids is best described as an upside-down brittle star with the ambulacral grooves and mouth directed upward. The mouth and larger organs are restricted to a small, cup-shaped body from which the arms radiate. Some crinoids have only five arms but most have many—up to 200—because of branching of the initial five. The arms also have small side branches. Tiny tube feet along these side branches secrete mucus, which aids in catching food particles. Food makes its way into the mouth by way of ciliated ambulacral grooves. These appendages tilt the body, so that the extended arms orient to currents for efficient suspension feeding.

Biology of Echinoderms Echinoderms are important members of bottom communities from the poles to the tropics. Radial symmetry is associated with sedentary lifestyles, and with the exception of limited swimming in some feather stars and some deep-water sea cucumbers, adult echinoderms are relatively slow bottom crawlers. But they do not need to be fast to be successful.

Feeding and Digestion The digestive system of echinoderms is relatively simple. Most sea stars are carnivorous. Many feed by extending, or everting, part of their stomach inside out through the mouth to envelop the food. The stomach then secretes enzymes produced by digestive glands that extend into the arms (see Fig. 9.44c). Digested food is carried into the glands for absorption

Figure 9.49 A sea cucumber (*Thelenota rubralineata*) from Papua New Guinea.

Figure 9.50 A feather star, or crinoid, photographed at night in the Marshall Islands. The outstretched arms of feeding individuals can reach a diameter of 70 cm.

and the stomach pulled back inside the body. The intestine is short or missing. Brittle stars and crinoids also have simple, short guts. Brittle stars lack an anus.

The gut of sea urchins and sea cucumbers is long and coiled (see Fig. 9.48). In sea urchins this is an adaptation for the lengthier digestion needed for the breakdown of plant material. A long gut also is advantageous to sea cucumbers because they need to process large amounts of sediment to obtain enough organic matter.

In echinoderms nutrients are transported in fluid that fills the body cavity. The fluid, called **coelomic fluid** because the body cavity is a coelom, also transports oxygen because most echinoderms lack a distinct circulatory system. In sea stars and sea urchins gas exchange occurs across small, projections of the body wall connected at the base to the coelomic cavity. Gas exchange in sea cucumbers occurs by drawing water through the anus into a pair of thin, branched tubes called **respiratory trees**. The respiratory trees are extensions of the gut and are suspended in the body cavity. They provide a large surface area close to the coelomic fluid, allowing considerable gas exchange.

Nervous System and Behavior The presence of a nerve net in echinoderms is reminiscent of cnidarians. The nervous system coordinates movements of tube feet and spines in the absence of a brain. Nevertheless, the righting of the body after being turned over, light-sensitive organs that allow some sea stars to recognize their home reef, and camouflaging with bits of debris by sea urchins, are evidence that the nervous system could not be as simple as it looks.

Reproduction and Life History The sexes are separate in most echinoderms. The gonads, 5, 10, or more in number, are usually located in the body cavity and open to the outside by way of a duct (see Figs. 9.44c and 9.48). They shed sperm or eggs directly into the water. Gametes do not survive long in the water, so in many species individuals spawn all at once to ensure fertilization.

Development of the fertilized egg proceeds in the plankton and typically results in a ciliated larva characteristic of each group. Echinoderm larvae are bilaterally symmetrical and it is not until metamorphosis that radial symmetry develops. Some echinoderms do not have planktonic larvae and brood their offspring in special pouches or under the body.

Some sea stars, brittle stars, and sea cucumbers regularly reproduce asexually by the separation of the central disk or body into two pieces. The resulting halves then grow into complete individuals. **Regeneration**, the ability to grow lost or damaged body parts, is highly developed in echinoderms. Sea stars, brittle stars, and crinoids regenerate lost arms. In some species of sea stars a severed arm can grow into a new individual (see Fig. 9.45b). Efforts to control sea stars that prey on oyster beds or corals by cutting them up and throwing the pieces back into the water have failed miserably. In most sea stars, however, only arms that include portions of the central disk can regenerate.

· ·

Section Review

1. Summarize What are the unique characteristics of echinoderms?

2. Describe What is the process of regeneration, and why is it important?

9.6 Hemichordates and Invertebrate Chordates

DIVE IN!

What do we really know about the fossil record of lancelets? Lancelets are small invertebrate chordates, about 7 cm long. These shallow-water filter feeders have all four chordate features, which makes them significant to evolutionary biologists. Lancelets have no hard parts, so they are uncommon in the fossil record. The earliest known fossil lancelet was from the Early Cambrian, 525 million years ago, and before vertebrates evolved.. The discovery of fossils like this strengthens scientists' hypothesis that vertebrates descended from a lancelet-like ancestor. Embryological evidence and gene sequence comparisons have also shown a close relationship between lancelets and vertebrates.

The search for links between the chordates—our own phylum—and other groups of animals has been a provocative challenge and several explanations have been proposed. As strange as it may seem, echinoderms and chordates share several features related to the development of our embryos.

Hemichordates

The wide evolutionary gap between echinoderms and chordates, however, could be filled by a small and infrequently seen group of worms, the hemichordates (phylum Hemichordata). Hemichordates share the same basic developmental characteristics of chordates and echinoderms. Some hemichordates also have a larva similar to that of some echinoderms. Hemichordates share with us chordates some of the features used to define our phylum. These characteristics—a dorsal, hollow nerve cord and openings along the anterior part of the gut—will be discussed in the following section.

The hemichordates include some 120 known species. Most of these are acorn worms, or enteropneusts, worm-like deposit feeders that live free or in U-shaped tubes. Some acorn worms have been found around hydrothermal vents, often in large numbers. They generally range in length from about 8 to 45 cm but some reach 2.5 m. Like sea cucumbers, practically all acorn worms ingest sediment, using a thick, mucus-secreting proboscis to collect organic material, which is then swept toward the mouth.

Invertebrate Chordates

The chordates (phylum Chordata) are divided into three major groups, or subphyla. Two of these lack a backbone and are discussed here with the invertebrates. These invertebrate chordates are collectively called protochordates. The third and by far the largest chordate subphylum comprises the vertebrates, the subject of Chapters 10 and 11.

The estimated 44,000 to 49,000 living species of chordates share many characteristics, but four stand out. During at least part of their development, all chordates, including humans, have (1) a single, hollow **nerve cord** that runs along the dorsal length of the animal, (2) **Pharyngeal slits**, small openings along the anterior part of the gut (or pharynx), (3) a **notochord**, a flexible rod for support that lies between the nerve cord and the gut, and (4) a **post-anal tail**—that is, a tail that extends beyond the anus (Figs. 9.51b). The notochord originally served for muscle attachment but in vertebrates it became surrounded or replaced by a series of articulating bones, the **backbone**. All chordates also have a ventral heart.

Tunicates The largest group of protochordates is the tunicates (subphylum Urochordata). All 3,000 known species are marine. Those we are most apt to see are the sea squirts, or ascidians (class Ascidiacea) (Fig. 9.51). Their sac-like bodies are attached to hard surfaces, often as fouling organisms (Fig. 9.51a), or anchored in soft sediments. They are the only sessile, or permanently attached, chordates. Some sea squirts can be confused with sponges because of their general appearance. The body of the sea squirt, however, is protected by a **tunic**, a leathery or gelatinous outer covering different from that of sponges, and the internal structure is completely different (Fig. 9.52b).

Sea squirts are filter feeders. Water typically flows through the mouth, or **incurrent siphon**, and is filtered by a ciliated, sieve-like sac. This sac represents the pharynx, and the openings are derived from the gill slits. Food is filtered from the water and passed into a U-shaped or coiled gut. Filtered water is expelled through a second opening, the **excurrent siphon**. When disturbed or expelling debris, ascidians force a jet of water out of both siphons, giving them their common name of sea squirts. Some sea squirts are colonial, some consisting of clumps of separate individuals (Fig. 9.52a) and others of a circular, flower-like arrangement of individuals having a common tunic and excurrent siphon.

Were it not for their planktonic larvae, the headless, attached sea squirts could easily pass as anything but chordates. The adults possess neither a notochord nor a dorsal nerve cord (Fig. 9.52b). Ascidian larvae are known as tadpole larvae because of their superficial resemblance to the tadpoles of frogs (Fig. 9.51b). Tadpole larvae clearly display the fundamental chordate traits. They have gill slits, a dorsal nerve cord, a notochord, and a well-developed post-anal tail. They also have an eye. Tadpole larvae do not feed; their only purpose is to find a suitable surface on which to settle. During metamorphosis of a tadpole larva into a juvenile ascidian, the notochord and tail are reabsorbed and free existence is no more.

Some tunicates lead a planktonic existence throughout their lives. Salps (class Thaliacea) have a transparent, barrel-shaped body with muscle bands for

Figure 9.51 (a) *Clavelina picta* is a colonial sea squirt, or ascidian. (b) The tadpole larva of ascidians exhibits all the distinguishing characteristics of chordates. Some of these characteristics are missing in the adults.

(a)

Incurrent siphon Sensory structure with eye Excurrent siphon Anus Dorsal nerve cord

Gill slits in pharynx Heart Stomach Notochord Tail

(b)

©Lynn Samuel

Cliona intestinalis, fouling species worldwide, and *Botryllus schlosseri*

(a)

(b)

Labels in (a): *Cliona intestinalis*; *Botryllus schlosseri*

Labels in (b): Incurrent siphon; Excurrent siphon; Tunic; Genital duct; Pharynx; Anus; Heart; Intestine; Gonad (ovary and testes); Stomach

Figure 9.52 (a) *Cliona intestinalis* is a shallow-water sea squirt that has become a common fouling animal in many parts of the world. (b) Adult tunicates may look like sponges externally, but possess an entirely different internal structure.

locomotion. Water enters through the anterior mouth, or incurrent siphon, and is forced out through the excurrent siphon on the posterior end. Salps can be extremely abundant, particularly in warm water. Some are colonial, much like floating colonies of sea squirts, and can reach several meters in length. Larvaceans, or appendicularians (class Larvacea) comprise another group of planktonic tunicates. They retain the body of a tadpole larva through life. Each tiny individual secretes a complex but delicate gelatinous "house" for protection and to filter water for food.

Lancelets The second group of invertebrate chordates consists of about 30 species of lancelets (subphylum Cephalochordata). The body, up to 7 cm long, is laterally compressed and elongated like that of a fish (Fig. 9.53). The basic chordate characteristics are well developed throughout life. Only the lack of a backbone separates lancelets from vertebrates. Lancelets are inhabitants of soft bottoms. They are filter feeders, using the gill slits to capture and concentrate organic particles.

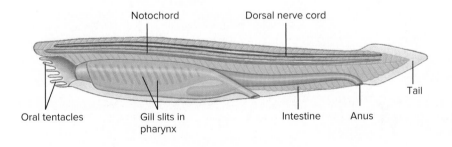

Labels: Notochord; Dorsal nerve cord; Oral tentacles; Gill slits in pharynx; Intestine; Anus; Tail

Figure 9.53 Though it looks like a fish, the lancelet (*Branchiostoma*) is an invertebrate because it lacks a backbone. The notochord, dorsal nerve cord, and gill slits show it is a chordate.

Section Review

1. **Identify** What are the characteristics of chordates?

2. **Describe** What are protochordates? What are the characteristics of organisms in each subphylum?

Table 9.1 Some of the Most Important Characteristics of the Major Animal Phyla

Phylum	Representative Groups	Distinguishing Features	General Habitat
Porifera (sponges)	Sponges	Collar cells (choanocytes)	notochord
Cnidaria (cnidarians)	Jellyfishes (sea jellies), sea anemones, corals	Nematocysts	Benthic, pelagic
Ctenophora (comb jellies)	Comb jellies	Ciliary combs, colloblasts	Mostly pelagic
Platyhelminthes (flatworms)	Turbellarians, flukes, tapeworms	Flattened body	Mostly benthic, many parasitic
Nemertea (ribbon worms)	Ribbon worms	Long proboscis	Mostly benthic, many parasitic
Nematoda (nematodes)	Nematodes, roundworms	Body round in cross section	Mostly benthic
"Chaetognatha (arrow worms)"	Arrow worms	Transparent body with fins	Mostly pelagic
Annelida (segmented worms)	Polychaetes, oligochaetes, leeches	Segmentation	Mostly benthic
Sipuncula (peanut worms)	Peanut worms	Long, retractable anterior end	Benthic
Mollusca (molluscs)	Snails, clams, oysters, octopuses, chitons	Foot, mantle, radula (absent in some groups)	Benthic, pelagic,
Arthropoda (arthropods)	Crustaceans (crabs, shrimps), insects	Exoskeleton, jointed legs	Benthic, pelagic, some parasitic
Ectoprocta (bryozoans)	Bryozoans	Lophophore, colonial	Benthic
Phoronida (phoronids)	Phoronids	Lophophore, worm-like body	Benthic
Brachiopoda (lamp shells)	Lamp shells	Lophophore, clam-like shells	Benthic
Echinodermata (echinoderms)	Sea stars, brittle stars, sea urchins, sea cucumbers	Tube feet, five-way radial symmetry, water vascular system	Mostly benthic
Hemichordata (hemichordates)	Acorn worms	Dorsal, hollow nerve cord, gill slits	Mostly benthic
Chordata (chordates)	Tunicates, vertebrates (fishes, reptiles, birds, mammals)	Dorsal, hollow nerve cord, gill slits, notochord	Benthic, pelagic

Ecology	Level of Organization	Symmetry	Segmentation	Body Cavity	Digestive Tract	Respiratory Exchange	Circulatory System
Filter feeders	Cellular	Asymmetrical	No	None	None	Body surface	None
Predators, passive suspension feeders; corals are important reef builders	Tissue	Radial	No	None	Incomplete	Body surface	None
Predators	Tissue	Radial	No	None	Incomplete	Body surface	None
Predators, many parasitic	Organ system	Bilateral	No	None	Incomplete or absent	Body surface	None
Many parasitic, deposit feeders	Organ system	Bilateral	No	Reduced (proboscis cavity)	Complete	Body surface	None
Predators	Organ system	Bilateral	No	Pseudocoelom	Complete	Body surface	Closed
Mostly pelagic	Organ system	Bilateral	No	Coelom	Complete	Body surface	None
Predators, deposit feeders, passive suspension feeders	Organ system	Bilateral	Yes	Coelom	Complete or absent	Gills or body surface	Closed
Predators	Organ system	Bilateral	No	Coelom	Complete	Body surface	None
Predators, grazers, filter feeders, some parasitic	Organ system	Bilateral	No	Coelom	Complete	Body surface	Open or closed
Predators, grazers, filter feeders, some parasitic	Organ system	Bilateral	Yes	Coelom	Complete	Gills	Open
Filter feeders	Organ system	Bilateral	No	Coelom	Complete	Gills (in many crustaceans)	None
Filter feeders	Organ system	Bilateral	No	Coelom	Complete	Body surface	Closed
Filter feeders	Organ system	Bilateral	No	Coelom	Complete	Body surface	Open
Predators, deposit feeders, passive suspension feeders	Organ system	Radial (adults) Bilateral (larvae)	No	Coelom	Complete	Body surface	None
Deposit feeders	Organ system	Bilateral	Reduced	Coelom	Complete	Body surface	Part closed, part open
Predators, grazers, filter feeders	Organ system	Bilateral	Reduced	Coelom	Complete	Gills, lungs	Closed

Chapter 9 Review

REVIEW QUESTIONS

Multiple Choice

1. All of the following invertebrate groups have tissues except
 - **a.** annelids
 - **b.** echinoderms
 - **c.** molluscs
 - **d.** sponges

2. In which phylum do organs and organ systems first appear?
 - **a.** Annelida
 - **b.** Hemichordata
 - **c.** Nematoda
 - **d.** Platyhelminthes

3. In which phylum does a coelom first appear?
 - **a.** Annelida
 - **b.** Mollusca
 - **c.** Nemertea
 - **d.** Phoronida

Short Answer

4. Compare and contrast suspension feeding and deposit feeding.

5. Choose two groups of invertebrates from this chapter and describe each group's commercial importance.

Critical Thinking

6. If bilateral symmetry were to evolve among cnidarians, in which group or groups would you expect it to occur? Why?

7. Cephalopods show a much higher degree of structural and behavioral complexity than other groups of molluscs. What factors triggered the evolution of these changes? A rich fossil record shows that once they were very common and even dominant in some environments. Now there are only about 650 living species of cephalopods, far fewer than gastropods. Were cephalopods successful? What do you think happened along the way?

DATA ANALYSIS LAB

What are the roles of jellyfishes in a marine ecosystem? In 2013, robotics designers created three prototypes of underwater jellyfish-killing robots. Although being stung by a jellyfish can be unpleasant and sometimes even life-threatening, many scientists are not enthusiastic about the idea of removing jellyfishes from marine ecosystems.

Data and Observations

The table shows facts and data about jellyfishes collected in a variety of ways.

Fact	Explanation
A	Reviewing studies that contained data on the stomach contents of fish showed that at least 69 species of fish eat jellyfish as part of their diet.
B	Leatherback sea turtles migrate several thousand kilometers from breeding grounds in tropical waters to feed on jellyfish in northern latitudes in the summer. Based on data collected by video cameras attached to 19 leatherbacks, scientists found that these turtles eat 330 kg (73% of the turtles' body weight) or about 261 jellyfish per day.
C	By collecting samples of jellyfish in the open ocean in clear glass jars, scientists found that many organisms, including juvenile fish, juvenile octopus, worms, and sea anemone larvae, cling to jellyfish, along for the ride as jellyfish float and move through the ocean.
D	The females of one group of organisms, hyperiid amphipods, live in jellyfish and use them as habitat to safely raise their juveniles.
E	Some species of tropical jellyfish, such as *Cassiopea* spp., take in species of yellow-brown algae. The algae photosynthesize, and the sugar produced by the algae is used by the jellyfish as a source of energy and nutrition.
F	Jellyfish eat copepods, other jellyfish, and fish. One study showed that barrel jellyfish sank an average of 619 m per day. The pattern of the jellyfishes' movement matched a pattern in which jellyfish could search for prey.

Think Critically

1. What roles do jellyfishes occupy in the oceans?

2. Predict what might happen to leatherback turtles if numbers of jellyfishes decreased significantly?

3. Describe the type of relationship the *Cassiopea* have with the algae they take in. Explain your reasoning.

Data obtained from: Milius, S., 2014. Seeing past the jellyfish sting. Science News, vol. 186, 22 August, pp. 16–20.

Marine Fishes

Chapter
10

THEME Patterns

Marine fishes all share defining features of their shared ancestry, as well as unique adaptations for their particular environment.

BIG IDEA

Marine fishes are a diverse assemblage of species, with many adaptations to the marine environment.

About the photo A dusky grouper (*Epinephelus marginatus*), surrounded by smaller fish, in the Mediterranean Sea.

Purestock/SuperStock

10.1 Fishes: the First Vertebrates

VOCABULARY

vertebrates

vertebrae

cartilage

placoid scales

caudal fin

dorsal fins

pectoral fins

demersal

cycloid

ctenoid

operculum

fin rays

swim bladder

Main Idea

Scientists categorize fishes into three main groups based on body structure.

Key Questions

1. Which characteristics of fishes separate them from the rest of phylum Chordata?
2. What are the three traditional groups of fishes?

DIVE IN!

What characterizes vertebrates? Most chordates are vertebrates. While vertebrates are traditionally defined by possessing vertebrae, or a spine, this is not always the case. Hagfishes, which are a type of jawless fish, do not have true vertebrae. They do, however, share several morphological and molecular characteristics with vertebrates. Because of their genetic similarities to lampreys (another jawless fish that DO have vertebrae), most biologists have placed hagfishes in the subphylum Vertebrata.

Vertebrates (subphylum Vertebrata) share four fundamental characteristics of the phylum Chordata with protochordates, the invertebrate chordates like lancelets and sea squirts. Vertebrates differ from these other chordates in having a backbone, also called the vertebral column or spine, which is a dorsal row of hollow skeletal elements, usually bone, called **vertebrae**. The vertebrae enclose and protect the nerve cord, also called the spinal cord, which ends in a brain that is protected by a skull made of cartilage or bone. Vertebrates also are characterized by a bilaterally symmetrical body and the presence of an internal skeleton, or endoskeleton.

Fishes are the oldest and structurally the simplest of all living vertebrates. They also are the most abundant vertebrates in terms of both species and individuals. About 32,000 species of fishes are presently known to science. They make up about half of all species of vertebrates on Earth. Most species of fishes, about 58%, are marine. Many new species are being discovered every year, but a good number is on the brink of extinction.

There is some disagreement as to how to classify the major groups of fishes. Three groups are traditionally recognized. One of the schemes proposed to show the relationships of these three groups with the rest of the vertebrates is given in Figure 10.1.

Jawless Fishes

The most primitive fishes living today are the jawless fishes (Agnatha). Because they lack jaws, they feed by suction with the aid of a round, muscular mouth and rows of teeth. The body is cylindrical and elongated like that of eels or snakes (Fig. 10.2). They lack the paired fins and scales of most fishes, but, like fishes and other vertebrates, the brain is protected by a skull. Some biologists do not even consider jawless fishes to be vertebrates because they lack true vertebrae.

Hagfishes Hagfishes, or slime eels (*Myxine, Eptatretus*), are jawless fishes that feed mostly on dead or dying fishes (Fig. 10.2, p. 268). They are known as the slime eel because of the abundant mucus produced by glands on the skin. The mucus is used by hagfishes to protect themselves while feeding with their

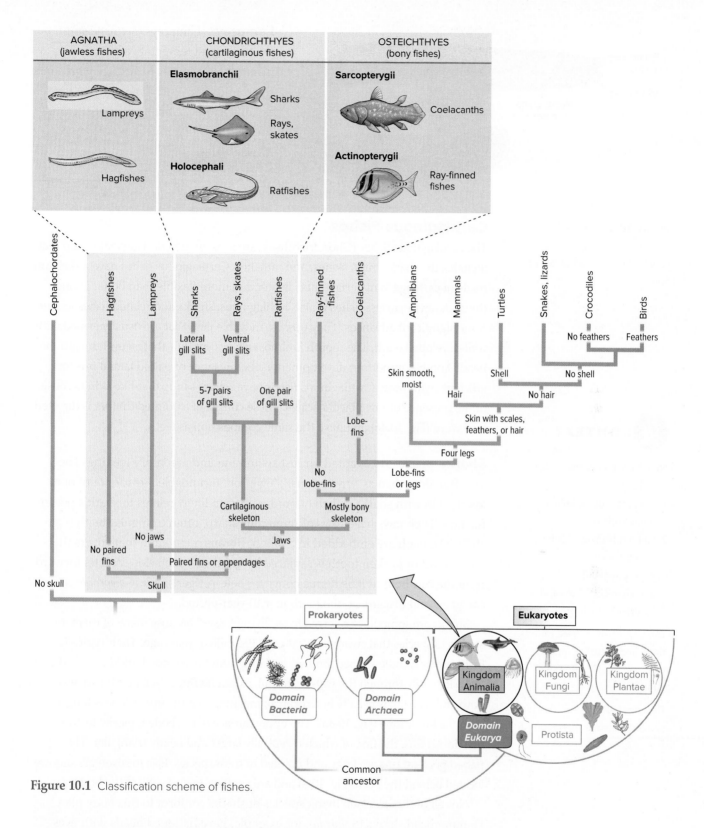

Figure 10.1 Classification scheme of fishes.

heads buried in dead fishes and whales. They sometimes bore into their prey and eat them from the inside out. Hagfishes live in burrows they dig in muddy bottoms, mostly at moderate depths in cold waters. Over 70 species are known. They reach a maximum length of approximately 80 cm. Their skin is used in the manufacture of leather goods, but they are mostly known for attacking bait or fishes on fishing lines, nets, and traps.

Gill slits

Paired tentacles

Mouth

Mucus gland pores

Figure 10.2 The Pacific hagfish (*Eptatretus stoutii*). Notice the absence of paired fins.

IN CONTEXT

Four basic characteristics of chordates:

1. A single, dorsal, hollow nerve cord
2. Gill or pharyngeal, silts
3. A notochord
4. A post-anal tail
• 9.6, Hemichordates and Invertebrate Chordates

The term "fish" is used for a single individual or for more than one individual of the same species.

The term "fishes" is used to refer to more than one species of fish.

Lampreys Lampreys (*Petromyzon*), found in most temperate regions, are primarily freshwater fishes. They breed in rivers and lakes, but some move to the sea as adults. They attach to other fishes and suck their blood or feed on bottom invertebrates. There are an estimated 30 species of lampreys.

Cartilaginous Fishes

The cartilaginous fishes (Chondrichthyes) are a fascinating and ancient group that includes the sharks, rays, skates, and ratfishes. Cartilaginous fishes have a skeleton made of cartilage, a material that is lighter and more flexible than bone. Though the skeleton of jawless fishes is also cartilaginous, sharks and related fishes feature some significant advances. They possess movable jaws that are usually armed with well-developed teeth. The mouth is almost always ventral—that is, underneath the head. Another important development is the presence of paired lateral fins for efficient swimming. Cartilaginous fishes have rough, sandpaper-like skin because of the presence of tiny placoid scales. These consist of a pointed tip that is directed backward (Fig. 10.4a) and have the same composition as teeth.

Sharks Sharks are adapted for fast swimming and predatory feeding. They are often described as "mysterious," "evil," or "formidable"—evidence of our fascination with sharks. Much of our fascination likely relates to shark's powerful jaws. Their jaws have rows of numerous sharp, often triangular teeth (Fig. 10.3). The teeth are embedded in a tough, fibrous membrane that covers the jaws. A lost or broken tooth is replaced by another, which slowly shifts forward from the row behind it as if on a conveyor belt. A tiger shark (*Galeocerdo cuvier*) can grow and discard 24,000 teeth in a 10-year-period.

Sharks are sometimes referred to as "living fossils" because many of them are similar to species that swam the seas over 100 million years ago. Their typically fusiform, or spindle-shaped, bodies, tapering from the rounded middle toward each end, slip easily through the water. The tail, or caudal fin, is well developed and powerful. The tail is usually heterocercal, meaning that the upper lobe is longer than the lower lobe (Fig. 10.4a). The upper surface of the body typically features two dorsal fins, the first of which is typically larger and nearly triangular. The paired pectoral fins are large and pointed in most species. Five to seven gill slits are located behind the head (Fig. 10.4) and are present on each side of the body.

Not all of the over 500 living species of sharks conform to this body plan. Hammerhead sharks (*Sphyrna*), for example, have flattened heads with eyes and nostrils at the tip of bizarre lateral extensions (Fig. 10.5a, p. 270). The head serves as a sort of rudder and separates the eyes and nostrils, which gives the shark binocular vision and sensory perception in practically all directions. The head of sawsharks (*Pristiophorus*) extends in a long, dorsoventrally flattened blade armed with teeth along the edges. The upper lobe of the tail is very long in the thresher sharks (*Alopias*; Fig. 10.5k). They use the tail to herd and stun

schooling fishes, which they then eat. Some sharks, like angel sharks (*Squatina*) and sawsharks (*Pristiophorus*), also have flattened bodies (Fig. 10.5b and e).

The size of fully grown adults also varies. The spined pygmy shark (*Squaliolus laticaudus*; Fig. 10.5l) grows to no longer than 25 cm. At the other extreme, the whale shark (*Rhiniodon typus*; Fig. 10.5i and 10.6) is the largest of all fishes. This huge animal, found in tropical waters around the world, can be as long as 110 m, though specimens longer than 12 m are rare. Whale sharks pose no danger to swimmers because they are filter feeders that feed on plankton. Another giant, second in size only to the whale shark, is the basking shark (*Cetorhinus maximus*; Figs. 10.5g and 10.6a, p. 272), also a filter feeder. There are reports of basking sharks 15 m long, but most do not exceed 10 m. The great white shark (*Carcharodon carcharias*; Fig. 10.5g), which is considered the most dangerous to humans, can exceed 6 m in length.

Figure 10.3 Shark jaws have several rows of triangular teeth. Individual teeth often have even smaller serrations, or denticles, along their edges. These are the jaws of a sandtiger shark (*Carcharias taurus*).

(a) **Cartilaginous fishes**

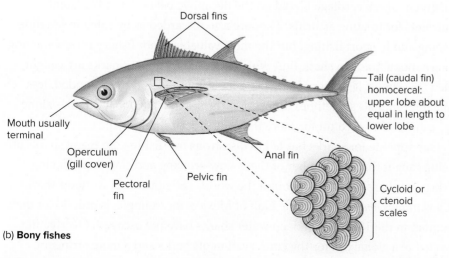

(b) **Bony fishes**

Figure 10.4 The most important external features of (a) cartilaginous and (b) bony fishes.

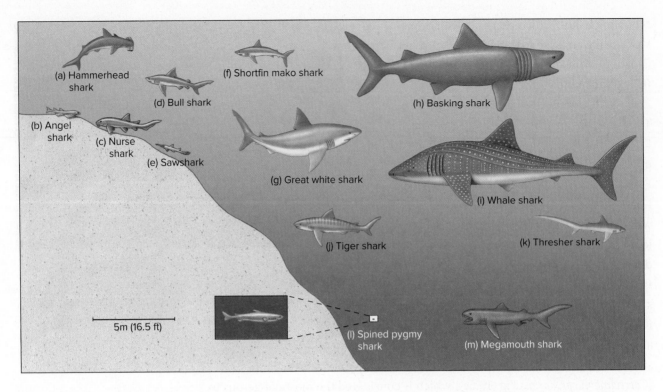

Figure 10.5 A sample of shark diversity: (a) hammerhead shark (*Sphyrna zygaena*), (b) angel shark (*Squatina californica*), (c) nurse shark (*Ginglymostoma cirratum*), (d) bull shark (*Carcharhinus leucas*), (e) sawshark (*Pristiophorus cirratus*), (f) shortfin mako shark (*Isurus oxyrinchus*), (g) great white shark (*Carcharodon carcharias*), (h) basking shark (*Cetorhinus maximus*), (i) whale shark (*Rhiniodon typus*), (j) tiger shark (*Galeocerdo cuvier*), (k) thresher shark (*Alopias vulpinus*), (l) spined pygmy shark (*Squaliolus laticaudus*) (not drawn to same scale), and (m) megamouth shark (*Megachasma pelagios*).

Sharks are found throughout the oceans at practically all depths, but they are more prevalent in tropical coastal waters. Sharks are primarily marine, but a few species travel far up rivers. The bull shark (*Carcharhinus leucas*; Fig. 10.5d) can be permanently established in some rivers and lakes in the tropics. Several sharks are found mostly in deep water (Fig. 10.6b).

Shark meat is eaten around the world. Many have tried shark without knowing because it is often illegally sold as "regular" fish or scallops. Sharks are also fished for their oil and skin, which is processed into a leather called shagreen. Shark cartilage is sold on the deceptive belief that it is a "joint nutrient" for treating arthritis. They are also discarded as by-catch in longline fishing and in sport fishing. But the most valuable shark fishery is the booming international trade in shark fins, which are used for soup. Sharks are caught, their fins cut off (a practice known as shark finning), and the wounded, less valuable, animals sometimes dumped in the sea to die. Global fin sales alone are estimated to account for more than the official shark catches.

Our appetite for sharks has led to disastrous overfishing. So many sharks are being caught that the number of these slow-growing and slow-reproducing fishes has sharply declined around the world. The number of whitetip sharks (*Carcharhinus longimanus*) in the Gulf of Mexico, for example, is only 1% of their number in the 1950s. Even deep-water sharks have not escaped. Overfishing has led to a global call for the conservation of sharks and a more stringent management of their fisheries.

Shark!

Most sharks are harmless—at least to humans. Nevertheless, 25 species of sharks are known to have attacked humans, and at least 12 more are suspected of doing so. Three are particularly dangerous: the great white, tiger, and bull sharks (see Fig. 10.5). The number of unprovoked attacks around the world was 75 in 2011, with 12 of those attacks fatal. The annual average for the decade from 2001–2010 was 64.3 (4.3 fatal). Even with an increase in tourists visiting pristine areas, shark attacks are rare. The chances of being attacked by a shark are lower than those of being struck by lightning.

Killer sharks? Shark attacks do happen: a diver in Southern California was last seen protruding from the mouth of a shark, and a series of attacks by tiger sharks has occurred in Hawai'i. Many grim stories about bleeding bodies surrounded by sharks spread during and after World War II, when the crews of torpedoed ships and downed airplanes were exposed to shark attacks. They prompted research on the aggressive behavior of sharks and the circumstances leading to shark attacks.

Great white shark attack humans wearing wetsuits because they look like their usual prey. Before attacking, sharks may display distinctive aggressive behaviors—pointing downward of the pectoral fins, pointing upward of the nose with back hunched, or the shivering of the body. Displaying sharks may attack if someone approaches.

Risk How can you decrease the risk of an attack? First, do not swim, dive, or surf in an area known to be frequented by dangerous sharks. Seal and sea lion colonies and coastal garbage dumps attract them. Blood, urine, and feces also attract sharks. Avoid murky water. Many sharks are more active at night, so avoid night swims. Sharks should not be provoked in any way. Even resting nurse sharks can turn and bite. Leave the water if fish suddenly appear in large numbers and behave erratically, which may be an indication that sharks are around. If you see a large shark, get out of the water with as little splashing as possible. If attacked, hit the shark's nose or gills with any object on hand, a spear or camera, but not with your bare hands! Do not panic.

The future of sharks We threaten the survival of sharks more than they threaten us. We should worry about the future of sharks rather than worrying about being attacked by them. They reproduce slowly, and their numbers are being depleted by overfishing. Around 100 million sharks are taken annually by commercial fisheries, mostly for their fins.

A negative attitude toward sharks is shortsighted, because they play an important role in marine communities. Overfishing of sharks in coral reefs has triggered an increase in the number of groupers and other small predators, which prey on parrotfishes and other algal feeders. Algae then thrive, growing over and killing reef-building corals.

Conservation efforts are slowly but surely being implemented. A shark sanctuary of nearly 2 milliion km2 was established in the Marshall Islands, central Pacific. CITES (Convention on International Trade in Endangered Species) approved the regulation, but not banning, of internationl trade f five shark species. A magnificent predator, the shark may soon be exterminated by humans, the fiercest predators of all.

The scalloped hammerhead shark (*Sphyrna lewini*) is not considered dangerous to humans.

(a)

(b)

Figure 10.6 (a) The great white shark (*Carcharodon carcharias*) is found along most coasts and in open water around the world. (b) The goblin shark (*Mitsukurina owstoni*) is typically found in water deeper than 100 m.

Rays and Skates The approximately 600 species of rays and skates have dorsoventrally flattened bodies and for the most part live on the bottom (Fig. 10.7). Fishes that live on the bottom are called **demersal**. Some true rays, like the guitarfishes (*Rhinobatos*), have a body that looks very much like a shark's. Only rays, skates, and related fishes, however, have their gill slits (always five pairs) on the underside of the body—that is, located ventrally (Fig. 10.7b)—rather than on the sides. The pectoral fins are flat and greatly expanded, looking almost like wings. They are typically fused with the head. The eyes are usually on top of the head.

The tropical sawfishes (*Pristis*) look very much like sawsharks but have ventral gill slits and are therefore grouped with the rays and skates. They feed by swimming through schools of fish and swinging their blades back and forth to disable their prey. They are known to grow up to 11 m long.

Many species of stingrays (Fig. 10.7a) and their relatives—the eagle, bat, and cow-nosed rays—have a whip-like tail usually equipped with stinging spines at the base for defense. Poison glands produce venom that can cause serious wounds to anyone who steps or falls on them.

Abdominal wounds from stingrays, which often occur when handling rays caught in nets, can result in death. Many stingrays cover themselves with sand, becoming nearly invisible. They feed on clams, crabs, small fishes, and other small animals that live in sediment. Stingrays have been known to damage valuable shellfish beds. They expose their food by excavating sediment with their pectoral fins. Their teeth are modified into grinding plates that crush their prey.

Electric rays (*Torpedo*) have special organs that produce electricity on each side of the head. They can deliver shocks of up to 200 volts that stun the fishes they eat as well as discouraging predators. The ancient Greeks and Romans used the shocks of electric rays to cure headaches and other ailments, the original shock treatment.

Not all rays spend their lives on the bottom. Eagle rays (*Aetobatus*) and the spectacular manta and devil rays (*Manta, Mobula*) "fly" through the water, using their pectoral fins like wings. Eagle rays return to the bottom to feed, whereas mantas feed in midwater on plankton. Both leap out of the water. The manta ray (*Manta birostris*; Fig. 10.7b) grows into a majestic giant. One individual was found to be almost 7 m wide.

Image Source/Getty Images

Figure 10.7 Rays and skates have a flattened body and ventral gill slits. Examples are (a) the yellow stingray (*Urolophus jamaicensis*) and (b) the manta ray (*Manta birostris*).

Skates (*Raja*) are similar to rays in appearance and feeding habits, but they lack a whip-like tail and stinging spines. Some have electric organs. Skates lay egg cases (similar to that shown in Fig. 10.30), whereas rays give birth to live young. Skates can be abundant, and the larger species are fished for food in some parts of the world.

Ratfishes About 30 species of strange-looking, mostly deep-water cartilaginous fishes are grouped separately because of their unique features. The ratfishes, or chimaeras (Fig. 10.8), have only one pair of gill slits covered by a flap of skin. Some have a long, rat-like tail. They feed on bottom-dwelling crustaceans and molluscs.

Bony Fishes

The great majority of fishes are bony fishes (Osteichthyes). As the name implies, they have a skeleton made at least partially of bone. There are over 31,000 species of bony fishes— about half of all vertebrates. Roughly 200 new species are described every year. A little more than half of all bony fishes live all or part of their lives in the ocean, where they are by far the dominant vertebrates.

Figure 10.8 The elephant fish (*Callorhinchus*), a ratfish, is caught commercially for food in the southern hemisphere.

The composition of the skeleton is not the only distinguishing feature of bony fishes. In contrast to the tiny, pointed placoid scales of cartilaginous fishes, bony fishes usually have **cycloid** or **ctenoid** scales, which are thin, flexible, and overlapping (see Fig. 10.4b, p. 269). Cycloid scales are smooth, whereas ctenoid scales have many tiny spines along their exposed borders. The scales are made of bone and are covered by a thin layer of skin as well as protective mucous. Some bony fishes, however, lack scales altogether. Another characteristic of bony fishes is the presence of an **operculum**, or gill cover, a flap of bony plates and tissue that protects the gills.

The upper and lower lobes of the tail, the caudal fin, are generally the same size, or homocercal (see Fig. 10.4b). The fins of bony fishes generally consist of thin membranes that are supported by bony spines, or **fin rays**, in contrast to the stiff, fleshy fins of cartilaginous fishes. Fin rays can consist of rigid spines that act as rudders or are used for protection, sometimes associated with venom glands. Some are flexible and used for propulsion and added maneuverability.

Whereas cartilaginous fishes have a ventral mouth, the mouth of most bony fishes is terminal—that is, located at the anterior end. Bony fishes have jaws with much more freedom of movement than those of sharks. The jaws of bony fishes are said to be protrusible because they can be projected outward from the mouth. The teeth are generally attached to the jawbones. Though they are usually replaced, the new teeth do not move forward in rows as do those in sharks.

Another important characteristic is the presence in many bony fishes of a **swim bladder**, a gas-filled sac just above the stomach and intestine. It allows the fish to adjust its buoyancy to keep from sinking or rising. This is a significant development that compensates for the relatively heavy bony skeleton.

Bony fishes are extraordinarily diverse in shape, size, color, feeding habits, reproductive patterns, and behavior. They have adapted to nearly every type of marine environment. All land four-legged vertebrates, or tetrapods, evolved from early bony fishes.

Section Review

1. **Determine** You've found an unfamiliar marine animal. How would you determine if it was a vertebrate or not without dissecting it?

2. **Describe** What are the main characteristics of the three major groups of fishes?

10.2 Fish Anatomy

Main Idea
Marine fishes are adapted to life in the marine environment in terms of their anatomy, including their organ systems.

. .

Key Questions
1. What are some of the specialized structures that help marine fishes live and reproduce in the ocean?
2. Why is the countercurrent system of flow important for fishes?

VOCABULARY

ichthyology

pyloric caeca

spiral valve

cloaca

gas exchange

spiracles

lamellae

countercurrent system of flow

chloride cells

urea

olfactory sacs

barbels

otoliths

lateral line

neuromasts

ampullae of Lorenzini

simultaneous hermaphroditism

sequential hermaphroditism

Discovering how fishes, cartilaginous and bony alike, have adapted to their environment so successfully is one of the objectives of **ichthyology**, the scientific study of fishes.

Body Shape

The body shape of a fish is directly related to its life-style. Fast swimmers like sharks, mackerels, marlins, and tunas (*Thunnus*, 10.9a, p. 276) have a streamlined body shape that helps them move through the water.

Laterally compressed bodies are good for leisurely swimming around coral reefs, kelp beds, or rocky reefs but are still efficient enough to allow for bursts of speed to escape from enemies or capture food. This body form is seen in many inshore fishes like snappers, wrasses, damselfishes, and butterflyfishes.

Many demersal fishes, like rays, skates, and sea moths (*Pegasus*; Fig. 10.9b), are dorsoventrally flattened. Flatfishes such as flounders, soles, and halibuts are flat and beautifully adapted to live on the bottom, but their bodies are actually laterally compressed (*Platichthys*, Fig. 10.9c). They lie on one side, with both eyes on top. They begin life with one eye on each side like other fishes, but as they develop one eye migrates up to lie next to the other one.

Distinctly elongated bodies are characteristic of freshwater eels (*Anguilla*, Fig. 10.9d , gunnels (*Pholis*, Fig. 10.9e) and snipe eels (*Nemichthys*, Fig 10.9f) among others. Eel-like fishes often live in narrow spaces in rocks or coral reefs, or among vegetation.

Many bony fishes, like seahorses depart from these generalized shapes (*Hippocampus*; Fig. 10.9g). Trunkfishes have truncate, or relatively short, bodies (*Ostracion*, Fig. 10.9h). Porcupine fish (*Diodon*, Fig. 10.9i) can inflate their bodies to almost spherical proportions when threatened.

Body shapes can be especially useful for camouflage. For example, some pipefishes live among the eelgrass they resemble. The long, thin trumpetfishes often hang vertically among gorgonian corals or tube-like sponges, or even sneak behind other fish when approaching prey. An irregular shape is often an excellent means of concealment. Bottom fishes such as blennies and sculpins have their outline broken up with irregular growths, particularly on the head, that

Figure 10.9 Body shape among bony fishes varies as an adaptation to habitat. Examples include (a) streamlined in tunas, (b) dorsoventrally flattened in the sea moth , (c) laterally flattened in flounder (d) elongated in eels, (e) ribbon-shaped in the gunnel (f) threadlike in snipe eel, (g) vertically elongated in the seahorse (h) triangular in the trunkfish, and (i) round in the porcupine fish.

(a) Tuna

(e) Gunnel

(b) Sea moth

(f) Snipe eel

(c) Flatfish (flounder)

(g) Seahorse

(h) Trunkfish

(d) Freshwater eel

(i) Porcupine fish

DIVE IN!

Can sharks really smell blood in the ocean? It is a common misconception that sharks can smell a single drop of blood in the "ocean." What is true is that sharks do have a powerful sense of smell. Unlike mammals, shark nostrils are only used for "tasting" the water, not for breathing. The olfactory area in a shark's brain is also large compared to other areas of the brain. When water containing chemicals such as oils or blood passes over the sensitive cells that line the inside of the nostrils, messages are sent to the brain for interpretation. Sharks also use their sense of smell to find a mate. Males can smell pheromones released by females, helping them to find a mate.

resemble seaweeds. The body of the stonefish (*Synanceia verrucosa*) resembles a rock so closely that it is almost invisible to both potential prey and humans (see Section 10.3). Unfortunately for humans, this shallow-water fish from the tropical Indian and Pacific oceans possesses the most potent venom known in fishes. Stepping on a stonefish is excruciatingly painful and potentially fatal.

Digestive System

After being swallowed, food passes through the pharynx and the esophagus into the stomach (Fig. 10.10). The stomach is where the process of digestion usually begins. It is typically J-curved or elongated but can be modified into a grinding structure or even lost altogether. The food passes from the stomach into the intestine. In most bony fishes the anterior portion of the intestine has many slender, blind tubes, the **pyloric caeca**, which secrete digestive enzymes. Other digestive enzymes are secreted by the inner walls of the intestine and the pancreas. Another organ important in digestion is the liver, which secretes bile needed for the breakdown of fats. The liver is particularly large and oil-rich in sharks, sometimes making up as much as 20% of their body weight.

A few fishes lack a stomach and tend to have a portion of the intestine expanded for the digestion of food. Carnivorous fishes have short, straight intestines. Fishes that eat seaweeds, which are more difficult to digest, have coiled intestines, which can be much longer than the fish itself. The intestines of cartilaginous fishes and a few primitive bony fishes contain a spiraling portion called the **spiral valve**, which increases the internal surface area of the intestine (Fig. 10.10a). The intestine is responsible for absorbing the nutrients resulting from digestion. These nutrients pass into blood to be distributed through the body. Undigested material exits through the anus, or the **cloaca**, the common passage for the digestive, excretory, and reproductive systems in cartilaginous fishes (Fig. 10.10a).

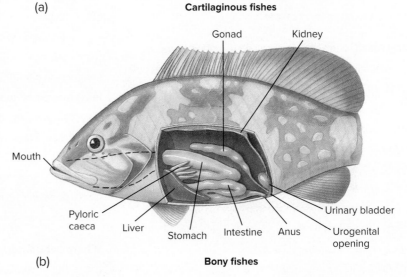

Figure 10.10 The digestive systems of (a) cartilaginous and (b) bony fishes display many of the basic features found in all vertebrates.

Circulatory System

All fishes have a two-chambered heart located below the gills (Fig. 10.11). Deoxygenated blood enters the first chamber of the heart from the body. The blood is then pumped to the second chamber, which then pumps it to the gills, where **gas exchange** takes place.

The oxygenated blood is then carried back to the body by blood vessels called arteries. The arteries branch out into thin-walled capillaries, which allow oxygen and nutrients to reach every cell. Other capillaries then collect into larger blood vessels, veins, which carry deoxygenated blood, along with dissolved carbon dioxide, back to the heart to complete the cycle.

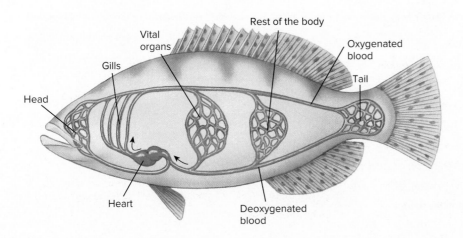

Figure 10.11 The circulatory system of fishes consists of veins that carry deoxygenated blood (in blue) from the body, a two-chambered heart that pumps blood to the gills for oxygenation, and arteries that carry oxygenated blood (in red) to the rest of the body.

A Fish Called Latimeria

In December 1938 the skipper of a fishing trawler operating in deep water off the Chalumna River in South Africa found a very strange fish in his catch. He took it to Marjorie Courtnay-Latimer, a young curator at the local museum, who recognized the fish as something special. She sent a sketch of the 1.5 m specimen to Dr. J. L. B. Smith at nearby Rhodes University, and history was made.

A living fossil The fish was a big catch, indeed. It was a coelacanth, a type of fish thought to have become extinct 60 million years ago. Previously, coelacanths were only known from fossils, some at least 400 million years old. They belong to the crossopterygian fishes, a group of fishes that might have given rise to the first land vertebrates. About 375 million years ago a crossopterygian fish with lobe-fins (paddle-like bony fins) crawled out of the water and changed life on Earth forever. There is ample evidence that all present-day tetrapods evolved from crossopterygian fishes (see Chapter 6, "When Fishes Stepped on Land,").

Dr. Smith officially described the fish and named it *Latimeria chalumnae* in honor of its discoverer and the river near which it was caught. Unfortunately, the internal organs of the fish had been discarded by the time Dr. Smith got to it, so nothing was known of its internal structure. It was not until 1952 that a second specimen was caught near the Comoro Islands, between Madagascar and mainland Africa. Ironically, the fish was well known to the locals - they dry and eat its oily flesh and use its rough skin for sandpaper!

Latimeria in the wild The fish, which is nocturnal in habits, is still very rare. None of the captured specimens has survived more than 20 hours, so little is known about its habits.

Latimeria is a large fish, up to 1.8 m in length and weighing as much as 98 kg. The body is covered by large scales. It feeds on fishes, squids, and octopuses. This living fossil is unique in many ways. The fish appears to stand on its stocky fins, but does not crawl over the bottom with them, as once thought. The pectoral fins can rotate nearly 180 degrees,

The coelacanth (Latimeria chalumnae), a living fossil.

allowing the fish to swim slowly over the bottom, sometimes standing on its head or with its belly up. Jelly-filled organs on the head could be used to detect electrical fields and thus help in prey location. Little is known about its reproduction. Females bear live young, as many as 26, and the huge eggs (about 9 cm) develop in the reproductive tract, apparently for at least a year.

Coelacanths have been found to be genetically closer to mammals, including humans, than to fishes. Evidence suggests that lungfishes, freshwater fishes with lobe fins and lungs, are closer to tetrapods than coelacanths are. Although truly "living fossils," the coelacanth genes have not remained unchanged. They have evolved but at a slower rate than in other vertebrates. There is perhaps less pressure to change adapt to changes in their very stable deep-water habitat. The close similarity of *Latimeria* to their fossil ancestors is perhaps best explained by this slow genetic change.

Where is *Latimeria* now? *Latimeria* remains a priceless catch, and several aquaria around the world would like very much to capture live specimens. As their value in dollars soars, so do the concerns about the 200 individuals that could still survive in the Comoros. International trade has now been officially outlawed. More recent catches along the Indian Ocean coast of East Africa as far north as Kenya, however, indicate a wider distribution than originally thought. The South and East African specimens are genetically similar, so they are members of the same species.

In 1997 *Latimeria* surprised everybody by turning up in a fish market on the island of Sulawesi, Indonesia, almost 10,000 km from the Comoros! Though very similar in appearance to the Indian Ocean specimens, DNA evidence showed that the Indonesian coelacanth belonged to a species different from that in the Indian Ocean. The Indonesian coelacanth was described as a new species, Latimeria menadoensis. Additional Indonesian coelacanths were filmed in deep water in 2006.

Are there any new coelacanths waiting to be discovered somewhere else?

Respiratory System

Fishes obtain oxygen dissolved in water and release carbon dioxide from their blood through paired gills. The gills lie in the pharynx, a chamber just behind the mouth that represents the front part of the gut.

Irrigation of the Gills Fishes get the oxygen they need by extracting it from the water. To do this, they must make sure that water flows over the gills—that is, they must irrigate, or ventilate, the gills.

Most sharks swim continuously. Swimming, plus the opening and closing of the mouth, forces water through the mouth, over the gills, and out through the gill slits (Fig. 10.12a). When caught in fishing nets, sharks cannot swim to force the water in and therefore "drown." Not all sharks need to swim to breahte, however. Nurse sharks and several other sharks rest on the bottom during the day. They force water over the gills by opening and closing the mouth.

Expansion and contraction of the walls of the pharynx and the gill slits assist in the irrigation of gills in sharks. Every gill lies in its own chamber, and each gill chamber opens to the outside by a separate gill slit. The first pair of gill slits of cartilaginous fishes is modified into **spiracles**, a pair of round openings just behind each eye. The spiracles are located on the dorsal surface of rays and skates (see Fig. 10.7b). They allow these fishes, many of which live on the bottom, to take in water even when the ventral mouth is buried in the sediment. When lampreys and other jawless fishes feed by sucking on other fishes, the passage of water through their mouth is blocked and they pump water in as well as out through the gill slits.

Most bony fishes have a more efficient mechanism to bring in water to the gills. The gills on each side share a gill chamber, which opens to the outside through an opening on each side of the head. Each opening is covered by an operculum (Fig. 10.12b). When the mouth opens, the opercula close and the pharynx expands, sucking water in. The fish does the reverse to pump water out: the mouth closes, the pharynx contracts, and the opercula open. Some fast swimmers simplify things by just opening their mouths to force water into the gills.

Structure of the Gills Fish gills are supported by cartilaginous or bony structures, the gill arches (Fig. 10.13, p. 281). Each gill arch bears two rows of slender, fleshy projections called gill filaments. Gill rakers project along the inner

Figure 10.12 The mechanism behind the irrigation of the gills. (a) In cartilaginous fishes, the rising of the floor of the mouth causes water to be pumped through the gills and out by way of the separate gill slits. The spiracles also allow water to move in. (b) In bony fishes, the closing of the mouth and the opening of the opercula force water through the gills and out.

(a)

(b) **Inhalation**

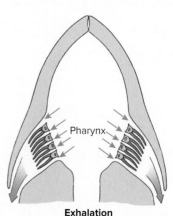
Exhalation

surface of the gill arch. They prevent food particles from entering and injuring gill filaments or can be specialized for filtering the water in filter-feeding fishes.

The gill filaments have a rich supply of capillaries (Fig. 10.13c), the oxygenated blood of which gives them a bright red color. Each gill filament contains many rows of thin plates or disks called **lamellae**, which contain capillaries. The lamellae greatly increase the surface area through which gas exchange can take place. The number of lamellae is greater in active swimmers, which need large supplies of oxygen.

Diffusion The movement of molecules from areas of high concentration to areas of low concentration.

Osmosis The diffusion of water across a cell membrane

Osmoregulation The internal control of salt and water balance

Gas Exchange Oxygen dissolved in the water diffuses into the capillaries of the gill filaments to oxygenate the blood. Diffusion will take place only if oxygen is more concentrated in the water than in the blood. This is usually true because the blood coming to the gills has already traveled through the rest of the body and is depleted of oxygen (see Fig. 10.11). As oxygen diffuses from the water to blood in the capillaries, the amount of oxygen in the surrounding water decreases as it increases in the blood. This can reduce the efficiency of gas exchange, which depends on the water having more oxygen than the blood. Fishes, however, have evolved a clever adaptation called a **countercurrent system of flow** to increase efficiency. The blood in the gills flows in the opposite direction to the water passing over them (Fig. 10.13d). When the water has passed over the gill and given up much of its oxygen, it meets blood that has just come from the body and is "hungry" for what oxygen remains in the water (Fig. 10.13f). By the time the blood has flowed most of the way through the gill, picking up oxygen, it encounters water that is just entering the gill chamber and is rich in oxygen. Thus, the oxygen content of the water is always higher than that of the blood. This system makes the gills very efficient at extracting oxygen. Without this countercurrent system, blood returning to the body would have less oxygen.

The blood disposes of its carbon dioxide using the same mechanism. Blood flowing into the gills from the body has a high concentration of carbon dioxide, a product of respiration. It easily diffuses out into the water.

Once oxygen enters the blood it is carried through the body by hemoglobin, a red protein that gives blood its characteristic color. Hemoglobin is contained in specialized cells called erythrocytes, or red blood cells. The hemoglobin releases oxygen to the tissues as it is needed. After it gives off its oxygen, the hemoglobin picks up carbon dioxide from the body and carries it to the gills, where it diffuses into the water.

Muscles use a lot of oxygen during exertion. They have a protein, called myoglobin, similar to hemoglobin, that can store oxygen. Hard-working muscles tend to have a lot of myoglobin, which makes them dark red. Strong swimmers, such as open-water sharks and tunas, have a high proportion of red muscle, as opposed to white muscle. Many other fishes have concentrations of red muscles at the base of heavily used fins.

Regulation of the Internal Environment

In contrast to most marine organisms, the blood of marine bony fishes is less salty than seawater. As a result, they lose water by osmosis. Marine bony fishes

therefore need to osmoregulate to prevent dehydration. To replace lost water, they swallow seawater (Fig. 10.14b, p. 282). Seawater contains excess salts, some of which pass straight through the gut without being absorbed. Salts that are absorbed are excreted by the kidneys, the most important excretory organs of vertebrates, and specialized **chloride cells** in the gills. The kidneys conserve water by producing only small amounts of urine.

Cartilaginous fishes use a different approach to salt balance (Fig. 10.14a). They reduce osmosis by increasing the amount of dissolved molecules, or solutes, in their blood, making the blood concentration closer to that of seawater. They do this by retaining a chemical called urea, a waste product that results from the breakdown of proteins. The amount of **urea** in the blood is controlled by the kidneys. In most animals urea is toxic and is excreted, but sharks and other cartilaginous fishes excrete only small amounts. Urea and related compounds are less toxic to cartilaginous fishes. They retain most of the urea in their blood; their gills help in this process by blocking the loss of urea.

Cartilaginous fishes also absorb water to prevent dehydration, mostly through the gills and from food. Excess salts are excreted by the kidneys, intestine, and a special gland near the anus called the rectal gland (Fig. 10.14a).

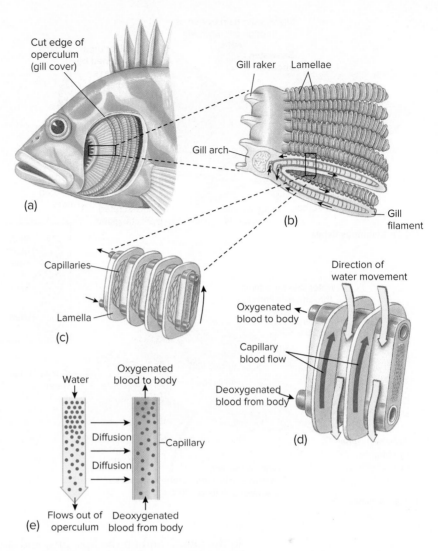

Figure 10.13 The gills of fishes are very efficient structures for gas exchange. Bony fishes have four pairs of gills (a), each containing two rows of numerous gill filaments (b). Lamellae in the gill filaments (c) increase the surface area for gas exchange. (d) Countercurrent system of flow increases oxygen diffusion into the blood. (e) The concentration of oxygen (indicated by dots) is always higher in the water than in the blood.

Nervous and Sensory Systems

Vertebrates have the most complex and advanced nervous systems of any animal group. At the heart of the system is the central nervous system, consisting of the brain and spinal cord. The central nervous system coordinates and integrates all body activities and stores information. The brain is divided into several regions known to serve as centers for particular functions such as olfaction and vision. It is protected by a cartilaginous or bony skull. Nerves connect the central nervous system with various organs of the body and with sense organs that receive information from the surroundings. This information is sent to the brain in the form of nerve impulses.

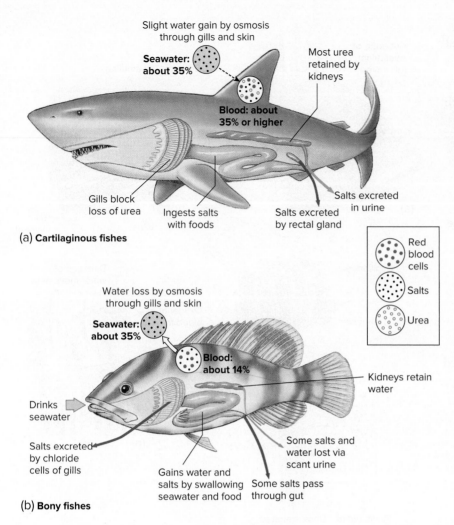

Slight water gain by osmosis
through gills and skin

**Seawater:
about 35%**

Most urea
retained by
kidneys

**Blood: about
35% or higher**

Gills block
loss of urea

Ingests salts
with foods

Salts excreted
by rectal gland

Salts excreted
in urine

(a) **Cartilaginous fishes**

Red
blood
cells

Salts

Urea

Water loss by osmosis
through gills and skin

**Seawater:
about 35%**

**Blood:
about 14%**

Kidneys retain
water

Drinks
seawater

Salts excreted
by chloride
cells of gills

Gains water and
salts by swallowing
seawater and food

Some salts pass
through gut

Some salts and
water lost via
scant urine

(b) **Bony fishes**

Figure 10.14 Marine fishes live in water that is saltier than their body fluids. (a) To prevent dehydration, cartilaginous fishes concentrate urea, absorb seawater through their gills and skin, and excrete excess salts through waste and glands. (b) Marine bony fishes conserve water with their kidneys, drink seawater, and excrete salts.

Olfaction Most fishes have a highly developed sense of smell, which they use to detect food, mates, and predators, and sometimes to find their way home. Fishes do this with special sensory cells located in **olfactory sacs** on both sides of the head. Each sac opens to the outside through one or two openings, the nostrils, or nares. The sense of smell is particularly well developed in sharks.

They can detect blood and other substances in concentrations as low as fractions of one part per million. Salmon, which live as adults at sea but reproduce in fresh water, use olfaction to find the stream where they were born years earlier. They accomplish part of this remarkable feat by memorizing the sequence of smells on their way out to sea.

Fishes detect other chemical stimuli with taste buds located in the mouth and on the lips, fins, and skin. Taste buds also are found on **barbels**, whisker-like organs near the mouth of many bottom feeders such as marine catfishes. Fishes that have barbels use them to detect food on the bottom.

Vision Fish eyes are not very different from those of vertebrates that live on land. One important difference, though, is the way they focus. Whereas the eyes of most land vertebrates focus by changing the shape of the lens, the round lens of the fish eye focuses by moving closer to or farther away from the subject. This is partially why fish eyes tend to bulge. Bony fishes appear to rely on vision more than most cartilaginous fishes. Many bony fishes—particularly shallow-water species—have color vision, but most cartilaginous fishes have little or none. Some sharks have a distinct nictitating membrane that can be drawn across the eye to reduce brightness and to protect the eye during feeding.

Hearing Fishes can also perceive sound waves with their inner ears, paired hearing organs located to the sides of the brain just behind the eyes. The inner ears are a set of fluid-filled canals that contain sensory cells similar to those in the lateral line canals. In some fishes the swim bladder amplifies sound by vibrating and transmitting sound waves to the inner ear. The inner ear is also

involved in equilibrium and balance. Many fishes detect changes in body position from the movement of calcareous **otoliths** (ear stones), that rest on sensory hairs, a mechanism similar to the statocysts of invertebrates.

Fishes have a unique sense organ called the **lateral line** that enables them to detect vibrations in the water. The lateral line consists of a system of small canals that run along the head and body (Fig. 10.15a, p. 283). The canals lie in the skin and in the bone or cartilage of the head. They are lined with clusters of sensory cells, or **neuromasts**, that are sensitive to vibration. The canals usually open to the surface through pores that are quite visible.

The lateral line system picks up vibrations resulting from the swimming of other animals as well as water displacements caused by sound waves. It allows fishes to avoid obstacles and predators, detect prey, orient to currents, and keep their position in a school.

Cartilaginous fishes also have sense organs in the head called the **ampullae of Lorenzini** (Fig. 10.15b) that can detect weak electrical fields. This system helps them locate prey. It can also help in navigation as a sort of electromagnetic compass or perhaps a detector of currents.

Reproductive System

The reproductive system of fishes is relatively simple. Both sexes have paired gonads located in the body cavity (see Fig. 10.10, p. 277). In cartilaginous fishes, ducts lead from the ovaries and testes into the cloaca, which opens to the outside (see Fig. 10.10a). Jawless and bony fishes have a separate opening for urine and gametes, the urogenital opening, which is located just behind the anus (see Fig. 10.10b).

In many marine fishes the gonads produce gametes only at certain times. The timing of gamete production is crucial. Both sexes must be ready to spawn at the same time. Spawning, as well as larval development, must take place during the period with the most favorable conditions. The exact timing of reproduction is especially critical for fishes that make long migrations to breed.

The timing of reproduction is controlled for the most part by sex hormones. Sex hormones are produced in the gonads and released in small amounts into the blood. They stimulate the maturation of gametes and can cause changes in color, shape, and behavior before breeding. The release of sex hormones is triggered by environmental factors such as day length, temperature, and the availability of food. Fishes can be artificially induced to spawn when these environmental factors are controlled or when hormones are injected. This technique is used to increase reproduction in fishes grown for food.

A few marine fishes are hermaphrodites. Some, like the hamlet, experience **simultaneous hermaphroditism** where they can produce sperm and eggs at the same time. Though able to fertilize their own eggs, simultaneous hermaphrodites usually breed with one or more other individuals, ensuring cross fertilization. One member of a pair of mating hamlets, for example, acts as a male and its sperm fertilizes the eggs released by the other individual. Roles are reversed after fertilization, and the "female" releases sperm to fertilize eggs released by

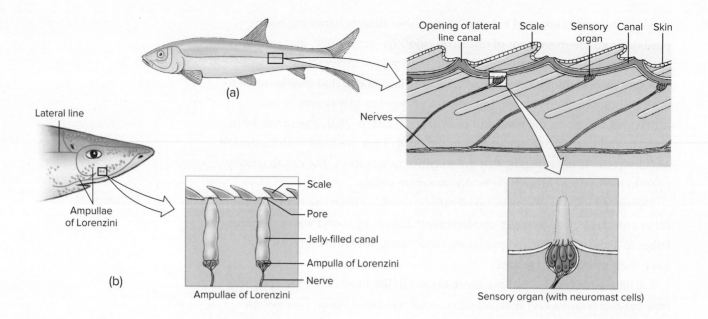

Opening of lateral line canal — Scale — Sensory organ — Canal — Skin

Nerves

Lateral line

Ampullae of Lorenzini

(a)

(b)

Scale
Pore
Jelly-filled canal
Ampulla of Lorenzini
Nerve

Ampullae of Lorenzini

Sensory organ (with neuromast cells)

Figure 10.15 (a) Cross section through the skin of a fish, showing the lateral line sensory system. The nerves collect nerve impulses from the sensory organs and transmit the information to the brain. (b) In addition to the lateral line, cartilaginous fishes have ampullae of Lorenzini, a network of jelly-filled canals capable of detecting electrical fields in the water.

Hormones Molecules that act as chemical messengers within the body.

• 6.1, The Ingredients of Life

the "male." Hermaphroditism also is found among several deep water fishes, an adaptation to the depths of the ocean where it can be difficult to find members of the opposite sex.

A variation of hermaphroditism among fishes is sex reversal or **sequential hermaphroditism**, in which individuals begin life as males but change into females (protandry) or females change into males (protogyny). These changes are controlled by sex hormones but triggered by social cues such as the absence of a dominant male. Sex reversal occurs in several groups of marine fishes, but it is most prevalent among sea basses and groupers, parrotfishes, and wrasses. Some rather complicated reproductive strategies have been discovered among these fishes.

In at least some species of anemone-fishes, all individuals begin as males. Each sea anemone is inhabited by a single, large female that mates only with a large, dominant male.

All the other fishes that live on the anemone are small, non-breeding males. If the female disappears or is experimentally removed, her mate changes into a female and the largest of the non-breeding males becomes the new dominant male. The new female can start spawning as soon as 26 days after her sex change. Males of some wrasses form harems of many females. If the male disappears, the largest, dominant female immediately begins to act as a male and within a relatively short period of time changes color and transforms into one that is capable of producing sperm.

· ·

Section Review

1. **Explain** How do marine fishes stay hydrated in such a salty external environment?

2. **Compare** What adaptations are common across groups of marine fishes?

10.3 Fish Adaptations

Main Idea
Fishes have several adaptations that have allowed them to thrive in a variety of marine habitats and become the most abundant type of vertebrate on the planet.

. .

Key Questions
1. What are some of the adaptations that help marine fishes live and reproduce in the ocean?
2. What are some of the behavioral adaptations that increase the fitness of fishes?

Marine fishes have anatomical and systematic adaptations to the ocean environment. Their abilities to regulate their internal environment and sense their external environment are vital to their survival. In addition, marine fishes have adaptations in coloration, movement, and behavior that have allowed them to radiate into the majority of marine niches.

Coloration
Some bony fishes use color for camouflage, but others, particularly those living in the tropics, are among the most brightly colored animals in the sea. The colored pigments in bony fishes are mostly found in special cells in the skin called **chromatophores**. These cells are irregular in shape and have branches radiating from the center. The amazing variety of colors and hues observed among marine fishes results from combinations of chromatophores with varying amounts of different pigments. Many fishes can rapidly change color by contracting and expanding the pigment within the chromatophores. Fishes can also have **structural colors** that result when a special surface reflects only certain colors of light. Most structural colors in fishes are the consequence of crystals that act as tiny mirrors. The crystals are contained in special chromatophores called **iridophores**. The iridescent, shiny quality of many fishes is produced by structural colors in combination with pigments (Fig. 10.16b, p. 286).

Colors can tell us a lot about fishes. Some change color with their mood or reproductive condition. They can also use color to advertise the fact that they are dangerous, are poisonous, or taste bad—a phenomenon known as **warning coloration** (Fig. 10.16c). **Cryptic coloration**, blending with the environment to deceive predators or prey, is a common adaptation (Fig. 10.16a). Flatfishes and some blennies, sculpins, rockfishes, and others can change color to match their surroundings. Another use of color is **disruptive coloration**, the presence of color stripes, bars, or spots that help break up the outline of a fish. These and other uses of color are especially common among coral reef fishes .

Figure 10.16 The variety of forms and habits among marine bony fishes is spectacular. (a) The stonefish (*Synanceia verrucosa*) uses cryptic coloration to look like stone. (b) The queen parrotfish (*Scarus vetula*) is an example of a fish with dramatic structural colors. (c) A colorful bar-tailed lionfish (*Pterois radiata*) advertises its venomous properties through warning coloration.

(a) (b) (c)

DIVE IN!

Why are fishes so successful? From an evolutionary perspective, marine fishes are a great success - they are the most abundant vertebrates on the planet. Fishes have numerous adaptations that allow them to fill many niches in the marine environment. The remoras and the sharks they live in association with are good examples. Sharks have fins and tails designed to propel them through the water and give them lift, and they are largely carnivorous, with sharp teeth and large jaws. In contrast, many species of remoras attach themselves to a host (often sharks, whales, or billfishes), catching a ride and feeding on the parasites, food scraps, and feces of their host.

Open-water fishes and shallow-water predators are rarely as colorful. Most of them have silver or white bellies in sharp contrast to dark backs. This distinctive color pattern, known as **countershading**, is a form of disguise in open water. Deep-water fishes also use color for concealment. They tend to be black or red, both of which are hard to see in the ocean depths.

Locomotion

Swimming is obviously a major part of the life of fishes. Fishes swim to obtain food, escape from predators, and find mates. Many cartilaginous and some bony fishes must also swim to flush their gills with water to obtain oxygen.

Most fishes swim with a rhythmic side-to-side motion of the body or tail. S-shaped waves of contractions moving from head to tail push against the water and force the body forward. Variations on this theme are illustrated in Figure 10.17.

The rhythmic contractions are produced by bands of muscle called myomeres, which run along the sides of the body (Fig. 10.18). The distinctive bands of muscle are easily seen in fish fillets. Myomeres are attached to the backbone for support. Muscles make up a large percentage of the body weight of a fish— as much as 75% in tunas and other active swimmers.

Sharks tend to sink because they lack the buoyant swim bladder of most bony fishes. To compensate, they have large, stiff pectoral fins that provide lift somewhat like the wings of a plane (Fig. 10.18a). The longer upper lobe of the tail tends to tilt the body upward, also generating some lift. The large amount of oil in the huge liver also provides buoyancy because it is less dense than water. In rays and skates the tail is greatly reduced and the pectoral fins are the main source of both thrust and lift, as in bird wings.

Because most bony fishes have a swim bladder, they do not have to rely on their pectoral fins to provide lift. The pectorals are thus free to serve other purposes.

Figure 10.17 Locomotion strategies in marine fishes include (a) full-body undulations as in eels, (b) flexing the caudal portion of the body as in tuna, (c) using just the fins as in surgeonfishes, or (d) using just the tail as in trunkfishes.

This gives bony fishes great maneuverability. Some can hover in the water or even swim back-ward, things that sharks cannot do. The other fins of bony fishes also help provide maneuverability. The dorsal and anal fins (see Fig. 10.4b) are used as rudders, at least part of the time, to steer and provide stability. The paired pelvic fins (see Fig. 10.4b, p. 269) also help the fish turn, balance, and "brake."

The flexibility of their fins has allowed many bony fishes to depart from the standard undulating style of swimming. Some open-water bony fishes, such as tunas, emphasize sheer speed. Many fishes, particularly those living around coral reefs, rocks, or kelp beds, swim mainly by moving their fins rather than their bodies, especially for the precise movements needed in feeding. Their tails are used almost solely as rudders. Wrasses, surgeonfishes, and parrotfishes, which all live on coral reefs, and the California sheephead (*Semicossyphus pulcher*), common in kelp beds, swim mainly with their pectoral fins. Triggerfishes undulate their dorsal and anal fins to swim. This style is perfect for hovering over the bottom while hunting for crabs and sea urchins. Flying fishes have greatly expanded pectoral fins, which they use to glide through the air. A large variety of bottom fishes (gobies, sculpins, and many others) crawl or rest on the bottom on their pectoral and/or pelvic fins. Clingfishes are small fishes that have their pelvic fins modified into part of a sucker that allows them to attach to rocks. Remoras attach to sharks, whales, turtles, and many types of large fishes, using a large sucker on top of their heads. This sucker is derived from part of the dorsal fin.

Feeding

Much like body shape can tell a great deal about where fishes live, the mouth shape and position offers many clues about how and what fishes eat.

Figure 10.18 Cartilaginous and bony fishes use different adaptations to maintain their position in the water. (a) Sharks use their fins for lift. (b) Many bony fishes have evolved a gas-filled swim bladder to compensate for their heavier bony skeleton.

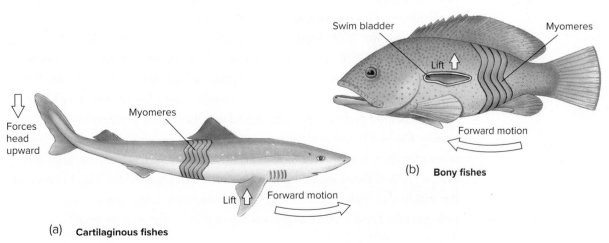

(a) **Cartilaginous fishes**

(b) **Bony fishes**

Cartilaginous Fishes Most sharks are carnivores, but in contrast to typical carnivores, which capture smaller prey, some sharks feed by taking bites from prey larger than themselves. For this they use their formidable jaws coupled with shaking of the head. Some sharks are not particular; almost anything can be found in a shark's stomach. A tiger shark that was caught off South Africa was found to contain the front half of a crocodile, the hind leg of a sheep, three gulls, and two cans of peas. Nurse sharks feed mostly on bottom invertebrates, including lobsters and sea urchins. Some deep-water sharks subsist mainly on squids. Cookie-cutter sharks are small, slow-moving, deep-water sharks that attack larger fishes and dolphins and cut out chunks of flesh with their razor-sharp teeth and sucking lips. They also eat whole squids. Scientists have suggested that they use bioluminescence to lure their prey. Even the rubber sonar domes of nuclear submarines have not escaped their bites.

Several species of cartilaginous fishes are filter feeders: whale sharks, basking shark, manta and devil rays, and the megamouth shark, a gigantic deep-water shark discovered off Hawai'i in 1976. Like other filter-feeding fishes, basking sharks filter the water with their gill rakers. Whale sharks have filter plates made of modified placoid scales. The large mouths of the three filter-feeding sharks have many small teeth and, excluding the megamouth, very long gill slits. The width of the spaces between the gill rakers or filter plates determines the size of the food captured. Water is strained through the gill rakers, and the shark swallows the food that is left behind. Whale sharks feed in warm water on small schooling fishes, squids, and planktonic crustaceans. Basking sharks, which live in colder water, feed on plankton by opening their mouths and slowly swimming through the water (see Fig. 10.5h, p. 270).

Mantas feed on plankton and small fishes, filtering them from the water with their gill rakers. Two fleshy, horn-shaped projections on the sides of the manta's mouth help channel food into the cavernous mouth (see Fig. 10.7b).

Bony Fishes Bony fishes are very diverse in the ways they feed. Their protrusible jaws allow them much more flexibility in feeding habits than sharks and rays have. Most bony fishes are carnivores; most groups of animals in the ocean are subject to being eaten by some bony fish. Bony fishes capture their prey from sediments, the water column, the surface of rocks, or other organisms, including other fishes, or combinations of these. Some chase their prey; others sit and wait.

Carnivorous bony fishes, such as the barracuda (*Sphyraena*, Fig. 10.19a) typically have well-developed teeth for catching, grasping, and holding their prey, which is usually swallowed whole. The roof of the mouth, gill rakers, and pharynx can also have teeth to help hold the prey. Deep-water fishes often have huge mouths and teeth, and a few capture and swallow prey larger than themselves.

Unusual food preferences have evolved in fishes. Some prefer sponges, infrequently touched by other carnivores. Others prefer sea urchins, sea squirts, or other seemingly tough-to-eat food items. Reef corals, skeleton and all, are eaten by several types of fishes including some butterflyfishes (*Chaetodon*, Fig. 10.19b) and parrotfishes, though only the living tissue is digested. Many species, however, are non-specialists and capture a wide variety of prey. Some feed on small

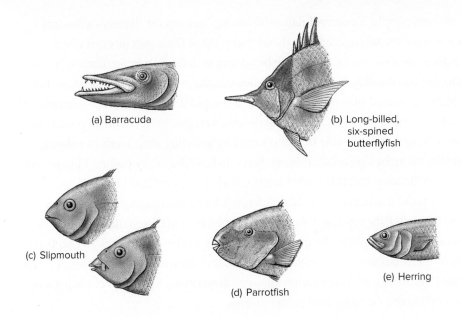

Figure 10.19 The shape of the mouth of bony fishes tells much about their diets. (a) Barracudas are carnivores that takes bites out of prey. (b) Butterflyfishes feed on very small prey. (c) Slipmouths feed on relatively small prey. (d) Parrotfishes graze on small algae and coral. (e) Herrings filter plankton out of the water.

(a) Barracuda

(b) Long-billed, six-spined butterflyfish

(c) Slipmouth

(d) Parrotfish

(e) Herring

invertebrates and dead animal material from the bottom. These bottom feeders have a downward-oriented mouth adapted to suck food from the bottom. Frog-fishes and anglerfishes use a modified spine on the head to lure small fishes.

Fishes that feed primarily on seaweeds and plants are known as grazers. Parrotfishes (*Scarus*, Fig. 10.19d), for example, graze on small algae growing on hard surfaces. Their front teeth are fused to form a beak-like structure. Some species use the beak to scrape off bits of live coral.

Fishes such as herrings (*Clupea*, Fig. 10.19e), sardines, anchovies, and menhadens filter plankton with their gill rakers. They typically strain their food by swimming with their large mouths open. These plankton feeders are small, in contrast to the huge plankton-feeding sharks. They usually occur in large, often immense, schools. Plankton feeders are the most abundant fishes in the ocean, and they are an important food source for many types of carnivores. They also account for a large share of the world's fish catch.

Behavior

Nearly all aspects of the lives of fishes involve complex behavior to adapt to light and currents, to find food and shelter, and to avoid predators. Behavior is also an important part of fish courtship and reproduction, which are described in the next section. This section provides only a glimpse of some aspects of fish behavior; some other highlights are discussed in other chapters.

Territoriality Many marine fishes, particularly open water species, do not reside in any particular area. Others, however, are known to establish territories, home areas that they defend against intruders. Some fishes defend territories only during reproduction. Many others have more or less permanent territories that they use for feeding and resting or as shelter. It is thought that fishes often guard territories to ensure that they have enough food and other resources. Territoriality is therefore most common in crowded environments like kelp beds and coral reefs, where resources are most likely to

be in short supply. Coral reef damselfishes are famous for fiercely defending their territories, often attacking fishes many times their size or even divers.

Fishes use a variety of aggressive behaviors to defend their territories. Actual fights are surprisingly rare. Instead, fishes usually prefer to avoid risking injury by bluffing. Raised fins, an open mouth, and rapid darting about are examples of such threatening postures. Territorial defense can also involve sound production. Marine bony fishes are able to make sound by grinding their teeth or rubbing bones or fin spines on another bone. Some fishes "drum" by pulling muscles on the swim bladder, and this sound is amplified by the air-filled bladder.

Sometimes a solitary individual defends a territory. In other species, like some butterflyfishes, territories are established by a male-female pair. Territories can also be inhabited by groups that belong to the same species. This is the case in many damselfishes, which inhabit spaces between the branches of corals (Fig. 10.20), and anemonefishes, or clownfishes. Members of such groups often divide the territory into subterritories.

Schooling Many fishes form well-defined groups, or **schools**. Some fishes, including herrings, sardines, mullets, and mackerels, school throughout their lives. Others are part-time schoolers, especially as juveniles or during feeding. Most cartilaginous fishes are solitary, but a few, such as hammerhead sharks, mantas, and other rays, sometimes travel in schools. It has been estimated that at least 4,000 species, including marine and freshwater species, school as adults. Schools can be huge, as large as 4,580 million m^3 in the Atlantic herring (*Clupea harengus*). The members of a school are typically all about the same size. The stationary schools that are common around coral reefs, kelp beds, rocks, and shipwrecks, however, can include members of different sizes or even different species.

Schools function as well-coordinated units, though they appear to have no leaders (Fig. 10.21). The individual fishes tend to keep a constant distance between themselves, turning, stopping, and starting in near perfect unison. Vision plays an important role in the orientation of individuals within a school. In some species, though, blinded fish can school in a coordinated way. These

Figure 10.20 Some damselfishes, such as the blue chromis (*Chromis cyanea*) from the Caribbean, inhabit group territories among corals.

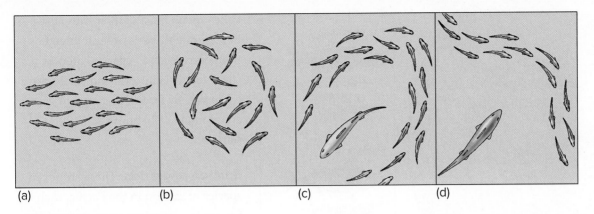

Figure 10.21 Fishes follow different recognizable patterns when schooling. Some common patterns include (a) traveling, (b) feeding on plankton, (c) encirclement of a predator, and (d) streaming to avoid a predator.

fish probably use the lateral line, olfaction, and sound they emit to keep track of each other. The tight coordination of schooling fishes can break down when they are feeding or attacked by a predator.

Why do fishes school? One explanation is that schooling offers protection against predation. Predators can be confused if, for instance, the school circles a predator (Fig. 10.21c) or splits into several groups. It is also difficult for a predator to aim for just one fish in a cloud of shifting, darting individuals (Fig. 10.21d). On the other hand, some predators, such as jacks, are more efficient when they attack schools of prey rather than individuals. Schooling could increase swimming efficiency because the fish in front form an eddy that reduces water resistance for those behind. There is experimental evidence, however, that this is not always the case and that fish do not always align in a hydrodynamically efficient way. In at least some fishes, schooling is advantageous in feeding or mating. There is probably no single reason that fishes school, and the reasons probably vary among species.

Migration Migration is the regular mass movements from one place to another once a day, once a year, or once in a lifetime. Schools of parrotfishes and other fishes migrate daily onshore and offshore to feed. Many open-water fishes migrate several hundred meters up and down the water column every day. The most spectacular migrations, however, are the transoceanic journeys made by tunas, salmon, and other fishes. We still know little about why fishes migrate, but most migrations seem to be related to feeding or reproduction.

Feeding is one reason behind the migration of open-water species like tunas. Recaptures of tagged fish have provided much information on how far, how fast, and when tunas migrate. Though essentially tropical, some species migrate long distances to feed in temperate waters. Such is the case in the skipjack tuna (*Katsuwonus pelamis*; Fig. 10.22, p. 292) and other tunas. Some sharks, like the ones discussed in the "Marine Science in Action" feature above, are now known to undertake long migrations.

Also amazing are the migrations between the sea and freshwater undertaken by some fishes that are dependent on fresh water for reproduction. **Anadromous**

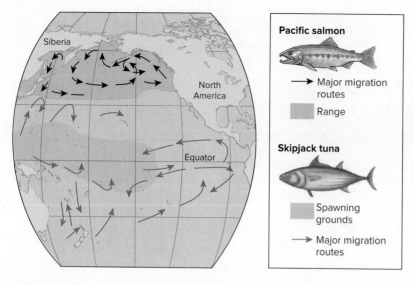

Pacific salmon

→ Major migration routes

Range

Skipjack tuna

Spawning grounds

→ Major migration routes

Figure 10.22 The skipjack tuna (*Katsuwonus pelamis*) undertakes extensive migrations that every year take it almost halfway across the globe. The Pacific salmon (*Oncorhynchus*) migrates to spawn only once in a lifetime. Approximate limits of distribution and migration routes indicated are for all seven species of salmon.

fishes spend most of their lives at sea but migrate to fresh water to breed. Sturgeons, whose eggs are eaten as caviar, some lampreys, and smelts are examples. By far the best known anadromous fish, however, is the salmon.

Salmon migration There are seven species of salmon in the Pacific, each known by several common names. They spend their adult lives in the North Pacific, traveling thousands of miles in vast sweeps along the coast, the Aleutian Islands, and the open sea (Fig. 10.22). Some even venture into the Bering Sea and the Arctic Ocean. How they navigate at sea is not known. It has been hypothesized that they use land features at least part of the way. Currents, salinity, temperature, and other water characteristics might provide clues. Other possibilities include orientation to polarized light, the sun, or the earth's magnetic field.

After several years at sea, salmon mature sexually and start migrating into rivers. They are probably guided at first by Earth's magnetic field. They do not feed once they enter fresh water, living instead on stored fat. Their kidneys must adjust to the change from salt to fresh water. Eventually, they reach the exact stream of their birth, sometimes far upstream (Fig. 10.23). The chinook, salmon (*Oncorhynchus tschawytscha*) and chum salmon (*O. keta*) reach as far inland as Idaho and the headwaters of the Yukon River. Salmon find their home streams with remarkable accuracy, using a type of chemical memory. They recognize not only the "smell" of their own stream but also those of other

Figure 10.23 The journey of Pacific salmon such as these sockeye salmon (*Oncorhynchus nerka*) is an arduous one, indeed. The fish have to swim far upstream, often leaping over rapids and waterfalls. Many fall prey to hungry predators or to the nets and lines of fishers.

©Frank Huber

Great White Shark Migrations

It was long thought that great white sharks (*Carcharodon carcharias*) spend a good part of their lives prowling along coasts hunting sea lions, seals, and large fishes. Current research in several parts of the world has nevertheless shown a new insight: great white sharks surprisingly travel long distances in regular, frequent migrations across the open ocean.

Following sharks Electronic tags and satellite technology allow scientists to record the long-term movements of large marine animals such as the great white shark. Individual sharks are tagged using a long pole while they swim along a boat, and photos of the dorsal fin are taken to record notches and other markings for individual identification. Photos are also taken to determine size and sex. The use of "pop-up" tags is a major advantage. They remain attached to a shark before being released at a pre-programmed time and float to the surface to transmit data on location, depth, and water temperature at previously determined intervals along the route. The data are sent to Argos (Advanced Research and Global Observation Satellite), a satellite-based system that supplies positions around the globe, which then relays the information to a lab on land.

Where do sharks go? Great whites tagged with pop-up tags off the central California coast were tracked to an area 2,400 km east of the Hawaiian Islands. The migratory route from California to the central Pacific followed a similar route during five years of tracking. The start of the migration was timed with a decrease in the abundance of sea lions and seals along the coast, suggesting that food was a factor in triggering migrations. Individual sharks remained in the central Pacific for as long as four months before returning to California. It is speculated that sharks gather in the area to feed on marine mammals, deep-water squids, and large fishes. A different study tracked great whites from off northern Baja California, Mexico, to the same central Pacific location where the California sharks gathered. Sharks returned to the same areas in California and Mexico, showing homing behavior. Long-lived batteries more recently allowed the tracking of females along the same route, suggesting that they give birth to their young upon returing to Mexico after an 18-month gestation period spent traveling to the central Pacific. Males, on the other hand, return to Mexico every year.

A similar study recorded the transoceanic migration of a shark (called "Nicole" after actress Nicole Kidman, known for her love of sharks) from South Africa to western Australia and back, a total of about 20,000 km, a trip that took under nine months to complete. The estimated minimum speed of this migration was 4.7 km/hr, not much different from the fastest tunas. Tagging has also shown that great whites undertake long migrations along the coasts of Australia as well as patrolling along the coast.

Future research One unanswered question is the role of long-distance migrations in the reproduction of great whites. So far there does not seem to be a connection with reproduction, but DNA analysis should help identify the genetic makeup of great white populations in different continents for determining the genetic parentage of individuals. How great whites orient during their transoceanic migrations without any clues from landmasses or islands is another provocative question. Of particular importance is that the discovery of long-distance migrations in the great white shark reveals that the species, which is listed as vulnerable in the IUCN Red List of threatened species, is particularly menaced while away from coasts where it is legally protected.

streams they pass along the way. There also is evidence that they respond to chemicals released by other members of their own species. The ability of an animal to find its way back to a home area is known as homing behavior.

Most salmon undergo dramatic changes before spawning. Both sexes of the sockeye salmon (*Oncorhynchus nerka*) turn from silver to bright red, giving rise to one of its common names, the red salmon. The jaws grow into vicious looking hooks in male sockeye and pink, or humpback, salmon (*Oncorhynchus gorbuscha*). Males of the latter species also develop a large hump. Salmon spawn on beds of clean gravel in the shallows. The female digs out a shallow nest, or redd, into which she deposits her eggs. The eggs are fertilized by the male and covered with gravel. After defending the nest for a while, the salmon die.

After hatching, the young salmon, or smolt, can return to sea immediately, as in the pink salmon, or remain in fresh water for as long as five years as in the sockeye salmon. The kokanee salmon, a race of this last species, is landlocked and does not migrate to sea at all. This is also the case of the rainbow trout, the landlocked variety of the steelhead salmon (*Oncorhynchus mykiss*). Unlike other salmon, the steelhead and rainbow trout can spawn many times before dying.

One of the biggest hazards Pacific salmon endure is the destruction of their natural environment by humans. Migration routes have been blocked by dams, spawning grounds filled with silt as a result of logging and cattle grazing, and rivers polluted by pesticides (which, among other toxic effects, are suspected of disrupting salmon's sense of smell), fertilizers, and animal waste. Around 11 to 15 million salmon once spawned in the Columbia River system in the northwestern United States. The number has now decreased by 90%, and most of the returning adults were born in hatcheries. Another problem is the severe drought in California that has left some rivers too dry for migrations. Drought has even forced the transport of Chinook smolt by truck. Some species, like the sockeye, Chinook, and chum salmon, have been declared endangered in some rivers. Sharp restrictions have been imposed on the harvest of salmon along the Pacific coast of the United States.

The Atlantic salmon (*Salmo salar*) breeds on both sides of the North Atlantic. It migrates across the ocean, mostly off Greenland, before returning to rivers from New England to Portugal. Atlantic salmon sometimes survive after spawning and return to the ocean. Some females have made as many as four round trips. Like their Pacific cousins, populations of the Atlantic salmon in the wild are in serious decline, though they are farmed for food in some temperate regions besides the Atlantic.

Catadromous fishes have a migratory pattern opposite that of salmon. They breed at sea and migrate into rivers to grow and mature. Several catadromous fishes are known, but the longest migration of any of them is that of freshwater eels. There are at least 16 species, including the American and European eels, which were once thought to belong to the same species.

Eel Migration Both American and European eels spawn in the Sargasso Sea north of the West Indies at depths of at least 400 to 700 m and then die (Fig. 10.24). They spawn at different, though overlapping, periods and locations in

the sea. The eggs hatch into tiny, transparent larvae that gradually develop into, leaf-shaped leptocephalus larvae. American eel larvae drift in the plankton for at least a year. The juveniles then move into rivers along the Atlantic coast of North America. The leptocephalus larvae of the European eel are believed to spend at least an extra two or three years drifting in the Gulf Stream to reach rivers throughout western Europe. Adults of both species eventually grow to more than 1 m in length. Both juveniles and adults are highly valued as food, particularly in Europe. After 10 to 15 years in fresh water, adults turn silver and their eyes become larger. Soon after this, they head out to sea.

The migration of eels back to the Sargasso Sea is not completely understood. There is experimental evidence that they use Earth's magnetic field as a cue for navigation. It has been assumed that European eels follow favorable currents that take them first along the coast of northwest Africa (Fig. 10.24)

Reproduction and Life History

Reproduction in fishes involves a number of adaptations of the reproductive system and the behavior that brings the two sexes together to ensure successful reproduction.

Reproductive Behavior Potential mates must get together at the right time to breed. Many species migrate and congregate in specific breeding grounds, as in the salmon and freshwater eels previously discussed. Sharks are usually loners but come together during the breeding season. Some of them appear to stop feeding at spawning time. Many bony fishes change color to advertise their readiness to breed, as in the dramatic case of salmon. Color changes also can be observed in tropical fishes. Many male wrasses, colorful all the time, appear even more spectacular before breeding.

The first step in reproduction is courtship, a series of behaviors that attract mates. These behaviors involve an exchange of active displays such as "dances," special postures that display colors, and upside down swimming (Fig. 10.25, p. 296). Each species has a unique courtship behavior. This is thought to help keep fishes from mistakenly mating with members of the wrong species.

Leptocephalus larva

Adult eel

American eel
→ Larva
⟹ Adult

Prevailing ocean currents
→

European eel
→ Larva
⟹ Adult

Figure 10.24 Two species of freshwater eels, the American (*Anguilla rostrata*) and European (*A. anguilla*) eels, breed in the Sargasso Sea and migrate to rivers in North America and Europe. The return trip of larvae is not precisely known, so the arrows indicate only the most likely routes.

Figure 10.25 The Atlantic cod (*Gadus morhua*) spawns in large aggregations in the North Atlantic. The two sexes follow a strictly choreographed series of behaviors.

Some fishes have **internal fertilization** of the eggs, in which the sperm is directly transferred from males to females through the act of copulation. **External fertilization**, the release of gametes into the water, or broadcast spawning, is more common in fishes, however.

Cartilaginous Fishes Internal fertilization occurs mainly in cartilaginous fishes. Unfortunately, not much is known about their sex life. Male sharks, rays, and skates have a pair of copulatory organs, called claspers, located along the inner edge of the pelvic fins (Fig. 10.26). The typical approach of an amorous male shark consists of biting his potential mate on the back. The male copulates by inserting his claspers into the female's cloaca. He bites and hangs from her or partly coils around her middle. Some skates mate after the male bites the female's pectoral fins. He presses his ventral surface against hers and inserts the claspers.

Bony Fishes Open-water fishes (sardines, tunas, jacks, and others) and those living around coral reefs and other inshore environments (such as surgeonfishes, parrotfishes, wrasses, and the capelin) spawn directly into the water after courtship (Fig. 10.27). Females typically release many eggs. In the Atlantic cod (*Gadus morhua*), for example, a female 1 m long can release up to 5 million eggs. The Atlantic tarpon (*Megalops atlanticus*) releases more than 100 million eggs every time it spawns.

Some fishes, like butterflyfishes, spawn in pairs (Fig. 10.28, p. 298), others in groups. Individual males can establish territories or aggregate into groups. Groups of males can be approached by single females or by females in groups. Usually, males seek out the females and entice them to spawn via courtship.

(t)Pixtal/age fotostock, (b)Gavin Parsons/Ardea.com, (inset)©Michael Huber

Figure 10.26 Fertilization in cartilaginous fishes is internal, so males possess claspers, which are located on the inner edge of the pelvic fins. The photo shows a male blacktip shark (*Carcharihinus limbatus*); the claspers of a scalloped hammerhead shark (*Sphyrna lewini*) are shown in the inset.

Figure 10.27 The capelin (*Mallotus villosus*) typically spawns on gravel beaches.

Two individuals can pair only during spawning time, as in butterflyfishes, or can establish long-lasting bonds.

Life History Eggs fertilized in the water column drift in currents and develop as part of the plankton. Most eggs contain oil droplets and are buoyant. Other eggs sink to the bottom. Herrings deposit their eggs on the surface of sea-grasses, seaweeds, and rocks. Lampreys and salmon bury their eggs after spawning. The California grunion (*Leuresthes tenuis*) buries its eggs on sandy beaches during high tides; they don't hatch until the next high tide.

Most of the eggs that are released into the plankton don't survive. Fishes and other marine animals that spawn into the plankton are broadcast spawners that release as many eggs as possible to ensure that at least some hatch and make it to adulthood. Eggs require a lot of energy to produce because they must contain enough yolk to nourish the young until they hatch and can feed.

Fishes that spawn fewer and larger eggs have evolved ways to take care of them. In many damselfishes males establish and defend breeding sites or

Figure 10.28 (a) In the oval butterflyfish (*Chaetodon trifasciatus*), pairs establish territories around coral colonies. (b) Experiments using transparent plastic cages can determine the sex of the wild fish,which is normally impossible in the field, based on their territorial response.

(a)

(b)

nests (Fig. 10.29) in holes among rocks or coral, empty mollusc shells, and other shelters—even discarded tires. After spawning, the eggs are retained in the nest and guarded by the males; the females leave after spawning. Males are promiscuous and will guard eggs they have fertilized from different females. Nests also are guarded by males in some gobies, blennies, and sculpins. In the Antarctic plunderfish (*Harpagifer bispinis*), the female prepares a breeding site and guards it for four to five months after spawning. If she disappears or is removed, her job is taken over by another plunderfish, usually a male.

Some fishes go even further and physically carry the eggs after they have been fertilized. Male pipefishes carry the eggs attached in neat rows to their bellies. A male seahorse literally becomes pregnant after the female deposits eggs in a special pouch on his belly. In some cardinalfishes, marine catfishes, and other groups, males brood the fertilized eggs in their mouths.

Figure 10.29 The yellow tail damselfish (*Stegastes arcifrons*) from the Galápagos Islands, like many other damselfishes, establishes and actively defends breeding sites where females lay their eggs.

Oviparity Most fishes spawn eggs and are known as **oviparous**. In oviparous sharks, skates, and other cartilaginous fishes, the embryo is enclosed by a large, leathery egg case (the "mermaid's purse" of skates) that drops to the bottom after spawning (Fig. 10.30). About 43% of all cartilaginous fishes are oviparous. The egg cases are rather large and often have thin extensions that attach them to surfaces. Only a few are laid at a time. The eggs have a large amount of yolk in a yolk sac that is attached to the embryo's belly. Yolk provides energy for several months of development, a long time by fish standards. As a result, the pup is well developed when it finally hatches. Development of the embryo proceeds rather quickly in most bony fishes. The transparent outer envelope of the eggs, the chorion, is thin, allowing oxygen to diffuse through. The eggs are usually spherical. The embryo is supplied with nutrient- rich yolk. After one or more days of development, the eggs hatch into free-swimming larvae, or fry. When they first hatch, the larvae still carry the yolk in a yolk sac. The yolk is eventually consumed, and the larvae begin feeding. Many larvae, like the leptocephalus larvae of eels, do not resemble their parents at all and undergo metamorphosis to a juvenile stage that resembles the adult.

Ovoviviparity In some cartilaginous fishes the female retains the eggs inside her reproductive tract for additional protection. The eggs develop inside the

(l)©Alex Kerstitch, Sea of Cortez Enterprises, (r)John M. Chase/E+/Getty Images

Figure 10.30 An egg of the swell shark (*Cephaloscyllium ventriosum*) containing a one-month-old embryo. The large, white structure is the yolk sac. The complete egg case of another shark is shown in the inset.

female, which gives birth to live young. The young are larger than in oviparous species. Such fishes are known as **ovoviviparous**. Most ovoviviparous fishes are cartilaginous. A total of 300 embryos, still in egg cases, were found ready to emerge inside the reproductive tract of a female whale shark. Some rockfishes are among the few marine bony fishes that can be classified as ovoviviparous. Most bony fishes are oviparous and spawn their eggs for external fertilization.

In some ovoviviparous sharks, the embryos rely on other sources of nutrition once they have consumed the yolk. In the sandtiger shark (*Carcharias taurus*; see Fig. 10.3) only two pups, which are large (up to 1 m) and active, are born. Each survived in one of the two branches of its mother's reproductive tract by eating its brothers and sisters. When that source of food is gone, they consume unfertilized eggs produced by their mother's ovaries.

Viviparity Some sharks and rays have embryos that actually absorb nutrients from the walls of the mother's reproductive tract. These sharks are said to be **viviparous**. The surfperches, which are bony fishes, are also viviparous. Their young have large fins that absorb nutrients from the walls of the mother's uterus. Some sharks go further and have evolved to possess what some call a placenta, tissue that provides nourishment to the embryo. It is not a true placenta, however, because it develops from the walls of the mother's uterus rather than from the embryo as in the case of the true placenta of practically all mammals, including humans. All viviparous fishes are therefore said to show aplacental viviparity.

There is now evidence of yet another method of reproduction among sharks. A female bonnethead shark (*Sphyrna tiburo*), a hammerhead shark, gave birth to a pup after three years in an aquarium together with two females of the same species but no males. Though it was possible that the female had stored sperm in her body as the result of sex with a male shark prior to capture, the DNA of the pup matched the DNA of one of the three females with no evidence of DNA from a male. Though very rare among vertebrates, **parthenogenesis**, the development of an egg into a normal embryo without fertilization by sperm, has also been reported in a few bony fishes.

Table 10.1 Most Important Characteristics of Marine Fishes

Group	Distinguishing Features	Skeleton	Feeding	Reproduction	Ecology
Hagfishes	No paired fins, no scales, xposed gill slits, marine	Cartilaginous skull, no vertebrae, no jaws	Suction by round, muscular mouth with teeth	Oviparous	Predators of dead or dying fishes and bottom invertebrates
Lampreys	Two dorsal fins only, no scales, exposed gill slits, fresh water or anadromous	Cartilaginous skull, no vertebrae, no jaws	Suction by round, muscular mouth with teeth	Oviparous	Suckers of fish blood or predators of bottom invertebrates
Rays, skates	Paired fins, placoid scales, pectoral fins greatly expanded, five ventral gill slits, mostly marine	Cartilaginous	Grinding plates to feed on bottom animals or gill rakers to filter plankton	Oviparous, viviparous	Predators of bottom animals or filter feeders
Sharks	Paired fins, placoid scales, 5–7 exposed and lateral gill slits, mostly marine	Cartilaginous	Teeth in jaws to capture prey or gill rakers to filter plankton	Oviparous, ovoviviparous, viviparous	Predators or filter feeders of plankton and small fishes
Ratfishes	Paired fins, placoid scales, one pair of gill slits covered by flap of tissue, deep water	Cartilaginous	Grinding plates to feed on bottom invertebrates	Oviparous	Predators of bottom invertebrates
Bony Fishes	Paired fins, cycloid or ctenoid scales (absent in some), gills covered by operculum, marine and fresh water	Bony	Teeth in jaws to capture prey or graze or gill rakers to filter plankton	Oviparous, ovoviviparous, viviparous	Predators, grazers, or filter feeders
Coelacanths	Paired lobe-fins, large scales, gills covered by operculum, deep water	Bony	Teeth in jaws to capture prey	Ovoviviparous	Predators of deep-water fishes, squids, and octopuses

Section Review

1. **Explain** What are some of the different ways fishes swim? What role do muscles and a swim bladder play in swimming?

2. **Describe** one reproductive strategy that increases the fitness of a marine fish.

Chapter 10 Review

Multiple Choice

1. Which is not a characteristic of jawed fishes?
 a. endothermy
 b. gills
 c. skin covered by scales
 d. two-chambered heart

2. Cartilagenous fishes and bony fishes are different in that only
 a. bony fishes have paired fins.
 b. bony fishes have a keen sense of smell.
 c. bony fishes have an operculum.
 d. cartilaginous fishes have a complete skeleton.

3. Bony fishes are divided into which two groups?
 a. hagfishes and lampreys
 b. sharks and ray-finned fishes
 c. ray-finned fishes and lobe-finned fishes
 d. jawless fishes and cartilaginous fishes

Short Answer

4. What is the evolutionary significance of lobe-finned fishes?

5. Compare and contrast cartilaginous fishes and bony fishes.

6. Describe an adaptation of a marine fish for feeding on small prey like tiny crustaceans.

Critical Thinking

7. Hagfishes and lampreys are the only living representatives of a very ancient group. Why do you suppose there are still some of these jawless fishes around?

8. A deep-water shark, new to science, is collected for the first time. The specimen is studied in detail, but its stomach is empty. How can you get a rough idea of its feeding habits? The specimen is female, and its reproductive tract is found to contain 20 eggs. Can you tell the type of development characteristics of this species?

How can we learn more about basking sharks?

Marine scientists know very little about the lives of basking sharks, including where they live, mate, and give birth. In an effort to learn more, scientists tagged 18 sharks off the coast of Massachusetts.

Data and Observations

The illustration shows the previously known ranges of basking sharks together with new information retrieved from the tags.

Atlantic Ocean

Key:
Previously known range of basking sharks
Migration path of tagged sharks

Think Critically

1. What did marine scientists learn from the data collected by the tags?

2. After analyzing the data from the tags in this study, the marine scientists hypothesized that the pregnant females and juveniles live in deep, tropical waters for months at a time. Describe a method in which this hypothesis could be tested.

Data obtained from: Hamady, L. L., 2011. Clues in shark vertebrae reveal where they've been. Oceanus, vol. 49, no. 1, Winter, pp. 10–13.

Chapter 11

Marine Reptiles, Birds, and Mammals

THEME Patterns

Reptiles, birds, and mammals show adaptations that allow them to occupy many environments both in the ocean and on land.

BIG IDEA

Studying the biology of marine tetrapods helps marine scientists identify how these predominately land animals have adapted to life in the oceans.

About the photo A group of common dolphins (*Delphinus*) hunt a school of sardines off the South African coast.

Dmitry Miroshnikov/Getty Images

11.1 Marine Reptiles

Main Idea

With the exception of the leatherback turtle, marine reptiles are poikilotherms and ectotherms, generally limiting their distribution to warm water.

. .

Key Questions

1. What are the main characteristics of reptiles?
2. Differentiate between the terms ectotherm, endotherm, and poikilotherm.

VOCABULARY

tetrapods

temperature-dependent sex determination

DIVE IN!

Where are marine iguanas found? Marine iguanas, *Amblyrhynchus cristatus*, are only found in the Galápagos Islands, off the west coast of South America. Scientists hypothesize that millions of years ago, iguanas that lived on mainland South America reached the Galápagos by drifting on wood or other floating objects. From these ancestors the marine iguanas we see today evolved. Due to relative geographic isolation, the iguanas on each island have particular colors, sizes, and behaviors. Their status is threatened and they are protected throughout the islands. Their eggs and juveniles, however, are still eaten by non-native rats, cats, and dogs that were introduced by humans.

As you learned in Chapter 10, fishes represent the first vertebrates, and it was a fish-like ancestor which gave rise to **tetrapods** – meaning "four-footed" animals with two pairs of limbs. Tetrapods arose from the oceans and populated the terrestrial environment. Having adapted to the land, various groups of reptiles (Class Reptilia), birds (Class Aves), and mammals (Class Mammalia) turned around and reinvaded the ocean. This chapter deals with these marine tetrapods. The marine animals to be covered in this chapter include some of the most fascinating and awesome creatures on the planet. Unfortunately, many are in danger of disappearing. Some already have become extinct.

Diversity of Marine Reptiles

There are around 7,000 living species of reptiles, including lizards, snakes, turtles, and crocodiles. Most reptiles are adapted to live on land: their dry skin is covered with scales to prevent water loss. Their eggs have a leathery shell that prevents them from drying out, so reptiles can lay their eggs on land. All reptiles are air-breathing. They first appeared more than 300 million years ago, and several different groups have invaded the seas. Many are long gone, like the ichthyosaurs (see Fig. 11.15) that thrived in the oceans during the so-called Age of Reptiles. Only a few reptiles still roam the seas. Some are rare and endangered; others, however, can be widely distributed in the tropics.

Sea Turtles Sea turtles belong to an ancient group of reptiles. Their bodies are enclosed by an armor-like shell, or carapace, that is fused to the backbone. Unlike land tortoises and turtles, sea turtles cannot retract their heads into the shell. Their legs, particularly the larger forelimbs, are modified into flippers for swimming.

There are only seven species of sea turtles (though some biologists recognize eight), which live primarily in warm waters. Green turtles (*Chelonia mydas*) were once common in coastal waters throughout the tropics, but like all species of sea turtles, their numbers have drastically declined due to overexploitation and increased mortality. Their shells can grow to 1 m in length. They feed mostly on seagrasses and seaweeds.

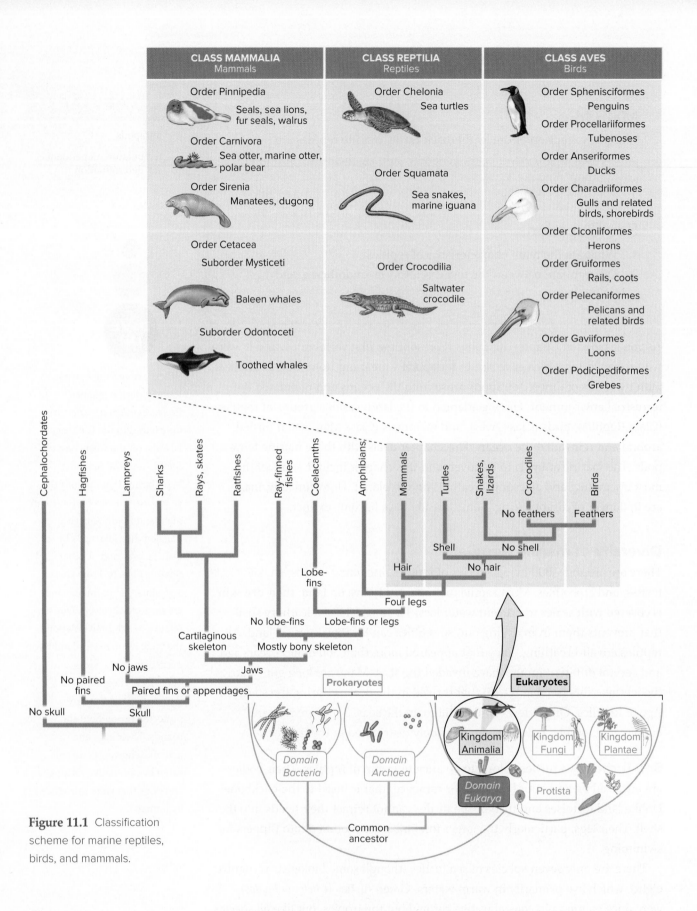

CLASS MAMMALIA
Mammals

Order Pinnipedia
Seals, sea lions, fur seals, walrus

Order Carnivora
Sea otter, marine otter, polar bear

Order Sirenia
Manatees, dugong

Order Cetacea
Suborder Mysticeti
Baleen whales

Suborder Odontoceti
Toothed whales

CLASS REPTILIA
Reptiles

Order Chelonia
Sea turtles

Order Squamata
Sea snakes, marine iguana

Order Crocodilia
Saltwater crocodile

CLASS AVES
Birds

Order Sphenisciformes
Penguins

Order Procellariiformes
Tubenoses

Order Anseriformes
Ducks

Order Charadriiformes
Gulls and related birds, shorebirds

Order Ciconiiformes
Herons

Order Gruiformes
Rails, coots

Order Pelecaniformes
Pelicans and related birds

Order Gaviiformes
Loons

Order Podicipediformes
Grebes

Figure 11.1 Classification scheme for marine reptiles, birds, and mammals.

Like all turtles, green turtles lack teeth, but they have strong biting jaws. The hawksbill turtle (*Eretmochelys imbricata*; Fig. 11.2*a*) is smaller, and the shell is reddish brown with yellow streaks. It uses its beak-like mouth to feed on encrusting animals (sponges, sea squirts, barnacles) and seaweeds.

The largest sea turtle is the leatherback (*Dermochelys coriacea*; Fig. 11.2*b*). Individuals can attain a length of nearly 3 m and weigh 900 kg. Instead of a solid shell, they have a series of small bones buried in the dark skin, forming seven distinct longitudinal ridges. Leatherbacks are the widest ranging of all marine reptiles, live in open water, and are rarely seen except on nesting beaches. They are deep divers—one individual was tracked diving to 640 m. Their diet consists largely of jellyfishes.

The remaining species of sea turtles have diverse diets, like the flatback sea turtle (*Natator depressus*) of northern Australia, which feeds on sea cucumbers, soft corals, and other invertebrates. Other sea turtles feed on soft, bottom invertebrates as well as hardier ones like crabs and molluscs. The jaws of most species are adapted for crushing. The leatherback jaws, however, are scissor-like to capture jellyfishes.

(a)

(b)

Sea Snakes About 70 species of sea snakes are found in the tropical Indian and Pacific oceans (Fig. 11.3). There are no sea snakes in the Atlantic, but a sea-level canal across Central America could allow their dispersal into the Caribbean. Most are 1 to 1.3 m long. Practically all sea snakes lead a totally marine existence, lacking the belly scales that allow land snakes to move. Like all snakes, sea snakes are carnivores. Most feed on eels and other bottom fishes, but a few specialize on fish eggs, one species on bottom invertebrates. They are closely related to cobras and their allies, the most venomous of all snakes. Sea snakes are among the most common of all venomous snakes, and their bites can be fatal to humans. Fortunately, they are rarely aggressive, and the mouth is too small to get a good bite. Most casualties occur in Southeast Asian and are from swimmers accidentally stepping on them and fishers removing them from

Figure 11.2 (*a*) The hawksbill turtle (*Eretmochelys imbricata*) takes its name from the shape of its jaw. It is the source of tortoiseshell. (*b*) The largest of all sea turtles, the leatherback turtle (*Dermochelys coriacea*), is found worldwide and sometimes ventures into cold waters as far north as Newfoundland and Alaska.

Figure 11.3 Sea snakes are found in the tropical Indian and Pacific oceans. The conspicuous coloration of sea snakes could be a warning to potential predators because many fishes learn to associate the bright colors with danger.

Figure 11.4 Though seemingly awkward on land, the marine iguana of the Galápagos Islands (*Amblyrhynchus cristatus*) is an elegant swimmer under water. It swims by undulating the body and the laterally flattened tail, the tip of which is shown on the bottom left.

nets. Sea snakes are also victims of overexploitation. They are hunted for their skins and meat, and some species have become rare in certain areas.

Other Marine Reptiles An unusual lizard is among the unique inhabitants of the Galápagos Islands off the Pacific coast of South America. The marine iguana (*Amblyrhynchus cristatus*; Fig. 11.4) spends most of its time basking in large groups on rocks along the coast, warming up after swimming in the cold water. It eats seaweeds and can dive as deep as 10 m to graze.

Another marine reptile is the saltwater crocodile (*Crocodylus porosus*), which inhabits mangrove swamps and estuaries in the eastern Indian Ocean, Australia, and some of the western Pacific islands. Saltwater crocodiles are also found in rivers but are known to venture into the open sea. There is a record of an individual 10 m long, but they are rarely over 6 m. They are among the most aggressive of all marine animals and do attack and eat humans. Where they occur, they are more feared than sharks.

The American crocodile (*Crocodylus acutus*) is also considered a marine species. Although it lives mostly in fresh water, it tolerates seawater and can be found in lagoons, river mouths, and other coastal areas in Florida, the West Indies, and on the Pacific and Atlantic coasts of Central America and northern South America. It has also been recorded in the open ocean.

Biology of Marine Reptiles

Most of today's reptiles are adapted to life on land. The oceans present an entirely different set of challenges and require different strategies and adaptations, particularly for locomotion and reproduction.

Thermoregulation Like most fishes and all land reptiles, marine reptiles are poikilotherms because their body temperature varies with that of the environment. With the exception of the leatherback turtle, marine reptiles are ectotherms because they lose most of their metabolic heat to the environment to

match their body temperature with the external temperature. Like other poikilotherms, their metabolic rate—and therefore activity level— varies with temperature so they get sluggish in the cold. This tends to keep reptiles out of cold regions, especially on land because the air temperature fluctuates more widely than does the ocean temperature.

A remarkable exception is the leatherback turtle, which can venture into cold water because of its ability to keep a body temperature higher than that of the surrounding water, which makes it an endotherm. Heat loss is reduced by transferring body heat back to the body by means of a network of blood vessels in the flippers as in the muscles of tunas and other fishes, by being large with a comparatively small surface area, and by storing fat for insulation.

Swimming and Diving Sea turtles, although slow on land, are fast swimmers in water. They use their flattened, flipper-like forelimbs to propel them through the water, while the smaller, paddle-shaped back limbs are mostly used for steering. The heavy shell is streamlined and flattened, but bones are spongy and the accumulation of fat further helps in buoyancy. While most sea turtles are shallow divers, leatherbacks have been tracked diving over 1,000 m.

The bodies of sea snakes, which are not particularly fast swimmers, are laterally flattened and the tail paddle-shaped for swimming. A similar adaptation is seen in the marine iguana, a stronger swimmer, where the tail is laterally flattened.

Reproduction All sea turtles must return to land to reproduce. They typically migrate long distances to lay their eggs on remote sandy beaches, and they have been doing so millions of years before humans appeared on the scene. Green turtles still gather to nest on beaches on the Caribbean coast of Central America, Northern Australia, Southeast Asia, Ascension Island (in the middle of the South Atlantic), and a few other locations.

Marine biologists have tagged adult sea turtles at Ascension and have found that the turtles regularly cross 2,200 km of open water from their feeding grounds along the coast of Brazil (see foldout map of the world at the end of the book), a journey that takes a little more than two months. Loggerhead turtles travel as much as 14,500 km between feeding and nesting grounds, and leatherbacks have been tracked as traveling as much as 17,614 km. Though we are still not completely sure how sea turtles find their way, there is evidence that they do it by sensing Earth's magnetic field.

Most of what we know about the reproduction of sea turtles is based on the green turtle. They return to their nesting areas every two to four years, often against prevailing currents. Copulating pairs of sea turtles are often seen offshore, but only females venture ashore, usually at night. Therefore, biologists have mostly tagged females because turtles can be tagged most easily on land. Evidence that females return to the beaches where they were born has been obtained by analyzing the DNA of breeding populations at sites around the world. The DNA of turtles breeding in one area differs from the DNA of turtles breeding at other sites. It thus appears that turtles keep returning to the same place generation after generation. The females congregate on the beach, and

IN CONTEXT

Poikilotherms Organisms where their internal body temperature changes with the temperature of their surroundings.

Ectotherms Organisms in which metabolic heat is lost to the environment and does not raise body temperature.

Endotherms Organisms in which metabolic heat is retained within the body and raises body temperature.

• 6.5, Challenges of Life in the Sea

The Endangered Sea Turtles

Humans are by far the most formidable and destructive enemies of sea turtles. Humans exploit turtles for a variety of reasons, from food to jewelry to cosmetics. Turtles are also suffering the effects of climate change, habitat destruction, and overfishing.

Direct collection The green turtle is especially esteemed for its meat, and its cartilage is used to make turtle soup. Sea turtles can live for months without food or water, and in the days before refrigeration sailors kept them alive aboard ship as a source of fresh meat, storing them on their backs for months. Coming ashore by the thousands, females were an easy catch. Thousands of sea turtles are illegally killed every year for meat in coastal regions of Latin America, Southeast Asia, and India. Sea turtle eggs, particularly those of leatherbacks, are said to be an aphrodisiac. There is still a flourishing, though technically illegal, market for sea turtle eggs in many coastal regions. Eggs are eaten or fed to pigs or cattle.

The polished shell of the hawksbill is the source of tortoiseshell used to make jewelry, combs, and other articles. Baby sea turtles are stuffed and sold as souvenirs, and sea turtle leather, which is soft and durable, is much prized for shoes, boots, handbags, and wallets. Sea turtle oil is also used in cosmetics.

Development and fishing Much more harmful is the fact that many nesting areas have been turned into resorts or public beaches. Females searching for nesting beaches avoid lights because dark areas along the horizon indicate land, and lights along a beach look like a starry horizon. Artificial lighting also disorients baby turtles after hatching, so they do not head to sea and therefore die, often eaten by seabirds and other predators.

Sea turtles also become entangled in gill and drift fishing nets, including abandoned ones, and drown because they cannot swim to the surface to breathe. Many choke to death after swallowing plastic bags they think are jellyfishes. Shrimp nets are estimated to kill thousands of sea turtles a year. They have been especially deadly to the Kemp's ridley turtle in the Gulf of Mexico, once very common but now so rare that only a few hundred breeding females remain. It is considered the most endangered sea turtle.

All seven species of sea turtles are classified as threatened worldwide because their numbers are low. The number of leatherbacks has decreased by 95% since the early 1980s, mostly victims of fishing nets, and only a few thousand nesting leatherbacks remain in the Pacific Ocean. The six species of sea turtles found in United States waters are protected under the Endangered Species Act of 1973. Four species are officially classified as endangered and are in danger of extinction: the leatherback, hawksbill, Kemp's ridley, and the green sea turtle (in Florida).

Green turtle (*Chelonia mydas*).

Conservation The numbers of green turtles, once very common, have been increasing in some areas, undoubtedly the result of protection. Shrimp nets in the United States are required to have turtle exclusion devices, or TEDs, to allow sea turtles to escape once caught in the nets. But sea turtles are not protected at all in many countries and, in those where they are, enforcement is difficult because it is impossible to protect all coasts and nesting grounds from fishers and egg hunters. Stricter worldwide enforcement of conservation practices, the establishment of reserves and national parks to protect nesting beaches, the control of pollution and fishing practices, the restocking of former nesting areas, and the regulation of trade in sea turtle products might help save them.

Georgette Douwma/Getty Images

each excavates a hole in the sand using both pairs of flippers (Fig. 11.5). They lay between 100 and 200 large, leathery eggs in each nest. Molecular studies have demonstrated the multiple paternities of offspring in single clutches of eggs in green turtles and other sea turtles. The female covers the eggs with sand before she returns to the sea. She can make several trips ashore during the breeding season, laying eggs each time.

Sea turtles, as in many other reptiles, show **temperature-dependent sex determination**. Almost all eggs incubated at lower temperatures, as in nests covered by vegetation or nests built in light-colored sand, hatch as males; most of those incubated at higher temperatures hatch as females. There are concerns that a global temperature rise could seriously affect sex ratios in sea turtles and other marine reptiles, increasing pressure on already threatened or endangered species. The eggs hatch after about 60 days of incubation in the sand. The baby turtles must then dig themselves out of the sand and crawl all the way back to the water, protected by darkness if they're lucky. Sea turtle eggs are often eaten by dogs, ghost crabs, wild pigs, and other animals. The hatchlings are easy prey for land crabs and birds, especially during the day. Even more, young turtles are lost in the water, where they are taken by a variety of fishes and seabirds.

Sea snakes mate in the ocean and are ovoviviparous, giving birth to live young. Males have two copulatory organs, so fertilization is internal. Evidence suggests that, in contrast to other marine reptiles, the single paternity of the offspring in each sea snake brood. Some form large aggregations, a behavior perhaps related to reproduction. A few species, however, still come ashore to lay their eggs. Female marine iguanas similarly burrow their eggs in sand, as many as six in the larger females and, like sea turtles, abandon their nests after a short time.

Figure 11.5 Egg laying in the olive ridley turtle (*Lepidochelys olivacea*) culminates a long and hazardous trip by females. It is at this time that they are most vulnerable. This photograph was taken on the Pacific coast of Costa Rica.

INCONTEXT

Ovoviviparous Live-bearing animals in which the eggs develop and hatch in the reproductive tract of females.
• 10.3 Fish Adaptations

• •

Section Review

1. **Summarize** Make a table like the one shown below to summarize the basic characteristics of the different groups of marine reptiles.

Group	Characteristics	Examples
Sea turtles		
Sea snakes		
Marine iguana		
Saltwater crocodile		
American crocodile		

2. **Discuss** What are the similarities and differences in reproduction in sea turtles, sea snakes, and marine iguana?

11.2 Seabirds

VOCABULARY

salt glands

erythrocytes

guano

Key Questions

1. What are the main characteristics of birds?

2. What is the definition of "seabird"?

Most living birds have the ability to fly, and in contrast to living reptiles, their bodies are covered with waterproof feathers that help conserve body heat. Waterproofing is provided by oil from a gland above the base of the tail, and birds rub the oil into their feathers with their beaks during preening. Flight is made easier by their light, hollow bones. Their eggs have hard shells that are more resistant to water loss than those of reptiles.

Seabirds are birds that spend a significant part of their lives at sea and feed on marine organisms. Most breed in large colonies on land, and often build their nests on the ground, sometimes traveling long distances to get food.

Diversity of Seabirds

Though comprising only about 3% of the estimated 9,700 species of birds, seabirds are distributed worldwide, and their impact on marine life is significant. Most are predators of fishes, squids, and bottom invertebrates, but some feed on plankton. Seabirds have amazing appetites, needing a lot of food to supply the energy required to maintain their body temperature.

Seabirds have evolved from several different groups of land birds. As a result, they differ widely in their flying skills, feeding habits, and ability to live away from land.

Penguins Of all seabirds, penguins (Fig. 11.6a) are the most radically adapted to live at sea. They are flightless, with wings modified into stubby flippers that allow them to "fly" underwater. All but one of the 18 species of penguins live primarily in Antarctica and other cold temperate regions of the Southern Hemisphere. The exception is the Galápagos penguin (*Spheniscus mendiculus*), which lives right on the Equator. Even so, this penguin is confined to regions in the Galápagos that are bathed by cold currents.

The larger penguins, like the imposing emperor penguin (*Aptenodytes forsteri*; Fig. 11.6a), hunt for fish and squid. The Adélie (*Pygoscelis adeliae*) and other small penguins feed mostly on krill. Penguins have strong beaks, a characteristic of seabirds that feed on fish and large plankton like krill (Fig. 11.7b, p. 312). Some species migrate seasonally between feeding grounds at sea and nesting areas on land or ice. They establish breeding colonies, which in Adélies can number more than a million pairs.

Tubenoses The tubenoses comprise a large group of seabirds with distinctive tube-like nostrils and heavy beaks that are usually curved at the tip (Fig. 11.7a). They spend months and even years on the open sea. Like other seabirds and

(a)

(b)

(c)

(d)

Figure 11.6 (a) An emperor penguin (*Aptenodytes forsteri*) and chick. (b) The brown booby (*Sula leucogaster*) nests in the Caribbean and Gulf of California. (c) The gannet (*Morus bassanus*) is the largest seabird in the North Atlantic. (d) The black-browed albatross (*Thalassarche melanophrys*) breeds mostly on the Falklands and on several islands that border the Southern Ocean. (e) The Atlantic puffin (*Fratercula arctica*) build their nests on cliffs and islands in the North Atlantic.

(e)

DIVE IN!

Which animal holds the record for the longest migration? In 2010, it was confirmed that the arctic tern, *Sterna paradisaea*, travels more than 71,000 km each year. The tern flies from its breeding grounds in Greenland to feeding grounds in Antarctica and back again. By attaching tracking devices to 60 terns, scientists discovered that their route is not as straight "as the crow flies." Rather, the birds flew out over the North Atlantic, then flew down to the coast of northwestern Africa. From there the birds split into two groups. One group continued south down the coast of Africa to Antarctica. The other group crossed the Atlantic Ocean and then flew south down the eastern coast of South America to Antarctica. After spending the summer feeding in Antarctica, all of the birds flew back in an "S" pattern, winding their way north first on the eastern side of the Atlantic, then near the equator, crossing over to the western side to finish the journey back to Greenland.

sea turtles, they have salt glands that empty into the nostrils and get rid of excess salts. Tubenoses include the albatrosses (*Thalassarche*; Fig. 11.6d), shearwaters, and petrels (*Pterodroma*; Fig 11.7a).

Tubenoses are very skillful fliers, often seen flying low over the water. Most catch fish at the sea surface (Fig. 11.7a), though some, like the giant petrel, scavenge on dead birds or whales. The whalebirds, or prions, feed on krill and other plankton. Albatrosses are magnificent gliders with huge wings that hardly

Figure 11.7 The shape of a seabird's beak is related to the bird's food and its feeding style. (a) In petrels (*Pterodroma*) the beak is relatively short, heavy, and hooked for holding and tearing prey that is too big to be swallowed whole. (b) The beak is more streamlined in razorbill (*Alca*) for diving deeper to feed on crustaceans and other prey. (c) Terns (*Sterna*) are plunge divers with a straight and narrow beak for feeding on fish that are swallowed whole. (d) Skimmers (*Rynchops*) are the only birds with a lower beak that is longer than the upper, which permits feeding while flying.

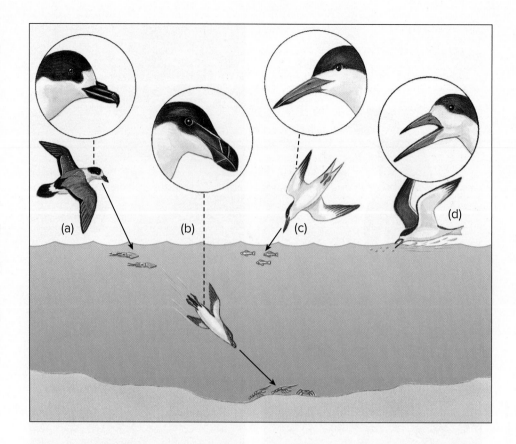

ever seem to flap. The wandering albatross (*Diomedea exulans*) has a wingspan of up to 3.4 m, the largest of any bird alive.

Tubenoses make some of the most spectacular migrations of any animal. Many breed on islands around Antarctica, then migrate across the open ocean to summer feeding grounds near the Arctic. The wandering albatross gets its name from the fact that it spends two years or more traveling, taking advantage of the strong winds that blow across the Southern Ocean, before returning to nesting sites near Antarctica. Some non-breeding individuals wander off and pay visits as far away as California and the Mediterranean. Albatrosses are monogamous because individuals have only one mating partner during most if not all of their lifetimes. They breed every two years since it takes almost a year to rear a single chick. Most species of albatrosses are threatened or endangered as a consequence of their slow reproductive rate and increased mortality resulting from longline fishing.

Pelicans and Related Seabirds Pelicans (*Pelecanus*) have a unique pouch below their large beaks that are used as nets for fishing. Some species, like the brown pelican, feed by plunging into the water and catching fish in the pouch. The brown pelican was once common along the coasts of the United States but was decimated by pesticide pollution. It has made a comeback as a result of restrictions on the manufacture and use of the pesticide DDT. Cormorants (*Phalacrocorax*; Fig. 11.8) are long-necked, usually black, seabirds that dive and pursue their prey. About 40 species are recognized, some known as shags. Cormorants can be easily identified by their low flights over water and the fact that they float low in the water, with only the neck above the surface. The Galápagos

cormorant (*P. harrisi*), however, is flightless. Cormorants are generally seen ashore in a spread-wing resting position; scientists believe this posture allows them to dry their feathers. Frigatebirds have narrow wings and a long, forked tail. They soar majestically along the coast, forcing other seabirds to regurgitate fish in midair or catching prey from the surface (Fig. 11.8). These agile pirates seldom enter the water, not even to rest, because their feathers are not very waterproof.

Gulls and Related Seabirds Gulls (*Larus*) and their kin make up the largest variety of seabirds. Common and widespread, gulls are predators and scavengers happy to eat just about anything (Fig. 11.8). They are very successful in the company of humans and congregate near piers, garbage dumps, or anywhere else we throw refuse. Jaegers and skuas are gull-like predators that steal fish from other birds (Fig. 11.8). They nest near the rookeries of penguins and other seabirds and eat their eggs and young.

Terns (*Sterna*; Fig. 11.7c) are graceful flyers that hover over their prey before plunging for it. Their slender beaks are specialized to catch small fish, which they swallow whole. The Arctic tern is another amazing wanderer. It breeds in the Arctic during the northern summer and can travel an average of 70,800 km, to Antarctica for the southern summer, and then returns to the Arctic.

Also related to gulls are several cold-water diving seabirds. Puffins (*Fratercula*, Fig. 11.6e, p. 311) have heavy beaks similar to those of parrots. The related razorbill (*Alca*) is a black and white bird reminiscent of penguins (Fig. 11.7b). In fact, these birds could fill the role of penguins, which are absent in the northern

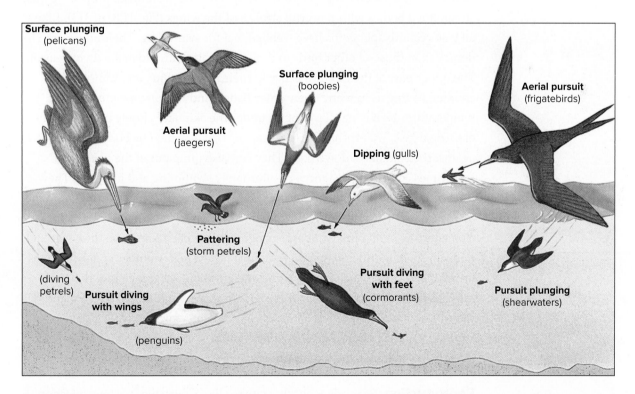

Figure 11.8 Feeding strategies vary widely among seabirds. Pelicans (*Pelecanus*) and boobies (*Sula*) plunge from the air into the water, jaegers (*Stercorarius*) pursue other seabirds and force them to regurgitate food, and frigatebirds (*Fregata*) take fish from the surface and steal fish from other seabirds. Gulls (*Larus*) rarely plunge from the air, and storm petrels (*Oceanodroma*) simply flutter over the waves. Divers like cormorants (*Phalacrocorax*) pursue prey under water.

Homeotherms Organisms able to keep their body temperature more or less constant regardless of the temperature of the environment.

Hemoglobin A blood protein that transports oxygen in many animals; it is contained in erythrocytes, or red blood cells, in vertebrates.

Myoglobin A muscle protein in many animals that stores oxygen.

• 6.5, Challenges of Life in the Sea

hemisphere. Like penguins, they use their wings to swim under water. Their extinct cousin, the great auk (*Pinguinus impennis*), looked and acted like a penguin. Great auks once lived in great numbers in the North Atlantic but were slaughtered for their eggs, meat, and feathers. The last great auk died in 1844.

Shorebirds Usually included among the seabirds are many species of wading shorebirds, most of which do not have webbed feet. Because they do not swim much, they are not seabirds in the strict sense. Many live in inland waters as well as the sea. Some are common in estuaries and coastal marshes. Plovers, sandpipers, and similar birds are related to gulls (see Fig. 11.1, p. 304). Many other shorebirds live on the coast: rails, coots, herons, egrets, and ducks.

Biology of Seabirds

Most biologists agree that birds are actually reptiles, having evolved from the same line that includes dinosaurs and crocodiles. However, birds do differ significantly from reptiles in several key adaptations, such as the maintaining a stable internal temperature, the presence of feathers, and flight.

Thermoregulation Birds are homeotherms, commonly referred to as "warm-blooded." They are also endotherms. This has allowed seabirds, in contrast to marine reptiles, to swim and dive in a wide variety of marine environments, mostly along coasts.

Swimming and Diving Penguins are spectacular swimmers, propelling their streamlined bodies with powerful strokes of the wings (Fig. 11.8, p. 313). Like all true seabirds, penguins have webbed feet for swimming. Their bones are denser than those of other birds to reduce buoyancy and to make diving easier. The upper part of their bodies is black, their bellies white, an example of countershading as in many open-water fishes, and they are adapted to cold temperatures by having a layer of fat under the skin and closely packed feathers. The dense, waterproof feathers trap air that, warmed by body heat, protects against the cold like a down coat. They can also jump out of the water and sometimes cover long distances by alternately swimming and jumping. They breathe air while jumping out of the water, with the efficiency of oxygen intake increased by air continuously flowing, in addition to the lungs, through air sacs characteristic of all birds. Emperor penguins have been known to dive over 200 m deep. As in marine mammals, penguins have a high number of **erythrocytes**, or red blood cells, which have a high concentration of hemoglobin, their muscles contain high amounts of myoglobin to store additional oxygen, and their heart rate is reduced to cut down oxygen consumption while diving. On land it is another story: they are clumsy and awkward. They are nearsighted, having eyes that are adapted for underwater vision.

Reproduction Seabirds typically mate for life. Courtship, partner selection, and recognition by the mated pairs involve elaborate display behaviors, which include calls.

The male emperor penguin incubates a single large egg during the dark Antarctic winter. The female leaves to feed as soon as she lays the egg. The male, standing on ice, must keep the egg warm by holding it on top of his feet and against his body for 64 days. Males huddle together in huge groups to protect themselves from the cold and winter storms. Emperors lay their eggs at the coldest time of the year so that the egg hatches during the productive Antarctic summer, when food is most plentiful. When the single egg hatches, the female finally returns and regurgitates food for the fuzzy chick. After that, both parents take turns feeding the chick. While the parents feed, the fast-growing young are herded into groups guarded by a few adult "baby sitters." Returning parents identify their chick among thousands by its voice and appearance. The parents continue to feed the chick for five and a half months, until it is strong enough to feed itself at sea (Fig. 11.6a, p. 311). In contrast, Adélies and other Antarctic penguins lay their eggs, typically two, during summer.

A threat to penguins is the reduction of Antarctic sea ice, a result of global climate change. There has been a reduction since 1986 in the populations of two species of penguins correlated with a decline in their food, krill, which depend on minute algae that grow under ice. Destruction of their breeding grounds in many areas of the Southern Hemisphere is another threat.

Albatrosses and other tubenoses take a long time to reach sexual maturity, around five years in albatrosses. They live for a long time, however, 50 years or more in albatrosses. Most return to nest on the same wind-swept, remote islands that are inaccessible to predators. Male and female remain faithful to each other and perform particularly elaborate courtship and greeting behaviors. Incubation of the egg in large nests and care of the single chick takes eight months or more. Incubation in the albatrosses lasts from 70 to 80 days, even longer in the large-size species. They feed the young regurgitated food and energy-rich oil produced by the parents' stomachs. The overall investment in reproduction is great so they typically do not lay eggs in successive years.

Pelicans nest in large colonies along the coast, where space, as in gannets (*Marus*; see Fig. 11.6c), can be limited. As in many other seabirds, they typically lay fewer eggs than land birds do, and their broods are reduced when less food is available. The messy nests are made of twigs and anything else they can find. The excrement of millions of pelicans, boobies (see Fig. 11.6b), cormorants, and other seabirds accumulates as **guano**. Guano deposits are particularly thick in dry coastal regions and islands near very productive waters like the coasts of Perú and southwest Africa. These deposits are mined for fertilizer.

· ·

Section Review

1. **Identify** What are the different groups of seabirds what characterizes each group?

2. **Describe** What are the general characteristics of reproduction in seabirds?

11.3 Marine Mammals

VOCABULARY

placenta

mammary glands

blubber

subspecies

convergent evolution

flukes

blowhole

ambergris

pods

Key Questions

1. What are the main characteristics of mammals?
2. How do the IWC and the Marine Mammal Protection Act of 1972 protect marine mammals?

DIVE IN!

How long can whales live? Scientists have determined that bowhead whales, a type of baleen whale that lives in the Arctic year round, may be the longest-living marine mammal. Bowheads can live as long as 200 years. Scientists studied the eye tissue of bowheads and estimated that about 5% of the population is over 100 years old, with some individuals reaching 160, 180, and even 200 years old. In comparison, blue whales are thought to live for about 80 years. Orcas live about 50 years. The smallest whale, the dwarf sperm whale, lives for about 25 years.

About 200 million years ago another major group of air-breathing vertebrates, the mammals (class Mammalia), evolved from now-extinct reptiles. For a long time the mammals were overshadowed by the dinosaurs, which were reptiles. About 65 million years ago, however, the dinosaurs became extinct. It was then that mammals thrived, taking the place of the dinosaurs. There are now roughly 4,600 species of mammals, including humans. Fishes, reptiles, and birds each outnumber mammals in number of species.

Like birds, mammals have the advantage of being endotherms and homeotherms. The skin of mammals, however, has hair instead of feathers to retain body heat. With few exceptions, mammals are viviparous. The embryo receives nutrients and oxygen through the **placenta**, or afterbirth, a membrane that connects it to the womb, thus an instance of placental viviparity. The newborn is fed by milk secreted by the mother's **mammary glands**. Instead of releasing millions of eggs, mammals produce few—but well-cared-for—young.

And then there is the brain. It is larger in relation to body size and far more complex than that of other vertebrates, allowing the storage and processing of more information. This accounts in part for the amazing adaptability of mammals. They live anywhere there is air to breathe and food to eat. This, of course, includes the ocean.

At least five different groups of land mammals succeeded in invading the oceans. They have followed different paths in adapting to the marine environment. Some are so fish-like that we have to remind ourselves that they have hair and bear live young nourished by their mother's milk.

Pinnipeds

Seals and related forms are marine mammals that have paddle-shaped flippers for swimming but still need to rest and breed on land. They make up one of the 19 or 20 major groups, or orders, of mammals, the pinnipeds (order Pinnipedia; see Fig. 11.1). Pinnipeds evolved from an early form of terrestrial carnivore (order Carnivora), which includes cats, dogs, bears, and their kin. The similarities are so close that many biologists classify them with the carnivores. Pinnipeds are predators, feeding mostly on fish and squid. Their streamlined bodies are adapted for swimming (Fig. 11.9).

Most pinnipeds live in cold water. To keep warm they have a thick layer of fat, called **blubber**, under their skin. Besides acting as insulation, it serves as a food reserve and helps provide buoyancy. Some pinnipeds also have bristly hair for added protection against the cold. Many of them are quite large, which also

helps conserve body heat because large animals have less surface area for their size than small animals and therefore lose less body heat.

Seals The largest group of pinnipeds, some 19 species, is the seals. Seals are distinguished by having rear flippers that cannot be moved forward (Fig. 11.9b). On land they must move by pulling themselves along with their front flippers. They swim with powerful strokes of the rear flippers.

Harbor seals (*Phoca vitulina*; Fig. 11.9b) are common in both the North Atlantic and North Pacific. Elephant seals (*Mirounga*; Fig. 11.10b, p. 318) are the largest pinnipeds. Males, or bulls, reach 6 m in length and can weigh as much as 3,600 kg. One unusual seal is the crabeater seal (*Lobodon carcinophagus*), which feeds on Antarctic krill. These seals strain krill from the water with their intricately cusped, sieve-like teeth. Unlike most seals, monk seals live in warm regions. The Mediterranean (*Monachus monachus*) and Hawaiian (*Monachus schauinslandi*) monk seals are now endangered. A third species, the Caribbean monk seal (*Monachus tropicalis*), was last seen in 1952.

Seals have been hunted for their skin and meat and for the oil extracted from their blubber. The Marine Mammal Protection Act of 1972 extends protection to all marine mammals and restricts the sale of their products in the United States. For some seals, however, this protection has not been enough.

Sea Lions Sea lions or eared seals, and the related fur seals are similar to the true seals except that they have external ears (Fig. 11.9b). They can also move their rear flippers forward, so they can use all four limbs to walk or run on land. The front flippers can be rotated backward to support the body, permitting the animal to sit on land with its neck and head raised. Sea lions are graceful and agile swimmers, relying mostly on their broad front flippers. Adult males are much bigger than females, or cows, and have a massive head with a hairy mane.

IN CONTEXT

Viviparous Live-bearing animals whose embryos develop within their mothers' bodies and are nourished by the maternal bloodstream.
• 10.3, Fish Adaptations

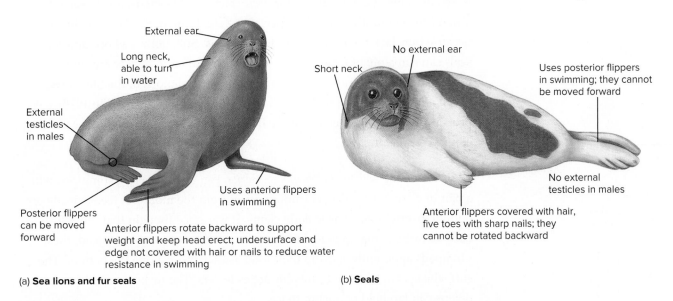

(a) **Sea lions and fur seals**

(b) **Seals**

Figure 11.9 Though they differ in some structural features and the ways in which they swim and move on land, (a) sea lions and fur seals and (b) seals are now thought to have evolved from the same group of land carnivores.

(a)

(b)

(c)

Figure 11.10 Seals. (a) The northern elephant seal (*Mirounga angustirostris*), is named because of the huge proboscis of the male. (b) The New Zealand fur seal (*Arctocephalus forsteri*), like the other fur seals, has a thick underfur. (c) Female harp seals (*Phoca groenlandica*), are threatened by hunting and decreased ice cover.

There are six species of sea lions and nine of fur seals. The most familiar of all is the California sea lion (*Zalophus californianus*) of the Pacific coast of North America. These sea lions are the trained barking circus "seals" that do tricks for a fish or two. Fur seals like the northern fur seal (*Callorhinus ursinus*; Fig. 11.10b) were once almost exterminated for their thick fur. Now most are protected around the world, though some species are still hunted. Sea lions are luckier because they lack the underfur of their cousins. Still, both sea lions and fur seals can run afoul of fishers. They sometimes drown in nets or are shot because of their notorious ability to steal fish. Populations of the Steller sea lion (*Eumetopias jubatus*) in Alaska have decreased by 80% since the 1970s, probably because of a decline in the fish they eat as a result of an increase in commercial fishing.

Walruses The walrus (*Odobenus rosmarus*; Fig. 11.11) is a large pinniped with a pair of distinctive tusks protruding down from the mouth. It feeds mostly on bottom invertebrates, particularly clams. It was once thought that the walrus used its tusks to dig up food, but there is no evidence for this. Instead, these pinnipeds apparently suck up their food as they move along the bottom. The stiff whiskers of the snout probably act as feelers. The tusks are used for defense and to hold or anchor to ice.

The walrus is restricted to the Arctic Ocean and adjacent waters. Two populations isolated from each other can be identified, one along the Arctic

Figure 11.11 The walrus (*Odobenus rosmarus*) lives at the edge of pack ice in the Arctic. They also crowd onto beaches on isolated islands and coasts, but a reduction of coastal sea ice has limited these resting areas. The walrus is still hunted legally by native Alaskans and Siberians.

coasts of Russia and Alaska extending to the North Pacific, and a second in northern Canada, Greenland, and the North Atlantic. Each population is considered as belonging to a geographically isolated group or **subspecies**.

Sea Otter, Marine Otter, and Polar Bear

The sea otter (*Enhydra lutris*; Fig. 11.12) is a member of the order Carnivora. The sea otter is the smallest truly marine mammal; an average male weighs 25 to 35 kg. It also differs from other marine mammals in lacking a layer of blubber. Insulation from the cold is provided by air trapped in its dense fur. This splendid, dark brown fur, unfortunately, attracted hunters. Sea otters were slaughtered to near extinction until they became protected by an international agreement in 1911. The California sea otter was then able to slowly expand from the few individuals that had managed to survive in some remote locations. Their numbers, how-ever, have been declining in recent years, particularly in the Aleutian Islands. In California they have slightly declined since peaking in

Figure 11.12 California sea otters (*Enhydra lutris*) live in and around kelp forests, and play an important role in ecosystem help by controlling grazers such as urchins.

2004, mostly as a result of disease. The sea otter is still a threatened species.

Three subspecies of sea otters are recognized. The Asian sea otter is found from northern Japan to Siberia, the Alaskan sea otter from the Aleutian Islands to the northwestern United States, and the California, or southern, sea otter in central California. They differ from one another in having slightly different sizes, in the amount of time they spend on land, and a few other characters.

Sea otters are playful and intelligent animals. They spend most or all of their time in the water, including breeding and giving birth. The furry pup is con-stantly groomed and nursed by its mother. Sea otters need to eat 25 to 30% of

Figure 11.13 The polar bear (*Ursus maritimus*) is beautifully adapted to live in the Arctic drifting ice, a habitat that is rapidly shrinking.

their body weight per day, so they spend a lot of time looking for food. They satisfy their ravenous appetites with sea urchins, abalone, mussels, crabs, other invertebrates, and even fishes. Sea otters help protect kelp beds from sea urchins. They are remarkable for their use of a tool—a rock for crushing sea urchins and molluscs. They carry a flat rock in side "pockets" of loose skin and fur. The sea otter floats on its back at the surface, places the rock on its chest, and crushes its toughest prey against it. Some have been observed carrying and using beer bottles as tools!

The rare marine otter, or sea cat (*Lontra felina*), is a small (up to 6 kg) otter that inhabits the Pacific coast of temperate South America. It feeds mostly on small invertebrates, but, unlike the sea otter, it spends most of the time on rocky shores. The marine otter is endangered because of hunting and habitat loss.

The polar bear (*Ursus maritimus*; Fig. 11.13) is the second member of the order Carnivora that inhabits the marine environment. Polar bears are semi-aquatic animals that spend a good part of their lives on drifting ice in the Arctic. They are threatened by the decrease in Arctic ice. They feed primarily on seals, which they stalk and capture as the seals surface to breathe or rest. Polar bears have legs, tail, and ears that are proportionally shorter than in other bears, adaptations to minimize heat loss in low temperatures.

Manatees and Dugong

It is hard to believe that relatives of the elephant, manatees and the dugong, live at sea. They are also known as sea cows, or sirenians (order Sirenia). Both have a pair of front flippers but no rear limbs (Fig. 11.14). They swim with up-and-down strokes of their paddle-shaped, horizontal tail. The round, tapered body is well padded with blubber. They have wrinkled skin with a few scattered hairs. The group is named after the mythical mermaids or sirens whose songs lured lonely sailors!

Sirenians are gentle creatures usually living in groups. They are the only strict vegetarians among marine mammals. Their large and prehensile lips are used to feed on seagrasses and other aquatic vegetation. All sirenians are large. Dugongs can reach 3 m and 420 kg, manatees 4.5 m and 600 kg. The largest sirenian of all was the Steller's sea cow, which supposedly grew to 7.5 m long. It was hunted to extinction.

Humans have exploited sirenians for their meat (which supposedly tastes like veal), skin, and oil-rich blubber. Like elephants and other large mammals, they reproduce slowly, typically one calf every three years. Seagrass beds, their main source of food, are being destroyed at an alarming rate by anchor and boat damage and by excess silt and nutrients from land, which often results from deforestation and intensive farming. All four species of sirenians are in danger of extinction. Three species of manatee live in the Atlantic Ocean; one is restricted to the Amazon and the other two inhabit shallow coastal waters and rivers from Florida to West Africa. The dugong (*Dugong dugon*) is strictly marine and survives from East Africa to some of the western Pacific islands. Its numbers are critically low throughout most of its range.

Figure 11.14 An estimated 5,000 West Indian manatees (*Trichechus manatus*) remain along the coasts and rivers of Florida. Manatees have been considered as a possible way to control weeds that sometimes block waterways.

Cetaceans

The largest group of marine mammals is the cetaceans (order Cetacea)—the whales, dolphins, and porpoises. No group of marine animals has captured our imaginations like the dolphins and whales, perhaps because they are mammals like us. They have inspired countless legends, works of art and literature, and films. The heated controversy surrounding the treatment of cetaceans, particularly of orcas, or killer whales, in theme parks and oceanaria is a good example of our emotional involvement with cetaceans.

Of all marine mammals, the cetaceans, together with the sirenians, have made the most complete transition to aquatic life. Whereas most other marine mammals return to land at least part of the time, these two groups spend their entire lives in the water. The bodies of cetaceans are streamlined and look remarkably fish-like (Fig. 11.15, p. 322). This is a dramatic example of **convergent evolution**, in which different species develop similar structures because they have similar lifestyles. Though they superficially resemble fishes, cetaceans breathe air and will drown if trapped below the surface.

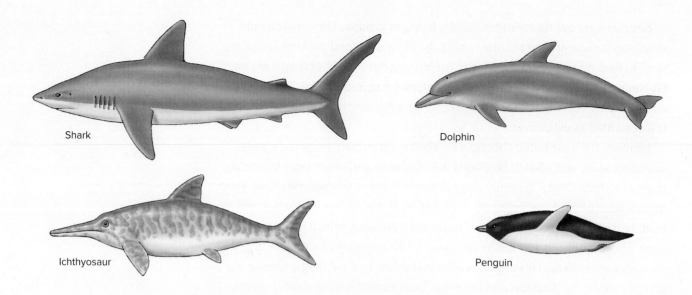

Figure 11.15 Streamlining to reduce water resistance evolved independently in different groups of fast-swimming marine animals: sharks, ichthyosaurs (extinct reptiles), dolphins, and penguins. Notice that dolphins lack posterior fins and that their flukes are horizontal, not vertical like the caudal fin of fishes.

They are endotherms, have hair (though scanty), and produce milk for their young.

Cetaceans have a pair of front flippers, but the rear pair of limbs has disappeared (Fig. 11.16). The rear limbs are actually present in the embryo but fail to develop (Fig. 11.17). In adults they remain only as small, useless bones. Like fishes, many cetaceans have a dorsal fin, another example of convergent evolution. The muscular tail ends in a pair of fin-like, horizontal **flukes**. Blubber provides insulation and buoyancy; body hair is practically absent. Cetacean nostrils differ from those of other mammals. Rather than being on the front of the head, they are on top, forming a single or double opening called the blowhole (Fig. 11.16). Barnacles, attached crustaceans, are often found living on the tough skin of whales.

There are some 90 species of cetaceans. They are all marine except for 5 species of river dolphins in Asia and South America. One South American species of river dolphin was formally described in 2014, but a Chinese species was declared extinct in 2006. Cetaceans are divided into two groups: the toothless, filter-feeding whales and the toothed, carnivorous whales, a group that includes the dolphins and porpoises.

Baleen Whales The toothless whales are better known as the baleen whales. Instead of teeth they have rows of flexible plates, named baleen, which hang from the upper jaws (Fig. 11.18, p. 325). Baleen is made of keratin, the same material as our hair and nails. The inner edge of each plate consists of hair-like bristles that overlap and form a dense mat in the roof of the mouth. The whale filter-feeds by taking a big mouthful of water and squeezing it out through the bristles by pushing up the tongue.

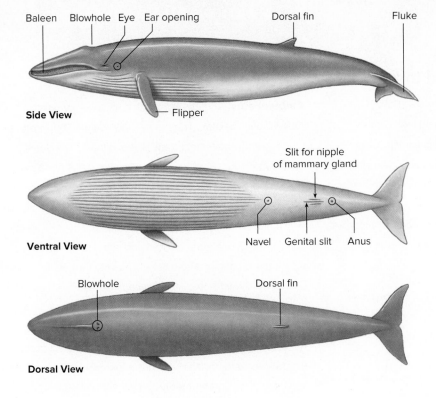

Baleen Blowhole Eye Ear opening Dorsal fin Fluke

Side View Flipper

 Slit for nipple
 of mammary gland

Ventral View Navel Genital slit Anus

 Blowhole Dorsal fin

Dorsal View

Figure 11.16 External morphology of a female blue whale (*Balaenoptera musculus*). Males have a genital slit halfway between the anus and navel, and they lack mammary slits.

Baleen whales are not only the largest whales but also are among the largest animals that have ever lived on Earth. There are 13 species of these majestic creatures. They were once common in all the oceans, but overhunting has brought many species to the brink of extinction. The blue whale (*Balaenoptera musculus*), which is actually blue-gray, is the largest of all (Fig. 11.19g, p. 326). Males average 25 m, and there is a record of a female 33.5 m. How do you weigh a blue whale? Very carefully—they average 80,000 to 130,000 kg, but the record is an estimated 178,000 kg.

The blue whale, the fin whale (*Balaenoptera physalus*; Fig. 11.19f, p. 326), and the minke whale (the common minke, *Balaenoptera acutorostrata*, and the Antarctic minke, *B. bonarensis*; Fig. 11.19b)—together with four other related species—are known as the rorquals. The rorquals and the humpback whale (*Megaptera novaeangliae*; Fig. 11.19c, p. 326), which is often included among the

Umbilical cord

 Rear limb

 Tail

Figure 11.17 The fetus of a white-sided dolphin (*Lagenorhynchus*) shows two distinct pairs of limbs; the rear pair will eventually disappear.

The Whales That Walked to Sea

Cetaceans are clearly identifiable as mammals. Their streamlined bodies and the absence of hind legs cannot disguise their affinities with land-dwelling mammals. Unlike the sea otter and seals, however, it is not easy to suggest what the first whales looked like and how they evolved into modern cetaceans. Extinct but fully marine cetaceans are found in the fossil record, but missing until recently were transitional fossils filling the gap between walking land mammals and swimming cetaceans.

Pakicetus An exciting discovery was made in Pakistan in 1979 when scientists found what proved to be the oldest known fossil whale. The fossil, officially described as *Pakicetus*, was found embedded in rocks formed from river deposits that were 52 million years old. The river was actually not far from the shores of the former Tethys Sea.

The fossil consists of a complete skull of an archeocyte, an extinct group of ancestors of modern cetaceans. Though limited to a skull, the *Pakicetus* fossil provides precious details on the origin of cetaceans. The skull is cetacean-like but its jawbones lack the enlarged space that is filled with fat or oil and used for receiving underwater sound in modern whales. *Pakicetus* probably detected sound through the ear opening as in land mammals. The skull also lacks a blowhole, another adaptation for diving in cetaceans. Other features show experts that *Pakicetus* is a transitional form between a group of extinct flesh-eating mammals, the mesonychids, and cetaceans. *Pakicetus* likely fed on fish in shallow water and was not yet adapted for life in the open ocean. It probably bred and gave birth on land.

Basilosaurus Several skeletons of another early whale, *Basilosaurus*, were found in 1989 in sediments left by the Tethys Sea and now exposed in the Sahara Desert in Egypt. This whale lived around 40 million years ago, 12 million years after *Pakicetus*. Many incomplete skeletons were found, including, for the first time in an archeocyte, a complete hind leg that features feet with three tiny toes. The legs are small, far too small to have supported the 15-meter-long *Basilosaurus* on land. *Basilosaurus* was undoubtedly a fully marine whale, possibly with non-functional, or vestigial, hind legs.

Ambulocetus Another remarkable find was reported in 1994, also in Pakistan. The now-extinct *Ambulocetus natans* ("the walking whale that swam") lived in the Tethys Sea 49 million years ago, 3 million years after *Pakicetus* but 9 million years before *Basilosaurus*. The fossil includes a good portion of the hind legs. The legs were strong and ended in long feet very much like those of a modern pinniped. The legs were certainly functional both on land and sea. The whale still retained a tail and lacked a fluke. The structure of the backbone shows, however, that *Ambulocetus* swam as modern whales do, by moving the rear portion of its body up and down, even if a fluke was missing. The large hind legs were used for propulsion in water. On land, where it probably bred and gave birth, *Ambulocetus* may have moved as a sea lion does. It was undoubtedly a whale that linked life on land with life at sea.

Ambulocetus natans, the walking whale that swam.

Other fossils found in Pakistan in 2001 link early cetaceans with ungulates, the group that includes cattle, pigs, and hippos. Some of the oldest bones (at least 50 million years old) were from land-living, wolf-like, hoofed animals. Another fossil, found in Australia and described in 2006, clearly belongs to a baleen whale, as evidenced by its skull. The small, big-eyed, 25-million-year-old whale used teeth to capture prey, unlike modern plankton-feeding baleens. It remains unknown if it used echolocation.

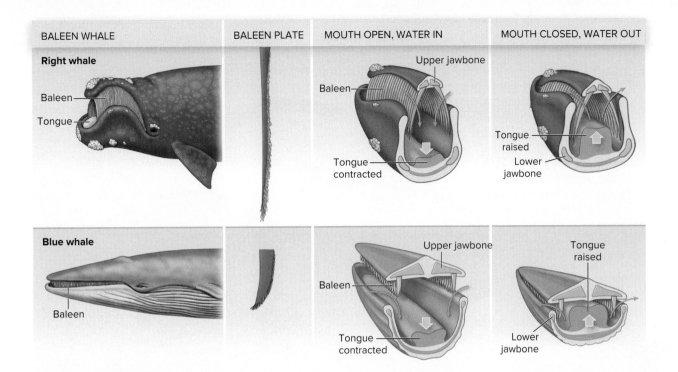

| BALEEN WHALE | BALEEN PLATE | MOUTH OPEN, WATER IN | MOUTH CLOSED, WATER OUT |

Right whale — Baleen, Tongue

Upper jawbone, Baleen, Tongue contracted

Tongue raised, Lower jawbone

Blue whale — Baleen

Upper jawbone, Baleen, Tongue contracted

Tongue raised, Lower jawbone

rorquals, feed by gulping up schools of fish and swarms of krill. The lower part of the anterior half of the body expands when feeding, hence the distinctive, accordion-like grooves on the underside of these whales (see Fig. 11.16). Krill is the most important part of the rorqual diet, especially in the southern hemisphere. Humpback whales often herd fishes like herring and mackerel by blowing curtains of bubbles around them and then surfacing after feeding with their mouth wide open (Fig. 11.20, p. 329).

The right whales (*Eubalaena, Caperaea;* Fig. 11.19a, p. 326) and the bowhead whale (*Balaena mysticetus;* Fig. 11.19e) feed by swimming along the surface with their huge mouths open. They have the largest baleen plates of the whales but the finest bristles (Fig. 11.18). This allows them to filter small plankton like copepods and some krill (Table 11.1).

Figure 11.18 Filtering in whales involves vertical baleen plates. The plates are up to 3 m in the northern right whale (*Eubalaena glacialis*) but only 1 m in the blue whale (*Balaenoptera. musculus*). Water is filtered as the mouth closes and the tongue (yellow arrow) pushes up, forcing the water out through the baleen.

Table 11.1 Diet of Whales

Whale Species	Bottom Invertebrates (%)	Large Zooplankton (%)	Small Squids (%)	Large Squids (%)	Small Pelagic Fishes (%)	Mesopelagic Fishes* (%)	Miscellaneous Fishes (%)
Blue	—	100	—	—	—	—	—
Bowhead	20	80	—	—	—	—	—
Bryde's	—	40	—	—	20	20	20
Fin	—	80	5	—	5	5	5
Gray	90	5	—	—	—	5	—
Humpback	—	55	—	—	15	—	30
Common and Antarctic minkes	—	65	—	—	30	—	5
Northern right	—	100	—	—	—	—	—
Southern right	—	100	—	—	—	—	—
Sei	—	80	5	—	5	5	5
Sperm	5	—	10	60	5	5	15

Fishes found at depths of around 200 to 1,000 m.

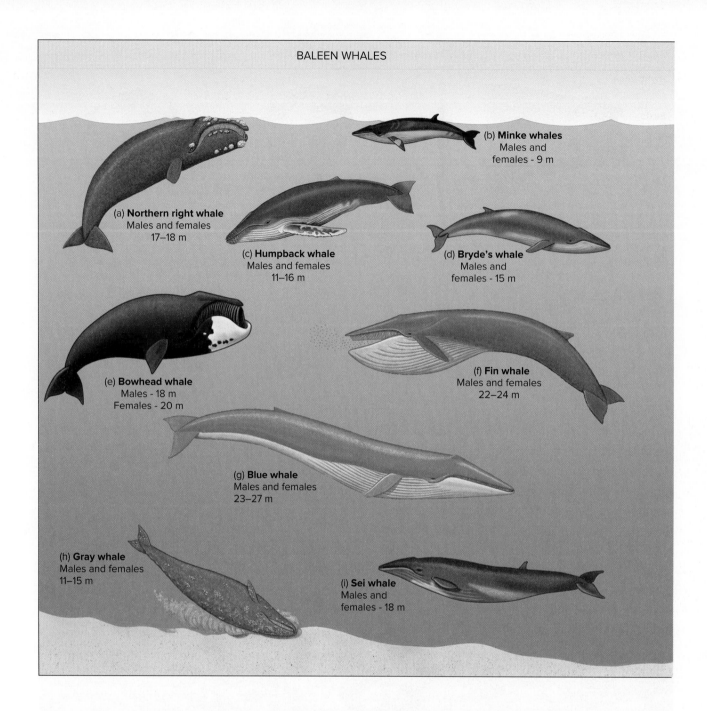

BALEEN WHALES

(a) **Northern right whale**
Males and females
17–18 m

(b) **Minke whales**
Males and
females - 9 m

(c) **Humpback whale**
Males and females
11–16 m

(d) **Bryde's whale**
Males and
females - 15 m

(e) **Bowhead whale**
Males - 18 m
Females - 20 m

(f) **Fin whale**
Males and females
22–24 m

(g) **Blue whale**
Males and females
23–27 m

(h) **Gray whale**
Males and females
11–15 m

(i) **Sei whale**
Males and
females - 18 m

Figure 11.19 Representative baleen and toothed whales.

IN CONTEXT

Amphipods Small crustaceans whose bodies are compressed from side to side.

• 9.4, Arthropods: The Armored Achievers

Gray whales (*Eschrichtius robustus*) are primarily bottom feeders. They feed mostly on amphipods that inhabit soft bottoms (Table 11.1, p. 325). Grays stir up the bottom with their pointed snouts and then filter the sediment (Fig. 11.19h), leaving characteristic pits on the bottom. Most appear to feed on their right side because the baleen on this side is more worn. Some, however, are "left-handed" and feed on the left side. A 10-week-old female kept in captivity in San Diego, California, was fed over 815 kg of squid every day, gaining weight at the rate of 1 kg an hour.

Toothed Whales The roughly 80 remaining species of cetaceans are toothed whales. Their teeth are adapted for a diet of fish, squid, and other prey. They use the teeth only to catch and hold prey, not to chew it, so food is swallowed whole. As in all cetaceans, food is ground up in one of the three compartments

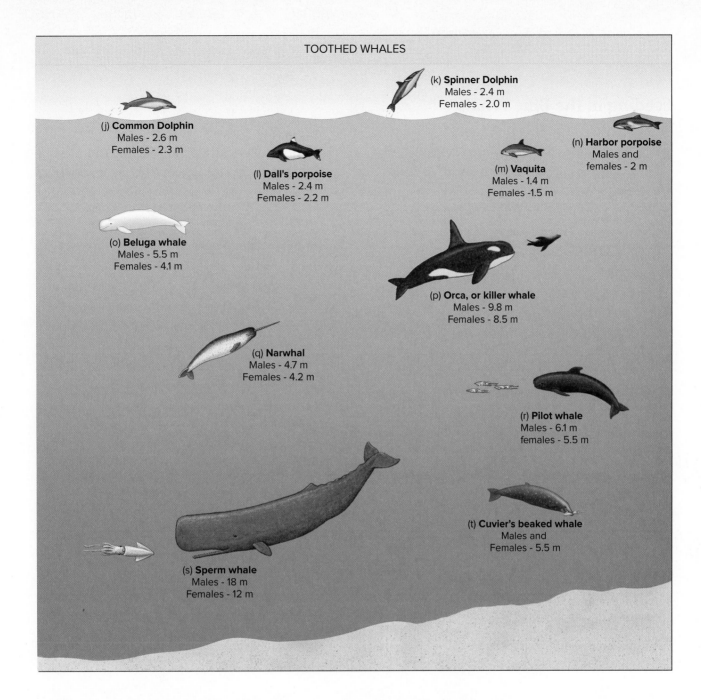

TOOTHED WHALES

(k) **Spinner Dolphin**
Males - 2.4 m
Females - 2.0 m

(j) **Common Dolphin**
Males - 2.6 m
Females - 2.3 m

(n) **Harbor porpoise**
Males and
females - 2 m

(l) **Dall's porpoise**
Males - 2.4 m
Females - 2.2 m

(m) **Vaquita**
Males - 1.4 m
Females -1.5 m

(o) **Beluga whale**
Males - 5.5 m
Females - 4.1 m

(p) **Orca, or killer whale**
Males - 9.8 m
Females - 8.5 m

(q) **Narwhal**
Males - 4.7 m
Females - 4.2 m

(r) **Pilot whale**
Males - 6.1 m
females - 5.5 m

(t) **Cuvier's beaked whale**
Males and
Females - 5.5 m

(s) **Sperm whale**
Males - 18 m
Females - 12 m

of the stomach. The blowhole has one opening, as opposed to two in the baleen whales.

The largest toothed whale is the sperm whale (*Physeter catodon*), the blunt-nosed giant of *Moby Dick* fame (Fig. 11.19s). Together, the sperm and baleen whales are often called the great whales. There is growing evidence that sperm whales, though toothed, are more closely related to baleen whales than to other toothed whales. The sperm whale is now the most numerous of the great whales, even though it was the mainstay of the whaling industry for centuries (Table 11.2). The largest on record weighed 38,000 kg.

Sperm whales are fond of squid, including the giant deep-sea ones. Undigested squid beaks and other debris accumulate in the gut as large globs of sticky material known as **ambergris**, an ingredient in fine perfumes. Sperm whales also eat a wide variety of fishes, lobsters, and other animals (Table 11.1).

Feeding in the Blue Whale

How blue whales (*Balaenoptera musculus*), the largest animals that have ever lived, maintain such a large size while having to dive for food, migrate long distances, maintain body temperature, and reproduce—all on a diet of krill—is a challenging question. Blue whales and other rorquals feed by lunging at high speeds, huge mouths open, into swarms of krill. They then filter and swallow the krill in the large volumes of water contained in their greatly inflated oral cavities by closing their mouths (see Figs. 11.18 and 11.19g). A unique adaptation in rorquals is the presence of longitudinal grooves along a good part of the lower surface of the body (see Fig. 11.16). This is how the rorquals got their name - from the French *rorqual*, originally from *röyrkval*, Norwegian for "furrow whale". The blubber along this area consists of tough ridges divided by deep channels of elastic tissue, which allow the whales to expand their massive oral cavities. The volume of the oral cavity is further increased by the retraction of the tongue along the floor of the mouth.

Energy balance We know that blue whales should get enough energy from plankton. Whales also have to expend energy, however, to accelerate their huge bodies, lunge, and feed. Expanding the oral cavity also increases drag, so scientists believed that they had to limit the time spent diving for food. Figuring out how blue whales cope requires estimating the energetic efficiency, or cost, of feeding. In other words, how much energy they obtain after subtracting the energy they burn while feeding. Measuring energy inputs and outputs in such large animals at sea seemed an impossible task. A team from the University of British Columbia, Canada, investigated this problem by tracking as many as 265 blue whales along the Pacific coast with several types of electronic sensors.

Energy out Whales were found to lunge to feed as many as six times during a dive, with each dive lasting from 3.1 to 15.2 minutes. The speed of whales as they lunged was calculated by correlating speed with the noise of water rushing by the animal, as measured by hydrophones attached to individual whales. Since the oral cavity inflates very much like a parachute, a parachute aerodynamics expert helped develop a mathematical model that allowed calculating the energy needed in each lunge. The team found that whales expend 3,226 to 8,071 kilojoules of energy per lunge.

Energy in Scientists then estimated the volume of the oral cavity from the size of jaw bones in museums, and the density of krill in the water from data in the scientific literature. They calculated that each mouthful of krill contained from 34,726 to 1,912,680 kilojoules, which is 240 times as much energy spent in each lunge! Even including the cost of diving, each dive could provide 90 times as much energy as whales use while diving. Blue whales certainly obtain more than enough energy from their food, and such extraordinary feeding efficiency helps explain their ability to survive despite their gigantic size.

Table 11.2 Status and Estimated Numbers of Great Whales

Whale Species	Status	Estimated Pre-exploitation Number	Current Estimated Numbers
Blue	Recovering populations	160,000–240,000	5,000-15,000
Bowhead	Aboriginal whaling	52,000–60,000	24,000
Bryde's	Aboriginal whaling	100,000	21,000-90,000
Fin	Hunted, including aboriginal whaling	300,000	100,000
Gray (eastern Pacific)	Recovering; aboriginal whaling	15,000–22,000	22,000–26,000
Gray (western Pacific)	Threatened with extinction	1,500–10,000	100–130
Gray (Atlantic)	Extinct	Unknown	0
Humpback	Aboriginal whaling	150,000	50,000-80,000
Common minke	Hunted, including aboriginal whaling	265,000	205,000
Antarctic minke	Hunted until 2013	670,000	460,000-690,000
North Pacific right	Threatened with extinction	unknown	60-80
North Atlantic right	Aboriginal whaling	Unknown	450-500
Southern right	Recovering populations	100,000	10,000
Sei	Hunted, including aboriginal whaling	100,000	60,000-80,000
Sperm	Recovering population	>2,000,000	300,000–500,000

The other toothed whales are much smaller than the great whales. One is the orca, or killer whale (*Orcinus orca*; Fig. 11.19p, p. 327), a magnificent black and white predator with a taste for seals, penguins, fishes, sea otters, and even other whales. They appear to use their white bellies to flash and frighten herring schools and their flukes to stun the fish. Orcas are most common in cold water but are found around the world. They have a nasty reputation, but there are very few confirmed cases of their attacking humans in the wild. DNA analysis has recently confirmed that there are at least three different species of orcas, each with a distinct appearance, hunting habit, and diet.

Figure 11.20 Humpback whales (*Megaptera novaeangliae*) often hunt in groups by surrounding schools of small fish with bubbles before lunging to the surface with open mouths and swallowing hundreds of fish in one gulp. Some of the whales help others by diving below the school to force the fish to the surface; others are known to flap the surface to herd the fish with their flukes before forming the bubbles. whales.

(a)

(b)

Figure 11.21 (a) Dolphins like the Atlantic sotted dolphin (*Stenella frontalis*) often ride the bow wave of boats or even that of whales. They ride without beating their tails, obtaining thrust from the pressure wave in front of the ship. (b) The Burrunan dolphin (*Tursiops australis*) from southern Australia was recognized as a new species in 2011. The Australian humpback dolphin (*Sousa sahulensis*) was also recognized as new in 2014.

One of the most enigmatic of whales is the narwhal (*Monodon monoceros*; Fig. 11.19q, p. 327), a small Arctic whale having a long, spiral tusk that can be as long as 2.7 m. Narwhal tusks washed ashore apparently gave rise to the legend of the unicorn. The tusk, a modified tooth found only in males, appears to be related to the establishment of a hierarchy of dominance. It has been recently shown, however, that the tusks have a network of small, fluid-filled tubes, which are connected to the narwhal's nervous system, and it has been suggested that they could sense the water for the detection of prey or other chemical cues.

Though they are all whales, most of the small toothed whales are called dolphins or porpoises. Technically, the six species of porpoises are blunt-nosed whales (Fig. 11.19m and n, p. 327) having flattened teeth in contrast to the conical teeth of dolphins. In some places, however, the name "porpoise" is given to some of the dolphins.

The many species of dolphins typically possess a distinctive snout, or beak, and a perpetual "smile." Playful, highly social, and easily trained, dolphins win people's hearts. They often travel in large groups called **pods**, herds, or schools. They like to catch rides along the bows of boats (Fig. 11.21a) or around great whales. The bottlenose dolphin (*Tursiops truncatus*) is the dolphin seen in marine theme parks and oceanaria around the world. An Australian population of bottlenose dolphins was recognized in 2011 as a new separate species, *Tursiops australis* (Fig. 11.21b). The spinner dolphin (*Stenella longirostris*; Fig.

11.19k, p. 327) is so named because of its spectacular twisting jumps in the air. It is one of the species of dolphins that get caught in the nets of tuna fishers. This happens because the tuna and dolphins eat the same fish and often occur together.

Whaling Whale hunting, or **whaling**, is an old tradition with a rich history. Stone Age people hunted whales as early as 6000 b.c. Native Americans hunted gray whales in prehistoric times; the aboriginal peoples of the Arctic still legally hunt them. Basques may have hunted them off Newfoundland before Columbus. It was not until the 1600s, however, that Europeans started to substantially exploit the great whales in the North Atlantic. Americans, who eventually dominated worldwide whaling, began hunting off New England in the late 1600s. Whales were harpooned from small, open boats (Fig. 11.22), a technique whalers learned from the natives. It was a rewarding fishery, though not one exploited primarily for food. Blubber provided "train oil," which was used to make soap and lamp oil. Baleen was used to make stays for corsets and other goods. Meat and other valuable products also were obtained. Whaling efforts rapidly increased after fast steamships and the devastating explosive harpoon were introduced in the 1800s. The largest and fastest whales, like the blue and fin whales, were then at the mercy of whalers.

Impacts of Whaling Whales are long-lived mammals with a low reproductive rate. The great whales generally give birth to one well-developed calf that has been carried by the mother for a year or more. Females usually don't become pregnant for one or two years after giving birth. As a result of this low reproductive potential, whale populations could not stand the intense whaling pressure, and many of the fisheries collapsed.

The first to be seriously depleted was the slow-swimming North Atlantic right whale (*Eubalaena glacialis*; Fig. 11.19a), the "right" species to be killed, according to whalers, because it floated after being harpooned. By the early 1900s whaling had moved to the rich feeding grounds around Antarctica, where other species of whales could be hunted. This location proved to be a

North Wind Picture Archives/Alamy

Figure 11.22 Long-boat crew harpooning a right whale in the 1800s.

bonanza. Whaling nations developed factory ships able to process whole carcasses. The Antarctic fishery reached its peak in the 1930s. The whales received a reprieve during World War II, but it was too late for saving the fisheries. It is estimated that more than a million whales were taken from Antarctica alone.

Blue whales, the largest of them all, were especially sought. A large specimen yielded more than 9,000 gallons of oil. It has been estimated that over 200,000 blue whales were taken worldwide between 1924 and 1971, close to 30,000 during the 1930–1931 whaling season alone.

Catches climbed way above the sustainable yield level. Catch per whaler-day's effort declined every year after 1936. As many as 80% of all blue whales caught by 1963 were sexually immature, so that there were even fewer individuals able to perpetuate the species.

Fin whales, the second largest of all whales, became the next major target as blue whales became more and more scarce. The 1950s and early 1960s saw annual catches of 20,000 to 32,000 fin whales per year, mostly from Antarctica. As catches dwindled, whalers shifted their target again in the mid-1960s, this time to the smaller sei whale (*Balaenoptera borealis*; Fig. 11.19i, p.326). The sei whale averages a length of around 13 m, whereas the fin whale averages 20 m.

Whaling Policy The abrupt disappearance of the more commercially valuable whales meant lower profits for the whaling industry. In 1946, 20 whaling nations established the International Whaling Commission (IWC) in an attempt to regulate whale hunting to stop overfishing. It collected data on the number of whales and set annual quotas for the number of whales to be caught each year, quotas that, unfortunately, were non-binding and could not be enforced. Furthermore, some whaling nations did not belong to the IWC. The blue whale was not completely protected by the IWC until the 1965–1966 season, long after its numbers had been drastically reduced; by then blue whales were so hard to find that the fishery for them was no longer profitable. Even under the protection of the IWC, blue whales were hunted at least until 1971 by countries that did not belong to the IWC.

Under mounting pressure from conservationists, the IWC gradually banned the hunting of other whales. Demand for whale products, mostly oil used in the manufacture of margarine and lubricants, was reduced because substitutes had been found for most of them. Whale meat, however, continued to be used as pet food and is still valued as human food, mostly in Japan. The lower quotas of the IWC were, unfortunately, not always accepted by all nations.

The United States Congress passed the Marine Mammal Protection Act of 1972, which bans the hunting of all marine mammals in U.S. waters (except in the traditional fisheries of Alaskan natives) and the importation of marine mammal products. By 1974 the IWC had protected the blue, gray, humpback, and right whales around the world, but only after their catches were no longer economically viable. Sperm, minke, fin, and sei whales were still hunted, but worldwide catches began to dwindle. Catches of these whales fell from 64,418 in 1965 to 38,892 in 1975 and to 6,623 in 1985. A moratorium on all commercial whaling was declared by the IWC in 1985, a move long sought

by conservation groups. The former Soviet Union halted all whaling in 1987. Japan, Iceland, and Norway, however, opted in 1988 to continue hunting minke, fin, and sei whales, as allowed by the IWC under the controversial title of "scientific whaling."

In 1994 the IWC created a vast sanctuary for all whales in the waters around Antarctica, the main feeding ground for 80% of the surviving great whales. Japan, however, decided to continue "scientific hunting" of whales in Antarctica. Starting with the 1997–1998 season, Japan took 440 minkes from Antarctica and 100 from the North Pacific each season until 2002, when it expanded its North Pacific catch to include 150 minke, 50 Bryde's (*Balaenoptera edeni;* Fig. 11.19d, p. 326), 10 sperm, and, for the first time since 1987, 50 sei whales. Disruptions from anti-whaling activists and low demand at home were responsible for the low Japanese catches, almost all minkes, in Antarctica in 2007–2008 (550), 2008–2009 (680), and 2009–2010 (507). The hunt was halted during the 2010–2011 season after catching 170 whales.

In 2006 the 70-nation IWC abandoned the conservationist policies of its 2003 Berlin Initiative on the belief that the populations of many of the whale species were large enough to support sustainable whaling. Norway had already resumed commercial whaling in 2005 by setting itself an annual quota of 796 North Sea minkes (Fig. 11.23). Iceland, which had not hunted whales since 1989, joined Norway in 2006 to further break the 1985 ban. South Korea announced in 2012 its intention to start scientific hunting of minkes but abandoned plans soon after.

The United Nations' International Court of Justice (ICJ) ordered Japan to stop "scientific" whaling in Antarctica in 2014, ruling that it was not for scientific purposes as claimed. Since 2005 and under the provisions of "scientific" whaling, about 3,600 minkes were hunted in Antarctica, a high of 680 during the 2005-2006 season to 251 in 2013-2014. The 2014-2015 hunt was called off. The IWC endorsed ICJ's ruling and approved stricter criteria for nations applying to undertake whaling for research purposes. Japan, however, will continue "scientific" whaling of minkes in the northwestern Pacific.

Small-scale whaling, or aboriginal subsistence whaling, remains part of the traditional fisheries of the native inhabitants of the Arctic region from Greenland to Siberia and in the Lesser Antilles in the Caribbean. One of the whales hunted in the Arctic, the bowhead, and another in the Lesser Antilles, the humpback, are endangered. Other smaller whales—the orca, narwhal, and beluga (*Delphinapterus leucas;* Fig. 11.19o, p. 327)—are also hunted in the Arctic.

Current Status of Whaling The current status of the great whale species according to population estimates varies from "vulnerable" to "critically endangered" depending on the classification of particular agencies or countries. The status may also vary among different populations of the same species. Collision with ships, entanglement with fishing nets, noise pollution, and illegal hunting remain as threats now that pressure from commercial whaling has almost disappeared. Nobody knows when the great whales will again roam the oceans

Figure 11.23 A harpooned common minke whale (*Balaenoptera acutorostrata*) being hauled on board a Norwegian whaling ship in the North Sea.

Figure 11.24 The blue whale (*Balaenoptera musculus*) has made a remarkable and unexpected comeback along the Pacific coast of North America. It is estimated that 2,200 individuals now live in the northeastern Pacific. These two individuals were photographed in the Gulf of California, Mexico.

in numbers approaching those before the start of large-scale whaling. Some experts are afraid that a few critically endangered species will never recover completely. Recovery is nevertheless under way in some species. The California gray whale, protected since 1947, has made a comeback. It was removed from the endangered species list in 1994. Recent research, however, estimates that 78,000 to 117,000 grays lived before exploitation, much higher than most estimates (see Table 11.2, p. 329). In 1997 the IWC allowed the catching of 600 gray whales by native hunters in Siberia and 20 by the Makah Indian tribe in Washington State. The tribe only took a single whale, in 1999. Resumption of the hunt has been stopped awaiting the result of court litigation. Even the blue whale, whose reproduction is severely limited by its restriction to small populations scattered around the world, is making a comeback. It has returned to the Arctic Ocean north of Norway, a region where they flourished before their near extermination by whalers. Sightings off the Pacific coast of North America have increased sharply (Fig. 11.24). Their numbers in Antarctica, however, are even lower than first estimates: around 500 individuals, or only 1% of those feeding there before whaling began.

Dolphins are also at great risk, having replaced the larger whales as the most threatened of all cetaceans. Small cetaceans are not under IWC regulation, and as many as 28 species are in immediate danger of extinction. Less than 600 *vaquitas*, or "little cows" (*Phocoena sinus*; Fig. 11.19m, p. 326), are left. This shy, shovel-nosed porpoise, known only from the northern Gulf of California, remained unknown to science until 1958. Everywhere, fishers are depleting stocks of fish and squid on which dolphins feed. Dolphins themselves are being hunted for human food. Most of the 5 species of river dolphins are critically endangered. China's Yangtze River dolphin, or *baiji* (*Lipotes vexillifer*), a freshwater dolphin found only in that Chinese river, was declared

extinct in 2006, the first cetacean to become extinct as a result of human activity. A second freshwater cetacean, the Yangtze finless porpoise (*Neophocaena phocaenenoides asiaorientalis*), is in danger of also becoming extinct.

Tuna fishers using giant purse seine nets sometimes trap and drown the many dolphins that often swim above schools of tuna, mostly yellowfin tuna (*Thunnus albacares*) in the eastern Pacific. During the early 1970s an estimated 200,000 dolphins died annually, mostly at the hands of American fishing fleets. The slaughter induced such public outrage that the United States, through the Marine Mammal Protection Act of 1972, called for a reduction in the accidental deaths of dolphins. It imposed a quota of 20,500 for the number of dolphins that could be killed by American fleets. The use of special nets was enforced, and observers were placed on board vessels to verify compliance with the ruling. By 1990 it was estimated that the number of dolphins killed by the American tuna fleet, by then operating in western Pacific waters, had reached zero. The same year the three biggest tuna packers in the United States pledged not to buy or sell fish that was caught using methods that injure or kill dolphins. Tuna cans began to display "dolphin-safe" labels, and imported tuna caught without the use of dolphin-safe methods were banned from sale in the United States.

Dolphins have also been entangled and killed by the thousands in drift nets, which also threaten sharks, sea turtles, seabirds, and other marine life. The nets, some as large as 60 km long and 15 m deep, have been used to catch fish and squid, but they actually catch practically anything that tries to swim by. Not only do they deplete valuable commercial fisheries like albacore tuna and salmon, but they also trap many noncommercial species. Hundreds of fishing boats outfitted for drift netting were used to catch tuna in the South Pacific. The United Nations called for a moratorium on drift-net fishing in 1992, and Japan and Taiwan, with the largest fleets of drift-net boats in the Pacific, ended the practice in 1993. Still drift-net fishing, some of it illegal, continues in some areas.

· ·

Section Review

1. **Summarize** Make a table like the one below to summarize the adaptations that the different groups of marine mammals have evolved to live in the marine environment.

Group	Examples	Adaptations
Pinnipedia		
Carnivora		
Sirenia		
Cetacea		

2. **Compare and contrast** baleen whales and toothed whales.

11.4 Biology of Marine Mammals

VOCABULARY

spout

bends

apneustic breathing

bradycardia

echolocation

melon

spermaceti organ

spermaceti

breaching

beaching

harems

bachelor groups

ovulation

gestation

delayed implantation

Main Idea
Marine mammals have adaptations and complex behaviors (particularly communication) which allow them to survive in the marine environment.

Key Questions
1. How are cetaceans adapted for swimming?
2. Why do pinnipeds and cetaceans migrate?
3. What is stranding?

It is surprising how little we know about marine mammals. Most are difficult or impossible to keep in captivity or even to observe for long periods at sea. Some whales and dolphins are rarely seen, so what little we know about them comes from captive or stranded individuals and information gathered over the years by whalers.

Swimming
Streamlining of the body for swimming is a hallmark of marine mammals. Seals, sea lions, and the walrus swim mostly by moving their flippers. Sea otters swim by using their hind feet, which are broad and webbed. Polar bears are strong swimmers and can cover long distances using all four legs for swimming, very much like a dog. Sirenians and cetaceans, in contrast, move their tails and flukes up and down (Fig. 11.25), in contrast to fishes who move their tails from side to side. Sea lions have been timed at speeds of 35 kph, orcas and blue whales at 50 kph. A group of common dolphins (*Delphinus delphis*; Fig. 11.19j, p. 327) was recorded bowriding at a speed of 64 kph.

Cetaceans have the advantage of having the blowhole on top of the head. This allows them to breathe even though most of the body is under water, although many pinnipeds and dolphins jump clear out of the water to take a breath when they are swimming fast. It also means that cetaceans can eat and swallow without drowning. The blowhole is absent in all other marine mammals and therefore must use their frontally located nostrils to breathe air just like us. To avoid inhaling water, all marine mammals take

Figure 11.25 Cetaceans swim with strong up-and-down movements of the tail and flukes.

	SURFACING AND BLOWING	START OF DIVE	END OF DIVE
Blue whale			
Fin whale			
Gray whale			
Right whale			
Sperm whale			
Humpback whale			

Figure 11.26 Great whales can be identified from a distance by their blowing pattern, their outline on the surface, and the way they dive.

very quick breaths. A fin whale can empty and refill its lungs in less than two seconds, half the time we take, even though the whale breathes in 3,000 times more air!

In the large whales, the moisture in their warm breath condenses when it hits the air. Together with a little mucus and seawater, this water vapor forms the characteristic **spout**, or blow (Fig. 11.26). The spout can be seen at great distances, and its height and angle can be used to identify the whale (Fig. 11.26). The blue whale, for instance, has a spout some 6 to 12 m high.

To keep warm in cold water, the great whales depend on a thick layer of blubber. Feeding, however, leaves their huge mouths exposed to low temperatures, a major problem in the very cold polar waters where they normally feed. A network of blood vessels in their tongues reduces heat loss by transferring heat from warm blood into vessels that carry it back to the body core.

Diving Marine mammals have mastered the art of diving, and most make prolonged dives to considerable depths for food. There is a wide range in diving ability. Sea otters can dive for only 4 or 5 minutes, to depths of perhaps 55 m. Pinnipeds normally dive for up to 30 minutes at maximum depths of 150 to 250 m. Female northern elephant seals (*Mirounga angustirostris*), however, are capable of continuous deep dives of up to 400 m. One individual was recorded diving to a depth of 1,500 m. The Weddell seal (*Leptonychotes weddelli*) has been recorded diving for as long as 1 hour 13 minutes and as deep as 575 m. Sirenians spend most of their time in relatively shallow water, but a dugong was recorded diving to 20 m. Polar bears have been estimated capable of diving as deep as 6 m.

The plankton-feeding habits of baleen whales do not require them to dive too deeply for their food, and they seldom venture below 100 m. Toothed whales, however, are excellent divers. Dolphins are known to dive as deep as 300 m. Though sperm whales are known to dive to 2,250 m for at least an hour, beaked whales are the champion divers. The Cuvier's beaked whale (*Ziphius cavirostris*; Fig. 11.19t, p. 327) regularly dives to more than 1,000 m to catch the deep-water squids that make up most of its diet. The deepest recorded dive of a beaked whale was close to 1,900 m for up to 85 minutes. Different species of toothed whales have evolved different hunting strategies. Short-finned pilot whales (*Globicephala*; Fig. 11.19r), for instance, sprint as fast as 9 m per second after single or several prey as deep as 1,000 m, very much like large wildcats do.

A potential problem faced by air-breathing, diving animals (including human divers) results from the presence of large amounts of nitrogen (70% of total volume) in the air. Nitrogen dissolves much better at high pressures, like those experienced at depth. The blood of scuba divers picks up nitrogen while they are below the surface. If the pressure is suddenly released, some of the nitrogen will not stay dissolved and forms tiny bubbles in the bloodstream. You can see a similar phenomenon when you open a bottle of soda pop. As long as the top is on, the contents are under pressure. The carbonation, actually carbon dioxide gas (CO_2), remains dissolved. When you open the bottle, the pressure is released and bubbles form. When nitrogen bubbles form in the blood after diving, they can lodge in the joints or block the flow of blood to the brain and other organs. This produces a painful condition known as the **bends**, or decompression sickness. To avoid the bends, human divers must be very careful about how deep they go, how long they stay under water, and how fast they come up.

Marine mammals dive deeper and stay down longer than human divers, so do they get the bends? Marine mammals do have adaptations that prevent nitrogen from dissolving in the blood. Human lungs work pretty much the same while scuba diving under water as on land. When marine mammals dive, on the other hand, their lungs actually collapse. They have a flexible rib cage that gets pushed in by the pressure of the water (Fig. 11.27). This, and increased water pressure with the increased depth, squeezes the air out of peripheral areas of the lungs where it readily dissolves into the blood. Air is moved instead into the central spaces of the lungs, where little nitrogen is absorbed. High concentrations of hemoglobin and myoglobin store the oxygen that marine mammals need in the blood and muscles during dives. Less air is thus needed in the lungs, which reduces the amount of nitrogen that can be absorbed. Apneustic breathing, exhaling before a dive, further reduces the amount of air—and therefore nitrogen—in the lungs. A reduction in the amount of air in the lungs has the additional advantage of decreasing buoyancy, which makes descents to deep water easier.

The long-held view that marine mammals don't get the bends may not be accurate after all. It has been suggested that deep-diving whales suffer from some of the effects of the bends, as evidenced by the degeneration of bone and cartilage in sperm whales, possibly caused by nitrogen bubbles.

Breathing without Gills The long, deep dives of marine mammals require several crucial adaptations. For one thing, they must be able to go a long time

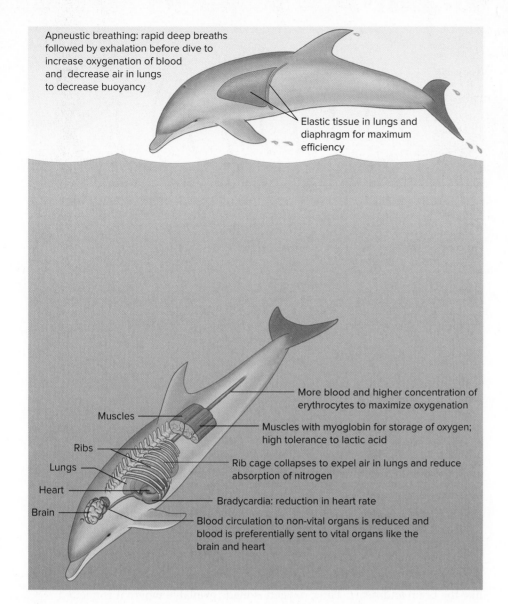

Apneustic breathing: rapid deep breaths followed by exhalation before dive to increase oxygenation of blood and decrease air in lungs to decrease buoyancy

Elastic tissue in lungs and diaphragm for maximum efficiency

More blood and higher concentration of erythrocytes to maximize oxygenation

Muscles with myoglobin for storage of oxygen; high tolerance to lactic acid

Rib cage collapses to expel air in lungs and reduce absorption of nitrogen

Bradycardia: reduction in heart rate

Blood circulation to non-vital organs is reduced and blood is preferentially sent to vital organs like the brain and heart

Muscles

Ribs

Lungs

Heart

Brain

Figure 11.27 Marine mammals have evolved a series of adaptations that allow them to undertake long, deep dives.

without breathing. This involves more than just holding their breath, for they must keep their vital organs supplied with oxygen. Being homeotherms means that marine mammals must use energy to keep their temperature constant— and the release of energy by aerobic respiration requires oxygen. To get as much oxygen as possible before dives, most marine mammals and some seabirds take several deep breaths, then rapidly exhale, a behavior known as **apneustic breathing** (Fig. 11.27). As much as 90% of the oxygen contained in the lungs is exchanged during each breath, in contrast to 20% in humans. Many marine mammals have elastic tissues in their lungs and diaphragms, which help lungs, fill and empty rapidly and efficiently.

Marine mammals are also better than other mammals at absorbing the oxygen from the air and storing it in their blood. To start with, they have relatively more blood than non-diving mammals. As in seabirds, their blood also contains a higher concentration of erythrocytes, and these cells carry more hemoglobin. Furthermore, their muscles are extra-rich in myoglobin, which means that the muscles themselves can store a lot of oxygen.

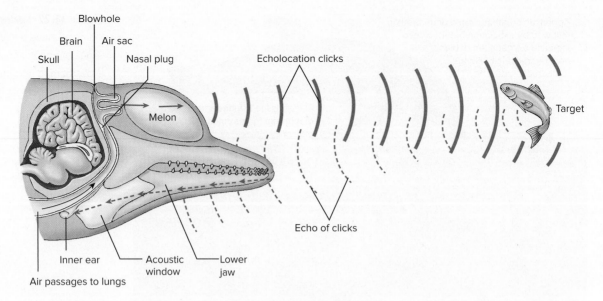

Figure 11.28 Dolphins echolocate by emitting bursts of sound waves, or clicks, as air is pushed through internal air passages. Two muscular nasal plugs act as valves, closing and opening the passages. Flaps of tissue on the plugs probably also produce sound by vibrating in the moving air. The clicks are focused into a beam by the melon. To cover a wider area, the dolphin moves its head from side to side. The melon is also known to receive the echoes and transmit them to the ears, but most are received by the lower jaw.

Marine mammals have adaptations that reduce oxygen consumption in addition to increasing supply. When they dive, the heart rate slows dramatically, an automatic reflex known as **bradycardia**. Diving bradycardia has been recorded in many diving vertebrates, including humans.

It is thought to have evolved as an oxygen-conserving mechanism. In the northern elephant seal, for example, the heart rate decreases from about 85 beats per minute to about 12. Blood flow to non-essential parts of the body, like the extremities and the intestine, is reduced, but it is maintained to vital organs like the brain and heart. Thus, oxygen is made available where it is needed most when oxygen supply is cut off during a dive.

The reduction or absence of oxygen triggers the muscle cells of many animals (including humans during exercise) to carry out anaerobic respiration, an "emergency" respiration that releases, though inefficiently, some energy without the need of oxygen. A product of anaerobic respiration is lactic acid, a chemical that produces cramps and is detrimental to muscular function. The tissues of marine mammals, however, have evolved a high tolerance to the accumulation of lactic acid, another adaptation to increase diving efficiency.

Echolocation

Marine mammals depend little on the sense of smell, which is so important to their terrestrial cousins. Their vision is excellent, but some have developed **echolocation**, another sensory system based on hearing. Echolocation is nature's version of sonar. Most if not all toothed whales, including dolphins and porpoises, and some pinnipeds are known to echolocate. At least some baleen whales could also use echolocation. There is no evidence of echolocation among sirenians, who use their sensitive whiskers when grazing. Echolocation is not

exclusive to marine mammals. Bats, for example, use it to find insects and other prey while flying at night.

Marine mammals echolocate by emitting sound waves, which travel about five times faster in water than in air, and listening for the echoes reflected back from surrounding objects (Fig. 11.28). The brain then analyzes the echoes. The time it takes the echoes to return tells the animals how far away the object is.

The sounds used in echolocation consist of short bursts of sharp clicks that are repeated at different frequencies. Low-frequency clicks have a high penetrating power and can travel long distances. They reflect from large features and are used to obtain information on the surrounding topography. Low-frequency sound waves can also be used by some toothed whales to stun their prey. It seems that sperm whales catch squids, including giant squids, without using their teeth, even if the teeth are actually of little help because they are present only in the lower jaw. In fact, live squids have been known to come out alive from the stomachs of freshly caught whales! To discriminate more detail and locate nearby prey, high-frequency clicks that are inaudible to humans are used. Experiments have shown that blindfolded bottlenose dolphins can discriminate between objects of slightly different size or made of different types of materials and can even detect wires.

We are not completely sure how echolocation operates in marine mammals. Cetaceans produce clicks, squeaks, and whistles as air is forced through the air passages and several associated air sacs while the blowhole is closed (Fig. 11.28). The frequency of the clicks is changed, or modulated, by contracting and relaxing muscles along the air passages and sacs. A fatty structure on the forehead of toothed whales, the **melon**, appears to focus and direct the outgoing sound waves.

The melon gives these whales their characteristic rounded foreheads. To accommodate the melon, the skull is modified to form a pointed, dish-shaped face. The skull is also asymmetric, the right side being slightly different from the left side. Belugas (Fig. 11.29) have a bulging forehead, which changes shape as the melon, moved by muscles, focuses the sound. The huge forehead of the sperm whale is filled in part with a massive melon called the **spermaceti organ**. Whalers originally thought this was the sperm sac of the whale, hence the peculiar name. This organ is filled with a waxy oil, **spermaceti**, once much sought after for making candles and still used as a lubricant for precision instruments. The actual function of the spermaceti organ is a controversial issue. It has been suggested

Figure 11.29 (a) The beluga (*Delphinapterus leucas*) is a white Arctic whale with a conspicuous melon. This individual was photographed under the ice in the White Sea, Russia. (b) In their natural environment belugas live in small groups.

(a)

(b)

that the deep-diving sperm whale might also use the spermaceti organ to regulate buoyancy or to absorb excess nitrogen, keeping it out of the bloodstream.

In toothed whales incoming sound waves are received primarily by the lower jaw (Fig. 11.28, p. 340). The ear canal that connects the outside with the inner ear is reduced or blocked in most cetaceans. The jawbones, filled with fat or oil, transmit sound to the two, very sensitive inner ears. Each ear receives sound independently. The ears are protected by a bony case and embedded in an oily mixture that insulates the ear but allows sound waves to pass from the jaws. Sound information is sent to the brain, which forms a mental "picture" of the target or surroundings. In fact, sight and sound information seems to be handled similarly by the brain. Captive dolphins can recognize by echolocation objects they have seen and recognize by sight those they have previously echolocated.

Behavior

The mammalian brain has evolved as an association center for complex behaviors in which learning, not instinct, dominates. In contrast to fishes, birds, and other vertebrates, mammals rely mostly on past experience, stored and processed by the brain, to respond to changes in the environment. Cetacean brains are as complex as those of humans as far as their anatomy is concerned. Their brains even have the same type of brain cell that is associated with abstract reasoning in humans.

Most marine mammals are highly social, living in groups at least part of the time. The polar bear is a notable exception. Many pinnipeds live in huge colonies during the breeding season. Most cetaceans spend their entire lives in highly organized pods of a few to thousands of individuals (Fig. 11.29b, p. 341). Some pods include smaller subgroups organized by age and sex. To keep in contact, many of their highly complex and sophisticated behaviors are directed toward members of their own species. It has been said that cetaceans, like humans, have complex cultures that vary from group to group.

Vocalizations Sounds, or vocalizations, play a prominent role in communication. Sea lions and fur seals communicate by loud barks and whimpers; seals use more sedate grunts, whistles, and chirps. The vocalizations of pinnipeds are especially important in maintaining territories during reproduction (see "Reproduction," below). Females and their pups or calves recognize each other partly by their sounds, partly by smell. Sea otters use body contact and several vocalizations to communicate with each other. Sirenians similarly use "chirps" and "squeaks" for communication.

Cetaceans produce a rich variety of vocalizations that are different from the sounds used for echolocation. Both types of sounds can be produced simultaneously, providing further evidence of the complexity of sound production in marine mammals. Social vocalizations are low-frequency sounds that humans can hear. The variety of sounds is amazing and includes grunts, barks, squeaks, chirps, and even "moos." Different sounds are associated with various moods and are used in social and sexual signaling. Whistles, emitted in many variations and tones, are characteristic of each species. Some of these sounds serve as a "signature," allowing individuals of the same species to recognize one another.

It was recently found that bottlenose dolphins use signature sounds to address each other. Male fin whales use high-intensity songs to attract females. Among the more than 70 calls that have been identified among orcas, some are present in all individuals, whereas others are "dialects" that identify certain pods.

Sounds are also used to maintain the distance between individuals and have an important role in the structure of the pod. Particular sounds are emitted during breeding, feeding, alarms, and birth. Mother gray whales grunt to stay in contact with their calves. Fin whales make a low-pitched sound thought to be involved in long-distance communication. Right whales have at least six distinct calls, each related to a specific function, and they can even call louder as background noise gets louder.

The humpback whale is renowned for its soulful songs. They are sung by breeding males to attract females by advertising their readiness to mate. The songs consist of phrases and themes repeated in a regular pattern for a half hour or longer. They may be repeated over and over for days. The songs change over time. Males also start each breeding season with the song they were singing at the end of the previous breeding season. New songs learned from immigrants have been shown to gain instant popularity among native whales. Researchers record and catalog songs to help track whales in their annual migrations. It has been found, for instance, that blue whales are enigmatically singing in a key that is lower than in previous records.

Communication among cetaceans is not restricted to vocalizations. Researchers have described a variety of postures and movements that can also indicate the animal's mood. Dolphins clap their jaws or turn around with their mouths open as a threat. The loud cracking sound made when some marine mammals slap their flukes or flippers on the surface is thought to be a warning signal.

Play Behavior Cetaceans are noted for their play behavior, seemingly pleasurable activities with no serious goal. Many species, including the great whales and orcas, play with food or floating objects like logs, kelp, and feathers, throwing them up in the air or holding and pushing them with their snouts. Individuals can swim head down or on their backs apparently just for the fun of it. Dolphins play with rings of air bubbles they create. Dolphins also like to surf, and pilot and right whales go sailing with their flukes out of the water to catch the wind. Sex play, the rubbing and touching of the genital opening, is also common.

The sight of a great whale **breaching**, leaping up in the air and loudly crashing on the surface, is breathtaking (Fig. 11.30). Breaching has been variously interpreted as a warning signal, for scanning the surface or the

Figure 11.30 A humpback whale (*Megaptera novaeangliae*) performing a full spinning breach.

©Bill Ober

(a)

(b)

(c)

Figure 11.31 Whale watchers can be rewarded with examples of the complex behavior of whales: (a) sperm whales (*Physeter macrocephalus*) surrounding an injured member of a pod, (b) "spying" behavior in orcas (*Orcinus orca*) in the wild, (c) two bottlenose dolphins (*Tursiops truncatus*) carrying a stunned companion to the surface to breathe.

shoreline, for of getting rid of external parasites or an ardent lover, and simply for fun. After a deep dive, sperm whales can breach, fall on their backs, and make a splash that can be heard 4 km and seen 28 km away! Many whales stick their heads out of the water to spy on their surroundings (Fig. 11.31b).

Social Behavior The complex behavior of cetaceans is evident in other ways. When one individual is in trouble, others can come to assist (Fig. 11.31a). Members of a pod refuse to leave a wounded or dying comrade. Whalers knew that a harpooned whale was a lure for others, which are drawn from miles around. Dolphins will carry injured individuals to the surface to breathe (Fig. 11.31c), and there are records of females carrying the body of a stillborn calf until it rots.

Many toothed whales work together when they hunt, some in coordinated pairs or groups. Sometimes whales take turns feeding while their partners herd a school of fish. An individual may investigate something strange lying ahead while the rest of the group waits for the "report" of the scout. Some humpback whales work in groups by producing bubbles to heard fish (see Fig. 11.20, p. 329). Studies of dolphins in the wild have shown that they belong to a complex society, one in which long-term partnerships of members of the same sex play an important role in sexual behavior, parental care, and other aspects of daily life. Social behavior in cetaceans could ultimately show many parallels with social behavior in large-brained mammals like apes and humans.

Beaching One of the mysteries of the behavior of cetaceans is the stranding, or beaching, of individuals, sometimes dozens, on beaches (Fig. 11.32). The animals refuse to move, and efforts to move them into deeper water usually fail. Even if they are pulled out to sea, they often beach themselves again. The whales die because their internal organs collapse without the support of the water. Stranding has been described in many species, but some, like pilot and sperm whales, strand themselves more often than others. It appears that whales become stranded when they follow one or more members of their group that have become disoriented by a storm, illness, or injury. This indicates the strong cohesiveness and herd instinct of the group.

Stranding of whales has been linked to high-intensity sonar that uses bursts of intense sound waves used by the navies of the United States and other nations to detect enemy submarines. Whales stranding after naval exercises where the sonar was used showed hemorrhages in the brain and inner ears, which cause disorientation and death. Tissue injury can also be caused by the formation of nitrogen bubbles, or the bends. Although the United States Navy has been sued, conflicts between the military and marine mammals will be an issue for years to come. Noise from ship propellers, depth sounders, and sonar used for scientific purposes has also been suggested as disturbing to whales. Dolphins spend more time on the surface and rest less, and orcas calls last longer, when whale-watching boats are close by.

Dolphins and Humans The relationship between dolphins and humans is a controversial one. Though exaggerations abound, there are authenticated cases of dolphins approaching human swimmers who appeared to be in trouble. For more than a century, fishers in southern Brazil have established a unique partnership with dolphins. The dolphins detect fish and deliver them to fishers waiting with nets. Fishers have learned to interpret cues given by the dolphins about the location and abundance of fish. Generations of dolphins have learned that a row of fishers holding a net in shallow water means an easy catch for themselves, even if it has to be shared with funny-looking, two-legged mammals.

Some people swear of experiencing "spiritual inspiration" while swimming among dolphins during the "dolphin encounters" offered by some resorts. Some even advocate "dolphin-assisted" deliveries of babies. Dolphins trained for military purposes by the former Soviet navy are being used to treat children suffering from behavioral disorders. Other people see this as outright exploitation of the captive animals, and interactions with humans could after all be harmful to dolphins.

Migrations

Many pinnipeds make seasonal migrations, often traveling thousands of miles from feeding grounds to breeding areas. Male southern elephant seals (*Mirounga leonine*) are known to travel as far as 8,000 km to mate. The walrus migrates following the coastal pack ice. Some migrate long distances to group in large numbers on isolated rocky beaches when the ice melts in the summer. The polar bear similarly follows the edge of the coastal pack ice, some making long annual migrations as the ice melts. The shrinking ice as a result of global climate change means that walruses and polar bears are keeping longer time on land as a result of the gradual disappearance of their normal habitat. Both are threatened by the scarcity of food; the huge numbers of walruses that seek refuge on shores are threatened by predators and stampedes that kill many, especially calves. Most toothed whales, the sea otter, and sirenians, on the other hand, do not migrate at all, though they can move about in search of food.

The migrations of the great whales are by far the most remark-able. Many baleen whales congregate to feed during the summer

Tony Ashby/AFP/Getty Images

Figure 11.32 These pilot whales (*Globicephala melas*) stranded themselves on a beach in Western Australia in 2009. About 80 pilot whales and bottlenose dolphins (*Tursiops truncates*) were stranded. Most perished, but 17 whales were saved by volunteer rescue workers.

How Intelligent *Are* Cetaceans?

We often hear that whales, dolphins, and porpoises are as intelligent as humans, maybe even more so. Are they really that smart? There is no question that cetaceans are among the most intelligent of animals. Cetaceans in captivity quickly learn tricks, and bottlenose dolphins have been trained to detect underwater mines in murky water.

Learning This type of learning is called conditioning. The animal simply learns that when it performs a particular behavior it gets a reward, usually a fish. Many animals, even invertebrates, can be conditioned to perform tricks. We certainly don't think of these animals as our mental rivals.

Unlike most other animals, however, dolphins quickly learn by observation and may spontaneously imitate human activities. Dolphins have also been seen imitating seals, turtles, and even water-skiers. Studies on discrimination and problem-solving skills in the bottlenose dolphin have concluded that its intelligence lies "somewhere between that of a dog and a chimpanzee", but such comparisons are unfair. It is important to realize that intelligence is a very human concept that we evaluate in human terms. After all, not many people would consider themselves stupid because they couldn't locate and identify a fish by its echo. Why should we judge cetaceans by their ability to solve human problems?

Cetacean brains Both humans and cetaceans have large brains with an expanded and distinctively folded surface, the cortex. The cortex is the dominant association center of the brain, where memory and sensory perception are centered. Cetaceans have larger brains than ours, but the body: brain weight ratio is higher in humans. In cetaceans, mainly the portions of the brain associated with hearing and sound processing are expanded. The enlarged portions of our brains deal largely with vision and hand-eye coordination. Cetaceans and humans almost certainly perceive the world in very different ways.

Can dolphins talk? Contrary to what is depicted in movies and television, there are not "talking" dolphins. Though they produce a rich repertoire of complex sounds, they lack vocal cords and their brains probably process sound differently than ours do. Bottlenose dolphins have been trained to make sounds through the blowhole that sound something like human sounds, but this is a far cry from human speech.

Captive bottlenose dolphins have been taught American Sign Language. These dolphins have learned to communicate with trainers who use sign language to ask simple questions. Dolphins answer back by pushing a "yes" or "no" paddle. Evidence also indicates that these dolphins can distinguish between commands that differ from each other only by their word order, a truly remarkable achievement. Nevertheless, dolphins do not have a real language like ours, and probably cannot convey very complex messages.

Bottlenose dolphins (*Tursiops truncatus*) participating in research to test acoustic communication. The devices on their heads, which are held in place by suction cups, light up every time a dolphin whistles.

Cetaceans in the wild Observations of cetaceans in the wild have provided some insights into their learning abilities. Several bottlenose dolphins off Western Australia, for instance, have been observed carrying large, cone-shaped sponges over their beaks. They supposedly use the sponges for protection against stingrays and other hazards on the bottom as they search for fish to eat. The use of sponges as tools is shown mostly by females, and it appears that this behavior is passed on by mothers to offspring. This is the first record of tool use and of its cultural transmission in a marine mammal in the wild.

Instead of "intelligence," some people prefer to speak of "awareness." In any case, cetaceans probably have a very different awareness and perception of their environment than do humans. Maybe one day we will come to understand cetaceans on their terms instead of ours, and perhaps we will discover a mental sophistication rivaling our own.

art.design-photography.com/Moment/Getty Images

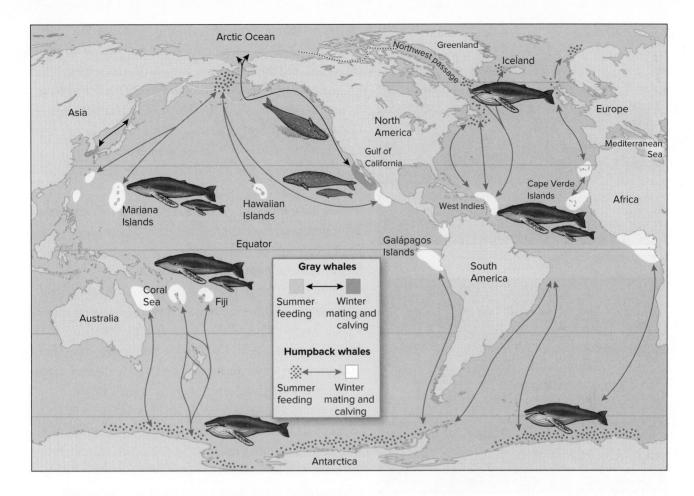

Gray whales

⬛↔️⬛
Summer Winter
feeding mating and
 calving

Humpback whales

⬚↔️⬚
Summer Winter
feeding mating and
 calving

in the productive waters of the polar regions of both hemispheres, where huge concentrations of diatoms and krill thrive in the long days. During the winter they migrate to warmer waters to breed. The seasons are reversed in the northern and southern hemispheres, so, when some humpback whales are wintering in the Hawaiian Islands or the West Indies, other humpbacks living in the southern hemisphere are feeding around Antarctica during the southern summer (Fig. 11.33).

The migratory route of the gray whale is the best known of any of the great whales (Fig. 11.33). From the end of May to late September, the whales feed in shallow water in the northern Bering, Beaufort, and East Siberian seas. They begin moving south in late September when ice begins to form. In November they begin crossing through the eastern Aleutian Islands. They eat less while on the move, burning off close to a quarter of their body weight. The whales cover about 185 km per day. They travel alone or in small groups along the coast of the Gulf of Alaska and down the western coast of North America en route to the Baja California Peninsula in Mexico. Migrating individuals often show spying behavior, pushing their heads out of the water. This raises the possibility that they navigate by using memorized landmarks. They reach Oregon around late November or early December and San Francisco by mid-December. Females generally migrate earlier. In late February pregnant females are the first to appear in shallow, quiet lagoons in southern Baja California and the southern mainland coast of the Gulf of California. It is here that females give birth and males mate with non-pregnant females.

Figure 11.33 Migration routes of humpback (*Megaptera novaeangliae*) and gray whales (*Eschrichtius robustus*). Both species tend to migrate and breed close to shore, where they were easily hunted. A gray whale was spotted in the Mediterranean Sea in 2010, apparently swimming there along the ice-free Northwest Passage in northern Canada. Another individual was spotted off the coast of Namibia, southern Africa, in 2013.

(a)

(b)

Figure 11.34 (a) A male Steller sea lion (*Eumetopias jubatus*) guarding his harem on a rocky island off the coast of Alaska. (b) A harem of female California sea lions (*Zalophus californianus*) on Santa Barbara Island, Southern California. The harem (center) is being guarded by a large, darker bull (top left).

The northbound migration begins by March, after the birth of the 700- to 1,400-kg calves. Females mate every two years, and the first to migrate north are the newly pregnant females that did not give birth. They will return 12 months later to give birth. Mothers with calves leave last. On the way north the whales tend to stay farther from the coast and move slower than on their southward migration, an average of 80 km per day because of the unfavorable currents and the newborn calves. The last whales leave the coast off Washington State by early May. They start reaching their feeding areas by late May, completing an amazing eight-month trip of up to 18,000 km, the longest migration of any mammal.

There is still much to be learned about the migrations of the gray and other whales. One vexing question is how whales navigate. It has been suggested that they use Earth's magnetic field, a possibility that implies that they carry some type of internal compass to orient themselves.

Scientists are using novel ways to investigate the migration of whales. Attaching small radio transmitters to whales and tracking their movements by satellite promises to uncover intriguing details. It has been found, for instance, that some isolated groups of gray whales along the migratory route do not migrate at all. One group resides in the Queen Charlotte Islands off the coast of British Columbia. Gray whales are also known to avoid cities by moving away from the coast. Females and young can slow down their migration back to the Arctic by taking shelter in kelp forests to avoid orcas. Analysis of the DNA of humpback whale populations in the Hawaiian Islands suggests that, as in the green turtle, individuals always return to the feeding grounds of their mothers.

Reproduction

The reproductive system of marine mammals is similar to that of land mammals. They have some unique adaptations to life in the water, however. To keep the body streamlined, male cetaceans and most other marine mammals have an internal penis and testes. The penis, which in blue whales can be over 3 m long, is kept rigid by a bone called the baculum. It is extruded just before copulation through the genital slit, an opening anterior to the anus (see Fig. 11.16, p. 323).

Reproduction in Pinnipeds Pinnipeds breed on land or ice, some migrating long distances to isolated islands to do so. In most species of seals each adult male breeds with only one female. Camcorders attached to animals in the wild have shown that male harbor seals make rumbling noises, quiver their necks, and release a stream of air bubbles, perhaps a display to attract females. In sea lions, fur seals, and elephant seals, however, a male breeds with many females. During the breeding season the males of these species, which are much bigger

and heavier than females, come ashore and establish breeding territories. They stop eating and defend their territories by constant, violent fighting. They herd **harems** of as many as 50 females onto their territories and keep other males away (Fig. 11.34). Only the strongest males can hold territories and breed.

The others gather into **bachelor groups** and spend much of their time trying to sneak into harems for a quick copulation. Defending the harem is exhausting work, and dominant males "burn out" after a year or two, making way for newcomers. It nevertheless pays off in the huge number of offspring they leave, compared with the males that never reach dominance, even though the subordinate males live longer.

Female pinnipeds give birth to their pups on shore. They seem to be indifferent to the birth process but soon establish a close relationship with the pup (Fig. 11.35). Because females continue to go to sea to feed, they must learn to recognize their own pups out of all the others by sound and smell. The pups generally cannot swim at birth. They are nursed for periods of four days to two years, depending on the species. Most pinnipeds have two pairs of mammary glands that produce a fat-rich milk ideal for the rapid development of the pup's blubber.

A female pinniped can become pregnant only during a brief period after **ovulation**, the release of an egg by her ovaries. This occurs just days or weeks after the birth of her pup. Females of most species return to the breeding grounds only once a year. By contrast, **gestation**, the length of time it takes the embryo to develop, is less than a year. This difference would cause the pup to be born too early, before the mother returns to the breeding ground. To prevent this, the newly formed embryo stops developing and remains dormant in the female's womb, the uterus. After a delay of as long as four months, the embryo finally attaches to the inner wall of the uterus and continues its normal development. This phenomenon, known as **delayed implantation**, allows pinnipeds to prolong the embryo's development so that the timing of birth coincides with the female's arrival at the safety of shore.

The walrus also shows delayed implantation as in other pinnipeds. Females give birth to one pup during the summer migration after a long gestation period of 15 to 16 months. They give birth every two years, the lowest reproductive rate among pinnipeds. Walruses mate both on land and in the water; nursing takes place in the water.

Reproduction in Sea Otters Sea otters reproduce year round, and males mate with multiple females. Mating takes place in the water, with males violently biting females on the nose. The gestation period, which includes delayed implantation, is of four to six months. Birth takes place in the water. Females are notoriously devoted to their pups, carrying and nursing them on their chests.

Reproduction in Polar Bears Polar bears are solitary except during the mating season in late spring or early summer. Mating induces females to ovulate so successive copulations are necessary to guarantee fertilization. Males can mate with several females, and DNA tests have shown that twins can be fathered by different males. Pregnant females build a den in the snow or ice in the fall

Figure 11.35 A California sea lion (*Zalophus californianus*) with pup.

(b)

(a)

Figure 11.36 Mating behavior in great whales. (a) Courtship in humpback whales (*Megaptera novaeangliae*) includes rolling, slapping of the flukes, and pairs surfacing vertically face to face. (b) Gray whales (*Eschrichtius robustus*) often mate with the help of a third party, another male that props the female against the mating male. Actual copulation is reported to last for just 30 to 60 seconds.

and hibernate, giving birth to one or two cubs after a gestation period of around eight months. Delayed implantation guarantees that the cub is born at the ideal time after winter. Cubs typically nurse for approximately two years.

Reproduction in Sirenians Manatees reproduce slowly, becoming mature after about five years and giving birth to a single calf every two to five years after a gestation period of about a year. Dugong females mature after at least six years and give birth to a calf every 2.5 to six years; gestation is from 13 to 15 months. Female sirenians, however, can mate any time of the year, typically with several males. They nurse their calves for a year or more from two mammary glands, each located behind a flipper.

Reproduction in Cetaceans We know relatively little about cetacean reproductive behavior. We do know that cetaceans are intensely sexual animals. Sex play is an important component of the behavior of captive dolphins. Like humans, they appear to use sex not only for procreation but for pleasure as well. Cetaceans reach sexual maturity relatively early, at age 5 to 10 in great whales. Sexual behavior appears to have a role in the establishment and maintenance of bonds among all individuals, not just potential mates. The sexes are typically segregated within the pod, and males perform elaborate courtship displays to catch the attention of potentially receptive females. Fights among rival males are common, but cooperation also occurs sometimes. Gray whales are known to copulate with the help of a third party, another male that helps support the female (Fig. 11.36b). Group matings have been observed in humpback and beluga whales. Considerable touching and rubbing precedes copulation (Fig. 11.36a). Actual copulation lasts less than a minute but is repeated frequently.

Gestation lasts for 11 or 12 months in most cetaceans. An exception is the sperm whale, which has a gestation period of 16 months. Development in most species of large baleen whales is relatively fast for a mammal of their size. It is synchronized with the annual migration to warm waters. It is remarkable that it takes 9 months for a 3-kg human baby to develop, but a 2,700-kg blue whale calf needs only about 11 months.

Cetacean calves are born tail-first (Fig. 11.37). This allows them to remain attached to the placenta, which provides oxygenated blood from the mother, for as long as possible to prevent oxygen deprivation. The calf immediately swims to the surface. In captive dolphins, the mother or an attending female can help the calf to the surface. Fat-rich milk is responsible for the rapid growth of calves, particularly in the great whales. They are born without their full complement of blubber and must gain weight before migrating with their mothers to feeding grounds in polar waters. It has been estimated that a blue whale calf gains 90 kg in weight and 4 cm in length every day for the first seven months of its life. The mother's milk is produced by two mammary glands with nipples located on both sides of the genital slit (see Fig. 11.16, p. 323). The milk is squirted into the calf's mouth, which allows the calf to drink under water. In at least some of the great whales, females do not seem to feed much while they are nursing. The calves are not weaned until they arrive at the feeding grounds. In some species they continue to nurse for more than a year after birth.

The relationship between mother and calf during the nursing period is close. Frequent contact and vocalizations are used in communication. Dolphin mothers modify their swimming behavior to allow calves to glide alongside. Mothers are known to defend their calves when there is danger. Sperm whale calves are watched by "babysitting" adults when mothers deep-dive for food. There is a report of a female gray whale lifting her calf onto her flipper to save it from the attacks of orcas. The bond between mother and calf probably lasts for several years. Captive young dolphins are known to return to their mothers for comfort in times of danger or stress.

Average estimated life expectancy in the wild is about 15-20 years in the polar bear, 40-45 years in the bottlenose dolphin, 40-50 years in manatees, and 20-30 years in male orcas, 30-50 in females. Lifespans of orcas in captivity is a controversial issue, some claiming that it is about the same as in the wild, others that it is lower. Great whales live at least 30 to 40 years, humpbacks at least 50 years, bowheads 150 years or more.

Figure 11.37 A bottlenose dolphin (*Tursiops truncatus*) giving birth in captivity. The gestation period lasted for about a year.

Section Review

1. **Describe** What adaptations do marine mammals have for diving?

2. **Explain** What is echolocation? How does it work?

3. **Explain** What is the role of vocalization in communication among marine mammals?

Table 11.3 Most Important Characteristics of the Marine Reptiles, Seabirds, and Marine Mammals

Group	Distinguishing Features	Temperature Regulation	Feeding	Reproduction	Ecology
Sea Turtles	Body covered by shell, scales on exposed parts of body, legs modified as flippers, found mostly in tropical seas	Poikilotherms, ectotherms	Toothless jaws adapted for crushing hard invertebrates o for picking soft invertebrates	Oviparious, laying eggs on sandy beaches	Predators of jellyfishes and bottom invertebrates, grazers of seagrasses and seaweeds
Sea Snakes	Skin with scales, no legs, laterally flattened for swimming, venomous, found only in tropical Indian and Pacific oceans	Pokilothermic, ectotherms	Small teeth adapted for capturing small prey	Ovoviviparous, giving birth at sea, or oviparous, laying eggs on land	Predators of bottom fishes; some feed mostly on fish eggs
Marine Iguana	Skin with scales, tail laterally flattened for swimming, found only in the Galápagos Islands	Pokilothermic, ectotherms	Three-cusped teeth adapted for grazing	Oviparous, laying eggs in nest on land	Grazer of seaweeds
Saltwater crocodile	Skin with scales, massive jaws and tail, found in coastal regions in Australia, Southeast Asia, and some western Pacific Islands	Pokilothermic, ectotherms	Heavily toothed jaws for capturing wide range of prey	Oviparous, laying eggs in nest of mud and vegetation on land	Predator of wide variety of coastal animals, including fishes, seabirds, sea turtles, and crabs
Seabirds	Feathers for insulation, webbed feet, light bones as adaptation for flight, found in all coastal regions	Homeotherms, ectotherms	Beachs adapted for capturing wide range of prey, including filtering	Oviparous, laying eggs in nest on land	Predators of fishes and many groups of surface-dwelling and shallow-water invertebrates, including plankton
Pinnipeds	Seals, sea lions, walrus: blubber, flippers, reduced hair (fur in some), found mostly in temperate and polar waters	Homeotherms, endotherms	Teeth for capturing and eating prey	Viviparous, giving birth on land	Predators of mostly fishes, crab-eater seal filters water for krill, leopard seal hunts mostly seabirds, walrus feeds on clams and bottom invertebrates
Sea Otter	Dense, dark fur, dorsoventrally flattened tail, flattened hind feet, found only in northern and northeastern Pacific Ocean	Homeotherms, endotherms	Flattened teeth for capturing and crushing prey	Viviparous, giving birth at sea or on land	Predator of sea urchins and wide range of bottom invertebrates and fishes in kelp forest
Polar Bear	Dense, white fur, proportionally short legs, tail, and ears; found in Arctic mostly on drifting ice	Homeotherms, endotherms	Powerful jaws and teeth for capturing and eating prey	Viviparous, giving birth on land	Primary predator of ring seals, other prey (birds, caribou) if available
Sirenians	Manatees, dugong: blubber, reduced hair, front flippers, paddle-shaped tail, tropical seas (one species only in fresh water)	Homeotherms, endotherms	Teeth as thick ridge pads for crushing vegetation	Viviparous, giving birth at sea	Grazers of seagrasses and other coastal vegetation
Baleen Whales	Blubber, streamlined body, reduced hair, front flippers, tail fluke, blowhole, found in all seas	Homeotherms, endotherms	Baleen on upper jaw for filtering small plankton	Viviparous, giving birth at sea in warm waters	Filter feeders of plankton and small fishes, mostly in polar water; grey whal feeds on small animals (mostly amphipods) in soft bottoms
Toothed whales	Blubber, streamlined body, reduced hair, front flippers, blowhole, found in all seas (some dolphins live in fresh water)	Homeotherms, endotherms	Conical teeth for capturing prey	Viviparous, giving birth at sea	Predators of fishes, squids, other marine mammals, and some large bottom invertebrates like lobsters

Chapter 11 Review

REVIEW QUESTIONS

Multiple Choice

1. Which of the following groups of organisms are homeotherms?

 a. reptiles and birds

 b. birds and mammals

 c. mammals and reptiles

 d. reptiles, birds, and mammals

2. Which describes reproduction in seabirds?

 a. ovoviviparous, giving birth at sea

 b. oviparous, laying eggs in nests on land

 c. viviparous, giving birth on land

 d. viviparous, giving birth at sea

Short Answer

3. How are penguins adapted for swimming and diving?

4. What has been done to try to curtail whaling? Have efforts been successful?

5. What is delayed implantation? What advantage does it give pinnipeds?

Critical thinking

6. Sea turtles have disappeared from many regions, and one way of trying to save them is to reintroduce them to areas where they have been wiped out. This is done by reburying eggs or by releasing newborn baby turtles on beaches. Why are eggs reburied or baby turtles released instead of fully-grown individuals?

7. Cetaceans give birth to few well-developed calves at well-spaced intervals. They also feed and protect the calves for long periods. This is in sharp contrast to most fishes, in which many eggs are spawned and the parents spend no time feeding and protecting the offspring. What do you think is the best strategy? Has this strategy paid off in the great whales?

DATA ANALYSIS LAB

How are seabird populations affected by introduced species? Scripp's murrelets (*Synthliboramphus scrippsi*) are small seabirds that only spend time on land to breed. One nesting site is Anacapa Island off the coast of southern California. Black rats, which feed on the eggs of Scripp's murrelets and other seabirds, were introduced to the island in the early 1900s. Between 2001 and 2002, the rats were eradicated from the island due to the use of a rodenticide.

Data and Observations

The graph shows the percentage of nesting success (which includes eggs hatching) of Scripp's murrelets on the island during the eradication year and the following 7 years.

Think Critically

1. Describe the change in nesting success between the two time periods.

2. How is this change in nesting success likely related to the eradication of rats from the island?

3. Scripp's murrelet is listed as a threatened species. Make an argument for or against the removal of introduced species, including plants, from the nesting sites of these birds. Explain your reasoning.

Data obtained from: Giller, G., 2013. Birds Return to a Rat-Free Island. Audubon, vol. 115, no. 5, September-October, p. 5.

354

Unit

3

Structure and Function of Marine Ecosystems

In a coral reef, organisms interact with each other and their environment constantly. Corals need the right combination of abiotic conditions to grow. They in turn provide a habitat for the most diverse communities in the oceans. Unit 3 looks at how organisms live and interact with each other and with the marine environment. We will explore the different habitats that make up the ocean world, from the water's edge to the darkest depths.

UNIT PROJECT

Evaluate a Solution

When working to solve a complex real-world problem, conflicting priorities and goals often present themselves. When evaluating a potential solution, these often conflicting goals should be considered, and prioritized relative to the overall goal of the project. Your task is to evaluate a solution to a marine science-related complex real-world problem based on prioritized criteria and trade-offs that account for a range of constraints, including cost, safety, reliability, and aesthetics, as well as possible social, cultural, and environmental impacts.

Chapter 12

An Introduction to Marine Ecology

THEME Energy and Matter: Flows, Cycles, and Conservation

The interaction between species, trophic relationships, and nutrient cycles are essential for understanding how energy and matter flow through an ecosystem.

BIG IDEA

The basic principles of ecology, such as population growth, interactions between species, food webs, and nutrient cycles all apply to marine ecosystems.

About the photo A cleaner shrimp (*Lysmata amboinensis*) removes parasites from a moray eel (Muraenidae)

AH Design Concepts/Getty Images

12.1 The Organization of Communities

Main Idea
Both abiotic and biotic factors affect the structure and organization of a marine community.

. .

Key Questions

1. Compare and contrast abiotic factors and biotic factors.
2. Which are some of the most important ways organisms interact?
3. What is an ecological niche? How does it differ from a habitat?

VOCABULARY

ecology
abiotic
biotic
evolutionary adaptations
exponential growth
logistic growth
carrying capacity
limiting resource
competition
competitive exclusion
resource partitioning
ecological niche
trophic cascade
stabilizing selection
directional selection
inducible defenses
coevolution
commensal
parasitism
mutualism
recruitment

The Science of Ecology

Here in Chapter 12 we first introduce a few basic principles that apply to all habitats. No matter where you go, for instance, algae and plants need light and animals need food. These are just two examples of the interactions among organisms and their environment. The study of these interactions and how they affect the abundance and distribution of organisms is called **ecology**. It is important to realize that ecology is a science, the branch of biology that studies how and why organisms interact with each other and with their environment. Contrary to the way the term is often used in the media, ecology is not the same as protecting the environment, as important as that may be.

The nature of life in a particular habitat is determined to a large extent by the non-living, or **abiotic**, environment, that is, the environment's physical and chemical features such as salinity or bottom type. Each environment has different characteristics that present both challenges and opportunities for the organisms living there. Organisms are also affected by other organisms—the living, or **biotic**, environment. Biological populations interact in complex ways that make the organisms in a community dependent on each other.

Organisms must adapt to both the biotic and abiotic features of their environment. Individual organisms can adapt to varying degrees by changes in behavior, metabolism, or other characteristics. Individual seagrasses, for example, adapt to low light levels by an increase in chlorophyll content, making them more efficient at capturing what little light *is* available, while corals often adapt through their growth form (Fig. 12.1, p. 358). In these physiological adaptations the organism's genes are unchanged, and the adaptation is not passed on to its offspring.

By contrast, genetic differences that make some individuals better adapted to the environment *are* passed on to these individuals' offspring. Those individuals whose traits are best suited to their abiotic and biotic environments are more successful at reproducing than less well-adapted individuals. The best-adapted individuals produce more offspring, and pass on the adaptive traits to the next generation. Because of this natural selection, the best-adapted individuals are better represented in each successive generation, and the population as

IN CONTEXT

Population A group of individuals of the same species that live together.

Community All the different populations of organisms that live in the same place.

• 6.2, Cellular Structure and Organization

Figure 12.1 These two corals belong to the same species, *Turbinaria mesenterina*. (a) The normal growth form occurs in relatively shallow water, where there is plenty of light. (b) In deeper water, with less light, the coral grows in a flattened shape. This flat shape is an adaptation that increases the coral's ability to capture the sunlight it needs. This variation in form reflects an individual colony's response to light, and not genetic differences among colonies.

a whole becomes more like the most successful individuals. In other words, the population evolves. Adaptations that result from natural selection and are passed on genetically are known as **evolutionary adaptations**.

How Populations Grow

Organisms can produce many more offspring than it takes simply to replace themselves. If every individual has more than one offspring—or each pair has more than two—the total number of offspring increases every generation (Fig. 12.2). The population grows faster and faster, and there is a population "explosion." Were its reproduction left unchecked, any species could grow to cover the earth in a relatively short time. This rapid, unchecked population growth is known as **exponential growth**.

Figure 12.2 All species are capable of undergoing population explosions. (a) When a single-celled dinoflagellate divides, it leaves two daughter cells. Each daughter cell produces two new daughter cells, which also divide, and so on. At each generation the number of cells doubles, and the population grows at an ever-increasing rate. (b) A graph of the number of cells at each generation produces a J-shaped curve, which is typical of populations undergoing unrestrained growth.

(a)

(b)

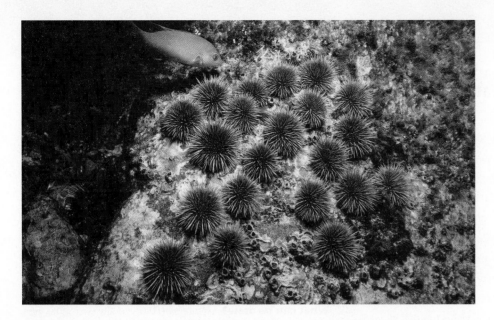

Figure 12.3 The purple sea urchin (*Strongylocentrotus purpuratus*) from the Pacific coast of North America sometimes undergoes dramatic population explosions, as do many other urchin species.

How can we study marine populations of the past? Many ecosystems were not studied until they had already been degraded by human activity. Marine scientists are teaming up with historians to analyze historical documents such as navigation charts and records from fishing and whaling ships for evidence of changing marine ecosystems. For example, a historical ecology project is underway in the Florida Keys, the largest reef system in the continental United States. Habitat destruction as well as overharvesting of marine aniamls in the Keys began long before biologists were studying the area. Natural history descriptions from historical documents give researchers a more accurate timeline of when commercial exploitation of marine resources began, and allows researchers a unique view of the ecosystem in a more pristine state.

Exponential growth sometimes occurs when conditions are just right for a population. Dinoflagellates and other algae, for example, sometimes form massive blooms known as red tides. Other organisms also occasionally undergo dramatic population explosions (Fig. 12.3).

Fortunately, we aren't up to our necks in dinoflagellates or sea urchins, because exponential growth cannot continue forever. Many factors can control, or regulate, the growth of populations. For one thing, explosive growth can occur only under favorable conditions. When the abiotic environment changes, the population may stop growing or decline. Some environmental fluctuations, like the seasons, are regular and predictable. Algal growth in temperate and polar regions, for example, often slows dramatically in winter because there is less light for photosynthesis. Other changes, like a sudden storm or a drifting log that smashes organisms on a rocky shore, are not so predictable. Predictable or not, large or small, changes in the environment can slow the growth of or even wipe out populations. Of course, conditions that wipe out one species might be just right for another.

There are also mechanisms that limit population growth even if the non-living environment doesn't fluctuate. Some animals slow down or stop reproduction when their habitat becomes too crowded. Others fight among themselves or even cannibalize each other. Natural enemies may be attracted when the population gets large, or diseases may spread

Figure 12.4 The kind of explosive population growth shown in Figure 12.2 would never go on forever. As the dinoflagellates get more and more numerous, they use up the nutrients and other resources that they need to grow and reproduce. Curves like this are often observed in laboratory experiments, but in the real world they are rarely so smooth.

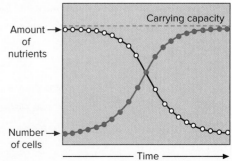

Fuse/Getty Images

Biodiversity: All Creatures Great and Small

From bacteria to baleen whales, our planet is home to tens of millions of life forms. Biologists can only guess at the true number of species. The richness and variety of life is referred to as biological diversity, or just biodiversity. In recent years biodiversity has been the focus of discussion in scientific circles, the news media, and the highest levels of government and international affairs. The reason for this attention, unfortunately, is that Earth's biodiversity is disappearing. Extinctions have occurred naturally since life first appeared on Earth, but pollution, habitat destruction, overexploitation, and other human activities are now driving species extinct at an unprecedented rate. Many species are likely to disappear from the face of the earth before we may even know they exist.

Conservation has often focused on organisms that we find beautiful or emotionally compelling. Whales and sea turtles evoke strong, positive reactions in many people. Worms, seaweeds, fungi, and millions of other seemingly insignificant species are much less popular, but are vital to Earth's—and our own—survival. There is an urgent need to reverse or at least slow the rate of mass extinction of all species, not just particular ones. So why should we care about how many different *kinds* of organisms there are, as long as the ones we care about are still around? For one thing, no organism lives in isolation. Organisms are bound together in complex food webs, nutrient cycles, symbioses, and other ecological interactions. The loss of even the "lowliest" species could have profound effects on many others, including "flagship species" like whales and sea turtles.

Benefits of biodiversity Another reason to conserve biodiversity is that it represents a hidden treasure trove. Most pharmaceuticals are derived from natural chemicals in organisms, but only a tiny fraction of species have been tested. New materials—perhaps a substitute for petroleum, useful chemicals, or better fibers for clothing—remain undiscovered. Many species have not even been identified, let alone evaluated for their usefulness. The next one that goes extinct might hold the cure for cancer, a solution to hunger, or maybe just the makings of an elegant new perfume—a secret that will be lost forever.

The loss of biodiversity is placing ecosystems in greater and greater jeopardy, not only because of the value of particular individual species but also because having a lot of species, whichever ones they are, is important. A diverse community, one with a lot of species, uses resources more efficiently, produces more food and oxygen, and breaks down waste faster. Diverse communities are more resilient to natural disasters and disease outbreaks, as well as to human pressures like pollution, fishing, and invasive species.

What can be done? Maintaining biodiversity is urgent. To some extent, governments have listened. At the 1992 Earth Summit in Rio de Janeiro, world leaders established the Convention on Biodiversity, an international agreement that recognizes that biodiversity conservation requires a coordinated global effort. At the World Summit on Sustainable Development in 2002, governments agreed to "significantly reduce" the rate of biodiversity loss by 2010. Unfortunately, there is little sign that this target was achieved. In 2012, biodiversity conservation progress was reviewed at the "Rio +20." The widespread view was "not enough" has been done, and the conference gave little hope for optimism in the future.

Important steps have already been taken, such as setting up global programs to monitor the health of the oceans, assist sustainable development projects, reduce pollution, and improve the management of natural resources but much more needs to be done. Ways must be found for poor countries to develop economically without destroying their natural biodiversity. Developed countries must support developing countries through scientific research and technical assistance, while becoming more sustainable themselves. Governments, scientists, and the world community at large must to work together to improve our understanding, management, and protection of the ocean's resources, including by establishing better networks of marine protected areas. Without strong, swift action the oceans are headed and rapid enough to prevent a catastrophic loss of marine biodiversity.

faster under crowded conditions. Large populations can pollute the environment with their own wastes.

As more and more individuals join the population, they use up their resources, the things they need to live and reproduce, like food, nutrients, and living space. Eventually there are just not enough resources to support any more individuals and the population levels off (Fig. 12.4, p. 359). This type of population growth is called **logistic growth**. The largest population size that can be sustained by the available resources is called the **carrying capacity**.

Organisms use many different resources. A lack of any one of these will slow or prevent growth and reproduction. A dinoflagellate living in the open ocean, for example, can be limited by a lack of nitrogen, even if it has plenty of light, water, carbon dioxide (CO_2), and other nutrients. During the winter in polar regions, there may be plenty of nutrients but not enough light. Seaweeds on a rocky shore, on the other hand, might have plenty of light and nutrients but be unable to grow because there is no space. A **limiting resource** is one whose short supply restricts the growth of a population.

Because of a drain on resources or other effects of crowding, populations do not grow forever. As the population becomes more crowded, its *growth rate* goes down. In this sense the population is self-regulating; that is, its growth rate depends on its own numbers. Whereas non-living factors such as the weather can affect populations of any size, self-regulation acts only when the population is large.

Ways That Species Interact

Species do not live in a vacuum, isolated from other species. Members of different species may have strong effects on each other. Species can interact in many ways, all of which influence the organization of communities.

Competition As resources are depleted, there are not enough to go around. Each individual has to vie with other organisms for what resources remain. **Competition** is the interaction that results when a resource is in short supply and one organism uses the resource at the expense of another (Fig. 12.5). When members of the same species compete it is called intraspecific competition.

When two species use the same resource and the resource is scarce, the species must compete just as if they were members of the same population. Competition between species is known as interspecific competition. One of the two species usually turns out to be better at the competition. If two species eat exactly the same food, for example, one of the two will be better at catching it. Unless something interferes, the inferior competitor loses out and the competitively superior species takes over. When one species eliminates another by outcompeting it, it is called **competitive exclusion**.

Sometimes a competitively superior species is prevented from excluding poorer competitors. Periodic disturbances such as severe storms, battering by drifting logs, or underwater landslides can reduce the population of a

Figure 12.5 The short supply of any resource can lead to competition. These left-handed hermit crabs (*Calcinus laevimanus*) are fighting over a shell they both want. The one on the left has evicted the one on top from its shell and will take over the shell.

Nutrients Raw materials other than carbon dioxide and water that are needed by primary producers for primary production.
• 6.1, The Ingredients of Life

Millard H. Sharp/Science Source

Figure 12.6 The whale shark (*Rhincodon typus*) feeds by straining plankton and small fish from huge mouthfuls of water. It has probably evolved its huge mouth and lost the large teeth of most other sharks not to avoid competition with other sharks but as adaptations for efficient plankton feeding.

dominant competitor and give other species a chance. Furthermore, which species is competitively superior sometimes depends on the conditions. On rocky shores in tropical Hong Kong, for example, foliose (leaf-like) algae are the dominant seaweeds during the relatively cool winter. In the summer heat these forms die out and are replaced by more resistant encrusting algae. Seasonal variation thus prevents either group from excluding the other and there is a shifting balance between the competing species.

Species can also avoid excluding each other if they share the limiting resource, with each species specializing on just part of the resource. For example, two species of fish that eat seaweeds might divide the resource by specializing on different kinds of seaweed. Animals that eat exactly the same thing might live in different places or feed at different times, staying out of each other's way. Ecologists call this sharing of a resource by specialization **resource partitioning**.

Resource partitioning allows species to coexist when they might otherwise exclude each other, but it does have its price. By specializing, each species gives up some of the resource. The fish that specializes on just one kind of seaweed, for example, will have less food available than if it ate all types. With fewer resources available, the size of the population tends to be smaller. On the other hand, the species might be able to use the resource more efficiently by being a specialist than if it were a "jack-of-all-trades." Different species of shorebirds, for example, have specialized bill shapes that make them more efficient at feeding on particular types of mudflat animals and may also reduce competition. Not all specializations, however, are a result of interspecific competition (Fig. 12.6).

To be successful in the long run, a species must find the right balance between specialization and generalization. There is no one best answer to the problem—just look at all the different kinds of organisms. Each species has its own special role, or **ecological niche**, in the community. A species' niche is defined by the combination of virtually every aspect of its lifestyle: what it eats, where it lives, when and how it reproduces, how it behaves, and so on.

Predator-Prey Interactions Species don't always compete for resources. Sometimes they use each other as the resource. In other words, they eat each other (Fig. 12.7). Predation is the act of one organism eating another. The organism that does the eating is the predator, and the organism that gets eaten

Figure 12.7 A purple cone shell (*Conus purpurascens*) engulfs its prey, a redhead goby (*Elacatinus puncticulatus*). It paralyzed the fish by injecting powerful venom through a hollow, harpoon-like "tooth" derived from the radula.

(t)Franco Banfi/WaterFrame/Getty Images, (b)©Alex Kerstitch, Sea of Cortez Enterprises

is the prey. The term "predator" is often reserved for carnivores, animals that eat other animals. A special case of predation occurs when organisms eat algae or plants. This is usually called herbivory rather than predation, and the organism doing the eating is called an herbivore. No matter what you call it, the alga or plant gets eaten just the same.

Predation obviously affects the individual organism that gets eaten, but it also affects the prey population as a whole by reducing prey numbers. If the predators don't eat too much, prey can reproduce to replace the individuals that have been eaten. Predation that is too intense, however, can wipe out a prey population.

Trophic Relationships The interaction between predator and prey is not a one-way street. After all, the predator depends on the prey for its food supply. If bad weather or disease decimates the prey, the predators suffer as well. And, when predators consume too many prey they deplete their own food supply and start to decline themselves.

Predators often affect species other than their prey. On temperate rocky shores, for example, sea stars that selectively prey on mussels make space for the mussels' competitors, who would otherwise be crowded out. Feeding, or trophic, interactions such as this, in which the effects of one species on another go on to affect a third species, are called indirect interactions. Indirect interactions sometimes result in relationships among species that are contrary to what might be expected. On Caribbean coral reefs, for example, a grazing chiton (*Choneplax lata*) actually *increases* the abundance of its primary prey, a coralline alga (*Porolithon pachydermum*). The chiton only scrapes a thin surface layer off the alga, which leaves most of the alga behind and actually stimulates its growth. The chiton also removes competitively superior algae that grow faster than the coralline alga. Furthermore, these fast-growing algae attract parrotfish grazers, whose deep bites do completely remove the coralline alga. There is often a web of indirect interactions, which means that an effect on one species may flow through an ecosystem in what is called a **trophic cascade**. In the northwest Pacific, for example, the effects of depleting fishes that are the prey of seals and sea lions flow through to killer whales, otters, sea urchins, and entire kelp beds.

Predator-prey interactions can also cause the effects of physical factors to cascade through a community. For example, a mud snail (*Hydrobia ulvae*) is the main grazer on seagrass epiphytes in North Sea seagrass beds. Epiphytes compete with the seagrass for light and nutrients, slowing seagrass growth. Where snail grazing keeps the seagrass leaves clean, the beds grow thick and lush. At exposed sites, however, currents wash the snails away, epiphytes grow in profusion, and the seagrass beds are sparser.

A tremendous variety of predatory strategies have evolved. Sharks and tunas are highly efficient killing machines, swift and powerful. Other predators sneak up on their prey, catching them unaware. Predators such as anglerfishes go a step further and lure their prey. Many snails can drill through their prey's shell, and many sea stars evert their stomachs to digest prey outside their bodies. Many predators eat only part of the prey, leaving the rest alive to continue growing. In the following chapters we will examine some of the many ways that predators catch their prey.

IN CONTEXT

Chitons Molluscs whose shells consist of eight overlapping plates on their upper, or dorsal, surface.

- 9.3, Molluscs: The Successful Soft Body

Epiphytes Photosynthetic organisms that live on the surface of algae or plants.

- 7.2, Prokaryotes

IN CONTEXT

Nudibranchs (sea slugs) Gastropods, or close relatives of snails, that lack a shell and have exposed gills.

- 9.3, Molluscs: The Successful Soft Body

Selection The most successful individual predators catch more food and thus tend to survive longer, grow faster, and produce more offspring. Natural selection, therefore, favors the most efficient predators in the population. This doesn't necessarily mean that predators always get bigger, stronger, and faster with each generation, because there are trade-offs. Getting bigger, for example, means needing to eat more, and perhaps getting slower or being more exposed to your own predators. In some situations, selection acts to weed out individuals that are too far from the optimal balance. This type of selection is called **stabilizing selection**. When conditions change—including not only the physical environment but the adaptations of competitors, predators, and prey—the optimal balance also changes and selection drives evolution toward a new optimum. This form of selection is called **directional selection**.

Natural selection also favors the prey individuals that are most successful at getting away. Organisms have evolved at least as many ways to escape from predators as there are predators. Some are fast and elusive; others use camouflage (Fig. 12.8). Many organisms have protective spines, shells, or other defensive structures. Many seaweeds and animals protect themselves with distasteful or even poisonous chemicals.

Organisms may also take advantage of the defenses of other species. Many sea slugs, or nudibranchs, for example, protect themselves from predators with noxious chemicals or stinging cells that they obtain from their prey. In North Carolina, a brown alga (*Sargassum filipendula*) competes with a red alga (*Hypnea musciformis*) and slows its growth when the two occur together. Nonetheless, the red alga *only* occurs together with the brown one when grazing fishes are abundant, because the brown alga is unpalatable and helps deter the fishes from eating the red one, which they relish.

Some organisms use their defenses only when they need them. The barnacle *Chthamalus anisopoma* from the Gulf of California, for example, normally grows upright, but when predatory snails are abundant it grows bent over. This makes it more difficult for the snails to get through the covering plates on the barnacles' upper surface. Some algae that use chemical defenses only produce the nasty chemicals after they have been damaged. Thus, they don't waste energy making the chemicals until they are needed. Blue mussels (*Mytilus edulis*) in southern New England can sense the presence of an introduced predator, the Asian shore crab (*Hemigrapsus sanguineus*), using an unknown cue, and grow thicker shells when the crabs are around. Defense mechanisms that are used only in response to predators are called **inducible defenses**.

Thus, there is a continual "arms race" between predators and their prey. The predator keeps getting better at catching the prey and overcoming its defenses. In response, the prey becomes more adept at escaping or develops better defenses. This interplay, with each species evolving in response to the other, is known as **coevolution**.

Figure 12.8 The frond-like projections of the leafy seadragon (*Phycodurus eques*) make these fish look like a bit of drifting seaweed. Not only does this hide them from predators, it also allows them to sneak up on their own prey, tiny shrimps. They also have sharp, protective spines in case a predator isn't fooled by the camouflage.

Paul A. Zahl/National Geographic/Getty Images

Symbiosis Coevolution becomes even more important when species interact more intimately. Members of different species may live in very close association, even with one inside another. Such close relationships are examples of symbiosis, which literally means "living together." The smaller partner in the symbiosis is usually called the symbiont and the larger one the host, though sometimes the distinction is unclear.

Biologists traditionally divide symbiosis into different categories according to whether the organisms involved benefit or suffer from the relationship. In **commensal** relationships, one species obtains shelter, food, or some other benefit without affecting the other species one way or the other. For example, certain barnacles live only on whales (Fig. 12.9). The barnacles get a place to live and a free ride. They feed by filtering the water, and as far as we can tell the whale is neither harmed nor helped by the barnacles. There are many other examples of symbiotic organisms that live on, or even in, a host without apparently either hurting or helping it.

On the other hand, sometimes the symbiont benefits at the expense of the host (Fig. 12.10). This is called **parasitism**. The giant tapeworms that derive food and shelter by living in the guts of whales are generally considered parasites because they may weaken their whale hosts. Marine parasites are very common. In fact, few marine species escape having at least one kind of parasite.

Not all symbiotic relationships are one-sided. In **mutualism**, both partners benefit from the relationship. In many places, small cleaner fishes and shrimps have mutualistic relationships with larger fishes known as cleaning associations.

In the case of cleaning associations, both partners can get by without the other if they have to, which is called facultative symbiosis. In obligate symbiosis one or both partners depend on the other. In the symbiosis that forms lichens, for example, both the fungus and the alga are obligate symbionts that cannot live without the other.

In practice, it can be difficult to tell whether the partners in a symbiotic relationship are helped or harmed. Some biologists prefer to emphasize the movement of energy or of food or other materials between the partners, without reference to possible harm or benefit.

Like other interactions among species, symbiosis can have indirect effects. In exchange for shelter and food, the crab *Trapezia flavopunctata* drives away the

Figure 12.9 Some barnacles live only on the skin of whales like in this humpback whale (*Megaptera novaeangliae*). Because they do not appear to affect their host at all, most such barnacles are considered to be commensals.

Figure 12.10 The isopod (*Anilocra*) just below the eye of this coney (*Cephalopholis fulva*) is a blood-sucking parasite of fishes.

Cleaning Associations

Some small fishes and shrimps form cleaning associations with larger and often predatory "clients." In these associations, the cleaner picks parasites and dead or diseased tissue off its clients. Cleaner fishes include wrasses, butterflyfishes, and gobies, while the banded coral shrimp (*Stenopus hispidus*) is the best-known cleaner shrimp. Usually the clients are fishes, but octopuses have also been seen to take advantage of the service. Cleaning behavior can be a full- or part-time job. Some species are cleaners all their lives, others only as juveniles. Cleaning associations are especially common in coral reef, kelp bed, and other shallow, hard-bottom communities.

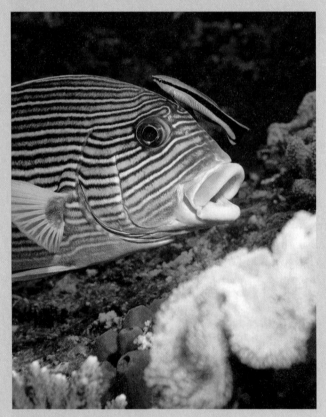

A cleaning wrasse (*Labroides dimidiatus*) doing its job on a yellow ribbon sweetlips (*Plectohinchus polytaenia*).

To attract clients, cleaners set up a "cleaning station," usually marked by a prominent feature like a large sponge or conspicuous rock. Cleaners also often have characteristic bright colors and distinctive stripes to advertise their services, which are popular: clients sometimes have to wait in line! Clients remember where the stations are and continue to visit for a while even after a cleaner is experimentally removed.

Client-cleaner communication Some clients are carnivores that could easily cheat the system and swallow a cleaner in one gulp. Perhaps to signal their good intentions, many clients swim unusually slowly, hovering in midwater. Some change color, or assume unusual postures. Cleaner fishes swim close to the client to inspect it. Some "dance" as part of the ritual, perhaps to remind the host that they are cleaners and not food. Cleaner shrimps use their antennae, tapping the host. Once the cleaning starts, the hosts remain still to allow the cleaners to pick at their skin and fins. They often open their gill covers to allow the cleaner to inspect and clean their gills. The cleaner can even swim inside the mouth without mishap, at least most of the time.

Cleaning associations benefit both sides. The cleaners get a reliable food supply that comes to them, although they are occasionally infected by the very parasites they are trying to eat. Cleaners effectively control their clients' parasites, which presumably benefits the client more in the long term than making the cleaner into an easy, short term meal.

Cheating Clients do sometimes cheat the system and eat a cleaner. Cleaners also cheat occasionally, biting off a scale or some protective mucus from a client. Cheating is rare, however, and both sides have strategies to keep the other side honest. Clients will punish cleaners who try to cheat, attacking them or taking their business to a different cleaning station. They will also move on if the cleaner keeps them waiting. It even appears that they watch how cleaners treat other clients and select cleaners that behave cooperatively. There is also evidence that cleaners behave better when they are being watched.

For their part, cleaners learn to recognize individual clients and may reject ones that didn't behave themselves during a previous encounter. They even give preferential treatment to clients that have access to a choice of cleaning stations and might seek better service elsewhere.

Some blennies have color patterns and behavior that mimic local cleaner wrasses. The blennies use this deception to get close to would-be clients, then dart in to take a bite of skin, scale, or fin before dashing to safety.

sea star predators of its coral host. This can protect nearby corals as well as the host coral, because the sea stars avoid crawling over the host to get to the other corals.

Larval Ecology Interactions among species and their environment occur over the organisms' entire lifetimes. Most marine animals have larval stages, and seaweeds have spores. What happens to the larvae and spores can be as important as interactions later in life—and for many species more so.

Many factors can affect how many larvae survive. Some larvae are sensitive to temperature, salinity, and other factors, so larval survival may depend on oceanographic conditions. Larvae may also have a critical period during which they must obtain the right food. The survival of haddock larvae off the coast of Nova Scotia, for example, is closely linked to the availability of particular plankton during a few critical weeks in early spring. In years when weather and oceanographic conditions favor an abundance of plankton during this period, the larvae survive particularly well and there are more adults of these fish in later years. Predators, pollution, and many other factors can also have a strong influence on larval survival.

If the larvae do survive, they still must reach their adult habitat for a successful **recruitment**, the addition of new members to a population. Some larvae remain near their parents, but the drifting larvae of many marine species depend on favorable currents to carry them to the right place. It was long assumed that currents scatter most larvae far and wide, but the larvae of at least some species are able to stay fairly close to the habitats where their parents produced them, making it easier for them to get back. The larvae of some coral reef fishes, for example, are surprisingly strong swimmers and are able to stay near their home reef, recognizing the reef by sound or perhaps smell. The larvae of other species tend to stay in water layers where there isn't much current, so they aren't carried far. If larvae do reach the right habitat, they have to recognize it and settle in the right place. The larvae of many bottom-dwelling species test the bottom before they settle. Others are attracted to chemicals produced by their hosts or adults of their own species.

In many marine communities it is unclear, and often controversial, whether the abundance of organisms is determined primarily by the supply of larvae to settle in the habitat (this school of thought has been dubbed "supply-side ecology") or by competition, predation, and other interactions after the larvae settle. As with many controversies in marine ecology, the answer probably lies in the middle.

· ·

Section Review

1. **Relate** Explain how the terms *exponential growth*, *logistic growth*, and *carrying capacity* are related.

2. **Define** the term *indirect interaction* and give an example.

3. **Identify** What are three types of symbiosis? Give one example of each.

12.2 Major Marine Lifestyles and Environments

VOCABULARY

benthic

pelagic

intertidal

subtidal

bathyal

abyssal

hadal

neritic

oceanic

epipelagic

mesopelagic

bathypelagic

abyssopelagic

hadopelagic

Main Idea

Marine scientists categorize marine communities based on where and how organisms live.

· ·

Key Questions

1. What are two main ways that marine communities can be classified?
2. What characteristics make an organism part of the plankton?
3. Which zones of the ocean do not have enough light to support photosynthesis?

It has already been noted that, because physical and chemical conditions change from place to place, different parts of the ocean harbor very distinct communities. Marine biologists therefore categorize communities according to where and how the organisms live.

Marine Lifestyles

Perhaps the simplest classification relates to the lifestyle of the organism: whether it lives on the bottom or up in the water column. **Benthic** organisms, or the benthos, are those that live on or buried in the bottom (Fig. 12.11). Some benthic organisms are sessile, or attached to one place; others move around. **Pelagic** organisms, on the other hand, live up in the water column, away from the bottom.

Figure 12.11 Three major groups, the benthos, nekton, and plankton, result when marine organisms are classified by lifestyle.

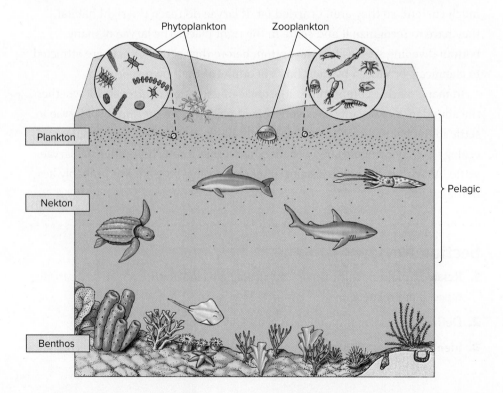

Pelagic organisms are further subdivided according to how well they can swim. Some marine organisms swim only weakly or not at all. These organisms, called plankton, are at the mercy of the currents and are carried from place to place. The term "plankton" comes from the Greek word for "drifters." Planktonic algae and other autotrophs are collectively called the phytoplankton and are the most important primary producers in many marine ecosystems. The heterotrophic plankton are called the zooplankton. The plankton can also be subdivided according to size, whether they live right at the sea surface or further below, or whether they spend all or only part of their lives in the plankton.

Animals that can swim well enough to oppose the currents are called nekton. Most nektonic animals are vertebrates, mainly fishes and marine mammals. There are a few nektonic invertebrates, however, such as squids. Not all nekton are pelagic. The ray in Figure 12.11, for example, is part of the nekton because it can swim. It is not considered pelagic, however, because it spends most of its time on or just above the bottom rather than up in the water column.

Zonation in the Oceans

Another way to classify marine communities is by where they live. Zonation in benthic organisms, for example, relates to depth and the continental shelf (Fig. 12.12). Right at the boundary between land and sea is the shallowest part of the shelf, the **intertidal** zone, sometimes called the littoral zone. This is the area between the tides that is exposed to the air when the tide is out but underwater at high tide. The intertidal zone makes up only a small fraction of the continental shelf. Most of the shelf is never exposed to the air, even at the lowest of low tides.

DIVE IN!

Can animals live in the deepest parts of the ocean? The abyssopelagic zone is found from 4000 m to 6000 m in the ocean, with no light, near freezing water, and pressures as high as 8,700 psi. Animals that live under these harsh conditions have adaptations that allow them to survive. Many animals are red or black in color or they are transparent, which means they cannot be seen in the darkness. Some animals do not have eyes, since they serve little purpose in the pitch black water. High pressure affects the function of proteins. Many animals of the abyssopelagic zone have increased amounts of a substance called trimethylamine oxide, which helps stabilize proteins. To deal with the cold temperatures, the cell membranes of many organisms contain unsaturated to increase flexibility.

Figure 12.12 The major subdivisions of the marine environment are based on distance from land, water depth, and whether the organisms are benthic or pelagic.

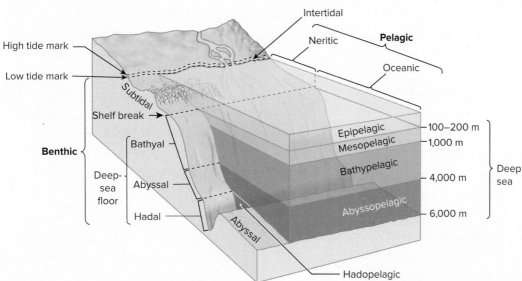

The Census of Marine Life

A decade ago, less than 1% of the oceans had been explored. We knew little about many cataloged species, and huge numbers of species remained unknown. In the year 2000, marine scientists around the world launched a 10-year Census of Marine Life (COML). Some 2,700 scientists, working at nearly 700 institutions in 81 countries, worked on COML. COML asked three questions: 1) What lived in the oceans before humans changed them? 2) What lives there now? 3) What will live there in the future?

The project's 540 expeditions collected over 6,000 candidates for new species. Some 1,200 of these have been formally described and documented, and work on the rest continues. Spectacular photographs of weird, wonderful, and beautiful new species from all corners of the ocean inspired both professional marine biologists and the public.

The Census was about more than collecting new species. There was a major effort to collate and make available existing information. COML created the Ocean Biogeographic Information System (OBIS) to make the vast amount of information—now over 30 million records—freely available online to anyone.

COML results Using historical records, genetic analysis, and data from sources like ancient fish scales, COML established the historical abundance of many fisheries species before intensive fishing began. This information is essential to determine how humans have changed the ocean. The project found, for example, that fishing had reduced Nova Scotian cod by 96% as early as 1850. Thus, a target to restore the cod to 1950s levels would hardly represent a historically healthy population. Some of COML's results are sobering. From a global decline in phytoplankton to evidence that most harvested fish species have become fewer and smaller, COML painted a picture of oceans even more profoundly affected by human activities than we thought. But COML also documented enough cases of population recovery to demonstrate that sound conservation measures work—if we choose to apply them.

COML also fostered development of new research tools. These include improved DNA "barcoding" to separate closely similar species or even confirm species' presence from DNA traces in the water. COML scientists developed acoustic

systems to track marine animals' movements, and improved electronic tags for satellite tracking of animals from juvenile salmon to seabirds. They invented Autonomous Reef Monitoring Structures that help coral reef scientists make comparable observations in widely separated parts of the world. COML's successes also motivated countries to invest in new research vessels.

COML strove to make its products widely accessible through books, the media, and the Internet. Anyone can search OBIS or learn more about particular species on the 80,000 species web pages that COML produced for the Encyclopedia of Life. Many of the thousands of COML's scientific publications can be freely downloaded. You can follow the movements of tagged sharks, seals, and sea turtles, and explore a global map of marine life in the past, present, and future. Some Census results are even on Google Earth.

COML's greatest achievements are perhaps intangible ones. COML scientists worked across borders, sharing knowledge, techniques, and above all data and information. This increased the scientific capacity of many countries, speeding the adoption of improved techniques and training of new scientists. The personal relationships and partnerships formed proved that scientists from around the world can work cooperatively toward common goals. The Census also generated great interest among governments and the public, fostering support for marine conservation.

What next? As the Census neared completion, marine biologists began asking "what next"? There is still are still to be discovered. Some components of COML will continue. OBIS will be maintained through the United Nations system. Tools like DNA bar coding, tracking networks, and satellite tags remain available. Marine biologists are laying the groundwork for COML's successor, the "Life in a Changing Ocean" project. The new project has three main themes about biodiversity: continuing to discover and map new species, understanding how marine ecosystems function, and assessing the impacts of human use of marine resources. Continuing the legacy of COML will keep marine biologists busy for at least another decade.

Benthic Zones Benthic organisms that inhabit the continental shelf below the intertidal live in the **subtidal** zone, sometimes called the sublittoral zone. Away from the shelf the benthic environment is subdivided by depth into the **bathyal**, **abyssal**, and **hadal** zones. For the sake of simplicity, we shall lump these different zones together and call them the "deep-sea floor."

Pelagic Zones The pelagic environment, too, is divided with reference to the continental shelf. The pelagic environment that lies over the shelf is called the **neritic** zone. Pelagic waters beyond the shelf break are the **oceanic** zone.

Besides being divided into the neritic and oceanic zones, the pelagic is divided vertically into depth zones that correspond to the amount of light. In the shallowest, the **epipelagic** zone, there is plenty of light for photosynthesis, at least for part of the year. Given a supply of nutrients, phytoplankton thrive, producing food for the rest of the ecosystem. Definitions vary, but this book defines the epipelagic zone as extending to a depth of 200 m. Because this is approximately the depth of the continental shelf, nearly all neritic waters lie in the epipelagic.

Other pelagic zones lie in the deep oceanic waters beyond the shelf. Below the epipelagic lies the **mesopelagic** zone. There is not enough light in the mesopelagic to support photosynthesis, so primary producers cannot grow. There is, however, enough light to see by. On even the brightest day, the mesopelagic is an area of twilight. This twilight zone typically extends to a depth of around 1,000 m.

No sunlight at all penetrates to the deepest parts of the ocean, the **bathypelagic**, **abyssopelagic**, and **hadopelagic** zones. Although each of these zones supports a different community of animals, they share many similarities. We refer to all these zones together as the "deep-sea" environment.

IN CONTEXT

Shelf Break The increase in steepness that marks the outer edge of the continental shelf.
• 2.3, Modern Ocean Basins

. .

Section Review

1. Construct Make a concept map that shows how marine organisms are classified based on their lifestyle. Include the following terms: *benthic, pelagic, plankton, nekton, phytoplankton,* and *zooplankton.*

2. Describe How are the different zones in the benthic and pelagic environments defined?

12.3 The Flow of Energy and Materials

Main Idea
The flow of energy and matter through an ecosystem can be traced through trophic relationships as well as nutrient cycles.

. .

Key Questions
1. What is a trophic level?
2. Why is the term *food web* more accurate than the term *food chain* when describing how energy and materials are passed from one trophic level to another?
3. What are the roles of decomposers in a food web?

All living things use energy to make and maintain the complex chemicals necessary for life. Autotrophs get this energy from the non-living environment, usually in the form of sunlight. They use the energy to make their own food

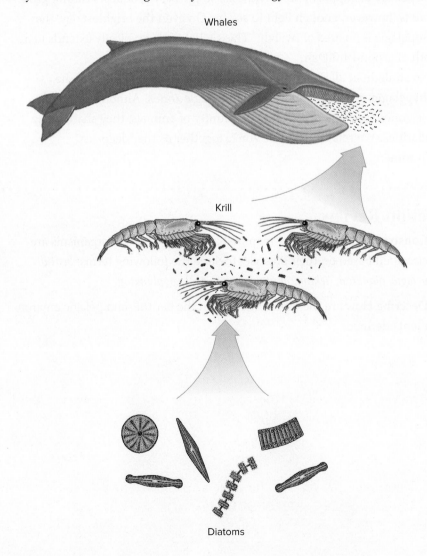

Whales

Krill

Figure 12.13 This three-step food chain is typical of Antarctic waters. Planktonic diatoms are the main primary producers during summer, but in winter diatoms and other algae growing in sea ice are more important. The energy is transferred from the diatoms to the krill that eat them, and then to blue whales that eat the krill.

Diatoms

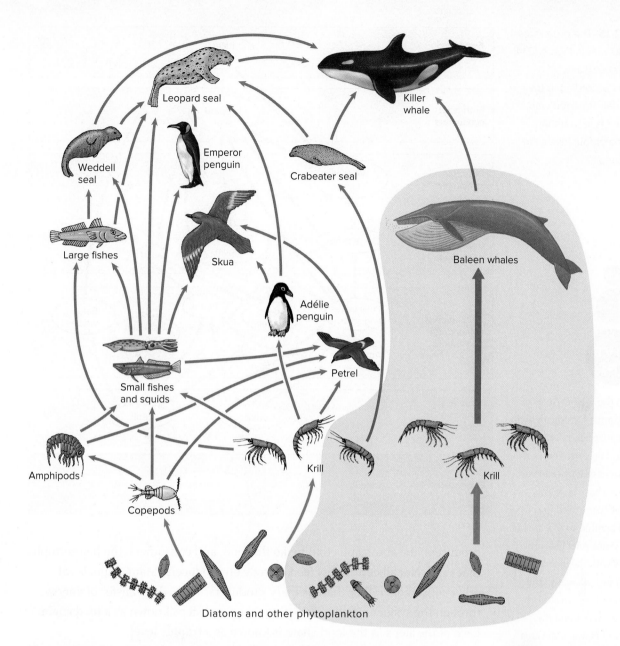

Figure 12.14 A simplified Antarctic food web. The simple diatom-krill-whale chain illustrated in Figure 12.13 (blue shading) is an important part of this food web.

IN CONTEXT

from simple molecules like carbon dioxide, water, and nutrients. As the autotrophs grow and reproduce they become food for heterotrophs (Fig. 12.13). When one organism eats another, both the organic material and the energy stored in it are passed from one to the other. Thus, energy and chemical substances flow from the non-living part of the ecosystem to organisms, and from organism to organism. The pathways taken by energy and materials tell us a lot about how an ecosystem works.

Trophic Structure

The flow of energy and matter through an ecosystem can be traced by observing the trophic relationships among its organisms: who makes the food and who eats it. The organisms can be divided into two broad components: primary

Autotrophs Organisms that can use energy (usually solar energy) to make organic matter.

Heterotrophs Organisms that cannot make their own food and must eat the organic matter produced by autotrophs.

Primary Production The conversion of carbon dioxide into organic matter by autotrophs, that is, the production of food.

• 4.1, The Ingredients of Life

Figure 12.15 The organisms in the food web in Figure 12.14 can be grouped into categories according to their trophic level. The food web shown here is longer than most; three or four levels are more common.

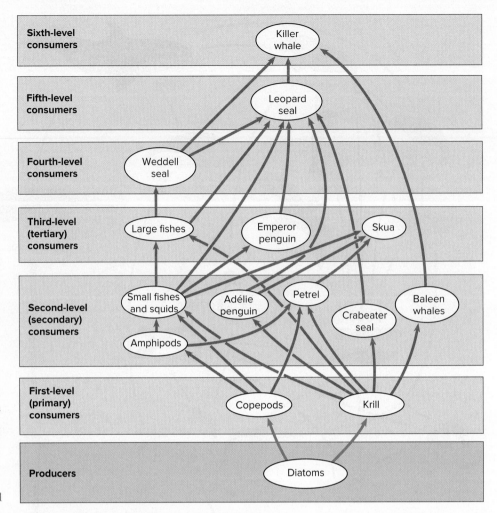

Sixth-level consumers	Killer whale
Fifth-level consumers	Leopard seal
Fourth-level consumers	Weddell seal
Third-level (tertiary) consumers	Large fishes · Emperor penguin · Skua
Second-level (secondary) consumers	Small fishes and squids · Adélie penguin · Petrel · Crabeater seal · Baleen whales · Amphipods
First-level (primary) consumers	Copepods · Krill
Producers	Diatoms

DIVE IN!

Deserts in the ocean? Areas of very low primary productivity in the ocean are often referred to as biological deserts. The oceanic zones of Earth's major ocean basins have very low productivity compared to coastal areas and benthic environments. In recent years, scientists are concerned with the increase in the size of desert areas in the ocean. According to a study by scientists at NASA and the University of Hawai'i, the area of ocean deserts increased by 15 percent from 1998 to 2007. Scientists note that the increase in the size of ocean deserts correlated with the increase in surface water temperature due to global climate change. As the temperature of surface water increases, fewer nutrients from deep water reach the surface, which limits primary productivity.

producers, the autotrophs that make the food, and consumers, the heterotrophs that eat it. Not all consumers feed directly on producers. Many animals eat other consumers rather than primary producers. Thus, the transfer of energy through the system usually takes place in several steps known as a **food chain**. Each of the steps in the food chain is known as a **trophic level**.

Most ecosystems have a number of different primary producers. Furthermore, many animals eat more than just one kind of food, and many change their diet as they get older and larger. For these reasons, an ecosystem's trophic structure is usually a complex, interwoven **food web** (Fig. 12.14) instead of a simple, straight-line food chain. Such food webs are difficult for biologists to unravel and understand, but their complexity is one reason for the tremendous diversity of life.

Trophic Levels Food chains and food webs are a little easier to understand if we consider the number of steps, or trophic levels, through which the energy passes (Fig. 12.15). The first step in the flow of energy, and therefore the first trophic level, is occupied by the primary producers that originally capture the energy and store it in organic compounds. Diatoms are the main primary producers in the food web illustrated in Figures 12.14 and 12.15. Consumers that feed directly on primary producers are called first-level, or primary, consumers and occupy the next trophic level. At the level above that are second-level, or secondary, consumers, predators that eat the primary consumers. Feeding on the

secondary consumers are the third-level, or tertiary, consumers, and so on. The same organism, however, may feed at different levels. Petrels, for example, act as second-level consumers when they eat copepods but as third-level consumers when they eat amphipods. Each trophic level relies on the level below for sustenance. At the end of the food web are top predators such as killer whales.

The concept of trophic levels helps us understand how energy flows through ecosystems, but the organisms themselves are not concerned about which level they feed at. Predators often feed on prey from different trophic levels, for example.

The Trophic Pyramid Instead of being passed on to the next higher level, much of the energy contained in a particular trophic level is used up by the activities of the organisms. Energy and organic matter are also lost as waste. Depending on the ecosystem, only about 5 to 20% of the energy in one trophic level is passed on to the next, on average about 10%. Suppose, for example, that the diatoms in the Antarctic food chain illustrated in Figure 12.13 (p. 373) contain a total of 10 million calories of energy. According to the 10% rule, only about 1 million calories will make it to the primary consumers, the krill, and the whales will get only about 100,000 calories. The trophic structure of ecosystems can be represented by a pyramid of energy (Fig. 12.16), with less energy contained in each succeeding level.

Because there is less energy available at each level, there are also fewer individual organisms. Thus, there are fewer primary consumers than

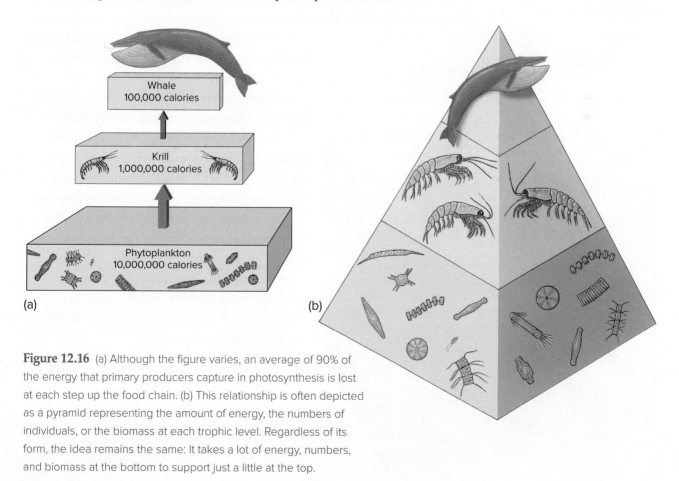

(a)

(b)

Figure 12.16 (a) Although the figure varies, an average of 90% of the energy that primary producers capture in photosynthesis is lost at each step up the food chain. (b) This relationship is often depicted as a pyramid representing the amount of energy, the numbers of individuals, or the biomass at each trophic level. Regardless of its form, the idea remains the same: It takes a lot of energy, numbers, and biomass at the bottom to support just a little at the top.

producers, and fewer secondary than primary consumers. Often, though not always, the **trophic pyramid** can be pictured in terms of numbers of individuals as well as energy and is then called the pyramid of numbers. The pyramid can also be expressed in terms of the total weight of tissue, or **biomass**, produced by the organisms at each level. In this case, the trophic pyramid is called the pyramid of biomass. To support a given biomass of primary consumers, primary producers must make about 10 times as much living tissue. To support 1,000 grams of copepods, for example, the copepods need to eat about 10,000 grams of phytoplankton. In turn, only about one-tenth of the primary consumer's biomass will make it to the secondary consumers.

At each step in the food web some organic matter is lost rather than being eaten by higher-level consumers. This material is not lost to the ecosystem, however. Decay bacteria, fungi, and other decomposers break down this non-living organic matter into its original components: carbon dioxide, water, and nutrients. Some of the organic matter is excreted as waste, is spilled during feeding, or leaks out of cells by diffusion. This material dissolves in the water, so it is known as dissolved organic matter (DOM). There is also a lot of dead organic matter in solid form, such as decaying seaweeds, cast-off seagrass and mangrove leaves, discarded exoskeletons, pellets of feces, and dead bodies. This material is called **detritus**. Detritus is an important energy pathway in marine ecosystems because many marine organisms feed on it. Large populations of microscopic decomposers live in association with detritus, and detritus feeders often derive as much or more nutrition from the decomposers as from the detritus itself. Similarly, many organisms eat the microorganisms that decompose DOM. Thus, decomposers play a vital role in the ocean by channeling dead organic matter back into the food web.

If it were not for decomposers, waste products and dead bodies would accumulate instead of rotting away. Not only would this make quite a mess, but nutrients would remain locked up in the organic matter. When decomposers break down organic matter, the nutrients incorporated into the organic matter during primary production are released, making the nutrients available again to photosynthetic organisms. This process is known as **nutrient regeneration**. Without it, nutrients would not be recycled and made available to autotrophs, and primary production would be greatly reduced.

Measuring Primary Productivity Because primary production supplies the food at the base of the trophic pyramid, it is useful to know how much production occurs in a given area. The rate of primary production, or **productivity**, is often

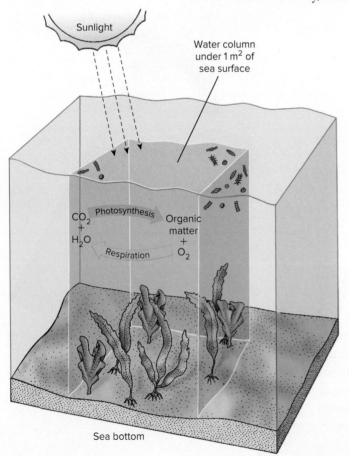

Figure 12.17 To describe primary productivity, marine biologists determine how much carbon ends up being fixed in the water column under 1 m² of sea surface in a given time. Gross primary production refers to the total amount of organic matter produced; net primary production is the amount left over after the respiration of the primary producers themselves. Primary productivity is expressed in units of grams of carbon per square meter per day or year (gC/m²/day or gC/m²/year).

Sunlight

Water column under 1 m² of sea surface

CO_2 + H_2O Photosynthesis Organic matter + O_2

Respiration

Sea bottom

expressed as the amount of carbon fixed under a square meter of sea surface in a day or in a year (Fig. 12.17). It includes the production of both phytoplankton in the water column and producers that live on the bottom.

The total amount of organic carbon manufactured by primary producers is called the **gross primary production**. Primary producers immediately respire some of the organic matter they make to meet their own energy needs, however, so it is not available as food to other organisms. It is the organic matter that is left over, or the **net primary production**, that forms the base of the trophic pyramid.

Net primary production by photosynthesis can be estimated by measuring either the amount of raw material used up or the amount of end products given off by photosynthetic organisms in the sunlight. Traditionally this was done by measuring the amount of oxygen (O_2) produced (Fig. 12.18) but now it is more common to measure the amount of carbon dioxide (CO_2) consumed. This method uses an isotope of carbon (^{14}C) that can be measured very accurately. To measure gross primary production, respiration by the primary producers can be estimated by keeping them in the dark where no photosynthesis can occur. Other methods to estimate primary production include measuring the amount of light absorbed by the producers and measuring the glow, or fluorescence, that chlorophyll emits when it absorbs sunlight.

IN CONTEXT

Photosynthesis

$CO_2 + H_2O$ + sun energy \longrightarrow organic matter (glucose)$+ O_2$

Respiration

Organic matter (glucose) + O_2 $\longrightarrow CO_2 + H_2O$ + energy

• 6.1, The Ingredients of Life

Isotopes Different atomic forms of an element.

• 2.2, How Plate Tectonics Formed the Oceans

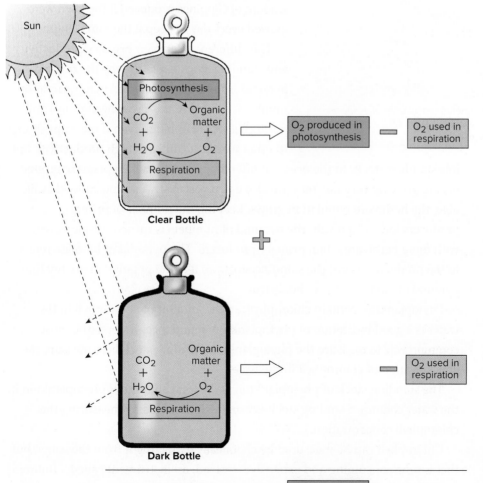

Figure 12.18 To measure primary production scientists put seawater samples in both clear and dark bottles. The clear bottle allows light to enter, so both photosynthesis and respiration take place. The amount of oxygen produced reflects the balance of photosynthesis over respiration, or the net primary production. To calculate the total production of organic matter by photosynthesis the amount of respiration must be known. This can be measured from the amount of oxygen used up in the dark bottle, where only respiration occurs because the light is blocked.

Table 12.1 Typical Rates of Primary Production in Various Marine Environments

Environment	Rate of Production (Grams of Carbon Fixed/m2/yr)
Pelagic Environments	
Arctic Ocean	<1–100
Southern Ocean (Antarctica)	40–260
Subpolar seas	50–110
Temperate seas (oceanic)	70–180
Temperate seas (coastal)	110–220
Central ocean gyres	4–40
Equatorial upwelling areas	70–180
Coastal upwelling areas	110–370
Benthic Environments	
Salt marshes	250–2,000
Mangrove forests	370–450
Seagrass beds	550–1,100
Kelp beds	640–1,800
Coral reefs	1,500–3,700
Terrestrial Enviroments	
Extreme deserts	0–4
Temperate farmlands	550–700
Tropical rain forests	460–1,600

Note: Production rates can be much higher at certain times or in specific locations, especially at high latitudes. Values for some selected terrestrial environments are given for comparison.

In the ocean the amount of primary production varies dramatically from one environment to the other (Table 12.1). Some marine environments are as productive as any on Earth. Others are biological deserts, with productivity as low as any desert on land. Productivity in the ocean depends largely on the physical characteristics of the environment, in particular on the amount of light and nutrients available. These characteristics will be explored in subsequent chapters.

Standing Stock The total amount of phytoplankton in the water, called the **standing stock** or standing crop of phytoplankton, is related to the primary productivity, but the two are not the same thing. The standing stock refers to how much phytoplankton is already in the water. Primary productivity refers to the amount of new organic material being created. In a lawn, for example, the standing stock would correspond to the length of the grass. The primary productivity would be the amount of clippings produced if the lawn were mowed every day to keep it the same length.

This difference between primary productivity and standing stock explains why the trophic pyramid is not always reflected in a pyramid of numbers. A lawn that is growing very fast will produce a lot of clippings even if it is kept short. In other words, it can have high productivity but low standing stock. In the ocean, it often happens that when primary producers are growing very fast they are also eaten very fast. With lots of food available, the herbivore population grows, keeping the number of primary producers low. As a result, the pyramid of numbers is turned upside down, with more herbivores than primary producers. The pyramid of biomass refers to the productivity, not the standing stock, so like the pyramid of energy the pyramid of biomass always holds true.

Phytoplankton contain chlorophyll, so the amount of chlorophyll in the water is a good indication of phytoplankton abundance. Indeed, the most common way to measure the phytoplankton standing stock is to measure the concentration of chlorophyll in the water.

The standing stock of phytoplankton is the total amount of phytoplankton in the water column. Standing stock is usually determined by measuring the chlorophyll concentration.

Chlorophyll can be measured by chemically extracting it from the water, but that is time-consuming and expensive. Instead, an instrument called a fluorometer is usually used to determine the amount of chlorophyll present from fluorescence it emits.

Figure 12.19 Global distribution of primary producer abundance.

Phytoplankton standing stock can also be measured from space. Satellites equipped with special cameras take color pictures of the sea surface and beam the images to Earth. Computer programs can estimate the concentration of chlorophyll from its characteristic green color. Phytoplankton standing stock can be estimated over large areas of the sea surface (Fig. 12.19) by applying various correction factors, taking account of the weather at the time the photo was taken, and relating the results to actual chlorophyll measurements made from ships. Satellites are the only means for assessing the large-scale distribution of phytoplankton; ships can measure standing stock at only one place at a time.

Cycles of Essential Nutrients

Once the energy stored in organic compounds is used in metabolism, the complex set of chemical reactions that take place in the cell, or given off as heat, it is lost to the system forever. Unlike energy, the *materials* that make up organic matter can be used over and over in a repeating cycle. These materials, like carbon, nitrogen, and phosphorus, originally come from the atmosphere, the earth's interior, or the weathering of rocks. Starting out as simple inorganic molecules, they are converted into other forms and incorporated into the tissues of autotrophs. When this organic material is broken down by digestion, respiration, and decomposition, the raw materials are released back to the environment, and the cycle begins again. Human alteration of these natural global cycles could have profound impacts on life on Earth that we are not yet able to predict.

The Carbon Cycle The carbon cycle provides a good example of this process (Fig. 12.20, p. 381). The carbon that forms the backbone of all organic molecules starts in the atmosphere as carbon dioxide, then dissolves in the ocean. CO_2 is

Drowning in Plastic

Humans appear to be moving into the ocean, or, at last their garbage is. The ocean is teeming with marine debris, anthropogenic solid material that is intentionally or unintentionally introduced into the marine environment. Much of this marine debris is composed of plastic. In 2012 alone, the U.S. generated 32 million tons of plastic waste, and an estimated 25% of that (8 million tons) entered the marine environment. Plastic products are so prevalent in the oceans because they are easily transported by wind and are buoyant. Common modes of introduction include littering, landfill escapement, wastewater runoff, and industrial runoff. Unlike organic materials, plastic does not decompose through mineralization into its basic chemical elements. Instead, plastic breaks down under light exposure, known as photodegradation, exponentially decreasing in size, but never truly going away. These microplastics can congregate in regions of the ocean where they will remain indefinitely and the magnitude of their ecological impacts is still unknown. Marine animals are literally drowning in our garbage.

A plastic food web Plastic debris can and is disrupting marine food webs. Floating plastic pieces can look a lot like squids or jellyfishes, common food items for birds, marine mammals, fishes, and sea turtles. This is particularly problematic for the Laysan albatross (*Phoebastria immutabilis*) in the Hawaiian Islands. Their chicks depend on the food regurgitated from their parents. As you can see in the photo to the right, the plastic fills their stomachs, falsely signaling a sensed fullness. Incapable of regurgitation, chicks ultimately die of starvation.

Sea turtles also lack the ability to regurgitate ingested plastic debris. Floating plastic bags look very similar to jellyfishes, the preferred prey of many species of sea turtle. Ingested plastics can cause esophageal and bowel obstructions or a false sense of fullness, all of which can lead to death. In addition, the partial digestion of plastic debris can trap gas in their bodies causing them to float at the surface. Unable to dive for food, they will die of starvation and will be unable to escape predators.

What can we do? To stop the onslaught of marine debris in the world's oceans, several organizations have developed cleanup initiatives. The Ocean Conservancy hosts an annual International Beach Cleanup Day to pick up and catalog beach litter. In 2013, they collected over 12.3 million pounds of garbage across 92 countries. The National Oceanic and Atmospheric Administration has developed a Marine Debris Tracker application for your smart phone. Anyone can pick up litter and upload their location to the database. Garbage cleanup is important anywhere in the world, because everything on land leads to a creek that leads to a river that leads to the ocean!

The stomach contents of a Laysan albatross (*Phoebastria immutabilis*) chick. Items include plastic fragment and a cigarette lighter. Adult albatrosses feed at sea and regurgitate food for their chicks.

Claire Fackler, NOAA National Marine Sanctuaries

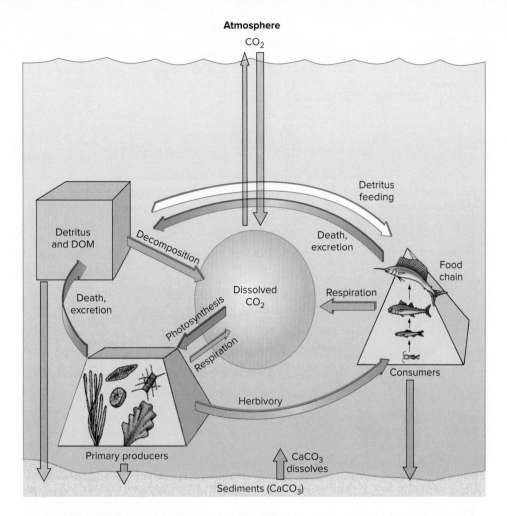

Atmosphere

CO_2

Detritus
and DOM

Decomposition

Death,
excretion

Photosynthesis

Respiration

Dissolved
CO_2

Detritus
feeding

Death,
excretion

Respiration

Food
chain

Consumers

Herbivory

Primary producers

$CaCO_3$
dissolves

Sediments ($CaCO_3$)

Figure 12.20 The carbon cycle in the ocean

highly soluble in seawater, so the ocean holds more than 50 times as much carbon as the atmosphere. This inorganic carbon is converted into organic compounds by photosynthesis. Respiration by consumers, decomposers, and the producers themselves breaks down the organic compounds and makes the carbon dioxide available to producers again. Some carbon is also deposited as calcium carbonate in biogenous sediments and coral reefs. Sediments contain 25 times again as much carbon as the ocean water. Under certain conditions, some of this calcium carbonate dissolves back into the water. Humans have greatly affected the global carbon cycle by burning fossil fuels and increasing the amount of CO_2 in the atmosphere.

Nitrogen, like carbon dioxide, is present in the atmosphere. Atmospheric nitrogen, however, is in the form of nitrogen gas (N_2), which most organisms are unable to use. Some cyanobacteria, other bacteria, and archaea can perform nitrogen fixation, and lightning also fixes atmospheric nitrogen. Without fixed nitrogen, algae and plants would not be able to get the nitrogen they need. In addition to nitrogen fixation that takes place in the ocean itself, fixed nitrogen enters the ocean from rivers and from the atmosphere in rain. Human activities, however, now produce more fixed nitrogen than natural sources, and most fixed nitrogen entering the ocean comes from agricultural fertilizers (either directly or as animal and human waste), cultivation of nitrogen-fixing crops, and fossil-fuel burning.

INCONTEXT

Biogenous Sediment Made of the skeletons and shells of marine organisms.

• 2.2, How Plate Tectonics Formed the Oceans

Nitrogen Fixation Conversion of nitrogen gas (N_2) into nitrogen compounds that can be used by primary producers as nutrients.

• 7.2, Prokaryotes

Figure 12.21 The nitrogen cycle in the ocean.

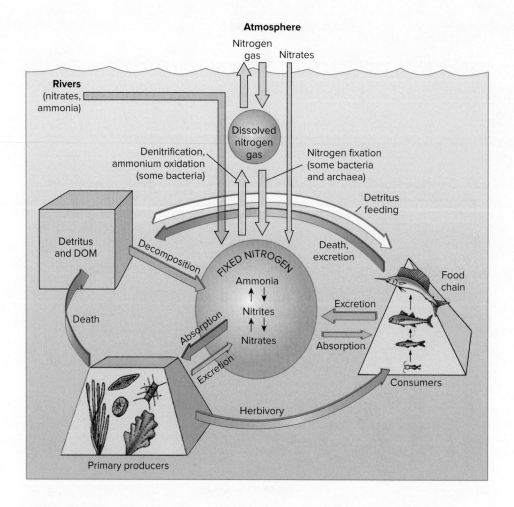

The Nitrogen Cycle Once nitrogen has been fixed, it cycles within the ecosystem as part of the nitrogen cycle (Fig. 12.21). There are several nutrient forms of nitrogen, the most important of which is nitrate (NO_3^{-1}). Nitrate and other nitrogen compounds are taken up during primary production and regenerated by decomposition. Most of the nitrogen-containing nutrients used in primary production come from this recycling of nitrogen compounds, not the original fixation of nitrogen. Until recent times natural nitrogen fixation was balanced by **denitrification**, which converts fixed nitrogen back to N_2, so the amount of fixed nitrogen in the global environment was more or less constant. Natural denitrification can no longer keep up with human activities, however, and fixed nitrogen is building up with potentially catastrophic consequences.

The Phosphorus Cycle Most of the phosphorus entering the ocean is brought in by rivers in the form of phosphate (PO_4^{-3}), although a small amount is deposited from the atmosphere. Much of this phosphate is derived naturally from the weathering of rocks and bird guano on land, but a great deal comes from fertilizers and other human sources as well.

On the other hand, humans sometimes reduce phosphate input to the ocean because phosphate gets trapped in sediment that builds up behind dams. After primary producers take up the phosphate and incorporate it into organic matter, the phosphorus cycles in organic form through the food web (Fig. 10.22). Phosphorus is deposited in ocean sediments either in biogenous sediments or

IN CONTEXT

Weathering The physical or chemical breakdown of rocks.

• 2.2, How Plate Tectonics Formed the Oceans

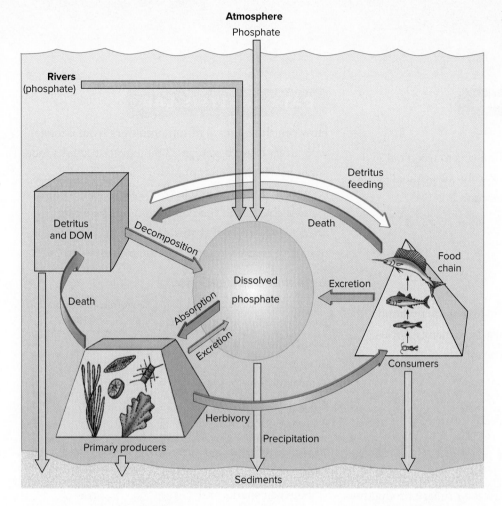

Figure 12.22 The phosphorus cycle in the ocean.

through **chemical precipitation**, in which chemical reactions convert phosphate into insoluble forms that settle on the sea floor. Some phosphorus is transported from the ocean back to land by seabirds and deposited as guano, but the flow of phosphorus through the ocean to the sediments is mostly a one-way street. Over millions of years, however, the phosphorus is recycled by geological processes such as the uplift of marine sediments onto land.

Other raw materials needed by organisms, like silicon and iron, undergo similar cycles. The tremendous abundance and diversity of living things depend on this recycling. Without it, the living world would soon run out of raw materials, and life on Earth would be scarce indeed.

. .

Section Review

1. **Summarize** Draw a sample food web in an open-water community, labeling each trophic level and example organisms in each level.

2. **Describe** What methods are used to measure net and gross primary production?

3. **Explain** Choose one nutrient cycle (carbon, nitrogen, and phosphorus) and explain how materials important for the production of organic matter move through the cycle.

Chapter 12 Review

REVIEW QUESTIONS

Multiple Choice

1. In some situations, selection acts to weed out individuals that are too far from the optimal condition. This type of selection is called

 a. directional selection.

 b. indirect selection.

 c. natural selection.

 d. stabilizing selection.

2. If the primary producers of an energy pyramid produce 10 million calories, how many calories will reach a fourth-level consumer?

 a. 1,000

 b. 10,000

 c. 100,000

 d. 1,000,000

Short Answer

3. What is the difference between primary productivity and standing stock?

4 Explain how human activities impact the nitrogen cycle.

Critical thinking

5. Two species of sea urchins live practically side-by-side on sandy bottoms. The two species appear to have the same diet. They are able to live in the same environment with minimal competition. How might they be able to share their habitat and food resources?

6. It is not always easy to categorize a particular case of symbiosis. Suppose a certain species of snail is always found living on a certain coral. No one has found evidence that the snail harms the coral, so the relationship is classified as commensalism. How would you go about testing this hypothesis? What kinds of observations might lead to the conclusion that the snail is a parasite, or that it has a mutualistic relationship with the coral?

DATA ANALYSIS LAB

How can the removal of top predators from a food web affect other organisms? Biologists studied a food chain in which the top predators, sharks, were essentially removed due to overfishing. The biologists studied changes in the population of 12 species of sharks, 14 species of skates and rays, and one species of bivalve in the western Atlantic from 1970 to 2005.

Data and Observations

The graphs show generalized population changes for each group of animals—sharks, skates and rays, and bay scallops—over the 25-year period.

Think Critically

1 What relationship between sharks and rays and skates could explain the changes to the population of skate and ray species?

2. What relationship between the skate and ray species and the bay scallops could explain the changes to the population of bay scallops?

3. What effect might the change in population of bay scallops have on populations of zooplankton and phytoplankton?

Generalized Population Changes

Data obtained from: Myers, R. A., J. K. Baum, T. D. Shepherd, S. P. Powers, and C. H. Peterson, 2007. Cascading effects of the loss of apex predatory sharks from a coastal ocean. Science, vol. 315, no. 5820, 30 March, pp. 1846–1850.

Between the Tides

Chapter
13

THEME **Structure and Function**

Intertidal zone organisms have structural and behavioral adaptations that help them survive the unique physical conditions associated with this habitat.

BIG IDEA

The intertidal zone is defined by changing conditions; organisms that live there have a range of adaptations that help them survive in this environment.

About the photo Bat sea stars (*Patiria minata*), rockweeds (*Fucus*), and other seaweeds on a rocky beach in British Columbia, Canada.

Chris Cheadle/All Canada Photos/Getty Images

13.1 Challenges of Life in the Intertidal

VOCABULARY

emersion

immersion

substrate

desiccate

tide pools

operculum

wave shock

dispersal

Main Idea

Organisms that live in the intertidal zone are well adapted to the harsh conditions found in this environment.

. .

Key Questions

1. What adaptations do intertidal organisms have that allow them to survive in this habitat?
2. How do organisms compete for space in the often crowded intertidal zone?

Of all the vast ocean, the intertidal zone, sometimes called the littoral zone, is the best known to marine biologists and the public alike. The intertidal zone is the narrow fringe along the shoreline that lies between the highest high and the lowest low tides. It is the only part of the marine world that we can experience firsthand without leaving our own natural element. We don't need a boat, and, at least at low tide, we can see it without a face mask and move through it without swim fins. More importantly from the scientist's point of view, we can work in the intertidal without cumbersome and expensive equipment and can easily return to the exact same place time after time. Intertidal communities are therefore among the most studied and best understood of all marine communities. Although the intertidal zone constitutes only a tiny fraction of the marine environment, lessons learned there have added immensely to our knowledge of ecology in general.

The intertidal is unique among marine environments in that it is regularly exposed to the air. Organisms that live in the intertidal must cope with this exposure, even if it means giving up characteristics that would be advantageous below the tides. Being out of the water and exposed to air is called **emersion**—the opposite of **immersion**, or being submerged.

The nature of intertidal communities depends critically on the type of bottom. The bottom—that is, the material on or in which an organism lives—is called the **substrate**. Hard, rocky bottoms are a very different habitat than soft bottoms made of mud or sand. In this chapter we consider intertidal communities on both kinds of substrate.

Exposure at Low Tide

Low tide presents problems for intertidal organisms. Left high and dry, they are exposed to the air, a harsher environment than water. The higher the organisms live in the intertidal, the more time they have to spend out of the water. The upper part of the intertidal zone is only immersed at the peak of high tide and then only briefly. The very upper edge may only be submerged during exceptionally high spring tides. The highest part of the "intertidal," in fact, is almost never immersed. It is kept wet by wave splash.

IN CONTEXT

Spring Tides: Tides with a large range that occur around times of the full or new moon.
• 4.3, The Tides

Organisms that live low in the intertidal, on the other hand, are submerged most of the time and have to cope with emersion only for short periods or, at the lowermost edge, only at the most extreme low tides.

Water Loss Marine organisms tend to dry out, or **desiccate**, when exposed to air. To survive, intertidal organisms must be able to prevent desiccation, tolerate it, or both. Most cope with the problem of drying out in one of two basic ways: They either run and hide or they "clam up."

The run and hide strategy is simple. When the tide goes out, the organism goes somewhere wet until the tide comes back in. Shore crabs, hermit crabs, snails, and other rocky shore denizens are often seen huddled in moist, shady cavities or crevices in the rocks (Fig. 13.1). **Tide pools**, depressions in the rocks that hold seawater after the tide goes out, are favorite places to hide (Fig. 13.2).

DIVE IN!

How are race cars, boats, and airplanes related to the rocky shore? The teeth of one of the hallmark organisms of the rocky shore—the small, domed-shaped limpet—may be the strongest known biological material on Earth. Limpets are molluscs that cling to rocks as they are constantly battered by waves. During high tide, limpets move along the rocks, using their teeth on their radulas to scrape algae from the rocky surface. In 2015, scientists determined that the teeth on the limpet's radula, made from a hard mineral called goethite, are up to 40% stronger than spider silk, another super-strong biological material. Engineers are interested in replicating the material in limpet teeth and applying it to the design of high-performance race cars, boat hulls, and aircrafts—all objects that need to be strong and durable, yet as lightweight as possible.

Figure 13.1 These periwinkles (*Littorina cincta*) get through low tide by clustering in a moist, shady crevice. They also seal against the rock to retain moisture.

Figure 13.2 When the tide goes out, a little bit of the ocean stays in this small tide pool in New Zealand. Organisms like these chitons (*Sypharochiton pelliserpentis*) and snails (*Diloma atrovirens*) move to such pools to stay moist.

(t&b)©Michael Huber

Figure 13.3 These seaweeds can't move to moist depressions at low tide, so they live there permanently. They don't grow on the surrounding rock because it dries out.

Even in tide pools, however, life can be tough. The water often undergoes drastic changes in salinity, oxygen content, and temperature. Some areas are kept moist by spray from waves or by water that slowly leaks out of tide pools. The hiding place may even be provided by other organisms: Dense clumps of mussels, algae, and other organisms provide a moist, shady habitat where smaller inhabitants of the intertidal can hide.

Seaweeds and sessile animals cannot run, but many are able to hide. Instead of moving to moist areas when the tide goes out, they live only in places that stay wet (Fig. 13.3). This may be because the larvae or spores settle only in moist, shady places or because those that settle elsewhere dry out and die.

Organisms that use the "clam up" strategy have some sort of protective covering, like a shell, that they can close to hold in water. Some, like barnacles and mussels, are completely enclosed and can seal in moisture simply by closing their shell. Others, like limpets, have an opening that cannot be completely shut. These organisms typically clamp themselves tightly to the rock to seal the opening. Some use mucus to make a better seal. They may also carve out shallow depressions, or "home scars," in the rock that make the seal more effective. They do this by scraping the rock with their shell or their radula. Every strategy has disadvantages, however: Clamming makes it harder for animals to obtain oxygen and food.

Some organisms use a combination of strategies. The periwinkles (*Littorina*) shown in Figure 13.1, for example, clamp themselves to the rocks to seal in moisture. They can also seal off the opening of their shell by closing the **operculum**, a stiff plate that fits the opening like a door. Periwinkles are still not immune to desiccation, so at low tide they congregate in moist, shady places, especially on hot days. Desiccation is a particular problem in hot, tropical climates, where intertidal organisms are often confined to tide pools or cool spots.

Finally, some intertidal organisms use neither the run-and-hide nor the "clam up" approach. Instead, they simply tolerate drying out. Some intertidal chitons can survive the loss of 75% of the water in their tissues. Some intertidal seaweeds, such as rockweeds, can withstand a water loss of as much as 90%, becoming almost completely dry and practically crunchy. They quickly recover when the tide comes in and wets their tissues.

Temperature and Salinity Emersion creates problems other than desiccation for intertidal organisms. Sea temperatures are relatively constant and mild because of the high heat capacity of water, but air temperatures can be much more extreme. At low tide, intertidal organisms are at the mercy of the sun's heat or the freezing cold of winter. Because tide pools are shallow, they too

IN CONTEXT

Heat Capacity: The amount of heat energy required to raise the temperature of a substance; also, its ability to hold heat. Water has one of the highest heat capacities of any natural substance.

• 3.1, The Unique Nature of Pure Water

©Michael Huber

experience extreme temperatures, though usually not as extreme as air temperatures.

Most intertidal organisms can tolerate a wide temperature range. Tidepool fishes, for example, are much more tolerant of extreme temperatures than related fishes that live below the tides. Certain species of periwinkle from high in the intertidal show remarkable heat tolerance. One is known to survive temperatures as high as 49°C in the laboratory.

Organisms deal with temperature extremes in ways besides simply being tough. Those that move to moist hiding places to avoid drying out, for example, also avoid high temperatures because such places are usually cool as well as damp. Some snails, especially in the tropics, have pronounced ridges on their shells. Like the fins on a car radiator, these ridges help the snail lose excess heat (Fig. 13.4).

The color of a snail's shell can also help it tolerate high temperatures. Snails that are regularly subjected to extreme high temperatures tend to be light in color. The light color helps reflect sunlight. On Atlantic shores, for example, the dog whelk or dog winkle (*Nucella lapillus*) has two color forms: whiteshelled and brownshelled. On shores with heavy wave action, brown whelks predominate. On sheltered shores the white form is most common. The difference appears to be related to temperature. Brown whelks absorb more heat than white ones and die in temperatures that don't harm the white snails. On the exposed coasts where brown snails live, there are dense mussel beds, which stay moist and relatively cool. Wave spray also helps cool the snails. Sheltered coasts have few mussels and little wave spray so they are hotter, and white whelks have an advantage.

Intertidal organisms also have to deal with extremes of salinity. When it rains, they have to endure fresh water, which is fatal to most marine organisms. Many simply keep the fresh water out by closing their shells—another benefit of the "clam up" strategy. Even so, rainstorms during low tide sometimes cause the mass mortality of intertidal organisms. Tidepool residents, too, face extreme variation in salinity. Rain dilutes the pools and lowers their salinity, while on hot days the salinity can increase markedly because of evaporation. To cope, most tidepool organisms tolerate a wide range of salinity as well as temperature. They may also burrow or reduce their activity to ride out extreme conditions and wait for high tide.

Restriction of Feeding Most of the sessile animals are filter feeders that have to be immersed in the water to filter it, and therefore can't feed at low tide. Even animals that are not filter feeders have trouble feeding at low tide. Many mobile animals in the rocky intertidal are grazers that scrape algae, bacteria, and other food from the rocks. Others feed on pieces of seaweed and other large detritus that drift in from deeper water. Some are predators and move over the rocks in search of prey. When mobile animals seek shelter or clamp to the rocks to avoid water loss at low tide they can't move around to find food.

Figure 13.4 The shell ridges on this tropical snail (*Nerita plicata*) help it stay cool by radiating heat. The snail's white color probably helps by reflecting sunlight.

CONTEXT

Deposit Feeders: Animals that eat organic matter that settles to the bottom.

Filter Feeders: Animals that actively filter food particles from the water. They are a common type of suspension feeder.
• 9.1, Sponges

Detritus: Particles of dead organic matter.
• 12.3, The Flow of Energy and Materials

Figure 13.5 Ocean waves expend tremendous energy when they break on the shore.

Being unable to feed when the tide is out is not much of a problem for animals low in the intertidal. They are immersed most of the day and have plenty of time to feed. Higher in the intertidal, however, animals may not be underwater long enough to get adequate feeding time. This can cause them to grow more slowly or even prevent them from living in the high intertidal at all.

The Power of the Sea

Even when the tide is in, life in the intertidal is not easy. Waves expend tremendous energy as they crash on the shore. Rocky intertidal organisms are exposed to the full power of the sea (Fig. 13.5).

Figure 13.6 Nearly all waves approach the coast at an angle. As a result, they refract, or bend toward the coast. By the time the waves break they are almost parallel to shore.

Figure 13.7 Wave refraction. Wave crests can be represented by lines on a diagram. Here the lines indicate the crest of a single wave at successive times. At time 1, the entire wave is in deep water. At time 2, the left end of the wave has entered shallow water and slows down, but the right end is still in deep water. The wave bends to become more parallel to the shore. This process continues so that the wave is nearly parallel to shore when it breaks (time 5).

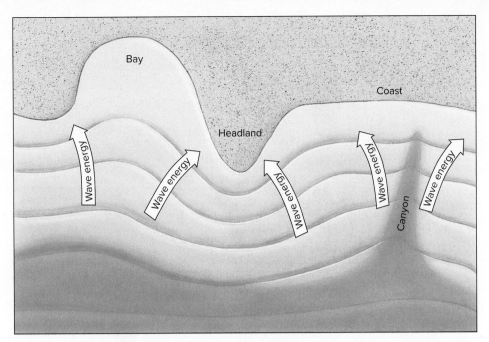

Figure 13.8 Both the coastline and bottom affect wave refraction. An incoming wave first hits shallow water in front of a headland. This part of the wave slows, lagging behind the sections on either side. As a result the wave wraps around the headland, which gets most of the wave shock. The opposite occurs in bays or over canyons: The central part of the wave is in the deepest water, and the wave turns off to both sides. This deflects wave energy away from the bay or the shore behind the canyon.

The Distribution of Wave Energy Along the Shore The impact of waves varies along the shoreline. Some areas are sheltered from the surf, others are fully exposed. Enclosed bays, for example, are usually protected from wave action, which is why they are used for harbors. It is not always easy to predict which areas will be sheltered and which exposed, however.

Recall from Chapter 4 that when a wave enters shallow water it "feels" the bottom and slows down. Thus, the same wave travels faster in deep than in shallow water.

Waves almost never approach the shore straight on, but instead come in at an angle (Fig. 13.6). One "end" of the wave reaches shallow water before the other. The end in shallow water slows down, but the end in deep water continues to travel at its original speed. As a result, the wave bends, just as a two-wheeled cart will turn to one side if one wheel sticks.

This bending, called refraction, causes the waves to become nearly parallel to the shore (Fig. 13.7). They never quite get perfectly parallel, however.

Refraction produces complex wave patterns when the coast is not a straight line. In particular, wave action tends to focus at headlands (Fig. 13.8). Bays, even if they are not physically sheltered from incoming waves, generally get less wave energy.

Offshore bottom features can influence the effect of waves on the coast. Submarine canyons, for example, may cause wave refraction. Also, waves often break on reefs or sand bars and expend their energy before they reach the shore.

Figure 13.9 On the Scottish coast the bladder rockweed (*Fucus vesiculosus*) occurs only in sheltered areas, the sea rockweed (*Fucus inflatus*) only at exposed sites. The spiral rockweed (*Fucus spiralis*) occurs in both, but lives higher up on the exposed coast. Such differences may be caused by the effects of waves on competitors or predators rather than the direct effects of wave shock. Diagrams like this are known as "kite diagrams." A wider blue "kite" indicates that more individuals of that species are at that elevation above low tide.

Holdfast: A structure used by many seaweeds to anchor themselves to the bottom.

• 8.1, Multicellular Algae: The Seaweeds

Swim Bladder: A gas-filled sac that provides many bony fishes with buoyancy.

• 10.1 Fishes: The First Vertebrates

Coping with Wave Shock The result of all the coastline variation is that there is tremendous variation in the intensity of wave impact, or **wave shock**, from place to place along the shore. Exposure to waves strongly affects intertidal organisms (Fig. 13.9). Some intertidal organisms simply can't withstand wave shock and are found only in sheltered locations. Such species are often delicate or unable to keep a firm grip on the rocks. Because sediment tends to accumulate in the sheltered spots where they live, these organisms often cope with sedimentation better than those that live on exposed shores.

Organisms exposed to the waves need some way to deal with wave shock. Sessile organisms anchor themselves firmly to the rocks to keep from being washed away. Seaweeds use their holdfasts or simply encrust on the rocks. Barnacles secure themselves with a glue so strong that several companies tried to duplicate it. Mussels attach with their byssal threads, strong fibers made of protein that the mussel produces with a special gland in its foot. These threads form the mussel's "beard." Though we usually think of them as stationary, mussels can slowly move around by putting out new threads and detaching old ones.

Mobile animals, too, often cling strongly to the rocks. Limpets and chitons, for example, use their muscular foot like a powerful suction cup. Gobies and clingfishes, which are common tidepool fishes, also have suction cups, formed by modified pelvic fins. The fishes don't stick as strongly as limpets, but unlike limpets they can swim home if dislodged. Intertidal fishes also tend to lack swim bladders, so they sink and stay on the bottom.

Rather than resist the waves, many intertidal animals find shelter when wave action gets too strong. This is especially true of highly mobile animals like shore crabs. In fact, there is a tradeoff between being able to move fast and having a strong grip on the rocks. Holding on tightly slows an animal

(a)　　　　(b)

Figure 13.10 (a) In sheltered locations the giant green sea anemone (*Anthopleura xanthogrammica*) anemone grows relatively tall. (b) Anemones from areas with heavy wave action are much shorter, reducing the exposed body area and reducing drag.

(a)　　　　(b)

Figure 13.11 Organisms like this brown alga, the kelp *Eisenia arborea*, withstand waves by being flexible. (a) In calm water the plant stands upright. (b) When a wave hits, the plant bends over and the fronds fold up.

(a)　　　　(b)　　　　(c)

Figure 13.12 Intertidal animals may get some protection from wave shock by growing in dense groups. (a) Mussels (*Mytilus*) growing as isolated individuals, for example, get the full impact of the waves. Larger mussels, with more exposed surface area, can be torn away. (b) When the mussels grow in dense clusters, the waves do not get a direct shot at any of them but slide over the tops. Eddies created as the water flows over the mussels may also hold plankton near the mussels, increasing their food supply, and the dense clumps retain moisture at low tide. (c) If the clumps get *too* dense, many of the mussels attach to other mussels instead of the rock. Waves can cause more strain than the mussels on the bottom can bear and rip away entire clumps.

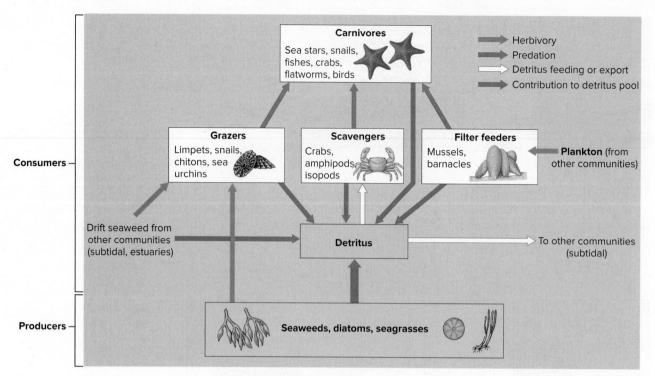

Figure 13.13 A generalized food web on a rocky shore.

down. Many intertidal snails are easier to knock off the rocks when they are moving around than when they stop and clamp down. Not surprisingly, they usually stop and hold on when wave action is heavy. This limits their feeding, which prevents some species from living in highly exposed areas.

Intertidal organisms have other adaptations to withstand wave shock. Animals in exposed locations tend to have thicker shells than those in sheltered ones. A compact shape can also help reduce the impact of the waves (Fig. 13.10). Many intertidal animals, including barnacles, mussels, limpets, and chitons, have low profiles close to the rocks. Colonies of the sandcastle worm *Phragmatopoma californica* provide another example of the importance of shape. These colonies are not especially strong and are easily crushed, but they can withstand considerable wave shock because of their domed shape. Some organisms, especially seaweeds, are flexible and can "go with the flow" (Fig. 13.11). There may also be safety in numbers (Fig. 13.12).

Even the firmest grip on the rocks is of little help if the wave smashes a boulder or drifting log onto the rocks, crushing whatever is there. Waves can also turn rocks over, even large boulders. When this happens the organisms on top of the rock may be smashed or buried, and sunlight for any algae is blocked. The underside of the rock supports different organisms than the top, and these fare no better. Once nestled snugly under the rock, they are suddenly exposed to sun and waves, not to mention hungry predators. The organisms on both the top and bottom of overturned rocks usually die. People can have the same effect. If you happen to go looking under rocks at the shore, don't forget to turn them back over the way you found them.

Because wave action varies from place to place, and organisms differ in how waves affect them, sheltered and exposed areas have very distinct

communities. Marine communities in quiet bays are much different from those at exposed headlands. The effect of wave action can also be seen on a smaller scale: The organisms in a sheltered crevice usually differ from those on exposed rocks nearby.

The Battle for Space

Despite the intertidal zone's physical extremes, there is usually plenty of food. The high light levels and nutrient rich coastal waters supply the basic requirements for photosynthesis, so algae grow vigorously. High tide brings in abundant plankton as well as drifting seaweed fragments and other detritus (Fig. 13.13), which is actually the most important food source in many intertidal communities. Limpets, snails, and other small, slow-moving animals that graze algae from the rocks are often limited by food availability because they are unable to range over a wide enough area. For most other intertidal animals, particularly sessile ones, food is not usually a limiting resource, though feeding time can be restricted near the upper end of a species' range.

Unoccupied space, on the other hand, is often in short supply. Sessile organisms need a permanent place to attach. Often there isn't enough room to go around, and the availability of space commonly limits rocky intertidal populations. Nearly all the space in the intertidal is occupied (Fig. 13.14), and

IN CONTEXT

Photosynthesis

$CO_2 + H_2O$ + sun energy → organic matter (glucose) + O_2
- 6.1, The Ingredients of Life

Limiting Resource: A resource that is so scarce that it restricts the growth of a population.
- 12.1, The Organization of Communities

Figure 13.14 Most organisms in the intertidal zone live attached to the rocks, and most of the available space is occupied. This little patch of rock supports two species of mussels and two of barnacles, plus limpets and sea anemones. What little open space there is doesn't last for long: A small bare patch (arrow) has already been colonized by newly settled barnacles.

©Michael Huber

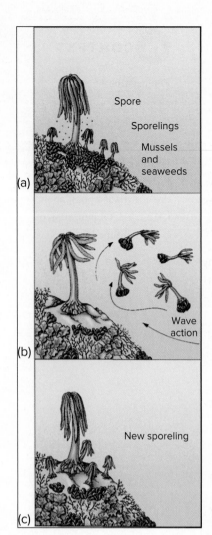

(a)

(b)

(c)

Spore

Sporelings

Mussels
and
seaweeds

Wave
action

New sporeling

Figure 13.15 The sea palm (*Postelsia palmaeformis*) takes advantage of wave action to compete for space. (a) Adults literally drip spores, which develop into young sporelings near the parent. (b) When the space below is occupied, the sporelings grow on these competitors, which are eventually pulled from the rocks. (c) This clears space for a new generation of young sea palms.

colonizers quickly take over space that opens up. Space is so scarce that organisms often attach to other organisms instead of the rocks.

Not surprisingly, competition for space is a dominant ecological factor in the rocky intertidal. There are many ways to compete for space. One is to be the first to get to open spots. For intertidal organisms, this means having an effective means of **dispersal**. Organisms that depend on being the first to occupy new patches of open space must be good at getting themselves or their offspring from place to place. Most rocky intertidal species disperse via larvae or spores, which settle on the rocks to colonize open space. Having taken over the space, the organism must either be good at holding on to it or be able to reproduce rapidly to disperse its young to the next opening. Both strategies are used in the intertidal.

Rather than colonizing open patches, some intertidal organisms take over space that is already occupied. Barnacles, for instance, undercut their neighbors, loosening them from the rock. Owl limpets (*Lottia gigantea*) maintain a territory by bulldozing out intruders. Less forceful methods also work. Many intertidal organisms simply grow over their competitors, making them vulnerable to waves (Fig. 13.16, p. 398), smothering them, or in the case of seaweeds, blocking precious sunlight. Some species grow in colonies, gradually increasing the amount of space they hold as the colony grows.

Two species of mussel provide an example of how competition and physical factors can interact to determine the distribution of organisms. On the west coast of North America the blue mussel (*Mytilus galloprovincialis;* also called the Mediterranean or bay mussel) is found mainly in sheltered locations, whereas the California mussel (*Mytilus californianus*) is found on the wave-exposed open coast. As you might expect, the California mussel has a thicker shell and attaches more strongly to the rocks than the blue mussel. On the other hand, the blue mussel does just fine on open coasts as long as the California mussel is absent. The blue mussel is apparently rare on open coasts not because it can't take the waves but because it is outcompeted by the California mussel. The California mussel, with its thicker shell, crushes the blue mussel when the two grow together. On the other hand, the blue mussel, unlike the California mussel, tolerates a lot of silt and can therefore survive in calm bays where silt settles out of the water.

. .

Section Review

1. **Explain** How do organisms in the intertidal zone cope with wave shock?

2. **Describe** How can some organisms solve the problem of competition for space in the rocky intertidal?

13.2 Rocky Shore Intertidal Communities

Main Idea

Zonation of organisms living in the rocky intertidal is determined by both physical and biological factors.

. .

Key Questions

1. What is vertical zonation?
2. What are the three main intertidal zones? Name three groups of organisms that can be found in each zone.

VOCABULARY

subsiding

epifauna

vertical zonation

pneumatocysts

keystone predators

ecological succession

climax community

Rocky shores generally occur on steep coasts without large amounts of sediment (see foldout map). Such areas have often been recently uplifted or are still rising as a result of geological events. These uplifted coasts have not had time to erode or accumulate sediment. The west coast of the Americas, for example, is largely rocky because its active margin has been uplifted by geological processes. Eastern Canada and New England were covered by huge sheets of ice during the last ice age. The ice sheets scraped sediment from the continental shelf, exposing the bare rock underneath. Under the tremendous weight of the ice, the coast actually sank partially down into the mantle. When the ice melted, the coast slowly rose, or rebounded, leaving a coastline of exposed rock. Sea level began to rise, however, and eventually the sea level rise overtook the rebound of the coast. The sea flooded the rocky coast, forming a deeply sculptured shoreline north of Cape Cod. This is the only part of the Atlantic coast of North America that has much rocky shore. The southern Atlantic and Gulf coasts of North America are slowly sinking, or **subsiding**, weighed down by huge amounts of sediment that are accumulating on the passive continental margin. Rocky areas are uncommon on these coasts.

Not all rocky shores are formed by uplift. Waves and currents can carry sediments away, leaving bare rock behind. Similarly, outcrops of hard, erosion-resistant rock may be left behind after softer surrounding rocks erode away. Much of the coast of the island of Hawai'i is rocky because it is geologically very young. Formed by successive flows of lava into the sea, the coast hasn't had time to accumulate sediment. In fact, the coast is still being formed by the periodic eruptions of the Kilauea volcano (Fig. 13.16, p. 398).

It is difficult to burrow through rock, though rockboring clams can do it in soft rock like sandstone. Most rocky intertidal organisms live on the rock's surface. Animals that live on the surface of the substrate—be it rock, sand, mud, or even other organisms—are called **epifauna**. Some epifauna move around on the rocks, but many are sessile and stay attached to the rock. Because little sediment accumulates in the rocky intertidal, deposit feeders are uncommon. Living on the rock surface, organisms in the rocky intertidal are fully exposed to the elements. This subjects them to considerable physical stress.

DIVE IN!

Are barnacles molluscs or crustaceans? Barnacles, one of the dominant organisms of the middle intertidal zone, are crustaceans. They nevertheless look like molluscs. In their larval stage, they resemble small shrimps. After they settle on a spot and attach themselves on their backs, they secrete calcium carbonate plates that surround their bodies. Like other crustaceans, barnacles have jointed appendages and an exoskeleton (not to be confused with the calcium carbonate plates).There are about 1,000 species of barnacles worldwide, ranging in size from 1 to 7cm. Barnacles filter feed on plankton and are preyed on by whelks, a snail also found in the intertidal zone.

IN CONTEXT

Active Continental Margin:
One that is colliding with another plate and therefore has a lot of geological activity.

Passive Continental Margin:
One that is on the "trailing edge" of a continent and has little geological activity.
• 2.3, Modern Ocean Basins

Mantle: One of the three main internal layers of Earth. The mantle lies between the crust, the outermost layer, and the core, the innermost layer.
• 2.1, Structure and Formation of Earth

Figure 13.16 Lava flows are creating new rocky shorelines on the island of Hawai'i.

Rocky intertidal communities vary markedly from place to place, but there is one feature that is remarkably common. The rocky intertidal community is usually divided into distinct bands, or zones, at characteristic heights in the intertidal (Figs. 13.17 and 13.18). Thus, a given species is usually found only within a particular vertical range, and not throughout the intertidal. Biologists call this pattern of banding **vertical zonation**.

When the shore consists of an evenly sloping rock face, the zones are often sharply defined belts that can be easily distinguished by the colors of the organisms (Fig. 13.17). Many shores are much more uneven, with scattered boulders, channels, and gullies. Zonation may not be as obvious in such areas, but it usually exists nonetheless.

Figure 13.17 Intertidal organisms often form distinct bands, or zones, at different heights on the shore.

Upper intertidal
Periwinkles dominant; limpets,
lichens, encrusting algae also common

Barnacles dominant

**Middle
intertidal**

Mussels,
barnacles,
seaweeds

Lower intertidal
Seaweeds,
surf grass

(a) Pacific

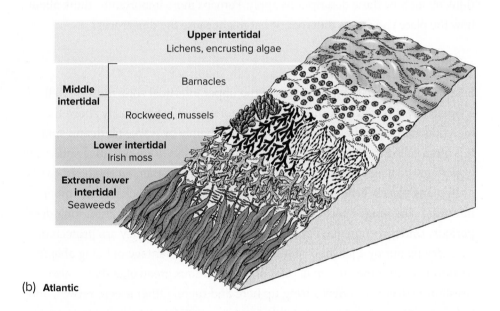

Upper intertidal
Lichens, encrusting algae

Barnacles

**Middle
intertidal**

Rockweed, mussels

Lower intertidal
Irish moss

**Extreme lower
intertidal**
Seaweeds

(b) Atlantic

Figure 13.18 Generalized patterns of zonation on temperate rocky shores on the (a) Pacific and (b) Atlantic coasts of North America. At any particular place on the coast the exact pattern will probably be different. Only organisms that characteristically dominate the various zones are shown. Hundreds of other species live in the intertidal.

Marine biologists have spent a lot of time and energy studying the causes of vertical zonation. Though there is still much to be learned, we know that zonation results from the complex interaction of physical and biological factors. A general rule is that the upper limit at which a species occurs is usually determined mainly by physical factors, whereas the lower limit is usually determined by biological factors, especially predation and competition.

This rule is a useful starting point in understanding the causes of vertical zonation. As with all generalizations, however, there are exceptions. For example, zonation can be produced if the larvae of a species settle only at a particular height on the shore. One cannot just *assume* that the upper limit of a certain species is set by physical factors. Instead, this hypothesis must be *tested* by experiment. Furthermore, different factors often interact to determine the limits, and sometimes the line between physical and biological factors is fuzzy. As noted previously, for example, some filter feeders do not live high in the

intertidal because they are not submerged long enough to feed. Is this upper limit due to a physical factor—emersion by the tide—or a biological one—feeding?

Many rocky shores around the world show a pattern of vertical zonation that is quite similar in its general nature but highly variable in its particulars. Biologists studying a certain area sometimes name the zones after the dominant organism, as in the "limpet zone" or the "mussel zone." These zones actually contain a variety of organisms. The mussel zone, for example, is also home to worms, crabs, snails, seaweeds, and many other species. Also, the specific organisms found in the intertidal vary from place to place. On the east coast of North America, for example, the mussel zone is often replaced by a rockweed zone.

Instead of naming the zones after organisms, we will simply divide the intertidal into the *upper, middle,* and *lower zones.* While not especially creative, at least these names apply everywhere! Bear in mind that the following sections are general descriptions of "typical" intertidal communities and don't apply exactly to any particular place. Furthermore, the boundaries between zones are not absolute and some species occur in more than one zone. The next time you go to the shore, think about how these descriptions apply. Perhaps more importantly, think about how the place you visit is different. What might cause the differences?

The Upper Intertidal

Seldom submerged, the inhabitants of the upper intertidal zone must be well adapted to withstand exposure to air. This zone actually lies mostly above the high tide mark, and the organisms are wetted mainly by wave splash and spray. It is often called the "splash zone." On exposed coasts with more wave splash the upper intertidal extends farther above the high tide line than on sheltered coasts.

In many places, lichens form black, tarlike blotches on rocks in the upper intertidal. The fungus part of the lichen soaks up water, storing it for long dry periods. Dark green mats of cyanobacteria are abundant. They are protected from drying out by a jellylike coating and have the advantage of being able to fix nitrogen from the air. Small tufts of a filamentous green alga that is also resistant to drying out may spring up here and there. Other green, brown, and red algae are occasionally found in the upper intertidal, usually in moist spots.

Large numbers of periwinkles graze on the algae, scraping it from the rocks with their radulas. Periwinkles (*Littorina*) are so abundant that the upper intertidal is often called the "*Littorina* zone." Not all of the many species of periwinkles live high in the intertidal, but those that do are well adapted to it. They can breathe air like land snails, live out of water for months, and tolerate extreme temperatures. They do have limits, however, and may have to seek shelter on hot days (see Fig. 13.1, p. 387).

Limpets also live in the high intertidal. Like periwinkles, limpets are hardy grazers. Shore crabs occasionally venture into the upper intertidal, scurrying over the rocks. They mostly scrape algae off the rocks with their claws, but also eat animals and detritus. Sea lice or sea roaches breathe air and live above the intertidal, but move down into the upper intertidal at low tide.

Few marine predators can reach the upper intertidal. Shore crabs occasionally eat periwinkles and limpets. From time to time predatory snails, such as

IN CONTEXT

Nitrogen Fixation: Conversion of nitrogen gas (N₂) into nitrogen compounds that can be used by primary producers as nutrients.
• 7.2, Prokaryotes

Lichens: Close, mutualistic associations between fungi and microscopic algae; mutualism is a type of symbiosis.
• 7.3, Fungi; and 12.1, The Organization of Communities

unicorn shells, also venture into the splash zone. On the other hand, land predators often visit the upper intertidal. Birds such as oystercatchers eat large numbers of limpets and snails. Raccoons, rats, and other land animals are also known to enjoy these tasty molluscs, as do people in most of the world.

The Middle Intertidal

Unlike the upper intertidal, the middle intertidal is regularly submerged and then uncovered by the tides. A diurnal tide exposes the organisms once a day, a semidiurnal tide twice. If the tide is mixed, the lower of the two successive high tides may not submerge the high part of the zone, and the higher of the two lows may not expose the bottom part. Thus, even in places with a semidiurnal tide, parts of the middle intertidal zone may be submerged or exposed only once a day. With so much variation in emersion time, different heights within the middle intertidal often support different organisms. In other words, the middle intertidal is often made up of several separate vertical zones.

Competition and Zonation The upper boundary of the middle intertidal is almost always marked by a band of acorn barnacles. In many places at least two species of barnacles, such as little gray and rock barnacles, form distinct bands, with little gray barnacles living higher up.

Experiments show that zonation results from a combination of the barnacles' larval settlement pattern, desiccation tolerance, competition, and predation (Fig. 13.19). The larvae of both species settle over a wider range than adults

IN CONTEXT

Semidiurnal Tide: A tidal pattern with two high and two low tides each day.

Diurnal Tide: A tidal pattern with one high and one low tide each day.

Mixed Tide: A tidal pattern in which two successive high tides are of different heights.

• 4.3, The Tides

Figure 13.19 Larvae of both little gray barnacles (*Chthamalus stellatus*) and rock barnacles (*Semibalanus balanoides*) settle out of the plankton over a wider range of the intertidal zone than is occupied by adults. Little gray barnacles are more resistant to desiccation so they can live higher on the shore. Little gray barnacles that settle where the rock barnacles can survive, however, are outcompeted. The lower edge of the rock barnacle zone occurs where predatory snails, called dog whelks (*Nucella lapillus*), can easily reach them. In some places, competition from mussels also helps determine the lower limit.

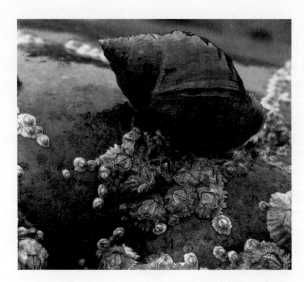

Figure 13.20 A dog whelk (*Nucella lapillus*) preying on rock barnacles (*Semibalanus balanoides*) on a rocky shore in Maine. In moist conditions like these the whelk is able to feed at low tide, but it will cease activity if the rock dries out.

occupy. The upper limit of both species appears to result from emersion, because larvae that settle too high in the intertidal die. Little gray barnacles tolerate drying better than rock barnacles and can therefore live higher up. At lower levels, where the rock barnacles can survive, they outcompete the little gray barnacles. Adult little gray barnacles, then, live perched in a narrow band above their competitors where they can survive emersion.

The lower limit of rock barnacles, like that of little gray barnacles, is determined by biological factors, in this case both predation and competition. Dog whelks are major predators of barnacles (Fig. 13.20). These snails use their radula to drill holes in barnacles' shells. They also secrete chemicals that soften the shells. If protected from the snails by wire cages, rock barnacles flourish below their normal limit. Unprotected barnacles, however, suffer heavy losses to snail predation. The only thing that saves the rock barnacles is that the dog whelks do most of their feeding during high tide. When emersed they lose a lot of water if they move around hunting barnacles, so to avoid desiccation at low tide they cease feeding or move lower in the intertidal. In the zone where rock barnacles are common, the snails don't have enough time at high tide to eat the barnacles. Sea stars, which also eat barnacles, have the same problem as the snails. Rock barnacles also face competition from mussels. In some places the lower limit of barnacles is set not by dog whelk predation but by competition with mussels. Like the little gray barnacles, the rock barnacles persist in a narrow zone between the level where they desiccate and die and the level where their enemies can reach them.

Many other organisms live in the middle intertidal among or below the barnacles. Which organisms are present, how many there are, and where they occur all depend on the unique combination of physical and biological factors at that site. The pattern of the tides, steepness of the shore, exposure to waves, and local weather all have an influence. Predation, competition, and larval settlement patterns are almost always involved. Considering all the possible ways in which these factors might interact gives you some idea of how complicated the story can be.

Depending on all these factors, mussels, gooseneck barnacles, and brown seaweeds, particularly rockweeds, often dominate the middle intertidal below the barnacles. Many rockweeds have air bladders called **pneumatocysts** that float their fronds closer to the light. They sometimes form a thick canopy that shelters many smaller organisms. If the rockweeds are experimentally removed, the algae and animals under the canopy typically bleach in the sun and die back, but small filamentous and branching alga often form a thick algal turf. The turf makes it more difficult for new rockweeds to settle, and recovery of the canopy and its associated community may take as long as a decade.

Transplantation, Removal, and Caging: Oh my!

In the intertidal zone, physical factors often determine how high up the shore a species can live. Biological factors often set the lower limit. Where does this rule of thumb come from? How can you tell why a certain species of barnacle or seaweed is found where it is? Like many scientific questions, this one is best answered by experiment. Three general types of experiments have been particularly effective.

Transplantation In transplantation experiments, organisms are transplanted from one place to another to see if they can survive under the new conditions. For example, barnacles were grown on panels that were placed at various heights on intertidal pier pilings. The barnacles died if they were moved higher than where they normally lived. Measurements of their body fluids showed the cause of death was drying out at low tide. The experiment showed that the barnacles cannot survive the conditions above their normal range. Thus, their upper limit is set by a physical factor—desiccation.

Transplantation can also tell us when physical factors are not important. When barnacles are transplanted lower in the intertidal than they normally live, they do just fine if protected from enemies. Their lower limit, therefore, is not set by the physical environment.

Removal experiments Experiments suggest that other organisms, either competitors or predators, often set a species' lower boundary. In removal experiments, one species is removed from an area, which is then compared to a control area. A classic experiment examined competition between two species of barnacles: *Semibalanus balanoides*

and *Chthamalus stellatus*. *Semibalanus* were removed from some of the rocks, while other contained both species.

In the lower zones, *Chthamalus* can survive on its own, but when the two species grow together the bigger, faster-growing *Semibalanus* crowd out or destroy *Chthamalus*. Thus, the lower limit of *Chthamalus* is determined by a biological factor: competition. Removal experiments have also shown the importance of grazing and predation. Another removal experiment, involving a sea star, is summarized in Figure 13.21.

Caging experiments It may be time consuming, difficult, or downright impossible to remove all the organisms of interest from an area, or they may quickly move back after removal. Instead, caging experiments can be used to keep them out. Scientists studied the role of spiny lobsters in the intertidal communities of the Channel Islands, off the coast of California. These intertidal habitats are dominated by thick mats of seaweed. They installed plastic cages to keep the lobsters out, as well as control cages that had roofs but no sides, to rule out the effect of reduced light penetration. Seaweeds continued to flourish in these control areas and in adjacent open areas, but within the fully enclosed cages the seaweeds were soon replaced by mussels. This demonstrated that seaweeds dominate because the lobsters eat the mussels, which compete with seaweeds for space.

Transplantation, removal, and caging experiments have been extraordinarily effective in unraveling the web of physical and biological interactions in the intertidal zone—so effective, in fact, that they have become an important tool for ecologists not only in the intertidal but in many other environments.

These cages were used in California to test the effects of spiny lobsters (*Panulirus interruptus*) in the intertidal zone. The lower cages have closed ends to keep lobsters out. The upper cages have open ends to allow lobsters free access and are controls for the effects of the cages themselves.

The mussels in this dark patch have taken over the space inside a closed cage (removed for the photo). The surrounding rock is dominated by red algae. The conclusion of the experiment is that mussels would dominate the community if not eaten by lobsters.

Predation and Zonation Mussels are especially common on stormy, exposed shores. They cannot live as high on the shore as barnacles, because they dry out and also because at high levels they don't spend enough time underwater to filter feed. At heights where they *can* live, however, mussels are dominant competitors, smothering or crowding out other organisms.

Sea stars are voracious mussel predators. Sea stars do not force open a mussel's shell to eat it. Instead, they insert their stomach into the shell and begin digesting the mussel from the inside. Sea stars need only a tiny crack to insert their stomach—only 0.2 mm for the ochre sea star (*Pisaster ochraceus*; see Fig. 13.25). The shell opens after the mussel is weakened.

Sea stars don't tolerate desiccation very well and need to be underwater to search for food though they often finish a meal after the tide goes out. In the lower part of the middle intertidal this is no problem, and sea stars take a heavy toll on mussels. Thus, mussels are only found above the level of heavy sea star predation. As with other intertidal species, the lower limit of mussels is set by a biological factor: predation by sea stars.

In some places spiny lobsters (*Panulirus*) instead of sea stars are the main mussel predators. Lobsters can almost completely eliminate mussels from the intertidal zone.

Sea stars and other mussel predators can have a strong effect even above the level where they completely wipe out the mussels. They venture into the mussel bed at high tide and eat some of the mussels before retreating with the tide. This opens up space for other organisms—such as acorn barnacles, gooseneck barnacles, and seaweeds—that would otherwise be crowded out. The sea stars also eat dog whelks, doing the barnacles another favor. If the barnacles live long enough, they get too big for the dog whelks to eat. By reducing the number of dog whelks, the sea stars give the barnacles more of a chance to grow big before a whelk eats them.

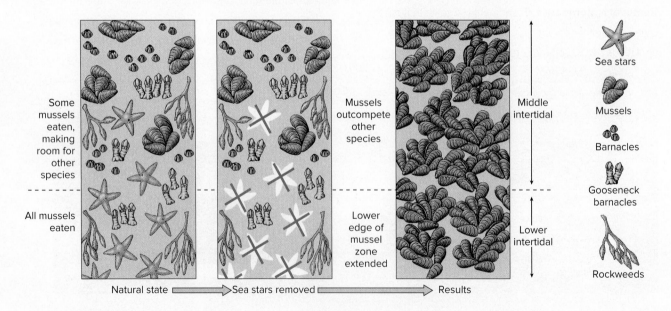

Figure 13.21 Removing predatory sea stars—or keeping them out with cages—shows how they affect the middle intertidal community. When there are no sea stars around, the mussels are able to live lower, and they monopolize the available space by overgrowing and crowding out other species. Sea stars thus maintain the diversity of the middle intertidal community.

The sea stars are absent from areas with extremely strong wave action. In such areas, or if the predators are experimentally removed (Fig. 13.21), mussels take over. They outcompete and crowd out other sessile organisms. Thus, many species can live in the middle intertidal only if sea stars keep mussel numbers under control.

Note that many intertidal species are more abundant than the sea stars, but the sea stars profoundly affect the entire community. This demonstrates a common feature of biological communities: Just because a species is relatively uncommon does not mean it is unimportant. Predators whose effects on their communities are proportionately much greater than their abundance are known as **keystone predators**.

Disturbance and Succession Natural disturbances can have effects similar to those of predation. When mussel beds get too dense, clumps of mussels can be torn away by waves (see Fig. 13.12, p. 393), exposing the bare rock. Drifting logs that batter the rocks and, in cold places, scouring ice and severe freezes, have the same effect of opening up new space. This prevents mussels from monopolizing the available space, allowing other organisms to persist.

When a patch of space is cleared, new organisms often move into the patch and get replaced by others in a regular sequence (Fig. 13.22, p. 406). The term **ecological succession** is used to refer to such regular patterns of regrowth. In the rocky intertidal the first stage is usually a thin film of bacteria and microscopic algae, such as diatoms, that covers the rock.

This film may condition the rock surface, since the settling larvae of some species prefer surfaces with the film over bare rock. Seaweeds often move in next, followed by barnacles and finally mussels, the dominant competitor. The final stage in an ecological succession is called the **climax community**.

What actually happens usually deviates, to varying degrees, from the idealized pattern shown in Figure 13.22. Grazers like limpets and chitons may remove newly settled larvae and spores.

If such grazers move into a patch early, the succession may never get past the bacterial and algal film stage and in some places grazers limit even the film.

On the other hand, stages in the succession may be skipped. Seaweeds may never colonize a patch, for example, if barnacles get there first. Timing is critical: If the patch opens up when barnacle larvae are abundant and seaweed spores are rare, barnacles will have the upper hand. Thus, there is always an element of chance in the development of the community. Stages may also be skipped if the bare patch is small. A small patch cleared in the middle of a mussel bed may be taken over by other mussels that move in from the sides before the succession gets started. Furthermore, the physical environment can have more subtle influences on the community than smashing it with waves or logs. For example, the sea star *Pisaster ochraceus* eats less in cold conditions, so its influence as a keystone predator can vary with the seasons. Other seasonal factors may come into play. On the Pacific Northwest coast of North America, low tides tend to occur during the day in spring and summer, but at night in autumn and winter. Rockweeds higher in the intertidal have difficulty growing in spring because of desiccation during the daytime tides, but in fall actually

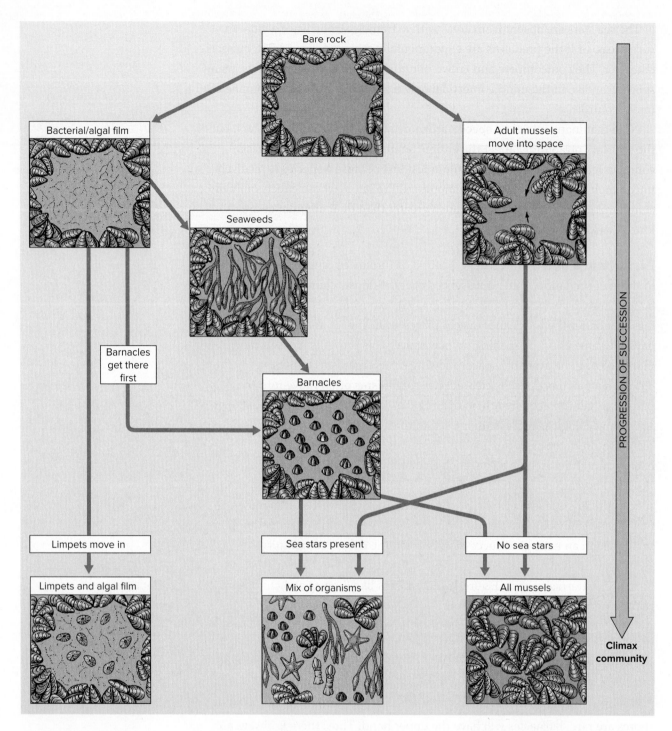

Figure 13.22 Ecological succession following the clearing of a patch in a mussel bed. The pathway taken and the end result depend on the size of the patch, when it opens up, and just plain luck—which organisms get there first.

Competitive Exclusion The elimination of one species by another as a result of competition.

• 12.1, The Organization of Cummunities

grow faster than rockweeds lower in the intertidal because they don't dessicate during the nocturnal low tides and are less exposed to predators and waves.

Without predation and other disturbances, the best competitors take over and many species disappear from the area. Disturbance can prevent this. Mussels, for instance, are often unable to completely cover middle intertidal rocks because sea stars eat them or waves tear them away. This gives other species a chance. Disturbance can thus increase the number of species that live in an area, that is, the diversity, by interfering with competitive exclusion. The

rocky intertidal in some places can be thought of as a mosaic of patches that were cleared at different times and have developed differently. Each patch supports a different set of organisms, so the number of species in the area as a whole increases. On the other hand, if predation and other disturbances happen too often, the community keeps getting knocked back to the starting point. It never gets a chance to develop, and not many species are able to live there. The highest diversity occurs when there is enough disturbance to prevent the dominant competitors from taking over, but not so much that the community is unable to develop (Fig. 13.23).

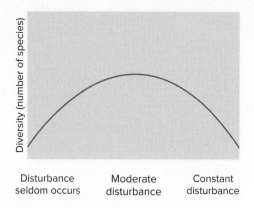

Figure 13.23 Other things being equal, the number of species depends on how often disturbances occur. When disturbance is rare, the dominant competitors take over and exclude other species. Intermediate levels of disturbance prevent this and give other species a chance; the number of species is highest when disturbance is moderate. If disturbance happens too often, most species cannot get a foothold and the number of species drops.

The Lower Intertidal

The lower intertidal is immersed most of the time. This makes it easy for predators like sea stars and dog whelks to feed, so mussels and barnacles are rare. The lower intertidal is dominated by seaweeds, which form a thick carpet on the rocks. These seaweeds, including species of red, green, and brown algae, can't tolerate drying out but in the lower intertidal they grow in profusion. Not surprisingly, grazing and competition are important in the lower intertidal. Light, as well as space, is an important resource. Seaweeds often compete by overgrowing each other, blocking the light.

A good example of the interaction of competition and grazing comes from the coast of New England. Two common seaweeds there are a green alga, *Enteromorpha*, and a red alga (not actually a moss) called Irish moss (*Chondrus crispus*). When the two seaweeds grow together in tide pools, *Enteromorpha* dominates by overgrowing the Irish moss (Fig. 13.24). Surprisingly, however, Irish moss is the most common seaweed in many tide pools.

The reason for this is that grazing sometimes prevents *Enteromorpha* from taking over. An abundant grazer in the area is the common periwinkle (*Littorina littorea*), which is closely related to the periwinkles that live in the upper intertidal. While *Enteromorpha* is one of its favorite foods, the snail hardly touches Irish moss. If a tide pool has a lot of snails, they eat the *Enteromorpha* and make way for Irish moss. If snails are transplanted into an *Enteromorpha* pool,

Subtidal Zone The part of the continental shelf that is never exposed by low tide.
- 12.2, Major Marine Lifestyles and Environments

Coralline Algae Red algae that deposit calcium carbonate (CaCO3) in their tissues.
- 8.1, Multicellular Algae: The Seaweeds

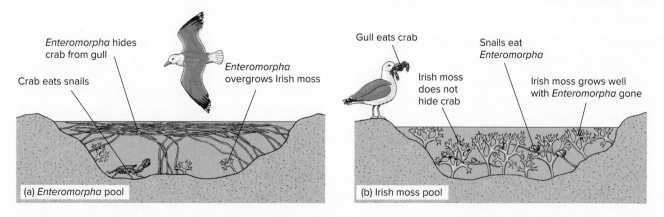

Figure 13.24 Biological interactions determine what type of seaweed predominates in New England tide pools. Unless something interferes, both types of pool are selfsustaining.

PISCO Studies in the Rocky Intertidal

Community interactions in the rocky intertidal are complex, and work across multiple scales. Experiments at one site help reveal the detail of how individual organisms interact, while studies of a bay or even an entire coastline provide a broader perspective. Timescale is critical, too: A one or two year study can't detect long-term changes. Similarly, while studies of single factors, sea star predation for example (see Fig. 11.21), reveal much about intertidal ecology, we also need to understand how many different factors interact—for example, how sea star predation, mussel and algal growth rates, and larval settlement might affect competition between mussels and seaweeds, and how oceanographic conditions and climate cycles influence the system.

Large-scale studies in space and time need a group effort. The Partnership for Interdisciplinary Studies of Coastal Oceans (PISCO) is a consortium of marine scientists doing integrated studies at many sites over many years along the Pacific coast of North America. In collaboration with government agencies, conservation groups, and other organizations, PISCO is monitoring more than 130 rocky intertidal sites from southeastern Alaska to Mexico.

"Robomussel" is an electronic temperature recorder that not only looks but also absorbs heat like real mussels.

Monitoring program Much of PISCO's research aims to understand how rocky intertidal ecosystems are changing, and predict how they will change in the future. Some sites have been monitored for over 25 years. At each site, scientists measure the abundance of major species annually or every few years. At many sites they also measure larval settlement and growth rates.

In California, many of the sites are inside marine protected areas to determine whether the protected areas are effective. They also assess whether the protected areas benefit surrounding non-protected areas. This helps agencies decide what management rules are needed. For example, when off-limits areas are opened up to people, food collecting and trampling cause declines in the abundance and diversity of intertidal organisms, so rules about what people are allowed to take and do in the intertidal are needed. The research also improves the design of protected areas by providing information on the best size for protected areas and the best places to put them.

Variation in the intertidal PISCO's research has shown largescale variation in rocky intertidal ecosystems on both short and long time scales. This variation is more closely linked to offshore oceanography than scientists previously thought. Scientists knew the importance of coastal upwelling along the continental shelf to plankton and fisheries, but PISCO scientists revealed that upwelling also influences predation, grazing, and other processes in the rocky intertidal.

On a longer time scale, larval recruitment of barnacles and mussels was higher on the Oregon coast in the 2000s than the 1990s. PISCO scientists found that much of this variation was related to large-scale ocean cycles such as the Pacific Decadal Oscillation and North Pacific Gyre Oscillation.

Experiments in the intertidal PISCO scientists conduct experiments at their monitoring sites on the intensity and effects of predation, competition, and physical factors. To study how climate change will affect intertidal communities, they placed "Robomussels" in mussel beds from Washington to Southern California to determine how splashing waves, tide, and latitude affect the mussels' body temperatures. They discovered "hot spots" along the coast where mussel temperatures were higher than expected for how far north they were. It had been assumed that global warming will mainly affect species at the southern end of their distribution as temperatures rise, but this work indicates that mussels could die off in patches all along the coast.

Enteromorpha quickly disappears and Irish moss takes over. On the other hand, *Enteromorpha* soon moves in after periwinkles are removed from an Irish moss pool.

The story doesn't end there. If the periwinkles like *Enteromorpha* so much you'd think that they would move into *Enteromorpha* pools, but they don't. For one thing, large periwinkles do not move around a lot and prefer to stay in one tide pool. Young periwinkles are eaten by green crabs (*Carcinus maenas*), which also live in tide pools. The crabs, in turn, are eaten by gulls. *Enteromorpha* provides cover for the crabs, protecting them from gulls. The crabs then eat young periwinkles, preventing *them* from colonizing the pool. In Irish moss pools, on the other hand, the crabs lack cover and are eaten by gulls. Without the crabs, young periwinkles can survive in the tide pools. Thus, both types of tide pool community perpetuate themselves as a result of both positive and negative interactions among species.

Figure 13.25 The bright green color of these giant green sea anemones (*Anthopleura xanthogrammica*) comes from zooxanthellae, symbiotic algae that live within the anemones' tissues and provide them nutrition, although the anemones also catch small prey with their tentacles. The anemones are surrounded by the sea star *Pisaster*, a keystone predator in the rocky intertidal.

The lower intertidal supports many other seaweeds besides *Enteromorpha* and Irish moss. These range from delicate reds and greens to large, tough kelps. The kelps mark the lower limit of the intertidal and continue down into the subtidal zone. There is no precise boundary between the lower reaches of the intertidal and the upper edge of the subtidal. Coralline algae can also be abundant. Surf grass, a flowering plant, is common in the lower intertidal on the Pacific coast of North America.

A host of small animals live among the seaweeds, hiding from predators and staying moist during very low tides. Sea urchins are common grazers on the seaweeds. Also present in the lower intertidal are sea anemones (*Anthopleura*; Fig. 13.25), polychaete worms, snails, sea slugs, and many others.

Most rocky intertidal fishes live in the lower intertidal or in tide pools. Gobies, clingfishes, sculpins, pricklebacks, and gunnels are among the most common. These are all small fishes adapted to the environmental extremes of the intertidal. Most are carnivorous.

. .

Section Review

1. Describe What are the main physical characteristics of each intertidal zone?

2. Identify Which factors determine the type of organisms that are present, how many there are, and where they are in the middle intertidal zone?

13.3 Soft-Bottom Intertidal Communities

VOCABULARY

infauna

interstitial water

bioturbators

meiofauna

DIVE IN!

What just touched me? If you were digging through the sand at the beach, it might have been a sand crab. Sand crabs, like *Emerita analoga*, moves up and down the beach with the tides, so that it can feed in the swash zone, (the area of breaking waves). These crabs are less than 35 mm in size, and unlike most crabs, have no claws on their first pair of legs. The carapace is smooth, and its color matches that of the sand. Sand crabs can only move in one direction: backwards. They feed by burrowing in the sand with their heads facing seaward. They expose only their eyes and first antennae above the sand. As receding water from a wave flows over them, they use featherlike antennae to filter plankton and other organic particles out of the water.

Main Idea

Adaptations for obtaining oxygen, burrowing, and feeding allow organisms to survive in the soft-bottom intertidal.

Key Questions

1. What is the relationship between grain size and conditions in bottom sediments?
2. What are three adaptations organisms have to live in intertidal soft sediments?

Any bottom that is composed of sediment, as opposed to rock, is considered a soft bottom. The dividing line isn't always clear. Boulder fields are usually thought of as rocky, but how small the rocks have to be before the bottom is considered soft is not precisely defined. In this text we refer to a bottom as soft if organisms are able to burrow in it easily.

Soft bottoms occur where sediments accumulate. In North America, soft bottoms dominate on the east coast south of Cape Cod and virtually all of the Gulf coast (see foldout map). The rocky west coast is often interrupted by sandy beaches and mudflats, especially in and near river mouths. Whether and what kind of sediment accumulates in an area depends on how much water motion there is and on the source of the sediment. In turn, the type of sediment strongly influences the community.

The Shifting Sediments

Soft bottoms are unstable and constantly shift in response to waves, tides, and currents. Thus, soft-bottom organisms do not have a solid place for attachment. Very few seaweeds have adapted to this environment. Seagrasses are the most common large primary producers on soft bottoms and often form dense intertidal beds. In this chapter we deal only with intertidal soft-bottom communities that lack seagrasses. Seagrass communities are discussed in Chapter 15 (see "Seagrass Meadows," in 15.2, Soft-Bottom Subtidal Communities).

The animals that live on soft bottoms also lack solid attachment sites. Though a few of them are epifauna and live on the sediment surface, most burrow in the sediment for protection and to keep from being washed away. Animals that burrow in the substrate are called **infauna** because they live *in* the sediment.

Types of Sediment The kind of sediment on the bottom, especially the size of the grains, is one of the most important physical factors affecting soft-bottom communities. Most people use the terms "sand," "silt," "clay," and "mud" without much thought. To a geologist these terms refer to sediments with specific particle sizes (Fig. 13.26).

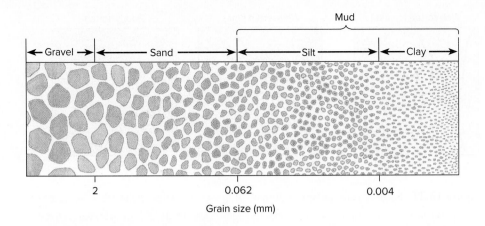

Figure 13.26 Sediments are classified by the size of the particles. Sand is relatively coarse; clay is very fine. Together, silt and clay are called mud.

Careful analysis is needed to determine grain size precisely, but a "quick and dirty" test will literally give you a feel for the different sediment types. Rub a pinch of sediment or soil between your fingers. Any grittiness is caused by sand. Silt and clay, which together are called mud, feel smooth. Also, the finest sand particles are the smallest you can see. To tell silt and clay apart, you can rub a little between your teeth. Silt feels gritty, but clay is still smooth.

Most sediments are made up of a mixture of different particle sizes. They are described according to the most common grain size. Sand, for instance, has mostly sand-sized grains but may also have small amounts of mud or a few large rocks.

Sediment and Water Motion Sediment composition is directly related to the degree of water motion. Imagine, or better yet actually perform, the following experiment. Take a handful of sediment or soil that contains a wide range of particle sizes, from large pebbles down to clay. Put it in a glass container with some water and shake it up. As soon as you stop shaking, the pebbles will sink to the bottom. As the water gradually stops moving, first the coarse sand and then finer particles will begin to sink. Very fine material will remain suspended for a long time, making the water look muddy. You might have to let the container sit for weeks before the finest clay settles out completely. If the container is stirred even gently, the fine material will never settle out.

This demonstrates a general rule: Fine sediments remain suspended with even a small amount of water motion, whereas coarse sediments settle out unless there is considerable flow. Bottom sediments thus reflect prevailing conditions. Calm, sheltered areas have muddy bottoms because fine sediment can settle out. Organic particles sink at about the same rate as clay particles, so the two tend to settle out together. Some organic material also binds to clay particles. Clay sediments are therefore usually organic-rich. Places with stronger waves and currents have coarser bottom sediments. If the water motion is strong enough, it may carry away *all* the loose material, leaving bare rock or large boulders.

In our sediment experiment, all the particles would eventually settle out if the container were kept still. A constant flow of water through the container, however, would carry away the fine material and only certain particles, the coarse ones, would remain. The coarse particles would be sorted out from the fine ones.

Well-sorted (coarse)

(a) Water drains quickly

Well-sorted (fine)

(b) Water drains slowly

Poorly sorted

(c) Water blocked

Figure 13.27 "Sorting" refers to the uniformity of grain size in the sediment. In well-sorted sediments (a and b) the grains are all about the same size. There is a lot of space between the grains, so water can flow through. The grain size, as well as sorting, affects how well water circulates; larger grains have larger spaces and are thus more porous. Poorly sorted sediments (c) have many differentsized grains. Small grains fill in the spaces between the larger ones, making water flow difficult.

Living in the Sediment

Living in sediment has advantages in the intertidal. Soft bottoms stay wet after the tide is out, so the problem of desiccation is not as critical as in the rocky intertidal. This depends on the grain size, however. Coarse sands drain, and therefore dry out, quite rapidly (Fig. 13.27). Partially because of this, coarse sand beaches have relatively little animal life.

Oxygen Availability The amount of organic matter in bottom sediments is particularly important to deposit feeders. Because there are relatively few primary producers, detritus is the main source of food for intertidal soft-bottom communities. Deposit feeders extract this organic matter from the sediments. The amount of detritus depends on grain size. Coarse sands contain very little organic matter. That is why we tend to think of sand as clean. Silt and clay, on the other hand, are usually rich in detritus, which is why they are often smelly.

The grain size also affects the amount of available oxygen in the sediments. Oxygen in sediments is used up by the respiration of animals and, more importantly, decay bacteria. Below the sediment surface there is no light and therefore no photosynthesis, so the infauna depend on the circulation of water through the sediments to replenish the oxygen supply. Grain size and the degree of sorting strongly affect how porous a sediment is (Fig. 13.27). Every backyard gardener knows that water flows more easily through sand than clay. Water circulation through fine sediments is greatly restricted.

Muddy bottoms, then, have a double problem. First, they have more organic material to decay and use up oxygen. Second, the flow of water that brings in new oxygen is reduced. Except for the upper few centimeters of mud the **interstitial water**, or water between the grains, is deficient in oxygen. If you go very far down at all, in fact, the oxygen is completely used up. Sediments with absolutely no oxygen are said to be anoxic.

Anoxic conditions are no problem at all for bacteria that are capable of anaerobic respiration, in which they break down organic matter without oxygen. A noxious gas called hydrogen sulfide (H_2S) is produced as a byproduct.

Figure 13.28 The black layer in the sediments shows where the oxygen has been completely depleted and hydrogen sulfide (H_2S) is produced by anaerobic bacteria.

If you have ever had to unclog the drain of your kitchen sink, you have probably run into hydrogen sulfide. It turns things black and smells like rotten eggs. In the sediments, just as in your drain, hydrogen sulfide is produced when bacteria continue to decompose organic material after all the oxygen has been used up. A distinct black layer appears where anoxic conditions begin in the sediments (Fig. 13.28).

Infauna must adapt to oxygen shortage, especially in muddy bottoms. Many animals pump oxygen-rich water from the sediment surface with siphons or through their burrows. Such animals are never actually exposed to low oxygen levels. Their burrows, and the effects of **bioturbators**—infauna that burrow through, turn over, or otherwise disturb the sediment—help oxygenate the sediment.

Nevertheless, muddy sediments are often oxygen-deficient. Animals living in them have adapted to low oxygen environments. They often have special hemoglobins and other adaptations to extract as much as possible of what little oxygen there is in the interstitial water. Some are sluggish, which reduces their oxygen consumption. They may even have a limited capability to carry out anaerobic respiration. A few animals have symbiotic bacteria that help them live in low oxygen sediments. Even so, hydrogen sulfide is toxic to most animals, and anoxic sediments have relatively little animal life.

Getting Around Soft bottom animals use a variety of methods to burrow through the sediments. Clams and cockles take advantage of being able to change the shape of their muscular foot. First, they make the foot thin and reach forward with it (Fig. 13.29). Then, the end of the foot thickens and acts as an anchor while they pull the rest of the body along. Many soft-bodied animals do something similar, using their body the way clams use the foot (Fig. 13.30, p. 414). Some worms force their pharynx into cracks in the sediment to expand the crack before pulling their bodies along behind. Heart urchins burrow or plough through the sediment with their spines and tube feet. The many crustaceans that live on soft bottoms—such as amphipods; sand, or mole, crabs; and ghost and mud shrimps—use their jointed appendages to dig.

IN CONTEXT

Respiration Organic matter (glucose) + O_2 → CO_2 + H_2O + energy
• 6.1, The Ingredients of Life

Hemoglobin Red protein in the blood of vertebrates and some invertebrates that transports oxygen.
• 10.2, Biology of Fishes

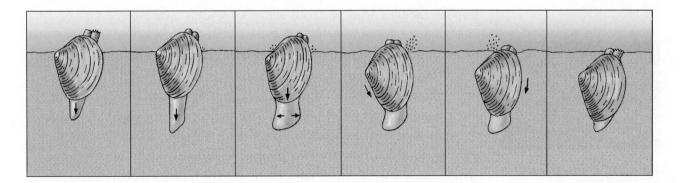

Figure 13.29 Burrowing in clams and cockles. The foot does all the work. The clam first pushes its foot down through the sediment. Then the clam expands the end of the foot, which anchors the clam as it pulls its body down.

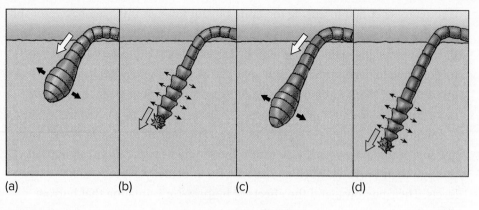

(a) (b) (c) (d)

Figure 13.30 Burrowing in the lugworm (*Arenicola*). (a) The worm expands the end of its body (black arrows). The expanded end acts as an anchor and the rest of the body is pulled along behind (white arrow). (b) The worm then flares its segments (black arrows), which prevents it from sliding backward when the contraction of the circular muscles in the anterior body extends the proboscis forward (blue arrow). As the process is repeated (c and d), the worm moves forward through the sediment.

Quite a few deposit feeders solve two problems at once by eating their way through the sediment. Sea cucumbers, certain worms, and other animals use the same inchworm style of locomotion as do the clams and worms described previously, with one important exception. Rather than push through the sediments, they push the sediments *through them*. They digest the organic matter and leave the rest of the sediment behind (Fig. 13.31).

Some soft-bottom animals are so small that instead of burrowing through the sediments they live in the interstitial spaces between the grains. These animals are collectively called the **meiofauna**. There are many different types of animals in the meiofauna, but most have independently evolved a worm-like shape.

Feeding As already noted, detritus is the main source of food in the soft-bottom intertidal (Fig. 13.32). Diatoms sometimes form highly productive mats on the sediment surface, but most of the time they don't account for much primary production. In any case, most animals don't distinguish between the diatoms and detritus. Plankton brought in by the tides also make a small contribution to the food supply.

Deposit feeders Many methods of deposit feeding have evolved. One of the most common has already been described: Animals such as sea cucumbers and various worms take in sediment as they burrow, digest the detritus and small organisms, and leave the rest of the sediment in their wake. This technique is more common on mud bottoms than sand, probably because mud contains more organic matter. Sand is also abrasive and hard on the digestive system.

Sand dollars are more selective. They use their tube feet to pick up organic particles. Sand dollars burrow near the sediment surface, where detritus is still accumulating and is therefore more abundant. The bentnosed clam also concentrates on the upper layer of sediment. It uses a long siphon to suck up food particles from the surface. Other animals catch particles as they settle out and are thus on the line between suspension feeding and deposit feeding. Some polychaete

Figure 13.31 After digesting out the organic matter, this tropical sea cucumber (*Holothuria edulis*) has deposited a pile of undigested sand pellets.

Figure 13.32 A generalized food web on a sandy beach.

Legend:
- Herbivory
- Predation
- Detritus feeding or export
- Contribution to detritus pool

Consumers

Carnivores
Fishes, birds, moon snails

Suspension feeders
Clams, sand crabs, polychaetes

Deposit feeders
Polychaetes, clams

Detritus

From other communities (rocky intertidal, estuaries, subtidal)

Producers

Plankton (from other communities)

Bottom diatoms

worms have long, sticky tentacles that they spread out on the bottom to gather food. Other polychaetes produce a mucus net that they use instead of tentacles.

Suspension feeders Suspension feeders, which don't wait for the detritus to settle, also inhabit intertidal soft bottoms. An olive shell from Central America makes a mucus net, but instead of spreading the net on the bottom the snail holds the net up in the water to filter food.

Sand, or mole, crabs hold a pair of large, bushy antennae up into the water to trap food. By quickly burying and uncovering themselves, they combine feeding and locomotion. When the tide is coming in and they want to move up the beach, they uncover just before a wave hits, ride the wave up, and raise their antennae. Then they burrow to keep from being washed back down. As the tide recedes, they do the opposite, burrowing before the wave arrives, then popping out to ride the backwash down the beach.

Predators Soft bottoms also have their share of predators. Moon snails burrow through the upper sediments looking for clams. When a snail finds a clam, it drills a hole in the clam's shell and eats it. Several polychaetes and other worms are also important predators. Birds can be major predators during low tide. At high tide, fishes can come in. Fishes often don't eat an entire animal. Instead, they just nip off clam siphons or other bits that stick out.

Zonation Because the organisms live in the sediments and can't be seen, zonation on intertidal soft bottoms is not as obvious as in the rocky

IN CONTEXT

Plankton Primary producers (phytoplankton) and consumers (zooplankton) that drift with the currents.

- 12.2, The Major Marine Lifestyles and Environments

Diatoms Singlecelled algae with a shell, or test, made of silica.

- 7.3, Unicellular Algae

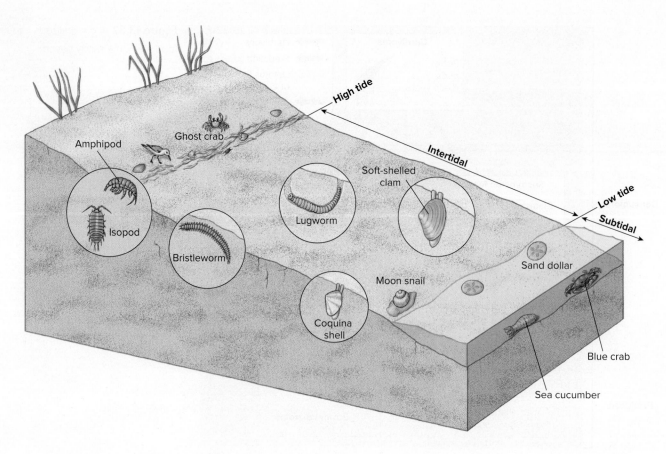

Figure 13.33 Generalized zonation pattern on sandy beaches of the Atlantic coast of North America. Sheltered and exposed beaches have different communities, and the animals shown here don't usually all occur on the same beach.

intertidal. Zonation does exist, however, especially on sandy beaches (Fig. 13.33). Water drains rapidly from the sand, and because the beach slopes, the upper part is drier than the lower part.

The upper beach is inhabited by beach hoppers, or sand fleas, which are really amphipods, and by isopods. In warmer areas, these small crustaceans are replaced by ghost and fiddler crabs that scurry about catching smaller animals, scavenging bits of dead meat, and gathering detritus. Polychaetes, clams, and other animals appear lower on the beach.

Zonation is even less obvious in muddy areas. The bottom in such places is flat, and fine sediments retain water. The habitat therefore does not change very markedly between the high and low water marks, and intertidal communities on muddy bottoms are much like the subtidal ones that are described in Chapters 14 and 15 rather than here in Chapter 13.

· ·

Section Review

1. Describe How do organisms such as infauna and bacteria deal with low oxygen conditions?

2. Identify What is the main food source in the soft-bottom intertidal? What is one way in which this food source is obtained?

Chapter 13 Review

REVIEW QUESTIONS

Multiple Choice

1. Competition for which resource is a dominant ecological factor in the rocky intertidal?

 a. food

 b. light

 c. oxygen

 d. space

2. Which is considered a keystone predator in the middle intertidal zone?

 a. birds

 b. crabs

 c. limpets

 d. sea stars

3. Which name is given to infauna that burrow through, turn over, or disturb sediment, helping to oxygenate the sediment?

 a. bioturbators

 b. meiofauna

 c. keystone predators

 d. suspension feeders

Short Answer

4. How is the color of a snail's shell related to its ability to tolerate high temperatures?

5. What factors can lead to stages of succession in the rocky intertidal being skipped?

6. Describe zonation on intertidal soft or muddy bottoms.

Critical Thinking

7. There are marked differences in the type of organisms found at four different locations at the same tidal height along a rocky shore. What might account for this? Offer at least three possible explanations.

8. Most marine biologists hypothesize that space, not food, limits populations in the rocky intertidal. What kind of experiments could be performed to test this hypothesis?

DATA ANALYSIS LAB

What are the feeding habits of the sea anemone *Anthopleura nigrescens*? Biologists collected data on the diet and feeding habits of the tropical sea anemone *Anthopleura nigrescens*. The anemones have been described as "opportunistic, polyphagous predators," meaning they feed on whatever food is available.

Data and Observations

The graphs show how many of each type of prey the anemones consumed at two different sites during high tide, both during the day and at night. Cyprids are a larval stage of barnacles.

(a)

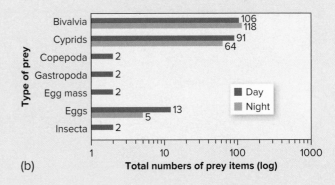

(b)

Think Critically

1. How does the data differ between the two sites? How does it differ between day and night?

2. How do these data support the description of anemones as "opportunistic, polyphagous predators"?

Data obtained from: Quesada et al., 2014. Diet of the sea anemone Anthopleura nigrescens: composition and variation between daytime and nighttime high tides. Zoological Studies, vol. 53, no. 26, June, pp. 1–7.

Estuaries: Where Rivers Meet the Sea

THEME　Patterns

By studying the structure of estuaries, scientists have classified four main types of estuaries, as well as four main estuarine communities.

BIG IDEA

Estuaries have unique physical and chemical properties and are of great importance to both the ecology of oceans and to human society.

About the photo　The red mangrove (*Rhizophora mangle*) is common in estuaries in tropical and subtropical regions. Their prop roots provide habitats for fishes and invertebrates.

Vilainecrevette/Alamy

14.1 Origins and Types of Estuaries

Main Idea
The four main types of estuaries are categorized by how they formed.

Key Questions
1. What are estuaries?
2. How do estuaries differ between the Atlantic and Pacific coasts of North America?

VOCABULARY

estuaries

drowned river valleys

coastal plain estuaries

bayous

bar-built estuary

tectonic estuaries

fjords

A unique environment develops where fresh water from rivers enters the sea. **Estuaries** are semi-enclosed areas where fresh water and seawater meet and mix. They therefore represent a close interaction between land and sea. Estuaries are typically inhabited by fewer species, and are therefore lower in biodiversity, than rocky shores. Nevertheless, they are among the most productive environments on Earth and are the breeding ground of many food species.

Estuaries also rank among the environments most affected by humans. Most natural harbors are estuaries, and many of the world's great cities—New York, London, and Tokyo, among others—developed along them.

Estuaries are scattered along the shores of all the oceans and vary widely in origin, type, and size. They may be called lagoons, sloughs, or even bays, but all share the mixing of fresh water with the sea in a partially enclosed section of the coast. Some oceanographers go as far as classifying enclosed seas with restricted circulation, like the Baltic and Black seas, as estuaries.

Broad, well-developed estuaries are particularly common in regions with flat coastal plains and wide continental shelves, a feature typical of passive margins. This is the case along the Atlantic coast of North America. The opposite is true for the steep coasts and narrow continental shelves of the Pacific coast of North America and other active margins. Here narrow river mouths carved along the steep coast have restricted the formation of estuaries.

Drowned River Valley and Coastal Plain Estuaries
Many estuaries were formed when sea level rose because of the melting of ice at the end of the last ice age during the Pleistocene, about 18,000 years ago. The sea invaded lowlands and river mouths in the process. These estuaries are called **drowned river valleys** or **coastal plain estuaries**. They are the most common type of estuary. Examples are Chesapeake Bay and the mouth of the Delaware River on the east coast of the United States (Fig. 14.1, p. 420), the mouth of the St. Lawrence River in eastern Canada, and the mouth of the River Thames in England. Also considered estuaries are the extensive **bayous** that are affected by tides along the low-lying mouth of the Mississippi River and adjacent coastal areas on the northern Gulf of Mexico.

DIVE IN!

Where are the world's largest estuaries? Many scientists consider the St. Lawrence River, a 1,200 km waterway that connects the Great Lakes to the Atlantic Ocean, to be the world's largest estuary. The largest estuary in the United States is Chesapeake Bay, which covers over 165,000 km². The Rió de la Plata is the world's widest estuary. The mouth of the estuary is about 200 km wide, and the length of the estuary forms the border between Uruguay and Argentina.

IN CONTEXT

Passive Continental Margin One that is on the "trailing edge" of a continent and therefore has little geological activity.

Active Continental Margin One that is colliding with another plate and therefore has a lot of geological activity.

• 2.3, Modern Ocean Basins

Pleistocene Geological period that began around 2.5 million years ago and during which sea levels fluctuated during several ice ages.

Bar-Built and Tectonic Estuaries

A second type of estuary is the **bar-built estuary**. Here the accumulation of sediments along the coast builds up sand bars and barrier islands that act as a wall between the ocean and fresh water from rivers. Bar-built estuaries are found, for instance, along the Texas coast of the Gulf of Mexico, the section of the North Carolina coast protected by the Outer Banks and Hatteras barrier islands (Fig. 14.1), and along the North Sea coast of the Netherlands and Germany.

Other estuaries, like San Francisco Bay in California, were created not because sea level rose but because the land sank, or subsided, as the result of movements of the crust. These are known as **tectonic estuaries**.

Fjords

Another type of estuary was created when retreating glaciers cut deep, often spectacular, valleys along the coast. The valleys were partially submerged when sea level rose, and rivers now flow into them. These estuaries, or **fjords**, are common in southeastern Alaska, British Columbia, Norway, Greenland, southern Chile, and the South Island of New Zealand (Fig. 14.2). Large glaciers, currently melting faster as a result of global climate change are still found in many fjords, particularly in Alaska, Greenland, and southern Chile.

Figure 14.1 Two types of estuaries are found along the eastern coast of the United States. Chesapeake Bay and the mouth of the Delaware River are drowned river valley estuaries; the Cape Hatteras islands form a bar-built estuary.

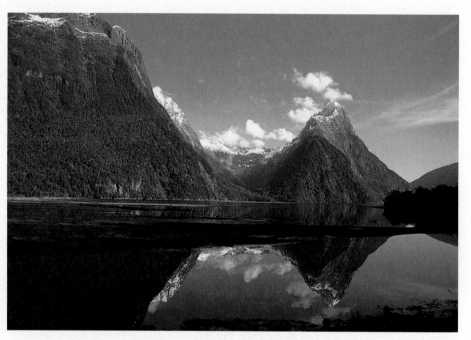

Figure 14.2 Milford Sound, on the southwestern coast of New Zealand's South Island, is a fjord. As in other fjords, the shallow entrance restricts the exchange of water between the fjord and the open sea, resulting in stagnant, oxygen-depleted, deep water.

Section Review

1. **Classify** Make a table to classify the different types of estuaries.

2. **Contrast** What are the differences in estuaries on passive margins versus those on active margins?

(l)NASA/Goddard Space Flight Center, The SeaWiFS Project and ORBIMAGE, (r)©Peter Castro

14.2 Physical Characteristics of Estuaries

Main Idea
Estuaries have wide fluctuations in salinity and dissolved oxygen, and contain various types of sediments.

. .

Key Questions

1. How does salinity change with depth and distance from the ocean in an estuary?
2. What is a salt wedge?
3. How do suspended sediments in estuaries affect water quality and the type of organisms living there?

DIVE IN!

Influenced by the tides and the mixing of fresh and salt water, estuaries have a unique combination of physical and chemical characteristics. These factors govern the lives of the organisms that live there.

Salinity

The salinity of estuaries fluctuates dramatically both from place to place and from time to time. When seawater, averaging about 35‰ salinity, mixes with fresh water (nearly 0‰) from rivers or melting glaciers, the mixture has a salinity somewhere in between. The more fresh water that is mixed in, the lower the salinity. Salinity therefore decreases as one moves upstream (Fig. 14.3).

Salinity also varies with depth in the estuary. The salty seawater is more dense and stays on the bottom. It flows in along the bottom in what is frequently known as a **salt wedge**. Meanwhile, the fresher, less dense water from the river flows out on the surface.

The salt wedge moves back and forth with the daily rhythm of the tides (Fig. 14.4, p. 422). It moves up the estuary on the rising tide, then recedes as the tide falls. This means that organisms that stay in one place are faced with

Can seagrasses survive in hypersaline water? Seagrasses are commonly found in subtidal marine habitats, where salinities average 35‰. In Laguna Madre, along Texas's Gulf of Mexico coast, however, salinities routinely reach over 80‰! It is one of only about six coastal hypersaline lagoons in the world. Despite this seemingly inhospitable salinity, Laguna Madre is home to 80% of Texas's seagrass habitat. The shallow lagoon receives ample sunlight year round. The limited connection to the ocean and low fresh water inflow result in low turbidity, a major benefit for seagrasses. The seagrass beds are vital nursery grounds for many fishes, crabs, and shrimps.

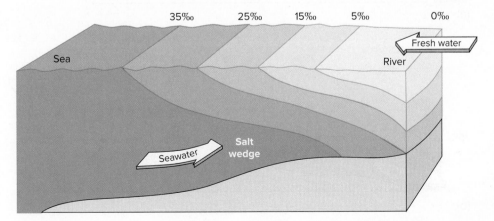

Figure 14.3 Profile of an idealized estuary. The lines across the estuary connect points of similar salinity and are known as *isohalines*.

dramatic fluctuations in salinity. They are submerged under the salt wedge at high tide and under low-salinity water at low tide. If the area has a diurnal tide, the organisms are subjected to two shifts in salinity every day: one as the tide moves upstream and a second as it retreats. In an estuary with semidiurnal tides, salinity changes four times a day.

The behavior of water masses in estuaries is not always this simple. The shape of the estuary and its bottom, the wind, evaporation of water from the surface, and changes in the tide all influence the distribution of salinity. Also of importance are seasonal variations in freshwater runoff from rivers as a result of rainfall patterns or snowmelt. Currents are especially important. Because most estuaries, including fjords, are long and narrow, the tide doesn't just rise; it rushes in, often creating strong **tidal currents**. In a few places the tide actually comes in as a nearly vertical wall of water known as a tidal bore. Tidal bores in the Qiantang River, China, can generate waves as high as 6 m! Such strong water movements greatly affect the pattern of salinity in an estuary.

Another factor that affects circulation in estuaries is the Coriolis effect. North of the equator, the fresh water that flows from rivers toward the sea is deflected toward the right. South of the equator, the flow is to the left. This means that in estuaries located in the northern hemisphere marine organisms can penetrate farther upstream on the left side when one faces seaward. In the southern hemisphere they extend up the right side.

In regions of little freshwater runoff and high evaporation, the salinity of an estuary can be higher than in the surrounding ocean. An example is Laguna Madre, a shallow bar-built estuary with limited access to the open ocean that

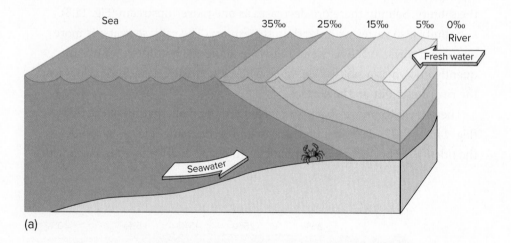

(a)

Figure 14.4 The salt wedge in a typical estuary moves in and out with the tide. (a) At high tide the crab is covered by water with a salinity of 35‰. (b) At low tide it is covered by water with a low salinity, between 5‰ and 15‰.

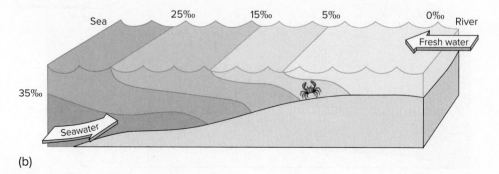

(b)

parallels the Texas coast for 185 km. The average salinity is over 50‰ in some areas, and it can reach 100‰ or more during dry spells. These high-salinity estuaries are called **negative estuaries**.

Substrate

Rivers carry large amounts of sediment and other materials, including pollutants, into most estuaries. Sand and other coarse material settle out in the upper reaches of the estuary when the river current slows. The fine, muddy particles, however, are carried further down the estuary. There, many of them settle out when the current slows even more, though the finest particles can be carried far out to sea. The substrate, or type of bottom, of most estuaries is therefore sand or soft mud.

Mud, which is actually a combination of silt and clay, is rich in organic material. As in other organic-rich sediments, respiration by decay bacteria uses up oxygen in the interstitial water, the water between sediment particles. Water cannot easily flow through the fine sediments to replenish the oxygen supply. As a result, the sediments in estuaries are often devoid of oxygen, or are anoxic, below the first few centimeters. They have the black color and rotten-egg smell typical of anoxic sediments, in which hydrogen sulfide (H_2S), which is toxic to most organisms, accumulates. Anoxic sediments are not completely devoid of life. Anaerobic bacteria, which do not need oxygen to carry out respiration, thrive under these conditions.

In estuaries that have unimpeded tidal flow, which includes most shallow ones, there is plenty of oxygen dissolved in the water. Some deep-water estuaries like fjords (see Fig. 14.2, p. 420), however, have a shallow "sill" at the entrance that restricts water circulation. Low-salinity water flows out unimpeded on the surface. The sill, however, prevents seawater from flowing in along the bottom. The stagnant deep water can become depleted in oxygen because of bacterial respiration associated with the decomposition of organic matter that sinks and accumulates on the bottom.

Other Physical Factors

Water temperature in estuaries, except fjords, varies markedly because of their shallow depths and large surface area. Organisms that are exposed at low tide have to face even more drastic daily and seasonal temperature fluctuations.

Large amounts of suspended sediments are typical of estuaries, greatly reducing the water clarity. Very little light thus penetrates through the water column. The particulate material in the water can also clog the feeding surfaces of some filter feeders and even kill organisms, like some sponges, that are sensitive to sediment.

· ·

Section Review

1. Relate How are salt wedges and tidal rhythms related to fluctuations in salinity in estuaries?

2. Describe the substrate in estuaries and the organisms found there.

14.3 Estuaries as Ecosystems

Main Idea

There are four main types of estuarine communities, each with organisms specially adapted to live there.

Key Questions

1. What challenges do organisms living in estuaries face?
2. What are pneumatophores?
3. Within estuaries, what is the main form of organic material that is available to consumers?

DIVE IN!

Can invasive cordgrass be biologically controlled? Smooth cordgrass (*Spartina alterniflora*) is an invasive species on the Pacific coast of North America, where it outcompetes native salt marsh plants. In an attempt to combat the spread of smooth cordgrass, scientists released a leaf-hopping insect (*Prokelisia*) that feeds on the sap of Spartina. Some studies have shown that *Prokelisia* is effective in reducing smooth cordgrass populations, but worryingly, other studies have found that certain genetic strains are quite tolerant of *Prokelisia*. Biologists and coastal managers must combine biological control with other measures (chemical or manual removal) to reduce smooth cordgrass populations.

An estuary may at first look bare, but this is far from the truth. Estuaries are tremendously productive and are home to large numbers of organisms, many of which are of commercial importance. Estuaries also provide vital breeding and feeding grounds for many animals. Estuarine ecosystems consist of several distinct communities, each with a characteristic assemblage of organisms.

Living in an Estuary

Life in an estuary revolves largely around the need to adapt to extremes in salinity, temperature, and other physical factors. Though other marine environments can be more extreme—they can be colder or more saline, for instance—none changes so rapidly or in so many ways as an estuary. Accumulation of sediment or the opening up of barrier islands as a result of storms can bring even faster changes. Living in an estuary is not easy, so relatively few species have successfully adapted to estuarine conditions.

Coping with Salinity Fluctuations Maintaining the proper salt and water balance of cells and body fluids is one of the greatest challenges facing estuarine organisms. Most estuarine organisms are marine species that have developed the ability to tolerate low salinities as a consequence of fresh water from rivers or melting glaciers (Fig. 14.5). How far they can move up the estuary depends on just how tolerant they are. Most estuarine organisms are **euryhaline** species, that is, they tolerate a wide range of salinities. The relatively few **stenohaline** species, those that tolerate only a narrow range of salinities, are limited to the upper or lower ends of the estuary and rarely penetrate into the estuary proper. Stenohaline species can be either marine or freshwater in origin. Some species are adapted to live in **brackish** water, or water of intermediate salinity. Some of these species are stenohaline; others are euryhaline.

Salinity adaptations in animals Because of their marine background, most estuarine organisms face the problem of the water in estuaries being diluted with fresh water. Those having an internal salt concentration higher than that of the surrounding water tend to take on water through osmosis. Some animals

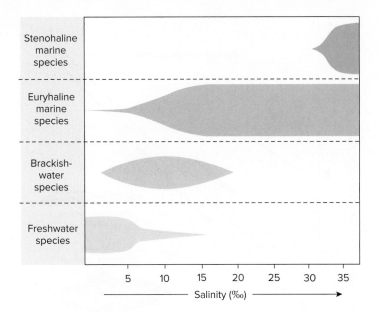

Figure 14.5 Types of species living in an idealized estuary in relation to salinity. The width of the bars represents the relative number of species, or biodiversity.

Stenohaline marine species

Euryhaline marine species

Brackish-water species

Freshwater species

5 10 15 20 25 30 35

⟶ Salinity (‰) ⟶

adapt by simple changes in behavior. They hide in their mud burrows, close their shells, or swim away if the salinity drops. These strategies are not widespread in estuaries, however, and most organisms rely on other mechanisms.

Soft-bodied estuarine animals, like many molluscs and polychaete worms, often maintain osmotic balance simply by allowing their body fluids to change with the salinity of the water. They are called osmoconformers (Fig. 14.6, p. 426). Many fishes, crabs, molluscs, and polychaetes are instead osmoregulators. They keep the salt concentration of their body fluids more or less constant regardless of the water salinity. They adapt to different salinities by regulating the concentration of their body fluids. For example, when the salinity of the water is lower than that of the blood, osmoregulators get rid of excess water and, via active transport, absorb some solutes from the surrounding water to compensate for those lost in the elimination of water. The gills, kidneys, and other structures accomplish this.

Bony fishes that inhabit estuaries also need to osmoregulate, since their blood is less salty than seawater. Salmon and freshwater eels migrate back and forth between rivers and the sea and still maintain a stable internal environment, thanks to the active transport of solutes by their kidneys and gills.

Few animals can be neatly classified as perfect osmoconformers or perfect osmoregulators. Many invertebrates, for instance, osmoregulate at low salinities and osmoconform at higher salinities. Even efficient osmoregulators like salmon and freshwater eels do not keep *exactly* the same concentration of salts and other solutes in their blood as salinity changes.

Salinity adaptations in plants Estuarine plants must also handle salinity variations. Grasses and other salt-marsh plants are land plants that have developed high salt tolerance. Some of these plants actively absorb salts and concentrate harmless solutes like sugars to match the outside concentrations and prevent water from leaving their tissues. Notice that this is opposite to the situation in marine organisms that live in estuaries, which have to adapt to reduced, not increased, salinities.

IN CONTEXT

Osmosis The movement of water from high to low concentrations across a membrane.

Active Transport The transfer of substances across membranes by a cell against a concentration gradient.

• 6.5, Challenges of Life in the Sea

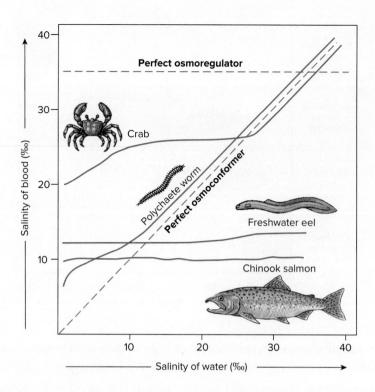

Figure 14.6 The salinity of the body fluids of estuarine animals responds in various ways to the salinity of the surrounding water. In a perfect osmoconformer, the salinity of the blood exactly matches that of the water. In a perfect osmoregulator, blood salinity stays the same no matter what the water salinity is. The important point is not the actual salinity of the blood but the fact that it remains relatively constant. Notice that some organisms, like the crab in the diagram, can only osmoregulate within a certain range of salinity; they are osmoconformers outside this range.

Several types of adaptations have evolved in some estuarine plants. Cordgrasses (*Spartina*; Fig. 14.7), other salt-marsh plants, and some mangroves actually excrete excess salts by way of salt glands in their leaves. Some estuarine plants, like pickleweed (*Salicornia*), accumulate large amounts of water to dilute the salts they take up (Fig. 14.8). Such fleshy plants are known as **succulents**.

Adapting to the Mud As discussed in Chapter 13, living in mud has its problems. There is nothing to hold on to, so most animals either burrow or live in permanent tubes beneath the sediment surface. Clams do well because they can extend their siphons through the mud to get water for food and oxygen. Because it is difficult to move through mud, the inhabitants tend to be stationary or slow-moving. Living in mud, however, has a benefit: Salinity fluctuations are less drastic than in the water column.

The depletion of oxygen caused by the decay of organic matter in the mud presents another challenge. This is no problem to burrowers that pump oxygen-rich water into their burrows. Burrowers without this luxury have special adaptations to low-oxygen environments. Some have blood that contains hemoglobin, which has a particularly high affinity for oxygen. It can hold and carry oxygen even when only minute quantities are available. Some clams and a few other mud-dwellers can even survive for days without oxygen.

Figure 14.7 Cordgrass (*Spartina*) is an important component of salt marshes on both the Atlantic and Pacific coasts of North America and other temperate shores worldwide.

Figure 14.8 Pickleweed (*Salicornia*) is a common succulent plant in salt marshes around the world.

Types of Estuarine Communities

Several distinct communities are associated with estuaries. One consists of the plankton, fishes, and other open-water organisms that come in and leave with the tide. Several other communities are permanent parts of the estuarine ecosystem.

Estuarine communities show low biodiversity. The relatively low number of estuarine species, however, are typically represented by many individuals. The low biodiversity is therefore compensated by a high biomass, the amount of living material measured as weight. A surprising number of estuarine species, particularly those inhabiting temperate estuaries, are widely distributed around the world. Humans have introduced many of this species, often with undesirable consequences.

Open Water The type and abundance of plankton inhabiting estuaries vary tremendously with the currents, salinity, and temperature. The murky water restricts the penetration of light and can limit primary production by phytoplankton, even when the density of plankton is typically very high. Most of the phytoplankton and zooplankton in small estuaries are marine species flushed in and out by the tides. Larger estuaries can also have their own, strictly estuarine, species.

One reason many of the world's great cities developed around estuaries is the rich supply of fish and shellfish in or near estuaries. Many species of commercially important fishes and shrimps use estuaries as nurseries for their young, taking advantage of the abundant food and relative safety from predators. About 90% of the marine commercial catch in the northern Gulf of Mexico, for example, is of species that depend on estuaries at some point in their lives.

IN CONTEXT

Plankton Primary producers (phytoplankton) and consumers (zooplankton) that drift with the currents.
- 12.2, Major Marine Lifestyles and Environments

Primary Production The conversion of carbon from an inorganic form, carbon dioxide, into organic matter by autotrophs—that is, the production of food.
- 6.1, The Ingredients of Life

Figure 14.9 The Atlantic menhaden (*Brevoortia tyrannus*) is one of the most important commercial fishes in the United States. Adult fish (one to three years old) spawn offshore and the larvae drift with the tides and currents into estuaries, moving into shallow areas in the salt marshes to grow.

IN CONTEXT

Anadromous Fishes Those that migrate from the sea to spawn in fresh water.

Catadromous Fishes Those that migrate from fresh water to spawn at sea.

• 10.3, Fish Adaptations

A relatively rich variety of fishes live in most estuaries. Many are the juveniles of marine species that breed at sea but use estuaries as nurseries. Examples are the menhaden (Fig. 14.9), anchovies, mullets, croakers, and many species of flatfishes. Some fishes move through estuaries during their migrations. Such fishes are either anadromous—like salmon, smelts, and shad—or catadromous—like freshwater eels. Relatively few fishes spend their entire lives in estuaries. Killifishes are one example.

Shrimps and crabs are often common in estuaries, and many commercially valuable species of invertebrates as well as vertebrates use estuaries as nurseries.

Mudflats The bottoms of estuaries that become exposed at low tide often form **mudflats** (see Fig. 14.14, p. 433). Mudflats are especially extensive in estuaries where there is a large tidal range and a gently sloping bottom. It all looks pretty much the same, but the mudflat can vary widely in particle size. Sand can accumulate to create sand flats near river mouths and in the tidal creeks that form as the tide changes. The calmer central part the mudflat contains more fine, silty material. Mudflats can also be found in fjords (see Fig. 14.2), where they occur mostly on the inner margin because the sides of fjords are steep and the bottom rapidly slopes to deep water.

Mudflat communities in estuaries are similar to those on muddy shores. Low tides expose organisms to desiccation, wide variations in temperature, and predation, just as in any other intertidal community. For these reasons mudflat communities could be classified as intertidal communities. In estuaries,

however, mudflat organisms must also withstand regular, often drastic, variations in salinity that define estuaries.

Primary producers Although not always evident, primary producers do flourish on mudflats. A few seaweeds—like the green algae *Entero-morpha* (Fig. 14.10) sea lettuce, and some red algae—grow on bits of shell. These and other primary producers can be particularly common during the warmer months. Large numbers of benthic diatoms grow on the mud and often undergo extensive blooms, forming golden-brown patches. In tide pools left by the receding tides these patches become coated with oxygen bubbles as intense photosynthesis takes place in the sunlight.

Bacteria and archaea are extremely abundant on mudflats. They decompose the huge amounts of organic matter brought in by rivers and tides. When the oxygen is used up by decay, some bacteria produce hydrogen sulfide. This in turn is used by sulfur bacteria, chemosynthetic bacteria that derive energy by breaking down sulfur compounds like hydrogen sulfide (see Table 5.1). Diatoms and bacteria, including photosynthetic bacteria, actually account for most of the primary production on mudflats.

Infauna The dominant animals on mudflats burrow in the sediment and are known as infauna (Fig. 14.11, p. 430). Burrowers, like the fat innkeeper (*Urechis caupo*; Fig. 14.11), actively excavate burrows in the sediment. It gets its common name because it shares its U-shaped burrow with a polychaete (*Hesperonoe adventor*), a crab (*Scleroplax granulata*), one or more fish (*Clevelandia ios*), and other guests. Though there are not many species of these burrowing animals, they often occur in immense numbers. At low tide their presence is often revealed only by small sediment mounds topped by a hole or piles of feces and other refuse. They feed on the abundant detritus in the sediment and water. Most of the food for these animals is brought in by the rivers and tides and is not actually produced on the mudflat.

Mudflat inhabitants that feed on detritus are deposit and suspension feed-ers, including filter feeders. Deposit feeders are more common than suspension feeders on mudflats and other muddy bottoms. On the Pacific coast of North America the fat innkeeper secretes a funnel-shaped net of mucus. It pumps water through this net to filter out food. Suspension feeders are at a disadvan-tage because their filtering mechanisms tend to get clogged by the high amounts of sediment that rains on soft bottoms. Furthermore, deposit feeders actually exclude many suspension feeders by disturbing the sediments, which clogs the suspension feeders' feeding structures and buries their newly settled larvae.

Suspension feeders, on the other hand, dominate where the sediment is more sandy. The wider interstitial spaces between the larger sand particles hold less detritus for deposit feeders to eat, and the abrasive sand is hard on their digestive systems.

Many deposit feeders and some other members of the infauna are known as bioturbators because they move and mix sediment when burying or digging.

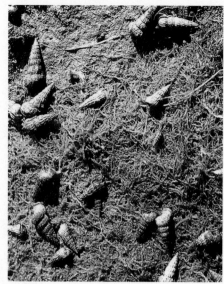

Figure 14.10 The California horn snail (*Cerithidea californica*), a deposit feeder, is abundant on mudflats. The green alga is *Enteromorpha*, which tolerates wide fluctuations in salinity and temperature as well as pollution.

IN CONTEXT

Chemosynthetic Bacteria
Autotrophic bacteria that use energy contained in inorganic compounds rather than sunlight to make organic matter.
• 7.2, Prokaryotes

Detritus Particles of dead organic matter.
• 12.3, The Flow of Energy and Materials

Deposit Feeders Animals that feed on organic matter that settles in the sediment.

Suspension Feeders Animals, including filter feeders, that feed on particles suspended in the water column.
• 9.1, Sponges, Cnidarians, and Comb Jellies

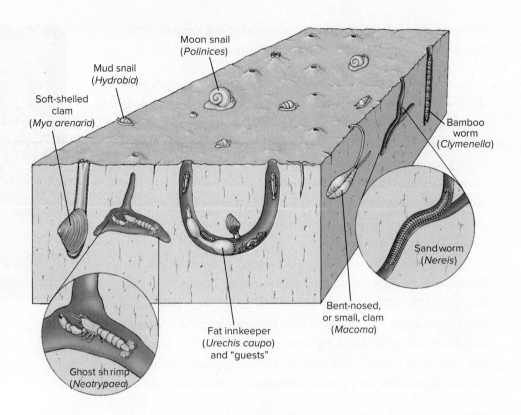

Figure 14.11 Some representative inhabitants of mudflats in temperate estuaries. Many mudflat organisms also can be found on muddy bottoms outside estuaries.

Moon snail (*Polinices*)

Mud snail (*Hydrobia*)

Soft-shelled clam (*Mya arenaria*)

Bamboo worm (*Clymenella*)

Sandworm (*Nereis*)

Bent-nosed, or small, clam (*Macoma*)

Fat innkeeper (*Urechis caupo*) and "guests"

Ghost shrimp (*Neotrypaea*)

Bioturbation helps in the oxygenation of sediment, a most important role in the case of the typically anoxic sediments of estuaries.

Meiofauna Protozoans, nematodes, and many other minute animals that compose the meiofauna also thrive on detritus. The meiofauna are also known as interstitial animals. The larger burrowing animals, or infauna, include many polychaetes. Most are deposit feeders. Other polychaetes are suspension feeders that filter water or extrude tentacles to collect the detritus that falls from the water column. Yet another detritus-feeding strategy among some polychaetes is to switch back and forth between suspension and deposit feeding, depending on the amount of suspended material in the water.

Epifauna Very few mudflat animals can be classified as epifauna, those that either live *on* the sediment surface or are attached to a surface as sessile forms. Crabs and some other animals can temporarily bury by covering themselves with sediment. Bivalves often abound on mudflats. Many are filter feeders that are also found on muddy and sandy shores outside estuaries. Examples from temperate waters are the quahog, or hard clam (*Mercenaria mercenaria*), the soft-shelled clam (*Mya arenaria*; Fig. 14.11), and razor clams. Some of these are of considerable commercial importance. Bent-nosed, or small, clams (*Macoma*; Fig. 14.11) are deposit feeders that use their long incurrent siphons to vacuum the surface.

The ghost and mud (*Upogebia*) shrimps make elaborate burrows that, as bioturbators, help oxygenate the sediment. These shrimps feed on detritus that they filter from the water and sift from the mud. Fiddler crabs (*Uca*) also live in burrows but are active on the mudflats at low tide. They process the mud and extract the detritus, which they eat.

Fiddler on the Mud

What lives in mud, feeds on mud, and finds mates by calling and waving across the mud? Of course—fiddler crabs!— best known as the ultimate mud aficionados.

The many species of fiddler crabs (*Uca*) are inhabitants of mud and sand flats in estuaries and other sheltered coasts. They are found mostly in the tropics and subtropics, but can be found as far north from the equator as Massachusetts and as far south as Argentina. Fiddlers are deposit feeders, which consume detrital food sources that have been passively deposited in the mud. Fiddler crabs feed at low tide when tidal flats are exposed to air. The detritus in the mud is extracted with the help of brush-like mouthparts. Water is pumped from the gill chambers into the mouth to make the lighter detritus float and thus help separate it from the mud. The detritus is swallowed, and the 'clean' mud is spat out on the substrate in neat little balls.

The importance of tides Fiddlers are active at low tide and retreat into their burrows at high tide. Each burrow has an entrance (revealed by the neat little balls of mud around it) that the occupant can plug when the tide comes in. At the next low tide, crabs emerge from home to feed and do whatever healthy, active fiddlers like to do. The tidal cycle is so ingrained in the crabs that if crabs are taken away and isolated in an environment of constant light and temperature, they will continue to be active at the times that correspond to low tide and sit quietly when they were expected to be in their burrows at high tide! Not only activity patterns but color changes that normally take place in their natural environment (darker color during the day, lighter at night) continue to be observed in isolation. These patterns are examples of biological clocks, repeated rhythms that are synchronized with time. In the case of fiddlers, activity patterns and color changes are synchronized with tides and with day-night changes.

The claw Fiddlers have an exciting sex life. Males feature one tremendously enlarged claw. It is brightly colored or highlighted with markings in many species. Males use their big claw to advertise their availability to females. They also use the claw to threaten away any other males that may be around. They wave the claw at low tide on territories established around their burrows. Claw-waving gets so heated sometimes that a whole mudflat seems to move up and down with hundreds of displaying males. Males entice females into their burrows, and a female may visit a few pads before deciding on a particular one. Males often fight for prospective mates but are also known to establish partnerships with neighbor males to defend their territories against intruders. Fighting with a stronger neighbor may cause a fiddler to end up with a lost claw, a more costly alternative than peaceful cooperation. A lost claw means disaster – the male can neither attract a mate nor defend himself. It takes many molts to regenerate one. A fiddler with a lost claw is also not as cool as his neighbor- literally. The enlarged claw is also used in temperature regulation, and crabs with an enlarged claw cool faster than those without it. On the other hand, while a large claw does have its advantages for attracting potential mates, it can also attract potential predators. A large claw would be easier for a bird to grab than a small one.

In those areas co-inhabited by several species of fiddlers, waving is used to prevent a male from attracting females of the wrong species. Some species wave up and down, others sideways. The angle and frequency of waving also vary, and bowing, fancy steps, and other body movements may form part of the ritual. Some beat the claw on the ground, and males of many species even produce sound by vibrating a joint of the large claw. It pays to advertise!

Signaling in a male fiddler crab (*Uca*)

Figure 14.12 Differences in the bill length of wading shorebirds from the west coast of North America allow them to feed on particular mudflat animals.

Some animals live on the surface of the mud or move in and out with the tide. These include deposit feeders like mud snails (*Cerithidea, Hydrobia;* Figs. 14.10 and 14.11 pp. 429-430), amphipods, and shrimps. Carnivores include polychaete worms (*Nereis;* Fig. 14.11), moon snails (*Polinices;* Fig. 14.11) and other predatory snails and swimming crabs.

Predators By far the most important predators in the mudflat community are fishes and birds. Fishes invade mudflats at high tide, whereas birds congregate at low tide to feed. Estuaries are important stopover and wintering areas for many species of migratory birds. The open spaces offer them safety from natural enemies, and food is plentiful. The most significant predators on mudflats are wading shorebirds (Figs. 14.12 and 14.13). These include the willet, godwits, dowitchers, and many species of plovers and sandpipers such a curlews and phalaropes. They feed on polychaetes, ghost shrimps and other small crustaceans, clams, and mud snails. Oystercatchers specialize on clams and other bivalves.

Figure 14.13 Feeding behavior also varies among shorebirds. (a) Sandpipers use their bills to search for food and follow a roughly straight path. (b) Plovers rely mostly on their vision, turning their heads sideways as they move.

These birds do not all exploit the same type of prey. The varying lengths of their bills represent a specialization in prey because different types of prey live at different depths in the mud (see Fig. 14.12). In addition, shorebirds use different strategies to locate their food. Birds like sandpipers rely mostly on their bills, probing in the mud as they walk around (Fig. 14.13a). Others, like plovers, use their eyesight to detect slight movements on the surface of the mud (Fig. 14.13b). Biologists think that these differences in feeding habits are an example of resource partitioning.

Herons and egrets compose yet another group of wading birds. They specialize in catching fishes, shrimps, and other small swimming prey. Birds that feed by swimming or diving in the estuary, like ducks, terns, and gulls, are often seen resting on mudflats.

Salt Marshes Estuaries in temperate and subarctic regions worldwide are usually bordered by extensive grassy areas that extend inland from the mudflats. These intertidal areas are partially flooded at high tide and are known as **salt marshes**, or sometimes tidal marshes (Fig. 14.14). Sometimes they are

IN CONTEXT

Resource Partitioning The sharing of a resource by two or more species to avoid competition.

• 12.1, The Organization of Communities

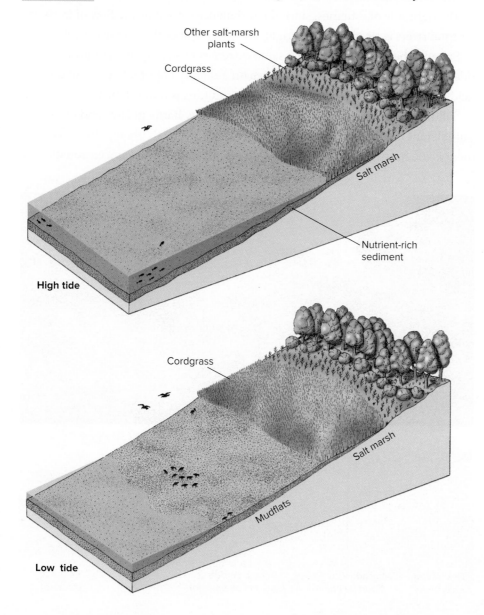

High tide

Low tide

Figure 14.14 The daily tides play a crucial role in salt marshes. They help circulate detritus and nutrients and expose mudflat organisms to predation by shorebirds and other animals.

grouped with coastal environments flooded at high tide and with freshwater marshes and collectively called **wetlands**. Though mostly associated with estuaries, salt marshes can also develop along sheltered open coasts. They develop as long as disturbance from wave action is minimal to allow the accumulation of muddy sediments. Salt marshes may also develop along the inner margins of fjords as long as there is an accumulation of muddy sediments free from the ice of glaciers. Tidal creeks, freshwater streams, and shallow pools frequently cut through the marsh. Salt marshes are typically replaced by mangrove forests along sheltered coasts in the tropics.

In North America, salt marshes are particularly extensive along the Atlantic and Gulf of Mexico coasts (see foldout map of North America at the end of the book). The broad estuaries and shallow, protected bays of these gently sloping coastlines provide optimal conditions for the development of salt marshes. The Pacific coast of North America, on the other hand, is generally steeper, exposed to wave action, so that most of the estuaries have formed along narrow river valleys. This has resulted in less extensive development of salt marshes. In the northern hemisphere salt marshes tend to be more extensive on the left side than on the right side of estuaries, since the turbulence caused by the flow of freshwater from rivers is stronger on the right side as a result of the Coriolis effect.

Salt marshes are subject to the same extremes in salinity, temperature, and tides that affect mudflats. They also live on a muddy bottom, but the mud is held together by the roots of marsh plants and thus is more stable.

Salt-marsh communities are dominated by a few hardy grasses and other salt-tolerant land plants. These plants thrive in the marsh, though the environment is too harsh for most other land plants. There is a pronounced zonation of

Figure 14.15 A salt marsh near Atlantic Beach, North Carolina. As in other Atlantic marshes, the smooth cordgrass (*Spartina alterniflora*) occupies the edge of the marsh that is flooded the most by seawater. It is replaced higher up in the marsh by saltmeadow grass, or salt-marsh hay (*Spartina patens*), a shorter and finer grass that is less salt tolerant and can form extensive meadows. It grows where the marsh is flooded only at high tide.

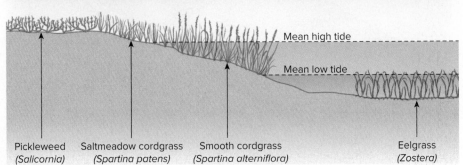

Pickleweed
(*Salicornia*)

Saltmeadow cordgrass
(*Spartina patens*)

Smooth cordgrass
(*Spartina alterniflora*)

Mean high tide

Mean low tide

Eelgrass
(*Zostera*)

plants in salt marshes (Fig. 14.15). The location of a given zone is related to the height relative to the tide, but it varies according to geographical location, type of substrate, and other factors. For instance, soil salinity can become particularly high at intermediate heights as a result of higher evaporation in marshes closer to hot, dry regions. This can result in areas bare of vegetation.

Salt marsh plants Cordgrasses (smooth cordgrass, *Spartina alterniflora*, on the Atlantic coast; California cordgrass, *Spartina foliosa*, on the Pacific coast; see Fig. 14.7) are typically the most common plants along the seaward limit of the salt marsh, where it meets the mudflats (Figs. 14.14 and 14.15). These grasses invariably occupy the fringe above the mean low tide level. Plants do better here because the soil is well drained and therefore richer in oxygen and less salty. The tops of their tall leaves remain exposed to the air even when the bottom is covered at high tide. The plants have extensive horizontal stems, or **rhizomes**, that stretch out underground. The leaves and roots develop from the stems. Rhizomes spread horizontally, farming extensive growths of genetically identical plants, or clones. Roots just below the soil also take in oxygen from the air, a crucial adaptation for living anoxic sediments.

Cordgrass can gradually invade mudflats because the plants slow down the tidal flow and thus increase the amount of sediment trapped among the roots. The landward extension of the salt marsh is eventually limited by the height of the highest tide. In addition to providing a significant portion of the high primary production of estuarine communities because such a high density of plant biomass, salt-marsh plants like cordgrass help stabilize soils by decreasing the effects of wave action. Smooth cordgrass from the Atlantic has been introduced in other parts of the world for the protection of shorelines, sometimes with negative.

Cordgrass gives way to other plants in the higher parts of the marsh. On the Atlantic coast, a second species of cordgrass (the salt-meadow, or salt-marsh hay, *Spartina patens*) dominates (Fig. 14.15), but rushes, pickleweed, and several other plants often form distinct zones. The higher levels of salt marshes on the Pacific coast are usually dominated by pickleweed (see Fig. 14.8, p. 427). The landward limit of salt marshes is a transition zone with adjacent terrestrial, or land, communities. It is characterized by a large variety of plants resistant to salt spray, like salt grasses and several species of pickleweed. It appears that in most salt marshes zonation is determined not only by the effects of flooding by tides but also by the combined effect of other factors. They include competition for space among salt-marsh plants, increased soil salinity in warm areas due to evaporation, and even the effect of burrowing animals.

Bacteria The muddy salt-marsh substrate is home to decay bacteria, diatoms, and thick mats of filamentous green algae and cyanobacteria. Bacteria play a crucial role by decomposing the large amounts of dead plant material produced in the salt marsh. These bacteria and the partially broken-down organic matter are a major source of the detritus that feeds many of the inhabitants of the estuary. Some bacteria are nitrogen fixers that enrich the sediment.

Salt marsh animals Some burrowing animals of mudflats also inhabit salt marshes. In addition, nematodes, small crustaceans, larvae of land insects, and

IN CONTEXT

Nitrogen Fixation Conversion of nitrogen gas (N_2) into nitrogen compounds that can be used by primary producers as nutrients.
• 7.2, Prokaryotes

Figure 14.16 The Atlantic ribbed, or horse, mussel (*Geukensia demissa*) lives half buried in the mud. It has the unusual ability to slightly open its valves, or gape, at low tide and take in oxygen from the air.

other small invertebrates live among the algal mats and decaying marsh plants. Crabs are conspicuous inhabitants of salt marshes. Fiddler crabs build burrows along the mudflat edges, where they increase the oxygenation of sediments, another example of bioturbation. Other marsh crabs are scavengers that eat dead organic matter. Some of these species live in burrows.

Marsh plants provide shelter and food to many marine and land animals. Coffee bean snails and marsh periwinkles are air-breathing snails that feed on detritus, minute algae, and fungi that grow on marsh plants. They move up the plants as the high tide moves in. Though they are air-breathers, they lay their eggs in the water and the larvae that hatch develop in the plankton. The ribbed, or horse, mussel (*Geukensia demissa*) is a suspension feeder that lives half buried in the mud among the cordgrass (Fig. 14.16). Killifishes and juvenile silversides are examples of fishes that inhabit tidal creeks and pools in the marsh at low tide. They move into the salt-marsh grass at high tide to escape predators like crabs and larger fishes, which enter the creeks with the rising tide. Crustaceans and small fishes retreat to pools or tidal creeks at low tide. Rails and American coots are among the birds that feed and nest here. Many other land birds and mammals, from ospreys to raccoons, are common visitors.

Mangrove Forests Mangrove forests are not limited to estuaries, but in some ways they are the tropical equivalents of salt marshes. Mangroves are flowering land plants adapted to live in the intertidal. These trees and shrubs often form dense forests called **mangals** to distinguish them from, the actual plants themselves, or mangroves. Mangroves are typical of tropical and subtropical regions, where they replace the temperate salt marshes. The two communities nevertheless coexist in some areas, as on the northern coast of the Gulf of Mexico, southeastern Australia, and northern New Zealand.

Global distribution It has been estimated that around 75% of all sheltered tropical shores were at one time fringed with mangroves, a figure that suggests their tremendous importance. Mangroves are killed by frost so their worldwide northern and southern limits are determined by low temperatures. A change in the distribution of mangroves as a result of global climate change has already been noted. Florida mangroves have been growing father north along the Atlantic coast, those in southern Australia have been expanding into salt marshes. Similar extensions of the geographical range of mangroves have also been recorded in Perú and on the Pacific coast of Mexico. Mangrove forests are been directly affected by humans in other ways.

Mangroves grow on protected coasts where muddy sediments accumulate. Though mangroves usually grow in tropical and subtropical estuaries, they are not restricted to them. Nevertheless, all mangroves require fresh water. As in salt-marsh plants, the various species of mangroves have different tolerances to immersion by high tide. Partly as a result of these differences in tolerance, they show a distinctive zonation in the intertidal, from a marine to a progressively terrestrial environment.

©Peter Castro

The Indo-West Pacific region has the world's most extensive mangrove forests with by far the largest number of mangrove species, which include species of *Aegiceras*, *Avicennia*, *Ceriops*, *Rhizophora*, and *Sonneratia*. Mangrove forests extend as far as 320 km inland in southern New Guinea and some islands in Indonesia, where the influence of the tides extend far up estuaries. Atlantic and eastern Pacific mangrove forests, though not as extensive and diverse as those in the Indo-West Pacific region, are nevertheless of great ecological significance.

Diversity The most common species of mangrove along the shores of southern Florida, the Caribbean, and the gulfs of California and Mexico is the red mangrove (*Rhizophora mangle*). It lives right on the shore (Fig. 14.17) and is easily identified by its peculiar prop roots, which support the trees very much like stilts. Flexible aerial roots drop down from the higher branches, helping extend the tree laterally. The trees can be as tall as 9 m. Under optimal conditions they form dense forests noted for their high primary production as a result of such high plant biomass. Other species of *Rhizophora* are found on tropical coasts around the world.

Along the Atlantic and Pacific coasts of the western hemisphere and the Atlantic coast of Africa, the black mangrove (*Avicennia germinans*) and the white mangrove (*Laguncularia racemosa*) live inland from the red mangrove (Fig. 14.17). Black mangrove seedlings can survive the high salinity of the water that remains standing after high tide flooding. As a result, the black mangrove grows higher in the intertidal than the red mangrove. White mangrove seedlings do not easily tolerate flooding by seawater, and as a consequence, the white mangrove is found only along the landward edge of the mangrove forest.

As land plants living at the sea edge, mangroves live under very stressful conditions. They are nevertheless adapted to the marine environment in several ways. Mangroves must get rid of salts from the water that is taken in by the roots.

IN CONTEXT

Indo-West Pacific Region The tropical Indian and western and central Pacific oceans.

Figure 14.17 Aerial view of a mangrove forest on the southern coast of Puerto Rico. The outer seaward edge of the forest is dominated by the red mangrove (*Rhizophora mangle*). As in Florida and most of the Western Hemisphere, the inner margin of the red mangrove is bordered by a broad belt of black mangrove (*Avicennia germinans*). Farther away from the black mangrove is the white mangrove (*Laguncularia racemosa*).

©Charlie Arneson

Most salts are actually not taken in by the roots of the red mangrove, a process that requires energy. Salt glands on the leaves of the black and white mangroves and several other mangrove species excrete salts, with salt crystals usually visible on the leaves as the drops containing the salts dry up (Fig. 14.18). The salt glands of the white mangrove are clearly visible as two bumps at the base of each leaf. Living in anoxic mud is also stressful, and the roots of all mangrove species have different types of root extensions, like the prop-roots of the red mangrove, for obtaining oxygen from the atmosphere. Several species of mangroves, like the black mangrove of the tropical western hemisphere and the Atlantic coast of Africa and the Indo-West Pacific species of *Sonneratia,* develop **pneumatophores**, conspicuous, unbranched extensions of the roots that grow upward from the anoxic mud to help aerate the plant tissues (Fig. 14.19).

Mangrove forest communities Many marine and land animals live in mangrove forests. Crabs are particularly common. Many feed on the abundant leaf litter that accumulates below the mangrove trees. Some species of crabs also feed on leaves, flowers, and seedlings still attached to mangrove trees. Mangrove crabs spend most of their lives on land, but when eggs are ready to hatch, females must release the larvae at sea.

Many types of organisms attach to, or take shelter among, the submerged mangrove roots (Fig. 14.20). These include cyanobacteria, seaweeds, sponges, polychaetes, gastropods, oysters, barnacles, and tunicates. Particularly rich in attached organisms are mangrove roots subject to a wide tidal range. Large sponges living on the roots have been found to provide significant amounts of

Figure 14.18 Leaves of the Indo-West Pacific black mangrove (*Aegiceras corniculatum*) from Australia, showing the salt crystals that form as the salty secretions of the salt glands dry up under the sun.

Figure 14.19 Pneumatophores, the vertical extensions of shallow roots, obtain additional oxygen in mangroves like *Sonneratia alba* in Palau.

nitrogen compounds to mangrove plants. They also help protect the roots from burrowing isopods, which otherwise can cause considerable damage.

The muddy bottom around mangroves is inhabited by a variety of deposit and suspension feeders, as on mudflats. These include polychaetes, mud shrimps, and clams. Several species of fiddler crabs excavate burrows in the mud. As in temperate mudflats and salt marshes, burrowing crabs help oxygenate the sediment. Mudskippers (*Periophthalmus;* Fig. 14.21) are unique fishes found in Indo-West Pacific mangrove forests and mudflats. They have burrows in the mud but spend most of their time out of the water, skipping over the mud and crawling up mangrove roots to catch insects and crabs. When out of the water, mudskippers can take oxygen from the air through blood-rich surfaces in the mouth and on the skin as long as the surfaces remain moist.

The channels that cross mangrove forests are rich nurseries for many species of commercially important shrimps, spiny lobsters, crabs, and fishes (Fig. 14.22, p. 440). They are also nurseries for coral reef fishes. Several species of oysters and other bivalve molluscs that live on or even within the roots of mangroves are used as food in some places around the world.

Many species of land animals use mangrove forests for shelter and food. Insects feed on leaves, flowers, and the attached seedlings, while bees and bats feed on the pollen and nectar of some mangrove flowers. Birds make their homes in the branches and feed on fishes, crabs, and other prey. Snakes, frogs, lizards, bats, and other land animals also live in the mangrove trees.

Figure 14.20 Seaweeds, sponges, oysters, sea anemones, barnacles, sea squirts, and many other types of organisms live attached to the roots of the red mangrove (*Rhizophora mangle*).

Figure 14.21 A mudskipper (*Periophthalmus*) from the mudflats of a mangrove forest in Singapore. Its protruding eyes are adapted to see in air. Each eye can be retracted into a moist pocket to keep it from drying out.

Figure 14.22 Small fishes, many of which are juveniles of commercially important species, seek shelter from predators and find food among the roots of mangrove. Shown here is the red mangrove (*Rhizophora mangle*) in a Florida mangrove forest.

Large amounts of leaves and other dead plant material accumulate in the mangrove forest. Considerable amounts are eaten by crabs, and some is exported to other ecosystems. Much of the detritus, however, is broken down by bacteria and fungi. This makes the mud among the mangrove roots black and oxygen-deficient, much like that in salt marshes. Because of the oxygen-poor sediments, and toxic substances released by the leaves, life on muddy bottoms in mangrove forests is less abundant than on similar bottoms elsewhere.

Besides its ecological importance, mangrove forests have an incalculable value to humans. They contribute significant portions of the economy of many developing nations, not only in terms of fishery resources but as a source of timber, charcoal, and revenue from the tourist industry. Mangrove forests improve water quality by removing pollutants and particulate matter and help protect coasts from wave action and increasing sea levels. There is evidence that damage by the 2004 Indian Ocean tsunami would have not been as severe if mangrove forests had not been cleared in some of these areas.

Other Estuarine Communities The bottom of estuaries beyond the intertidal mudflats and salt marshes is typically muddy and inhabited by organisms similar to those characteristic of muddy subtidal communities. Brackish-water species are gradually replaced by euryhaline marine species and ultimately by stenohaline marine species (see Fig. 14.5). Fjord bottoms are typically anoxic and support communities there are characterized by low biodiversity but high biomass. The abrasive effect of drift ice from glaciers is also an important factor in some fjords. Deep-water coral "reefs" high in biodiversity have nevertheless been discovered in some fjords.

Seagrasses The muddy bottoms below low tide levels are sometimes covered by beds, or meadows, of grass-like flowering plants known as seagrasses. They include eel grass, which is restricted to temperate waters, and turtle grass, a warm-water species often found around mangrove forests. The roots of seagrasses help stabilize the sediment, and their leaves provide shelter to many organisms as well as being an additional source of detritus. Seagrass meadows are not restricted to estuaries and are discussed in more detail in Chapter 15.

Oyster reefs Oysters can form extensive beds on the muddy bottoms of estuaries in temperate waters in North America, Europe, and Australia. These oyster reefs, sometimes known by the more general name of shellfish reefs, gradually develop as successive generations of oysters grow on the shells of their predecessors (Fig. 14.23). They provide a complex, three-dimensional surface for many organisms. The oyster-reef community includes seaweeds, sponges, tubeworms, barnacles, and other organisms that attach to the hard shells. Other animals take shelter among or even inside the shells. A similar estuarine community develops in association with mussel beds, though the mussels require a hard substrate for attachment.

Oyster reefs, once an important feature of many estuaries along the Atlantic coast of North America, have sharply declined as a result of pollution, overharvesting, disease, and disturbance by boats. This decline has had a major impact on the biodiversity in these estuarine communities. In addition to providing a habitat to many species, oysters increase the clarity and light transmission of water for seagrasses and other plants by filtering the water; an adult oyster filters as much as 270 liters of water daily. Oyster reefs also help protect shores from erosion by waves.

Figure 14.23 An oyster reef formed by the eastern oyster (*Crassostrea virginica*) on the Atlantic coast of the United States, exposed at low tide.

Figure 14.24 Generalized food webs in estuarine ecosystems. Salt marshes (left) occur in temperate regions, mangroves (right) in the tropics.

Feeding Interactions Among Estuarine Organisms

Though estuaries are low in biodiversity when compared to rocky shores, they reap the benefits of living in a very productive ecosystem. The generalized food webs shown in Figure 14.24 summarize the feeding relationships among different organisms in estuarine ecosystems.

Productivity Why do estuaries have a high primary production? There are several reasons. Nutrients brought in by the tide and rivers, together with those generated by nitrogen-fixing organisms and the decomposition of detritus, are used by plants, algae, and bacteria, the primary producers. Primary production is especially high in the communities that surround estuaries. The high biomass of cordgrass and other salt-marsh plants (or mangroves in the tropics) are particularly adapted to live on the mud and thus take advantage of the high concentration of nutrients in the sediments. The diatoms and bacteria in the mud and the phytoplankton in the water also contribute significantly to primary production.

Detritus also tends to sink to the bottom. Bottom water, which has a higher salinity and density than shallower water, thus acts as a nutrient trap in deep estuaries. Some phytoplankton are known to migrate to deep water at night to take in nutrients and move up to shallow, sunlit water the next day to carry out photosynthesis.

Primary production by estuary plants and other organisms varies geographically and seasonally, as does their relative contribution to the ecosystem as a whole. Estimates of primary production range from 130 to nearly 6,000 grams dry weight/m²/year for cordgrasses in salt marshes on the Atlantic coast of the United States. See Table 12.1 for a summary of the typical rates of primary production for salt marshes, mangroves, and seagrass meadows.

Detritus-based Food Web The organic material manufactured by primary producers is made available to consumers mainly in the form of detritus. A distinctive feature of estuarine ecosystems is that most of the animals feed on dead organic matter. Except for insects, geese, and some land animals on the fringes, relatively few herbivores actually graze on salt-marsh plants. Many detritus feeders obtain more energy from the bacteria and other decomposers in the detritus than from the dead organic matter itself. They excrete any detritus that remains undigested, however, returning it to the detritus pool. The surplus detritus is exported to the open ocean and neighboring ecosystems in a process known as **outwelling**. The exported detritus serves as a valuable source of food and nutrients to other ecosystems. The amount of exported detritus varies among estuaries, and some are actually net importers. Nonetheless, outwelling is an important role of estuaries, an additional reason that they should be preserved and protected.

· ·

Section Review

1. Identify What are some of the adaptations estuarine plants and animals have to manage the fluctuations in salinity?

2. Summarize What are the four main types of estuarine communities and the dominant organisms in each?

Estuary	Substrate	Producers	Consumers
Open water			
Mudflats			
Salt marshes			
Mangroves			

3. Explain Why do estuaries have such high primary productivity?

14.4 Human Impact on Estuarine Communities

DIVE IN!

How might rising sea level affect estuaries? The current average rate of sea level rise is about 3.2 mm per year. Scientists are concerned about the potential impacts of rising sea level on estuaries. As the sea level rises, ocean water will reach further upstream, increasing salinity in the upper reaches of estuaries. This could change the composition of estuarine communities, resulting in the local extiction or migration of species that cannot withstand higher salinities. The types of habitats within an estuary could also shift. With rising sea levels, mudflats, salt marshes, and mangrove forests that are regularly exposed at low tide would be covered and destroyed by exposure to wave action.

Main Idea

Estuaries provide many important functions, both to the marine environment and to humans, which are lost and when humans impact estuaries by dredging or filling them.

Key Questions

1. What are the ecosystems services of estuaries?
2. How much of estuaries have been lost worldwide due to human activities?

The worldwide economic value of marine wetlands as sources of food and as nurseries of many food species, in recreation and tourism, as sources of wood and other materials, in protection against storms and erosion, as well as in the maintenance of biodiversity and water quality, is enormous. Estuaries also serve as critical stopovers for migratory seabirds around the world. Furthermore, mangrove forests and seagrass meadows that often develop in tropical estuaries help reduce terrestrial runoff that protects coral reefs from sedimentation. Yet, the environmental consequences of human intrusion in marine wetlands, particularly in highly productive salt marshes and mangrove forests, have been disastrous. Countless have been obliterated, and many surviving ones are in danger of disappearing. What's more, oyster reefs are now considered by many as the world's most imperiled marine habitat.

Threats to Estuaries

All around the globe estuaries are being dredged to make marinas, artificial harbors, and seaports. Others are filled to create everything from industrial parks and urban development to garbage dumps. The dredging of navigation channels increases the exposure of estuaries to wave action, which often results in the destruction of salt marshes.

Another problem in some estuaries is the reduction or elimination of normal freshwater input when rivers are dammed or diverted. Sediment input from rivers is decreased so that erosion of sediments by tides is not replenished. The opposite can also be a problem. Deforestation and agriculture brings about an increase in sediment, which in estuaries like Chesapeake Bay, decreases water quality and increases pollution.

Threats to Salt Marshes Oil spills can be particularly harmful to salt marshes, as evidenced by the 2010 *Deepwater Horizon* spill in the Gulf of Mexico. Wetlands along the Louisiana coast of the Gulf of Mexico have also

Restoration of Salt Marshes

The destruction of salt marshes by human activities is a major worldwide problem. The restoration of disturbed salt marshes to their natural condition is being carried out as part of efforts to protect and preserve coastal wetlands. Some biologists encourage the natural regeneration of disturbed salt marshes, but others prefer to accelerate recovery by replanting dominant species, particularly cordgrass (*Spartina*). For example, the University of Southern Mississippi operates a nursery that grows salt marsh plants that are transplanted in damaged marshes.

Does restoration work? The success of replanting depends on factors such as wave action and tides, water chemistry (salinity, dissolved nutrients), and the substrate (slope, sediment size, oxygen content). Recovery can take many years, and the length of the recovery period depends on the degree of disturbance, environmental factors, and the relative maturity of marshes. The more mature marshes take the longest to recover. Restoration does not always lead to a community where species diversity matches that of a targeted natural state. Invasive plants, feral animals, unusual weather, and other unpredictable disruptions often disrupt restoration. Systematic monitoring is vital to restoration efforts.

Why restore marshes? The technology behind salt-marsh restoration is also being applied other habitats. Flooded areas previously cut off from the ocean, construction mitigation projects, and salt marshes affected by oil spills are habitats that can benefit from restoration. Restoring salt marshes can defend against erosion, provide flood management, and create habitats for seabirds and other wildlife.

Restoration is also undertaken in salt marshes impacted by invasive species of cordgrass, as in the case of smooth cordgrass (*Spartina alterniflora*). *S. alterniflora* is native to the Atlantic coast of North America and has become a pest on the Pacific coast. Seeds were probably accidentally introduced when oysters were transplanted to the Pacific, but some may have come in ships' ballast water. In some areas of the Pacific coast the plant was intentionally introduced for erosion control. Smooth cordgrass has spread over mudflats, tidal creeks, oyster grounds, and eelgrass beds. Its roots accumulate additional sediment, which turns natural habitats into swampy land within a decade or so. Smooth cordgrass can hybridize with the native Pacific cordgrass (*Spartina foliosa*), and the hybrids are taller, spread into deeper water, and release more pollen and seeds than the native cordgrass. Spreading is accelerated because smooth cordgrass, unlike the native species, can self-pollinate. These hybrids are by now the dominant cordgrass in many Pacific salt marshes. Efforts to control invasive cordgrass are being undertaken in places such as San Francisco Bay and Willapa Bay on the west coast of the United States.

been gradually eroded away as the result of the cutting of canals to install pipelines and haul oil drilling equipment. Another threat, particularly along the Pacific coast of North America, is the effect of cordgrass **invasive species**.

About one-third of all salt marshes in the United States have disappeared altogether; close to 70% of those in California have been lost. Similar or even higher losses have been documented in many countries. Erosion and flooding as a result of sea level rise brought about by global climate change is another threat to salt marshes.

Threats to Mangrove Forests The same factors that threaten salt marshes are menacing mangrove forests in the tropics. Mangrove forests once fringed around 75% of all sheltered tropical coastlines, but 35 to 50% of these forests have been destroyed. The figure is higher for the very diverse mangrove forests of Southeast Asia. It has been estimated that mangrove forests are disappearing at an annual rate of at least 1 to 2% worldwide, possibly higher than coral reefs and tropical rain forests. The United Nations' Environment Program estimated in 2014 that about 20% of mangrove forests had been lost worldwide between 1980 and 2005.

Mangrove forests have been cleared at this alarming rate to provide space for agriculture, urban and industrial development, roads, and garbage dumps. Shrimp aquaculture, a booming operation and an important source of revenue and employment in the developing nations of Southeast Asia, South America, and other tropical and subtropical regions, has nevertheless been particularly destructive. Forests have been destroyed to build the ponds to raise the shrimp; many others have been built to raise fish (Fig. 14.25). Water from the ponds, which contains large amounts of waste, excess nutrients, and uneaten food, is often flushed out to sea, causing serious pollution. Excess nutrients are responsible for **eutrophication**, the increased growth of algae as a consequence of excess nutrients.

The growing use of mangrove wood as fuel and timber in some areas is another cause for concern. Mangrove forests have also been impacted by increased boat traffic, oil spills, and pollution by plastics and solid waste

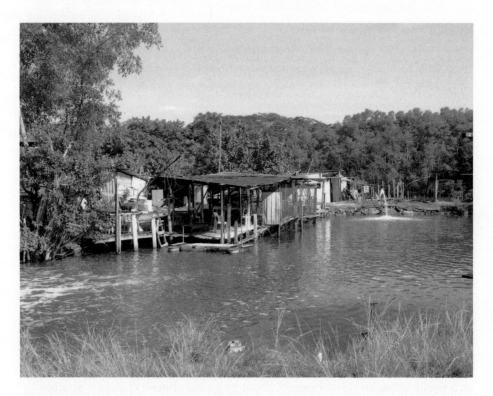

Figure 14.25 Shrimp and fish aquaculture has been responsible for the destruction of many mangrove forests, particularly in Southeast Asia and South America. Waste that is released into coastal waters from ponds like this, where fish are farmed, is an important source of pollution.

Figure 14.26 Solid waste, especially plastics and other durable materials, often accumulate in mangrove forests. Some of the waste is brought in by high tide, but forests are sometimes used as convenient garbage dumps.

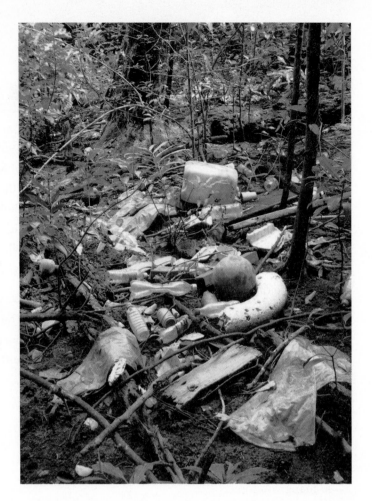

(Fig. 14.26). Together with salt marshes, it has been predicted that mangrove forests will be among the first marine communities to be adversely affected by sea level rise due to global climate change. Increased temperatures, however, are promoting the extension of mangroves further north or south from the Equator.

Mangrove forests remove carbon dioxide (CO_2) from the atmosphere and trap an unusually high proportion of their primary production as organic carbon in the leaves and detritus that are deposited in the sediments. When the mangroves are destroyed, much of this trapped CO_2 is released back to the atmosphere when the organic matter decays. Mangrove deforestation accounts for an estimated 10% of all CO_2 emissions caused by deforestation, even though mangrove forests occupy only 0.7% of the area of tropical forests.

Conservation Efforts

Conservation efforts through protective legislation, restoration projects involving the mass replanting of mangrove seedlings, proper management, and the creation of marine protected areas and reserves have nevertheless resulted in a reduction in the rate of loss of mangrove forests in some areas. Of particular importance is the development of sustainable exploitations of fishery resources in developing countries, where economic development is often in conflict with the conservation of natural resources. Many conservation programs thus emphasize the education of local communities on the economic and ecological values of mangrove forests, and on providing economic alternatives to practices that lead to mangrove destruction.

Section Review

1. **Describe** How have human activities led to the reduction of estuaries worldwide?

2. **Relate** How might the reduction of mangrove forests affect global climate change?

Chapter **14** Review

Multiple Choice

1. Estuarine organisms that are euryhaline

 a. tolerate only a narrow range of salinities.

 b. tolerate a wide range of salinities.

 c. are limited to the upper end of an estuary.

 d. are limited to the lower end of an estuary.

2. Some estuarine plants accumulate large amounts of water to dilute the salts they take up. Fleshy plants such as these are known as

 a. anadromous.

 b. catadromous.

 c. osmoregualtors.

 d. succulents.

3. Which type of estuarine community consists of extensive beds of filter feeders on the muddy bottom?

 a. mangrove forests

 b. mudflats

 c. oyster reefs

 d. seagrass beds

Short Answer

4. What factors affect the behavior of water masses and circulation in estuaries?

5. What are the ecosystem services of mangroves? What conservation efforts help protect mangroves?

Critical Thinking

6. A proposal is made to deepen the entrance and main channel of an estuary. What do you think will happen to the salt marshes that surround the channel? What do you predict will happen to the primary production of the estuary as a whole?

7. Some of the organic material manufactured in estuarine communities is exported to other ecosystems. What type of ecosystems receive this material? How is this material transported?

Can an oil spill affect salt marshes decades later? In 1969, 189,000 gallons of oil was spilled onto the coast of Massachusetts. In 2000, scientists found around 100 kg of oil remaining in salt marsh sediments. Biologists collected data on the depth and shape of fiddler crab burrows in two salt marsh locations—one where oil from the spill was still present and one which had never been affected by the spill.

Data and Observations

The illustration shows the depth and shape of the burrows at Location 1 (never affected by the oil spill) and Location 2 (oil from spill still present in sediment).

Think Critically

1. What are some differences in the fiddler crab burrows between the two locations?

2. What effect do you think a shallower burrow would have on predation on fiddler crabs?

3. Fiddler crabs are bioturbators who move sediments around as they dig their burrows. Why would any affects on this process be important to salt marsh communities?

Data obtained from: Reddy, C., 2007. Still toxic after all these years. Oceanus, vol. 45, no. 3, July, pp. 26–29.

Life on the Continental Shelf

Chapter
15

THEME Patterns

Scientists have defined different regions of the continental shelf and described the conditions and organisms that exist in each.

BIG IDEA

Continental shelf communities develop on soft-bottom and hard-bottom environemts, where different organisms adapt to different physical and biological factors.

About the photo This Eyed Flounder (*Bothus ocellatus*) is adapted for life on the unvegetated continental shelf off the coast of Florida.

Pixtal/age fotostock

15.1 Physical Characteristics of the Subtidal Environment

turbulence

stratification

sedimentation

Benthos Organisms that live on the bottom.

Plankton Those that drift with the currents.

Nekton Those able to swim against the currents.

- 12.2, Major Marine Lifestyles and Environments

Main Idea

The subtidal zone consists of the sea floor from the low tide level to the outer edge of the continental shelf.

Key Questions

1. What is the continental shelf? What is the subtidal zone?
2. Why is the continental shelf the most biologically rich part of the ocean?

The submerged edge of the continents, the continental shelf, was once considered the beginning of the unknown blue sea. But when the shelf came under scrutiny during the 1800s, the unknown began to give up its secrets. We have learned even more because the use of scuba and underwater habitats has allowed firsthand study and observation (Fig. 15.1). Coral reefs, seagrass meadows, and kelp forests, among the most productive ecosystems in the ocean, are found on the shelf. The shelf also includes the world's most important fishing grounds, which yield about 90% of the total marine global catch. Oil and minerals have been found on the shelf, and nations have extended their international borders to protect their new-found, or yet-to-be-found,

Figure 15.1 Scuba diving has dramatically expanded the possibilities for the study of the subtidal environment. These two marine biologists are using a grid to determine the density (number per area) and spacing, or distribution, of bottom organisms.

©Bill Ober

resources. The continental shelf, being close to shore, is also profoundly affected by pollution and other activities of humans on land.

Geology of the Subtidal Zone

The part of the continental shelf that is never exposed at low tide constitutes the subtidal zone of the marine environment. The subtidal zone, sometimes called the sublittoral zone, extends from the low tide level on shore to the shelf break, the outer edge of the continental shelf. The depth of the shelf break varies, but it averages around 150 m. The width of the shelf is also highly variable. The continental shelf is much wider on the trailing edge of plates (passive margins) than on the leading edge of plates (active margins), where the plates are colliding with other plates. The shelf is therefore much wider on the Atlantic margins of North and South America than on the Pacific side of the same continents. The width can also vary in a particular stretch of the coast, as in the case of the coast of California. It ranges from less than 1 km to over 750 km, with an average of approximately 80 km.

The benthos of the continental shelf live in the subtidal zone, which is discussed in this chapter. Chapter 17 covers the plankton and nekton of the water column over the continental shelf, which constitutes the neritic zone.

Defining Physical Characteristics

The physical factors that affect subtidal organisms are linked to two of the shelf's fundamental characteristics: its relatively shallow water and its proximity to land. Because the bottom is shallow, temperature varies more from place to place in the subtidal zone than on the deeper bottom beyond the shelf. This is significant because temperature is one of the most important factors influencing the distribution of marine organisms. The kinds of organisms found in the subtidal change dramatically from the Equator to the poles. There are also *more*, not just different, species in the tropics than in temperate or polar waters. The sea floor beneath the Arctic and Antarctic ice, however, is far from being a desert.

The bottom in shallow water is also much more affected by waves and currents than in deep water. The rise and fall of tides can produce particularly strong tidal currents on the shelf, especially in bays and narrow straits. Wind waves can affect the bottom even at depths of 200 m. Water motion, or **turbulence**, stirs up the water column and prevents **stratification**. Thus, except for seasonal or short-term variations, the temperature and salinity of the water at the bottom of the subtidal are not much different from those at the surface. More importantly, nutrients do not concentrate in the bottom layer and become unavailable to primary producers at the surface, as they do in deep water. Nutrients are also brought in by rivers, sometimes as by-products of the highly productive salt marshes or mangrove forests that fringe the shore. As a result, the water over the continental shelf is far more productive and plankton-rich than the open ocean, so much more food is available. The high concentration of phytoplankton, plus decaying organic matter brought in by rivers, gives coastal water a greenish tint as opposed to the deep blue of open-ocean water.

Rivers, obviously, also bring in large amounts of fresh water. Freshwater runoff dilutes the seawater, reducing its salinity. The effects of large rivers like the Mississippi and the Amazon extend many miles offshore. Away from the influence of rivers, however, the salinity of coastal water is about the same as that beyond the shelf.

The combination of proximity to land and shallow water greatly influences **sedimentation**, the settling of sediment particles from the water. Most of the sediment on the shelf is lithogenous, and rivers bring in huge quantities of sediment from the continents. Water motion sorts out the particles by size and density, especially on the shallower parts of the shelf. Large-grained material like gravel and sand settles out even in areas with strong waves and currents. On the other hand, turbulence keeps fine particles like silt and clay in suspension. Fine particles are deposited only in quiet areas or in deeper water when turbulence does not reach the bottom.

The abundant phytoplankton plus the sediment that is brought in by rivers and stirred up by waves and currents make shelf water murkier than that in the open ocean. Light does not penetrate as deeply, which reduces the depth at which primary producers—phytoplankton and attached seaweeds and seagrasses—can carry out photosynthesis and grow.

Other factors affecting subtidal communities, particularly those close to land, are disturbance by ice in polar and subpolar regions, trawling by commercial fishing, and pollution by fertilizers, sewage, oil, and other chemicals.

· ·

Section Review

1. **Identify** Which abiotic factors influence life on the continental shelf?

2. **Explain** How is sediment is sorted out on the continental shelf?

15.2 Soft-Bottom Subtidal Communities

Main Idea
Soft-bottom subtidal communities can be unvegetated or vegetated.

. .

Key Questions
1. What factors influence the distribution of soft-bottom subtidal communities?
2. Compare and contrast unvegetated soft-bottom communities and seagrass meadows.

VOCABULARY
patchy

unvegetated

passive suspension feeders

scavengers

The type of bottom, or substrate, is most important in determining which particular organisms inhabit the floor of the continental shelf. It also dictates how they are sampled by marine biologists (Fig. 15.2). In fact, subtidal communities are often classified according to the type of substrate. The most common of these are communities associated with soft bottoms.

Defining Characteristics of Soft-Bottom Communities
Sandy and muddy substrates dominate the continental shelves. Large areas covered by soft sediments stretch from the shore to the edge of the shelf. Even along rocky shores, rocks eventually give way to sand or mud in deeper water.

Figure 15.2 Soft bottoms at greater depths can be studied by using various sampling gears. Bottom grabs (*a*) like the Van Veen grab are lowered from the surface. The jaws close, grabbing a sample of bottom sediment, as the grab is retrieved. Different types of bottom trawls (*b*) are used to collect the larger inhabitants of soft bottoms. *Dredges* (*c*) with heavy metal frames are used to scrape organisms off rocky bottoms.

DIVE IN!

Why is turbidity important? Turbidity is the measure of how clear water is. It is based on how much particulate matter, such as soil particles and plankton, is suspended in the water. As you might guess, turbidity greatly influenecs the amountof light that can penetrate the water, and thus the types of plants and seaweeds found there. Marine scientists track turbidity changes over longterm studies, as it can be a good indicator of how a community might change. Tracking changes in turbidity can indicate increases in soil erosion, runoff, algal blooms, and pollution.

The organisms that inhabit these often flat and seemingly homogeneous bottoms (Fig. 15.3) are not necessarily distributed by chance. It has long been recognized that these organisms form distinct communities whose distribution is greatly influenced by factors like the particle size and stability of the sediment, light, and temperature.

The shelf's soft-bottom communities share many traits with the communities of sandy beaches and mudflats covered in Chapters 13 and 14. Infauna, animals that burrow or dig in the sediment, predominate in all these communities. There are also, however, some epifauna, animals that live on top of the sediment. Some members of the epifauna can temporarily bury by covering themselves with sediment. Because the soft bottom offers nothing to hold on to, sessile, or attached, forms are rare.

The number of species living on subtidal soft bottoms is usually higher than the number of species on soft bottoms in the intertidal zone. This is largely because of the relatively more stable physical conditions below the low tide mark. Desiccation is not a problem. Furthermore, there are none of the drastic temperature changes caused by exposure at low tide or the salinity variations of estuarine mudflats.

The close relationship between particle size and the distribution of organisms is particularly evident in the infauna, which for the most part are very selective as to where they live. Different species can also partition their environment and thus reduce competition by living at different depths in the substrate. Because only the upper layer of sediment contains enough oxygen, however, the depth at which the infauna can live is limited. Oxygen is quickly used up by decomposition. In contrast, sand usually contains less organic matter in mud that is rich in organic matter. It is also more porous, so water can circulate through the sediment and replenish the oxygen. Thus, the infauna can burrow deeper in sand than in mud. Even in mud, however, the presence of burrows aids circulation, helping organisms obtain oxygen.

Soft substrates often contain patches of different types of sediment. As a result, the distribution of organisms along the bottom is typically **patchy**—that is, the organisms occur in distinct clumps or patches rather than in random or regular pattern (Fig. 15.4). Some species have a patchy distribution because their planktonic larvae select a particular environment on which to settle and undergo metamorphosis. Some larvae are known to postpone metamorphosis and "taste" the bottom until they find a certain type of substrate. Metamorphosis can be triggered by highly specific chemical, physical, and biological factors (Fig. 15.5). The larvae of some species can sense adults and prefer to settle near them. This can cause the species to be distributed in clumps even if there is a uniform bottom. The arrival of new individuals to a population, either by settling larvae or by migrating juveniles or adults, is known as recruitment. The presence and abundance of a particular

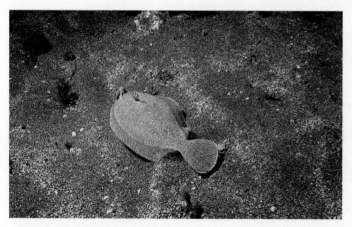

Figure 15.3 Conspicuous life on shallow sandy bottoms includes occasional clumps of seaweeds that manage to grow on small rocks or shells and large predators like flatfishes.

©Charlie Arneson

(a) Random (b) Regular (c) Patchy

species on the bottom is nevertheless not always dependent on the supply of larvae. Interactions with other species after larval settlement, factors like predation, competition, and bioturbation, most probably enter into the equation.

Unvegetated Soft-Bottom Communities

The great majority of the shelf's soft-bottom communities lack large amounts of seaweeds or seagrasses. They are known as **unvegetated** communities. The absence of large seaweeds and plants is the defining feature of these communities. Sea lettuce (*Ulva*) and other seaweeds can grow in the shallows as long as there is enough light, but only on hard surfaces like rocks and shell fragments.

IN CONTEXT

Metamorphosis The transformation of one stage of a life cycle into another, as when a larva changes into a juvenile.
- 9.1, Sponges, Cnidarians, and Comb Jellies

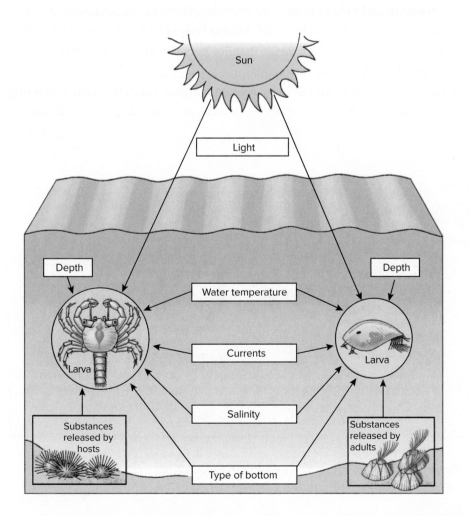

Figure 15.5 Many factors are known to influence the settlement and metamorphosis of planktonic larvae. The larva illustrated on the left, a megalopa, belongs to a crab (*Echinoecus pentagonus*) that as an adult lives on tropical sea urchins. The larva on the right is the cypris larva of a barnacle. Many barnacle larvae settle near adults of their species. Both larvae are tiny when compared to the adults. They are enlarged to show their morphologies.

Primary Producers Organisms
that can create organic matter
from carbon dioxide (CO_2),
usually by photosynthesis.
• 6.1, The Ingredients of Life

Detritus Particles of dead
organic matter.
• 12.3, The Flow of Energy and
Materials

The main primary producers are diatoms and a few other microscopic algae
and bacteria that grow on sand or mud particles in shallow water. This scarcity
of seaweeds and plants means that, as on sandy beaches, primary production
by benthic primary producers is usually very low. Nearly all primary production
is by phytoplankton that are not part of the bottom communities.

Infauna The benthic animals that are part of subtidal soft-bottom communi-
ties vary widely across the world, but four groups of large-size invertebrates
dominate for the most part: polychaetes, molluscs, crustaceans, and echino-
derms. Though there are exceptions, temperate soft bottoms, which constitute
roughly 45% of the planet's continental shelves, are inhabited by communities
that are typically dominated by molluscs and polychaetes. Tropical soft-
bottoms, about 30% of all continental shelves, are dominated by communities
dominated for the most part by crustaceans and molluscs. Tropical subtidal
communities, in contrast to temperate ones, can be particularly influenced by
sharp seasonal variations in salinity and sedimentation due to heavy seasonal
rainfall typical of the tropics. Polar (Arctic and Antarctic) subtidal communities
are unique and very different from each other.

Because there is little primary production by benthic organisms, detritus is a
very important food source for many inhabitants. Detritus is brought in by
currents from estuaries, rocky shores, and other, more productive coastal
communities. It is also present as feces, dead individuals, and other debris from
the plankton and nekton in the water column. Detritus is also generated by the
bottom inhabitants themselves. The detritus is used by bacteria and by the
many types of microscopic animals that live between the sediment particles,
the interstitial animals, or meiofauna.

Larger benthic invertebrates also feed on detritus. They are mostly burrowing
deposit feeders (Fig. 15.6). Polychaetes are the most diverse group of deposit

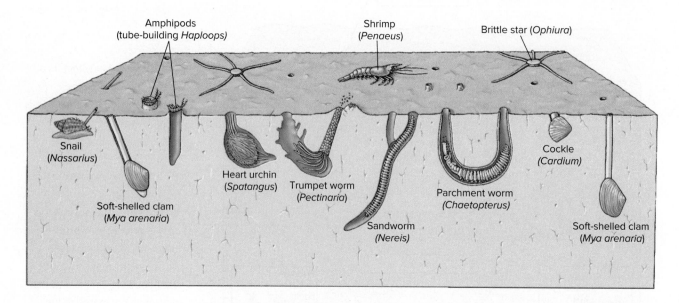

Figure 15.6 Representatives of the infauna and epifauna of subtidal soft bottoms in
different parts of the world.

Figure 15.7 Heart urchins (*Spatangus purpureus*) burrow below the surface and use modified tube feet for feeding. This photo was taken in the Mediterranean Sea at night, when the urchin emerged from the sediment.

Figure 15.8 These sand dollars (*Dendraster excentricus*) form dense beds on sandy bottoms along the California coast. They lie partially buried in the sand, angled into the current.

feeders in the soft sediments of the shelf. Trumpet (*Pectinaria*; Fig. 15.6) and bamboo worms are deposit-feeding polychaetes that inhabit tubes they build from sediment particles. Lugworms (*Arenicola*) and others live in burrows. These polychaetes and other worms eat detritus and other organic matter by collecting it with their tentacles or by ingesting sediment and extracting food from it.

Some sea urchins are uniquely adapted to living as deposit feeders on soft bottoms. Heart urchins (*Spatangus*) have abandoned the round shape of most urchins, becoming more streamlined with shorter spines that lie flat (Figs. 15.6 and 15.7). Sand dollars (*Dendraster*) are almost completely flat and have very short spines (Fig. 15.8). They typically feed on detritus and use mucus to carry particles to the mouth. Other deposit feeders among the infauna include echiurans (Fig. 15.9), peanut worms, sea cucumbers, and ghost.

The infauna include suspension feeders that eat drifting detritus and plankton from the water column. Many of these suspension feeders are filter feeders, which actively filter the water to obtain suspended particles by pumping water or sweeping filtering structures through the water. In contrast, **passive suspension feeders** only use mucus or cilia to move suspended particles into the mouth. Soft-bottom suspension feeders include many types of clams (Fig. 15.6): razor clams, the quahog (*Mercenaria mercenaria*), cockles, the soft-shelled clam (*Mya arenaria*), and others. Most clams are filter feeders, but some—like the bent-nosed, or small, clams and the file yoldia—are deposit feeders. They collect detritus and microscopic organisms with their siphons or specialized appendages (Fig. 15.10, p. 458). Also among the filter feeders are some amphipods (Fig. 15.6) and polychaetes like parchment worms (Fig. 15.6) and terebellids (Fig. 15.11, p. 458).

Figure 15.9 *Thalassema steinbecki* is a deposit-feeding echiuran that inhabits soft bottoms along the Pacific coast of North and South America from Baja California to the Galápagos Islands.

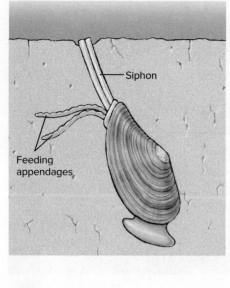

Siphon

Feeding appendages

Figure 15.10 The file yoldia (*Yoldia limatula*) inhabits mud and muddy sand from shallow water to depths of 30 m in the North Atlantic. This clam uses two grooved, tentacle-like appendages to feed. Each groove contains cilia that bring small organisms to the mouth.

Figure 15.11 *Thelepus* is a terebellid polychaete that burrows in soft bottoms. Its long tentacles are extended over the bottom for suspension feeding.

IN CONTEXT

Deposit Feeders Animals that feed on organic matter that settles in the sediment.

Suspension Feeders Animals, including filter feeders, that feed on particles suspended in the water column.

• 9.1, Sponges, Cnidarians, and Comb Jellies

The distribution of burrowing deposit and suspension feeders is influenced by several factors. Already mentioned in the previous section is the type of substrate. Deposit feeders tend to predominate in muddy sediments because more detritus settles in areas of low turbulence and is ultimately retained in the smaller interstitial spaces between mud particles (Fig. 15.12). Suspension feeders, on the other hand, are more common on sandy bottoms. Suspension feeders like sea pens (*Stylatula*; Fig. 15.13) and sea pansies (*Renilla*) can often be found in dense stands.

The types of organisms present also affect the establishment of others. Deposit feeders, for instance, are known to exclude suspension feeders. Some

	Less turbulence ⟶ More turbulence	
Size of sediment particles	Mud	Sand
Oxygen and detritus concentration	Less oxygen, more detritus	More oxygen, less detritus
Type of animals	Deposit feeders	Suspension feeders

Figure 15.12 The distribution of suspension and deposit feeders in soft bottoms is largely influenced by the size of the sediment particles. This relationship, however, is not clear-cut. Animals already established in the sediment, for instance, influence colonization by others.

deposit feeders, other members of the infauna, as well as some burying animals are known as bioturbators because they move sediment while burrowing, burying, or feeding. Bioturbation uncovers and oxygenates deeper sediments and buries those on the surface. Sea cucumbers, echiurans (see Fig. 15.9), peanut worms, some fishes, and even gray whales are important bioturbators.

Many tube builders are deposit feeders, but they help stabilize the substrate, making it more suitable for suspension feeders. Their tubes can also interfere with the activities of burrowing deposit feeders. The tubes slow down the flow of water over the sediment, which decreases the rate of particle suspension. The modification of the environment by deposit feeders, especially of bioturbators, and the effects of tubes are examples of how some species can indirectly influence the recruitment and survival of others.

Epifauna Many epifaunal invertebrates are deposit feeders. These include most amphipods and other small crustaceans as well as many species of brittle stars (see Fig. 15.6). Most of these brittle stars use their tube feet to collect detritus from the bottom. Other brittle stars are suspension feeders that raise their arms up into the water to catch suspended particles with their tube feet. Still others are **scavengers** and feed on dead animals. Shrimps, including commercially valuable species, are another type of epifaunal scavenger.

Some members of the soft-bottom epifauna are predators. They can burrow through the sediment to get their prey or catch it on the surface. As in other communities, predators are important in regulating the number and type of bottom inhabitants. They remove individuals and cause disturbances in the sediment that allow recolonization by different types of organisms. Snails like whelks (*Nassarius*; see Fig. 15.6) and moon snails feed on clams by drilling a hole in the shell and rasping away the flesh. Sea stars (*Astropecten*; Fig. 15.14) prey on clams, brittle stars, polychaetes, and other animals. Predatory amphipods influence subtidal communities by preying on the settling larvae of other species. Crabs are common predators and scavengers on soft bottoms. Examples are the blue crab (*Callinectes sapidus*) and other swimming crabs like the lady crab (*Ovalipes ocellatus*), and many others that partially bury themselves in the sand or mud. Other predatory crabs, hermit crabs, lobsters, and octopuses take shelter on rocky bottoms but move to soft bottoms to feed.

Fishes Many fishes are common predators. Most bottom-dwelling, or demersal, fishes in these soft-bottom communities are carnivores. Rays and skates scoop up clams, crabs, and other infauna and epifauna. Flatfishes like flounders, halibuts, soles, and turbots lie camouflaged or covered on the bottom and forage for a wide variety of prey. Pelagic fishes and squids also feed on the inhabitants of the shelf's soft bottoms. Some fishes only nip off the

Figure 15.13 Sea pens are suspension-feeding cnidarians. Each individual is a colony that includes numerous minute polyps and a fleshy stalk that is anchored in soft substrates. This is *Stylatula elongata* from the Pacific coast of North America.

Figure 15.14 Sea stars like *Astropecten* prey on clams and other burrowing invertebrates.

Under the Antarctic Ice

Antarctica, one of the last unspoiled wildernesses on Earth, is for many of us a vision of awesome whales, seals, and penguins. Life below the ice on the Antarctic continental shelf is as fascinating and pristine as in the cold waters above it.

Bottom under the ice, McMurdo Sound, Antarctica. Benthic invertebrates include sea urchins, Antarctic scallops, and a brittle star.

Life under the ice Ice is a major physical factor for shelf organisms in the Antarctic. The organisms that live here are so adapted to the extreme environment that there are even diatoms that can live in the ice. While extreme, the physical environment in Antarctic waters is remarkably constant. Year-round low temperatures mean that life processes, including growth, proceed at a slow pace.

The nutrient-rich waters result in very high primary production when sunlight becomes available during the long summer days. Rich blooms of phytoplankton provide food for the many suspension feeders. Sessile organisms are rare at the shelf's shallower depths because of the scouring effect of winter ice. Below this depth, however, benthic invertebrates are plentiful and diverse. Many of the invertebrate species are endemic to the continent because they evolved over a long period of isolation from the continental shelves of other continents.

Diversity of life Sponges in particular are so common that they cover most of the bottom in many places. These sponges provide refuge and perches for many fishes and other mobile animals, much as kelps or corals do elsewhere. Sponges are so common that the bottom below 30 m or so is nothing but a thick mat of siliceous, or glass, sponge spicules—truly a glass bottom.

Glass spicules play a key role in some Antarctic sponges that harbor symbiotic algae. The green algae may provide some nutrients to the sponges, but must obtain light in order to carry out photosynthesis. But how can algae photosynthesize from deep within the sponge tissue, on the floor of the ocean covered by ice? The answer seems to be

in the glass spicules. They efficiently transmit what little light reaches the sponges to the algae, a biological analog to fiber-optic cables!

Other invertebrates abound: sea anemones, soft corals, seastars, nudibranchs, and sea urchins. Predators are largely absent in Antarctica as a result of the low temperatures. To protect themselves against the relatively few predators, many sponges produce toxic substances. These chemical defenses are especially common in those sponges with bright colors. Not all of these subtidal invertebrates are unique to Antarctica. Some are found elsewhere, but in progressively deeper water as one gets closer to the Equator.

Antarctic fishes are remarkable in that they can remain active at temperatures that are close to the freezing point of seawater. Many have a chemical "antifreeze" in their blood. The cold water holds so much dissolved oxygen that some fishes have colorless blood that lacks hemoglobin. The Antarctic cod (*Notothenia*) actually lowers its metabolism and goes into a sort of hibernation during winter.

An emerald notothen (*Trematomus bernacchii*), an icefish with a chemical "antifreeze" in its blood, on an Antarctic sponge.

The future of the Antarctic Sadly, life below and above the Antarctic ice is threatened. The Antarctic Treaty, signed by 48 nations to administer and manage the continent, now forces polluters to repair damage that they cause or be fined. However, an increase in cruise-ship traffic and fishing vessels increases the risk of garbage, oil, and sewage pollution. Larger scale problems, such as an ozone hole the effects of global climate change, may have drastic consequences for Antarctic life. Average water temperatures west of the Antarctic Peninsula have already increased by 1 °C since the 1950s. King crabs (*Neolithodes*), previously unrecorded in Antarctica because of the low temperatures, have been recently found in large numbers at the edge of the continental shelf. The potential effects of these predators on Antarctic shelf life are unknown.

siphons of molluscs instead of eating the entire animal. A very large predator indeed is the gray whale (*Eschrichtius robustus*), which filters the sediment for amphipods and other small animals. Large predators like whales, walruses, and rays have important effects on soft-bottom communities. They not only remove their prey but also are bioturbators that dig pits in the bottom while feeding, killing large numbers of infauna that aren't eaten and changing the characteristics of the sediment.

Seagrass Meadows

Soft bottoms along the coast are occasionally carpeted by seagrasses. These flowering plants, grass-like in appearance but unrelated to true grasses, have become established in the marine environment. Seagrass meadows, or beds, develop best in sheltered, shallow water. They are also found in estuaries and in association with mangrove.

Seagrasses can grow into thick, luxuriant meadows, which often consist of more than one species of seagrass and many species of seaweeds (Fig. 15.15). Their roots and a network of underground stems keep them anchored in the face of turbulence. The roots and stems also help stabilize the soft bottom, and the leaves cut down wave action and currents. This decrease in turbulence causes more and finer sediment to be deposited, which in turn affects colonization by other organisms. Turtle grass is particularly important in Caribbean coral reefs, where it helps stabilize the landward, shallow sides of the reefs. Reduced turbulence also improves water clarity because less sediment remains suspended in the water column.

Productivity Seagrass meadows contain a very high plant biomass, as high as 1 kg dry weight/m² for turtle grass. With such a high density of photosynthesizing plants, seagrass meadows have a higher primary production than anywhere else on soft bottoms. In fact, they rank among the most productive communities in the entire ocean. Values as high as 8 gC/m²/day (where gC is grams of carbon fixed) have been recorded for seagrasses alone. Part of the reason for such high primary production is the fact that seagrasses have true roots, unlike seaweeds, and are therefore able to absorb nutrients from the sediment. Phytoplankton and seaweeds, by contrast, must depend on nutrients dissolved in the water column. Some typical rates of primary production in seagrass meadows and other marine environments are summarized in Table 12.1.

Figure 15.15 A thick meadow of neptune grass (*Posidonia oceanica*) in the Mediterranean.

Many small algae grow on the surface of seagrass leaves. These algae, known as epiphytes, further increase primary production in seagrass communities, although they can actually shield seagrasses from some of the light needed for photosynthesis. Microscopic diatoms are particularly abundant. Some epiphytic cyanobacteria also carry out nitrogen fixation and release nutrients in the form of nitrogen compounds.

IN CONTEXT

Demersal Living on or near the bottom
• 10.1 Fishes: The First Vertebrates

Pelagic Pertaining to the water column away from the bottom or other structure.
• 12.2, Major Marine Lifestyles and Environments

iStockphoto/Getty Images

Biomass The total weight of living organisms.
- 12.3 The Flow of Energy and Materials

Nitrogen Fixation Conversion of nitrogen gas (N_2) into nitrogen compounds that can be used by primary producers as nutrients.
- 7.2, Prokaryotes

The number of grazers, or herbivores, varies with geographical location. Herbivores, not many of which eat seagrass leaves, directly consume much less than half of seagrass primary production. Herbivores that do eat seagrasses include sea turtles, manatees, sea urchins, and some parrotfishes. Seagrasses are also important in the diet of some birds, like ducks and geese in the Arctic and other regions.

Animals take advantage of the high primary production of seagrasses in several ways, even when they do not actually eat the seagrasses themselves (Fig. 15.16). Many feed on the large amounts of decaying leaves and seaweeds. Deposit-feeding polychaetes, clams, and sea cucumbers live in or on the detritus-rich sediment. Detritus is also exported to other communities, like the unvegetated soft bottoms of deeper water. Some detritus, however, is buried in the sediment.

Seagrass Meadows as Habitat The dense seagrasses also offer shelter to many animals that do not feed on detritus. More animals live in or on the sediments in seagrass meadows than in the surrounding unvegetated sediments. A variety of small sessile or crawling animals live on the leaves: hydroids, snails, tiny tube-dwelling polychaetes, amphipods, shrimps, and the like (Fig. 15.17). Larger animals, like the queen conch (*Strombus gigas*, Fig. 15.18), live among the plants. Filter feeders that live in the sediment include clams and the pen shell (*Pinna carnea*). Several species of fishes and the West Indian sea star (*Oreaster reticulatus*) are examples of carnivores. Seagrass meadows also serve like nurseries for commercially valuable species such as the Atlantic bay scallop (*Argopecten irradians*) and shrimps.

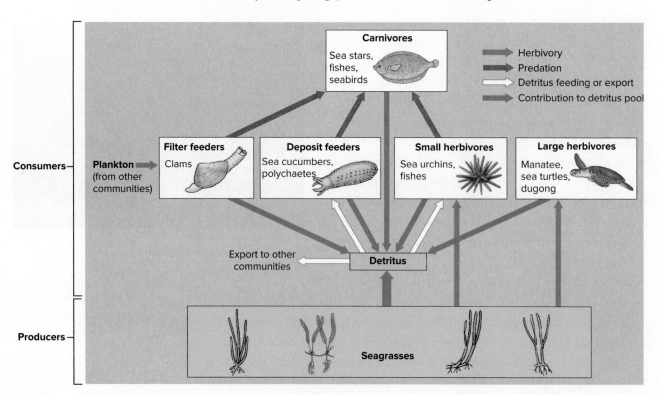

Figure 15.16 Generalized food web in a seagrass community. As in estuaries, detritus generated from dead plants is an important source of food for many animals.

Figure 15.17 Eelgrass (*Zostera marina*) leaves are often inhabited by sessile, or attached, organisms.

Green algae (*Enteromorpha*)

Trumpet-stalked jellyfish (*Haliclystus*)

Tube-dwelling polychaete (*Spirorbis*)

Colonial sea squirts

Eelgrass (*Zostera marina*)

Bryozoan (*Membranipora*)

Human Impact on Seagrasses The total area covered by seagrass meadows around the world has decreased by about a third since the late 1800s—and the loss is speeding up. Seagrasses have been adversely affected by sediment deposition from dredging, damage by boat propellers, and pollution, most especially excess nutrients from agriculture and urban runoff. Excess nutrients stimulate the growth of epiphytes, which cut down available light and thus photosynthesis by the seagrasses. Overfishing can further stimulate the growth of epiphytes because fishes normally keep small grazers like crabs and snails under control (see Fig. 15.28, p. 478).

©Charlie Arneson

Figure 15.18 The queen conch (*Strombus gigas*) is a large snail commonly associated with turtle grass meadows in Florida and the Caribbean. It is highly prized as food.

Toxoplasma: From Cats to Sea Otters

A growing number of marine mammals, notably the sea otter, are coming down with diseases found in land animals. One of these diseases is caused by *Toxoplasma gondii*, a single-cell protozoan parasite that is known to cause brain infections in humans and other animals on land.

Who is succeptable to *Toxoplasma*? *Toxoplasma* infects the tissues of many land mammals and birds, but it typically doesn't cause any symptoms. It is estimated that about 20% of the United States population is infected with *Toxoplasma* without causing any ill-effects. *Toxoplasma* can be lethal in the immunocompromised, or immune-deficient, individuals. The human fetus is also susceptible. Pregnant women infected with the parasite can be symptomless but pass the parasite to the fetus and cause congenital toxoplasmosis, which can lead to blindness or the death of the fetus.

Mammals get infected with *Toxoplasma* by ingesting food or water infected with the non-sexual stage of the parasite, or contaminated by a resistant stage which is found only in the feces of the members of the cat family. *Toxoplasma* reproduces sexually only in the intestines of cats. *Toxoplasma* infections are readily diagnosed by a blood test.

***Toxoplasma* and sea otters** The California sea otter has been particularly affected by *Toxoplasma*. The parasite is responsible for approximately 17% of sea otter deaths. It has been found mostly in the sea otter brains, where it causes inflammation, or encephalitis. The analysis of antibodies in sea otter blood between 1998 and 2004 in California showed positive infection with *Toxoplasma* in 52% of 305 recently dead animals and 38% of 257 live individuals. The parasite has been pointed out as the main reason of the slow recovery of the California sea otter. *Toxoplasma* has also been found in several species of seals, sea lions, and dolphins on the Pacific and Atlantic coasts of North America, Australia, and the Mediterranean, as well as in the walrus and the beluga whale. Congenital toxoplasmosis was discovered in a one-day-old harbor seal.

How do marine mammals become infected? How a cat parasite can infect sea otters and other marine mammals remains a mystery. The search for an answer is the subject of current research. The most obvious source of infection is fresh water runoff contaminated with parasites from cat feces and cat litter. Areas of high marine mammal infection are near urban centers with fresh water coastal runoff, and parasites in the water could be consumed by filter feeders that are then eaten by sea otters and other marine mammals. There is experimental evidence for this: *Toxoplasma* DNA was found in the gills of anchovies exposed to parasites under laboratory conditions. Cats, domestic and living in the wild, are undoubtedly the main source of infection. Domestic cats that are allowed outside the house are easily infected by eating rodents or birds. The resistant stage of the parasite can survive sewage treatment. They can also survive in landfills for as long as 18 months. The best recommendation to cat owners is to bag and dispose of cat feces in approved landfills where runoff is strictly controlled.

Disease also has had devastating effects. An epidemic of seagrass wasting disease resulted in about 95% loss of meadows on the Atlantic coasts of North America and Europe in the 1930s. Some meadows never recovered and disappeared forever. More recent but less dramatic outbreaks have taken place among Atlantic, Mediterranean, and Pacific seagrasses. Wasting disease, as well as other outbreaks of diseases among marine organisms, seems to be somehow related, directly or indirectly, to the impact of humans on the marine environment. It was not until relatively recently that the cause of seagrass wasting disease was discovered: *Labyrinthula*, a microscopic, ameba-like organism with unclear affinities to the protozoans. It produces enzymes that break down the

contents of cells, causing dark-brown to black spots, streaks, and eventually large patches on the leaves. *Labyrinthula* is known to reduce the efficiency of photosynthesis by the plant, which ultimately kills it. *Labyrinthula* produces a network of filaments along which its cells move and absorb nutrients from seagrasses and seaweeds. It spreads through direct contact between attached plants and by drifting leaves.

The rate of loss of seagrass meadows parallels that of other highly productive but endangered marine ecosystems: mangrove forests and salt marshes and coral reefs. The loss of seagrass meadows is more widespread, however, because they extend into both tropical and temperate regions in contrast to the tropical coral reefs and mangroves and the temperate salt marshes.

As a result of the high primary production of seagrass meadows, a large amount of organic carbon is stored not only in the plants themselves but buried as detritus in the substrate below. The loss of seagrasses is therefore of considerable significance because much of this organic carbon, some trapped over thousands of years, will be decomposed and released back to the atmosphere as CO_2, the most important of the greenhouse gases responsible for global climate change.

· ·

Section Review

1. **Identify** Which organisms are the main primary producers in soft-bottom subtidal communities? What is an important source of food for soft-bottom animals?

2. **Describe** Make a table like the one below to describe the different types of feeders in soft-bottom unvegetated communities.

Feeding Type	Definition	Example Animal
Deposit feeder		
Suspension feeder		
Scavenger		
Predator		

3. **Summarize** What are the benefits and services provided by seagrass meadows?

4. **Describe** What are the human impacts on seagrass communities?

15.3 Hard-Bottom Subtidal Communities

VOCABULARY

calcareous algae

perennial

fronds

canopy

monsoons

understory

DIVE IN!

What ecosystem services do kelp forests provide? Kelp forests provide habitat for many species of other algae, fishes, invertebrates, and marine mammals. They are also a nursery for commercially important fishes, such as kelp bass and rockfishes. Kelp is a food source for certain animals, such as sea urchins and kelp snails. Kelp is also a source of food and pharmaceutical compounds for humans. The thick growth of kelp forests reduces wave action, which can reduce shoreline erosion. Kelp forests are also a source of recreation for humans, who enjoy fishing, kayaking, and scuba diving through the fronds.

Main Idea

Two types of hard-bottom subtidal communities are submerged rocky bottoms and kelp forests.

Key Questions

1. Which are the dominant organisms in hard-bottom subtidal communities?
2. What is the basic structure of giant kelp communities?
3. What physical factors influence kelp communities?

Hard bottoms make up a relatively small portion of the continental shelf. They usually represent the submerged extensions of rocky shores. There are also some subtidal rocky reefs of varying size. In some cases a significant component of the hard substrate is provided by **calcareous algae**, the tubes of polychaete worms, and oyster shells. These three-dimensionally complex hard bottoms are often called reefs, but they should not be confused with the coral reefs in tropical waters or deep-water coral reefs.

Rocky Bottoms

Unlike those in the intertidal, the communities that develop on subtidal rocky bottoms are never subject to desiccation. This means that a wider variety of organisms can live there than in the intertidal.

Seaweeds Rocky-bottom communities in shallow water are rich and productive as long that enough light penetrates through the water column. Oil and gas rigs, although human-made, provide a hard bottom that can support a wide variety of organisms. Seaweeds are the most conspicuous inhabitants, especially on flat or gently sloping bottoms. They occur in an awesome array of colors, growth forms, textures, and sizes. Most are brown and red seaweeds. They can be filamentous, branched, thin and leafy, or encrusting. Many of these species can also be found in the intertidal.

As in the intertidal, one of the main problems for seaweeds and sessile animals in the subtidal is to find a place to attach. Practically all seaweeds have a holdfast (see Fig. 15.21, p. 470) to anchor them to the substrate. Every bit of space is occupied, even if it seems clean to our naked eye. Thus, there is intense competition for living space on the rocks. Different species of seaweed have different competitive abilities. They also vary in their tolerance to wave action, grazing by sea urchins and other herbivores, temperature, light, and the stability of the substrate.

The effect of light is particularly intriguing. As depth increases, the amount of light available for photosynthesis decreases, making it harder for seaweeds to

live there. Some species are better adapted to deep water than others, however. Several can live at depths of over 200 m in clear water. Deep-water seaweeds have increased levels of chlorophyll and other pigments that capture light energy. These adaptations can be found in all groups of seaweeds and are not restricted to the red algae, once thought to be the best adapted to deep water. In any case, depth zonation is not determined by light alone. It is affected by grazing, competition for space, and other factors.

Seaweeds also vary in their life history strategies. Some are **perennial** and the holdfast survives more than a year, whereas others are found only at certain times of the year. Though some grow fast and are short-lived, others grow slowly and are sturdy and long-lived. Fast growers are typically the first to colonize surfaces recently disturbed by grazers, turbulence, or other phenomena. Some seaweeds use both strategies by alternating a fast-growing, short-lived stage with a slow-growing, perennial one. Many have different growth forms at different times of their lives. Growth forms belong to the same species but look and function differently.

Some seaweeds have evolved defenses against grazing. These include chemicals, like sulfuric acid and phenols, that make them unpalatable and the ability to rapidly regrow tissue lost to grazers. Some seaweeds even have life history stages that bore into mollusc shells, a good deterrent to being eaten by a hungry sea urchin! Upright seaweeds tend to be tough and leathery. Calcareous algae, which include coralline and calcareous green algae, deposit calcium carbonate in their cell walls, making them particularly resistant to grazing.

Grazers and Carnivores Seaweeds must compete for space not only with each other but also with many sessile animals. Invertebrates tend to dominate on steeply sloping bottoms and vertical walls, where light limits seaweed growth. Hard substrates provide a good place for organisms to attach but are much harder to burrow through than sand or mud. As a result, hard-bottom communities tend to have a rich epifauna and a poor infauna, the reverse of soft-bottom communities. Sponges, hydroids, sea anemones, soft corals, bryozoans, tube-dwelling polychaetes, barnacles, and sea squirts are among the most common groups. Many form colonies that outcompete other species by simply growing over them. A few, like the rock-boring clam (*Pholas*), live embedded in rocks.

IN CONTEXT

Encrusting Seaweeds Those that grow as a thin layer over rocks, like some of the coralline algae.
- 8.1, Multicellular Algae: The Seaweeds

©Charlie Arneson

Figure 15.19 Sea urchins, like *Strongylocentrotus purpuratus* from the Pacific coast of North America, are important in influencing the composition of hard-bottom subtidal communities. They graze on seaweeds and on some of the encrusting animals and thus uncover new surfaces where other sessile organisms can settle.

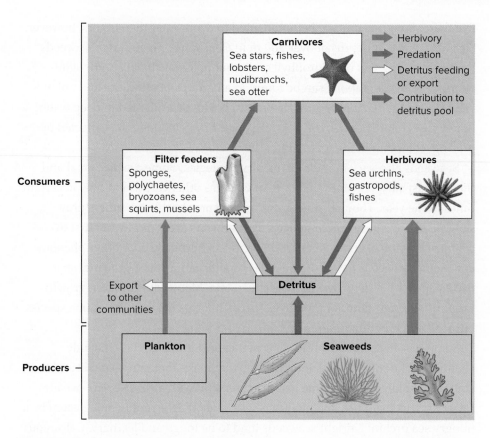

Figure 15.20 Generalized food web in a hard-bottom subtidal community. Most of the detritus generated is in the form of drift seaweeds.

Grazers on subtidal rocky bottoms are usually small, slow-moving invertebrates. Perhaps most important are sea urchins (*Strongylocentrotus*; Fig. 15.19). Chitons, sea hares, limpets, abalones, and other gastropod mollusks, as well as several groups of fishes, are also important grazers (Fig. 15.20).

Carnivores, including nudibranchs and other gastropods and sea stars, feed on attached invertebrates. Sea urchins feed not only on seaweeds but also on some of the flimsier attached invertebrates. Other carnivores, like crabs, lobsters, and many types of fishes, prey on grazers and smaller carnivores.

Grazers and predators strongly influence the composition of hard-bottom communities. As in the rocky intertidal, they remove residents from the rocks, opening up space for other organisms. Settling larvae and seaweed spores colonize the patches cleared by grazers and carnivores. Because these planktonic stages are often seasonal, patches formed at different times can be colonized by different species. This increases the total number of species, or biodiversity, in a given area. Carnivores can also act as a check on the grazers, helping maintain a relatively stable situation. This balance can be easily upset, however. For example, explosive increases in sea urchin populations have been reported in several subtidal communities, as have mass mortalities. Like all biological communities, subtidal communities are constantly changing.

Oil Rigs as Artificial Reefs

Oil rigs pepper the continental shelves in many parts of the world. These offshore rigs extract oil and natural gas from below the sea floor, and may be fixed to the sediment, attached to an artificial island, or float. While drilling for oil and natural gas presents many risks to the marine environment, the rigs also provide a unique habitat for marine life.

The rigs serve as important hard substrate in an otherwise mostly flat and muddy underwater landscape. A typical four-legged drilling structure provides two to three acres of habitat for reef-dwelling organisms to colonize. Sessile organisms, such as algae, sponges, sea urchins, and barnacles, attach to the legs of the platforms, providing a readily available food source for fishes, sea turtles, and many invertebrates. Corals even colonize rigs that are farther offshore, away from coastal runoff and turbidity. As many as 30,000 fish can inhabit one oil rig, including federally managed species. Rigs are popular destinations for recreational fishing and scuba diving. Endangered sea turtles are often seen at the artificial rig reefs. The structure of the rigs provides turtles a place to rest after feeding.

Natural reefs vs. artificial reefs Coral reefs serve many functions in marine systems, and artificial reefs are no different. Pharmaceutically important species like algae and sponges inhabit both natural and artificial reefs. Bryozoans host symbiotic bacteria that have been found to produce a compound that inhibits the growth of cancerous cells. Other coral reef organisms are sources of new medicines being used to treat arthritis, bacterial infections, heart disease, and viruses. As the demand for natural marine products increases, and natural coral reefs decrease, artificial reefs could provide necessary habitat for these organisms.

When the rigs are no longer in use there is debate over whether or not they should be removed or kept in place as artificial reefs. Proponents of removing the rigs argue that a rig kept in place is more likely to leak or have a significant impact than a removed rig. They also argue that overpopulating the continental shelf with such structures can have deleterious effects. Advocates for keeping the rigs in place argue that removing the rigs cause much more destruction than keeping them in place. Some say that removing artificial rigs can cause more damage to fisheries than the actual act of fishing does. The Rigs to Reef program in the Gulf of Mexico and the California Artificial Reef Enhancement Program on the Pacific coast are two programs fighting to keep rigs in place after they have been decommissioned. They either cut the rig down to 85 ft. below the surface so boat traffic is not affected by the structure, or they tow the entire rig to another location and sink it well below the water line so it does not impede navigation. The Rigs to Reef program is just one of several programs around the world dedicated to conserving this important marine habitat.

An oil platform in the Gulf of Mexico forms an artificial reef.

Greg McFall, Gray's Reef NMS, NOS, NOAA

Kelp Communities

Among the most fascinating of marine communities, kelp communities are also one of the most productive. Kelps are a group of large, fast-growing brown seaweeds that live in relatively cold water and are restricted to temperate and subpolar regions. They are true giants compared with other subtidal seaweeds or seagrasses, creating luxuriant forests that are home to a vast assortment of organisms.

Figure 15.21 Examples of North Atlantic kelps. The hollow-stemmed kelp reaches lengths of 12 m in deeper water.

Hollow-stemmed kelp
(Laminaria longicruris)

Edible kelp
(Alaria esculenta)

Horsetail kelp
(Laminaria digitata)

Kelp Diversity and Distribution There are several species of kelp. In the North Atlantic and on the Asiatic coast of the North Pacific, various species of *Laminaria* predominate (Fig. 15.21). Their simple or cleft blades can be 3 m long. The giant kelp, *Macrocystis*, the planet's largest benthic organism, dominates kelp communities on the Pacific coasts of North and South America and other parts of the southern hemisphere (see Figs. 15.22, 15.24). Each kelp individual is attached to the rocky bottom by a large holdfast. Several long stipes, intertwined to form a trunk-like foundation, grow from a single holdfast. The **fronds**, leaf-like blades, grow from the stipes. A single stipe can reach 20 to 30 m or more. In some southern hemisphere communities, *Ecklonia* rather than *Macrocystis* is the main species.

Figure 15.22 The geographic distribution of kelps is greatly affected by surface temperatures, which are influenced by the surface circulation of the ocean.

Figure 15.23 The elk kelp (*Pelagophycus porra*) grows on the outer, deeper edges of some giant kelp forests. It has two impressive, antler-like branches from which large blades hang at the mercy of the currents.

Kelp communities on the Pacific coast of North America are particularly diverse. A number of species, including the bull kelp (*Nereocystis*), the elk kelp (*Pelagophycus;* Fig. 15.23), *Alaria*, and *Pterygophora*, can be important or even replace *Macrocystis*. The bull kelp consists of many blades, each up to 5 m long, that hang below the surface suspended by a gas-filled pneumatocyst located at the end of a long stipe. *Alaria* (Fig. 15.21), which like many other species of kelp is edible, is also important in the North Atlantic.

Large, dense patches of kelp are known as kelp beds. They are called kelp forests when the fronds float on the surface in a thick mat (Fig. 15.24). This floating **canopy** is characteristic of Pacific kelp forests dominated by the giant and bull kelps. *Ecklonia* and some species of *Laminaria* and *Alaria* also form a canopy. Kelp forests can be extensive enough to be visible from space!

Water Temperature Physical factors have a major influence on kelp communities. Temperature is of particular importance because kelps are restricted to cold

Figure 15.24 Giant kelp (*Macrocystis pyrifera*) forest at Cedros Island, Baja California, Mexico.

Upwelling The upward flow of
cold, nutrient-rich deep water
to the surface.
• 5.2, Surface Circulation

water. This is partially because kelps don't do well in warm water and partially
because warm waters tend to lack the rich supply of nutrients that kelps need.
This dependence on cold water is reflected in the geographic distribution of kelps
(see Fig. 15.22). The surface waters of the oceans flow in great clockwise gyres in
the northern hemisphere and counterclockwise gyres in the Southern Hemi-
sphere. The currents that flow toward the poles on the western sides of the oceans
carry warm water from equatorial regions. Because of this, kelps are restricted to
high latitudes on the western sides of the oceans. On the other hand, kelps extend
well down eastern shores, where cold, nutrient-rich currents flow down from high
latitudes. *Ecklonia* occurs in low latitudes not far from the Equator along the
southern coast of the Arabian Peninsula (see Fig. 15.22). It is found there only
during the summer, when the summer **monsoons**, strong winds that blow north-
eastward across the Arabian Sea, cause strong upwelling that lowers water tem-
perature. The disappearance of the monsoons when the winds reverse direction
during the winter causes an increase in water temperature, which kills the kelp.

Substrate All but a few species of kelp are limited to hard surfaces, which can
include substrates like worm tubes and the holdfasts of other seaweeds. Given a
suitable place to attach, kelps will grow in water as deep as light allows. This
can be quite deep, up to 40 m in some species. Their fronds float at the surface,
basking in sunlight, while their stipes connect them to the bottom far below.

Wave Action Kelps are actually quite fragile largely *because* of their size. The
fronds cause a lot of drag, but the long, thin stipes can break even if they are
adapted to stretch in storm surge. Holdfasts are more often torn from the
bottom as the result of storm waves and damage by grazers. Drifting kelps
cause more damage by entangling other fronds. Kelps don't do well where
there is heavy wave action, and storms can be disastrous. Though they must
extend their blades to the surface, many species prefer to attach to the bottom
in relatively deep water, where wave action is reduced. Thus, the very adapta-
tions that allow these species of kelp to live in deep water, large size and
floating fronds, also tend to restrict them there.

Kelp Life History Kelps can grow very fast, and, not surprisingly, kelp
communities are very productive. Primary production reaches 1,000 gC/m^2/
year in *Ecklonia* in Australia and South Africa, close to 2,000 gC/m^2/year in
Laminaria in the North Atlantic, and as high as 5,400 gC/m^2/year in giant kelp.

The large kelps we see are only one stage in their life history. All kelps go
through two stages: a large, spore-producing sporophyte and a microscopic
gametophyte, which produces male and female gametes (Fig. 15.25). The
sporophyte is the organism we see.

In some kelps the sporophyte is an annual. In contrast, the giant kelp and others
are perennial and can live for several years. These long-lived kelps often lose fronds
to grazers, storms, and waves. They are able to grow them back because, unlike
many seaweeds, growth takes place at the holdfast as well as the end of the stipes.

Kelp forests along the Pacific coast of North America are arranged in distinct
depth zones, each made up of species that grow at a characteristic height above

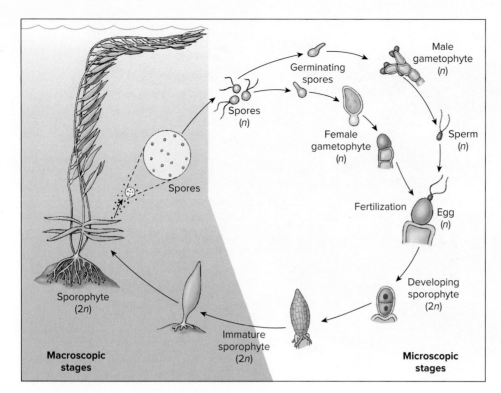

Figure 15.25 The life history of the giant kelp (*Macrocystis*) and other kelps includes a large, spore-producing sporophyte. Spores settle on the bottom and develop into minute male or female gametophytes. Each gametophyte releases male or female gametes, which, after fertilization, develop into the sporophyte, thus completing the cycle.

the bottom (Fig. 15.26, p. 474). This structure results from the interaction of several physical and biological factors. The floating canopy of the giant kelp, for instance, develops only where the water is deep enough to reduce wave action but shallow enough for light to reach the bottom, permitting growth from the holdfasts. Other kelps can contribute to the surface canopy. These include the bull kelp, found closer inshore, and the feather-boa kelp (Egregia), which lives in shallower water subject to wave action. The elk kelp forms a midwater canopy in deeper water along the outer edge of the giant kelp canopy.

The dense canopy of giant kelp forests cuts down the amount of light underneath. Diving under the canopy is like walking into a dense forest on land: It takes a few moments for your eyes to adjust to the dim light. The kelps that live there also have to adapt to the reduced light level, but life under the canopy is rich and varied. Smaller kelps exploit the **understory**, the area below the canopy. They include *Laminaria*, *Pterygophora*, and other kelps that have either erect fronds that stand above the bottom or fronds that lie right on it.

Various shorter algae, mostly red algae, live on the bottom under the two overlying layers, even when light is greatly reduced. Both branching and encrusting coralline algae are common. Some seaweeds are more abundant in shallow water, where increased wave action reduces the canopy.

North Atlantic kelp forests are dominated by species of *Laminaria* that typically do not form a canopy. They are similar to Pacific kelp forests, however, in that they include many species of seaweeds and are arranged in depth zones.

Diversity within Kelp Forests The complex, three-dimensional structure of kelp forests is exploited by many different animals, which explains the high species diversity of kelp communities. Assortments of polychaetes, small crustaceans, brittle stars, and other small invertebrates live on the kelp

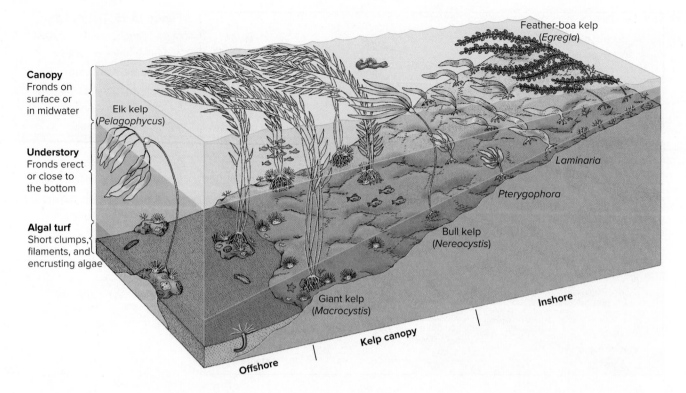

Canopy
Fronds on
surface or
in midwater

Understory
Fronds erect
or close to
the bottom

Algal turf
Short clumps,
filaments, and
encrusting algae

Elk kelp
(*Pelagophycus*)

Feather-boa kelp
(*Egregia*)

Laminaria

Pterygophora

Bull kelp
(*Nereocystis*)

Giant kelp
(*Macrocystis*)

Inshore

Kelp canopy

Offshore

Figure 15.26 Distribution of the major types of kelps and other seaweeds in a generalized giant kelp forest on the Pacific coast of North America. The complex distribution of algae results from the effects of factors like light, type of substrate, wave action, depth, number and type of grazers, and even time of year because some of the seaweeds are annuals.

IN CONTEXT

Ecological Niche The combination of what a species eats, where it lives, how it behaves, and all the other aspects of its lifestyle.

• 12.1, The Organization of Communities

holdfasts, particularly those of the giant kelp. Tube-dwelling polychaetes and other sessile organisms are common on the blades and stipes. Like the animals that live associated with the holdfast, they are mostly suspension feeders. One conspicuous inhabitant of the blades is a bryozoan (*Membranipora*) that forms thin, lace-like colonies. Its calcareous encrustations weigh the blades down and cover photosynthetic tissues, but their effect appears to be minimal. The rocky bottoms around kelps are inhabited by sponges, sea squirts, lobsters, crabs, hermit crabs, sea stars, abalones, and octopuses, among others.

Fishes Fishes are very common in kelp communities. They use the food resources and habitats provided by the kelp community in many ways, thus occupying many different ecological niches. For instance, fishes often use the available resources by feeding and taking shelter in different areas within the forest. Some species feed close to the bottom. In kelp forests along the Pacific coast of North America, bottom feeders include many species of rockfishes and the kelp bass (*Paralabrax clathratus*). The California sheephead (*Semicossyphus pulcher*) uses its dog-like teeth to crush sea urchins, crabs, and other bottom invertebrates. Surfperches and others can feed in different parts of the canopy, around the holdfasts, or in the open water among the kelp. Topsmelt are plankton feeders that take advantage of large swarms of opossum shrimps, or mysids, and other planktonic animals found around kelps. Fishes can define additional ecological niches by being active at different times of the day or night.

Invertebrate grazers Small algae are grazed by snails, crabs, sea urchins, and fishes, but surprisingly few grazers eat the large kelps. One giant kelp grazer, the Steller's sea cow, is now extinct. Isopods are small crustaceans that burrow into the holdfast, weakening it. A few fishes graze on kelps, but they do not appear to cause much mortality. Instead of feeding on the attached, actively growing

individuals, animals use most of the huge production of kelps in the form of drift kelp, pieces that break loose and sink to the bottom or are washed ashore. Many species of filter feeders (sponges, polychaetes, bivalves, sea squirts) feed on the small particles of detritus generated from drift kelp (see Fig. 15.20). As with seagrasses, salt marsh plants, and mangroves, however, much of the drift kelp is exported to other communities. Sea urchins are by far the most important grazers in kelp communities. Of special importance are the red (*Strongylocentrotus franciscanus*) and purple (*S. purpuratus*) sea urchins on the Pacific coast of North America and the green sea urchin in both the North Atlantic and North Pacific.

Threats to Kelp Forests Sea urchin populations sometimes "explode". These explosions have had devastating impacts on kelp communities in different parts of the world. Sea urchins normally feed on drift kelp. During population explosions, or "plagues," however, the urchins eat live kelps and other seaweeds. When the sea urchins eat the holdfasts or stipes, the kelps break loose, float away, and die. The sea urchins can even climb up the kelp, weighing down the fronds and allowing other urchins to reach them. The urchins can completely clear large areas, which are then known as "urchin barrens" or "urchin deserts." Encrusting coralline algae are virtually the only seaweeds left on these barrens. It is likely that a combination of factors causes urchin plagues, which probably are natural fluctuations in population size.

Trophic cascades The reasons for such outbreaks of urchins remain unclear. A possible cause in the North Pacific is the decline in sea otters, an urchin predator that has disappeared from most of its former range. Fluctuations in the number of urchins could be caused by the higher survival of their planktonic larvae, perhaps the result of more favorable temperatures or more abundant food.

Another possibility is that a decrease in the amount of drift algae in Southern California might have caused the urchins to switch to feeding on live kelps. Pollution from sewage and stormwater runoff and temperature increases are factors that could have caused a drop in the availability of drift kelp. Some of these factors can also stimulate the growth and survival of sea urchins. Some organic compounds released in sewage, for example, are used as nutrients by juvenile sea urchins.

A study of kelp forests in the Aleutian Islands, Alaska revealed that kelp forests were healthier where sea otters were common. In contrast, there were many sea urchins and few kelps on islands where sea otters were absent (Fig. 15.27, p. 476). There is evidence that otter populations have declined because of predation by orcas, or killer whales, which normally prefer seals and sea lions. Seals and sea lions have become increasingly scarce since the late 1980s, perhaps because overfishing has reduced their food supply. Orcas also appear to have changed their feeding habits when an important food source, the carasses of harpooned whales tied to whaling ships, dwindled when most commercial whaling was banned. Overfishing may have affected kelp forests through a chain of events that includes seals and sea lions, orcas, sea otters, and sea urchins. This interaction is an example of a trophic cascade, when changes in the abundance of an organism, usually a predator, extend throughout the food chain.

IN CONTEXT

Trophic cascade A process in which an effect on one species flows through an ecosystem because of indirect interactions through the food web.
• 12.1, The Organization of Communities

Figure 15.27 The effect of predation of sea otters (*Enhydra lutris*) by orcas, or killer whales (*Orcinus orca*) on the populations of sea urchins (*Strongylocentrotus*) and kelps (*Alaria, Laminaria*) in kelp forests in the Aleutian Islands, Alaska. The years without bars in the graphs indicate years without data and not zero values.

HEALTHY KELP FORESTS

OVERGRAZED KELP FORESTS

Observations in the Aleutians, however, do not help explain the situation in Southern California, where sea otters were wiped out almost 200 years ago, but the destruction of giant kelp forests by sea urchins was not observed until the 1950s. Heavy fishing on other urchin predators, including lobsters, crabs, and fishes, probably plays a role in the increase in urchin populations in Southern California and other places. The harvesting of Pacific abalones for food could have been a cause of sea urchin plagues. These large molluscs compete with sea urchins for shelter in rock crevices. Removal of the abalones, now rare, can provide more space for urchins.

To appease fishers, who were angry at the possibility of sea otters feeding on valuable shellfish, it was agreed that sea otters were to be kept from naturally migrating from central to Southern California. The "otter-free" zone was abandoned in 2012 and sea otters were allowed to roam free in Southern California. An estimated 16,000 sea otters lived from Oregon to Baja California, Mexico in

the nineteenth century. A 2014 survey showed that 2,944 sea otters inhabited California, slightly higher than the 2,939 recorded in 2013 but down from a 2007 peak of 3,026. A total of 68 sea otters were counted in San Nicolas Island off Southern California in 2014, where 140 individuals were relocated there from central California between 1987 and 1990. The 89 sea otters transplanted from Alaska to Vancouver Island, British Columbia, Canada between 1969 and 1972 now number about 5,000. Sea otters are being infected by bacteria that thrive in water polluted by nutrients and by parasites like *Toxoplasma*.

Sea otters also appear to have helped in the recovery of seagrass meadows in Elkhorn Slough, a large estuary in Monterey Bay, central California. Fertilizer runoff carrying nutrients from the many farms nearby has stimulated the growth of algae that live as epiphytes on the seagrass leaves, shading the leaves from sunlight. Seagrass meadows nearly disappeared from the estuary. Marine biologists have recently provided evidence that the return of sea otters to Elkhorn Slough in the mid-1980s signaled the recovery of the seagrass meadows. Crabs are a major part of the sea otter's diet in the estuary. Increased predation by sea otters have decreased the number of crabs, which feed on gastropod molluscs and other small invertebrates like isopods that graze on the epiphytes. A smaller number of crabs triggers an increase in the abundance of grazers, which then eat more of the seagrass epiphytes, resulting in an increase in the amount of sunlight reaching the leaves (Fig. 15.28, p. 478). Such interaction is another example of a trophic cascade involving sea otters.

Another example of a trophic cascade has been described in the Gulf of Maine on the Atlantic coast of the United States. Overfishing of the Atlantic cod (*Gadus morhua*) and other bottom-feeding fishes reduced pressure on young American, or Maine, lobsters (*Homarus americanus*), crabs, and sea urchins. The number of sea urchins steadily increased and they grazed on kelp, which is used by lobsters as shelter. Heavy harvesting of sea urchins for export to Japan led to a reduction in their numbers in the early 1990s. This development and continued overfishing of bottom fishes have resulted in a sharp increase in the number of lobsters. The 2014 catch in the state of Maine alone totaled 56,099 metric tons worth around $456.9 million, more than twice the 2000 catch of 25,900 metric tons.

Climate In addition to grazing by sea urchins, another factor affecting the health of kelp forests is climate. A strong El Niño in 1983 produced severe storms and unusually warm currents, which caused high kelp mortality in California and other regions along the eastern Pacific. The 1997–1998 El Niño was similarly destructive. A La Niña that followed the 1997–1998 El Niño, however, brought cold, nutrient-rich currents to the same region, stimulating the recovery of kelp forests. Recovery proved to be slow in kelp forests as well as in subtidal seaweed communities. The recovery of Southern California kelp forests has nevertheless progressed relatively well because of the absence of strong El Niños and effects of cool, nutrient-rich water during the last few years. It has been aided in part by the relative success of forests developing on artificial rocky reefs. Increased water temperatures during recent summers, most probably associated with global climate change, may nevertheless slowdown recovery.

IN CONTEXT

El Niño A warming of the surface water in the Eastern Pacific, part of large-scale changes in atmospheric and ocean current patterns, or *ENSO*.

La Niña A cooling trend of the surface water in the Eastern Pacific.

- 5.4, Interactions between Oceans and the Atmosphere

NO SEA OTTERS

MANY EPIPHYTES ON
SEAGRASSES

HEALTHY OTTER POPLULATION

HEALTHY SEAGRASSES

Sea otter abundance

Crab biomass

Grazer biomass

Algal epiphytes

Eelgrass biomass

Figure 15.28 Sea otters (*Enhydra lutiris*) are also part of a trophic cascade in the eelgrass (*Zostera marina*) meadows of Elkhorn Slough, CA. An increase in the number of sea otters reduced the number of crabs which in turn increased the number of epiphyte grazers, helping the recovery of seagrass meadows.

Evidence from other parts of the world, however, seems to indicate that catastrophic disturbances of kelp communities are usually followed by a recovery, all in recurring cycles. Sea urchins and kelps are apparently kept in a delicate "balance" that can be tipped one way or the other by factors like climate fluctuation, the effects of nutrient pollution on the survival of urchin larvae, and the removal of urchin predators.

Section Review

1. **Summarize** What abiotic and biotic factors affect the species composition of a hard-bottom subtidal community?

2. **Describe** What defenses have seaweeds evolved against grazing?

3. **Describe** the trophic cascade in kelp forests in the Aleutian Islands, Alaska.

Chapter 15 Review

REVIEW QUESTIONS

Multiple Choice

1. The continental shelf is wider on Atlantic margins than Pacific margins because the

 a. shelf is narrower on the trailing edge of plates.

 b. shelf is wider on the trailing edges of plates.

 c. plates on the Atlantic margin are colliding with other plates.

 d. plates on the Atlantic margin are subducting other plates.

2. Microscopic animals that live between sediment particles are called

 a. epifauna.

 b. epiphytes.

 c. infauna.

 d. meiofauna.

3. A bottom-dwelling fish is referred to as

 a. demersal.

 b. deposit feeder.

 c. filter feeder.

 d. pelagic.

Short Answer

4. What factors affect the distribution of deposit and suspension feeders in soft-bottom communities?

5. Why is there high species diversity in kelp communities?

Critical Thinking

6. The settling and metamorphosis of planktonic larvae may be triggered by the type of bottom. What kind of bottom features might be involved?

7. The life history of kelps consists of a very large sporophyte and a tiny gametophyte. Sea lettuce and some other seaweeds, however, have a gametophyte that is as big as their sporophyte. Do you see any advantages for kelps having such an inconspicuous, puny gametophyte?

DATA ANALYSIS LAB

What role do humans play in the illnesses of marine organisms? In the last ten years, the incidence of terrestrial diseases being found in marine organisms has increased. Scientists have come up with a new term, *pollutagen,* to refer to the land-based bacteria, fungi, and parasites reaching ocean waters. Scientists are seeing infected animals, including seals, sea lions, dolphins, and sea otters, worldwide.

Data and Observations

The flowchart shows one example of how a terrestrial parasite reached marine animals.

Opossums are the primary host of *Sarcocystis neurona*, a parasite similar to the organism that causes malaria.

Sarcocystis sporocysts are excreted in opossum feces.

In April 2004, a rainstorm transported a lot of material from land into the waters of Morro Bay, California.

Sarcocystis sporocysts were taken up by filter-feeding clams, which are a main food source for sea otters.

Shortly after the rainstorm, 40 sea otters washed ashore, infected and for the first time dying from *Sarcocystis*.

Think Critically

1. How might an outbreak of *Sarcocystis* in sea otters impact kelp communities along the California coast?

2. Scientists did not start testing for pollutagens in marine organisms until about 10 to 15 years ago. How can scientists measure changes in the amounts of pollutagens from this point on?

Data obtained from: Solomon, C., 2013. How kitty is killing the dolphins. Scientific American, vol. 308, no. 5, May, pp. 72–77.

Chapter 16

Coral Reefs

16.1 The Organisms that Build Reefs

16.2 Kinds of Coral Reefs

16.3 The Ecology of Coral Reefs

THEME Stability and Change

By studying coral reef ecosystems, which are built by living organisms, scientists can learn how high diversity is maintained despite complex changes in abiotic and biotic factors, including human intrusion.

BIG IDEA

Coral reefs are complex ecosystems supporting many organisms that interact in a multitude of ways.

About The Photo A queen angelfish (*Holacanthus ciliarus*) in a coral reef in the Caribbean.

16.1 The Organisms that Build Reefs

Main Idea

Cnidarians and other organisms, such as encrusting coralline algae, build coral reefs in shallow, warm waters.

. .

Key Questions

1. What are coral reefs?
2. What physical conditions do reef-building corals need to build reefs?

VOCABULARY

hermatypic

ahermatypic

planula

mesenterial filaments

algal ridge

bioerosion

bleaching

There is something special about tropical coral reefs. The warm, clear water, spectacular colors, and multitude of life captivate almost everybody. Coral reefs rival that other great tropical ecosystem, the rain forest, in their beauty, richness, and complexity. Tropical rain forests and coral reefs are also similar in that the basic physical structure of both communities is produced by organisms. Like the giant trees of a rain forest, reef-building corals create a three-dimensional framework that supports a vast assortment of organisms. Coral reefs are massive—in fact, they are considered not only biological communities but geological structures, the largest geological features built by organisms.

Coral reefs are made of huge amounts of calcium carbonate ($CaCO_3$), or limestone, deposited by organisms. Of the thousands of species in coral reef communities, only a fraction produce the limestone reef. The most important of these reef-building organisms, not surprisingly, are corals.

Reef Corals

"Coral" is a general term for several different groups of cnidarians, only some of which help build reefs (Table 16.1). In reef-building, or **hermatypic**, corals the polyps produce calcium carbonate skeletons. The skeletons built by billions of these tiny polyps form a massive reef. The most important reef builders are a group known as scleractinian corals, sometimes called the stony or "true" corals.

Nearly all reef-building corals contain symbiotic zooxanthellae (Fig. 16.1) that help the corals make their calcium carbonate skeletons. Corals can produce their skeletons without zooxanthellae but only very slowly, not nearly fast enough to build a reef. It is the zooxanthellae as much as the corals themselves that help

Robert Sisson/National Geographic Creative

IN CONTEXT

Cnidarians Animals in the phylum Cnidaria; they have radial symmetry, tissue-level organization, and tentacles with *nematocysts*, specialized stinging structures. Life stages of cnidarians:

Polyp The sessile, sac-like stage, with the mouth and tentacles on top.

Medusa The swimming, bell-like stage, with the mouth and tentacles underneath.

- 9.1, Sponges, Cnidarians, and Comb Jellies

Figure 16.1 The microscopic golden-brown zooxanthellae that pack these polyp tentacles help corals build reefs. They also occur in giant clams, some sea anemones, and other invertebrates. Cells in the white, bulbous tips of the tentacles are packed with nematocysts.

What is a reef? The word "reef" can mean different things in different situations. Seamen and surfers use the term "reef" to mean submerged hazards that could make navigation or surfing dangerous. To marine scientists, a reef is a rigid, or hard, structure built by a living organism. Corals are not the first or the only organisms to build reefs. Cyanobacteria built reefs as far back as 3.5 billion years ago. The fossil record shows that many marine animals built reefs at one time, including sponges, bryozoans, and bivalves. The coral reefs that we are familiar with today only developed during the last 60 million years.

Table 16.1 Major Groups of Corals and Related Cnidarians on Coral Reefs

Common Name	Importance as Reef Builders	Notes
ANTHOZOANS (Lack medusa stage and live only as polyps; same group as sea anemones)		
Scleractinian corals	The main reef builders	Produce a calcareous skeleton; nearly all have zooxanthellae.
Soft corals	Not reef builders	Common on coral reefs; most have zooxanthellae but do not produce a rigid calcareous skeleton.
Organ-pipe corals (*Tubipora*)	Minor	Named for pipe-like structure of skeleton; have zooxanthellae and produce a calcareous skeleton.
Blue coral (*Heliopora*)	Significant in some places	Named for distinct blue color of skeleton; has zooxanthellae and produces a calcareous skeleton.
Gorgonians	Not reef builders	Sea fans and sea whips; hard skeleton is made mostly of protein and not calcareous; some have zooxanthellae.
Precious corals	Not reef builders	Gorgonians with a pink, red, or gold calcareous skeleton used for jewelry; occur mainly in deep water rather than on coral reefs and do not contribute to reef growth; lack zooxanthellae.
Black corals	Not reef builders	Used to make jewelry and sometimes called "precious corals" but are a different group from gorgonian precious corals; hard skeleton is made of protein and is not calcareous; commonly occur in deep and cold surface waters as well as on coral reefs.
HYDROZOANS (Medusa and polyp stages; only polyps on reef; same group as some jellyfishes)		
Fire corals (*Millepora*)	Significant in some places	Named because touching them produces a burning sensation from their powerful nematocysts; have zooxanthellae and produce a calcareous skeleton.
Lace corals (*Stylaster*)	Insignificant	Named for delicate, branched colonies; lack zooxanthellae but produce a calcareous skeleton; commonly occur in deep and cold surface waters as well as on coral reefs.

IN CONTEXT

Zooxanthellae Dinoflagellates (single-celled, photosynthetic algae) that live within animal tissues.
• 7.3, Unicellular Algae

Symbiosis The living together in close association of two species, often divided into parasitism, where one species benefits at the expense of the other; commensalism, where one species benefits without affecting the other; and mutualism, where both species benefit.
• 12.1, The Organization of Communities

construct the reef framework, and without zooxanthellae there would be no reefs. **Ahermatypic** corals, such as precious corals and black corals, do not help build reefs and often lack zooxanthellae.

The Coral Polyp You have to look close to see the little polyps that build coral. Coral polyps are not only small but deceptively simple in appearance. They look much like little sea anemones, consisting of an upright cylinder of tissue with a ring of tentacles on top (Figs. 16.2 and 16.3b). Like anemones and other cnidarians, they use their nematocyst-armed tentacles to capture food, especially zooplankton. The tentacles surround the mouth, the only opening to the sac-like gut.

Most reef-building corals are colonies of many polyps, all connected by a thin sheet of tissue. The colony starts when a planktonic coral larva, called a **planula**, settles on a hard surface. Coral larvae generally do not settle on soft bottoms. The larva transforms, or metamorphoses, into a single "founder" polyp that, if it survives, divides over and over to form the colony. The digestive systems of the polyps usually remain connected, and they share a common nervous system (Fig. 16.4). A few reef-building corals consist of only a single large polyp (Fig. 16.3a).

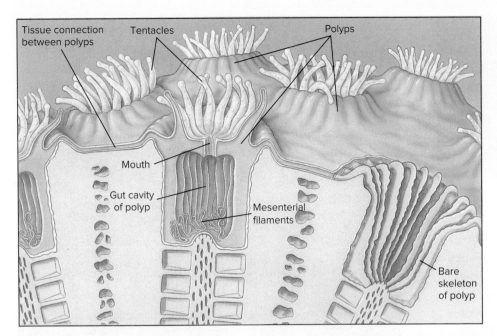

Figure 16.2 Cutaway view of one of the polyps in a coral colony and of the calcium carbonate skeleton underneath. The polyps are interconnected by a very thin layer of tissue.

Tissue connection between polyps

Tentacles

Polyps

Mouth

Gut cavity of polyp

Mesenterial filaments

Bare skeleton of polyp

(a)

(b)

(c)

Figure 16.3 Coral polyps. (a) A few corals consist of a single polyp, which as in this mushroom coral (*Fungia*) may be free-living, or not attached to the bottom. Most corals are colonies of many individual polyps. (b) In some of these, like the large-cupped boulder coral (*Montastrea cavernosa*) from the Caribbean, each polyp has its own individual cup, called a corallite. (c) In coral colonies like this brain coral (*Diploria*) from the Caribbean, meandering lines of polyps lie in a common corallite.

Figure 16.4 The polyps in a coral colony are interconnected. If you touch an extended polyp, it will contract, and so will its neighbors, and so on, so that a wave of contraction passes over the colony. In this colony of the coral *Goniopora lobata* the wave of contraction is moving up from the bottom.

Figure 16.5 The calcium carbonate skeleton makes up most of a coral colony. A live colony (*Pocillopora verrucosa*) is shown on the right. On the left is a colony with the living tissue removed. The main difference is the color; the live part is only a thin layer of tissue on the surface.

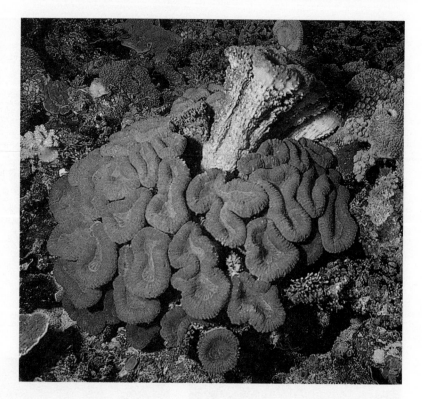

Figure 16.6 This coral (*Lobophyllia hemprichii*) shows particularly well how corals build up their skeleton. Each of the irregular rings (arrows) is a single polyp that has built up a column of calcium carbonate skeleton.

IN CONTEXT

Plankton Primary producers (*phytoplankton*) and consumers (*zooplankton*) that drift with the currents.

- 12.2, Major Marine Lifestyles and Environments; Figure 10.11

Photosynthesis $CO_2 + H_2O +$ sun energy → organic matter (glucose) + O_2

- 6.1, The Ingredients of Life

Detritus Particles of dead organic matter.

- 12.3, The Flow of Energy and Materials

Coral polyps lie in a cup-like skeleton of calcium carbonate that they make themselves. The polyps continually lay down new layers of calcium carbonate, building up the skeleton so that it grows upward and outward (Fig. 16.6). The skeleton forms nearly all the bulk of the colony (Fig. 16.5) and can take many different shapes (Fig. 16.7). The actual living tissue is only a thin layer on the surface. It is the calcareous coral skeletons that form the framework of the reef.

Coral Nutrition Zooxanthellae nourish the host coral as well as help it deposit its skeleton. They perform photosynthesis and pass some of the organic matter they make on to the coral (Fig. 16.8). Thus, the zooxanthellae feed the coral from the inside. Many corals can survive and grow without eating, as long as the zooxanthellae have enough light.

Although corals get much of their nutrition from their zooxanthellae, most eat when they get the chance. They prey voraciously on zooplankton. The billions of coral polyps on a reef, along with all the other hungry reef organisms, are very efficient at removing zooplankton brought in by currents. Indeed, the reef has been called a "wall of mouths."

Coral polyps catch zooplankton and detritus with their tentacles, or in sheets of mucus that they secrete on the colony surface. Tiny, hair-like cilia gather the mucus into threads and pass them along to the mouth. Some corals hardly use their tentacles and rely on the mucus method. A few have even lost tentacles altogether.

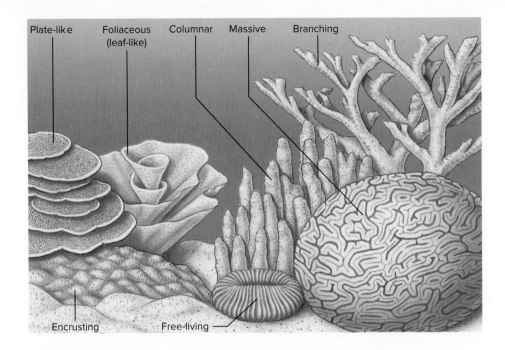

Plate-like | Foliaceous (leaf-like) | Columnar | Massive | Branching

Encrusting | Free-living

Figure 16.7 Corals come in a multitude of shapes, or growth forms.

Coralline Algae Red algae that deposit calcium carbonate ($CaCO_3$) in their tissues.
- 8.1, Multicellular Algae: The Seaweeds.

Corals have still other ways of feeding themselves. Attached to the wall of the gut are a number of long, coiled tubes called **mesenterial filaments** (see Fig. 16.2). The mesenterial filaments secrete digestive enzymes. The polyp can extrude the filaments through the mouth or body wall to digest and absorb food particles outside the body. Corals also use the mesenterial filaments to digest organic matter from the sediments. In addition, corals can absorb dissolved organic matter (DOM).

Other Reef Builders

Although they are the chief architects, corals don't build a reef alone. Algae, for example, are essential to reef growth. In fact, some marine biologists prefer to call coral reefs "algal reefs" or "biotic reefs." One reason for this is that zooxanthellae, which are algae, are essential to the growth of corals. There are other algae, however, that have key roles in building the reef. Encrusting coralline

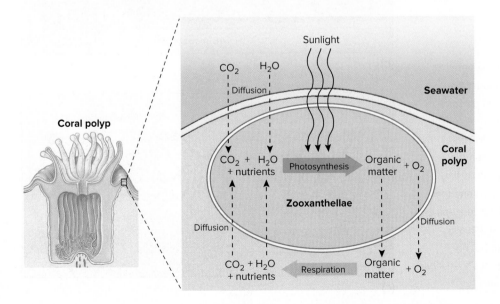

Coral polyp

Sunlight

CO_2 H_2O

Diffusion

Seawater

$CO_2 + H_2O$ + nutrients → Photosynthesis → Organic matter $+ O_2$

Coral polyp

Zooxanthellae

Diffusion

Diffusion

$CO_2 + H_2O$ + nutrients ← Respiration ← Organic matter $+ O_2$

Figure 16.8 Zooxanthellae in a coral's tissue perform photosynthesis and transfer some of the organic matter they manufacture to the coral host. Carbon dioxide and nutrients are continually recycled between the coral and its zooxanthellae.

IN CONTEXT

Bryozoans Small, colonial, encrusting animals that make delicate, often lace-like, skeletons.

• 9.2 Marine Worms and Some Small Invertebrate Phyla

algae grow in rock-hard sheets on the reef. They deposit considerable amounts of calcium carbonate, sometimes more than corals, and thus contribute to reef growth. Coralline algae are more important on Indian and Pacific Ocean reefs than Atlantic ones.

Encrusting coralline algae not only help build the reef, they also keep it from washing away. The stony pavement they make is tough enough to withstand waves that would smash even the most rugged corals. The algae form a distinct ridge on the outer edge of many reefs, especially in the Pacific. This **algal ridge** absorbs the force of the waves and prevents erosion from destroying the reef.

Cementing the Reef Encrusting algae do yet another job that is vital to reef growth. Coral skeletons and fragments create an open network, full of spaces, that traps coarse carbonate sediments (Fig. 16.9). Sediment, especially fine

(a)

(b)

Figure 16.9 Reef growth involves several processes. (a) The framework is made when reef-building corals settle and grow on some hard surface, usually a preexisting reef. (b) The spaces in this framework are partially filled in by coarse carbonate sediments. (c) When the sediments are glued together by encrusting organisms, new reef "rock" is formed and the reef has grown. On a real reef all three steps go on at the same time. Reefs are not actually completely solid as depicted here but are porous, with many holes and crevices that serve as home to a multitude of organisms.

(c)

sediment, damages corals when it settles directly on them, but the buildup of coarse sediment in the reef framework is an essential part of reef growth. The structure of a reef is formed as much by the accumulation of calcareous sediment as by coral growth. Encrusting algae grow over the sediment as it builds up, cementing it in place. Thus, encrusting coralline algae are the glue that holds the reef together. Some invertebrates, notably sponges and bryozoans, also help bind sediments.

Nearly all the sediment that accumulates to help form the reef comes from corals and the shells or skeletons of other organisms. In other words, nearly all the sediment is biogenous. Coral rubble, fragments of broken coral, is one important source of sediment on reefs. Another important sediment-forming organism is a calcareous green alga called *Halimeda* (Fig. 16.10). *Halimeda* deposits calcium carbonate within its tissues to provide support and discourage grazers—a mouthful of limestone is pretty unappetizing. The remnants of *Halimeda* accumulate on reefs in huge amounts, to be bound together by encrusting organisms.

Many other organisms make calcium carbonate sediments and contribute to reef growth. The tests of forams, as well as snails, clams, and other molluscs, are very important. Sea urchins, bryozoans, crustaceans, sponges, bacteria and a host of other organisms add or help bind carbonate sediments. Reef growth is truly a team effort.

Bioerosion A great deal of the sediment on reefs comes from the activities of organisms that don't deposit calcium carbonate but instead break down the solid reef structure. Many animals scrape or bite their food off the reef with some sort of hard structure. Parrotfishes, for example, are named for their fused teeth that form a parrot-like beak (see Fig. 8.8b). Sea urchins scrape algae off the reef with their Aristotle's lantern. In the process, these and many other grazers remove bits of calcium carbonate from the reef to form sediment. The erosion caused by such organisms, or **bioerosion**, tends to wear the reef away, although some of the sediment they create does get reincorporated into the reef structure. Many other organisms—including sponges, clams, polychaete and other worms, and algae—cause bioerosion by burrowing into or through the reef limestone, either scraping it away to form sediment or dissolving it. A reef grows only if corals, coralline algae, and sediment-forming and binding organisms build it up faster than bioeroders wear it away.

Bioeroders have another important role in reef communities. Their burrowing creates spaces in the reef framework, nooks and crannies where other organisms can hide. More species live inside the reef framework, in fact, than on its surface.

Conditions for Reef Growth

Other organisms are important, but coral reefs do not develop without reef-building corals. Corals have very particular requirements that determine where reefs develop. Reefs are rare on soft bottoms, for example, because coral larvae need to settle on a hard surface.

Figure 16.10 This calcareous green alga (*Halimeda*) is one of the main sediment-forming organisms on most reefs. About 95% of the plant's weight is calcium carbonate, and there is only a thin layer of live tissue on the outside. When the tissue dies the segments separate, each leaving a piece of limestone.

IN CONTEXT

Foraminiferans (Forams)
Protozoans, often microscopic, with a calcium carbonate test.
• 7.3, Single Celled Eukaryotes

Aristotle's Lantern A complicated set of calcium carbonate ($CaCO_3$) teeth and associated muscles that is found in sea urchins.
• 9.5 Echinoderms

Light and Temperature Corals can grow only in shallow water, where light for their zooxanthellae can penetrate. Calcareous algae need sunlight, too. Different corals and algae have different depth limits—some can live deeper than others—but reefs rarely develop in water deeper than about 50 m. Because of this, coral reefs are found only on the continental shelves, around islands, or on top of seamounts. Many types of coral live in deep water and do not need light, but these corals do not contain zooxanthellae or build massive reef structures. Corals tend to prefer clear water, since water clouded with sediment or plankton reduces light penetration.

Reef-building corals are limited to warm water and can grow and reproduce only if the average water temperature is above about 20 °C. Most reefs grow in considerably warmer areas. Figure 16.11 illustrates the relationship between coral reefs and water temperature.

Water that is *too* warm is also bad for corals. The upper temperature limit varies, but is usually around 30 to 35 °C. The first outward sign of heat stress, or stress of other kinds, is **bleaching**, in which the coral expels its zooxanthellae. It is called bleaching because the golden-brown or greenish zooxanthellae give corals most of their color; without zooxanthellae the coral's tissue is almost transparent and the coral looks white. Corals also slough off large amounts of slimy mucus when stressed. Corals often recover from bleaching, regaining their symbionts either from the regrowth of the few that remain or by acquiring new ones from the water. If the warm conditions last too long or the temperature gets too high, however, the coral dies.

Temperature limits vary among coral species and from place to place. Corals from warm locations, for example, tolerate higher temperatures than corals from cooler waters (Fig. 16.12). Corals can also adapt to fluctuating

Figure 16.11 The distribution of reef coral communities, including those that do not form structural reefs. Note that the distribution of reef corals, like that of kelps, is related to temperature. Reef corals require warm water, however, while kelps need cold water.

● Coral reefs ▨ Reef-building corals --- Average 20 isotherm ← Warm current ← Cold current

488 Chapter 16 *Coral Reefs*

conditions. For instance, reefs grow in parts of the Persian Gulf where the water temperature ranges from 16 to 40 °C.

One way that corals adapt to temperature may be to have different species or strains of zooxanthellae that vary in temperature tolerance. There is even evidence that corals bleach in order to expel zooxanthellae that can't take the heat and acquire different strains that are better adapted to warm conditions. On the Great Barrier Reef, for example, corals have been found to harbor more high-temperature zooxanthellae after recovering from temperature-induced bleaching. This hypothesis, however, is controversial.

In any case, the upper thermal tolerances of corals are usually not far above the normal temperature range where they live. Corals suffer in temperatures above the normal range. This sometimes happens during extreme low tides, when shallow pools on the reef may be cut off from circulation. Heated by the sun, the water can warm up to fatal temperatures. By discharging heated water, power stations and other industrial plants can also kill corals.

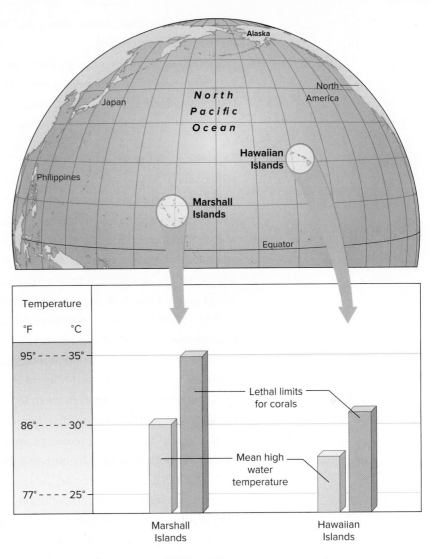

Figure 16.12 The upper temperature limit of corals is related to the temperature at their home sites. The water in the Marshall Islands is warmer on average than that in Hawai'i. Corals from the Marshalls can tolerate correspondingly higher temperatures. Why do you think the highest temperature that the corals can tolerate would be higher than the average high temperature at their location?

El Niño brings unusually warm water to many parts of the ocean, and with it coral bleaching and mortality. During the unusually strong El Niño of 1997–98, there was severe bleaching on many reefs around the world. In some places, including the Caribbean and parts of the Great Barrier Reef, bleaching did not kill many corals, and reefs quickly recovered after the water cooled. In other areas, such as the Indian Ocean, Southeast Asia, and the far western Pacific, many corals died after bleaching and some reefs were severely damaged. Many of these reefs have recovered slowly or not at all.

High water temperatures have led to other, generally more localized, bleaching events almost every year since the 1997–98 El Niño. Bleaching is becoming more frequent and more intense as a result of global climate change. We must remember that coral bleaching and El Niño are natural events, and some of the apparent increase in bleaching may just reflect better monitoring and reporting. Temperature records stored in oxygen isotopes of fossil corals indicate that the strongest El Niño events of the past millennium probably occurred in the mid-seventeenth century, long before the industrial revolution began

intensifying the greenhouse effect, and long before scientific records of coral bleaching were recorded.

Nevertheless, the sea has steadily warmed over the last 50 to 100 years, exposing reefs more and more often to stressful temperatures, and global warming will continue. Coupled with ocean acidification, pollution, overfishing, and other human-induced stresses the world's coral reefs are in serious trouble.

Salinity, Sediments, and Pollution Most corals are quite sensitive to reduced salinity and don't do well near river mouths, for example, where there is a lot of freshwater input. This is not only because of the lowered salinity but also because rivers bring in a lot of silt, which is generally unfavorable to corals. It clouds the water, cutting down light for the zooxanthellae. Sediment on the colony surface can smother the coral or cause disease, though corals can clean themselves to some extent by sloughing off mucus that carries the sediment away. Sediment also covers up hard substrate that coral larvae need to settle on, and even when they do find a place to settle the new coral polyps are especially susceptible to sediment because they are so small, and therefore easily smothered.

Some corals do thrive in silty environments and even feed on organic-rich sediment particles. Sediment-tolerant corals can build diverse, productive reefs in silty environments. Reefs in clear, low-sediment waters, however, are dominated by species that are sensitive to sediment unless there is enough wave or current action to wash the sediment away (Fig. 16.13). Many reefs around the world have been damaged by human activities like mining, logging, and agriculture that greatly increase the flow of sediment onto the reef.

Corals are also sensitive to pollution of many kinds. Even low concentrations of pesticides and toxic wastes can harm them. Again, larvae are especially sensitive. Large amounts of nutrients can also be harmful. Humans release tremendous quantities of nutrients in sewage, in fertilizers that are washed from farmland to the sea, and in emissions to the atmosphere from burning fossil fuels. There is some evidence that the nutrients can harm the corals

Figure 16.13 Corals often flourish where there is plenty of wave action. The water motion keeps sediment from settling on the corals and brings in food, oxygen, and nutrients.

Is there a Doctor on the Reef?

Coral reefs face many challenges to survival: bleaching due to increased temperatures, ocean acidification, destructive fishing practices, sea level rise, and negative impacts from shoreline development. On top of all of these, corals can get sick, just like humans. Coral diseases are poorly understood despite being first reported in the 1970s, but they are an avenue of ongoing research that will continue to help marine scientists understand the resiliency of coral reefs.

The symptom of black-band disease is exactly what it sounds like. A black band, or ring, forms on the coral and moves outward, leaving in its trail bare coral skeleton. Typically, black band disease infects brain corals (*Diploria*) and star corals (*Montastraea*), but it can infect many other species found in Florida and the Caribbean. Researchers have determined that infection is caused by a group of bacteria, including cyanobacteria, sulfur-reducing, and sulfur-oxidizing bacteria. Fungi are also involved Experiments have revealed that unless multiple species of the bacteria consortium are present, the disease does not occur. Even more perplexing is that the bacterial infection does not appear to attack the coral polyps directly. Instead, it asphyxiates the coral by competing for oxygen. In the anoxic microenvironment of the infection, hydrogen sulfide (H_2S), a toxic chemical at even low concentrations, is produced and might further harm the corals.

Why do corals get sick? Are coral diseases random occurrences or are there triggers for the infection? Some research has shown that disease transmission occurs more readily between coral colonies close to each other than in corals spaced further apart. This is not surprising since that idea parallels our experience in the terrestrial world. Other studies have found a trend that disease outbreaks usually occur after a reef is exposed to stressful conditions. For example, a period of low salinity may impair the corals' natural defensive response to a disease-causing microorganism, giving the microbe a brief opportunity to cause an infection. A period of increased water temperature can harm a coral, including impairing its defenses against disease, while at the same time it boosts the growth rate of a bacterium. Likewise, eutrophication is known to harm corals via poor water quality, but it might also "fertilize" bacteria, helping them grow more quickly.

Coral diseases are a serious threat to the fragile reef ecosystem, especially if additional environmental stressors increase the virulence or frequency of disease. How is the disease transmitted between individuals? Why is the Caribbean is a global "hot spot" for coral diseases? Why do disease outbreaks not occur after all environmental stresses experienced by a reef, but only after some? Researchers continue to investigate these important questions.

Coral with black band disease.

directly by interfering with skeleton formation. More importantly, increased nutrients alter the ecological balance of the community. Most coral reefs grow in water that is naturally nutrient-poor. Nutrient limitation slows the growth of seaweeds, which are also kept under control by grazers. This allows corals to compete successfully for space and light. When nutrients are added, seaweeds may grow much faster, and shade and choke out the slow-growing corals. This is a particular problem when, as is often the case, fishing has reduced populations of herbivorous fishes, urchins, and other grazers.

©Diane Nelson

The Kāne'ohe Bay Story One of the best-known examples of the harmful effects of nutrient enrichment occurred in a partially enclosed bay in the Hawaiian Islands. Kāne'ohe Bay, located on the northeast shore of the island of O'ahu, once had some of the most luxuriant reefs in Hawai'i. Until the 1930s the area around the bay was sparsely populated. In the military buildup of O'ahu before World War II, the population increased. The increase continued after the war with residential development around the bay.

Treated sewage from the expanding population was discharged into the bay. By 1978, about 20,000 m³ of sewage were entering the bay every day. Long before then, by the mid-1960s in fact, marine biologists began to see changes in the middle part of the bay. Loaded with nutrients, the sewage was a great seaweed fertilizer. One seaweed in particular, a green alga called bubble weed (*Dictyosphaeria cavernosa*), found the conditions particularly to its liking and exploded, literally covering the bottom in many parts of the bay. Bubble weed overgrew and smothered corals. Phytoplankton also bloomed, clouding the water. Kāne'ohe Bay's reefs began to die. Such accelerated algal growth due to nutrient input is called eutrophication.

For a time it seemed the story had a happy ending. As the once-beautiful reefs smothered, scientists and the general public cried out. It took a while, but in 1978 public pressure finally forced the diversion of the treated sewage offshore, greatly reducing nutrient input to the bay.

The response was dramatic. Bubble weed died back in much of the bay, and the corals began to recover faster than anyone expected. By the early 1980s, bubble weed were fairly scarce and corals were growing back. The reefs weren't what they once were but seemed to be on track to recover.

Then the ghost of pollution past reared its ugly head. In November 1982, Hurricane Iwa struck Kāne'ohe Bay. Layers of coral skeleton that formed during the years of pollution were fragile and crumbly. When the hurricane hit, this weak layer collapsed and many reefs were severely damaged. Fortunately, the corals were already beginning to recover, and some of the broken pieces were able to grow back. If the hurricane had hit during the years of pollution, the reefs of Kāne'ohe Bay—and their benefits to fishing, tourism, and recreation—might have disappeared forever.

The rapid recovery of Kāne'ohe Bay's reefs that was seen during the early 1980s did not continue. By 1990 the recovery leveled off, and bubble weed came back. Nutrients are still a problem. A lot of nutrients from the old sewage outfalls were stored in the sediments, and are still being released. New nutrients enter from lawn and golf course fertilizers, boats, and the septic tanks of private homes. There is also evidence that fishing has reduced populations of fishes that graze on bubble weed. Furthermore, the grazing fishes that remain prefer to eat other species of seaweed introduced from outside Hawai'i, and have shifted away from eating bubble weed. Bubble weed started to dominate again, but then in 2006 declined steeply. No one is sure why, though the decline followed a period of unusually rainy weather, or if bubble weed will make another comeback. But bubble algae are no longer the only problem. While some species of algae introduced to Ka-ne'ohe Bay are favorite foods of

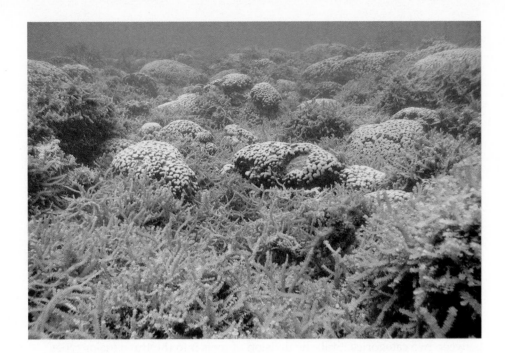

Figure 16.14 The red alga *Eucheuma denticulatum* is one of several species of red algae introduced to Kāneʻohe Bay for experimental mariculture that have gotten out of control. The algae are overgrowing and smothering corals in many parts of the bay.

grazing fishes, others are not. These species are also able to grow without a lot of nutrients, and are proliferating and smothering the bay's reefs (Fig. 16.14). Trying to fight back, biologists and conservationists came up with "Super Sucker," an underwater vacuum cleaner that divers use to suck up the dense mats of invasive algae, hoping to give corals and other native species a chance. Biologists are also trying to increase numbers of a native urchin (*Tripneustes gratilla*) that, unlike the grazing fishes, does like to eat the invasive algae. Clearly, it is going to be a lot harder to restore the coral gardens of Kāneʻohe Bay than it was to spoil them.

The Kāneʻohe Bay story, unfortunately, is far from unique. Most of the world's tropical coasts are undergoing increasing development and population growth. More and more nutrients are ending up in waters that support coral reefs, and there are many reports of reefs being threatened by eutrophication. New research, however, is showing that the effects of added nutrients on coral reefs are more complicated than we thought. Experiments indicate that algal growth is not nutrient-limited on at least some reefs. There are even suggestions that added nutrients may sometimes be good for zooxanthellae and help corals grow faster. The overwhelming weight of evidence, though, is that eutrophication is damaging, especially when populations of algal grazers are decimated by overfishing or disease.

· ·

Section Review

1. **Explain** How do corals build reefs? Include in your answer organisms other than corals involved in building reefs.

2. **Discuss** coral sensitivity to water temperature, sediments, and pollution.

3. **Construct** a timeline to describe the major event in the history of coral reefs in Ka-neʻohe Bay in Hawaiʻi.

16.2 Kinds of Coral Reefs

VOCABULARY

fringing reefs

reef flat

reef slope

reef crest

lagoon

back-reef slope

fore-reef slope

sand cays

spur-and-groove formations

windward

leeward

DIVE IN!

Who's guarding this reef? *Pocillopora* and related corals in the Indo-West Pacific and tropical eastern Pacific have a symbiotic relationship with several species of "coral guard crabs" (*Trapezia*). In exchange for shelter and food from their coral hosts, the crabs defend the coral against predators such as the crown-of-thorn sea star. The crabs also help to remove sediments from the coral. Coral guard crabs usually live in male-female pairs, and each species has a characteristic bright color pattern. Several species of crab may live together peacefully on a host coral. The crabs have also been observed to cooperate with other coral symbionts, such as shrimp.

Main Idea

Marine scientists categorize coral reefs based on how they form and on their structure.

· ·

Key Questions

1. What are the three main types of coral reefs?
2. Due to the fact that they are close to land, which factors can affect health of fringing reefs?

Coral reefs are usually divided into three main categories: fringing reefs, barrier reefs, and atolls. Many reefs do not fit neatly into any particular category or fall between two categories. Still, the division of reefs into the three major types works well for the most part. The three main types of reefs are fringing reefs, barrier reefs, and atolls.

Fringing Reefs

Fringing reefs are the simplest and most common reefs. They develop along the shore throughout the tropics, wherever there is a hard bottom where coral larvae can settle. Rocky shorelines provide the best conditions for fringing reefs. Fringing reefs also grow on soft bottoms if there is even a small hard patch where corals can get a foothold. Once they get started, the corals create their own hard bottom and the reef slowly expands.

As their name implies, fringing reefs grow in a narrow band or fringe along the shore. Occurring close to land, they are especially vulnerable to sediment, freshwater runoff, and human disturbance. Under the right conditions, however, fringing reefs can be impressive. In fact, the longest reef system in the world (though not the one with the largest coral area) is not the famous Great Barrier Reef in Australia but a fringing reef that runs some 4,000 km along the coast of the Red Sea. Part of the reason this reef is so well developed is that the climate is dry and there are no rivers that bring in sediment and fresh water.

Fringing Reef Structure The typical structure of a fringing reef is shown in Figure 16.15. Depending on the place, the shore may be steep and rocky or have mangroves or a beach. The reef itself consists of an inner **reef flat** and an outer **reef slope**. The reef flat is the widest part of the reef. It is shallow, sometimes exposed at low tide (Fig. 16.16), and slopes very gently toward the sea. Being closest to land, it is the part of the reef most strongly affected by sediments and freshwater runoff. The bottom is primarily sand, mud, or coral rubble. There are some living corals, but neither as many colonies nor as many

(t)©Michael Huber, (b)©L. Newman & A. Flowers/Science Source

Figure 16.15 Typical structure of a fringing reef. Fringing reefs, like this one in the Bismarck Archipelago in the southwest Pacific (photo), can grow right up to the shore.

Figure 16.16 The upward growth of most reefs is limited by the tides. When extreme low tides occur, shallow areas like this reef flat on the Great Barrier Reef are exposed. If the corals are exposed for only a short time, they can survive, but they will die if exposed for too long. Photograph courtesy of the Great Barrier Reef Marine Park Authority.

species as on the reef slope. Seaweeds, seagrasses, and soft corals may also occupy the reef flat, sometimes in dense beds.

How Fringing Reefs Form The reef slope can be quite steep, even vertical. It is the part of the reef with the densest cover (Fig. 16.17, p. 495) and most species of coral, because the slope is not exposed at low tide and is away from shore and the effects of sediment and fresh water. Also, the waves that bathe the slope provide good circulation, bring in nutrients and zooplankton, and wash away fine sediments. The **reef crest** is the shallow upper edge of the reef slope. The crest usually has even more luxuriant coral growth than the rest of the reef slope. If there is intense wave action, however, the crest may consist of an algal ridge, with the richest coral growth just below. Because there is less light in deep water, the deepest part of the reef slope usually has less live coral and fewer coral species.

Figure 16.17 The elkhorn coral (*Acropora palmata*) is a dominant coral on the reef slopes of fringing reefs in the Caribbean and Florida. Its broad branches rise parallel to the surface to collect light. This colony suffers from a disease known as white-band disease.

Large amounts of sediment and coral rubble tumble down the reef slope and settle at the base. As this material builds up, reef organisms may begin to grow on it, depending on the water depth and other factors. Thus, the reef can grow outward as well as upward. Beyond the base of the slope, the bottom is usually fairly flat and composed of sand or mud. On many Caribbean reefs, turtle grass (*Thalassia testudinum*) dominates the bottom beyond the slope.

Barrier Reefs

The distinction between barrier reefs and fringing reefs is sometimes unclear because the two types grade into one another. Like fringing reefs, barrier reefs lie along the coast, but barrier reefs occur considerably farther from shore, occasionally 100 km or more. Barrier reefs are separated from the shore—which may also have a fringing reef—by a relatively deep **lagoon** (Fig. 16.18). Largely protected from waves and currents, the lagoon usually has a soft sediment bottom. Seagrass beds often grow in shallow parts of the lagoon. Scattered coral formations variously known as patch reefs, coral knolls, or pinnacles depending on their size and shape may grow up nearly to the surface.

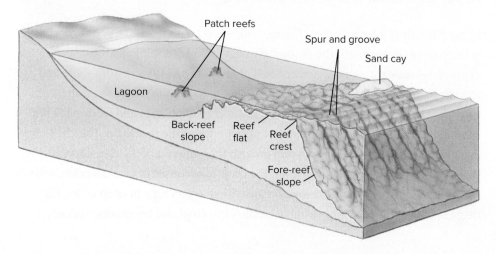

Figure 16.18 Typical structure of a barrier reef.

Courtesy of John C. Halas, Resource Manager with the Florida Keys National Marine Sanctuary

Barrier Reef Structure The barrier reef consists of a **back-reef slope**, a reef flat, and a **fore-reef slope**, which corresponds to the reef slope of a fringing reef and has a reef crest. The back-reef slope may be gentle or as steep as the fore-reef slope. It is protected from waves by the rest of the reef, but waves wash large amounts of sediment from the reef down the slope. As a result, coral growth is often not as vigorous on the back-reef slope as on the fore-reef slope. This isn't always true, though, and some back-reef slopes, especially gentle ones, have luxuriant coral growth (Fig. 16.19).

The reef flat, like that on fringing reefs, is a shallow, nearly flat platform. Sand and coral rubble patches are interspersed with seagrass or seaweed beds, soft corals, and patches of dense coral cover. Waves and currents may pile up sand to form small sand islands called **sand cays** or, in the United States, keys.

The richest coral growth is usually at the outer reef crest. There may be a well-developed algal ridge if there is strong wave action, with coral growth most luxuriant just below the crest. Exposed fore-reef areas often have a series of finger-like projections alternating with sand channels (Figs. 16.18 and 16.20). There is still debate about what causes these formations, known as **spur-and-groove formations** or buttresses. The wind, waves, or both are definitely involved, because spur-and-groove formations develop primarily on reef slopes that are exposed to consistent strong winds. These formations are found on atolls and some fringing reefs as well as barrier reefs.

Fore-reef slopes vary from relatively gentle to vertical. The steepness depends on the action of wind and waves, the amount of sediment flowing down the slope, the depth and nature of the bottom at the reef base, and other factors. As with other types of reefs, the abundance and diversity of corals on the fore-reef slope generally decrease with depth. Corals' growth form also changes down the slope. At the crest, under the pounding of the waves, the corals are mostly stout and compact; many are massive (see Fig. 16.7). Below the crest there is great variety in form. Whether they form branches, columns, or whorls, corals in this zone often grow vertically upward. This may be an adaptation for competition. Corals that grow upward like

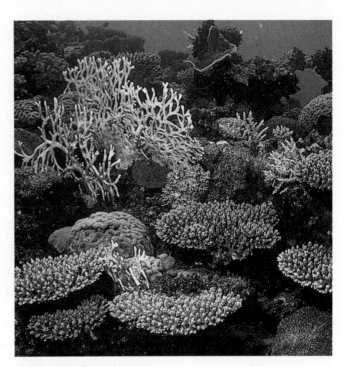

Figure 16.19 A rich back-reef slope on a Pacific barrier reef.

Figure 16.20 Spur-and-groove formation on Guam, one of the Mariana Islands in the western Pacific Ocean (also see Fig. 16.21 below).

skyscrapers rather than outward need less space to attach. They are also less likely to be shaded, and if they spread out at the top, can shade out neighboring corals and algae. Deeper on the reef slope, corals tend to grow in flat sheets to collect light.

Large Barrier Reefs The largest and most famous barrier reef is the Great Barrier Reef. It runs more than 2,000 km along the northeastern coast of Australia, varying in width between about 15 and 350 km and covering an area of over 225,000 km². Though not the longest reef in the world, it covers such a large area and is so complex and well developed that it is generally regarded as the largest reef structure. Actually, the Great Barrier Reef is not a single reef but a complex system of more than 2,500 reefs and associated lagoons, channels, islands, and sand cays (Fig. 16.21).

The largest barrier reef in the Caribbean lies off the coast of Belize, Central America. Other major barrier reefs include the Florida Reef Tract and barrier reefs associated with the islands of New Caledonia, New Guinea, and Fiji in the Pacific. There are many other smaller barrier reefs, especially in the Pacific. Like the Great Barrier Reef, these are usually not single reefs but complex systems of smaller reefs.

Figure 16.21 The Great Barrier Reef is a complicated system of thousands of small reefs, sand cays, and lagoons. Shown here are two reefs, one with a sand cay. The spur-and-groove formation (Fig. 16.20) is clearly visible along the outer margin of the barrier reef in the foreground..

Atolls

An atoll is a ring of reef, and often islands or sand cays, surrounding a central lagoon (Figs. 16.22 and 16.23). The vast majority of atolls occur in the Indo-West Pacific region, that is, the tropical Indian and western Pacific oceans. Atolls are rare in the Caribbean and the rest of the tropical Atlantic Ocean. Unlike fringing and barrier reefs, atolls can be found far from land, rising up from depths of thousands of meters. With practically no land around, there is no river-borne silt and little or no freshwater runoff. Bathed in clean blue ocean water, atolls display spectacular coral growth and breathtaking water clarity. They are a diver's dream.

Figure 16.22 A small atoll in the Republic of Belau (Palau), western Pacific.

Atoll Structure Atolls range in size from small rings less than a mile across to systems more than 50 km in diameter. The world's largest growing atoll, with an area of some 12,600 km², is the Great Chagos Bank in the Indian Ocean. Mostly submerged, it has only a few tiny islands and is now uninhabited, though in the past is supported a small plantation population and was

Figure 16.23 Typical structure of an atoll.

used by sailors. The largest inhabited atolls are Suvadiva, in the Maldive Islands of the Indian Ocean, and Kwajalein, one of the Marshall Islands in the central Pacific. These atolls cover areas of more than 1,200 km². Atolls can have dozens of islands and be home to thousands of people.

An atoll's reef flat is much like the reef flat on a fringing or barrier reef: flat and shallow, with low coral abundance and diversity. The fore-reef and back-reef slopes can now be thought of as outer and inner slopes, respectively, since they extend all the way around the ring-shaped atoll.

Atoll reef crests are strongly influenced by wind and waves. Because most atolls lie in the trade winds belt, winds come from a consistent direction. Consequently, the wind affects various parts of the atoll in different ways. Encrusting coralline algae, which can endure the constant pounding of waves, build a distinct algal ridge on the reef crest of the **windward** side of the atoll, the side that faces the prevailing wind. On the few Caribbean atolls, coralline algae may be replaced by especially wave-resistant corals. The algal ridge is less prominent or absent on the **leeward**, or sheltered, side of the atoll. Spur-and-groove formations, too, are better developed on the windward side.

The fore-reef, or outer, slope is vertical or nearly so, usually with ledges and overhangs. The rocky wall of the reef extends down to great depths, far beyond the limit of living corals themselves. The water may be hundreds, even thousands, of meters deep just a stone's throw from the reef. The lagoon, on the other hand, is relatively shallow, usually no more than about 60 m deep. The bottom of a lagoon is uneven, with depressions and coral pinnacles. Some pinnacles rise almost to the surface, where they may form "mini-atolls," small rings of coral within the lagoon.

IN CONTEXT

The trade winds, the steadiest winds on Earth, blow from latitudes of about 30° toward the Equator.

• 5.1, Atmospheric Circulation

How Atolls Form When European explorers discovered atolls, scientists were at a loss to explain them. They knew corals can only grow in shallow water, yet atolls grow right from the depths of the open ocean. Therefore, the atoll could not have grown up from the ocean floor. If the reefs grew on some kind of shallow structure that was already there, like a seamount, why is there no sign of it? The islands on atolls are simple sand cays that have been built by the accumulation of reef sediments and would not exist without the reef. They are *products* of the reef and could not provide the original foundation for reef growth. Finally, why do atolls always form rings?

The puzzle of atoll formation was first addressed by Charles Darwin in the mid-nineteenth century. Darwin is most famous, of course, for proposing the theory of evolution by natural selection, but his theory of atoll formation was another of his many important contributions to science.

Darwin reasoned that atolls are formed by reef growth on a subsiding island. The atoll gets its start when a deep-sea volcano erupts to build an island or seamount. Corals soon colonize the shores of the new island, and a fringing reef develops (Fig. 16.24a). As with most fringing reefs, coral growth is most vigorous at the outer edge of the reef. The inner reef is strongly affected by sediment and runoff from the island.

Surprisingly, Darwin's explanation of atoll formation was pretty much ignored for a century while scientists proposed various other hypotheses, none of which held up. That is the strength of the scientific method. Eventually, scientists found conclusive evidence that Darwin was right. Unlike other hypotheses for atoll formation, Darwin's hypothesis predicted that, below the thick calcium carbonate cap formed by the reef, there should be volcanic

Subsidence The slow sinking into the mantle of a part of the earth's crust that contains a landmass.

• 13.2, Rocky Shore Intertidal Communities

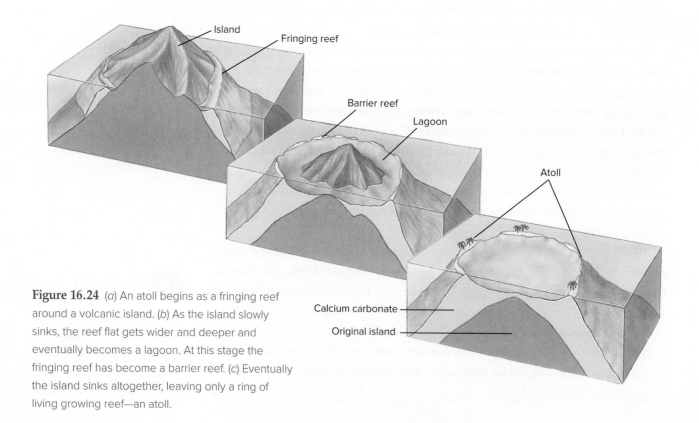

Figure 16.24 (*a*) An atoll begins as a fringing reef around a volcanic island. (*b*) As the island slowly sinks, the reef flat gets wider and deeper and eventually becomes a lagoon. At this stage the fringing reef has become a barrier reef. (*c*) Eventually the island sinks altogether, leaving only a ring of living growing reef—an atoll.

Caves in the Reef

Caves are unique ecosystems that are often found within coral reefs. Some are formed by wave action, others were formed through erosion thousands of years ago when sea levels were lower and reefs were exposed. Whether formed recently or in the geologic past, caves in coral reefs are home to organisms that are different from those on the reef just outside. Even at shallow depths, coral reef caves house organisms that closer resemble deep-sea coral communities because of the low light levels inside of the cave.

Cave corals Just like reefs have zones of different coral development, reef caves have zones as well. The entrance of caves have corals and invertebrates typical of the surrounding reef. There is enough light available for zooxanthellae to photosynthesize. Deeper within the caves light levels are reduced so corals living here are similar to deep-sea red and black corals that do not have zooxanthellae. These corals do not receive enough sunlight to support zooxanthellae. Instead, they depend entirely on their consumption of zooplankton to survive. Reef caves are typically very porous, so there is a large amount of exchange between water inside and outside the cave. Caves that have limited water exchange usually have less biodiversity than caves with a large amount of exchange.

Cave diversity While they might be less diverse than the surrounding reefs, coral caves nevertheless support an intriguing mix of organisms. Sponges and hydroids are common inside reef caves. Japanese biologists recently discovered two species of *Palythoa*, a colonial cnidarian related to corals that like corals harbor zooxanthellae. These new species, however, lack zooxanthellae and are adapted to life in the dar crevices of coral caves. Coral caves also provide refuge for nekton like nocturnal fishes, including sharks. In particular, whitetip reef sharks (*Triaenodon obesus*), which actively forage for food at night, spend the day resting in coral caves. These sharks are not terratorial, and it is common to find several sharing a cave. Coral caves tend to have higher dissolved oxygen levels and lower salnities than exposed reefs. Marine scientists will undoubtedly discover new species and new adaptations as they explore additional coral caves.

rock—the original island. In the 1950s the United States Geological Survey drilled several deep holes on Enewetak atoll in the Marshall Islands. These cores revealed exactly what Darwin predicted: volcanic rock far beneath the calcium carbonate of the reef. The thickness of the carbonate cap is impressive: The volcanic island underlying Enewetak is covered by more than 1,400 m of calcium carbonate!

There were, of course, a few details to be added to the picture, in particular the effects of changes in sea level. When sea level is low, atolls may be stranded above the surface. The reef dies and is eroded by wind and rain. If sea level rises rapidly, the atoll may be drowned, unable to grow in deep water. There are a number of such atolls, drowned and with only a few small living reefs, in the Indo-West Pacific.

Section Review

1. **Compare and contrast** fringing reefs and barrier reefs.
2. **Construct** a flowchart to show the steps involved in the formation of an atoll.

16.3 The Ecology of Coral Reefs

VOCABULARY

sweeper tentacles

sclerites

pre-settlement hypothesis

post-settlement hypothesis

microherbivores

facultative

obligate symbionts

Main Idea
Coral reefs are productive and diverse ecosystems where inhabitants interact with each other and with their environment in complex ways.

. .

Key Questions
1. For which factors do sessile coral reef organisms compete?

2. What are the many roles of symbiosis in coral reef communities?

Coral reefs are impressive to geologists, but to biologists they are simply awesome. They are easily the richest and most complex of all marine ecosystems. Thousands of species may live on a single reef. How do all these species live? How do they affect each other? What is their role in the reef ecosystem?

Our ability to answer these questions is surprisingly limited. This is partly because reefs are so complex. Just identifying the vast number of different organisms is hard enough; the task of figuring out what they all *do* is mind-boggling. Furthermore, until the last half century or so most marine biologists lived and worked in the temperate regions of the Northern Hemisphere, far from the nearest reef. As a result, relatively few biologists studied reefs. Tremendous progress is now being made, but there is still much to learn. The rest of this chapter summarizes the ecology of coral reefs and a few of the many important questions that remain.

The Trophic Structure of Coral Reefs
The tropical waters where coral reefs are found are usually nutrient-poor and therefore have very little phytoplankton and primary production. In these barren waters, coral reefs are oases of abundant life. How can such rich communities grow when the surrounding sea is so unproductive?

Part of the answer lies in the mutualistic relationship between corals and their zooxanthellae. We have already learned what the zooxanthellae do for the coral: They provide food and help make the calcium carbonate skeleton. In return, the zooxanthellae get a protected place to live and a steady supply of carbon dioxide and nutrients such as nitrogen and phosphorus. Most of the coral's nitrogen and phosphorus waste products are not released into the water. Instead, they are taken up and used as nutrients by the zooxanthellae. Using sunlight, the zooxanthellae incorporate the nutrients into organic compounds, which are passed on to the coral. When the coral breaks down the organic matter, the nutrients are released and the whole process begins again. The nutrients are recycled, so that far fewer new nutrients are needed than would otherwise be the case.

Nutrient Recycling Nutrient recycling occurs not just between corals and their zooxanthellae, but among all the members of the coral reef community. Sponges, sea squirts, giant clams, and other reef invertebrates have symbiotic algae or bacteria and recycle nutrients like corals do. Recycling also takes place outside the bodies of reef organisms. When fishes graze on seaweeds, for example, they excrete nitrogen, phosphorus, and other nutrients as waste. These nutrients are quickly taken up by other algae. Many corals provide shelter to schools of small fish. The fish leave the coral at night to feed and return during the day. The waste products of the fish can be an important source of nutrients and help the coral grow faster. In this way nutrients are cycled from the fish's prey to the fish to the coral. Nutrients pass through the community again and again in this cycle of feeding and excretion.

Nutrient Sources Coral reef communities use nutrients very efficiently as a result of recycling. The recycling isn't perfect, however, and some nutrients are lost, carried away by the currents. Thus, the reef still needs a supply of new nutrients. Recycling alone is not enough to account for the high productivity of reefs.

The reef provides some of its own nutrients. Coral reefs have among the highest rates of nitrogen fixation of any natural ecosystem. The main nitrogen fixers are cyanobacteria, especially a free-living one called *Calothrix* and another group that lives symbiotically in sponges. There is evidence that corals, too, have symbionts that fix nitrogen, providing nutrients for the zooxanthellae. Just what the symbionts are is not known. Because of this nitrogen fixation, nitrogen probably does not limit most coral reef communities, though not all reef biologists agree.

Ocean currents bring in additional nitrogen and, more importantly, phosphorus and other nutrients that are not produced on the reef. Corals, bacteria, algae, and other organisms absorb nutrients directly from the water. Even though the water is nutrient-poor, if enough water washes over the reef the nutrients add up. In business terms, this is a low-margin, high-volume proposition. More importantly, the water carries zooplankton, a rich nutrient source. When zooplankton are captured by the "wall of mouths," the nutrients in the zooplankton are passed on to the reef community. In fact, many biologists think that corals eat zooplankton to get nutrients for their zooxanthellae as much as to feed themselves. Animals like seabirds, dolphins, and large fishes that spend part of their time around reefs but venture out into the open ocean to feed also transfer nutrients from oceanic waters to the reef community.

Primary Production The production and efficient use of nutrients by coral reef communities result in high primary productivity. This is reflected in the richness of the community. Scientists aren't sure, however, just how much primary production there is on coral reefs, or which particular organisms are the most important producers. There is no doubt that zooxanthellae are very important, but because they live inside corals, it is hard to measure exactly how much organic matter they produce. For a time it was thought that very few animals eat coral, since there is so little live tissue on a coral colony. It was therefore believed that, even though zooxanthellae produce a lot of organic

DIVE IN!

How do parrotfishes improve reef health? Although it may not seem like a big job, parrotfishes and other grazers help keep coral reefs from being overgrown by algae. Many reefs in the Caribbean have suffered due to the overfishing of parrotfishes. Biologists have found that in areas where parrotfishes are protected, such as the Flower Garden Banks National Marine Sanctuary in the Gulf of Mexico, coral reefs are healthier than in non-protected areas. Other countries have restricted or banned fish trapping or spearfishing to protect healthy reefs. The Global Coral Reef Monitoring Network, the International Union for the Conservation of Nature, and the United Nations Environmental Programme stated in 2012 that the overfishing of parrotfishes and other algae grazers has caused more harm to Caribbean reefs than climate change.

IN CONTEXT

Nitrogen Fixation The conversion of nitrogen gas (N_2) into nitrogen compounds that can be used by primary producers as nutrients.
• 7.2, Prokaryotes

Figure 16.25 A number of animals eat coral directly, and many others feed on the mucus that corals produce or on coral eggs and larvae. The primary production of coral zooxanthellae is thus passed on to coral feeders, and then to the animals that eat them.

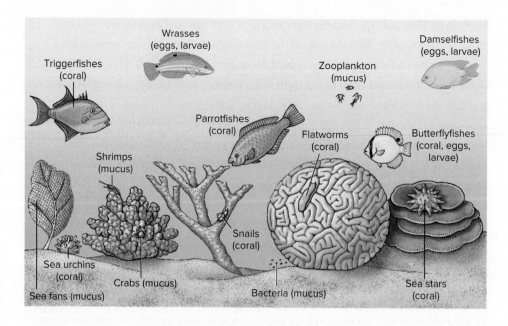

Wrasses (eggs, larvae)

Triggerfishes (coral)

Zooplankton (mucus)

Damselfishes (eggs, larvae)

Parrotfishes (coral)

Flatworms (coral)

Butterflyfishes (coral, eggs, larvae)

Shrimps (mucus)

Snails (coral)

Sea urchins (coral)

Crabs (mucus)

Bacteria (mucus)

Sea stars (coral)

Sea fans (mucus)

Figure 16.26 A generalized coral reef food web. Coral reefs are extremely diverse, and most components include many organisms in addition to those listed here.

matter, most of it is consumed by the coral and not much passed on to the rest of the community. As biologists looked closer, however, they found more and more animals that eat corals or their products (Fig. 16.25). Primary production by coral zooxanthellae therefore may be important not only to corals but to the community at large. Exactly how much production corals and their zooxanthellae contribute is still unknown.

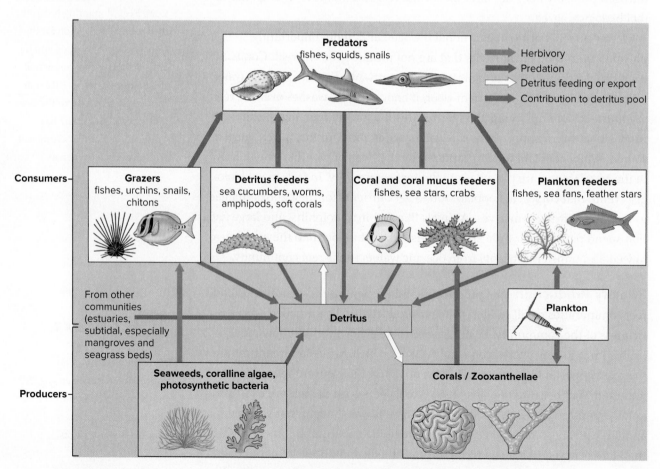

Predators
fishes, squids, snails

Herbivory
Predation
Detritus feeding or export
Contribution to detritus pool

Consumers

Grazers
fishes, urchins, snails, chitons

Detritus feeders
sea cucumbers, worms, amphipods, soft corals

Coral and coral mucus feeders
fishes, sea stars, crabs

Plankton feeders
fishes, sea fans, feather stars

From other communities (estuaries, subtidal, especially mangroves and seagrass beds)

Detritus

Plankton

Seaweeds, coralline algae, photosynthetic bacteria

Corals / Zooxanthellae

Producers

Seaweeds are also important primary producers on the reef (Fig. 16.26), especially small, fleshy or filamentous types that are called turf algae because they often grow in a short, thick turf on the reef flat. A great many fishes, sea urchins, snails, and other animals graze on these seaweeds. Turf algae may perform more photosynthesis on the reef than zooxanthellae, but biologists are not sure. Cyanobacteria and coralline algae are also primary producers on coral reefs. They probably account for less primary production than do zooxanthellae and turf algae.

Coral Reef Communities

With so many species on coral reefs, the interactions among them are exceedingly complex. What is known about these interactions is fascinating; what remains to be learned will be even more so.

Competition Space is at a premium on coral reefs, as it is in the rocky intertidal. Corals, seaweeds, and many others need to attach to a hard surface. Corals and seaweeds need not just space but space in the sunlight. The reef is crowded, and most of the space is taken. As a result the sessile organisms, those that stay in one place, must compete for space.

Corals Corals compete for space in different ways. The fast-growing ones tend to grow upward and then branch out, cutting their neighbors off from the light. Other corals take a more direct approach and actually attack their neighbors (Fig. 16.27). Some use their mesenterial filaments for this. When they contact another coral, they extrude the filaments and digest away the tissue of the other coral. Still other corals develop special long tentacles called sweeper tentacles that are loaded with nematocysts and sting neighboring colonies. Corals differ in their aggressive abilities. The most aggressive corals tend to be slow-growing, massive types, whereas the less aggressive forms are usually fast-growing, upright, and branching. Both strategies have their advantages, and both kinds of corals thrive on the reef.

Corals compete for space and light not only with each other, but also with seaweeds and sessile invertebrates. Like corals, encrusting algae have to produce a calcium carbonate skeleton and therefore grow relatively slowly. They tend to be most abundant in places where corals don't do well because of sedimentation, wave action, or predation.

Under the right conditions seaweeds, except for encrusting forms, can grow much faster than either corals or encrusting algae. Even with the nitrogen fixation and nutrient cycling that occur on reefs, seaweeds are probably somewhat nutrient-limited most of the time. Hence, they usually grow fairly slowly. The reef also has an abundance of hungry grazers that eat seaweeds. The combination of nutrient limitation and grazing keeps seaweeds in check. If the nutrient levels increase or the grazers are removed, seaweeds may rapidly take over, overgrowing and choking out corals and other organisms.

Figure 16.27 The pink band separating the brown (*Porites lutea*) and blue (*Mycedium elephantotus*) corals is a dead zone where the blue coral has killed the brown one in the process of overgrowing it. The width of the pink band corresponds to the length of the blue coral's tentacles. To the upper left of the brown coral is a soft coral (*Sarcophyton*), which may be attacking the brown coral by releasing poison. The brown coral seems to be stuck between a rock and a soft place!

©Michael Huber

"Must Have Been Something I Ate"

Ahhhhh! You're relaxing on a warm tropical night after a delicious fish dinner. The palm trees sway, a warm tropical breeze blows, and suddenly your bowels begin to churn. Before long your lips burn and tingle, while your hands and feet feel like pins and needles. You go into the kitchen for a drink, but the cold water feels warm. The floor and the cold compress you put on your forehead are also warm to the touch; the sensations of hot and cold are reversed. As you stagger to the bathroom, your arms and legs are heavy and weak. Now feeling really sick, you mumble, "Must have been something I ate."

So what did you eat? What you have is a case of ciguatera—tropical fish poisoning. You're not alone. Tens or perhaps hundreds of thousands of people get ciguatera every year. And relax, you probably won't die, though you might be a little sick (if you're unlucky, *very* sick) for months or even years. Although there is no cure, various folk remedies and the intravenous administration of an inexpensive sugar, mannitol, can sometimes alleviate symptoms. Although fish dinners are the most common cause, you can also get ciguatera from eating molluscs, and perhaps sea urchins.

So what *is* ciguatera, anyway? Ciguatera is caused by toxic dinoflagellates that live on the surface of seaweeds. Herbivorous fishes like surgeonfishes and parrotfishes ingest the dinoflagellates, and their toxins, along with the seaweeds. Predatory fishes pick up the toxin when they eat the herbivores. An intense, localized bloom of the dinoflagellates can result in herbivorous fishes feeding in that spot becoming toxic, while fishes a short distance away are unaffected. If the dinoflagellates are widespread at relatively low levels, herbivorous fishes may remain safe, while predators that range over larger distances and eat many fish steadily accumulate toxins, known as bioaccumulation, and become poisonous.

Staying safe It is hard to predict which fish are safe to eat, although certain species are known to have a higher risk than others. Also, risky species vary from place to place; local knowledge is essential to avoid poisoning. It is possible to tell if a fish is affected—all you need is a laboratory. The toxins can be identified through chemical analysis, or by exposing test organisms such as mice or mosquito larvae to extracts. Test sticks that change color when rubbed into the flesh of toxic fish have been developed but are not very reliable, sometimes indicating that toxic fish are safe and vice versa. This may be because there are at least 20 toxins involved, not just one, and it is hard to develop a test to reliably pick them all up in any combination. For now, at least, the best strategy is to avoid eating fish species that are known to be risky.

Soft corals Soft corals are also important competitors for space on reefs (Fig. 16.28), and in some places they make up almost half the living tissue. Like most seaweeds, soft corals lack a calcium carbonate skeleton and are able to grow faster than hard corals. Some soft corals contain sharp little calcium carbonate needles, or **sclerites**, that discourage predation. Many of them also contain chemicals that are toxic or taste bad to predators. Because of these defense mechanisms, only a few specialized predators eat soft corals. The defensive chemicals can also be released into the water, where they kill hard corals that come too close (see Fig. 16.27). Another competitive advantage enjoyed by some soft corals is that they are not completely sessile. Though they stay in one place most of the time, they can slowly move about. This helps them invade and occupy available space on the reef.

With all these competitive weapons at their disposal, why don't soft corals take over? Not much is known about how soft corals compete with reef-building corals and other reef organisms or what determines the winner. Many soft corals appear to have shorter lives than reef-building corals, although some live

for decades. They are also more easily torn away by storm waves. They have symbiotic zooxanthellae like reef-building corals but are much less efficient photosynthesizers. They also seem to depend on favorable physical conditions. It is not known how these factors interact to determine when and where soft corals successfully compete for space on the reef.

Sponges Like soft corals, sponges often have needle-like structures, spicules, and nasty chemicals that protect them from predators. They can be important users of space on reefs, much more so in the Caribbean than in the Pacific and Indian oceans. Part of the reason for this seems to be that there are fewer species of coral in the Caribbean than in the Indo-West Pacific, a result of geological history. During the most recent series of ice ages, the sea surface was cooler. Corals survived in the heart of the Indo-West Pacific region, around Indonesia and New Guinea, but many species became extinct in other parts of the ocean. When the ice age ended, corals spread out again across the Pacific. The Isthmus of Panamá emerged in the meantime, however, and blocked the dispersal of corals back into the Caribbean. It is thought that the Caribbean contains only those coral species that managed to survive the ice ages there. Sponges are even more abundant in the Caribbean now than in the past, however, because Caribbean corals have been hard-hit by pollution, disease, and bleaching, making way for the sponges.

Figure 16.28 Soft corals can form dense patches on coral reefs, as on this reef in Papua New Guinea.

Figure 16.29 Diving on a coral reef is like taking a swim in a tropical fish aquarium—brightly colored fishes seem to be everywhere. The competitive relationships among the various species of fishes are poorly understood.

Fishes Coral reef fishes are another group in which competition may be important. Along with corals, fishes are probably the most conspicuous and abundant animals on the reef (Fig. 16.29). Many of these fishes share similar diets; for example, many species eat coral, many graze on algae, and many are carnivores. Different species of fishes of the same feeding type at least potentially compete with each other.

There is debate about how competition affects coral reef fishes. One hypothesis is that competition is relatively unimportant, and that the abundances of individual species are determined mainly by how many of their larvae settle out

from the plankton. With favorable currents and other conditions, many larvae of a given species will settle out and the species becomes abundant. If larval supply is low, for example if currents carry larvae away from the reef, the species becomes rare. This is called a pre-settlement hypothesis because it holds that the nature of reef fish communities is determined by the availability of different species' larvae before they settle.

An alternative post-settlement hypothesis is that most species have plenty of settling larvae. The abundant species are those that compete successfully for space, food, and other resources after the larvae settle, that is, as juveniles and adults. This hypothesis holds that so many different fish species can live on the reef because each does something a little different from other species, thereby avoiding competition and the risk of competitive exclusion. In other words, each species has its own particular ecological niche. According to this view, the structure of the fish community on a reef is determined by the range of resources available on that reef, and the fish communities of reefs vary because different reefs offer different resources.

It is still not clear whether the supply of settling larvae or post-settlement competition is more important in structuring reef fish communities. Their relative importance probably varies from species to species and from reef to reef, and other factors such as predation and natural disturbances are also important.

Predation on Corals As in other communities, predation and grazing are important in structuring coral reef communities. A variety of animals eat corals, but instead of killing the coral and eating it entirely, most coral predators eat individual polyps or bite off pieces here and there (Fig. 16.30). The coral colony as a whole survives and can grow back the portion that was eaten. In this respect, coral predation is similar to grazing by herbivores.

Coral predation affects both the number and type of corals that live on a reef, as well as how fast the reef as a whole grows. In Ka-neʻohe Bay, for example, a butterflyfish (*Chaetodon unimaculatus*) slows the growth of a particular

Figure 16.30 The chevron butterflyfish (*Chaetodon trifascialis*) is one of many reef animals that feed on corals without killing the entire coral colony. Its mouth is adapted for nipping off individual coral polyps.

Borut Furlan/WaterFrame/Getty Images

coral (*Montipora verrucosa*) that it likes to eat. When the coral is protected from the fish by a cage, it grows much faster. If it were not for the fish, this fast-growing coral would probably dominate other corals in the bay. Coral-eating snails have similar effects on some reefs.

The crown-of-thorns sea star Another example of the effect of coral predators is the crown-of-thorns sea star (*Acanthaster planci*; Fig. 16.31). The crown-of-thorns feeds by pushing its stomach out through the mouth, covering all or part of the coral colony with the stomach, and digesting away the live coral tissue. The crown-of-thorns has distinct preferences for certain types of coral and avoids others—some corals must taste bad. Other corals harbor symbiotic crabs, shrimps, and fishes that discourage the sea stars by pinching and biting their tube feet.

The crown-of-thorns has had a major impact on some reefs. Beginning in the late 1950s, people began to notice large numbers, sometimes thousands, of the sea star on reefs scattered across the Pacific (Fig. 16.32). The sea stars in these large aggregations move en masse across the reef, consuming almost every coral in their path. Reefs recover in 10 to 15 years, though it may take longer for slow-growing coral species to grow back.

The first response to the problem was panic. Coral reefs are valuable resources: They support fisheries, tourism, and recreation and protect coastlines from tsunamis, storms, and erosion. With the crown-of-thorns apparently threatening reefs, people decided to take action and control the sea star. The first attempt backfired. With limited knowledge of the animal's biology, some people cut the sea stars into pieces and dumped them back in the sea. Because sea stars can regenerate, the pieces grew into new sea stars!

Figure 16.31 The crown-of-thorns sea star (*Acanthaster planci*) is an important and controversial coral predator.

Tube Feet Water-filled tubes possessed only by echinoderms, many of which end in a sucker and can be extended and contracted to grip things and to move around.
• 9.5 Echinoderms

Figure 16.32 An outbreak of crown-of-thorns sea stars (*Acanthaster planci*) in the Hawaiian Islands. The white colony of *Pocillopora* shows the bare calcium carbonate skeleton left behind after being fed upon by the sea star.

(t)©Charlie Arneson, (b)NHPA/Franco Banfi

More sophisticated methods, such as poisoning the sea stars, were tried, but these methods were time consuming and expensive, did not work that well, and sometimes did more harm than good. Fortunately, the outbreaks mysteriously went away by themselves. Did the sea stars starve? Did they move away? No one knows. Crown-of-thorns outbreaks are still appearing, and disappearing, without explanation.

Outbreak causes The debate over what causes these outbreaks has been controversial and sometimes emotional. At first it seemed obvious that the outbreaks must be unnatural, with humans to blame. After all, the plagues had never occurred before. Or had they? People had only begun using scuba a short time before the sea star plagues were noticed. Even if the plagues had been occurring for a long time, there were no scientists around to see them. Geologists found fossil evidence of crown-of-thorns outbreaks dating back thousands of years, and in some places there are old stories and other historical evidence for past outbreaks. The outbreaks, then, may be a natural part of the reef ecosystem. A leading hypothesis is that in unusually wet years river runoff naturally brings more nutrients into the sea than normal. According to this hypothesis, the extra nutrients increase the growth of phytoplankton that are food for sea star larvae. It may also be that periodic population explosions are a natural part of the sea star's biology.

The evidence for past outbreaks of the crown-of-thorns, however, is hotly disputed. Some scientists are convinced that the plagues result from some human activity that has altered the ecological balance of reefs. Even if outbreaks did occur in the past, they seem to be happening more often. If occasional peaks in nutrient inputs cause natural outbreaks, nutrients from fertilizers, sewage, and other human sources may be making them worse. Another hypothesis is that fishers have caught too many fishes that eat young sea stars, allowing more to survive to adulthood. Some biologists have suggested that plagues occur because shell collectors have removed the triton shell, a large snail that preys on adult crown-of-thorns. Others argue that even without shell collectors the triton shell has always been naturally rare and unlikely to control sea star populations. They also point out that crown-of-thorns outbreaks continue to occur in areas where collecting triton shells has been banned for years.

The question of crown-of-thorns outbreaks has practical implications for the management and protection of reefs. If the outbreaks are caused by humans and threaten reefs, then we should probably try to do something to stop them. On the other hand, the "plagues" may be a natural, potentially important, part of the ecosystem. We might do more harm than good by interfering in a system we don't understand. The managers of the Great Barrier Reef have taken a middle road. They have developed an effective method to kill sea stars by injecting them with a poison. They use the poison, however, only if a crown-of-thorns outbreak is threatening a particularly valuable part of the reef, such as a popular dive spot or a scientific research site. Otherwise, they allow

IN CONTEXT

Trophic Cascade A process in which an effect on one species flows through an ecosystem because of indirect interactions via the food web.

- 12.1, The Organization of Communities

the outbreak to run its course. They are re-thinking this policy, however, as crown-of-thorns outbreaks have become unusually bad in recent years.

Grazing Grazing on algae by herbivores is at least as important in coral reef ecosystems as is predation on corals. Many fishes, especially surgeon-fishes, parrotfishes, and damselfishes, graze intensively on reefs. Among the invertebrates, sea urchins are especially important. Many **microherbivores**, small invertebrates like snails, chitons, crustaceans, and polychaete worms, also eat algae.

Many seaweeds grow rapidly and can potentially outcompete and overgrow corals. Under natural conditions they are common on the reef, but are kept in check by grazers and nutrient limitation. Caging experiments have been used to demonstrate the importance of grazers. For example, seaweeds are abundant on sand flats next to many Caribbean reefs but relatively scarce on the reef itself. To test the hypothesis that reef fishes were responsible for this, biologists transplanted seaweeds from the sand flat to the reef. If left unprotected, they were soon eaten by fishes. When protected by cages, they grew even faster than on the sand flat! The seaweeds are perfectly capable of living on the reef, therefore, but are rare there because they get eaten. Caging experiments on the Great Barrier Reef have had similar results.

If grazers are removed, seaweeds can flourish and take over space from corals and other organisms. In many parts of the Caribbean, for example, grazing reef fishes have become less common because of fishing. When this happened another important grazer, a sea urchin (*Diadema antillarum*; Fig. 16.33), became more common. The urchin apparently benefited from the reduced competition. For years, the urchin seems to have picked up the slack for the fishes, and seaweed populations remained more or less stable. In 1983, however, a disease wiped out populations of the urchin over much of the Caribbean. Seaweeds, released from grazing pressure, became much more abundant on many reefs, at the expense of corals. In Jamaica they took over many reefs almost completely, and the formerly rich coral reefs are now more seaweed bed than coral reef.

There is growing evidence that fishing can reduce fish grazing on reefs—even if the grazers themselves aren't fished—through a trophic cascade. On many reefs, sharks, the top fish predators, have been severely depleted by fishing because they reproduce slowly and are highly targeted, mostly for their valuable fins. Without shark preda-tion, large predatory fishes such as groupers become more abundant, and in turn eat smaller grazing fishes. It is not yet clear how wide-spread this phenomenon is, because the predatory fishes are usually targeted as well as sharks. It is probably most common in remote areas where there isn't a lot of local fishing pressure but fishing vessels targeting sharks come in from far away.

Such changes are becoming more and more common as fishing pressure on reefs escalates because of increasing populations and demand. There is often a double whammy. Coastal development

Figure 16.33 The long-spined black sea urchin (*Diadema antillarum*) is one of the most important grazers on Caribbean reefs. Closely related species are found on reefs in other parts of the world.

Reef Check

So you want to try your hand at marine science? Have we got a deal for you! Scientists are increasingly concerned about the future of the world's coral reefs, which need extensive monitoring. But there are tens of thousands of reefs around the world, many in remote locations. There are only so many reef biologists, and like most scientists, they are consistently short of research funding. How can a limited number of scientists, with limited resources, monitor the global status of coral reefs?

Since 1996, the Reef Check program has provided part of the answer by enlisting the services of an enthusiastic volunteer "army." Every year millions of recreational divers journey to the world's coral reefs—the more remote the better—to explore and experience their wonders. Most divers lack a formal scientific background, but are keenly interested in marine life, know quite a bit about it, and want to help protect it. Best of all (from a cash-strapped scientist's point of view), they pay for their own travel, food, and lodging, and even bring their own equipment!

Citizens as scientists To take advantage of this untapped resource, Reef Check designed methods that are simple enough to be used by citizen scientists, and still produce reliable, scientifically valuable information for assessing the condition of reefs. After a short training session, Reef Check volunteers don their dive gear to survey a site. After recording a basic site description based upon a predetermined set of questions, they stretch a measuring tape at a given depth along the bottom of the sea floor. Then they record the type of bottom (live coral, dead coral, algae, etc.) at specified points under the tape and count designated "indicator organisms." The indicator organisms are selected to be easy to identify and distinguish from other species by sight, without professional training. They are a good gauge of human impact on the reef. For example, population declines of popular game species, such as spiny lobsters and groupers, are a good indicator of fishing pressure since they are preferentially caught by fishers.

What did they find? Reef Check provided the first-ever global snapshot of the state of coral reefs. Unfortunately, it confirmed the widespread degradation of reefs. Reef Check surveys since have indicated that most reefs have been negatively impacted by humans. The research also suggests that, given wise use and management, reefs are able to recover. In addition to its regular surveys Reef Check now helps support improvements in the management and protection of reefs, and recently extended its approach to rocky temperate reefs on the California coast. So if you're a diver looking for a holiday that's a little different, and would like to get a taste of being a marine biologist while helping protect coral reefs and having a lot of fun in the process, check out ReefCheck.org.

For more information, explore the links provided on the Marine Biology *Online Learning Center.*

often leads not only to more intensive fishing, but also to greater release of nutrients from sewage and agricultural fertilizer. Thus, seaweed growth may be artificially enhanced by eutrophication at the same time that the grazers that keep the seaweeds under control are removed. This is one of the main global threats to coral reefs.

In addition to controlling how *much* algae there is, grazers affect which particular *types* of seaweed live on the reef and where. Coralline algae, for example, are abundant because the calcium carbonate in their tissues discourages grazers. Other seaweeds produce noxious chemicals that are poisonous or taste bad; these seaweeds also tend to be abundant. Seaweeds that lack such defenses are most heavily grazed and thus tend to be rare. Even so, they generally grow rapidly and are an important food source.

Damselfishes provide some interesting examples of the effects of grazing on reefs. Many damselfishes graze on seaweeds inside territories that they vigorously defend, chasing away other fishes. Many such damselfishes actually "farm" their territories. They weed out unpalatable algae, pulling them up and carrying them outside the territory. What is left in the territory is a dense mat of tasty seaweeds, usually fine, filamentous types. Protected by the damselfish, these algae grow very rapidly and outcompete corals and coralline algae. Outside the territory, parrotfishes and surgeonfishes gobble up the algae, clearing space for other organisms. Thus, the community inside the territory is very different from that outside. One interesting point is that cyanobacteria, which are nitrogen fixers, are more common inside damselfish territories than outside. Thus, damselfishes may indirectly have an important role in the nutrient balance of the reef.

Symbiosis Among the vast number of species that live on coral reefs, many have evolved special symbiotic relationships. There are far too many instances of symbiosis on reefs to describe here. In fact, coral reefs probably have more different symbiotic relationships than any habitat on Earth. The few examples discussed here will give you some idea of how fascinating these relationships are.

Photosynthetic symbionts We have already seen how mutualism between corals and their zooxanthellae is the essential feature of reef formation. Many other organisms also have photosynthetic symbionts. Sea anemones, snails, and giant clams all harbor zooxanthellae. The "deal" between the two partners is the same as in corals: The zooxanthellae get nutrients and a safe place to live, and the host gets food. Giant clams grow so large (Fig. 16.34) because their zooxanthellae provide a constant food supply.

There are primary producers other than zooxanthellae that live inside reef animals. As mentioned previously, some sponges have cyanobacteria that fix nitrogen in addition to performing photosynthesis. Certain sea squirts house a photosynthetic bacterium called *Prochloron*. *Prochloron* is especially interesting because it may be similar to the symbiotic organisms that eventually became the chloroplasts of plants.

Protective symbionts Another important example of mutualism on the reef is the relationship between corals and the crabs, shrimps, and fishes that help protect them from the crown-of-thorns sea star and other predators. Most corals host a number of **facultative** and **obligate symbionts**, especially crustaceans. Some of the obligate symbionts are parasites and harm the coral, some are commensals and apparently have no effect, and some are mutualists that benefit the coral. It is often hard to tell which is which, because for most symbionts the nature of the relationship between coral and symbiont is not well understood. The crabs that protect their coral from predators, for instance, were once thought to be parasites.

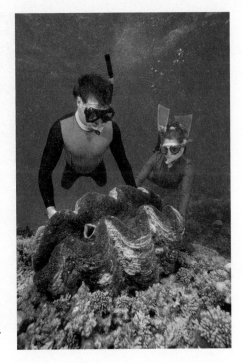

Figure 16.34 Symbiotic zooxanthellae give the upper surface of the giant clam (*Tridacna gigas*) its blue and green colors, and provide food that allows the clam to get so big. The clam also filter-feeds. Stories of people getting trapped in them and drowning are myths.

Figure 16.35 Orange-fin anemonefish (*Amphiprion chrysopterus*) and their sea anemone host.

Anemonefishes (*Amphiprion*; Fig. 16.35) have an interesting relationship with several kinds of sea anemones. The anemones inhabited by anemonefishes have a powerful sting and are capable of killing the fish. The fish have a protective mucus, however, that keeps the anemone from stinging. It is not known whether mucus is produced by the fish themselves, by the anemone, or both. When anemonefish are newly introduced to an anemone, they typically rub against the anemone's tentacles. They may be coating themselves with the anemone's mucus. Anemones have no eyes and no brain, but lots of tentacles. If they simply stung everything they touched they would sting themselves. The anemone recognizes itself by the "taste" of its own mucus, and the anemonefishes may be taking advantage of this. On the other hand, the fishes' own mucus may protect them from the anemone, and the fish sometimes safely enter an anemone without rubbing on the tentacles first.

Anemonefishes are protected by the anemone's stinging tentacles and brood their eggs under the anemone. The anemone also benefits. Anemonefishes drive away other fishes that eat anemones, and may benefit their hosts nutritionally. Anemones with anemonefish survive longer, grow faster, and reproduce by asexual division more often than anemones without anenomefish.

Section Review

1. **Explain** How does nutrient recycling occur within a coral reef community? Why is this of particular importance in coral reefs?

2. **Identify** The four main types of interactions among organisms in coral reef communities, giving an example of each.

Chapter 16 Review

REVIEW QUESTIONS

Multiple Choice

1. Corals obtain food using all of the following except:
 a. mutualistic relationship with zooxanthellae.
 b. capturing zooplankton and detritus.
 c. absorbing DOM from water.
 d. scavenging the bottom for dead animals.

2. When corals expel zooxanthellae it is called
 a. bleaching.
 b. eutrophication.
 c. nitrogen fixation.
 d. subsiding.

3. Which is the outer, steep margin of a coral reef?
 a. reef crest
 b. reef flat
 c. reef slope
 d. lagoon

4. Which are the most important primary producers in a coral reef?
 a. diatoms and seaweeds
 b. dinoflagellates and seaweeds
 c. zooxanthellae and seagrasses
 d. zooxanthellae and algae

Short Answer

5. What factors affect growth on the fore-reef slope of a barrier reef? What adaptations do corals have to deal with these factors?

6. Give an example of a trophic cascade in a coral reef community.

7. Which is a leading hypothesis that explains crown-of-thorns outbreaks? Why is it important to understand why these outbreaks occur?

Critical thinking

8. What factors might account for the fact that the vast majority of atolls occur in the Indian and Pacific oceans and that atolls are rare in the Atlantic?

DATA ANALYSIS LAB

Are Invasive Lionfishes a Threat to Caribbean Reefs?

Lionfishes (*Pterois*) are invasive and have few natural predators on Caribbean coral reefs. Biologists are concerned about this invasive species and the effects of its presence on Caribbean reefs.

Data and Observations

The graphs show changes in abundance of lionfishes and other fishes on nine coral reefs in the Bahamas.

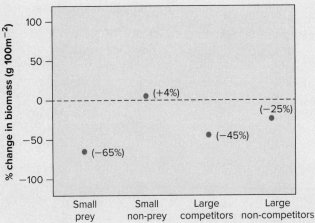

Think Critically

1. Describe the trends in the abundance of lionfishes and other fishes from 2004 to 2010.

2. Predict what will happen to the biodiversity of reefs due to the introduction of the lionfishes if nothing is done to control their populations.

Data obtained from: Green SJ, Akins JL, Maljković A, Côté IM (2012) Invasive Lionfish Drive Atlantic Coral Reef Fish Declines. PLoS ONE 7(3)

Chapter 17

Life Near the Surface

THEME Patterns

Pelagic organisms are adapted in many ways to the unique conditions of life in the water column.

BIG IDEA

The epipelagic consists of a huge volume of water and its inhabitants face unique challenges to life.

About The Photo A Portuguese man o' war (*Physalia physalis*) drifts at the surface of the ocean. Part of its body, a float, extends above the water.

Stephen Frink/The Image Bank/Getty Images

17.1 The Organisms of the Epipelagic

Main Idea
Plankton and nekton are the two overarching categories of organisms found in the epipelagic.

Key Questions
1. Name and describe the two main divisions of the epipelagic.
2. What is the difference between plankton and nekton?

"The ocean," for most of us, conjures up images of beaches and cliffs, breaking surf, or quiet bays. Such familiar nearshore waters, however, make up only a small fraction of the world ocean. The rest is the vast open sea, the pelagic realm. Though distant and unfamiliar, the open ocean affects us all. It regulates our climate and atmosphere, and provides food and other resources. The pelagic realm holds practically all the liquid water on Earth, the water planet.

The Epipelagic Environment

The pelagic environment is the water column itself, away from the bottom or the shore. Pelagic organisms live suspended in their liquid medium. With few exceptions, the pelagic lacks the solid physical structure provided either by the bottom and other geological features or by large organisms like corals and kelps. There is no place for attachment, no bottom for burrowing, nothing to hide behind. Imagine spending your life floating in the air, never touching the ground. This is what life is like for pelagic organisms.

Chapter 17 deals with the surface layers of the pelagic environment: the epipelagic, or upper pelagic, realm. The epipelagic is often defined as the zone from the surface down to 200 m depth. Being the shallowest part of the pelagic realm, the epipelagic is the warmest, and, of course, the best lit. The epipelagic is therefore much the same as the **photic zone**, the zone from the surface to the depth where there is no longer enough light to support photosynthesis. The depth of the photic zone varies, depending on water clarity and the amount of sunlight.

The epipelagic is divided into two main components. Epipelagic waters that lie over the continental shelf are referred to as neritic. The neritic environment is only a small part of the epipelagic, but it is important to humans because it is near the shore and supports most of the world's marine fisheries. Surface waters beyond the continental shelf are known as the oceanic part of the epipelagic.

Like nearly all ecosystems, the pelagic realm is fueled by solar energy captured in photosynthesis. Epipelagic ecosystems differ from many shallow-water ones in that nearly all the primary production takes place within the epipelagic system itself. Coastal ecosystems often receive large amounts of food from elsewhere. The intertidal zone, for example, gets plankton and drifting seaweeds from

DIVE IN!

What is a continuous plankton recorder (CPR)? CPRs are devices that filter plankton from the water for marine scientists to analyze in the lab. They can be towed behind boats, allowing researchers to efficiently collect plankton data over large areas of the open ocean. Understanding the biodiversity within the plankton in the pelagic environment is an important baseline for many ecological studies, as plankton are the base of open ocean food webs. Recently, marine ecologists have been working with toxicologists to use CPRs to study microplastics, tiny particles of plastic that are becoming distressingly common in the oceans. Toxicologists are testing the plastic for surface contaminants so that ecologists can better understand the ways in which plastic debris can impact marine ecosystems.

offshore, and rivers carry organic material into estuaries. The pelagic realm, far from the shore and bottom, gets little external input of organic matter.

The epipelagic does supply food to other communities. Large amounts of organic matter sink out of the epipelagic to feed the organisms below. Ocean currents carry epipelagic plankton into shallow water, where it is consumed by a profusion of suspension feeders. Epipelagic fishes and zooplankton provide food not only to other marine communities, but also to land-dwelling birds and mammals, including humans.

Because there is no bottom where organic-rich sediments can build up, the epipelagic lacks deposit feeders. Suspension feeders, on the other hand, abound. After all, the food is suspended in the water column. There are also many large predators, like fishes, squids, and marine mammals.

The Plankton: A New Understanding

For more than a century the main way that scientists studied the plankton was to catch them by towing nets through the water (Fig. 17.1), or sometimes by pumping water through a filter. This reliance on plankton nets profoundly affected our understanding of the epipelagic. Imagine if our only way to study flying organisms was with nets. If our nets were too small or too slow, we might never know there were birds. Nets with too coarse a mesh would leave us completely ignorant of mosquitoes. Imagine organisms with the consistency of clouds that slipped through any net! Similar misunderstandings resulted from oceanographers' dependence on plankton nets to study the epipelagic.

In recent decades, new techniques including better microscopes, underwater photography, satellites, chemical procedures, and, yes, even improved nets and

(a)

(b)

Figure 17.1 (a) A typical phytoplankton net is small and has fine netting that prevents many microscopic plankton from passing through. (b) A bongo net, so called because of the two rings that hold the mouth of the net open, is used to sample zooplankton. Zooplankton nets are usually larger and have coarser mesh than phytoplankton nets. The coarse mesh lets phytoplankton pass through but retains the larger zooplankton.

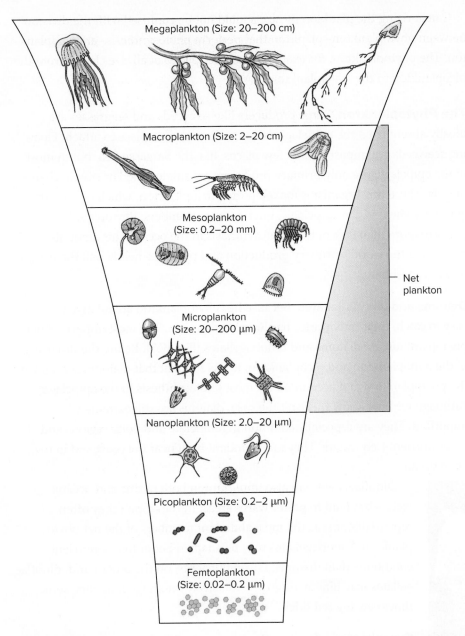

Megaplankton (Size: 20–200 cm)

Macroplankton (Size: 2–20 cm)

Mesoplankton
(Size: 0.2–20 mm)

Microplankton
(Size: 20–200 µm)

Nanoplankton (Size: 2.0–20 µm)

Picoplankton (Size: 0.2–2 µm)

Femtoplankton
(Size: 0.02–0.2 µm)

Net
plankton

Figure 17.2 Plankton are often separated into categories based on their size (also see Appendix A). Note that the size of the organisms in each category increases logarithmically, that is, the size range of each category is 10 times that of the next smaller category.

filters, have revolutionized our understanding of the epipelagic. New discoveries have raised even more questions, and it has become clear that we don't understand the plankton nearly as well as we once thought.

Of particular importance was the discovery of vast numbers of plankton that are too small to be caught in standard plankton nets. These are often subdivided on the basis of the size (Fig. 17.2). The **femtoplankton** is composed almost entirely of viruses and is also called the virioplankton. Most of the **picoplankton** consists of archaea and bacteria. The larger **net plankton**, so called because they *can* be caught in nets, are also subdivided on the basis of size into the micro-, meso-, macro-, and megaplankton, but these categories are not stressed in this book.

Categorizing the plankton by size should not be confused with the division between phytoplankton—plankton that perform photosynthesis—and zooplankton. The phytoplankton, for example, include species of all size classes from the picoplankton to the megaplankton.

The Phytoplankton Large producers like seaweeds and seagrasses are mostly absent from the epipelagic because they have no place to attach. Floating seaweeds are important in a few places, like the Sargasso Sea, but in most of the epipelagic the only primary producers are single-celled or simple chains of cells. These tiny organisms make up the phytoplankton, which occur virtually everywhere in the epipelagic, often in huge numbers. Phytoplankton perform more than 95% of the photosynthesis in the ocean. This amounts to nearly half the world's primary production and produces nearly half the oxygen in our atmosphere.

Diatoms and dinoflagellates Because net phytoplankton are relatively easy to catch, their importance has long been known. The most important net phytoplankton are diatoms and dinoflagellates (Fig. 17.3). Before the discovery of the nanoplankton and picoplankton, it was thought that diatoms dominated the phytoplankton and accounted for most photosynthesis in the epipelagic. Although we now know this isn't the case, diatoms are still extremely important. They are especially common in temperate and polar regions and other nutrient-rich waters. They are abundant both near the coast and in the open ocean.

Dinoflagellates are also important in both neritic and oceanic waters but tend to prefer warm areas. In the tropics they often replace diatoms as the most abundant members of the net phytoplankton. Dinoflagellates may be adapted better to low-nutrient conditions than diatoms. Given nutrients, on the other hand, dinoflagellates may **bloom**, or grow explosively into huge numbers, sometimes causing **red tides**.

Cyanobacteria Because they are so small and hard to catch, or even to see under a microscope, photosynthetic picoplankton and nanoplankton are not as well known as net phytoplankton, and their great importance has only been appreciated in the last few decades. Photosynthetic picoplankton and nanoplankton are far more abundant than net phytoplankton and contribute most of the epipelagic's photosynthesis, 90% or more in many places.

Cyanobacteria are the most abundant picoplankton and account for at least half the ocean's total primary production. One group, the single-celled *Prochlorococcus*, is the most abundant of all marine phytoplankton and is especially dominant in nutrient-poor tropical and subtropical waters. The closely related *Synechococcus* is also very abundant in all but polar waters.

A group of cyanobacteria, *Trichodesmium,* grows in filamentous colonies that can be caught in nets. They can be very abundant,

(a)

(b)

Figure 17.3 Diatoms have sculptured shells, or frustules, made of silica. (a) *Thalassiosira allenii* (top and side views) consists of a single cell. (b) *Fragilaria* is a diatom that forms chains of cells. Each division in the long, green filaments is an individual cell.

sometimes forming red tides. They are particularly abundant in nutrient-poor waters, perhaps because they can fix nitrogen.

Protists Although cyanobacteria dominate the photosynthetic picoplankton, various protists are also important. These eukaryotes typically make up less than 5% of the total picoplankton cells in the water, but have been known to account for over 30%. An ample supply of nutrients appears to favor an increased abundance of protists relative to cyanobacteria. Many of the photosynthetic protists in the picoplankton have not been isolated and are known only from characteristic nucleic acid sequences in the water. Their ecological role in the phytoplankton is poorly understood.

Although protists have an uncertain role in the picoplankton, they dominate the nanoplankton, and many are photosynthetic. Coccolithophorids are one of the more abundant groups, and probably the best known. They do best in the open ocean but also occur in coastal waters.

Another group of minute phytoplankton known as cryptophytes are plentiful in neritic waters. They appear to be very important in epipelagic food webs but not much is known about them. Silicoflagellates occasionally bloom and become important primary producers. Several other groups of protistan nanoplankton appear to be important primary producers in the epipelagic. Most diatoms and dinoflagellates belong to the net plankton, but a few unusually small species are part of the nanoplankton.

Table 17.1 shows the major groups of marine phytoplankton. Various other groups may form blooms, but relatively rarely, and are of secondary importance in the overall ecology of the seas.

The Zooplankton Phytoplankton form the base of the food web. The solar energy that they capture and store in organic matter is passed on to the other creatures of the epipelagic, from minute zooplankton to gigantic whales.

IN CONTEXT

Archaea Single-celled, prokaryotic organisms once thought to be bacteria but now known to be as different from bacteria as from humans.
• 7.2, Prokaryotes

Dinoflagellates Single-celled organisms that use two flagella to swim.
• 7.3, Single-celled Eukaryotes

Eukaryotes Organisms with cells that contain a nucleus and other organelles that are enclosed by membranes.
• 6.2, Cellular Structure and Organization

Protozoans Single-celled organisms that are animal-like in that they ingest food and usually can move around.
• 7.3, Single-celled Eukaryotes

Table 17.1 The Major Groups of Marine Phytoplankton

Size Category	Group	Neritic or Oceanic	Latitude/Temperature	Notes
Net plankton	Diatoms	Both	Everywhere, but most common in temperate and polar waters	Important primary producers
	Dinoflagellates	Both	Everywhere, but most common in warm waters	Common red tide organisms
	Colonial cyanobacteria (*Trichodesmium*)	Oceanic	Mainly tropical	Can fix atmospheric nitrogen; causes red tides in the Red Sea and elsewhere
Nanoplankton	Coccolithophorids	Oceanic	Everywhere, but most commonin tropical waters	Important primary producers in nutrient-poor waters
	Cryptophytes	Both	Everywhere, but most common in temperate and polar waters	Very important primary producers; poorly known
	Silicoflagellates	Neritic	Temperate and polar waters	Sometimes form blooms
Picoplankton	Unicellular cyanobacteria (*Prochlorococcus*, *Synechococcus*)	Both	Everywhere, but most important in warm waters	Dominant primary producers, especially in nutrient-poor waters
	Various protists	Not known	Not known	Presence of many groups recently discovered

Red Tides and Harmful Algal Blooms

From time to time, especially near the coast, the sea becomes bright red literally overnight. This phenomenon, known as a red tide, has occurred for thousands of years. The Old Testament may contain the earliest known reference to a red tide when it described the waters of the Nile turning to blood. Red tides are unrelated to tides; they are massive phytoplankton blooms. Furthermore, they aren't always red. The sea may instead turn orange, brown, or bright green. Red tides are usually blooms of a single species, and are unpredictable because we don't fully understand what causes them. At least 200 species of phytoplankton are known to cause red tides. About half are dinoflagellates, but many other organisms, including cyanobacteria, diatoms, and chrysophytes are involved.

Harmful Algal Blooms When red tides cause ecological problems, they are classified harmful algal blooms (HABs). Harmful algal blooms occur when biotoxin-producing phytoplankton grow rapidly, and can cause a variety of problems: blocking out sunlight, depleting oxygen, and poisoning marine organisms (and humans!). Crustaceans and bivalves avoid poisoning by storing the toxin in their digestive glands, kidney, liver, or other tissues. When humans consume contaminated shellfish they can become very ill or even die. Toxic algae cause up to 200,000 cases of serious poisoning a year, of which as much as 10% lead to death. Additionally, swimming or boating in affected water, or even breathing sea spray, can result in minor poisoning.

Bloom of a dinoflagellate called *Noctiluca* caused this red tide in Mexico. The organism is bioluminescent, and blooms like this produce spectacular blue-green light at night.

What causes HABs? HABs have been around since long before biblical times. There is fossil evidence of red tides from over 100 million years ago. Scientists are concerned because HABs are becoming more frequent, widespread, intense, and persistent than in the past. This increased occurrence could partially be the result of improved monitoring technology, but, there is strong evidence that HABs are expanding globally.

Coastal eutrophication is a strong driver of increased algal bloom frequency, including HABs. Nitrogen pollution, primarily from primarily non-point sources such as agricultural run-off, overwhelms nitrogen-limited marine systems. When nitrogen is added to a system, phytoplankton grow uncontrollably until they run out of the next limiting item. Other factors, including currents, temperature, and the abundance of zooplankton grazers and disease-causing viruses, also play a role.

HABs are also spreading to new places. Ships can transport hundreds of millions of HAB-causing organisms in their ballast water. Natural events can also spread HABs. Shellfish poisoning was unknown in the Gulf of Maine until 1972, when a tropical storm carried in a seed population of toxic dinoflagellates. There have since been outbreaks almost annually in the area.

What can we do? Coastal nations are taking steps to control HABs. The U.S. estimates an economic loss of $82 million annually due to HABs shutting down fishing and recreational activities. In their 2001 proposal to Congress, The National Sea Grant College Program identified the need for further research to understand the relationship between land-use changes that lead to nutrient loading and HABs. Shipping nations have agreed on an international ballast water convention to reduce the spread of HAB organisms and ballast water disinfection systems are being tested. Scientists are beginning to develop models to predict HABs, but in most places it is still unknown as to what triggers HABs well enough to predict them. Utilizing the constant tracking of the Integrated Ocean Observing System, with satellites, buoys, and ocean floor sensors collect scientists at least are able to quickly detection HABs and predict their spread as early warning.

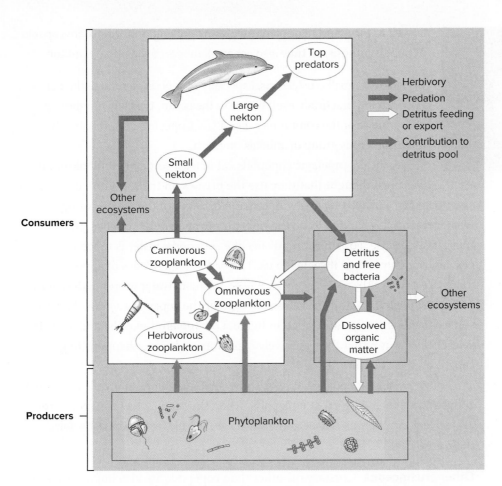

Figure 17.4 A simplified epipelagic food web.

Herbivory
Predation
Detritus feeding or export
Contribution to detritus pool

Consumers

Producers

Top predators

Large nekton

Small nekton

Other ecosystems

Carnivorous zooplankton

Omnivorous zooplankton

Herbivorous zooplankton

Detritus and free bacteria

Dissolved organic matter

Other ecosystems

Phytoplankton

The first step in this flow of energy occurs when herbivores eat phytoplankton. Herbivores are the vital link between the primary producers in the phytoplankton and the rest of the community. Large epipelagic animals cannot feed directly on the tiny phytoplankton. They rely instead on herbivorous zooplankton, which can. Thus, a fundamental part of the epipelagic food web is the flow of energy from phytoplankton to herbivorous zooplankton.

Very few zooplankton are strict herbivores and most that eat phytoplankton also eat other zooplankton when they can. Most zooplankton species, however, are primarily carnivorous and hardly eat phytoplankton at all. These carnivores may feed directly on herbivorous zooplankton and thus reap the production of phytoplankton with only one intermediate step, or trophic level, in the food web. They may also eat other carnivores, adding links to the food web (Fig. 17.4).

Protozoan zooplankton The tiny picoplankton and nanoplankton are too small for most multicellular animals to catch and eat. Protozoans, however, can catch them. When it was discovered that pico- and nanoplankton are the most abundant phytoplankton, the importance of protozoans was also recognized. Without protozoans, much of the primary production in the epipelagic would go unused. The most important of these protozoans are various flagellates, protozoans that move around by means of flagella. Ciliates, foraminiferans (Fig. 17.5), and radiolarians are also important protozoan grazers, on net plankton

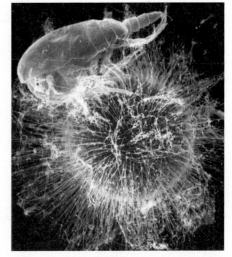

Figure 17.5 A planktonic foramaniferan (*Orbulina universa*) captures a copepod on its long, sticky pseudopodia. The copepod will be drawn to the surface of the foram's shell, where it will be digested. The shell is 0.5 mm in diameter.

Figure 17.6 An epipelagic copepod. The orange balls attached to the tail are eggs.

Detritus Particles of dead organic matter.
- 12.3, The Flow of Energy and Materials

as well as pico- and nanoplankton. Many grazing protozoans are capable of photosynthesis, so they also act as members of the phytoplankton.

Copepods Small crustaceans, especially copepods, dominate the net zooplankton practically everywhere in the ocean, typically numbering 70% or more of the community (Fig. 17.6). Copepods are probably the most numerous group of animals on Earth.

Nearly all epipelagic copepods eat at least some phytoplankton. It was once thought that they use the bristles on their mouthparts and antennae (Fig. 17.7a) to sieve phytoplankton cells out of the water, eating whatever particles happen to be the right size to snag in the bristles. In fact, this is not the case. For small organisms like copepods, water is much more viscous, or syrupy, than it seems to us. The bristly appendages of copepods act more like paddles than sieves, and are used to create water currents that draw individual phytoplankton cells close enough for the copepod to grab. At least some copepods can see and smell individual phytoplankton cells, and actively select cells to eat rather than passively consuming whatever happens to get caught on their bristles.

Though most copepods eat at least some phytoplankton, many also eat other zooplankton, including other copepods, when they can. A few copepods are exclusively carnivorous. These generally have claw-like appendages to seize their prey (Fig. 17.7b).

Other crustaceans Crustaceans other than copepods are also important in the zooplankton. Though not as abundant globally as copepods, shrimp-like krill often aggregate into huge, dense swarms. Preferring cold oceanic waters, krill sometimes dominate the zooplankton in polar seas. They are efficient filter

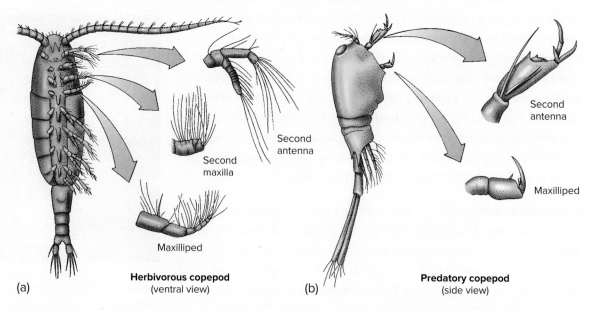

Second maxilla

Second antenna

Maxilliped

Herbivorous copepod
(ventral view)

(a)

Second antenna

Maxilliped

Predatory copepod
(side view)

(b)

Figure 17.7 (a) The mouthparts and appendages of copepods that feed mainly on phytoplankton, such as *Calanus*, have many long bristles. (b) Copepods that prey on other zooplankton, such as *Corycaeus*, have appendages with fewer and shorter bristles. Their appendages are better adapted for grasping. All appendages are paired, but (a) shows only those on one side.

Kjell B. Sandved/Science Source

feeders, capturing phytoplankton, especially diatoms, with their bristly appendages. Krill also eat small zooplankton and detritus, including **fecal pellets**, the solid waste excreted by other zooplankton.

Copepods are so small that most large animals can't catch them. Krill, by contrast, are relatively big, up to 6 cm in Antarctic krill (*Euphausia superba*). Fishes, seabirds, and even great whales eat them.

Although copepods and krill are by far the most abundant, there are many other crustaceans, both herbivorous and carnivorous, in the zooplankton. At certain times and places, these other crustacean groups can be very abundant. Amphipods and most other planktonic crustaceans are small, like copepods. A few larger crustaceans, however, also occur in the net zooplankton. These include a number of decapods—crabs, shrimps, and their relatives. Although relatively few of these large crustaceans inhabit the epipelagic realm as adults, many do so during their larval stages. The decapods are almost exclusively carnivorous.

Non-crustacean zooplankton The zooplankton includes many animal groups other than crustaceans. Salps are transparent, plank-tonic relatives of the sea squirts, or tunicates, that live on the bottom (Fig. 17.8). Salps catch phytoplankton by filtering water through a sieve-like sac or a fine mucus net.

Larvaceans are also relatives of sea squirts, though they hardly look it—except for their larvae (Fig. 17.9). Sea squirt larvae, known as tadpole larvae, are very similar to larvaceans. Larvaceans live inside a "house" they make of mucus. By beating its tail, the larvacean pumps water in through passages in the house. Food particles are caught in a mucus net that is secreted inside the house. At least some larvaceans also secrete a net that can extend as far as 2 m outside the house. The use of sticky mucus nets allows larvaceans to capture extremely small food particles. Larvaceans and some salps are among the few animals that can catch pico- and nanoplankton. They are therefore important in linking primary production by pico- and nanoplankton to the rest of the epipelagic community.

The openings through which water enters the house filter out particles too large for the larvacean to eat. These filters eventually get clogged. Some species can reverse the water flow to clear the filters. If the filters get too clogged or a predator threatens, the larvacean simply abandons the house and swims away. It can produce a new house within minutes and resume feeding. When disturbed, some species can build and discard houses every 10 minutes! Even under normal circumstances, most species change houses every four hours or so. Discarded houses are an important source of detritus and transport large amounts of carbon to the deep sea as they sink.

A group of molluscs called pteropods also includes phytoplankton grazers. Pteropods are small snails in which the foot has been modified to form a pair of "wings" that they flap to stay afloat (Fig. 17.10, p. 527). Some feed by

Figure 17.8 A pelagic salp (*Tethys vagina*). These gelatinous animals feed by filtering phytoplankton. Like larvaceans, some can capture bacteria and other very tiny plankton. They sometimes provide a home to other animals.

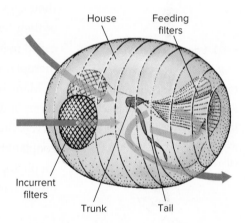

Figure 17.9 A larvacean spends most of its time inside a mucus "house." By beating its tail, the larvacean draws water into the house (blue arrows) through two intake filters. Food particles are caught on the mucus feeding net, and the filtered water flows outward.

Arrow Worms or
Chaetognaths Small, worm-like
predators with fins along the
body and spines on the head.

• 9.2, Marine Worms and
 Some Small Invertebrate
 Phyla

Siphonophores Drifting
colonial cnidarians in which
different members of the
colony are specialized for
different tasks.

• 9.1, Sponges, Cnidarians,
 and Comb Jellies

Comb Jellies or **Ctenophores**
Radially symmetrical animals
that resemble jellyfishes, or sea
jellies, but have eight rows of
cilia and no nematocysts.

• 9.1, Sponges, Cnidarians,
 and Comb Jellies

capturing phytoplankton, including picoplankton and nanoplankton, in mucus nets or threads; others are carnivorous.

Arrow worms, or chaetognaths, feed mostly on copepods. This may not be because they actually prefer copepods but because copepods are so abundant. Arrow worms consume an assortment of other prey when it is available. They can be very abundant, and have a major role in epipelagic food webs.

Not all carnivorous zooplankton are small. Jellyfishes, or sea jellies, and siphonophores, for example, can be large but are weak swimmers and drift with the currents as part of the plankton. Sea jellies are carnivorous, and many eat small fishes as well as zooplankton. The somewhat similar comb jellies are also carnivorous (see Fig. 17.18b, p. 534). The ocean sunfish (*Mola mola*), which grows to 2,300 kg, swims weakly and is sometimes considered a member of the zooplankton.

Meroplankton The zooplankton discussed so far, except for a few sea jellies, spend their whole lives in the plankton and are called **holoplankton**. In addition to these permanent members of the zooplankton, there are a vast number of organisms that have planktonic larvae. These animals—whether from estuaries, the rocky intertidal, kelp beds, coral reefs, or even the deep sea—release their eggs or young into the water column, and the larvae spend the early part of their lives in the plankton. Such temporary members of the plankton are called **meroplankton**. Coastal waters are particularly rich in meroplankton, but meroplankton are found throughout the ocean.

Many invertebrates have a characteristic larva (Fig. 17.10). Some have a whole series of different larval stages. Nearly all marine fishes also have planktonic larvae. Small larvae tend to feed mostly on phytoplankton; larger ones eat zooplankton. If there is a series of stages, as in crustaceans, the larvae may feed on phytoplankton initially and later switch to zooplankton. Fish larvae, too, often change from herbivores to carnivores as they grow (see Fig. 17.23, p.540).

The Nekton

Plankton are by far the most abundant organisms in the sea, and they form the base of the epipelagic food chain. Most of us, however, are more familiar with the nekton, the large, strong swimmers. Fishes, marine mammals, and squids are the most abundant nekton. Turtles, sea snakes, and penguins are also nekton.

Nearly all nekton are carnivorous. **Planktivorous** nekton, those that eat plankton, include small fishes like herrings, sardines, and anchovies. They also include the world's largest fishes, the whale shark (*Rhinocodon typus*) and basking shark (*Cetorhinus maximus*). The largest nekton of all, baleen whales, also eat plankton, mostly krill. Seals, penguins, squids, and an assortment of fishes, including salmon, tunas, and flying fishes, also eat krill.

Fishes like lanternfishes that stay in deeper water during the day and swim up into the epipelagic at night also consume large amounts of zooplankton. These fishes are important foods for some large epipelagic nekton, such as tunas and dolphins, that dive into deeper water to feed.

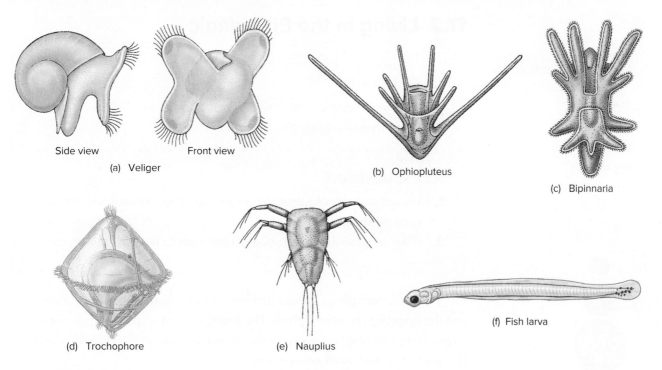

Figure 17.10 There is tremendous variety of larvae in the meroplankton. Some examples are (a) the veliger of molluscs, (b) the ophiopluteus of brittle stars, (c) the bipinnaria of sea stars, (d) the trochophore of polychaete worms and some molluscs, (e) the nauplius of many crustaceans, and (f) fish larvae.

Side view Front view
(a) Veliger

(b) Ophiopluteus

(c) Bipinnaria

(d) Trochophore

(e) Nauplius

(f) Fish larva

Most species of nekton, however, eat other nekton rather than plankton. Fishes, squids, and large crustaceans are the main foods. Epipelagic predators are usually not fussy and will eat many different prey, provided they are the right size. In general, the larger the predator, the larger the prey, though there are exceptions like baleen whales and whale sharks. Thus, small fishes like herrings have small prey: zooplankton. Bigger fishes eat these small fishes and in turn are eaten by even bigger predators. At the top of the food chain are the largest predators of all, the top predators, which eat the largest prey. The sperm whale, the largest of all nekton except baleen whales, eats giant squid more than 10 m long. There are other top predators that eat massive prey. Among predators, killer whales (*Orcinus orca*) are second in size only to the sperm whale. They not only eat porpoises and seals, but by hunting in groups can also attack and kill baleen whales. Large sharks such as the great white (*Carcharodon carcharias*) and mako (*Isurus*) sharks eat seals, billfishes like marlins and sailfishes, and even other sharks.

Section Review

1. **Construct** a Venn diagram to compare and contrast phytoplankton and zooplankton.

2. **Describe** What are the most important characteristics of nekton?

17.2 Living in the Epipelagic

VOCABULARY

drag

neuston

pleuston

rete mirabile

vertical migration

DIVE IN!

Are there advantages to not having a swim bladder? Sharks do not have a swim bladder and, although this means that they have to swim constantly to avoid sinking to the bottom, there are some advantages to not having one. Sharks can move from deep water to shallow water and back quickly without the risk of their bodies imploding or exploding as the pressure changes. Fishes that have a swim bladder and live very deep in the ocean often do not survive when brought to the surface. As the pressure decreases, the swim bladder expands. In some cases it expands enough enough that other internal organs are forced out of the mouth of the fish.

Main Idea

Organisms in the epipelagic have adaptations that allow them to stay afloat and feed while living in the water column.

· ·

Key Questions

1. The adaptations of epipelagic organisms center around which two main needs?
2. What adaptations do epipelagic fishes have to be able to swim fast?

Every environment places special demands on the organisms that live there, and the epipelagic is no exception. The adaptations of epipelagic organisms largely center around two main needs: the need to stay in the epipelagic and the need to eat and avoid being eaten.

Staying Afloat

To live in the epipelagic, organisms must *stay* in the epipelagic. Epipelagic organisms face a fundamental problem: Cells and tissues are denser than water. Shells and skeletons are even more dense. Thus, organisms sink unless they have some adaptation to prevent it. This is not a problem for organisms on the bottom, but organisms that live near the surface must somehow avoid sinking to stay in their habitat.

Phytoplankton have to stay in relatively shallow water to get enough light for photosynthesis. Once a phytoplankton cell sinks or is carried by water movement out of the photic zone it is doomed, unless it can somehow get back into the sunlit layer. Animals, too, whether zooplankton or nekton, must be able to stay near the surface, not because they need sunlight but because their food is found in the shallow waters.

Organisms that can't swim have two basic ways to keep themselves from sinking out of the epipelagic. One is to increase their water resistance so they sink slower. The other is to be more buoyant so they don't sink in the first place. Buoyancy also benefits organisms that swim because it allows them to stay at a given depth with less work. This applies not only to the nekton, which by definition are strong swimmers, but also to zooplankton and even some phytoplankton, like dinoflagellates. They swim up and down in the water column even though they can't swim strongly enough to move against the currents.

Increased Resistance How fast an organism of a given weight sinks depends on how much water resistance, or **drag**, it encounters. The drag depends on how much water a moving organism pushes in front of it and the amount of friction between the water and the organism's surface. For small organisms like most plankton, drag depends mostly on surface area: the more surface area, the higher

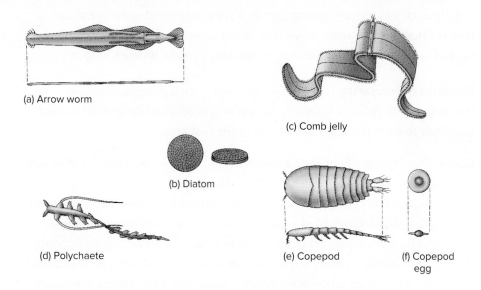

(a) Arrow worm

(c) Comb jelly

(b) Diatom

(d) Polychaete

(e) Copepod

(f) Copepod egg

Figure 17.11 Planktonic organisms with flat shapes include (a) some arrow worms (*Sagitta*), (b) many diatoms, (c) *Cestum*, a highly modified comb jelly, (d) *Tomopteris,* a planktonic polychaete worm, (e) certain copepods, such as *Sapphirina,* and (f) some copepod eggs.

the water resistance and the slower the organism sinks. This is one reason plankton are usually small: Small organisms have more surface area per unit volume than large organisms and thus a slower sinking rate.

An organism's shape can also increase its surface area to make it sink slower. The parachute-like shape of many sea jellies helps slow sinking. Many plankton have extremely flat shapes (Fig. 17.11). You can see how this helps slow sinking with a simple experiment. Take two sheets of paper and crumple one into a ball. Drop both pieces from the same height. The flat sheet will take much longer to hit the ground because it offers more air resistance. Exactly the same principle applies in water. A flat shape has another advantage for phytoplankton. Note how the flat sheet of paper zigzags from side to side as it falls. Flat phytoplankton cells do the same thing. The back-and-forth motion keeps seawater flowing over the cell's surface, renewing the layer of water with which the cell exchanges gases, nutrients, and waste products.

Long projections or spines are another common adaptation to increase surface area in planktonic organisms (Fig. 17.12). Spines have the added advantage of making the organism harder to eat. Many phytoplankton form chains. Chains sink slower than the individual cells would if the chain were broken up.

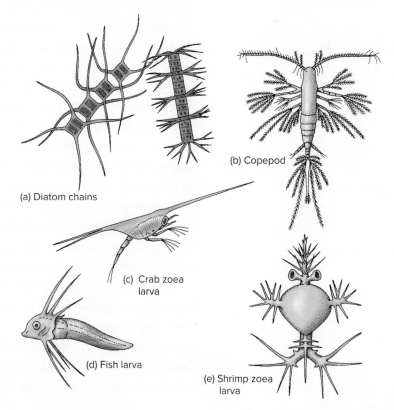

(a) Diatom chains

(b) Copepod

(c) Crab zoea larva

(d) Fish larva

(e) Shrimp zoea larva

Figure 17.12 Some planktonic organisms have long spines or projections. They may also form chains. Examples are (a) diatoms, (b) copepods like *Augaptilus*, (c) zoea larvae of porcellanid crabs, (d) fish larvae (*Lophius*), and (e) zoea larvae of *Sergestes,* a shrimp.

Epipelagic nekton, by contrast, rarely have pronounced spines or other features that increase water resistance. Since swimming burns a lot of energy, nekton usually have adaptations to *reduce* drag to make it easier to swim.

Increased Buoyancy The second way that epipelagic organisms stay near the surface is to be more buoyant. As opposed to adaptations to slow sinking, buoyancy reduces the tendency to sink in the first place.

Lipids One common buoyancy adaptation is to store lipids, such as oils or fats. Lipids are less dense than water and float. Many plankton—notably diatoms, copepods, and fish eggs and larvae—contain oil droplets. Many adult epipelagic fishes also gain buoyancy by storing lipid. This is especially true of sharks, tunas and their relatives, and other fishes in which the swim bladder is poorly developed or missing altogether. Epipelagic sharks, for example, have enlarged livers with a very high oil content. Whales, seals, and other marine mammals have a thick layer of fatty, buoyant blubber under the skin.

Oils and fats also serve other functions. They are the most efficient way to store energy. In marine mammals the blubber provides insulation from the cold as well as buoyancy. Insulation is especially important to warm-blooded animals, which must burn energy to maintain their body temperatures.

IN CONTEXT

Swim Bladder A gas-filled sac that lies inside the body cavity of most bony fishes.
• 10.1, Fishes: the First Vertebrates

Gases Pockets of gas are another adaptation for buoyancy. Cyanobacteria have tiny gas bubbles, or vacuoles, inside their cells. They are able to change the number and buoyancy of the vacuoles. This allows them to increase buoyancy to move up in the water column or decrease it to sink. Some other plankton, especially large ones, have special gas-filled floats. Such floats are especially common in organisms that live right at the sea surface, as described in the next section. Most epipelagic bony fishes have swim bladders for buoyancy.

A gas-filled bladder or float provides much more buoyancy than lipid, but it has a major disadvantage. Gas expands and contracts as the animal moves up and down in the water column and the pressure changes. When the volume of gas changes, so does the amount of buoyancy. To control its buoyancy, a fish must be able to regulate the amount of gas in its swim bladder. Most fishes can do this, but the mechanism varies. Some fishes pump gas in and out of the bladder through a special duct. They can adjust fairly rapidly to pressure changes pressure. Other fishes regulate the amount of gas in their swim bladders much more slowly and can't cope well with depth changes. Many fishers have pulled up fish with bulging eyes or the stomach protruding from the mouth. This happens when the gas in the swim bladder expands as the fish is brought to the surface. In normal life the fish would never come to the surface or would surface much more slowly than it did on the end of the line.

The swim bladder is often poorly developed or absent in active fishes that frequently change depth, such as tunas. It never evolved in sharks. Such fishes must compensate for the loss of buoyancy. In addition to their large oily livers, sharks have large, stiff fins and asymmetrical tails that provide lift when the shark swims. Members of the tuna family also rely on stiff, almost wing-like fins and must swim constantly to provide lift.

Figure 17.13 The by-the-wind sailor (*Velella*).

Body fluids Yet another way to gain buoyancy is to control the composition of the body fluids. Organisms can reduce their density by excluding heavy ions, like sulfate (SO_4^{-2}) and magnesium (Mg^{+2}), and replacing them with lighter ones, especially ammonium (NH_4^+) and chloride (Cl^-). Some dinoflagellates do this, and there is evidence that other phytoplankton do too. So do various zooplankton, including salps, comb jellies, and some squids.

The Floaters The plankton include an unusual and highly specialized group of organisms that live at the very surface of the ocean. Organisms that live right at the sea surface but remain underwater are known as **neuston**, and those whose bodies project through the sea surface into the air are called **pleuston**.

All plankton need to keep from sinking, but the neuston and pleuston have to go one step further and actually float. The most common method of doing this is to have some sort of gas-filled structure to provide buoyancy. There are many variations on this seemingly simple theme.

The upper surface of the by-the-wind sailor (*Velella*), a colonial, jellyfish-like cnidarian, is specialized as a float (Fig. 17.13). The float protrudes into the air and acts like a sail, pushing the colony along in the wind. The Portuguese man-of-war (*Physalia*), a siphonophore, is famous for its powerful sting. Like the by-the-wind sailor, part of the colony acts as a sail. Several other cnidarians live at the sea surface. Some can even control the buoyancy of their floats and sink below the surface for shelter during storms.

Despite their powerful stinging cells, these relatives of sea jellies have enemies. The violet snail (*Janthina*) makes a raft of mucus filled with bubbles from which it hangs upside down (Fig. 17.14, p. 532). These inch-long snails drift about, eating Portuguese men-of-war and other delicacies when they find them. Another mollusc, a sea slug (nudibranch) called *Glaucus*, stays afloat by swallowing a bubble of air. It feeds on the by-the-wind sailor and its relative, the small, disk-shaped *Porpita* (Fig. 17.15, p. 532). *Porpita*'s stinging cells, still capable of firing, end up in the flaps that protrude from the sea slug's back and will sting animals that disturb it. Thus, *Porpita* provides not only food but also protection for the sea slug.

Figure 17.14 A violet snail (*Janthina janthina;* left), covered with gooseneck barnacles, eats a Portuguese man-of-war (*Physalia physalis;* right).

Figure 17.15 The pelagic sea slug (*Glaucus*) feeding on *Porpita,* a relative of jellyfishes, or sea jellies. The sea slug not only gets a meal but also uses *Porpita*'s stinging cells for defense.

One unusual inhabitant of the surface doesn't have a float. In fact, it doesn't actually float at all. The water strider (*Halobates*), the only insect that lives in the open ocean, skims over the water surface (Fig. 17.16). It is closely related to the water striders common in lakes and ponds. The water strider isn't marine in the strict sense because it cannot swim and drowns if it falls through the surface. Still, it manages to live all over the oceans.

Predators and Their Prey

Many of the adaptations of epipelagic animals, or for that matter all animals, are related to the need to find food and at the same time avoid being eaten.

Sense Organs Predators and their prey play a continual game of hide-and-seek. The predator tries to find its prey and attack before the victim can escape, while the prey wants as much warning of approaching enemies as possible. Most animals in the epipelagic therefore have highly developed sense organs to help them detect their prey and enemies.

Figure 17.16 The marine water strider (*Halobates sericeus*) relies on greatly elongated legs and the water's surface tension to skate on the surface.

The epipelagic has plenty of light to see by, at least during the day, so it is no surprise that vision is important to many epipelagic animals. Many zooplankton have well-developed eyes. Most can't see actual images, but they can detect motion, shapes, and shadows. Copepods and other zooplankton probably use vision to locate their prey and avoid predators.

Squids, fishes, and marine mammals all have good eyesight. Vision is especially important to nekton in the epipelagic because there are no solid structures that can be used for concealment. In other environments, animals can hide in the sediments, under rocks, or behind seaweeds. There are no such hiding places in the epipelagic. The large, well-developed eyes of most epipelagic fishes (Fig. 17.17) and other nekton are used not only to find prey and watch for predators, but also to find mates and stay together in schools.

Fishes have another remote sensing system: the lateral line. Like all fishes, those in the epipelagic are extremely sensitive to vibrations in the water. Fishes use their lateral lines to stay in touch with schoolmates and to detect predators. Even blind fishes can school and react to the lunge of an enemy. Most predatory fishes—including sharks, tunas, and billfishes—also have well-developed hearing and are strongly attracted to splashes on the surface and irregular vibrations in the water, the kinds of vibrations made by an injured fish. Dolphins and many other cetaceans have another remote sensing system, echolocation. This sophisticated built-in sonar allows them to locate their prey at a distance.

Coloration and Camouflage Just because there are no hiding places doesn't mean that epipelagic organisms cannot conceal themselves. Indeed, protective coloration, or camouflage, is nearly universal among epipelagic animals. This isn't surprising in a world where the hunters rely so heavily on vision to find their victims, and the prey use vision to get away.

One way to be inconspicuous in the vast, watery world of the epipelagic is to be transparent. This adaptation is seen in a variety of zooplankton, some of which are almost perfectly clear (Fig. 17.18, p. 534). Gelatinous zooplankton like some sea jellies, salps, larvaceans, and comb jellies are probably the best at being invisible. Many other groups of zooplankton are at least partially transparent, with perhaps only the eyes, a few spots of pigment, or the internal organs visible. Some nearly transparent zooplankton have a faint, bluish tinge, possibly helping them blend in with the surroundings.

CONTEXT

Lateral Line A system of canals on the head and sides of fishes that helps them detect vibrations in the water.
• 10.2, Fish Anatomy

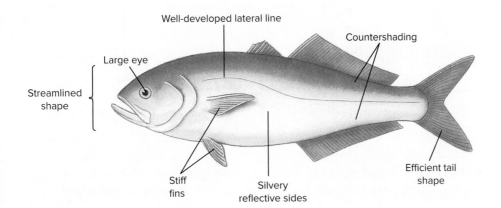

Well-developed lateral line
Countershading
Large eye
Streamlined shape
Stiff fins
Silvery reflective sides
Efficient tail shape

Figure 17.17 Typical adaptations of epipelagic fishes.

Figure 17.18 Many species of zooplankton are transparent or nearly so. Examples include (a) hyperiid amphipods, (b) comb jellies, and (c) squids. These photos have been taken to make the organisms more visible than they actually are.

Another common form of protective coloration, particularly in nekton, is countershading, where the back (dorsal surface) is dark, usually green, blue, or black, and the belly (ventral surface) is white or silver. Countershading is particularly well suited to the epipelagic. To a predator looking down, the ocean depths are a dark blue. Viewed from above, the dark back of a countershaded fish blends in against the dark background. Looking up, on the other hand, the predator sees the bright light filtering down from above and the silvery-white sea surface. Against this bright background a dark object stands out like a sore thumb, but the silvery-white undersides of epipelagic nekton reflect light and match the bright surface. Laterally compressed bodies, also common, reduce the size of the silhouette, whether viewed from above or below (Fig. 17.19).

Most epipelagic fishes have silvery sides that reflect light. This helps them blend in when viewed from the side as well as from above or below. It is also common for them to have vertical bars or irregular patterns that help break up their outline in the dappled underwater light.

Flying fishes have evolved a distinctive defense that is not exactly camouflage but does make them hard to see, at least temporarily. When threatened, flying fishes launch out of the water and glide through the air on their greatly

Figure 17.19 Like many epipelagic fishes, the sailfish (*Istiophorus platypterus*) has a laterally compressed body; that is, it is high and narrow when viewed in cross section.

enlarged pectoral fins (Fig. 17.20). To a predator, the flying fish seems to have disappeared, because it is difficult to see beyond the reflective sea surface. This strategy is only partly effective, however, and flying fishes are a favorite food of many epipelagic predators.

Swimming: The Need for Speed Protective coloration or not, the keen senses of epipelagic predators enable them to find their prey. When this happens the prey's only hope is to flee, because there are no hiding places. Whether the prey gets away or the predator gets a meal often depends on which swims faster. The emphasis is on sheer speed, as opposed to maneuverability, or being able to burrow in the bottom or hide in crevices in the rocks. It is no coincidence that the epipelagic is home to the ocean's most powerful swimmers.

Body shape As noted above, whereas plankton are typically adapted to increase drag to slow their sinking, epipelagic nekton typically reduce drag to swim fast. Practically all epipelagic nekton have streamlined bodies for more efficient swimming. They rarely have bulging eyes, long spines, or other projections that increase resistance. Instead, they are sleek and compact. They often have smooth body surfaces to help them slide through the water. Epipelagic fishes usually have small scales, or none at all, and dolphins and whales have lost nearly all their hair. Fishes produce mucus that lubricates the skin to slip through the water even more easily.

Epipelagic nekton are also firm and muscular, another adaptation for swimming. The force produced by the muscles is delivered almost entirely by the tail. Epipelagic fishes rarely swim with their pectoral fins, which are used only for steering and lift. Most epipelagic fishes have a high, narrow tail, which in engineering terms is the most efficient shape for high-speed swimming. Tunas and sharks have independently evolved specialized long tendons, not found in other fishes, that efficiently transfer the muscles' power back to the tail.

Bony fishes in other habitats typically have fins with movable spines connected by a thin membrane, making them dexterous and flexible. By contrast, the fins of epipelagic bony fishes tend to be stiff, giving them the strength to provide maneuverability and lift at high speed. These stiff fins aren't much good for hovering, swimming backward, poking into holes and crevices, or other activities that require tight maneuvers at slow speed. Epipelagic fishes don't do such things often anyway.

Muscles Epipelagic nekton not only have a lot of muscle, but their muscles are strong and efficient. Fishes have two kinds of muscle, red and white—the equivalent of the light and dark meat on a turkey. Red muscle gets its color from a high concentration of myoglobin and can store a lot of oxygen. It is best suited for long, sustained effort. White muscle, on the other hand, provides short bursts of power. Epipelagic fishes use their red muscle for sustained cruising, and typically have more red muscle than coastal fishes, which usually don't do as much sustained swimming. Epipelagic fishes still have white muscle that kicks in when they need an extra burst of speed.

Figure 17.20 Flying fishes (*Cypselurus*) can't actually fly, but their expanded pectoral fins allow them to glide for long distances.

CONTEXT

Myoglobin A red muscle protein in vertebrates that stores oxygen.
• 10.2, Fish Anatomy

WaterFrame/Alamy

Swimming Machines

Tunas, mackerels, and billfishes swim continuously. Feeding, courtship, reproduction, and even "sleep" are carried out in constant motion. As a result, practically every aspect of their body forms is adapted to enhance swimming ability.

Tunas and billfish routinely cover vast distances in their annual migrations. One Pacific bluefin tuna (*Thunnus orientalis*) tagged southeast of Japan was recaptured off the Pacific coast of Baja California, Mexico, 10,800 km away. Tagged Atlantic bluefin (*T. thynnus*), have crossed the Atlantic in 119 days, averaging a straight-line distance of over 65 km per day. The fish actually cover a much greater distance as they dash here and there in search of food. Tunas could cross the Atlantic in less than two months at their slowest cruising speed if they took a direct route.

Tunas and billfishes are primarily endurance swimmers, adapted for sustained cruising, but they are also accomplished sprinters and can make blistering high-speed bursts. The fastest fish of all is the Indo-Pacific sailfish (*Istiophorus platypterus*), which can exceed 110 km/h for short periods. Several large species can reach 75 km/h. Their phenomenal speed has probably evolved not so much to catch prey—they could do that at slower speeds—but to

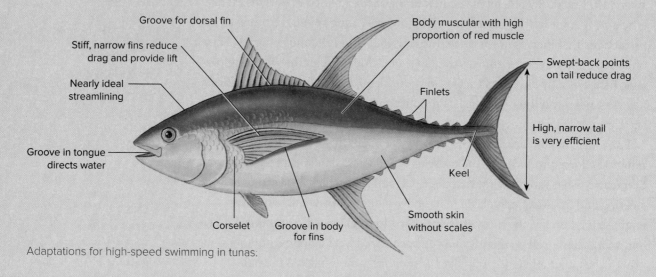

Adaptations for high-speed swimming in tunas.

Rete mirabile Muscles work more efficiently at warm temperatures. Epipelagic sharks, tunas, and billfishes have evolved a system to conserve the heat their muscles generate to keep their internal temperature warmer than the surrounding water. These fishes can almost be considered "warm-blooded." In most fishes, heat generated in the muscles is carried by the blood to the skin and then lost to the water. "Warm-blooded" fishes have a special arrangement of the blood vessels known as the **rete mirabile**, or "wonderful net." In this arrangement (Fig. 17.21), body heat carried outward in the blood is transferred to inward-flowing blood and taken back into the body core. This greatly reduces heat loss. The fish's core temperature remains higher than the surrounding water, while the skin is nearly at water temperature.

Vertical Migration The surface layers of the ocean contain by far the most food in the pelagic realm. They can be a dangerous place, though, with voracious predators zooming around and nowhere to hide. Some zooplankton

compete with schoolmates. The first one to the prey, after all, gets the meal.

Hydrodynamic adaptations Many adaptations of these fishes reduce water resistance, and may resemble aerodynamic features of high-speed jets. Human engineers are new to the game—tunas and their relatives evolved their "high-tech" designs long ago.

Tunas, mackerels, and billfishes have made streamlining an art form. The sleek, compact body shapes of tunas are nearly ideal from an engineering perspective. Most species lack scales over most of the body, making it smooth and slippery. The eyes lie flush instead of protruding, and are covered with a slick, transparent lid that reduces drag. The fins are stiff, smooth, and narrow, also cutting drag. When not in use, the fins tuck into special grooves or depressions so they lie flush with the body and don't break up its smooth contours. Jets retract their landing gear in flight for the same reason.

These fishes have even more complex adaptations to improve their hydrodynamics. The long bill of marlins and billfish probably helps them slip through the water. Many supersonic aircraft have a similar needle at the nose. Most tunas and billfishes have a series of keels and finlets near the tail. Tunas and mackerels retain a patch of coarse scales near the head called the corselet. The keels, finlets, and corselet help direct the flow of water over the body surface in such a way as to reduce resistance.

Respiration Because they are always swimming, tunas can force water over their gills by simply opening their mouth. Accordingly, they have lost most of the muscles used to suck water in and push it past the gills. Tunas must swim to breathe. They must also keep swimming to avoid sinking, since most have largely or completely lost the swim bladder.

A potential problem is that opening the mouth to breathe detracts from the streamlining of these fishes and tends to slow them down. Some tunas have specialized grooves in their tongue. The grooves help channel water through the mouth and out the gill slits to reduce water resistance.

Drag Tunas and billfishes have adaptations to increase forward thrust as well as to reduce drag. Their high, narrow tails with swept-back tips are almost perfectly adapted for propulsion with the least possible effort. Perhaps most important of all to these and other fast swimmers is their ability to sense and make use of swirls and eddies in the water. They can slide past eddies that would slow them down and then gain thrust by "pushing off" the eddies. Scientists and engineers are beginning to study this ability of fishes in the hope of designing more efficient propulsion systems for ships.

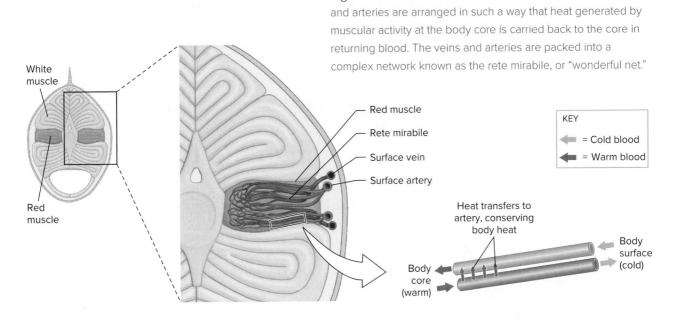

Figure 17.21 In "warm-blooded" epipelagic fishes, small veins and arteries are arranged in such a way that heat generated by muscular activity at the body core is carried back to the core in returning blood. The veins and arteries are packed into a complex network known as the rete mirabile, or "wonderful net."

White muscle

Red muscle

Red muscle
Rete mirabile
Surface vein
Surface artery

KEY
← = Cold blood
← = Warm blood

Heat transfers to artery, conserving body heat

Body core (warm)

Body surface (cold)

Figure 17.22 The depth distribution at different times of day of a vertically migrating copepod.

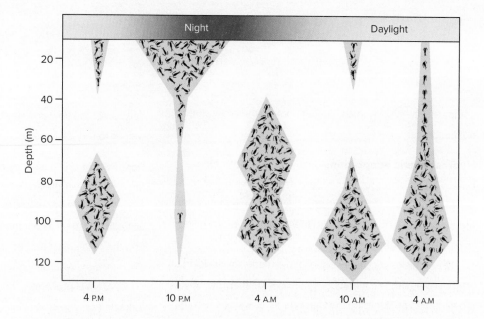

spend only part of their time near the surface and retreat to safer territory when not feeding. These animals undertake **vertical migration**. During the day they live at considerable depth, usually at least 200 m or so (Fig. 17.22). At these depths there isn't much light, so the zooplankton are relatively safe from predators that use vision. At night, the zooplankton swim up to the surface layers to feed on phytoplankton and other zooplankton. Experiments in deep tanks have shown that zooplankton vertically migrate only if predatory fish are in the tank; they stop migrating when the predators are removed.

These daily movements require a lot of energy. For an organism that is 2 mm long, a 200-m migration is the rough equivalent of a 200-km swim for a human! That is a long way to go to avoid predators when many other zooplankton survive predation without migrating. Various explanations for vertical migrations in addition to escaping predation have been proposed. Zooplankton may be able to slow their metabolism and conserve energy by spending part of their time in the deep water, which is cold and reduces their body temperature. This is somewhat like sleep or hibernation. It has also been suggested that zooplankton migrate to avoid toxins produced by some phytoplankton. The phytoplankton produce more toxins during the day, when they are actively photosynthesizing.

Many species of nekton, mainly fishes and some large shrimps, also make vertical migrations. These animals generally migrate to much greater depths than zooplankton.

Section Review

1. **Construct** a t-chart to list the adaptations epipelagic organisms have to either increase resistance (drag) or increase buoyancy to stay afloat.

2. **Discuss** why each adaptation is important in the epipelagic and give an example of each: vision, camouflage, ability to swim fast, and ability to migrate vertically.

17.3 Epipelagic Food Webs

> ## Main Idea
> Epipelagic food webs are an important part of how matter and energy flow through the ocean.
> ..
>
> ## Key Questions
> 1. What is the importance of DOM in the epipelagic?
> 2. What are two of the main factors that limit primary production in the epipelagic?

VOCABULARY

microbial loop

marine snow

light-limited

self-shading

nutrient-limited

regenerated

spring bloom

fall bloom

Epipelagic food webs are of great interest, especially because epipelagic fisheries provide food and employment to millions. Species that are caught commercially can be understood and managed only if we understand the rest of the food web on which they depend.

Trophic Levels and Energy Flow

The trophic structure of the epipelagic is extremely complex. The feeding habits of most epipelagic species are known poorly, if at all. It's hard to describe the trophic structure of a community if you don't even know what many of the animals eat!

Another difficulty in understanding epipelagic food webs is that most of the animals eat a variety of prey, often from different trophic levels. Zooplankton such as copepods often eat both phytoplankton and zooplankton; thus, they function as both herbivores and carnivores. Nekton, too, consume prey at different levels on the food web. A tuna, for example, may act as a secondary consumer when it eats krill, as a third-level consumer when it eats a sardine that ate zooplankton, and as a fourth-level consumer when it eats a mackerel that ate a flying fish that ate zooplankton. This makes it difficult to assign many animals to a particular trophic level.

Yet another complication is that most epipelagic animals consume different prey at different times in their lives. Most eat different food as larvae than they eat as adults, and even the larvae often switch from phytoplankton to zooplankton. Even after they leave the larval stage, many fishes eat different prey as they grow larger (Fig. 17.23, p. 540).

Food Web Complexity

The basic flow of energy in the epipelagic can be depicted as phytoplankton → zooplankton → small nekton → large nekton → top predators. The top predators, of course, are also nekton. This scheme is greatly simplified, and each level contains a mini-web. Within the zooplankton, for example, there are herbivores and several levels of carnivores, forming a food web among themselves. There is also an important side branch called the microbial loop, which is described in the next section.

DIVE IN!

We are what we eat? Epipelagic food webs are difficult to construct, largely because the feeding habits of most species are not well understood. To figure out what animals eat without observing them feed, marine scientists can use a tool known as stable isotope analysis. Stable isotope analysis uses the ratio of heavy to light isotopes as a chemical signature of what an animal has eaten. In food webs, the most useful isotopes to use are those of carbon and nitrogen. The isotopic signature of an animal is directly related to that of its prey, and values change in a predictable manner between successive levels in a food web. Stable isotopic analysis has many limitations, but it can help marine ecologists track the feeding and migratory behavior of open ocean animals.

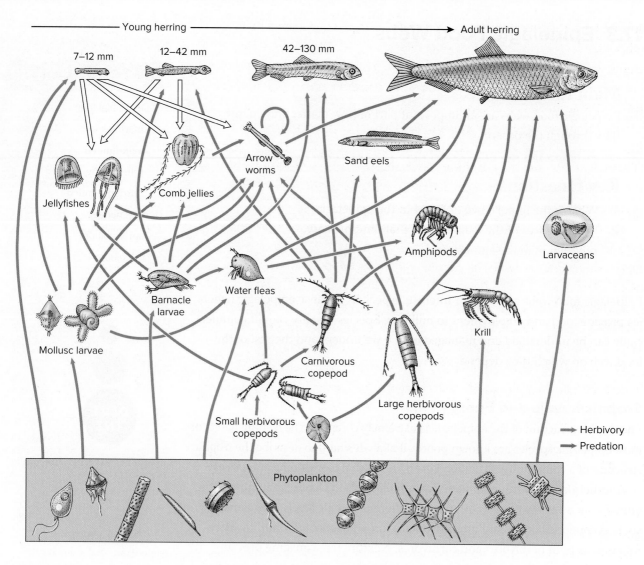

Figure 17.23 The North Sea herring (*Clupea harengus*) eats different foods at different stages of its life. Its larval stages are eaten by predators that do not feed on adult herring (open arrows). Many, if not most, epipelagic animals have similarly complex feeding relationships.

Epipelagic food chains usually have many steps and are generally longer than in other ecosystems. The number of steps varies: Tropical food webs tend to have more levels than those in colder waters. Even in cold waters there may be five or six steps between the primary producers in the phytoplankton and a top predator. There are exceptions, of course. The diatoms-to-krill-to-whales chain, for example, is among the shortest of food chains.

The epipelagic is an exception to the rule of thumb that only about 10% of the energy contained in one trophic level is passed on to the next. Epipelagic herbivores convert more than 20%, on average, of the energy derived from phytoplankton into growth. Epipelagic carnivores, too, are more than 10% efficient, though not as efficient as the herbivores.

Even with their high efficiency, the length of epipelagic food webs means that most of the energy fixed by phytoplankton is lost before it reaches the top carnivores. After all, even an efficiency of 20% means that 80% of the energy is lost in just one step. Large animals that feed directly on plankton have an

advantage in that they have more food available than do predators that eat other high-level carnivores. By feeding directly at low trophic levels, they eliminate many intermediate steps. It is no accident that the largest epipelagic nekton, baleen whales and whale sharks, and the most abundant epipelagic fishes, like anchovies and herrings, eat plankton.

Most of the primary production in the epipelagic was once thought to flow through the simplified food chain of phytoplankton → zooplankton → nekton. "Phytoplankton" and "zooplankton" referred to the relatively large net plankton. The discovery of vast numbers of picoplankton and nanoplankton in the ocean greatly changed our understanding of epipelagic food webs.

The ocean contains huge amounts of dissolved organic matter (DOM), organic material that exists not as particles but in dissolved form. Some of the DOM simply leaks out of cells. Zooplankton spill a lot of DOM when they eat, and excrete it as waste. The largest source of DOM, however, results from the activities of viruses. There are tens of millions of viruses per milliliter in the ocean that infect large numbers of organisms, particularly microbes. The viruses cause the infected cells to burst, releasing the cells' contents.

We may not think of such material as food in the ordinary sense, but it contains a lot of energy. Calorie counters who like sugar in their coffee are familiar with this fact. Very few animals, however, are able to use DOM as an energy source. Though some phytoplankton can use it, it was once thought that the huge pool of energy in DOM went largely unused.

The Microbial Loop

We now know that the DOM in the ocean isn't just sitting there, but is the basis of a major energy pathway called the **microbial loop** (Fig. 17.24, p. 542). The microbial loop starts with phytoplankton of all sizes. As much as *half* the organic matter manufactured by these primary producers becomes DOM. The DOM is not lost to the food web, but is used by bacteria and archaea. These microbes are so small that most animals can't eat them, but they are eaten by tiny protozoans in the nanoplankton. The energy from DOM is finally passed up the rest of the food chain when net zooplankton eat the protozoans.

The microbial loop channels not only DOM but also the energy in living pico- and nanoplankton up the food chain. Photosynthetic cyanobacteria like *Prochlorococcus* and *Synechococcus* are consumed directly by protozoan grazers that are able to eat such small cells. Biologists have discovered that the cyanobacteria live in close association, perhaps even symbiosis, with heterotrophic bacteria, so that the photosynthetic production is transferred to the heterotrophic bacteria before moving up the food chain through the microbial loop.

Detritus—particulate rather than dissolved organic matter—is also important in the epipelagic, as in other marine food webs. Two important sources of detritus have already been mentioned: fecal pellets and abandoned larvacean houses. Larvacean houses and other accumulations of mucus can be very abundant and support rich populations of bacteria. They are often called **marine snow** because underwater they look like snowflakes.

IN CONTEXT

Heterotrophs Organisms that cannot make their own food and must use the organic matter produced by autotrophs.
• 6.1, The Ingredients of Life

Figure 17.24 A modern view of epipelagic food webs. The left side of the diagram shows the flow of energy from the net phytoplankton to the nekton. This part of epipelagic food webs has been known for decades. The right side of the diagram shows the more recently discovered microbial loop. Much of the flow from organisms into dissolved organic matter (blue arrows) is caused by viruses.

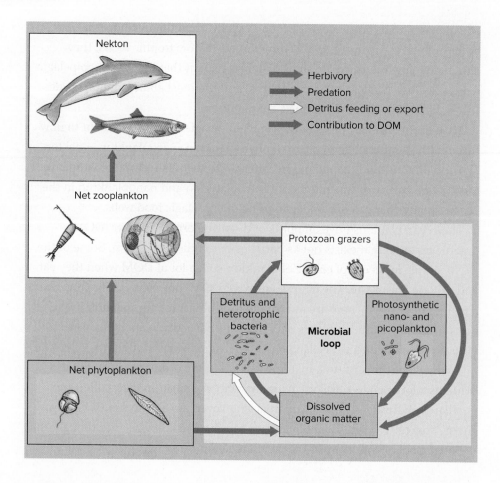

Many zooplankton and small fishes eat marine snow. Much of the detritus, however, sinks out of the epipelagic zone to deeper zones before epipelagic organisms can get to it. Furthermore, as previously noted, the epipelagic does not get much detritus from other systems.

Patterns of Production

Epipelagic food webs are complex, but they all share one simple feature: Primary production by phytoplankton, be they pico-, nano-, or net plankton, is the base. All other organisms, from the smallest zooplankton to the largest predators, depend on this primary production. Some parts of the epipelagic rank among the most productive ecosystems on Earth. There are also large areas that are among the least productive, the "deserts" of the ocean. The abundance of animals, from zooplankton to whales, generally follows the pattern of primary production (Fig. 17.25). It is therefore essential to understand the factors that control primary production by phytoplankton.

Limiting Factors Phytoplankton need two main things to perform photosynthesis. They need sunlight, the ultimate source of energy for the ecosystem, and they need a supply of essential nutrients. Without sunlight and nutrients, phytoplankton cannot grow and produce the food that fuels the food web.

Light The epipelagic is the sunlit layer of the ocean, but there is not always enough light for photosynthesis. In other words, primary production may be **light-limited**.

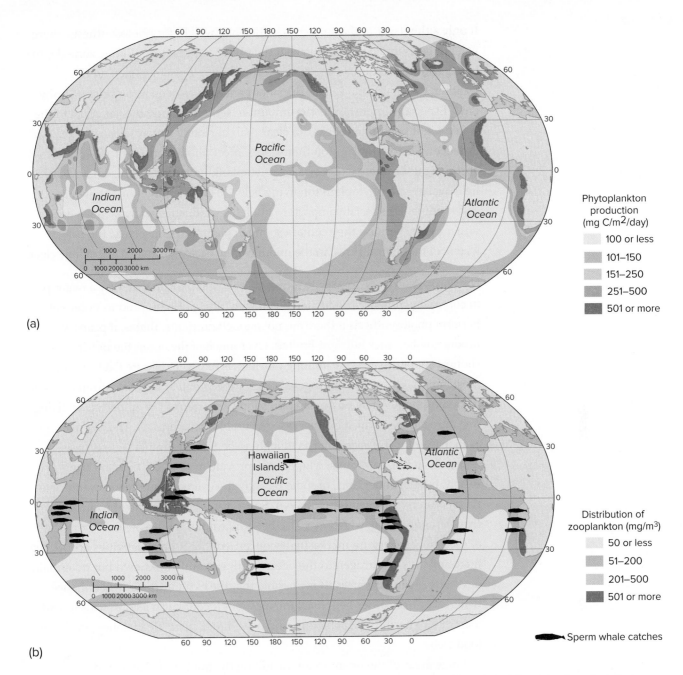

(a)

(b)

Figure 17.25 (a) The distribution of average daily primary productivity in the oceans. Note the relationship of productivity to current and upwelling systems. Because primary production is the base of the food web, this pattern is reflected at higher levels, for example, in the average distribution of zooplankton (b) and of sperm whales, as indicated by whale catches from 1760 to 1926. Some areas with low primary production, such as the Hawaiian Islands, show high catches because of whale migration and the proximity to land bases.

There is no light at night, of course, so phytoplankton must get enough light during the day to grow. At high latitudes, phytoplankton become light-limited during winter, when the days are short and the sunlight weak. In winter, phytoplankton at these latitudes store less energy in photosynthesis than they burn up in respiration. In tropical and subtropical waters, on the other hand, there is enough light to support photosynthesis throughout the year, at least at the surface.

The total primary production in the water column depends not only on the intensity of light at the surface, but also on how deep the light penetrates.

If only a thin layer at the surface is sufficiently well lit for photosynthesis, there will not be as much production as if the photic zone is deep. Photic zone depth varies with the season, being deepest in summer. It also depends on the weather. On cloudy days, light does not penetrate as deep as on bright, sunny days. Especially important is the amount of sediment and other material in the water. Light penetrates deeper in clear than in cloudy water, so the photic zone is deeper. The phytoplankton themselves affect photic zone depth. Because they absorb light for photosynthesis, the phytoplankton cut down the amount of light available to phytoplankton living below. This phenomenon, known as **self-shading**, is especially important in highly productive waters, which are rather murky because they contain so much plankton. Conversely, the barren waters of the central gyres, in the middle of the ocean basins, are incredibly clear.

Nutrients Nutrients, especially nitrogen, iron, and phosphorus, play a major part in controlling primary production. Even with plenty of light, producers cannot perform photosynthesis if there are not enough nutrients, that is, if primary production becomes **nutrient-limited**. Over much of the ocean the nutrient in shortest supply, that is, the limiting nutrient, is nitrogen. Nitrate (NO_3^{-1}) is the most important source of nitrogen. Some nitrogen enters the ocean from rivers or the atmosphere. A few phytoplankton supply nitrogen to the epipelagic by fixing free nitrogen (N_2), which is otherwise not usable by organisms. The colonial cyanobacterium *Trichodesmium* is thought to be the most important of these, especially in tropical waters, where it can provide most of the fixed nitrogen supply, but the importance of some recently discovered single-celled cyanobacteria that can fix nitrogen is not known. Some diatoms have symbiotic nitrogen-fixing cyanobacteria, which may also be significant in certain areas.

Oceanographers are increasingly concerned that human activities are adding huge amounts of nitrogen to the ocean, with currently unknown but potentially devastating consequences. It is possible that the extra nitrogen will enhance fisheries or have other beneficial impacts, but it could also lead to widespread increases in harmful algal blooms and oxygen-deficient "dead zones" or disrupt food webs.

Large areas of the open ocean, including the Southern Ocean and the equatorial south Pacific, are limited by iron instead of nitrogen. This has been demonstrated in dramatic experiments where phytoplankton grow explosively after iron is added, becoming so abundant they turn the water pea-soup green. There are highly controversial proposals to attempt to reduce global warming by fertilizing these parts of the ocean with iron, hoping that the resulting huge phytoplankton blooms will remove carbon dioxide (CO_2) from the atmosphere. Experiments to date, however, indicate that this "geoengineering" would not be effective. The ecological consequences can only be guessed at, and the risks to the planet are immense.

Phosphate (PO_4^{-3}) and various other nutrients can also limit phytoplankton growth at certain times and places. The supply of silicon sometimes limits diatom growth, for example, and a variety of substances that phytoplankton need in minuscule amounts, including vitamin B_{12}, cobalt, and nickel, may

sometimes affect productivity. Biologists are also learning that sometimes things are more complicated than having a single limiting nutrient. *Trichodesmium* needs phosphorus to fix nitrogen, for example, and when phosphorus shortage prevents its nitrogen fixation, other phytoplankton may become nitrogen-limited.

Most of the nutrients that support the ocean's primary productivity come not from new inputs but from nutrient recycling. The cycle begins when dissolved nutrients are incorporated into living organic matter by phytoplankton. Later, when the living material dies, the nutrients are **regenerated**, or released, as bacteria decompose the material.

The dependence on recycled nutrients profoundly affects production in the epipelagic. Much of the organic material produced by phytoplankton eventually ends up as detritus: fecal pellets, dead bodies, marine snow, and other organic particles. These organic particles sink, and many sink out of the epipelagic into deeper waters before they decay. The nutrients they contain are released below the photic zone (Fig. 17.26). The net result of this process is to continually strip nutrients out of the epipelagic zone into the dark, cold waters below. The epipelagic is often depleted in nutrients, and phytoplankton growth is nutrient-limited. Deep water, on the other hand, is nutrient-rich, but there is not enough light for photosynthesis, so the nutrients can't be used in primary production. The general nutrient profiles shown in Figure 17.27 (p. 546) are characteristic of most of the pelagic realm.

Because many organic particles sink out of the epipelagic before the nutrients they contain are regenerated, surface waters are nutrient-poor and phytoplankton growth is often nutrient-limited. Deep waters are nutrient-rich because of the rain of organic particles from the surface.

The phytoplankton that support epipelagic food webs, then, have a problem: At the surface there is sunlight but hardly any nutrients, whereas in deep water there are plenty of nutrients but not enough light. Neritic phytoplankton have it somewhat easier than oceanic forms. Neritic areas are relatively shallow, so the bottom traps sinking organic particles and some of the regenerated nutrients are returned to the water column by mixing. This is one reason the continental shelves are highly productive. Another reason is that rivers bring in fresh nutrients from land. Even so, neritic waters may have very low surface concentrations of nitrate, phosphate, and other nutrients.

Seasonal Patterns For primary production to occur in oceanic areas, the nutrients contained in the deep water must somehow get to the surface. The only effective way for this to happen is for the water itself to move to the surface, carrying the nutrients along with it. This sounds simple enough, but it occurs only at certain times and places. The ocean is usually stratified; that is, warm surface layers float on the denser water below (see 5.3, Thermohaline

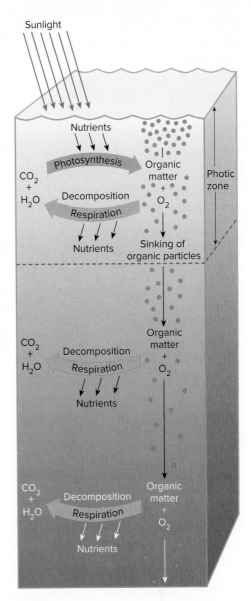

Figure 17.26 The amount of nutrients at different depths is controlled by photosynthesis, respiration, and the sinking of organic particles. In primary production, dissolved nutrients are removed from the water and incorporated into organic matter. Respiration breaks down the organic matter and regenerates the nutrients. The uptake of nutrients can occur only at the surface, where there is light. Because organic particles sink, however, much of the regeneration occurs below the photic zone.

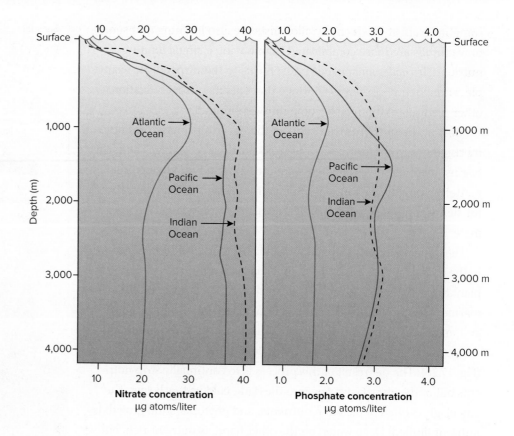

Figure 17.27 Typical profiles of nitrate and phosphate in the open ocean. Note that surface waters generally have very low levels of these nutrients.

Circulation and the Great Ocean Conveyor). A thermocline, a zone of transition, lies between the warmer, less dense layer at the surface and the colder, denser layer below. Just as energy is needed to push a cork below the surface or lift a weight up from the bottom, it takes energy to push surface water down across the thermocline into the denser water below, mixing it with the deep water, or to bring dense, nutrient-rich deep water to the surface. Much of the time this simply doesn't happen, and surface production is often nutrient-limited.

One way that deep-water nutrients get to the surface is for the water column to mix. On the continental shelf, strong wind and waves may mix the water column all the way to the bottom. Often, however, wind and waves do not mix the water deeply enough to bring nutrients to the surface, especially in the open ocean.

If the surface water gets denser, however, it loses its tendency to float. In the ocean this happens at high latitudes when winter cold and winds cool the surface water. If the surface water gets denser than the water below, overturn occurs. The thermocline breaks down, the surface water sinks, and nutrient-rich deep water is mixed up to the surface. Overturn depends on very cold conditions and is not usually widespread except in polar waters. Even in the absence of overturn, however, surface cooling can promote mixing. Since the cooler surface water becomes more dense, less energy is needed to mix it with the deep water. The mixing of deep nutrients to the surface is assisted by winter storms, which bring strong winds and large waves. Because of overturn and mixing during the winter, polar and temperate waters are highly productive (see Fig. 17.25, p. 543).

The productivity figures in Figure 17.25 refer to production over the entire year. This obscures the fact that primary production often varies during the year. In warm temperate and tropical waters such seasonal variation is relatively small.

The water column remains stable throughout the year, restricting the transport of deep nutrients up into the photic zone. For this reason, most tropical waters have low but fairly constant levels of production throughout the year (Fig. 17.28a), although there are occasional bursts of production as small-scale mixing events push localized plumes of nutrient-rich water up into the photic zone.

In temperate waters the effects of the seasons are profound. Overturn and mixing during the winter bring large amounts of nutrients to the surface. The phytoplankton are light-limited in winter, however, and unable to use the nutrients in photosynthesis. Primary production during winter months, therefore, is low at temperate latitudes (Fig. 17.28b).

Spring blooms When spring comes, the days get longer and the sunshine more intense. Light is no longer limiting, and the surface water contains plenty of nutrients because of overturn and mixing during the preceding winter. The availability of both light and nutrients provides ideal conditions for phytoplankton and produces a period of rapid growth called the **spring bloom** (Fig. 17.28b). In many temperate areas the spring bloom represents most of the primary production for the entire year.

As the spring bloom develops, two things occur, neither beneficial for phytoplankton growth. First, the increase in sunlight warms the surface water, making it less dense. This increases the stability of the water column, eliminating overturn and greatly reducing the mixing of nutrient-rich deep water to the surface. Second, the phytoplankton rapidly use up nutrients as they photosynthesize and grow. Thus, nutrients are removed from the water at the same time that the supply is blocked. The amount of available nutrients in the water falls, and the phytoplankton become nutrient-limited. Nitrate is usually the nutrient that runs out first.

During most of the summer, primary production is nutrient-limited. Productivity depends on nutrient recycling and the new input of nutrients. If the water column never becomes very stable, some mixing and therefore some production takes place. If the surface water warms considerably and the water column becomes stratified, nutrient input by mixing is low and primary production falls. Temperate waters generally have low productivity during the summer months because of stratification.

Figure 17.28 Generalized seasonal cycles of primary productivity (mg C/m²/day) in (a) tropical, (b) temperate, and (c) polar waters.

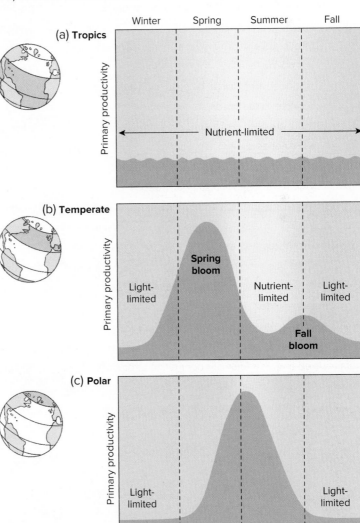

Biological Nutrient Pumps

The epipelagic tends to be low in nutrients (oligotrophic), so primary production depends heavily on nutrient transport into the photic zone. Oceanographers have traditionally looked at this as a physical process, with nutrients from deep waters supplied via upwelling, overturn, and mixing by wind and waves. Decades ago, a leading oceanographer suggested as a joke that animals might also pump nutrients up to the surface. Marine biologists are now taking this hypothesis seriously.

Animal currents When marine animals swim, their fins, bodies, and paddling appendages create turbulence in their wake. Nobody thought much about this in terms of ocean mixing until researchers calculated that the total turbulent energy that organisms could impart could be comparable to wind and waves. Shortly thereafter, research in a Canadian fjord concluded that turbulence from swarms of vertically migrating krill increased mixing by a factor of 100. Biologists soon realized that such biological mixing might be a major contributor to primary production in the ocean. A single krill probably can't do much, but there are billions of them. Larger animals like fishes and whales are not as numerous but are a lot more powerful. Unlike wind and wave mixing, this "biomixing" can occur throughout the water column, not just at the surface. Furthermore, marine animals often congregate near the boundaries of ocean layers, where mixing is most effective in pumping nutrients up from one stratified layer to another.

Like all scientific hypotheses, this one was tested. A study in Europe failed to detect significant biomixing by vertically migrating zooplankton. Theorists calculated that much of the turbulent energy that swimming animals generate is dissipated as heat instead of driving mixing. Then a research team realized that, in addition to generating turbulence, animals carry water along with them as they push through the water. This gives the water momentum, so it keeps moving for a while after the animal stops. Thus, in addition to stirring the water with turbulence, the animal creates a "mini-current." Experiments with swimming jellyfish confirmed this. Unlike turbulence generation, this mechanism doesn't lose energy through heat dissipation, and new calculations show that the total mixing energy imparted by the ocean's animals could indeed be more than that imparted by winds and waves.

Animals as nutrient pumps Animals also pump nutrients into the epipelagic by moving nutrients themselves instead of nutrient-rich water. Iron is the critically limiting nutrient in Southern Ocean around Antarctica. Antarctic krill (*Euphausia superba*) store large amounts of iron in their tissues, but the iron isn't available to support primary production by phytoplankton. In fact, krill were thought to strip iron from the epipelagic as their iron-rich fecal pellets sank to abyssal depths. Scientists recently discovered, however, that krill travel out of the epipelagic, feeding on iron-rich detritus and copepods in the abyssal depths. Thus, the krill scavenge this critical nutrient and carry it with them when they return to surface waters.

Of course, the nutrients mostly remain "locked up" in the animals' bodies, though some are excreted. This is where whales and large predators may come in. In the Antarctic, for example, baleen whales eat huge amounts of krill, and upon defecating they release iron in concentrations 10 million times that of the surrounding water. Similarly, whales and seals release more nitrogen via excretion into the Gulf of Maine than all of the inflowing rivers. Scientists were long perplexed by the decline in primary production in many parts of the epipelagic over the last century or so. The drastic reduction of baleen whale and top-predator populations by humans may provide part of the answer.

Fall blooms Production in autumn depends largely on which comes first: the short days and weak sunlight that bring light limitation, or surface cooling and storms, which increase mixing of nutrients. If the days are cool and sunny, for example, a strong wind can mix up nutrients while the phytoplankton still have enough light to grow. If so, there will be a burst of production known as the **fall bloom** (Fig. 17.28b, p. 547). If warm, mild conditions persist late into the fall the phytoplankton become light-limited before overturn and mixing begin. Even though production at temperate latitudes is light-limited during the winter and nutrient-limited during the summer, enough production takes place during blooms to give these waters a high total production over the year (see Fig. 17.25).

Cold polar waters may not become stratified at all and may continue to produce at very high levels throughout the short summer when the sun shines most of the day (Fig. 17.28c). Primary production falls as the days shorten and is greatly reduced by light limitation during the long winter months. Polar waters can be extremely productive, as is true in some parts of the Southern Ocean. Production in much of the Southern Ocean, however, is limited by a shortage of iron. The Arctic Ocean lies at such a high latitude and is so extensively covered by ice that it is generally light-limited and unproductive in spite of overturn and mixing.

Geographic Patterns Water depth and temperature, prevailing winds, and surface currents all affect the ocean's productivity, as we have seen in the previous sections. The geographic distribution of productivity (see Fig. 17.25) reflects these factors. Coastal waters are highly productive because the shallow bottom prevents nutrients from sinking out of the photic zone and because wind and waves mix the water column. Coastal areas that experience upwelling are especially productive. Equatorial waters, too, are productive because of upwelling, but not as productive as coastal upwelling zones. Polar and cold temperate regions have high productivity because overturn and upwelling bring nutrients up into the sunlit layer.

None of the processes that bring nutrients to the surface operate in the vast central gyres. These lie in relatively warm latitudes, so the surface waters never cool enough for overturn or mixing. Far from both the coast and the equatorial currents, the central gyres never experience upwelling, either. The gyre regions are permanently limited by nutrients; with the exception of the light-limited Arctic Ocean, they have the lowest productivity in the epipelagic. In fact, the central gyres are among the least productive ecosystems on Earth. As you will recall from Chapter 5, productivity patterns are also influenced by large-scale events like El Niño-Southern Oscillations.

. .

Section Review

1. **Describe** an epipelagic food chain that involves tuna.

2. **Describe** What is the microbial loop? Why is it important?

3. **Discuss** Use the graphs in Figure 17.28 to discuss seasonal variations in primary production in tropical, temperate, and polar waters.

Chapter 17 Review

REVIEW QUESTIONS

Multiple Choice

1. Which group of plankton consists mostly of viruses?
 a. femtoplankton
 b. nanoplankton
 c. net plankton
 d. picoplankton

2. Which zooplankton are transparent and related to sea squirts?
 a. copepods
 b. krill
 c. pteropods
 d. salps

3. Organisms that live right at the sea surface but remain underwater are called
 a. nekton.
 b. neuston.
 c. plankton.
 d. pleuston.

Short Answer

4. What is the significance of zooplankton in epipelagic food web?

5. How do fish regulate their swim bladders?

6. How does upwelling impact primary productivity?

Critical Thinking

7. Plankton are unable to swim effectively and drift about at the mercy of the currents. You might think that the currents would scatter planktonic organisms throughout the oceans, but many species are restricted to particular regions. What mechanisms might allow a species to maintain its characteristic distribution?

8. Spiny species of diatoms are found in both warm subtropical waters and colder areas. Because warm water is less dense than cold water, would you predict any differences between the spines of warm-water and cold-water individuals? Why?

DATA ANALYSIS LAB

How might seasons affect vertical migration in zooplankton? Marine scientists have known for many years that zooplankton migrate vertically in ocean waters—spending the daylight hours at depth and rising to the surface at night. They hypothesize that this behavior allows zooplankton to avoid predation during the day, while feeding on phytoplankton at the surface at night.

Data and Observations

The graph shows the average time spent at the surface each night for several species of copepods in the North Atlantic Ocean throughout the year.

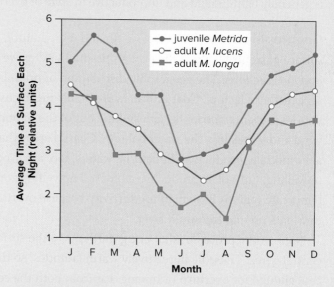

Think Critically

1. Describe the trend in the amount of time spent by the zooplankton at the surface at night throughout the year.

2. How does this pattern coincide with the change in seasons?

3. Juvenile *Metrida* are smaller than both adult species of *Metrida* shown in the graph. The adult *Metrida longa* is the largest of the three. What differences exist in length of time spent at the surface based on body size? Present a hypothesis to explain these differences.

Data obtained from: Hays, G., 2003. A review of the adaptive significance and ecosystem consequences of zooplankton diel vertical migrations. Hydrobiologia 503: 163–170, 2003.

The Ocean Depths

Chapter

18

THEME Structure and Function

Living in the deep ocean requires organisms to be adapted to a variety of harsh conditions including low or no light, high pressure, and low availability of food and mates.

BIG IDEA

Mesopelagic and deep-sea animals have adaptations that allow them to survive in a unique environments.

About The Photo The underside of this deep-water squid is lined with photophores which produce light and help to camouflage the squid.

Nathan Tia, Simplynate Photography/Moment/Getty Images

18.1 The Twilight World

VOCABULARY

midwater animals

photophores

bioluminescence

deep scattering layer

tubular eyes

retina

counterillumination

oxygen minimum zone

Main Idea

Organisms in the mesopelagic must deal with extremely low light conditions as well as a limited food supply.

Key Questions

1. Describe the mesopelagic zone.
2. Describe the relationship between depth and the number of pelagic organisms. What explains this relationship?

DIVE IN!

Vampires in the oceans? Despite their frightening name, vampire squid (*Vampyroteuthis infernalis*) do not suck blood and are not top predators. These small squid feed mainly on detritus and plankton. What is unique about vampire squid is their reproductive pattern. Most cephalopods reproduce just once during their life, a massive event that requires a great deal of energy. In contrast, vampire squid reproduce in several times throughout their life. The University of California, Santa Barbara researchers who discovered this hypothesize that multiple spawns are a strategy to conserve energy. The amount of energy a squid can put into reproduction is limited by the amount of food it can find. Since food is limited in the deep sea, multiple spawning events allow female squid to store up enough energy to produce eggs.

"Inner space" it's been called. Dark, cold, and inhabited by bizarre, fearsome-looking creatures, it *is* reminiscent of the outer space of science-fiction movies. As in outer space, humans can venture into this mysterious realm only in elaborate, specially designed craft. Even then there is an element of risk. But "inner space" is here on Earth: the ocean waters below the sunlit surface layer.

Several different habitats are discussed in this chapter; each supports a distinct community of organisms. These communities share one important feature: the lack of primary production by photosynthesis (Fig. 18.1). Without primary production to support the rest of the food web, most organisms below the photic zone depend for food on organic material produced in the surface layers of the ocean. Some of this surface production sinks into the dark waters below. Without a steady supply of food from above, there could be little life below the sunlit layer of the sea. Exceptions to this are discussed in section 18.4, Hot Springs, Cold Seeps, and Dead Bodies.

The Mesopelagic Habitat

Because of the dependence on surface production for food, life is much less abundant below the photic zone than in the sunlit surface layer. Most food particles get eaten before they sink into deeper water. With food in short supply, pelagic organisms become more and more scarce at greater depths. There are typically 5 to 10 times fewer organisms at 500 m, for example, than there are at the surface, and perhaps 10 times fewer again at 4,000 m.

Deep-water organisms depend on the surface not only for food, but also for oxygen (O_2). If the ocean were stagnant, the oxygen below the surface would be quickly used up by respiration and animal life would be impossible. Fortunately, however, thermohaline circulation and the great ocean conveyor constantly replenish the supply of oxygen to the deep sea. Oxygen still becomes depleted in some places, but by and large the ocean depths have enough oxygen to support respiration.

The mesopelagic is a world of twilight. In the upper part of the mesopelagic the dim light during the day is enough to see by, but there is not enough light to support phytoplankton growth. As depth increases, of course, the sea gets

darker. Eventually, typically at a depth of about 1,000 m, there is no light at all. The absence of light marks the bottom of the mesopelagic zone, which thus stretches from the bottom of the epipelagic, around 200 m deep, to about 1,000 m deep.

The temperature at a given depth in the mesopelagic generally varies much less than in the epipelagic. The mesopelagic, however, is the zone where the main thermocline occurs (Fig. 18.1), so organisms that move up and down in the water column encounter large changes in temperature. Mesopelagic organisms that stay at about the same depth experience more constant temperatures.

Figure 18.1 Life in the mesopelagic and deep sea is closely linked to the abundance of plankton and light intensity in the water column.

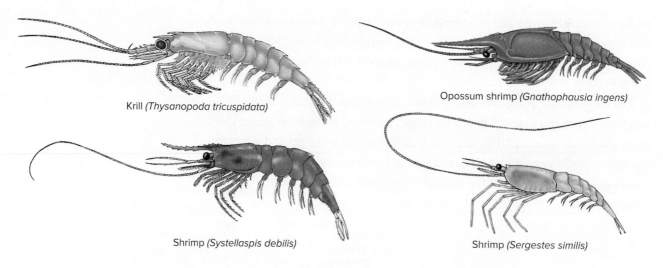

Krill *(Thysanopoda tricuspidata)*

Opossum shrimp *(Gnathophausia ingens)*

Shrimp *(Systellaspis debilis)*

Shrimp *(Sergestes similis)*

Figure 18.2 Some crustaceans that inhabit the mesopelagic zone include krill, opossum, or lophogastrid, shrimps, and true (decapod) shrimps like *Systellaspis debilis* and *Sergestes similis*.

IN CONTEXT

Krill Planktonic, shrimp-like crustaceans

- 9.4, Arthropods: The Armored Achievers

Copepods Tiny crustaceans that are most often planktonic

- 17.1, The Organisms of the Epipelagic

Arrow Worms or **Chaetognaths** Small, worm-like predators with fins along the body and spines on the head.

- 9.2, Marine Worms and Some Small Invertebrate Phyla

The Animals of the Mesopelagic

Though phytoplankton and other photosynthetic organisms cannot grow in the dim light, the mesopelagic supports a rich and varied community of animals, which are often called **midwater animals**.

Zooplankton The major groups of animals in the mesopelagic zooplankton are much the same as those in the epipelagic. Krill and copepods are generally dominant, as they are in surface waters. Several different kinds of shrimps are relatively more common in the mesopelagic than in the epipelagic (Fig. 18.2). Krill and most mesopelagic shrimps have a common adaptation of midwater animals: **photophores**, or light organs, which are specialized structures that produce light. The function of this "living light," called **bioluminescence**, is discussed later in the chapter.

A group of crustaceans known as ostracods can be very abundant in the mesopelagic. Ostracods have a characteristic shell, or carapace, that makes them look like tiny clams with legs (Fig. 18.3). They are crustaceans, however, and unrelated to clams. Like copepods, most ostracods are small, usually only a few millimeters long. One group (*Gigantocypris*), however, can reach 1 cm. Amphipods and other crustaceans are also part of the midwater plankton.

Other Invertebrates Arrow worms, or chaetognaths, are important midwater predators. At times they are among the most abundant components of the midwater zooplankton, especially in the upper parts of the mesopelagic. Jellyfishes, siphonophores, comb jellies, larvaceans, and pteropods are also common. Most of the other major groups that are part of the epipelagic zooplankton are also represented in the mesopelagic, though the particular species are usually different.

Cephalopods, especially squids, are also prominent members of the midwater community (Fig. 18.4). Some squids swim weakly and are planktonic, while strong-swimming squids are part of the nekton. Mesopelagic squids usually

have photophores, typically arranged in a species-specific pattern. The vampire squid (*Vampyroteuthis*; Fig. 18.5) looks something like an octopus but is neither a squid nor an octopus and is in its own separate group. Like true squids, the vampire squid has photophores. There are also a few species of octopuses in the mesopelagic zone. Bioluminescence is not as common in mesopelagic octopuses as in squids, but some species do have photophores. One (*Stauroteuthis*) has light-emitting suckers. Like the vampire squid, mesopelagic octopuses have evolved membranes that connect their tentacles like the web between a duck's toes. This may help them float like a parachute in the water or envelope prey.

Figure 18.3 A mesopelagic ostracod (*Gigantocypris*). This one is about 1 cm long. Most ostracods are much smaller.

Abraliopsis *Chiroteuthis* *Thelidioteuthis* *Pterygioteuthis*

Figure 18.4 Most mesopelagic squids have photophores (shown in black), usually arranged in a different pattern in each species.

Midwater Fishes Nearly all mesopelagic fishes are small, typically 2 to 10 cm long, although a very few species get considerably larger. Bristlemouths and lanternfishes (Fig. 18.6, p. 556) are the most abundant fishes in the mesopelagic, and may account for 90% or more of the fishes collected by midwater trawls (Fig. 18.7, p. 557). Bristlemouths are the most common of all. One species (*Cyclothone signata*) is the most abundant fish on Earth, surprising and impressive when you consider the huge schools of fishes like sardines and herrings in the epipelagic. Bristlemouths are named for their many sharp teeth. They have rows of photophores on their underside, or ventral surface.

Lanternfishes get their name from the photophores adorning their heads and bodies. Like squids, each species typically has a distinctive photophore pattern. Lanternfishes have blunt heads, relatively large mouths, and large eyes, probably an adaptation to help them see in the dim light. They are very general in their food habits, eating just about anything they can get down.

Bristlemouths and lanternfishes are the most numerous, but many other fishes live in the mesopelagic. Marine hatchetfishes look somewhat like the unrelated freshwater hatchetfishes sold in pet shops, except for their large eyes and mouths and ventral photophores. Viperfishes, dragonfishes, barracudinas, sabertooth fishes, lancetfishes, snake mackerels, and cutlassfishes are all long, eel-like fishes with large mouths and eyes. Many have photophores, usually in rows on the ventral surface. Most are less than 30 cm long, but

Figure 18.5 The vampire squid (*Vampyroteuthis infernalis*) differs from the closely related true squids in that 2 of its 10 tentacles are modified into long, retractable sensory tentacles, instead of being used to capture prey. The vampire squid only grows to about 30 cm.

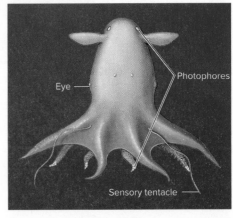

Eye

Photophores

Sensory tentacle

there are exceptions. One of the lancetfishes (*Alepisaurus ferox*; Fig. 18.6) grows to about 2 m. The black scabbard fish (*Aphanopus carbo*; Fig. 18.8), a kind of cutlassfish, can get well over 1 m long.

Adaptations of Midwater Animals

By the standards of the surface world, many of the animals that live below the photic zone seem bizarre, indeed. Though they look strange to us, they are well adapted to their unique environment. Species from the same depth often have very similar characteristics even though they are unrelated. On the other hand, closely related species from different depths can differ markedly in appearance and other characteristics.

Feeding and Food Webs

Most of the food produced in the epipelagic is used there; only about 20% of surface primary production sinks to the mesopelagic. This means that the mesopelagic is chronically short of food, which is why there are fewer organisms in the mesopelagic than in the epipelagic. The abundance of midwater organisms in a particular geographic area reflects the productivity of the waters above. There is more mesopelagic life under highly productive surface waters than under areas with low primary production.

Many of the characteristics of midwater animals are directly related to the lack of food in the mesopelagic. The small size of midwater fishes, for example, is thought to be an adaptation to the limited food supply. Animals need a lot of food to grow large, so there may be an evolutionary advantage to limited growth and small size.

Figure 18.6 Some typical mesopelagic fishes and their approximate maximum length.

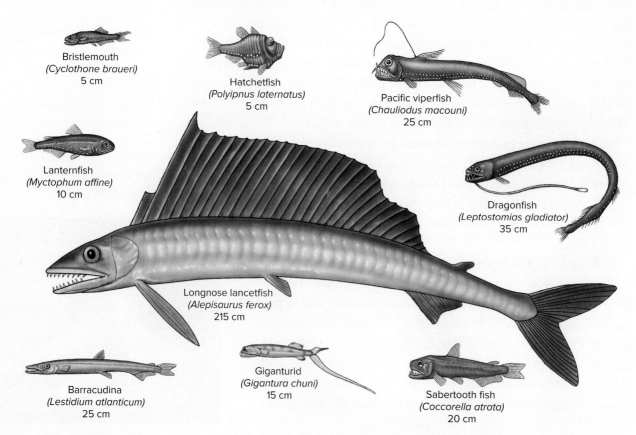

Bristlemouth
(*Cyclothone braueri*)
5 cm

Hatchetfish
(*Polyipnus laternatus*)
5 cm

Pacific viperfish
(*Chauliodus macouni*)
25 cm

Lanternfish
(*Myctophum affine*)
10 cm

Dragonfish
(*Leptostomias gladiator*)
35 cm

Longnose lancetfish
(*Alepisaurus ferox*)
215 cm

Barracudina
(*Lestidium atlanticum*)
25 cm

Giganturid
(*Gigantura chuni*)
15 cm

Sabertooth fish
(*Coccorella atrata*)
20 cm

(a)

(b)

Figure 18.7 Nets called rectangular midwater trawls are commonly used to collect mesopelagic organisms. The net can be opened and closed by remote control when it is at the desired depth. This prevents surface organisms from being caught while the net is being lowered and retrieved. In both (*a*) and (*b*) the mouth of the net is open.

Figure 18.8 A black scabbard fish (*Aphanopus carbo*) being sold by a Madeiran fisherman. The fish, known locally as *espada*, is a relatively rare example of a midwater fish that is caught for food.

Midwater fishes usually have large mouths, and many have hinged, extendible jaws equipped with fearsome teeth (Figs. 18.9 and 18.10, p. 558). Because food is scarce below the photic zone, most fishes cannot afford to be picky. They usually have very broad diets and eat just about anything they can fit into their mouths. The large, protrusible jaws allow them to eat a wide range of prey. The advantage of this is that they don't have to pass up a potential meal because it is too large to eat. Some mesopelagic fishes can even eat prey larger than themselves! The long, sharp teeth help keep prey from escaping.

Midwater animals fall into two major groups: those that stay in the mesopelagic and those that migrate to the surface at night. The non-migrators include a few species of small zooplankton, mainly copepods and krill, that filter out detritus and the small amount of phytoplankton that sinks out of the photic zone. The fecal pellets of epipelagic copepods and other surface grazers form an important part of the detritus eaten by mesopelagic filter feeders. Fecal pellets sink much faster than individual phytoplankton cells, so they stand a better chance of making it to the mesopelagic before being eaten.

Most non-migrating midwater animals, however, are not zooplankton but fishes, shrimps, and squids. They are sit-and-wait predators that lurk in the dim light, gulping down anything that comes within range. With food hard to come by, these animals are adapted to conserve energy. Instead of energy-consuming muscle, non-migrating midwater fishes have flabby, watery flesh. To fill their swim bladder, fishes must produce a gas pressure inside the bladder that matches the external water pressure. Because pressure is greater at depth than at the surface, filling the swim bladder requires more energy in the mesopelagic. As a result, most non-migrating midwater fishes have lost the swim

IN CONTEXT

Detritus Particles of dead organic matter.
- 12.3, The Flow of Energy and Materials

Swim Bladder A gas-filled sac in the body cavity of most bony fishes that helps them control their buoyancy.
- 10.2, Fish Anatomy

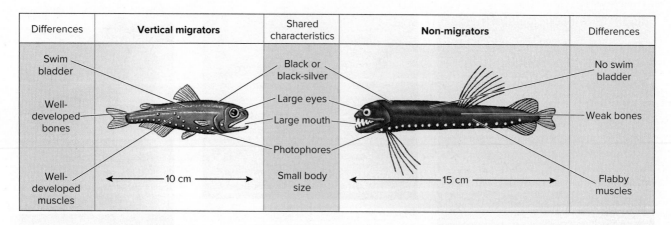

Differences	**Vertical migrators**	Shared characteristics	**Non-migrators**	Differences
Swim bladder		Black or black-silver		No swim bladder
Well-developed bones		Large eyes		Weak bones
		Large mouth		
		Photophores		
Well-developed muscles	← 10 cm →	Small body size	← 15 cm →	Flabby muscles

Figure 18.9 Some adaptations of typical mesopelagic fishes, including some differences between vertical migrators like lanternfishes (left) and non-migrators like dragonfishes (right). Compare these with the adaptations of epipelagic and deep-sea fishes.

bladder. To compensate, they have evolved soft, weak bones and lost defensive structures like spines and scales. These adaptations reduce weight and make them more neutrally buoyant, so they can float at a constant depth without wasting energy swimming. Since they don't swim much, they have little need for the streamlining that is so characteristic of epipelagic fishes.

Vertical Migration and the Deep Scattering Layer Instead of staying put and waiting for food to fall from above, many mesopelagic organisms make vertical migrations. They swim up at night to feed in the rich surface layers and descend during the day to depths of several hundred meters or more. In the dim light they are probably relatively safe from predators. Some vertical migrators spend the day in a lethargic stupor, conserving energy until their next foray to the surface. Vertical migration is also seen in many zooplankton that live in the deeper parts of the epipelagic.

Vertically migrating fishes differ in several important ways from those that stay in the mesopelagic (see Fig. 18.9). Well-developed muscles and bones are needed to make the daily swim up and down the water column. These structures are heavy, so mesopelagic fishes that vertically migrate have retained the swim bladder for buoyancy. As they move up and down, they experience dramatic changes in pressure. Some vertically migrating fishes can rapidly

Figure 18.10 (a) The viperfish (*Chauliodus*) has hinged jaws that can accommodate large prey. (b) The rattrap fish (*Malacosteus*) has a similar jaw arrangement.

(a)　　　　　　　　　　**(b)**

adjust the volume of gas in the swim bladder to prevent it from collapsing or exploding when they change depth. Many, however, have a swim bladder filled with fat instead of gas. Fat, unlike gas, does not expand and contract with pressure changes, making it much easier for the fishes to regulate their buoyancy. Vertically migrating fishes can also tolerate the temperature changes they experience as they move up and down across the thermocline.

Vertical migration by mesopelagic animals was discovered during World War II, when sonar came into use. Sonar soundings regularly showed a series of sound-reflecting layers, or "false bottoms." The real bottom gives a sharp, clear echo, but these layers, collectively dubbed the **deep scattering layer** (DSL), give a soft, diffuse echo that makes a shadowy trace on sonar plots. During the day the DSL lies at depths of 300 to 500 m, but at sunset it rises to the surface. The depth of the DSL is clearly related to light intensity: The DSL stays deeper when there is a full moon than on moonless nights and even moves up and down when clouds pass over the moon.

Net tows have shown that the DSL is composed of fishes—especially lantern-fishes—krill, shrimps, copepods, jellyfishes, squids, and other midwater animals. Most of these organisms, though they are found within the DSL, do not contribute to its sound-reflecting properties. Because organisms are made mostly of water, sound passes through them much as it does through seawater, without bouncing off. Air pockets, however, reflect sound quite strongly. The echo of the DSL comes mostly from sound waves bouncing off the gas-filled swim bladders of some vertically migrating fishes.

When lanternfishes, krill, and other animals return to the mesopelagic after feeding near the surface, they carry the products of surface production down with them. Thus, vertical migration greatly increases the food supply in the mesopelagic. Many non-migratory midwater predators feed heavily on vertically migrating species. The migrators are more muscular than non-migrators and provide a nutritious meal.

Figure 18.11 Many midwater fishes, like this barreleye (*Dolichopteryx longipes*), have tubular eyes that give them very acute upward vision. The large white lens at the top of the eye is very effective at gathering the dim light in the mesopelagic. This species has a second lens on the side of the eye, visible below the white lens that provides peripheral vision. In some species the tubular eyes point forward.

Sense Organs To help them see in the dim light, midwater fishes characteristically have eyes that are large and highly sensitive. Large, light-sensitive eyes also occur in squids, shrimps, and other groups. Some midwater fishes have developed **tubular eyes**, a complex visual system that is almost like having two pairs of eyes (Fig. 18.11). Tubular eyes give very acute vision in the direction that the eyes are pointed, either upward or forward, but are not good for lateral vision. Some fish species with tubular eyes can rotate the eyes to look in different directions. In addition, in many fishes with tubular eyes the **retina**, the light-sensitive part of the eye, extends partway up one side of the eye (Fig. 18.12a, p. 560). This allows the fish to see to the side and below as well as upward (Fig. 18.13, p. 560). In normal fish eyes the retina lies only at the back of the eye.

Dante Fenolio/Science Source

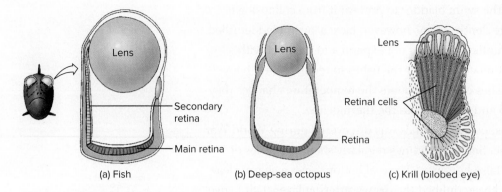

Figure 18.12 (a) The tubular eye of a midwater fish (*Scopelarchus*). Eyes with a similar structure have evolved in (b) octopuses (*Amphitretus pelagicus*) and (c) krill (*Stylocheiron suhmii*).

(a) Fish

(b) Deep-sea octopus

(c) Krill (bilobed eye)

Lens · Secondary retina · Main retina

Lens · Retina

Lens · Retinal cells

Upward visual field

Lateral visual field

Figure 18.13 The field of vision of a midwater fish (*Scopelarchus*). The fish has two main visual fields corresponding to the two parts of the retina shown in Figure 18.12a.

IN CONTEXT

Countershading A pattern of coloration common in open-water nekton in which the back is black or dark blue and the underside white or silver.

• 17.2, Living in the Epipelagic

Adaptations very similar to tubular eyes are found in at least one octopus (*Amphitretus;* Fig. 18.12b) and in the bilobed eyes of some krill (*Stylocheiron;* Figs. 18.12c). One squid, *Histioteuthis,* has taken a different path to the same apparent end. One of the squid's eyes is much larger than the other. The squid floats in the water column with the large eye pointed upward and the small eye downward.

Coloration and Body Shape Like their epipelagic counterparts, mesopelagic predators rely heavily on vision. Because midwater prey cannot afford the energetic costs of fast swimming or heavy defensive spines and scales, camouflage is perhaps even more important than in the epipelagic. The basic strategies, however, are very similar: countershading, transparency, and reduction of the silhouette. There are many variations on these basic strategies, especially in relation to depth and light levels.

Transparency is particularly common in the shallower and better-lit parts of the mesopelagic. Copepods, jellyfishes, shrimps, bristlemouths, and other animals that live in the upper mesopelagic tend to be transparent, some almost completely so. Deeper in the mesopelagic, fishes tend to be more silvery, and in the deepest, darkest part, black or red. Zooplankton from the deeper mesopelagic are often red, orange, or purple. These colors might be conspicuous at the surface, but colors change underwater. Our eyes see a fish as red at the surface because its skin absorbs most of the colors in sunlight but reflects the red. Red light doesn't penetrate to mesopelagic depths, however, and the red pigment absorbs the blue light that does, so the organism appears dark against the dark background of the surrounding water.

Mesopelagic fishes often have black backs and silvery sides, reminiscent of the countershading of epipelagic animals. In the mesopelagic, however, there is not enough light to reflect off the white or silver belly typical of epipelagic countershading to mask the animal's outline. Even a white object produces a silhouette in twilight. A silhouette makes the animal conspicuous and vulnerable to all those sharp eyes peering up from the depths. To reduce their silhouettes, some mesopelagic fishes, like hatchetfishes, have laterally compressed bodies. This reduces the size of the body outline whether the animal is viewed from above or below.

Some mesopelagic squids and octopuses can switch between camouflage strategies. Many cephalopods have a chameleon-like ability to change color, and

some midwater species can change from being practically transparent to a mottled reddish-orange depending on light level and the presence of predators.

Bioluminescence Most midwater animals have evolved an even more effective way to mask their silhouettes. Their bioluminescent photophores, found mostly on their undersides, produce light that breaks up the silhouette and helps the animal blend in with the background light filtering down from the surface (Fig. 18.14). This adaptation, which functions in a similar manner to countershading, is called **counterillumination**.

The light produced by midwater animals is closely matched to the background light. Like the natural light at these depths, nearly all midwater bioluminescence is blue-green. Furthermore, many mesopelagic animals can control the brightness of the light they produce to match the brightness of the background. This has been shown experimentally by placing special blinders on shrimps and other animals. The blinders allow experimenters to control how much light the animal sees (Fig. 18.15). When the animal is exposed to bright light, it produces bright bioluminescence; when the light is dim, so is the bioluminescence. When opaque blinders are put on the shrimp so that it sees no light, it turns off its photophores completely. This control of the bioluminescence is essential. It is easy to see an animal if it produces light at night or if the light is brighter than the background. On the other hand, the animal creates a silhouette during the day if the light isn't bright enough.

Figure 18.14 The value of having photophores on the ventral surface of midwater animals is shown in this simple demonstration. (a and b) Silhouettes of two organisms as they would appear without photophores. (c and d) The same animals with the photophores in white to match the background. This effect is more pronounced when the animals are a little out of focus (b and d), as they are when viewed through the water.

Figure 18.15 This experiment showed that a midwater shrimp (*Sergestes similis*) produces bioluminescence to match the background light intensity. Shrimp having both eyes covered with blinders did not produce any light (lower line), no matter how much outside light there was. Those with clear blinders produced more light as the outside level was increased (upper line).

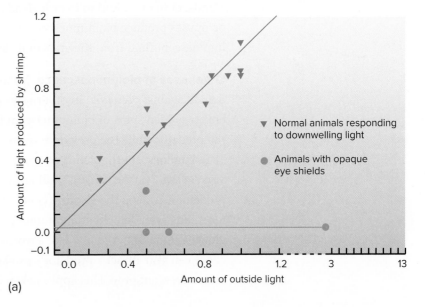

(a)

Cookie-cutter sharks (*Isistius*) appear to use a twist on counterillumination to attract prey rather than avoid predators. Cookie-cutter sharks cut out oval chunks of flesh from much larger fishes and even dolphins, and also eat whole squids. Cookie-cutter sharks swim only weakly, however, making it hard for them to attack their faster-moving prey. These sharks have diffuse photophores over almost all of their undersides, which is thought to blend in with light from the surface as in normal counterillumination. Just behind their mouth, however, is a black band without photophores that creates a false silhouette. The silhouette is thought to attract predators from below, which the shark attacks as they approach.

Counterillumination arms race Midwater organisms have evolved ways to defeat counterillumination. The bioluminescence of most mesopelagic organisms nearly matches the color of natural light coming from above, but is just a bit greener. Some fishes have visual pigments that allow them to distinguish subtle differences among colors in the blue-green range, and many others have yellow filters that serve the same purpose. Some midwater squids can also tell the difference between bioluminescence and natural light, but they do it by having the equivalent of polarized sunglasses rather than color discrimination—natural light in the mesopelagic and bioluminescence are polarized differently. The use of counterillumination to avoid predators, and even attract prey, and the evolution of measures to defeat counterillumination, is a classic example of the "arms races" that evolve between predators and prey.

Most midwater animals are bioluminescent, and midwater organisms have evolved many different ways to produce light. Photophores, as we have seen, are common. In some species the light is produced by the animal's own specialized tissue. In other species, symbiotic bacteria live inside the light organ and produce the light. In either case, the photophore can be quite complex (Fig. 18.16).

Bioluminescence is not always produced by specialized photophores. Many jellyfishes and other gelatinous animals produce light with cells that are scattered over the body surface. Some copepods, ostracods, shrimps, and other animals secrete bioluminescent fluids that they squirt out through special glands, either in addition to or instead of having photophores. Some squids and octopuses produce bioluminescent ink, and the vampire squid can release luminescent fluid from the ends of its tentacles.

Other uses of bioluminescence Mesopelagic animals use bioluminescence to defend against predators in other ways than counterillumination. Many flash or emit a sudden spurt of bioluminescent fluid when attacked, which may startle the predator while the prey darts away.

Bioluminescent fluid emitted by some squids, octopuses, and shrimps may also act as a "smoke screen," hiding the escape route, or as a decoy, with the predator attacking the glowing cloud of fluid instead of the prey. Some midwater squids even detach, or autotomise, a tentacle when disturbed, and the glowing tentacle, wriggling like a worm, may be an effective decoy. Bioluminescence can also attract a predator's predators, making it riskier to attack the bioluminescent prey. This applies even if the prey is eaten, since the prey, or its

body parts, may continue to glow inside the predator. Some mesopelagic animals that can regrow lost body parts, including the squid just mentioned, may sacrifice bioluminescent body parts for this purpose. As a counter-strategy, many mesopelagic predators that are otherwise transparent have an opaque lining around their gut to black out their glowing meals.

Bioluminescence has offensive as well as defensive functions. The cookie-cutter shark is not the only mesopelagic animal that uses bioluminescence to attract prey. Mesopelagic fishes like viperfishes and dragonfishes (see Figs. 18.6 and 18.10, p. 556 and 558) have a modified, elongated fin spine or a barbel under the chin, with a lure, a fleshy bit of bioluminescent tissue, on the end. The fishes dangle the glowing lure in front of their mouths. Anglerfishes use the same tactic, though they are more common in the bathypelagic zone than the mesopelagic. The mesopelagic squid *Chiroteuthis* (see Fig. 18.4, p. 555) has light organs on the end of long modified tentacles that hang down as lures. Even the bioluminescent suckers of the octopus *Stauroteuthis* are thought to attract prey. Some midwater fishes and squids seem to use bioluminescence to stun or confuse their prey. Some mesopelagic fishes have light organs around their eyes that may help them see their prey. In a few species, such photophores produce red light instead of the usual blue-green. The red light is invisible to most other midwater organisms, but these fishes can see it and probably use it to spy undetected on potential prey.

Mesopelagic animals probably also use bioluminescence for purposes unrelated to predation. The distinctive pattern of photophores on many species, which often differ between males and females, may help them identify potential mates. Some mesopelagic fishes, squids, and octopuses are also thought to use bioluminescence to communicate.

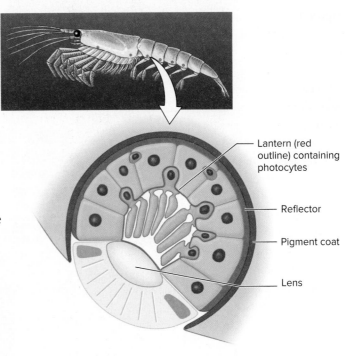

Figure 18.16 Cross section of a krill photophore.

Lantern (red outline) containing photocytes

Reflector

Pigment coat

Lens

The Oxygen Minimum Zone In many places, midwater organisms have to deal with a shortage of oxygen in the water. Oxygen enters the ocean in two ways: by gas exchange with the atmosphere and as a by-product of photosynthesis. Once a water mass leaves the surface and descends to mesopelagic depths, there is no way for it to gain oxygen. It is removed from contact with the atmosphere, and there is not enough light for photosynthesis. Respiration by decay bacteria, however, continues to use up oxygen (Fig. 18.17, p. 564). As a result, the water becomes depleted in oxygen, often in a fairly well-defined layer around 500 m deep known as the **oxygen minimum zone** (OMZ). The amount of oxygen in the OMZ can drop to practically nothing. Below the OMZ there is very little food and therefore very little respiration and decomposition, so oxygen is not used up so quickly. Thus, the water below the OMZ retains most of the oxygen it had when it left the surface.

Figure 18.17 Surface waters are rich in oxygen from the atmosphere and photosynthesis. In the mesopelagic zone neither source contributes oxygen, but bacterial decomposition of sinking organic matter sinking uses up oxygen. This results in an oxygen minimum zone. Below the oxygen minimum zone, most of the organic matter has already decayed, and oxygen remains dissolved in the water. Additional oxygen is brought in by the deep thermohaline circulation.

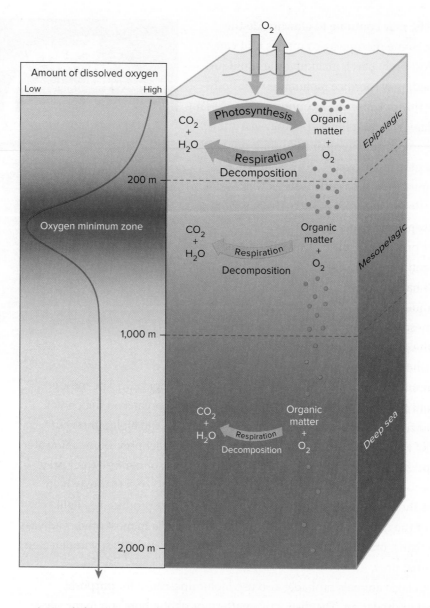

IN CONTEXT

Hemoglobin A blood protein that transports oxygen in many animals; in vertebrates it is contained in erythrocytes (red blood cells).

• 10.2, Fish Anatomy

Animals live in the oxygen minimum zone despite the low oxygen levels. Fishes, krill, and shrimps that live there usually have large, well-developed gills to help extract what little oxygen there is. They also tend to be relatively inactive, which lowers their oxygen consumption. Many also have complex biochemical adaptations, like hemoglobin that functions well at low oxygen concentrations.

Section Review

1. **Summarize** Make a table like the one below to summarize the adaptations of the two main groups of mesopelagic animals.

Group	Example Animals	Adaptations
Zooplankton		
Fishes		

2. **Explain** What is bioluminescence and what do midwater animals use it for?

3. **Describe** What is the oxygen minimum zone (OMZ)?

18.2 The World of Perpetual Darkness

Main Idea

Deep-sea animals live in conditions of complete darkness, high pressure, and a limited amount of food.

. .

Key Questions

1. What are the zones that the deep sea is divided into?
2. What are the general characteristics of deep-sea pelagic fishes?

VOCABULARY

pheromones

male parasitism

DIVE IN!

Below the mesopelagic lies the little-known world of the deep sea, where sunlight never penetrates. This alien environment is vast, indeed. It is the largest habitat on Earth and contains about 75% of our planet's liquid water. The deep sea can be divided into several pelagic depth zones. The bathypelagic zone includes depths between 1,000 and 4,000 m, and the abyssopelagic zone lies from 4,000 to 6,000 m. The hadopelagic, or hadal pelagic, zone consists of the waters of the ocean trenches, from below 6,000 m to just above the sea floor, as deep as 11,000 m. Each of the depth zones supports a distinct community of animals, but they also have much in common. Here we stress the similarities, rather than the differences, among the depth zones of the deep sea.

How deep in the ocean can fishes live? The greatest depth at which a fish has been found is 8,145 m, in the Mariana Trench. Found in 2014, scientists think it may be a new species of deep-sea snailfish. Part of the fish's body is transparent and its tissues appear to be very delicate. Organs are visible through the transparent tissues. The fish was captured on video, but scientists were unable to capture a live specimen. As a result, they cannot say for sure that it is a new species or learn more about its structure and physiology. Based on physiological evidence, scientists believe that no fish can survive below 8,200 m.

The Deep Pelagic Habitats

The conditions of life in the deep pelagic environment change very little. Not only is it always dark, it is always cold: The temperature remains nearly constant, typically at 1 to 2 °C. Salinity and other chemical properties of the water are also remarkably uniform. The deep sea also includes the ocean bottom beyond the continental shelf. Bottom-living organisms are covered separately (see "The Deep-Ocean Floor," below).

In the darkness of the deep sea there is no need for countershading. Many animals, especially zooplankton, are a drab gray or off-white. Deep-sea fishes are generally black or a reddish brown, which in the deep sea has the same effect as being black. A few deep-sea fishes are bright red, as are many deep-sea shrimps.

As in mesopelagic animals, bioluminescence is very common in animals that live in the upper part of the deep sea. Deep-sea animals do not use bioluminescence for counterillumination, however, since there is no sunlight to create a silhouette. They have fewer photophores than midwater species, and the photophores are usually on the head and sides of the body rather than on the ventral surface. In the deep sea the primary uses of bioluminescence are probably prey attraction, communication, and courtship. Bioluminescence becomes less common in the deeper parts of the deep sea, for reasons that are not understood.

IN CONTEXT

Trenches Deep depressions in the sea floor that are formed when two plates collide and one sinks below the other.

• 2.2, How Plate Tectonics Formed the Oceans

Figure 18.18 Some typical characteristics of deep-sea pelagic fishes.

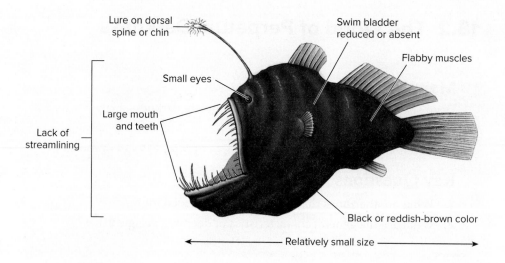

Lure on dorsal spine or chin

Swim bladder reduced or absent

Flabby muscles

Small eyes

Large mouth and teeth

Lack of streamlining

Black or reddish-brown color

Relatively small size

Mesopelagic (*Gonostoma denudatum*)

Brain

Gill filaments

Deep-sea (*Gonostoma bathyphilum*)

Brain

Gill filaments

Figure 18.19 Comparison of typical adaptations in mesopelagic and deep-sea pelagic fishes. Shown are closely related bristlemouths from the mesopelagic (*Gonostoma denudatum*) and the deep sea (*Gonostoma bathyphilum*). The deep-sea fish has smaller eyes, less muscle, and fewer light organs. It also has less-developed nervous and circulatory systems, as indicated by the smaller brain and gill filaments.

The large eyes of midwater animals are not needed in the deep sea, where not even dim sunlight penetrates. The deep sea is not *completely* dark, however, because of bioluminescence. Many deep-sea animals have functional eyes, especially in the upper parts of the deep sea, but the eyes are generally small (Figs. 18.18 and 18.19). Animals from the deepest regions tend to have even smaller eyes or be blind altogether. Deep-sea fishes that are blind are not bioluminescent, one indication that the main function of vision in the deep sea is to see bioluminescence.

The Lack of Food

Deep-sea organisms may not have to adapt to variations in the physical environment, but they face a continual shortage of food. Very little, only about 5%, of the food produced in the photic zone makes it past all the hungry mouths in

How to be Invisible

White light is a composition of all the colors of the rainbow, but not all parts of the light spectrum are absorbed by water the same way. As you will recall, red light is absorbed very near the surface of the ocean (see Fig. 3.8, p. 65) . If you were to carry a red apple down with you during a scuba dive, it would soon appear black! This phenomenon is no magic trick; it's the physics of how water absorbs light. Likewise, a red fish in the surface of the ocean would become black in appearance as it swam to the depths, out of range for red light to reflect off its body. In a sense, being red in the deep ocean is like wearing an invisibility cloak.

Who needs invisibility? Why do fishes and invertebrates that live this deep need to worry about being invisible if there's no sunlight? As you have learned, the predators and prey in the deep sea are in a constant arms race over bioluminescence. It's no coincidence that most bioluminescence is blue light since it's the wavelength that best travels through water. It's also no surprise that most animals in the depths, despite never seeing sunlight, have eyes tuned to seeing blue light. The animals of the deep sea use bioluminescence to blend in, attract prey, and ward off predators.

Ways to be invisible There are other ways to be invisible in the deep sea. Some animals are entirely black. Similar to how wearing black clothes at night makes you harder to see, the black body of a deep-sea animal provides excellent camouflage. Another way not to reflect the faint light of the deep sea is to be completely clear. Any light will pass through a transparent creature making it seem like it's not there.

One amazing adaptation for transparent animals is having a black gut. This allows them to consume bioluminescent prey while remaining invisible. It would be quite counterproductive to be invisible until you ate something, and then bioluminesce uncontrollably until your meal was digested!

Another fascinating adaptation using the properties of light in the deep sea belongs to a group of dragonfishes called loosejaws. They have adapted to emit and see red light, which most deep-sea creatures cannot detect. They search out prey using their red spotlights while their prey perceive perfect darkness. Without warning, an unsuspecting but illuminated animal becomes lunch for a loosejaws fish. The deep sea, seemingly dark, is in fact full of light, while the animals that live there constantly battle against each other to remain invisible.

A flashlight fish with a bioluminescent patch below its eye.

Danté Fenolio/Science Source

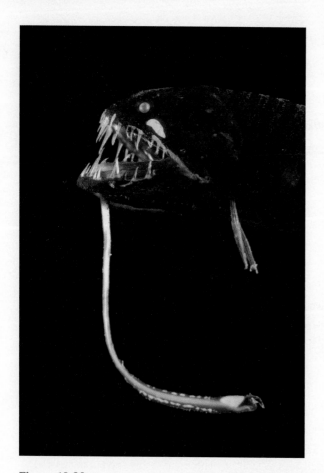

Figure 18.20 The fearsome jaws and teeth of a scaleless black dragonfish (*Melanostomias biseriatus*). The long barbel below the chin has a lure with bioluminescent tissue.

the waters above to reach the deep sea. Deep-sea animals do not make vertical migrations to the rich surface waters, probably because the surface is too far away and the change in pressure too great. With food critically scarce, deep-sea animals are few and far between.

Deep-sea fishes, the most common of which are bristlemouths and anglerfishes, are relatively small, generally 50 cm or less, but on average they are larger than mesopelagic fishes. It is somewhat surprising that deep-sea pelagic fishes tend to be larger than mesopelagic ones, since there is even less food available in the deep sea than in the mesopelagic. It is thought that deep-sea fishes put their energy into growth, reproducing slowly and late in life, while mesopelagic fishes spend less energy on growth and more on reproduction. In addition, vertically migrating mesopelagic fishes expend a lot of energy in their migrations, reducing the energy available for growth.

The energy-saving adaptations to food shortage seen in midwater organisms are accentuated in the deep sea. Deep-sea fishes are even more sluggish and sedentary than midwater fishes. They have flabby, watery muscles, weak skeletons, no scales, and poorly developed respiratory, circulatory, and nervous systems. Nearly all lack functional swim bladders. These fishes hang in the water column, expending as little energy as possible, until a meal comes along. Most deep-sea fishes have huge mouths and often fearsome teeth (Fig. 18.20), can consume prey much larger than themselves. This trend reaches its peak in the swallowers (*Saccopharynx*) and gulpers (*Eurypharynx;* Fig. 18.21), which look like swimming mouths. To go along with their large mouths, many species have stomachs that can expand to accommodate the prey once it has been engulfed.

Like their mesopelagic counterparts, bathypelagic anglerfishes have a first spine on their dorsal fin that is modified into a "pole" that dangles a bioluminescent bait to attract prey. In some species the bait even resembles a shrimp or a worm. In most species only the females have a pole and bait. Many other deep-sea fishes also attract prey with bioluminescent lures.

Sex in the Deep Sea

Food is not the only thing that is scarce in the deep sea. In such a vast, sparsely populated world, finding a mate can be difficult—even harder than finding food. After all, most deep-sea animals are adapted to eat just about anything they can get, but a mate has to be both the right species and the opposite sex! At least one deep-sea squid (*Octopoteuthis*) doesn't worry about that. Male squids mate by implanting a sperm packet on the female. Male *Octopoteuthis*

take no chances, and implant sperm packets indiscriminately on both males and females, presumably getting it right half the time.

Many deep-sea fishes have solved the problem by becoming hermaphrodites. After all, it would accomplish nothing if two members of the same species finally got together but were both the same sex. If every individual can produce both eggs and sperm, the ability to breed is guaranteed.

Deep-sea animals probably actively attract mates. Bioluminescence, for example, may send a signal that draws other members of the same species. The lure of female anglerfishes differs among species, so it may have a role in attracting mates as well as prey. Chemical attraction is also important. Male anglerfishes have a very powerful sense of smell, which they use to locate females. The females apparently release a special chemical that the male can detect and follow. Such chemicals are called **pheromones**.

Some anglerfishes have evolved an extreme solution to the problem of finding mates. When a male locates a female, who is much larger, he bites into her side, where he remains attached for the rest of his life (see Fig. 18.21). In some species the male's modified jaws fuse with the female's tissue. Their circulatory systems join, and the female ends up nourishing the male. This arrangement, sometimes called **male parasitism**, ensures that the male is always available to fertilize the female's eggs.

Hermaphrodites Individuals with both male and female gonads.
• 6.3, Perpetuating Life

Deep-sea anglerfish
(*Gigantactis macronema*)
35 cm

Deep-sea devilfish
(*Caulophryne pelagica*)
15 cm

Deep-sea bristlemouth
(*Gonostoma bathyphilum*)
20 cm

Attached male

Female deep-sea anglerfish
with attached male
(*Cryptopsaras couesi*)
45 cm

Swallower
(*Saccopharynx ampullaceus*)
160 cm

Gulper
(*Eurypharynx pelecanoides*)
100 cm

Figure 18.21 Some deep-sea fishes and their approximate maximum length.

Neither hermaphroditism nor male parasitism seems to be common in deep-sea invertebrates. The mechanisms that bring males and females together, if any, are unknown. There is evidence that some groups aggregate to breed, perhaps attracted by bioluminescent signals.

Living Under Pressure

Under the weight of the overlying water column, the pressure in the deep sea is tremendous. This is one reason so little is known about the deep sea. The submersibles and equipment needed to study the deep sea must withstand the pressure without being crushed and are very expensive. Only a very few submersibles can venture into the deepest trenches, where the pressure may exceed 1,000 atmospheres. It is just as difficult to bring animals up from the deep sea as it is to go down to them. Unable to endure the enormous change in pressure, they usually die when brought to the surface. A few scientists have succeeded in retrieving organisms from the deep sea in special pressurized chambers. Much has been learned from such work, but it is frustratingly difficult.

It is clear that pressure has important effects on deep-sea organisms. The absence of a functional swim bladder in most deep-sea fishes, for example, is probably due to the high energetic cost of filling the bladder with gas under extreme pressure. Along with the availability of food, pressure seems to be a major factor causing zonation in deep-sea pelagic organisms, that is, dividing the deep sea into bathypelagic, abyssopelagic, and hadopelagic zones.

In shallow-water organisms the enzymes that control metabolism, the vast set of chemical reactions that sustains life, are strongly affected by pressure, and would cease to function at the pressures of the deep sea. Deep-sea organisms, however, have much more pressure-resistant enzymes. Some also have high concentrations of a chemical that helps to stabilize the enzymes. Such molecular adaptations enable deep-sea organisms to survive under pressures that would kill any surface-dweller.

Nonetheless, pressure probably limits the depth range of most organisms, and the number of species declines going deeper and deeper. The deepest known fish was recorded at a depth of 8,370 m. Invertebrates and protozoa, however, have been found at the deepest depths.

Enzymes Proteins that speed up and control chemical reactions in organisms.
• 6.1, The Ingredients of Life

. .

Section Review

1. **Describe** What adaptations or strategies do deep-sea animals have to help ensure successful reproduction?

2. **Describe** What adaptations do deep-sea animals have to survive under high pressure?

18.3 The Deep-Ocean Floor

Main Idea

Despite living in the deepest parts of the ocean, there is a surprisingly high biodiversity in the deep-sea benthos.

Key Questions

1. What type of feeders dominate the deep sea benthos? Why?
2. How do scientists explain the fact that most deep-sea animals grow slowly and live for a long time?

VOCABULARY

benthos

macrofauna

deep-sea gigantism

psychrophilic

barophilic

DIVE IN!

The floor of the deep sea shares many characteristics with the pelagic waters immediately above: the absence of sunlight, constant low temperature, and great hydrostatic pressure. Nevertheless, the biological communities of the deep-sea floor are very different from pelagic communities because of one key factor: the presence of the bottom.

Marine biologists have learned a bit more about the **benthos**, or bottom-inhabiting organisms, of the deep sea than about deep-water pelagic communities, but much is still unknown. Of the 270 million km² of deep-sea floor, only about 500 m², the floor area of a large house, have been quantitatively sampled. What we do know has been learned using a variety of techniques. Devices called epibenthic sleds are dragged along the bottom, scooping up organisms, and corers bring a piece of the bottom to the surface. Deep-sea cameras photograph fast-swimming animals like fishes that are hard to catch in nets. Scientists have even developed a miniature transmitter that can be hidden in bait and used to track the movements of a fish that swallows it. Deep submersibles like *Alvin* have also been very useful, even more so than in the water column, as have remotely operated vehicles, or ROVs (Fig. 18.22), which are used to collect samples, take photos, and perform experiments. In 1995 one such craft succeeded for the first time in reaching the deepest part of the ocean. The Japanese-built *Kaiko* descended 10,991 m to the floor of the Challenger Deep in the Mariana Trench. This broke the long-standing record of the submersible *Trieste*, which in 1960 descended to 10,919 m, also in the Mariana Trench. Completely autonomous submersibles, which are pre-programmed to operate independently of direct human control, are also making contributions.

What is a "whale fall"? A whale fall occurs when a whale dies and sinks to the seafloor. This is like a "windfall" for the deep-sea benthos. First, scavengers consume the dead tissue. After several months, the soft tissues are gone. Within a year, the bones and surrounding sediment provide habitat and food for numbers of polychaete worms, molluscs, crustaceans, and other invertebrates. The animals obtain nutrients from the organic material in the bones and in the sediment. About two years after the whale hits the bottom, bacteria colonize the remaining bones and use the sulfur as a source of energy for chemosynthesis. The bacteria then become a source of food for other animals.

Figure 18.22 The ROV *Hercules*. Controlled from the surface, it is used to take samples, install equipment, and take photographs and videos to depths of 4,000 m.

Courtesy Mountains in the Sea Research Team, the IFE Crew, and NOAA/OAR/OER

The Nature of Life in the Deep-Sea Benthos

Food shortage is critical on the floor of the deep sea. Very little of the surface production makes it all the way to the bottom. Benthic organisms, however, have a major advantage over the deep-sea pelagic ones above. In the water column, food particles that are not immediately located and eaten sink away and are lost. Once food reaches the bottom, by contrast, it stays put until it is found. Thus, although pelagic animals may get first crack at food sinking out of the photic zone, benthic animals have more time to find and eat it. Food particles that reach the bottom tend to be those that sink fairly rapidly, minimizing the chance that they will be eaten on the way down. Fecal pellets, for example, are an important source of organic matter for the deep-sea benthos.

Feeding in the Deep Sea Still, the rain of organic matter to the sea floor is actually more like a drizzle. Very little food is available to the benthic community. Furthermore, much of the material that reaches the sea floor, like the chitinous remains of crustacean zooplankton, is not immediately digestible. On the sea floor, however, bacteria decompose the chitin and become food for other organisms.

Most of the deep-sea floor is covered in fine, muddy sediment. The meiofauna, tiny animals that live among the sediment particles, are the most abundant organisms on the deep-sea floor, typically outnumbering larger organisms by a factor of 10. The meiofauna graze on bacteria and absorb dissolved organic matter (DOM) from the water between the sediment grains. The energy in the bacteria and DOM is thus made available to larger animals, or **macrofauna**, that graze on the meiofauna.

Suspension feeders are rare among the larger organisms in the deep-sea benthos. Instead, deposit feeders dominate. Many of these are infauna, burrowing in the sediments. Others, the epifauna, rest on the sediment surface.

IN CONTEXT

Suspension Feeders Animals, including filter feeders, that eat particles suspended in the water column.

Deposit Feeders Animals that eat organic matter that settles to the bottom.

- 9.1, Sponges, Cnidarians, and Comb Jellies

Grenadier
(*Lionurus carapinus*)
45 cm

Tripod fish
(*Bathypterois viridensis*)
20 cm

Deep-sea spiny eel
(*Notacanthus bonapartei*)
20 cm

Hagfish
(*Eptatretus stouti*)
50 cm

Eelpout
(*Lycenchelys jordani*)
35 cm

Brotulid
(*Holcomycteronus profundissimus*)
30 cm

Figure 18.23 Some typical deep-sea bottom fishes and their approximate maximum length.

Deep Sea Diversity Polychaete worms are usually the most abundant macrofauna on the deep-sea floor, followed by crustaceans and bivalve molluscs, but there is considerable variation from place to place. Sea cucumbers are sometimes dominant, for example. Sea cucumbers from the deep sea often have strange, highly modified body forms. Some have leg-like appendages and walk across the bottom in search of organic-rich sediment. "Herds" of some species have been observed from submersibles. Other species can swim above the sea floor by undulating their bodies or squirting jets of water. Some parts of the deep sea are dominated by brittle stars, and sea stars can also be abundant.

There are predators in the deep-sea benthos, but they seem to be fairly rare. The main predators on deposit-feeding animals are probably sea stars, brittle stars, and crabs. Members of the nekton, like fishes and squids, are also important predators. Sea spiders, or pycnogonids, prey on other invertebrates by sucking out their soft parts.

Tripod fishes are another interesting group of deep-sea benthic predators. Nearly blind, these fishes sit on the bottom on their elongated fins (Fig. 18.23), facing into the current and snapping up passing plankton.

The slow rain of food to the bottom is interrupted by an occasional "storm." Large pieces of food that sink rapidly, like the dead bodies of large fishes or whales, are an important source of food to the benthos. Mobile deep-sea animals rapidly congregate around such "baitfalls" (Fig. 18.24). Among the most common of these are crustaceans, especially amphipods, which arrive soon after the bait touches down. Many deep-sea benthic amphipods are generalists that feed on detritus and perhaps prey on live organisms if nothing else is available. Some, however, seem to specialize as scavengers. They apparently have a well-developed sense of smell, which probably helps them find new baitfalls. When caught in traps, these amphipods often have nothing in their guts except the bait, which indicates that they have not fed for some time before. This, and the fact that they have an expandable gut, may mean that the amphipods are adapted to capitalize on large but infrequent meals.

While most deep-sea benthic animals are small, the deep-sea members of some groups are giants compared to their shallow-water relatives (Fig. 18.25). Most amphipods, isopods, and sea spiders are less than 1 cm in size, but deep-sea isopods can reach 36 cm and amphipods 34 cm in length, and deep-sea sea spiders can reach 80 cm across. This phenomenon, known as **deep-sea gigantism**, is thought to be an adaptation for fast, efficient locomotion so that the animals can cover a larger area in search of baitfalls. The giant size also provides for more energy storage between meals.

Various fishes also find freshly placed bait quickly. The most common are grenadiers, or rattails, brotulas and cusk eels, deep-sea spiny eels, and hagfishes (Fig. 18.23). These bottom scavengers tend to be larger on average than deep-sea pelagic fishes, relatively muscular, and active, unlike bathypelagic fishes (Table 18.1, p. 574). They are adapted for cruising along the bottom in search of the

Figure 18.24 A variety of fishes quickly locate large pieces of food on the deep-sea floor. These cusk eels (*Genypterus blacodes*), spiny dogfish sharks (*Centrophorus squamosus*), cutthroat eels (*Diastobranchus capensis*), and snubnose eels (*Simenchelys parasiticus*) have come to a large piece of fish placed at 1,500 m on the sea floor near New Zealand.

Figure 18.25 The deep-sea isopod (*Bathynomus giganteus*) is related to common garden pillbugs, but can grow to 36 cm long.

occasional bonanza. In several groups the parts of the brain devoted to olfaction are highly developed compared to close relatives from the mesopelagic, where visual areas dominate. Thus, deep-sea bottom fishes, like invertebrates, probably rely a lot on their sense of smell. At depths above 2,000 m, sharks may also show up at deep-sea bait falls, quickly putting an end to the feed.

Life Histories There is a growing realization that life in the deep sea proceeds at a very different pace than at the surface. Most deep-sea animals seem to grow very slowly, probably because of the lack of food. On the other hand, they live for a long time. Deep-sea clams have been estimated to be 50, 60, or even 100 years old. Some fishes live even longer (see "Biodiversity in the Deep Sea"). Perhaps the low temperature and high pressure slow down the processes of life in the deep sea.

It may also be that deep-sea animals need to live a long time to store up enough energy to reproduce. The larvae of deep-sea forms do not spend time in the food-rich photic zone. The chances of making it all the way to the surface and then back to the deep-sea floor are simply too small. Instead, deep-sea animals tend to produce large eggs, with enough yolk to see the larva through its early stages without eating. It takes a lot more energy to produce a large egg than a small one, so deep-sea animals produce only a few eggs. In at least some animals, reproduction may be tied to feeding. In some species of amphipods, individuals caught in baited traps are all sexually immature. Biologists have speculated that they do not reproduce until they manage to find a good meal.

Microbes in the Deep Sea

Bacteria, archaea, and viruses play many important roles in deep-sea benthic ecosystems that we are only beginning to understand. In 1968 a famous unplanned "experiment" gave an early indication of how much there is to learn.

Table 18.1 Characteristics of Fishes from Different Depth Zones in the Pelagic Realm

	Epipelagic	Mesopelagic (migrator)	Mesopelagic (non-migrator)	Deep Pelagic	Deep-sea Bottom
Appearance					
Size	Wide range	Small	Small	Small but larger than mesopelagic	Relatively large
Musculature	Strong muscles, fast swimming	Moderately strong muscles	Weak, flabby muscles	Weak, flabby muscles	Strong muscles
Eyes	Large	Large, sensitive	Very large, sensitive, sometimes tubular	Small, sometimes absent	Small
Coloration	Typical countershading — dark back, white or silver belly	Black or black with silver sides and belly; counterillumination	Black or black with silver sides and belly; counterillumination	Black or reddish brown, occasionally bright red, or lacking in coloration	Dark brown or black
Bioluminescence	Relatively uncommon	Common, often used for counterillumination	Common, often used for counterillumination	Common, often used to attract prey	Only found in a few groups

Biodiversity in the Deep Sea

With its immense pressure, near-freezing waters, absence of sunlight, and chronic food shortage, the deep sea seems like one of Earth's least hospitable environments. In the 19th century, it was hypothesized that no life existed below about 600 m. This "azoic hypothesis" was quickly proven wrong, nevertheless, animal abundance does decrease with depth.

It was once thought that in addition to low abundance there were relatively few animal species in the deep sea. Just the opposite is true: The deep sea may be the most biologically diverse environment on Earth. Biologists are unable to collect and identify every species present in most environments. They can estimate species numbers, however, from "species accumulation curves." This involves analyzing how often they find new species as they collect more and more individual organisms. The first organism collected will always be a species new to the study. The second organism could be from the same species, leaving the total number of species collected at one, or it might be a new species, bringing the species total to two. If the number of species in the habitat—one measure of its biodiversity—is small, then as more and more individuals are collected the number of new species increases relatively slowly and soon starts to level off near the total number of species that live in the environment. In highly diverse habitats, ones with many species, the number of species increases rapidly as more individuals are collected and takes longer to level off.

Rainforests of the deep? When this was done on the deep-sea floor, the species curve went through the roof, leading the researchers to estimate that certainly more than 1 million, and possibly more than 10 million, species live on the deep-sea floor, making it as or more diverse than rain forests and coral reefs. This was little short of astonishing because up until then most biologists had thought that no more than a few hundred thousand species lived in the entire ocean!

The estimate of 10 million species on deep-sea floor has been controversial, in part because it is extrapolated from samples of only a tiny fraction of the sea floor. Whatever the true number, it is clear that the deep-sea floor is among the planet's most diverse habitats.

Seamounts Seamounts, which like the rest of the deep sea have hardly been studied, also have very high diversity. Samples from only 24 seamounts in the southwest Pacific near Australia produced more than 850 species, and the species accumulation curve was still rising, indicating that many more species would have been found had more samples been taken. What is more, about a third of the species collected were new to science, and probably restricted to small seamount chains or even individual seamounts. Seamounts seem to be deep-sea islands, each holding a collection of organisms found nowhere else.

With some 30,000 seamounts in the deep sea, nearly all unexplored, there must be vast numbers of seamount species waiting to be discovered, but many may go extinct before that happens. Seamounts are home to the orange roughy (*Hoplostethus atlanticus*) and several other commercially valuable fishes. These fishes live for more than a hundred years, but take decades to reach reproductive age and reproduce very slowly, so they are vulnerable to overfishing. Numerous newly discovered populations on seamounts in the Indian Ocean, for example, were completely fished out less than three years after their exploitation began. With many seamounts located on the "high seas" outside the jurisdiction of any country, protecting the rich biodiversity of seamounts will require an international effort to control the damage inflicted by seamount fisheries.

Species accumulation curves from the deep-sea and shallow-water benthos off the coast of New England.

As the crew of *Alvin* was preparing for a dive the submersible was swamped by a wave, its supporting cable snapped, and *Alvin,* hatch open, dropped 1,540 m to the bottom. The crew escaped but left their lunch on board. It was to become the most famous lunch in marine biology.

When *Alvin* was recovered 10 months later, scientists discovered that the long-lost lunch was in amazingly good condition. Though soggy, the sandwiches looked almost fresh; the bologna was still pink inside. The rest of the lunch—apples and a thermos of soup—also looked good enough to eat. Once brought to the surface the food soon spoiled, even though it was refrigerated.

Why was the food preserved in the deep sea? Though the deep sea is cold, so is a refrigerator. Are bacteria absent from the deep sea? Does the pressure somehow inhibit bacterial decay? Is there some other explanation? These questions sparked a flurry of research.

Deep Sea Bacteria It is now known that bacteria do live in huge numbers in the deep sea, as in every other environment on Earth. Pressure and cold slows bacterial growth, however, and most shallow-water bacteria cannot grow at the pressure and temperature of the deep sea. Therefore, the bacteria that were already in the lunch probably died when *Alvin* sank.

Even if the surface bacteria died, there are still plenty of bacteria already living in the deep sea. Deep-sea bacteria don't just tolerate the prevailing cold and high pressure, many of them grow best or even *only* under these conditions. Even though these cold-loving (**psychrophilic**) and pressure-loving (**barophilic**) bacteria grow best under deep-sea conditions, they still grow slower than shallow-water bacteria and can take up to 1,000 times longer to decompose organic matter. It has also been suggested that deep-sea bacteria are adapted to use organic matter in low concentrations, and are overloaded by such unusually rich fare as a bologna sandwich. Furthermore, the lunchbox may have prevented many bacteria, such as those that live on or in amphipods and other deep-sea animals, from getting to the food in the first place.

Archaea and Viruses Vast numbers of microbes live in deep-sea sediments. Until the last few years it was thought that bacteria dominate, but new surveys show that archaea are more abundant in deep-sea sediments. The earlier studies used techniques that are not well-suited for detecting archaea. Other studies indicate that viruses are a key driver of the microbial ecosystem in deep-sea sediments, where they are extremely abundant. Researchers estimate that viruses kill some 80% of the bacteria and archaea in the sediment, causing their cells to burst and release large amounts of dissolved organic carbon (DOC). Other microbes take up the DOC as an energy source, so the viruses promote rapid carbon recycling, as well as recycling of nutrients such as nitrogen and phosphorus. It thus appears that viruses play a major role in the deep-sea carbon cycle.

Deep-sea sediments also contain large numbers of chemosynthetic prokaryotes, which are probably an important food source for deposit feeders. They may be involved in the formation of polymetallic nodules and other mineral deposits. Chemosynthetic bacteria are even gradually digesting the wreck of the *Titanic!*

IN CONTEXT

Chemosynthetic Prokaryotes
Autotrophic bacteria and archaea that use energy contained in inorganic chemicals rather than sunlight to make organic matter.
• 7.2, Prokaryotes

Alvin Reborn

The research submersible *Alvin* played a central role in some of the greatest discoveries in deep-sea biology, but she isn't getting any younger. Launched in 1964, she has more than 4,600 dives under her belt. Over the years *Alvin* has often had facelifts and upgrades, and not much of the original sub remains. Even so, moving into the 21st century *Alvin* seemed to have reached the end of her working life, and her depth limit of 4,500 m meant that much of the deep ocean was out of reach. In 2004, a plan was developed to retire *Alvin* and build her replacement.

Alvin rebooted Escalating costs and tight research funding meant scrapping the plan for a new *Alvin* replacement. Instead, *Alvin* was transformed through her largest upgrade ever, which was accomplished in two phases. The first phase replaced the heart of *Alvin*, the titanium sphere that houses the crew. The new sphere is capable of depths of 6,500 m, *Alvin* will eventually have access to 98% of the ocean floor. The new sphere will have more and larger windows to improve visibility. The interior is being redesigned to be more comfortable—or, perhaps more accurately, less uncomfortable—during the long descents into the abyss.

Other improvements in Phase I included better lighting and high-definition camera systems as well as an improved control system. Also improved was *Alvin's* launch system and storage in its support vessel, the research vessel *Atlantis*. *Alvin* made her last dive before her rebuild late in 2010, and the reborn *Alvin* was launched in 2014.

Deep ocean frontiers Even the new *Alvin's* won't be able to go to the very deepest trenches. There have been two manned submarines that have traveled to the bottom of the Mariana Trench: Jacques Piccard and Don Walsh in 1960 (*Trieste*) and the movie director James Cameron, in 2012. Some ROVs and "landers," equipment packages that are simply dropped to the bottom and retrieved, can reach the deepest parts of the oceans. In addition, private groups are currently developing glider-like submersibles that will be able to "fly" to the deepest parts of the ocean to help reveal the secrets of the least-explored places on Earth.

Section Review

1. **Explain** What is deep-sea gigantism? What are some possible advantages of this phenomenon?
2. **Describe** the role of viruses in the deep-sea carbon cycle.

18.4 Hot Springs, Cold Seeps, and Dead Bodies

DIVE
IN!

Where is the deepest
hydrothermal vent? The Beebe
Vent Field in the Cayman
Trough is 4,960 m below the
surface. It is in the Caribbean
trench between the Cayman
Islands and Jamaica. Not only
is it the deepest vent found to
date, but it is also thought to be
the hottest, registering
temperatures of 400°C.
Scientists believe that the vent,
discovered in 2010 by
researchers using a remote-
controlled submarine, is home
to several new species as well.

Main Idea

Hydrothermal vents are unique environments that support robust
communities of equally unique organisms.

. .

Key Questions

1. Which primary producers supports vent communities?
2. What are cold seeps?

The year 1977 marked one of the most exciting discoveries in biology, and it
wasn't even made by biologists! Marine geologists were using *Alvin* to look for
hydrothermal vents. When the scientists found the vents, they also found
something completely unexpected: a rich, flourishing community unlike
anything ever imagined (Fig. 18.26). In the days that followed, more vents were
found, each teeming with animals. The vents were like oases of life on the
barren deep-sea floor.

Hydrothermal Vent Communities

Over 400 new animal species and counting have been found in hydrother-
mal-vent communities, which vary considerably from place to place. Vestimen-
tiferan tubeworms, clams, crabs, and shrimps typically dominate the eastern
Pacific vent communities. Unusual snails and barnacles are most common at
vents in the western Pacific, while vents on the Mid-Atlantic Ridge are typically
dominated by a shrimp (*Rimicaris*).

Vent Microbes With no sunlight to support photosynthesis, the primary
producers in hydrothermal-vent communities are chemosynthetic archaea and
bacteria. Around the mid-ocean ridge, seawater trickles down through cracks
and fissures in Earth's crust and is heated to very high temperatures. Rich in
sulfide minerals, it emerges at hydrothermal vents to form "black smokers,"
"chimneys," and other mineral deposits. The hot water also contains large
amounts of hydrogen sulfide (H_2S), which is toxic to most organisms but is an
energy-rich molecule.

Chemosynthetic prokaryotes that use the energy in hydrogen sulfide and
sulfide minerals to make inorganic matter are the base of the food chain. Some
of these microbes are extremophiles that can live at temperatures over 120 °C,
the highest temperature at which life is known to occur. New studies are
expected to shatter this record.

Water near the vents contains so many microbes that they cloud the water.
Some vent animals feed by filtering them from the water, but this isn't the
principal mode of feeding. One of the dominant tubeworms in eastern Pacific
vent communities (*Riftia*) does not filter-feed. In fact, it doesn't even have a
mouth or digestive tract! Instead, these worms have a highly specialized organ

Figure 18.26 Rich, colorful
animal communities live at many
hydrothermal vents. Image
courtesy of Richard A. Lutz.

Image Courtesy of Richard A. Lutz

called a trophosome packed with symbiotic bacteria (Fig. 18.27). The bacteria pass much of the organic matter they produce on to the worm. The worm, in turn, supplies the bacteria with raw materials. The bright-red plume acts like a gill, exchanging not only carbon dioxide and oxygen, but also hydrogen sulfide. The tubeworm's blood has special hemoglobin that protects the worm from the toxic effects of hydrogen sulfide, and transports it to the bacteria. Other vent animals chemically process hydrogen sulfide to avoid being poisoned.

Vent Animals Numerous other vent animals, including mussels (*Bathymodiolus*) and large clams (*Calyptogena*), contain symbiotic bacteria, though they can filter-feed as well. Non-symbiotic microbes are also an important food source. For example, the shrimp (*Rimicaris*) that dominates vent communities on the Mid-Atlantic Ridge scrapes off and eats bits of microbe-covered mineral from the chimneys that form at vents. The microbes are digested and the mineral eliminated.

Rimicaris do not have recognizable eyes, but they do have two light-sensitive patches on their upper surface. The patches can detect much fainter light than humans can see. Before the discovery that these shrimp could "see," no one suspected there was any light at all at deep-sea vents. A special low-light camera similar to those used to study distant stars revealed a faint glow, invisible to the human eye, around the vents. The source of the glow is uncertain, although some of it can be explained by the heat of the emerging water. Biologists speculate that the shrimp use this dim light both to locate active vents and to avoid getting cooked by coming too close to the scalding water.

Boiling-hot black smokers are not the only hydrothermal habitat in the deep sea. **White smokers** occur a little further away from the active rift and are cooler, so that black sulfide minerals precipitate out before the water emerges from the sea floor. White minerals remaining in the water form the "smoke." **Snowblower vents** form where chemosynthetic bacteria living in rocks beneath the sea floor produce filaments of sulfur that coalesce into snowflake-like particles when they emerge from the vent into cold water. Cooler vents at greater distance from the mid-ocean ridge or associated with transform fault and trenches support snails, sponges, deep-sea corals, and other organisms, but not in the abundance seen at mid-ocean ridge vents. Thick mats of chemosynthetic archaea and bacteria also grow in some of these areas, using carbonate minerals rather than sulfide minerals as an energy source. "Blue smokers" have also been discovered.

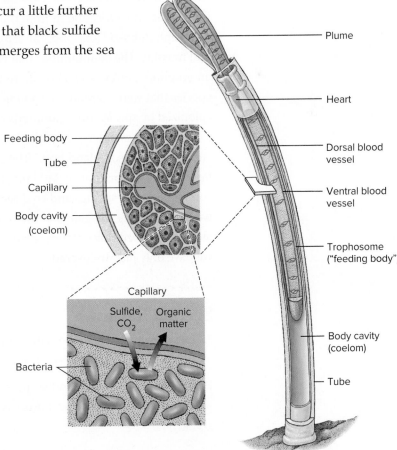

Figure 18.27 The anatomy of the giant hydrothermal-vent tubeworm (*Riftia pachyptila*). The plume at the end acts like a gill, except that it exchanges hydrogen sulfide (H_2S) as well as carbon dioxide (CO_2) and oxygen (O_2). The carbon dioxide and sulfide are carried in the blood to the feeding body, where symbiotic bacteria use them to make organic matter by chemosynthesis.

Plume

Heart

Dorsal blood vessel

Ventral blood vessel

Trophosome ("feeding body")

Body cavity (coelom)

Tube

Feeding body

Tube

Capillary

Body cavity (coelom)

Capillary

Sulfide, CO_2

Organic matter

Bacteria

Other Chemosynthetic Communities

Chemosynthetic communities aren't confined to hydrothermal areas. **Cold seeps** are places, mostly along continental margins or in sediment-rich basins like the Gulf of Mexico, where hydrogen sulfide and methane (CH_4) produced by the decay of organic matter seep out from the sea floor. Primary production by chemosynthetic prokaryotes that can use one or the other of these energy-rich molecules supports communities that are similar in many ways to those at hydrothermal vents, though the individual species are mostly different.

Communities based on chemosynthesis have even been found on deep-sea "graves." As previously discussed, occasional baitfalls like dead whales are an important source of food for deep-sea scavengers. When the scavengers are through, the decomposing remains produce hydrogen sulfide and methane, supporting a community similar to those at vents and seeps.

Unlike most deep-sea organisms, those at hot springs, cold seeps, and dead bodies enjoy an energy-rich environment and grow fast and large. On the other hand, their specialized habitats are tiny oases separated by vast distances. These oases are also unreliable. Hot springs and cold seeps can be wiped out by volcanic eruptions, undersea landslides, and other disturbances. As long as the flow of energy-rich molecules continues the communities recover fairly quickly as new larvae recolonize the vent but sometimes the flow is blocked or dries up. In 2002 scientists returned to the Rose Garden, a hydrothermal vent near the Galápagos Islands first discovered in 1979 and revisited several times since, to find no trace of the rich vent community—only fresh lava flows. To avoid extinction, vent and seep species must be able to disperse to new oases, and indeed they do. Only 200 m from the Rose Garden site scientists discovered a new vent, dubbed "Rosebud," already colonized by juvenile tubeworms, clams, and mussels. The communities that recolonize new vents or areas wiped out by an eruption, however, can be different from those before the eruption, with species that were common becoming rare and *vice versa,* and vents may be colonized by species from hundreds of kilometers away.

The carcasses of large organisms quickly rot away, so the species they support must also "island hop." The remains of whales, seals, sharks, and other large organisms are particularly tiny habitats in the expanse of the deep sea, but compared to hot springs and cold seeps there are a lot of them. Or at least there were. By hunting the giants of the sea to the brink of extinction, we may have already removed these stepping stones and doomed an entire ecosystem that we have only just discovered.

. .

Section Review

1. Describe the symbiotic relationship between chemosynthetic bacteria and tubeworms in hydrothermal vent communities.

2. Explain How do vent and seep species avoid extinction if their habitats are destroyed through natural processes?

Chapter 18 Review

REVIEW QUESTIONS

Multiple Choice

1. Deep-water organisms depend on the surface for

 a. food and nitrogen

 b. food and oxygen

 c. light and food

 d. light and nitrogen

2. What percent of food from the surface reaches the deep sea?

 a. 1 percent

 b. 5 percent

 c. 10 percent

 d. 15 percent

Short Answer

3. What is the deep scattering layer?

4. Compare and contrast the adaptations of mesopelagic fishes to those of deep-sea fishes.

5. What methods do scientists use to explore the deep-sea floor?

6. Compare and contrast two types of hydrothermal vents—black smokers and white smokers.

Critical Thinking

7. The deep-sea floor has been considered as a potential site for the disposal of toxic and radioactive wastes. What questions about the biology, geology, and chemistry of the deep-sea environment do you think should be answered before such plans are approved?

8. How do you think non-migratory midwater fishes, with their flabby muscles, are able to prey on vertical migrators, which have well-developed muscles?

9. Fats are highly energy-rich molecules that require a great deal of energy to produce, but in the mesopelagic, where food is scarce, many fishes have swim bladders filled with fat. How might this be explained?

DATA ANALYSIS LAB

Where does the main source of food for organisms in the deep sea come from? The deepest parts of the ocean have near freezing water temperatures, they are absent of light, and new sources of oxygen in the water come from water masses that sink from the surface. Despite these harsh conditions, a variety of organisms live in the deep sea.

Data and Observations

The illustration shows a generalized food web for the ocean.

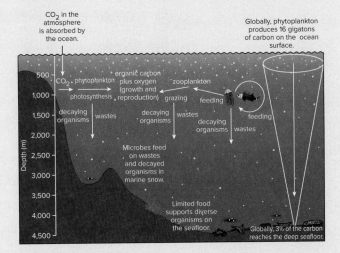

Think Critically

1. According to the illustration, where does the main source of food for organisms in the deep sea come from? In what form is the food?

2. How much of the carbon produced at the surface reaches the deep sea floor? What happens to it as it sinks?

3. Research has shown that diversity in the deep sea is comparable to diversity in temperate coastal zones and, in some cases when looking at macrofauna (small polychaete worms, mollucs, and crustaceans), may be close to that of diversity in tropical rain forests. Form a hypothesis as to why a limited amount of food still supports a diverse number of organisms in the deep sea.

Data obtained from: McClain, C., 2010. An empire lacking food. American Scientist, vol. 98, no. 6, November/December, pp. 470–477.

Unit 4

Humans and the Sea

People have used the many resources that the oceans offer ever since the first humans ventured to the shore. Evidence of the use of marine life as food goes back to our African ancestors more than 150,000 years ago. Unit 4 examines the resources we extract from the sea and the impacts that human populations have on the marine environments. In discussing these topics, we will draw on our knowledge of the sea floor and the chemical and physical features of the ocean (discussed in Unit 1 of the book) and of the types and distribution of marine life (discussed in Units 2 and 3). Many other disciplines, ranging from technology development to the marketing of new products, are also involved.

UNIT PROJECT

Model the Impact of a Solution

Computer simulations and mathematical models are often used when working to solve a complex real-world problem. These methods may be used when attempting to predict the effects of a design solution, and can be employed throughout the engineering design process. Your task is to use a computer or mathematical model to evaluate the impact of proposed solutions to a marine science-related real-world problem.

Resources from the Sea

19.1 Fisheries

19.2 Aquaculture

19.3 Commerce and Recreation

19.4 Non-Living Resources from the Sea Floor

19.5 Non-Living Resources from Seawater

THEME Stability and Change

It is important to study and monitor marine ecosystems from which we are removing living and nonliving resources.

BIG IDEA

Humans obtain many important resources from the oceans, including fish and shellfish for food, oil and natural gas for energy, and minerals for use in industry.

About The Photo A fisherman pulls a crab trap out of the ocean off the coast of Great Britain.

Monty Rakusen/Cultura/Getty Images

19.1 Fisheries

People have used the many resources that the oceans offer ever since the first humans ventured to the shore. Evidence of the use of marine life as food goes back to our African ancestors more than 150,000 years ago. Our exploitation of marine resources is now much more sophisticated, but, unfortunately, it has also become more destructive. Fishers cross the oceans assisted by satellites, marine chemists extract wonder drugs from marine organisms, genetic engineers develop faster-growing fish, and ocean engineers investigate better ways of harvesting energy from waves and tides.

Food from the Sea

The oceans provide food as well as employment for millions of people (Fig. 19.1). Many different types of marine organisms are harvested. Seaweeds and creatures as diverse as jellyfishes, sea cucumbers, and even polychaete worms are part of the diet in many cultures. The vast majority of the harvest, however, consists of fishes, which are known in fisheries terminology as **finfish**. Finfish constitute nearly 90% of the total world marine catch. Molluscs and crustaceans, which together are called **shellfish**, are less important than finfish in the size of their catches but their economic value is highly significant.

The Importance of Seafood Most of the world's food is grown on land, and food from the sea represents only about 1% of all the food eaten. Still, seafood is one of the world's most important foods. It is rich in protein, which is essential for normal growth. Finfish, ocean catches, and fish-farming figures combined, provides around 30% of the animal protein consumed by people around the world. Dependence on seafood for protein is much higher in most coastal nations, especially poor ones. The world's population is still growing (Fig. 19.2, p. 586), and it is increasing much faster in the places where it hurts the most: the poor, already crowded developing world.

Figure 19.1 Although threatened by overfishing and the high cost of fuel, marine fisheries provide food to millions. Recent food shortages in developing nations have increased their importance as a source of protein. Catches are consumed in many ways, and the by-products processed into food for farmed fish, chickens, and pets, as well as fertilizer and other products.

Michael Melford/National Geographic/Getty Images

Where did your dinner come from? People in the United States collectively consume more than 2.2 billion kilograms and spend 83 billion dollars on fish and shellfish per year. As a consumer, your knowledge and demand for sustainably harvested or farmed fish can benefit fisheries, the environment, and the industry as a whole. Organizations such as the Monterey Bay Aquarium Seafood Watch and the Safina Center produce lists of finfish and shellfish species that are best choices, those that are sustainably harvested or farmed, and those that should be avoided because the fishery is poorly managed and/or overexploited. Consumers with increased awareness can make informed choices and drive the market towards sustainability.

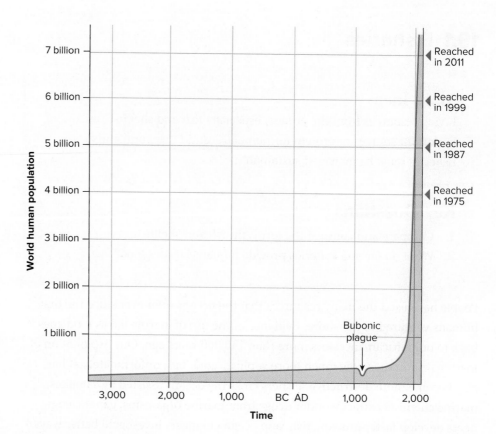

Figure 19.2 The curve that represents the growth of the world human population over the centuries looks like a giant "J": Very slow, almost non-existent population growth is followed by a sudden explosion of very fast growth. Global population is expected to be over 10 billion by 2085, and U.S. population is projected to reach 478 million by 2100, whereas that of Africa will rise from 1 billion in 2010 to 3.6 billion in 2100.

Marine fisheries provide much-needed food and employment. As we shall see, however, the world's most important fisheries are already overexploited, and some are exhausted. Worldwide catches have nearly stabilized despite increasing effort since the late 1980s (Table 19.1) after increasing almost fivefold since 1950s. Neither the world fish catch nor the world grain output can catch up with global population growth. The amount of grain and fish per person has already started to decrease. The increase in fish production needed to meet the demand is being met not by marine catches but by farming of mostly freshwater fishes. The world's oceans have actually been running out of fish because of overfishing.

Fishing is one of the oldest human endeavors. Small-scale fisheries that involve simple gear and methods are still practiced around the world, especially in developing countries. These fisheries, in fact, employ most of the world's fishing people. They also provide animal protein to people whose diets are often protein-deficient.

Demand for Seafood The increased demand for seafood in affluent countries and the expansion of the world's population have steadily increased the pressure on the food resources of the sea (Fig. 19.3, p. 588). The application of technology, from the development of more effective gear to the use of satellites to find fish, has greatly increased the efficiency with which we harvest these

Table 19.1 **World Commercial Catches of Selected Fishes, Molluscs, and Crustaceans in Millions of Metric Tons**

Catch	1975	1985	1995	2000	2005	2012
Herrings, sardines, etc.	13.43	21.10	22.01	24.90	22.27	17.55
Miscellaneous pelagic fishes	not listed	not listed	13.93	10.64	11.30	10.41
Cods, haddocks, hakes, pollocks	11.85	12.46	10.74	8.68	8.07	7.70
Tunas, bonitos, billfishes	2.06	3.18	4.89	5.82	6.51	7.18
Salmon, smelts	0.55	1.17	1.15	0.80	1.03	0.97
Flounders and other flatfishes	1.16	1.35	0.92	1.01	0.90	0.99
Sharks, rays	0.59	0.62	0.76	0.87	0.78	0.77
Total marine fishes	**51.93**	**64.40**	**72.00**	**71.84**	**69.60**	**65.49**
Freshwater fishes	**5.96**	**8.74**	**5.80**	**6.80**	**8.17**	**10.29**
Squids, octopuses	1.18	1.79	2.94	3.66	0.73	4.03
Scallops	0.29	0.60	0.54	0.66	0.73	0.75
Clams, cockles	0.94	1.51	0.96	0.80	0.65	0.61
Oysters	0.85	1.09	0.19	0.25	0.16	0.17
Mussels	0.53	0.97	0.24	0.26	0.13	0.10
Total marine molluscs	**4.03**	**6.18**	**6.38**	**7.25**	**6.49**	**6.55**
Shrimps	1.33	2.12	2.44	3.08	3.20	3.35
Crabs	0.75	0.89	0.95	1.09	1.23	1.50
Lobsters	0.10	0.20	0.22	0.23	0.23	0.29
Krill	0.04	0.19	0.12	0.11	0.13	0.19
Total marine crustaceans	**2.35**	**3.42**	**4.77**	**5.91**	**5.20**	**5.75**
World total catch in marine fishing areas	**61.22**	**75.68**	**85.04**	**86.80**	**92.49**	**91.3**

Note: Catches tabulated in groups defined by the Food and Agriculture Organization of the United Nations (FAO); only selected groups are listed and, when added, will not equal totals.

resources. Nations like China, Japan, South Korea, and the United States have developed high-tech fishing fleets that can remain on the fishing grounds away from the home port for long periods. Factory, or mother, ships process and store fish taken by smaller boats. The catch is processed, marketed, and sold in a multitude of ways: fresh, frozen, canned, dried and salted, smoked, marinated, as fish sticks, and as fish meal. Fishing operations have been known to diversify into businesses that develop products such as boats and fishing gear. Some even get involved in the management of fishing resources to prevent them from disappearing. Modern fishing ventures thus provide jobs not only to fishers but also to technicians, marine biologists, marketing specialists, economists, and other experts. Women and the elderly often undertake processing and marketing in developing countries, which contributes to social equity. Commercial fisheries provide a considerable portion of the export earnings of countries like Canada and Iceland and help sustain these nations' economies.

Global consumption of fish, marine and freshwater, reached 19.2 kg per capita in 2012, which provided about 16.7% of the animal protein intake and 6.5% of all protein, plant and animal. The amount of seafood consumed in various cultures, however, varies widely because of consumer tastes, price, and availability. Seafood consumption has been steadily increasing as consumers

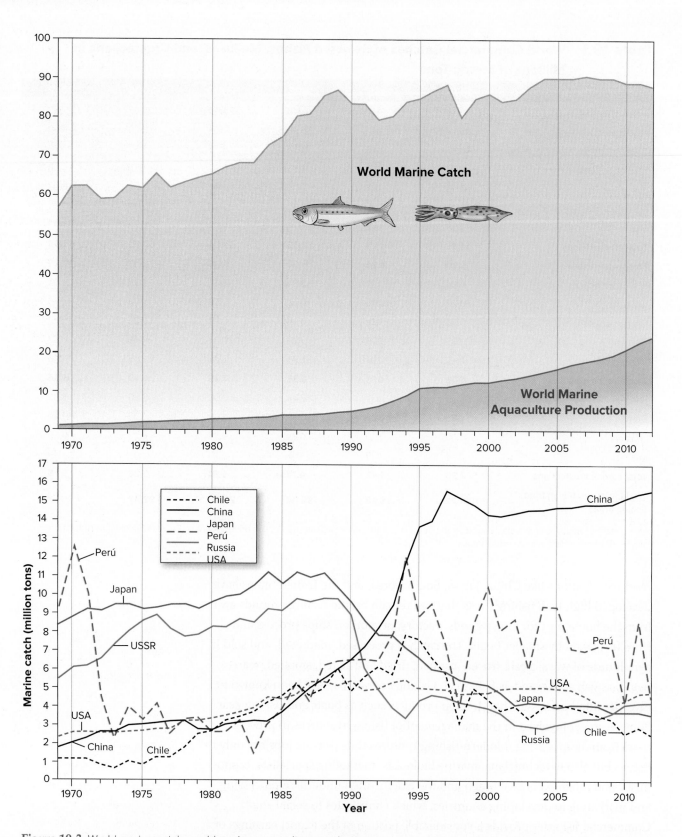

Figure 19.3 World marine catch, world marine aquaculture production, and marine catches by six of the major fishing nations of the world. Marine fishes constitute the bulk of the catches, but shellfish are also included; seaweeds and marine plants are not included. Russian catches included those from the former Soviet Union until 1990. The figures for Perú and Chile show the dramatic changes of their anchovy fisheries.

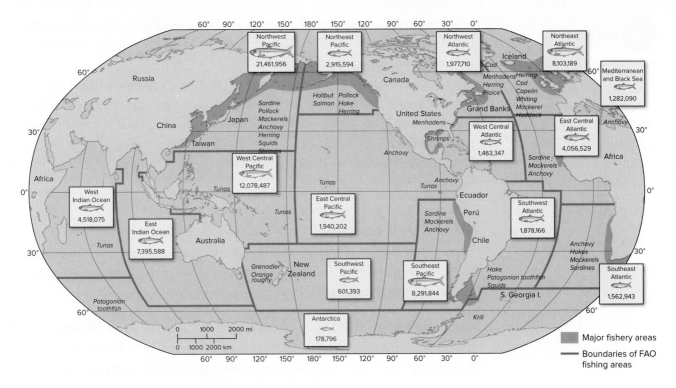

Figure 19.4 The major marine fishing areas of the world, based on boundaries established by the Food and Agriculture Organization of the United Nations (FAO). Statistics refer to the 2012 marine fish and shellfish catches (in metric tons) reported for each region. The figure for Antarctica incorporates catches for three separate FAO regions. Eight of 18 FAO regions showed lower catches in 2012 than in 2011. Not included is the 1 metric ton reported in 2012 for the Arctic.

broaden their tastes and look for healthy, low-fat sources of protein. China and Japan are the two leading consumers of seafood, followed by the United States, where annual consumption in 2012 was of 6.5 kg per person. The developed world is for the most part buying more fish rather than catching it. Japan, home to 2% of the world's population but consuming about 10% of the world's catch, buys more than twice what it catches.

Major Fishing Areas

An estimated 95% of the world's major fisheries are located on coastal waters and over half of the catches are taken from less than 7% of the ocean surface (Fig. 19.4). These coastal fisheries exploit the rich waters of the continental shelf. Because shelf waters are shallow compared with the open ocean, it is relatively easy to harvest bottom-dwelling, or demersal, species. Coastal fisheries also include pelagic catches: fishes, squids, and other animals that live in the water column. Primary production is higher over the shelf than farther offshore, supporting much more abundant life. Significant coastal fisheries are found where the continental shelf is very wide, as in the case of the Grand Banks of Newfoundland, off east Canada, the North Sea, and the Bering Sea.

Notice from the map in Figure 19.4 that the largest catches are taken from the western Pacific because they are closer to nations where large populations consume large amounts of fish. Most fishing grounds in the Atlantic, Pacific, and Mediterranean are in decline or exhausted. A few other areas are less heavily exploited and still have some underexploited resources. This is the case

Continental Shelf The shallow, gently sloping edge of the continents, extending from the shore to the *shelf break*, the point where the slope gets steeper.

- 2.3, Modern Ocean Basins

Primary Production The conversion by autotrophs of carbon dioxide into organic carbon, that is, the production of food.

- 6.1, The Ingredients of Life

in the Indian Ocean and around Antarctica. These areas are more remote and therefore more expensive to exploit. Fisheries in such remote areas, however, are rapidly increasing.

The open ocean, 92% of the total area of the ocean, supports fewer fisheries than the continental shelf but they are nevertheless very valuable. Some of the richest fishing areas of the world are located in upwelling areas just beyond the continental shelf, including those of Perú and northwest Africa (see Fig. 19.4, p. 589). Other fisheries, like those for migratory species like tunas, extend far into the open ocean.

Major Food Species

There are thousands of species of fishes, crustaceans, and molluscs, but relatively few support major fisheries. A summary of the major commercial catches for the world is given in Table 19.1 (p. 587).

Clupeoid Fishes The largest catches in the world are those for **clupeoid** fishes, small, plankton-feeding fishes that form huge schools. They include herrings (often sold as "sardines"), anchovies, sardines (sometimes called pilchards), menhadens, and shads (Table 19.2). Clupeoid fishes are usually concentrated over the continental shelf, but some species can also be found offshore in upwelling areas. They are caught with large **purse seines** that surround and trap the schools (Figs. 19.5).

Figure 19.5 The European anchovy (*Engraulis encrasicolus*) is an important food fish. It is canned, salted, or cooked fresh. Basque fishers from the port of Bermeo (a) in the Bay of Biscay, northern Spain, make bountiful catches using purse seines (b). Fish is unloaded (c) and sold in the fish market (d). Women workers, or *neskatillas*, play an important role in the fishing routine. They are in charge of carrying and selling the fish.

(t to b)©Peter Castro

Table 19.2 Examples of Commercially Important Fishes Around the World

Species	Distribution and Habits	Species	Distribution and Habits
Herring (*Clupea*)	North Atlantic and Pacific; schooling, plankton feeders; 38 cm	Hakes and whiting (*Merluccius*)	Temperate worldwide; demersal, feed on bottom invertebrates and fishes; 1 m
Sardines, or pilchards (*Sardinops, Sardinella, Sardina*)	Mostly temperate worldwide; schooling, plankton feeders; 30 cm	Flatfishes (*Platichthys, Hippoglossus,* etc)	Mostly temperate worldwide; demersal, feed on bottom invertebrates and fishes; 2 m in some halibuts
Anchovies (*Engraulis*)	Worldwide; schooling, plankton feeders; 20 cm	Tunas (*Thunnus, Katsuwonus*)	Tropical and temperate; schooling, carnivores; 4.3 m in the northern bluefin tuna
Menhadens (*Brevoortia*)	Temperate and subtropical Atlantic; schooling, plankton feeders; 38 cm	Mackerels (*Scomber, Scomberomorus*)	Tropical and temperate worldwide; schooling, carnivores; 2.4 m
Cods (*Gadus*)	North Atlantic and Pacific; demersal, feed on bottom invertebrates and fishes; 1.5 m in the Atlantic cod (no longer this large)	Salmon (*Oncorhynchus, Salmo*)	North Pacific and Atlantic; open ocean and rivers, carnivores; 1 m
Alaska/walleye pollock (*Theragra*)	Temperate North and Central Pacific; demersal, feeds mostly on bottom invertebrates and fishes; 90 cm	Haddock (*Melanogrammus*)	North Atlantic; demersal, feeds mostly on bottom invertebrates; 90 cm

Note: Fishes are not drawn to scale. The measurements given are of the approximate maximum length.

Herrings, anchovies, and sardines are eaten fresh, canned, or pickled. Fish flour, an odorless powder used as protein and dietary food supplements for human consumption, is also made from these and other fishes. About 90% of the clupeoid catch, however—particularly that of menhadens and herrings—is ground into fish meal to feed farmed fish, poultry, and livestock. The fish are also pressed to obtain fish oil, used in the manufacture of products like margarine, cosmetics, and paints as well as in omega-3 fatty acids supplements. Some of the catch finds its way into fertilizers and pet food. Industrial fisheries such as these, where the catch is used for purposes other than direct human consumption, accounted for 15.8% of fishery products in 2009.

Cods and Related Fishes Cods and related fishes—the pollock, haddock, hakes, and whiting—make up another important component of the worldwide catch (Table 19.2). These fishes are all demersal, cold-water species. They are mostly caught with **trawls** that are dragged along the bottom (Fig. 19.6d). The Alaska, or walleye, pollock (*Theragra chalcogramma*) is the largest fishery in United States waters and in 2012 second in the world after the Peruvian anchovy (see below). Overfishing, however, is threatening the fishery. The fall in the pollock catch is a reason why the number of Steller sea lions (*Eumetopias jubatus*) in the area has dropped by 80% since the 1970s.

The Atlantic cod (*Gadus morhua*) supported a most important fishery for centuries. European fishers were already fishing for it on the Grand Banks of Newfoundland a century before the arrival of the first colonists (and probably before Columbus). It was a vital source of inexpensive protein in many parts of the world. In an age before refrigeration, cod was salted and dried for preservation before being sent to Europe and the Caribbean. The dried fish was soaked in water before cooking, just as it is done today in the Mediterranean and Caribbean. Now, however, most cods and related fishes are sold fresh or frozen.

Atlantic cod catches in Canadian and other Western Atlantic waters peaked to close to 2 million metric tons in the 1960s before collapsing in the early 1990s. The cod fishery on the Grand Banks lasted until 1992 when a moratorium was declared in an attempt to save the fishery from extinction. The fishery had peaked in the late 1960s, almost 25 years before the moratorium. The Georges Bank fishery off New England was similarly closed in 1994, that off Newfoundland and Labrador, Canada in 2003. Cod was listed as an endangered species in Newfoundland, unthinkable a few decades ago. North Sea catches were also affected, dropping from 277,000 metric tons in 1971 to 59,000 metric tons in 2000. As a result, vast areas of the North Sea were closed to cod fishing in 2001.

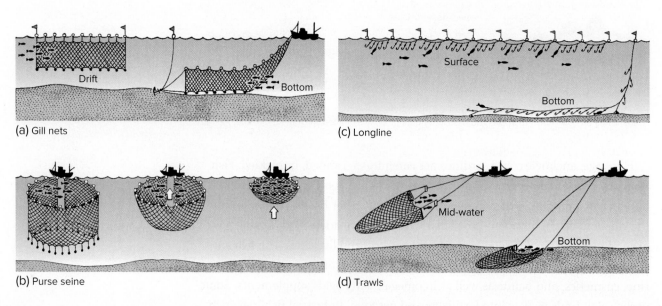

Figure 19.6 Examples of fishing methods. Gill nets (a) can be highly effective, and their mesh size and length are often regulated for taking particular catches or for reducing by-catch. Purse seines (b) are especially important for capturing schooling fishes like clupeoid fishes and tuna. The use of longlines (c) and trawls (d) is controversial due to the large, unintentional by-catch.

Closures caused high unemployment among fishers and related workers on both sides of the Atlantic, also threatening to destroy traditional fishing cultures.

A total of 1.1 million metric tons of cod were taken in 2012, but mostly on the northeastern Atlantic. Although there are signs of a slow recovery in Newfoundland, the Gulf of Maine cod fishery has nearly collapsed. It was closed for six months in late 2014. Recovery is slow because fishes like herring and mackerel, which are normally eaten by adult cod, are now more common and they happen to feed on juvenile cod. An increase in the North Sea catches was recommended in 2015.

Mackerels, Flatfishes, and Salmon Jacks, mullets, rockfishes (including the ocean perches, also known as redfishes), and mackerels are also important in terms of worldwide tonnage (Table 19.2). Canned mackerel from Southeast Asia, Japan, South America, and elsewhere has become an important source of cheap protein in some parts of the world. Flounders, halibuts, and other flatfishes are important catches in the United States, Canada, and other countries. Salmon, though now caught in much smaller numbers than formerly, are a very valuable catch, especially in the North Pacific. Salmon are caught both along the coast and in the open ocean. The fishery is threatened, however, by overfishing and environmental degradation, especially the destruction of salmon breeding sites. Most of the salmon consumed now is farmed (see below).

Tunas The most important open-ocean fisheries are those for several species of tunas (Table 19.2). The migration routes of tunas crisscross the oceans, mostly in tropical waters. Skipjack, yellowfin, albacore, bigeye, and bluefin tunas command high prices in world markets (see "The Aquaculture of Bluefin Tunas"). Most are eaten in affluent countries, either canned or raw as *sashimi*. They are caught by modern fleets that use sophisticated techniques to spot the schools in mid-ocean. The fishes are caught with large seines, surface **longlines**, or **gill nets** (Fig. 19.6). The fishing boats are equipped with freezers so that the fishes can be brought to port long after they are caught. Besides these high-tech fleets, some local, small-scale fisheries continue to exploit bonito, skipjack, and other tunas.

Molluscs and Crustaceans Molluscs are the second most valuable group of marine food species after fishes. Several types of cephalopods provide the largest catches of molluscs (Table 19.1). Squids, cuttlefishes, and octopuses are especially popular in Asian and Mediterranean countries. Fishers use lighted boats to catch squid at night. Other important food molluscs are clams, oysters, mussels, scallops, abalones, and a few other gastropods like conch in particular parts of the world.

Crustaceans are prized as food the world over. Major fisheries exploit many species of shrimps (or "prawns") and lobsters, all of which bring a high price. Crabs like the blue (*Callinectes sapidus*), king (*Paralithodes camtschaticus*), and Dungeness (*Cancer magister*) are also harvested. *Langostino* (some species of *Cervimunida* and *Pleuroncodes*), sometimes called "langostino lobster" or "squat lobster," has become an important food item. It is neither a true lobster nor a shrimp but a galatheid crab more related to hermit crabs.

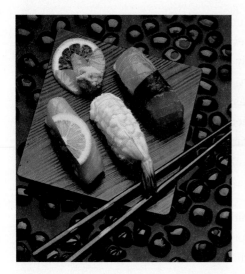

Figure 19.7 Raw fish, shrimp, seaweed, scallops, and sea urchin roe are used in *sushi* and *sashimi*, Japanese delicacies that have become popular around the world.

Other Food Species Other groups of marine organisms are also eaten, though they do not contribute much to global fisheries production (Fig. 19.7). Seaweeds, for instance, are consumed in many cultures, especially in Asia. Most seaweeds are farmed for carrageenan, which is used as an emulsifier; Japanese kelps come second. Sea urchins are esteemed for their gonads, or roe, which command an astronomical price, particularly in Japan. There is a fishery for the red sea urchin (*Strongylocentrotus franciscanus*) in California. Most of the roe is exported to Japan or sold to Japanese restaurants where, known as *uni,* it is eaten raw. Some species of sea cucumber are another common seafood. Called *trepang* or *bêche-de-mer,* they are dried, boiled, smoked, or eaten raw in Asia. Examples of other invertebrates used as food are jellyfishes in Asia, gooseneck barnacles in Spain, and polychaete worms in the South Pacific.

Sea turtles are still hunted and their eggs still gathered, even where they are officially protected. Seals and whales are also eaten in some places, and there are traditional fisheries for marine mammals in the Arctic, North Atlantic, West Indies, and South Pacific. These traditional fisheries are a source of considerable controversy because most of the species involved are endangered and modern technologies like powerboats are used. Should traditional fisheries be banned or left entirely unregulated? A compromise between these two extremes involves questions of the rights of indigenous cultures, the ethics of conservation, and international politics.

Optimal Yields and Overfishing

Like all living things, species we harvest for food can reproduce to replace the individuals that are lost to disease and predators, including humans. Because these living resources can replace themselves, they are known as **renewable resources**. Resources that are not naturally replaced, like oil and minerals, are called **non-renewable resources** (see more in 19.4 and 19.5). Experience has shown, however, that, though the food resources of the ocean are renewable, they are not inexhaustible.

Imagine that fishers discover a new, untapped population of sardines. The new fishery is very successful, and the fishers make handsome profits. The news spreads, and soon other fishers join in. After several years, **overfishing** results: Catches dwindle, and the fish caught are smaller. The fishery becomes unprofitable, and many, if not most, of the fishers go out of business.

What happened? The reproductive rate of sardines—or salmon, scallops, or sharpnose surfperches—depends, at least in part, on the size of the population, or **stock**, as it is known to fisheries biologists. Like dinoflagellates growing in a jar, fish stocks grow fastest when there are neither too few nor too many individuals in the population. If the population size is very small, the number of young being born is also small because there are not many potential parents. If the population is too large, competition, overcrowding, and diseases slow down the growth rate. The number of new young, that is, the population growth rate, is highest at some intermediate population density (Fig. 19.8).

Sustainable Yields To harvest the stock in the optimal way, one must consider this feature of population growth. Obviously, it would be foolish to harvest the entire stock at once, because no individuals would be left to reproduce, and the fishery would be destroyed forever. If the fishery is to last indefinitely, or to be sustainable, the number of fish caught can be no more than the number of new fish added through reproduction; if more are caught, the population will decline. The **sustainable yield** of a population is the amount that can be caught and just maintain a constant population size. In other words, the catch is just large enough to prevent the population from growing but not so large as to reduce it.

Because the growth rate depends on the population size, so does the sustainable yield. When the stock is large, it is held in check by natural mortality; even a small harvest causes a reduction in the population. Very small populations also have a low growth rate. In this case, fishing is more dangerous because it can threaten the seed stock and drive the species to the brink of extinction.

The highest catch that can be continued year after year without threatening the stock is called the **maximum sustainable yield**. It occurs at medium population size, when the natural growth rate is highest. From the point of view of profit and food production, this can be thought of as the optimal catch.

The size of the catch, and therefore of the stock, is directly related to **fishing effort**: how many boats and fishers there are, how long they stay at sea, and so forth. When little effort is put into the fishery, of course, the catch is small. Only a small fraction of the stock is removed by fishing, and the population grows until limited by natural factors. The population remains secure, but the fishery yields less food and less profit than it could safely sustain.

With more intense effort, such as larger, faster, and more numerous boats, the catch exceeds the maximum sustainable yield. Before long the stock declines and catches fall no matter how much fishing effort is expended: Overfishing has occurred.

A plot of the sustainable yield versus fishing effort gives us a **catch-effort curve**, which is very similar to a plot of the population growth rate versus the size of the stock (Fig. 19.9, p. 597). The maximum sustainable yield occurs at moderate levels of effort, which result in intermediate stock sizes.

Overfishing Another problem is that the biological properties of most fisheries stocks do not mix well with the principles of economics or with politics. Most fisheries are profitable when harvested at the level of the optimum catch. As long as there is money to be made, more fishers and more boats will be attracted to the fishery, and the catch will exceed the optimal level. In an open fishery, one that is completely unregulated, market forces inevitably lead to overfishing. A fishery can still be profitable to some fishers, even if catches are below the maximum sustainable yield. The total profit from the fishery, however, decreases.

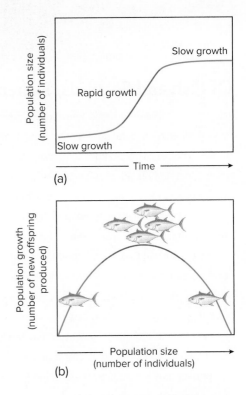

Figure 19.8 (*a*) The theoretical growth of a population that begins with just a few individuals. At first, with only a few adults to produce offspring, the population increases very slowly (blue line). As the population size gradually increases, so does the number of offspring, and the rate of growth begins to increase (red line). Eventually, the population reaches the point where limiting factors like food shortage and crowding prevent further growth (blue line). (*b*) Thus, the rate of growth is directly related to population size and is maximal at intermediate abundances.

Of Fish and Seabirds, Fishers and Chickens

The ocean has been good to Perú. As the cold Perú Current flows northward, coastal upwelling pumps nutrients from the depths. The resulting increase in primary production supports a rich harvest of plankton that in turn feeds immense schools of the Peruvian anchovy, or *anchoveta* (*Engraulis ringens*). This high productivity is a critical part of a complex cycle that has had tremendous consequences for the economy of Perú.

The productive fisheries support large populations of cormorants, boobies, and other seabirds, which roost and nest on small islands along the coast. With hardly any rain to wash them away, the birds' droppings slowly build up in the dry climate. The droppings have accumulated into layers as thick as 45 m. This thick crust, called guano, is rich in nitrogen and phosphorus transported through the food web as part of the nutrient-rich upwelled water.

Islas Ballestas, guano islands off Pisco, Perú.

An upwelling economy In what is now referred to as the "Guano Age," Peru experienced a major economic boom due to their export of guano as an agricultural fertilizer. The guano industry brought considerable wealth to Perú, and by the end of the nineteenth century millions of tons had been mined. Though it is still sustainably harvested today, what took thousands of years to deposit was depleted in a matter of decades.

The anchovies were next. A commercial anchovy fishery was established in the early 1950s and grew to unprecedented dimensions. It became the world's largest single fishery and made Perú the largest fishing nation. Catches continued to increase, except in years when El Niño, the warm currents that occasionally arrive at Christmastime, interfered with upwelling. Most of the catch was exported as fish meal for chicken feed.

An unsustainable haul By the late 1960s, the annual catch surpassed the maximum sustainable yield of around 10 million tons per year. After the 12.3 million ton catch of 1970, the fishery collapsed. The 1972 El Niño added to these woes and then the seabirds suffered a dramatic decline. Today, only 4 million seabirds live on the Peruvian coast, down from 20 million in the 1950s. The early 1970s collapse had widespread effects. Without the fish meal produced from anchovies, the price of other protein meals increased. In Brazil alone, several million acres of rain forest were cleared to grow soybeans.

So what caused the fishery to collapse? The reasons are not entirely clear but overfishing was not the only cause. Clupeoid fishes like anchovies and sardines naturally undergo population cycles. It may well have been that the combination of heavy fishing pressure and a dip in the natural cycle pushed these fishes over the brink. El Niño also enters into the picture. Harvest records since the beginning of the twentieth century show that when the Peruvian anchovy was thriving, the Pacific sardine occurred in low numbers, and vice versa. Populations of these two Eastern Pacific species thus appear to have cycles that are opposite to each other for reasons that are not clear. It has been speculated that warm water in the Pacific due to a strong El Niño increases sardine populations; cool water and a weak or absent El Niño increases anchovy populations. Management of fisheries for species like the Peruvian anchovy needs to be tuned to the fishes' biology, with oscillating harvesting intensities. The Peruvian government has now established annual fishing quotas, but the maximum sustainable yield remains unknown.

Fishers unloading a net full of Peruvian anchovies.

Overfishing is not just a theoretical problem. The Food and Agriculture Organization of the United Nations (FAO) has estimated that out of the 600 stocks of marine fishes it monitors, 52% are fully exploited and are close or have reached the maximum sustainable yield, 17% are overexploited (or overfished) and 7% are already depleted, both being below the maximum sustainable yield. Only 3% are underexploited.

Most of the top-10 commercial species, which account for almost a third of global catches, are fully depleted. Good examples are the stocks of cod, herring, halibut, salmon, Alaskan king crab, and other species in the North Sea and the North Atlantic and North Pacific oceans. As with the great whales, fishing effort in these fisheries has long surpassed the optimal catch. Overfishing is by far the biggest threat for fish species listed as endangered or vulnerable to extinction. The traditional major fishing areas of the world are not the only ones becoming exhausted. Small-scale local fisheries in the tropics, where most people live and where food is needed the most, are being rapidly decimated.

The danger of overfishing can be particularly serious in herrings, anchovies, and other clupeoid fishes. These species often undergo dramatic population cycles and can therefore be especially vulnerable to fishing pressure. This is the situation in the Peruvian anchovy, or *anchoveta* (*Engraulis ringens;* see "Of Fish and Seabirds, Fishers and Chickens").

Overfishing is threatening many other species, particularly large fishes like tunas, swordfish, and sharks. Recent studies have shown that around 90% of the biggest fishes in the world oceans are already gone, taken by big commercial fishing fleets. Only around 10% of the original stocks remain. The fish that now remain are also smaller than before the intensive fishing started, an average of half the size 20 years ago in the case of tunas. Small females produce less eggs than large females, and, at least in some species, larvae from small females are smaller and grow slower than larvae from larger females.

Fishing Down the Food Web The disappearance of the most valuable fishes, which happen to be higher in the trophic pyramid, has forced fishing lower on the pyramid, which is referred to as fishing down the food web. Such a shift can cause unknown changes in the whole ecosystem, like an increase in the number of non-utilized plankton or in undesirable predators like jellyfishes, or sea jellies, and comb jellies.

Bigeye tuna (*Thunnus obesus*) were about two times as heavy and eight times more abundant in the 1950s than at present. The northern, or Atlantic, bluefin tuna (*Thunnus thynnus*) is among the largest of bony fishes and the most valuable.

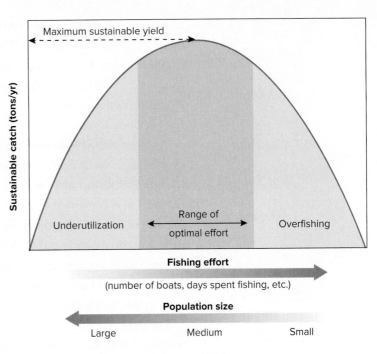

Figure 19.9 A generalized, theoretical curve showing that as fishing effort increases, so does the catch, but only up to a point, which is the biologically optimal catch. After this point the overexploited fishery will yield smaller and smaller catches as the effort increases. The curve is called a catch-effort curve.

Bluefin stocks are critically overfished, and the western Atlantic populations have fallen by 90% since 1970 (see "The Aquaculture of Bluefin Tunas").

Sharks, slow to grow and reproduce, have critically declined in many parts of the world. A 2002 study showed that populations of most species of sharks along the eastern coast of the United States had declined by at least 50% from the previous 15 years.

Another prized food fish, the swordfish (*Xiphias*), has also been victim of overfishing. North Pacific populations fell approximately 70% between 1960 and the late 1990s, and most of the fish were caught before they were old enough to reproduce. A campaign to reduce consumption of swordfish was very successful, and ultimately swordfish fisheries closed in large areas of United States waters in the Atlantic and Gulf of Mexico. Longline fishing for tuna was closed in some areas because many swordfish are caught and killed on longlines. There is evidence that these measures are working, and the number of swordfish, at least in the North Atlantic, is on the increase and perhaps on a road to recovery.

The potential effects of overfishing on whales, seals, and other marine mammals that feed on fish are the subject of much debate and controversy. Conservationists see overfishing as a threat to the already endangered marine mammals, while fishers see protected marine mammals as a threat to their diminishing stocks. A 2004 study suggested that on a global scale marine mammals and fishers rarely prey on the same stocks. Significant overlaps between feeding by marine mammals and fishing, however, do exist in some regions.

Overfishing could be responsible for the destruction of kelp forests by sea urchins in the Aleutian Islands in Alaska. Because fish-eating seals and sea lions have become less common, orcas, or killer whales that normally hunt them have turned instead to sea otters, which keep sea urchins under control.

A fishery is regarded as collapsed if catches fall below 10% of historic highs. Nearly one-third of all the species that are fished have already collapsed, and around 60% are predicted to collapse by 2025. A 2006 worldwide study by 14 marine scientists concluded that *all* of the world's fishing stocks will collapse by mid-century if no drastic measures are taken to curb overfishing, by-catch, destruction of habitats, and pollution. Good management, however, has yielded some signs of improvement in some regions.

Other Threats to Fisheries

Habitat destruction is another threat to fishery resources. Estuaries, mangrove forests, seagrass beds, and other threatened environments are critical breeding grounds and nurseries for fishes, lobsters, shrimps, scallops, and other valuable species. Three-quarters of the commercial catch of the United States consists of species that inhabit estuaries during part of their life. Bottom trawls are very destructive to life on the sea floor, another form of habitat damage. Seamounts, inhabited by many unique species, are particularly vulnerable.

By-catch, organisms that are caught unintentionally while fishing for other species, is another problem. It is estimated that around 25% of the worldwide catch, roughly 23 million metric tons, is by-catch. A good portion of the

by-catch, however, goes unreported. The swordfish caught and killed on longlines set for tuna are one example of by-catch; longlines also catch sharks, marlin, turtles, seabirds, dolphins, and other marine life. Trawling also produces a lot of by-catch. This is particularly true of shrimp fisheries, where as much as 95% of the total catch is by-catch because of the small size of the net's mesh. Some of the by-catch is kept but much of it is discarded. Even if it is released alive, most of the discarded by-catch dies. A large part of the by-catch in trawl fisheries is juveniles of valuable species sought in other fisheries, or worse, of protected species like cod in the North Atlantic. **Ghost fishing**, trapping by abandoned fishing gear like gillnets, longlines, and various kinds of traps, also kills uncounted numbers of marine life each year.

Another threat to fisheries is pollution from sewage, oil spills, and toxic chemicals. The high level of mercury in large fishes like tuna and swordfish is of particular concern to consumers.

Managing the Resources

Given the danger of overfishing, fisheries resources should be harvested in a way that does not deplete them beyond recovery. In other words, fisheries must be managed to ensure their long-term value. The wise management of fisheries stocks is much harder than it sounds. For one thing, the maximum sustainable yield can be very difficult to estimate. To do so, fisheries biologists need detailed information about the size of fisheries stocks, how fast the organisms grow and reproduce, how long they live, and what they eat (which often changes at different stages of the life cycle). Such information is rarely easy to obtain and is often unavailable. Biologists may have to rely on rough "guesstimates" or questionable assumptions, making their estimates of the optimal catch less reliable.

Fisheries are Complex Furthermore, real fisheries are much more complex than we have indicated. The smooth catch-effort curve shown in Figure 19.9 represents a greatly simplified model and does not take many natural features into account. For example, the harvested species could be competing with other species, and fishing pressure might alter the competitive balance. Schools can get smaller as a result of fishing. This might not only make the fish harder to catch but also adversely affect their behavior and reproduction. Simple models do not consider the size of the fish caught, but it makes a difference whether large, older individuals or small, younger ones are caught, for example, because large individuals typically produce many more eggs than small ones. Similarly, the time of year when fishing is done can be important, like whether the catch is taken before or after the breeding season. These natural complexities can have dramatic consequences and could cause a fishery to fail unexpectedly if not taken into account.

A major problem is that the actual population size of most commercial catches remains unknown. We are only assuming in many cases that fishing at an "intermediate" abundance is at the theoretical "optimal" level and this may not be the case in the real world. Given the uncertainties of determining optimal catches and the disastrous effects of overfishing, it can be argued that the catch should be set somewhat lower than the estimated optimum just to be on the safe side. A fisher,

however, with a family to feed and boat payments to make, might have different ideas. The same can be said of cannery and dockyard workers, gear merchants, bankers, and all the other people who depend directly or indirectly on the fishery. Thus, fisheries management is a complex, often controversial, matter that is affected by economic and political factors as well as biological ones. If the catch is harvested by more than one nation, international relations also come into play.

Management Strategies Once a desired catch level is agreed on, there are a number of ways to manage the fishery. The total catch could be limited by closing the season as soon as the target catch is reached or by restricting the length of the season, the number of boats or fishers, or the areas open to fishing. The size and power of boats can be limited or certain methods of fishing can be banned altogether. Limits can be placed on the catch of each boat or, if an international fishery, each nation. By regulating the mesh size of nets, fishers could be prohibited from taking individuals of certain sizes (Fig. 19.10). Leaving the largest individuals to reproduce can help in the recovery of overfished stocks. If sexes can be told apart, fishers could be prohibited from taking females.

The total collapse of once thriving fisheries has sometimes forced the use of drastic measures. Governments could reduce their fishing subsidies, they could pay fishers to give up fishing, or a fishery could be closed altogether. Complete closures, however, are in effect in less than 1% of the world's fishing grounds. Besides preventing overfishing, management measures can be undertaken to help restore stocks that have become depleted. These measures could involve the improvement and protection of essential habitats or the transplantation of artificially raised young.

Marine Protected Areas The establishment of **marine protected areas**, or MPAs, in many parts of the world (see foldout map of North America at the end of the book) is another way of restoring and protecting fisheries. Fishing in MPAs can be either banned or strictly managed. It has been demonstrated that both biodiversity (number of species) and biomass (abundance of particular species) in MPAs is higher than in non-protected areas. Non-protected areas open to fishing benefit by offspring and adults settling in from protected areas.

More than 90% of the ocean's present fisheries lie within an exclusive economic zone (EEZ) of one nation or another, where nations can completely control their fisheries, oil, and mineral resources within this zone. The high

Figure 19.10 Both the size and species of fishes caught by nets can be controlled by the net's mesh size. A fine net might catch both sardines and tuna (a). With a larger mesh size the sardines escape and only tuna are caught (b). With an even larger mesh size only large tuna are taken (c).

(a)

(b)

(c)

seas, however, are still considered common property. A number of potential new fisheries lie outside territorial boundaries and thus are open to all nations. Furthermore, fishing regulations are being increasingly ignored as catches diminish. Estimates are that a fifth of all worldwide catches are illegal.

UN Convention on the Law of the Sea Fishing rights over a nation's territorial waters are a complicating factor in the management of resources. A country's territorial waters traditionally extended 3 nautical miles offshore. Some nations, however, began to exclude fishing fleets from other nations by extending their territorial waters. Fishing disputes even led to wars. A general compromise was finally reached in 1982 with the United Nations Convention on the Law of the Sea (UNCLOS), which so far has been ratified by 189 nations. It allows nations to establish an EEZ 200 nautical miles offshore. Nations are required to use their resources in a sustainable way. They can keep foreign fishers out if they wish or sell fishing licenses to foreign fleets. EEZs cannot always be policed, of course, and foreign fishers can sneak in and fish without permission or with illegal gear.

Management in the United States In the United States, which has signed but not yet ratified UNCLOS, the Sustainable Fisheries Act of 1996 required federal fisheries managers to develop management plans to prevent overfishing, restore depleted stocks, and reduce by-catch, the first time protection has been extended to non-commercial species. Vessels of foreign nations that have signed fishing agreements with the United States are allowed to fish in United States waters only after receiving a fishing permit. The 2003 Pew Oceans Commission went further by calling for policies to protect marine ecosystems. Such policies are beginning to yield results. A record 34 commercially and recreationally important fisheries have returned to health since 2000, which means that 86% of near 250 monitored commercial fisheries are not being overfished and 79% are thought to be healthy.

Efforts toward the better management of resources have already brought about some positive results elsewhere. A limited number of fisheries in a few countries are now being successfully managed. Fish from these sustainable fisheries are "eco-labeled," so that consumers can select it over fish from overexploited catches (Fig. 19.11, p. 602). The role of consumer awareness is most important. We should minimize our impact by choosing fish whose fisheries are sustainable as well as those that generate less by-catch. DNA tests could soon be widely used to identify fish caught from unsustainable or illegal fisheries.

New Fisheries

Fisheries scientists have for decades been absorbed with the possibility of tapping new or underexploited fisheries. One possibility is to increase the use of by-catch. The main problem with much of the by-catch is not so much that there is anything wrong with it but that people simply will not buy it. Consumer tastes are fickle and vary from place to place. What is considered a delicacy in one place is junk in another. Hake (*Merluccius*), for instance, is prized in Europe but not in the United States. Croakers and sea robins are common throwaway fishes in the United States that are eaten elsewhere. Squids are caught in large numbers in California, but 90% of the catch is

exported, primarily to Asia. Some of it is processed in China, repacked, and sent back for consumption in California. Squids are a delicious source of lean protein, but American consumers were suspicious of the sucker-armed molluscs, though marketing squid by its Italian name of *calamari* has increased its appeal. A name change also helped overcome consumer resistance to eating dolphinfish (*Coryphaena hippurus*), certainly a fish and not a dolphin. The more seductive Hawaiian name of *mahimahi* helped enhance its marketability.

A potential use of the by-catch is to process it into products that have more market appeal. In the *surimi* process, for example, low-value fish is washed to remove fat, minced, and mixed with flavorings and preservatives. Appropriately shaped, it is sold as imitation crab, lobster, shrimp, or scallop. Alaskan pollock is now the main component, but less desirable fish species could also be used. Processed fish products like fish sticks could also open the market to less popular catches. Unwanted catches could also be converted into fish flour and used to protein-enrich all sorts of processed foods.

The highly productive waters around Antarctica are often mentioned as potential sources of increased commercial catches. Antarctic catches include various fishes, squids, the southern king crab, and krill (Fig. 19.12). Even in the harsh conditions of Antarctic waters, however, overfishing is taking its toll. The waters around South Georgia Island, for instance, are already overfished. The Patagonian toothfish (*Dissostichus eleginoides*), marketed as "Chilean sea bass" in the United States, has been seriously depleted, mostly by illegal fishing. Many illegal fishing vessels operate under flag-of-convenience countries that can't enforce international maritime law. Illegal longline fishing has also affected populations of seabirds, especially albatrosses.

There are other potential sources of food from the sea. Some of these are underutilized. Others are unconventional species like squid and other

IN CONTEXT

Deep Scattering Layer
Animals, mostly fishes and shrimps, that migrate vertically between the mesopelagic and epipelagic zones.
• 18.1, The Twilight World

BEST CHOICES	GOOD ALTERNATIVES	AVOID
Abalone	Basa/Pangasius/Swai	Abalone (China & Japan)
Arctic Char (farmed)	Cod: Atlantic (imported)	Cod: Pacific (Japan & Russia)
Bass: Striped (US hook & line, farmed)	Cod: Pacific (US trawl)	Crab: Canned (imported)
Catfish (US)	Crab: Blue & King (US)	Crab: Red King (Russia)
Clams, Mussels & Oysters	Grouper: Black & Red (US)	Lobster: Spiny (Belize, Brazil,
Cod: Pacific (US hook & line)	Halibut (US Pacific gillnet & trawl)	Honduras & Nicaragua)
Lobster: Spiny (Mexico)	Lobster (Bahamas & US)	Mahi Mahi (imported)
Prawn: Spot (AK & Canada)	Mahi Mahi (Ecuador & US)	Orange Roughy
Rockfish (US)	Monkfish (US)	Rockfish (Canada trawl)
Sablefish/Black Cod (AK & Canada)	Pollock (US)	Salmon: Atlantic (farmed)
Salmon (AK)	Salmon (CA, OR & WA wild)	Sardines: Atlantic (Medit. Sea)
Sanddab (CA, OR & WA)	Scallops (wild)	Sharks
Sardines: Pacific (Canada & US)	Shrimp (Canada & US wild,	Shrimp (imported)
Scallops (farmed)	Ecuador farmed)	Squid (imported)
Seaweed (farmed)	Snapper (US)	Swordfish (imported)
Shrimp (AK wild, US farmed)	Squid (US)	Tuna: Albacore/White canned
Tilapia (Ecuador & US)	Swordfish (US)	(except Canada & US troll, pole
Trout: Rainbow (US farmed)	Tilapia (China & Taiwan)	and US longline)
Tuna: Albacore/White canned	Tuna: Albacore/White canned	Tuna: Bluefin
(Canada & US troll, pole)	(US longline)	Tuna: Skipjack/Light canned
Tuna: Skipjack/Light canned	Tuna: Skipjack/Light canned	(imported longline and purse seine)
(FAD free, US troll, pole)	(imported troll, pole and US longline)	Tuna: Yellowfin (except troll, pole
Tuna: Yellowfin (US troll, pole)	Tuna: Yellowfin (imported troll, pole	and US longline)
	and US longline)	

Start with "Best Choices" then check the other columns—your favorite seafood could be in more than one.

Best Choices
Buy first, they're well managed and caught or farmed in ways that cause little harm to habitats or other wildlife.

Good Alternatives
Buy, but be aware there are concerns with how they're caught or farmed.

Avoid
Don't buy, they're overfished or caught or farmed in ways that harm other marine life or the environment.

Visit us online or download our app for a **full list** of our recommendations.

Figure 19.11 Consumers can decide which ocean-friendly seafood they should buy or order by using the Monterey Aquarium Seafood Watch pocket guide (www.seafoodwatch.org). It is available as a United States national guide (reproduced here) or as one of six regional guides. The Monterey Aquarium also offers a free smart-phone application for all of its guides.

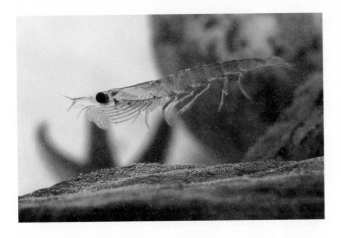

Figure 19.12 Krill (*Euphausia superba*) form dense schools in Antarctic waters and were considered potential human food. Predictions of annual catches in the millions of tons, however, have not yet materialized. It is one of the world's largest underexploited catches, although it is now harvested to feed farmed salmon and as a source of omega-3 fatty acids.

Figure 19.13 *Pleuroncodes planipes*, also called the lobster krill or tuna crab, is a pelagic crab that has been suggested as a potential source of human food. A possible problem with this plan is that *Pleuroncodes* is an important food for tunas.

cephalopods, flying fishes, and pelagic crabs (*Pleuroncodes*) (Fig. 19.13). Large swarms of the crab, each having a maximum length of 13 cm, are often washed ashore on the Pacific coast of California and Mexico, turning beaches bright red. Lanternfishes (*Myctophum*) represent another potential fishery. They are found in deep water but migrate to the surface at night in the deep scattering layer forming dense schools in most oceans. These and other species remain underexploited or completely unexploited mainly because of lack of consumer acceptance or the absence of efficient fishing methods. Some could eventually support successful fisheries, though probably for fish meal rather than for human consumption. The development of a new fishery must nevertheless be undertaken only after a careful investigation of the biology of the species in question, particularly on their interactions with other species in the community.

Section Review

1. **Summarize** Make a table like the one below to summarize information about some of the major marine food species.

Group	Examples	Fishery Technique	Uses	Status of Fishery
Clupeoid fishes				
Cod and related species				
Tuna				
Molluscs				
Crustaceans				

2. **Explain** Once a desired catch level is agreed upon, how can a fishery by managed to prevent overfishing?

19.2 Aquaculture

VOCABULARY

aquaculture

mariculture

seeding

salmon ranching

salmon farming

genetic engineering

biotechnology

Main Idea
Farmed fish and shellfish must be managed sustainably because they contribute an increasing amount of food to people around the world.

. .

Key Questions
1. What is the difference between aquaculture and mariculture?

2. List examples of organisms that are raised in aquaculture.

Instead of overfishing the dwindling natural food resources of the oceans, why not raise marine organisms as we do farm animals? This is not a new idea. The Chinese have been doing it with freshwater fishes for thousands of years, the Romans farmed oysters, and the early Hawaiians built walled fish ponds along the coast to raise mullet and milkfish. The application of farming techniques to the growth and harvesting of aquatic organisms is known as **aquaculture**. The specific farming of marine and brackish-water organisms is called **mariculture**.

Growth of Aquaculture
The worldwide culture of marine and brackish-water fish and shellfish alone reached 25.6 million metric tons in 2012, with an estimated value of $58.7 billion. These figures, more than twice the volume and more than three times their value in 2000 (11.9 million metric tons with a value of $16.7 billion), show the rapid growth of marine aquaculture (see Fig. 19.3 p. 588). The 2012 figures nearly double if the culture of seaweeds (23.7 million metric tons with a value of $6.3 billion) is included. The culture of freshwater animals in 2012 was 41.1 million metric tons, with a value of $79.0 billion. Freshwater fish accounts for 58% of all fish farmed worldwide in 2012. Aquaculture has also grown more than three times faster than livestock production on land, and according to FAO aquaculture is the world's fastest-growing food-production sector. Aquaculture has already passed worldwide beef production. In fact, farming salmon is about seven times more efficient than raising beef! Demand for seafood is expected to grow by 35% in the next 20 years as the result of the growth in population and incomes as well as in the status of fish as a healthy food.

It is clear that this increased demand must be met by aquaculture because of the stagnant growth in global wild-fish catches. Dwindling catches mean an even higher consumer price for wild-caught fish. Nearly half of all the fish and shellfish consumed around the world today is farmed, although consumption is mostly of freshwater fish. Aquaculture production showed an average annual growth rate of 6.1% between 2002 and 2012, from 36.8 to 66.6 million metric tons. It has increased 14-fold since 1980. Shrimp aquaculture shown tremendous growth, with farmed shrimp now accounting for around 25% of the estimated 3 million metric tons of shrimp consumed every year around the world.

Figure 19.14 Milkfish (*Chanos chanos*) is harvested from floating cages and brackish-water ponds in Southeast Asia. Some will be eaten, others allowed to mature and spawn to replenish the cages and ponds. The fish are shown in the insert.

Aquaculture has made China the largest producer and exporter of seafood in the world. Though aquaculture is used primarily to grow food species, it is also used to grow commodities like cultured pearls and fishes for the aquarium trade.

Aquaculture Operations

Traditional aquaculture is still practiced in several parts of the world, particularly in Asia. Milkfish is farmed in ponds filled with brackish water (Fig. 19.14). The immature fish, or fry, are captured from the wild, transferred to the ponds, and fed with agricultural by-products. In this way, wastes are converted to fish flesh that is cheaper to produce than pork or beef. These fish farms provide not only high-protein food but jobs as well.

With various modifications, similar operations are used to farm other fishes, molluscs, shrimps, and seaweeds in many parts of the world. The fry or spat (immature bivalve molluscs) are transplanted to favorable sites in coves (Fig. 19.15, p. 606), natural ponds, estuaries (including fjords), and mangrove forests. Already, shrimp aquaculture is a significant source of foreign exchange earnings for several developing countries. Salmon and other fishes are raised in large floating pens or cages (Fig. 19.16, p. 607). Oysters, mussels, and other molluscs are grown on racks, in baskets, hanging from rafts, and in other ingenious ways. Aquaculture of these filter-feeding molluscs can be particularly successful because it is a self-supporting system that does not involve adding food, fertilizers, or antibiotics.

Sometimes, however, the farming of high-value food species is a different story. In what can best be described as the domestication of marine organisms, growth is maximized by almost totally controlling the organisms and their environment (see "The Aquaculture of Bluefin Tunas").

Challenges in Mariculture Mariculture poses many problems, and relatively few marine food species can be farmed (Table 19.3, p. 606). Most species need specialized sites and equipment to be farmed successfully. Water pumped into tanks, pools, or other holding facilities must be free of pollutants and carefully

DIVE IN!

Genetically modified salmon at the market? Atlantic salmon that have been genetically modified (GM) to grow faster and larger than natural, non-GM salmon are being raised by an aquaculture operations. The GM salmon consume 25% less food compared to the non-GM salmon. In 2012, the U.S. Food and Drug Administration issued a draft review of the GM salmon and indicated that the GM salmon are safe to eat and would not have a significant impact on the environment. Some scientists are concerned that if the GM salmon were to escape, they would breed with wild salmon or brown trout, leading to unknown effects on the species and food webs. The company that makes the GM salmon has responded to concerns by explaining that the GM salmon that will be marketed are sterile females and, therefore, would not be able to reproduce in the wild. As of 2015, the GM salmon have not been approved for the marketplace and are not available to consumers in the U.S.

Table 19.3 Examples of Marine and Brackish-Water Species Used in Commercial Aquaculture

Species Cultured	Major Areas Where Cultured
FISHES	
Atlantic salmon (*Salmo salar*)	U.S., Canada, Chile, Europe
Pacific and southern bluefin tunas (*Thunnus*)	Japan, Mexico, Mediterranean, Australia
Cobia (*Rachycentron canadum*)	Central America, Caribbean, Southeast Asia
Cod (*Gadus*)	Europe
Flatfishes (*Solea* and others)	U.S., Europe
Horse mackerel (*Trachurus*)	Japan
Milkfish (*Chanos chanos*)	Southeast Asia
Mullets (*Mugil*)	Southeast Asia, Mediterranean
Pacific salmon (*Oncorhynchus*)	U.S., Canada, Chile, Japan
Puffer (*Fugu*)	Japan
Red seabream (*Pagrus major*)	Japan
Yellowtail (*Seriola*)	Japan
MOLLUSCS	
Abalone (*Haliotis*)	U.S., Japan, Australia, South Africa
Clams, cockles (*Anadara* and others)	Southeast Asia
Mussels (*Mytilus, Perna*)	U.S., Europe, New Zealand
Oysters (*Crassostrea, Ostrea*)	U.S., Europe, Japan, Taiwan, Australia, New Zealand
Fishers unloading a net full of Peruvian anchovies Scallops (*Pecten, Patinopecten*)	Europe, Japan
CRUSTACEANS	
Shrimps (*Penaeus* and others)	U.S., Ecuador, Mexico, Japan, China, Southeast Asia, India, Bangladesh
SEAWEEDS	
Red (*Porphyra, Eucheuma*) and brown (*Laminaria*) algae	Japan, China, Southeast Asia, South Pacific, U.S.

Figure 19.15 The farming of oysters (*Crassostrea glomerata*) in New Zealand.

checked for temperature, salinity, and other factors. Food can be a problem because the various stages of a species usually have different food requirements. This is a particular problem when trying to culture planktonic larval stages because they require specially cultured food. Still it takes less food to grow a pound of fish than it takes to grow a pound of chicken, pork, or beef. With a few exceptions, fishes are "cold-blooded" and don't need energy, and hence extra food, to keep a relatively constant body temperature as birds and mammals. Predators and disease-causing organisms must be periodically removed or killed. Parasites and disease can be devastating under the crowded conditions typical of culture operations. Viral infections are a serious problem in salmon and shrimp farming. Crowded conditions tend to cause cannibalism among fishes and crustaceans.

Commercial aquaculture, however, is a challenge that many enterprising, business-oriented fisheries scientists

Figure 19.16 Most salmon consumed worldwide is farmed, and introduced Atlantic salmon (*Salmo salar*) has made Chile the most important salmon-producing nation in the world after Norway. Fish-meal-fed salmon is raised in cages anchored offshore, and the harvested fish flown to consumers around the world. There are, however, serious environmental problems. Decomposition of feces and uneaten food depletes bottom oxygen in the once-pristine waters of southern Chile, and escaped salmon threaten local fish populations. A viral disease has also cased a drop in production.

are willing to accept. It is now possible, for example, to freeze shrimp sperm, ensuring year-round reproduction. We have learned to induce the settlement of the planktonic larvae of abalone, a valuable mollusc. Scientists have created oysters with three sets of chromosomes, instead of the normal two. This makes them sterile. Normal oysters develop a strong taste when their eggs mature, but sterile ones, which do not produce eggs, are always tasty. Different strains of oysters are being bred in efforts to develop faster-growing and larger hybrid oysters, very much as in plants.

Mariculture Strategies Rather than raising them all the way to harvest, some species are grown for only a short time, then released as fry or juveniles to enrich natural populations in a process known as **seeding**. Salmon are a good example of seeding. In some places, especially Europe, salmon are held in captivity their entire lives, from hatching to harvest. In other places the salmon are released as juveniles to feed and grow at sea. Several years later they return as full-grown adults to the stream where they were released. Not only can they be caught by commercial and sport fishers in the usual way, but they swim right back to the hatchery, where the ones not used for breeding can be caught and harvested. This practice, known as **salmon ranching**, originated in the Pacific Northwest of North America but has been transplanted to other locations. Salmon raised from young stages in offshore enclosures, which is known as **salmon farming** (Fig. 19.16), is responsible for making salmon the most important fish raised at sea.

Drawbacks of Aquaculture

Expensive equipment and specialized personnel further restrict the number of species that are cultured: Some species that could be farmed are not because it is not profitable. Species that are cultured are usually those that command high prices. Thus, instead of mass-producing cheap protein for a hungry world, aquaculture tends to provide variety in the diet of the affluent.

Pollution Another drawback of aquaculture is pollution. Fish farming in which thousands of fish are concentrated in ponds along the coast or in extensive

floating pens releases huge amounts of feces, urine, and uneaten food that ultimately degrade water quality. The decomposition of these wastes depletes oxygen in the water and releases nutrients that can trigger harmful algal blooms. Farmed fishes also concentrate toxic chemicals. Furthermore, the production of fish meal, often used as food in fish farming, is an energy-consuming process that releases greenhouse gases. Aquaculture now uses almost 70% of the fish meal and 90% of the fish oil produced world-wide. There is nevertheless a movement to use more plant protein from soybeans and less fish meal.

Impacts on Wild Fishes Some fish farms also use chemicals like antibiotics, pesticides, and synthetic pigments, which could be harmful to surrounding marine life. Another risk is the introduction of parasites and diseases to local marine species in the wild. More devastating can be the destruction of the natural environment. Clearing for shrimp and fish farms, has caused the destruction of mangrove forests, salt marshes, and other coastal communities in the tropics and subtropics. This has reduced the stocks of local species that use mangrove forests and salt marshes as nurseries. Though shrimp is more profitable, less food is actually produced than in the destroyed environments. Introduced farmed species that escape can be harmful to native species and the communities they live in, and catching fish from the wild to feed farmed fish, a common practice, can also deplete wild fisheries. Escaped Atlantic salmon breed with wild populations and dilute the genome of the wild populations, perhaps making the offspring more vulnerable to diseases and other stresses.

Human Health There are also major concerns about the harmful effects that harmful polychlorinated biphenyls, or PCBs, antibiotics, and other chemicals could have on human consumers. These contaminants accumulate on farmed fishes, particularly salmon. Still, some fisheries scientists and economists are optimistic. predicting that aquaculture will provide relatively cheap food. They pin these hopes on the development of technological improvements (Fig. 19.17) and fast-growing food species, and even turning carnivorous fishes into vegetarians, which are much cheaper to raise. Other options are pumping nutrient-rich water from the deep sea to provide better culture conditions or from sewage discharges and use it to grow the planktonic algae that are fed to the larvae of fish and other species. The application of **genetic engineering** and other forms of **biotechnology** to aquaculture is another possibility. Genetic engineering could allow scientists to alter the genetic information contained in a species' DNA to produce faster-growing, more disease-resistant, or better-tasting strains. There is great concern, however, about the effects of escaped organisms that breed with wild populations.

Figure 19.17 Mobile, wire-mesh pens of different shapes and sizes are being used experimentally or commercially in the offshore aquaculture of shrimps and of fishes like amberjack (*Seriola*), cobia (*Rachycentron*), and rockfish (*Sebastes*) in in many parts of the world. Food is piped in from barges or boats.

©Ocean Farm Technologies, Inc.

The Aquaculture of Bluefin Tunas

The Food and Agriculture Organization of the United Nations (FAO) reported in 2011 that about 60% of the 23 tuna stocks are fully exploited worldwide, up to 35% are overexploited or depleted, and only a few appear to be underexploited. The three species of bluefin tunas, which *sashimi* and *sushi* connoisseurs value for their dark-red meat, are fully exploited, and their value increases as the fish, particularly large ones, become scarcer. Bluefin tuna is a coveted luxury food; in 2012, individual fish sold in Tokyo for as much as $736,000.

Managing tuna There are three species of bluefin tunas: Atlantic/northern (*Thunnus thynnus*), Pacific (*T. orientalis*), and southern (*T. maccoyii*). Internationally, the Atlantic and southern bluefins are listed as critically endangered and the Pacific are listed as threatened by the International Union for Conservation of Nature and Natural Resources (IUCN). In 2011, however, the United States ruled that the western Atlantic populations of the Atlantic bluefin did not warrant the status of endangered species and named them a "species of concern." Meanwhile, the Mediterranean populations of the Atlantic bluefin have halved in size since the 1970s. Efforts to ban fishing or to stop international trade have failed. In addition, poaching has presented huge problems for the population management of bluefin tuna. In 2014 alone, 175 tons of illegally caught tuna was seized in Italy and Spain.

Is aquaculture the answer? High prices, high demand, and dwindling stocks have encouraged the farming of bluefin tunas. One type of tuna farming is known as "tuna ranching". Schools are caught in giant purse seines, slowly grown and fattened in floating pens, harvested, and flown refrigerated or frozen to fish markets. Tuna ranching unfortunately does not help in replenishing the dwindling natural populations because the captured bluefins typically do not reproduce in floating pens.

Attempts are consequently being made to raise bluefins from eggs to sexually mature adults through intensive, or closed, aquaculture. This domestication process could supplement fish stocks to satisfy the high demand for human consumption and ultimately help save bluefins from extinction. Researchers have had major breakthroughs with bluefin breeding in captivity. One technique involves implanting a device into adults that slowly releases synthetic gonadotropin-releasing hormone (GnRH), a hormone that triggers spawning. Bluefins raised in captivity appear to lack the natural production of GnRH. Using GnRH hormone replacement, Japanese scientists have successfully bred domesticated bluefin to a third generation. A Spanish team first reported in 2010 spontaneous spawning without the use of hormonal induction in the Atlantic bluefin, a major breakthrough in the domestication of bluefins.

These promising innovations in captive bluefin breeding, however, have been obtained on an experimental basis and not on a mass culture level. Furthermore, fish farming has many potential challenges to address before it can be commercially viable, such as the pollution released from the large pens, similar to salmon farming. Bluefin tuna biologists are hopeful that future advancements in breeding programs will result in the production of captive bluefin tuna populations in a sustainable environmentally safe method and thus relieve pressure on the dwindling wild stocks.

. .

Section Review

1. **Describe** a basic aquaculture operation.

2. **Summarize** the problems that can occur in mariculture.

3. **Discuss** how pollution and destruction of habitat that result from aquaculture impact marine environments.

19.3 Commerce and Recreation

DIVE IN!

What is ecotourism?
Ecotourism is responsible travel
to natural areas for leisure and
education. An increasing
number of companies in the
tourism industry are offering
ecotourism packages that allow
people to explore and learn
more about marine ecosystems
with the least amount of
impact as possible. These
packages include whale
watching and trips to
mangrove forests and coral
reefs. Studies have shown
mixed results on the success of
ecotourism. Corals near
viewing platforms on the Great
Barrier Reef in Australia show
a higher incidence of disease
than corals in nearby areas
where there are no platforms.
Studies also show that whale
watching may adversely affect
the animals by preventing
them from feeding as well as
expending energy to swim
away from boats full of tourists.
Ecotourism must be truly
eco-friendly and regulated in
order for it to achieve the goals
in its definition.

Main Idea
Humans use marine living resources in other ways besides food,
including for recreation.

Key Questions
1. What are the non-food living resources from the ocean?
2. What type of recreational activities associated with the oceans do
 people participate in?

The living resources of the sea are used in many ways other than food. Timber
is obtained from mangroves. Jewelry is made from pearls, shells, and corals;
leather from sharks. Chemical compounds that are obtained from marine
organisms, or **marine natural products**, have potential for use in medicine.
Many of these resources are being overexploited. Mangrove forests are being
rapidly cleared, and coral have disappeared in many areas.

Tourism
For millions, the living resources of the sea provide not a livelihood but relax-
ation. Marine recreational fisheries, diving, and surfing are the basis of a
multi-billion-dollar business worldwide. Recreational fishers, for instance, buy
boating and fishing equipment and travel long distances to pursue their hobby,
supporting the tourist trade. Whale-watching cruises and diving tours offer
opportunities for nature lovers, creating a tourist industry that provides jobs
and help many local and national economies. Many see marine life as a source
of income that needs to be protected and managed in a sustainable way with a
minimum of disturbance, something that unfortunately is not always the case.

The Aquarium Trade
Business and recreation are also combined in the thriving aquarium trade. The
export of colorful marine tropical fishes such as damselfishes is a significant
component of trade in countries like the Philippines and Fiji. The United States
is the largest importer of aquarium marine life, a trade that is estimated to be
worth over $100 million annually. The unrestricted collection of marine life for
the aquarium trade can be very destructive to the natural environment. Aquar-
ium fishes are often collected by using poisons or explosives that kill hundreds
or thousands of fish for every one that makes it to a pet shop. The management
of aquarium-species harvests, however, has helped reduce the impact of
collecting in Australia and a few other places by harvesting reef fishes in a
sustainable way. Live coral reef fishes are also removed to be sold as a luxury
food item. Most of the fishes are taken from reefs in Southeast Asia and sold in
China, Hong Kong, and other markets in the region.

Take Two Sponges and Call Me in the Morning

Obtaining drugs from the sea is not an easy task. It involves a sharp eye, good luck, and lots of frustration. It means diving in awesome coral reefs as well as working countless hours in the chemistry lab.

Colorful attached invertebrates and algae that live in coral reefs—sponges, soft corals, sea squirts, seaweeds—are good candidates for biomedical research. Though seemingly defenseless, these very visible and soft animals are rarely eaten by hungry fishes or crabs. This suggests that they have some type of chemical defensive mechanism. Many of these colorful invertebrates produce unique natural products that make them taste bad or allow them to get rid of hostile neighbors. These chemical substances may have properties that can be beneficial to humans, or rather, harmful to disease-causing pathogens.

Testing for new drugs Marine species of interest are collected, frozen, and sent to the lab for analysis. Crude extracts using alcohol and other solvents undergo trials to test their ability to kill or inhibit the growth of bacteria, fungi, and viruses. Extracts are also tested for their effect on cancer cells and for their potential use as anti-inflammatory agents. Any promising extracts are further separated and chemically identified. Chemicals with particularly promising properties are then used by pharmaceutical companies to develop new drugs. This process entails another long series of tests that ultimately includes rigorous clinical testing on humans.

Sponges and gorgonians are a rich source of natural products; some act as antibiotics, whereas others have

This sponge from coral reefs in Fiji (*Acanthella cavernosa*) contains a chemical that kills some human parasites.

anti-inflammatory or painkilling properties. Squalamine, which circulates in the blood of sharks, has antibiotic and antifungal properties. It is being tested for its ability to starve tumors by cutting off blood supply. Horseshoe crab blood is used to test for contaminant bacteria in vaccines, injections, and space probes. Deep-sea bacteria and archaea, dinoflagellates, and seaweeds are cultured to study their anti-tumor capabilities.

Toxins The toxins found in many marine organisms are also of potential value in medicine. Tetrodotoxin, the powerful toxin from certain pufferfishes and porcupine fishes, is used in Japan as a local anesthetic and as a painkiller in terminal cancer patients. Some conotoxins, a group of toxins from cone shells are used to treat chronic pain. Other conotoxins are undergoing clinical tests as painkillers and to treat epilepsy and depression.

Marine organisms are also used as remedies in Eastern medicine. Chinese traditional medicine, for instance, recommends red seaweeds for weight loss and for lowering cholesterol. Seahorse extracts are believed to "strengthen" the kidneys and the circulatory system, but as a result seahorses are rapidly becoming rare worldwide.

Although the use of marine natural products in medicine may sound promising, only relatively few of them have made their way into pharmacies and hospitals. Nevertheless, one never knows what's waiting to be discovered in the tissues of a sponge, sea cucumber, or sea anemone on a distant coral reef.

Section Review

1. **Identify** What problems are associated with the harvest of mangroves and corals?

2. **Identify** What are the impacts of the aquarium trade on marine ecosystems?

19.4 Non-Living Resources from the Sea Floor

VOCABULARY

methane hydrate

polymetallic nodules

massive sulfide deposits

DIVE IN!

Where did this sand come from? If you're standing on a beach that has been part of a beach nourishment program, it may have come from the outer continental shelf (OCS). To combat coastal erosion, communities and governments have established beach nourishment programs, which replenish beaches with sand from somewhere else. The United States Bureau of Ocean Energy Management (BOEM) has provided over 80 million cubic meters of sand for beach nourishment and coastal restoration projects. Most of this sand is dredged from the floor of the OCS. Maintaining sandy shorelines increases coastal resiliency, as beaches and barrier islands provide protection from storms and sea level rise. Many scientist are concerned, however, about the impacts of dredging and removal of sediments from the OCS. BOEM has partnered with researchers with the Navy and at the University of Florida to better understand the ecological function and recovery timeline of dredged coastal shelf habitats.

Main Idea

Non-living resources from the seafloor include those used for energy or other industries.

Key Questions

1. Why are humans now turning to the seafloor for resources?

2. Which energy resources can be extracted from the seafloor?

The ocean contains many resources like oil, natural gas, and minerals that are not naturally replaced. These non-renewable resources are much harder and more expensive to locate and exploit under the sea than they are on land. Not surprisingly, the ocean's non-renewable resources went largely unused while supplies on land were plentiful. Land reserves are running out, however, and we are turning to the sea floor for new sources.

Oil and Gas

Oil and natural gas were among the first of the sea's non-renewable resources to be used commercially. In the late 1800s it was discovered that some California oil fields extended well offshore beneath the seabed. The oil was extracted using oil rigs mounted on wooden piers that extended out from shore. Underwater oil and gas deposits have resulted from millions of years of decomposition under high pressure of buried sediments rich in diatoms and other organic matter.

Offshore Drilling The offshore oil industry has undergone tremendous expansion since the 1970s, because the high price of oil and gas makes offshore drilling immensely profitable. Oil and gas production on the continental shelf now occurs in the Persian Gulf, the Gulf of Mexico, the North Sea, in the South Atlantic off Brazil, and in other areas on all continents except Antarctica. A large percentage of the continental shelves around the world is considered a potential source of oil and gas, and drilling has now extended into deeper water beyond the shelf. The melting of the Arctic ice cap due to climate change has made the Arctic seabed particularly attractive. The enormous economic potential of these reserves was one of the factors that prompted the establishment of EEZs to protect national interests.

Exploiting offshore oil and gas can be a difficult and costly operation, especially when the deposits lie many miles offshore as they do in the North Sea. Powerful currents, huge waves, and miserable weather compound the problems.

Overcoming these difficulties has been a major accomplishment of ocean engineering. Exploratory drilling is done from drill ships or partly submerged or elevated platforms that can be towed from place to place and anchored in position on the sea floor. Once oil or gas is found, huge steel or concrete

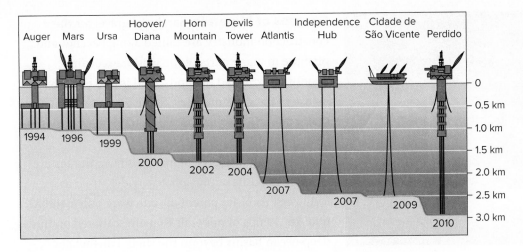

Figure 19.18 Offshore oil-drilling rigs and ship platforms have been built to get oil from deeper water farther offshore than ever before.

platforms are erected and secured to the bottom to extract it (Fig. 19.18). Remotely controlled submarines are being used to build facilities on the bottom, which include pipelines to transport the oil or gas to terminals on land.

Methane and Natural Gas Another potential source of energy from the sea floor is methane (CH_4), a major constituent of natural gas. Vast stores of methane, both as bubbles and in the form of frozen compounds, **methane hydrate** (or clathrates), have been found in frozen Arctic soil or deep under the ocean floor worldwide. Methane hydrates consist of methane molecules locked within ice. Methane is a product of microbial decomposition of organic matter but the gas can be locked into the ice due to low temperature and/or high pressure. If warmed, methane hydrates revert back to water and methane, which can be used as fuel. It has been suggested that the energy contained in all methane hydrate deposits could be higher than that contained in all other known fossil fuels. Japan was the first nation to successfully extract natural gas from methane hydrates taken 25 km off its central coast in 2013. Some methane hydrate deposits have been found at shallower depths, which could require scooping rather than drilling. It will be some time, however, before the technology needed to exploit this resource is developed. One drawback is that methane is a "dirty fuel," a greenhouse gas that may escape into the atmosphere and accelerate global climate change.

The potential threat of oil pollution, however, is a factor of great concern. Pollution from drilling operations can affect coastal fisheries and recreation. As a consequence, oil and gas drilling have been banned or strictly regulated in many coastal regions.

Ocean Mining

The seabed is a potential source of many types of minerals; in fact, some believe the bottom of the sea to be the planet's richest source. Though most marine deposits are not presently exploited, ocean mining will become more viable as we develop new technologies and as high-grade ores on land become exhausted.

Figure 19.19 ROVs (remotely operated vehicles) are being used to explore and sample the ocean floor off Papua New Guinea for minerals containing gold, silver, zinc, and, shown below in an underwater view, copper.

 CONTEXT

Hydrothermal Vents Undersea hot springs associated with the mid-ocean ridges encircling Earth, where new material is added to Earth's crust.

• 2.3, Modern Ocean Basins

Sources of Minerals Sand and gravel for the construction industry are mined along beaches and offshore in many parts of the world, exceeding quantities extracted on land. Sand is also extracted for its use in the glass industry. Some tin and even diamonds and gold are mined from offshore deposits brought to the seabed by rivers.

One potential source of minerals from the sea are offshore deposits of phosphates, widely used as fertilizers. Another potential source are **polymetallic nodules**, lumps of minerals that are scattered on the deep-ocean basins beyond the shelf. Polymetallic nodules contain manganese, nickel, copper, and cobalt, all of which are important to industry. Large amounts of "rare-Earth" minerals, needed for hybrid-car batteries and other types of electronic technologies, have been recently found as deep as 6,000 m in the Pacific near Hawai'i and French Polynesia.

Perhaps the most promising source of minerals are the seafloor **massive sulfide deposits** that are found along the mid-ocean ridges and hydrothermal vents. The minerals are extracted from deep in Earth's crust as hot water percolates through volcanic fissures and vents. They are deposited when the hot water emerges and hits the cold ocean water. These deposits are rich in copper, lead, and zinc.

Challenges of Ocean Mining There are many technical problems associated with mining the deep-sea floor. It is expensive, mainly because of the tremendous depths. The possible detrimental effects of mining on the surrounding marine environment, in some ways similar to the destruction of habitats caused by trawling, have also been raised. The ownership of mineral resources that lie outside the EEZ is another unresolved issue. As metal prices rise, however, deep-sea mining could become more attractive. Several agencies and private companies continue to investigate the prospects of deep-water mining (Fig. 19.19). Future prospects include the use of specially designed dredges or suction pipes suspended from ships.

Section Review

1. **Describe** What are the potential threats of offshore drilling to marine ecosystems?

2. **List** potential mineral resources that could be extracted from the seafloor. What problems are associated with their exploitation?

19.5 Non-Living Resources from Seawater

Main Idea
Non-living resources from seawater include energy from tides or waves, fresh water, and table salt.

. .

Key Questions
1. What resources can be extracted from seawater?
2. What challenges have affected the development of new technologies to harness energy from the ocean?

VOCABULARY

tidal energy

wave energy

ocean thermal energy conversion

desalination

DIVE IN!

Energy

The ocean is a vast reservoir of renewable energy that can be put to human use. Harnessing the sea's energy is the objective of several concepts that could help ease our dependency on fossil fuels and reduce carbon dioxide (CO_2) emissions. An indirect way of using the ocean as a source of energy are the offshore solar-panel plants that are already operational or planned in the Far East, offshore wind farms that are common in western Europe, and floating nuclear power plants being built or planned.

Tidal Energy Mill wheels have been used since ancient times to harness **tidal energy**, the tremendous energy contained in the normal ebb and flow of the tides. Modern schemes call for the construction of large barriers across narrow bays and river estuaries in areas where the tidal range is high, at least 3 m. Water moving in with the high tide is caught behind the barrier, and locks are opened to release the water at low tide. The flowing water drives turbines that generate electricity from mechanical energy, just as in the hydroelectric plants in river dams. One such electrical generating plant has been operational in northwestern France since 1966; another became operational in Northern Ireland in 2008 (Fig. 19.20, p. 616). A few other facilities have been built on an experimental basis in western Europe, Canada, and other countries.

The use of tidal energy is pollution-free and relatively efficient, but the resulting changes in the tidal patterns can be highly destructive to the nearby environment. The rich marshes in estuaries can be damaged or destroyed. Pollutants from other sources tend to accumulate upstream because normal tidal flushing is restricted. River flows can also be altered, increasing the risk of floods inland. The artificial lagoon created upstream, however, can be used in recreation, as in the case of the French plant.

Wave energy Another potential source of energy is **wave energy** from wind-generated waves and strong ocean currents, particularly in areas subject to powerful, regular waves. As with tidal energy, water flow can be converted into electricity by turbine generators. Wave energy is being used to power

Can ocean water become drinking water? The idea of desalinating ocean water for use as fresh water is being discussed more and more as a result of population growth and as droughts intensify in many parts of the world. In the past, desalination was considered expensive and inefficient. New technologies, however, have decreased the cost and made the process of reverse osmosis more efficient. The world's largest desalination plant, which began providing water to 20% of households in Israel in 2015, used larger pipes, which reduced the cost of materials, and increased efficiency by using high-efficiency pumps and recovering energy from the plant. A proposed desalination plant in Saudi Arabia will use solar energy to run the plant, a way to avoid the use non-renewable energy sources to power desalination plants.

Figure 19.20 The tidal power station at Strangford Lough, Northern Ireland, became operational in 2008.

Figure 19.21 Several experimental devices have been designed to tap energy from waves. The sea snake (a) consists of several long, floating segments, which rise and fall in relation to each other as the result of wave action. Internal hydraulic rams drive generators that produce electricity. Small turbines secured on the bottom (b) or paddles with buoyant flaps (c) have also been envisioned as ways of generating electricity.

navigation buoys. Several projects, mostly experimental, are being developed in western Europe and the United States (Fig. 19.21).

Thermal Energy Another way to harvest energy from the ocean is to take advantage of the temperature difference between surface and deep water, a process called **ocean thermal energy conversion** (OTEC). There is renewed interest in OTEC, with several pilot plants already in operation or planned throughout the Pacific and Indian oceans and the Caribbean. OTEC plants circulate ammonia, propane, or another liquid that boils at low temperature through pipes bathed in warm surface water (Fig. 19.22). The liquid evaporates and the gas is forced through turbine generators to produce electricity. The pipes then flow through cold water pumped up from the deep, condensing the gas back into a liquid, and the cycle repeats. Temperature differences of at least 20 °C between the surface and below are needed, a condition met in the tropics.

Electricity generated by OTEC could be sent ashore by power lines, or used at sea for various industrial operations, including the desalination of seawater into fresh water (see below). The cold, nutrient-rich water pumped from the deep could be used for aquaculture or to grow algae for animal feed, fertilizers, or biofuels.

Several challenges have affected the development of new technologies to harness energy from the ocean. Problems include not only corrosion and the effects of exposure to the elements, but biofouling, the unwanted growth of seaweeds, encrusting animals, and microorganisms.

Fresh Water

Desalination plants that convert seawater into fresh water were first built in coastal regions that lack a sufficient supply of fresh water, mainly in desert and semi-desert regions like the Middle East and North Africa. Widespread water shortages, however, have now made desalination a more feasible source of fresh water, so that the number of people using water from desalination plants has more than doubled in the past decade. More than 17,000 desalination plants had been built around the world.

(a)

(b)

(c)

Warm water

Ammonia converted to gas by warm-water current

Generator

Ammonia converted to liquid by cold-water current

Cold water

Warm-water intake

Warm-water intake

Warm water

Warm water

Cold water

Cold water

Cold-water intake

Figure 19.22 This ocean thermal energy conversion (OTEC) plant is an experimental model consisting of a platform containing generators and a long tube for obtaining cold water.

Several desalination systems are currently employed. The process of reverse osmosis, for example, uses a membrane that allows water to pass through but blocks the passage of ions, thus desalinating the water. Reverse osmosis plants are becoming common in small-scale projects like resorts as well as in large desalination plants. Desalination, however, requires a great deal of energy and is therefore expensive. It also produces a highly saline residue that can cause environmental problems. New methods that use a renewable energy source like wind or wave power should make desalination more practical and economical.

Minerals

Every element on Earth is present in seawater, but mostly in extremely small quantities. The chief product presently obtained from seawater is table salt, or sodium chloride (NaCl), which is composed of the two most abundant ions in seawater, sodium and chloride. Salt, a precious commodity to ancient civilizations, has been produced for centuries by evaporating seawater. Most of the other constituents of seawater are difficult to extract because they are present in such small amounts. Magnesium and bromine are also obtained from seawater but valuable trace metals like uranium could one day be commercially extracted.

. .

Section Review

1. Identify What are the advantages and disadvantages of tidal energy?

2. Identify What are the advantages and disadvantages of desalination?

Chapter **19** Review

REVIEW QUESTIONS

Multiple Choice

1. Which term refers to organisms caught unintentionally while fishing for other species?

 a. by-catch

 b. ghost fish

 c. stock

 d. sustainable yield

2. Oil and natural gas production on the continental shelf occurs on all continents except

 a. Antarctica.

 b. Europe.

 c. North America.

 d. South America.

Short Answer

3. What factors lead to overfishing?

4. How does overfishing affect other non-commercial species?

5. What are methane hydrates? Why might they become important?

6. What is ocean thermal energy conversion (OTEC)? Why might the process become important?

Critical Thinking

7. For the last three years the annual catches of a commercially important fish have been above the maximum sustainable yield. One option is to decrease the fishing effort by decreasing the number of fishers. This, however, would cause unemployment in a region where unemployment is already high. What other options might ensure a lower fishing effort? How could they be done?

8. It has been suggested that cheap electricity generated from tides, waves, or currents could be used to pump nutrient-rich water from the deep. The water could then be used to grow algae to feed the larvae of farmed fishes and invertebrates. How else could this energy be used to decrease the costs of aquaculture?

DATA ANALYSIS LAB

Is aquaculture a more efficient use of resources than raising other types of animals for food? The human population is over 7 billion and still growing. Many scientists are concerned with how the production of animal protein for food will feed the growing population. Many are looking to aquaculture as a solution.

Data and Observations

The table shows the feed conversion ratio for different farmed animals. The feed conversion ratio is an estimate of the amount of feed needed for a farmed animal to gain one pound of body mass.

Farmed Animals	Pounds of Feed Needed to Gain One Pound of Body Mass
Chickens	1.7
Cattle*	6.8
Fishes**	1.1
Pigs	2.9

*Based on Hereford breed (used mainly for meat production in temperate areas), ratios for other cattle breeds vary

**Based on salmon, ratios for other farmed fish are slightly higher

Think Critically

1. Which farmed organism needs the most pounds of feed to produce one pound of body mass? Which farmed organism needs the least?

2. Fishes are ectotherms and the buoyant force of water counteracts the force of gravity on their mass. How do these facts help explain the data in the table?

3. It would seem from the data in the table that aquaculture is the most efficient use of resources for producing animal protein as food for humans. What other variables may need to be considered before a final conclusion can be reached?

*Data obtained from: Bourne, J. K., 2014. How to farm a better fish. National Geographic, vol. 225, no. 6, June, pp. 92–113.

The Impact of Humans on the Marine Environment

THEME Cause and Effect

Human activities, both on land and at sea, affect marine environments and the organisms that inhabit them.

BIG IDEA

Human impacts on marine environments can be immediate, as is the case in an oil spill, or long-term, such as pollutants that bioaccumulate in the cells of organisms.

About the Photo Wildlife officers in Barataria Bay, Louisiana, prepare to net a brown pelican (*Pelecanus occidentalis*) that is covered in oil from the *Deepwater Horizon* spill in the Gulf of Mexico in 2010.

Coast Guard photo by Petty Officer 2nd Class John Miller

20.1 Modification and Destruction of Habitats

Main Idea

Modifying or destroying marine habitats though changing the abiotic or biotic characteristics of the habitats often negatively affects marine life.

Key Questions

1. Where are the marine environments that are most impacted by human activities located?
2. What is the difference between direct and indirect impacts on marine environments?

The media offer disturbing news about the health of the oceans. Stories about oil spills, global climate change, dying coral reefs, and vanishing sea life are all too familiar. This is, however, only a small sample of anthropogenic impacts, or the effects of human activities, on the marine environment (Fig. 20.1). More than 7 billion people now live on our planet, and more than half of these live within 100 km of the coast, which is more than the population of the *whole* planet in 1950! Everywhere, not just along industrial areas, humans are

Low Impact ← → Very High Impact

Figure 20.1 The impact of humans on the marine environment reaches practically every corner of the ocean planet, even the surface water farthest from land. The most impacted marine environments are along the coast near the densely populated regions of the northern hemisphere. Data was gathered from maps showing 17 different human activities like pollution, fishing, fertilizer runoff, and climate change, as well as maps for 14 marine ecosystems worldwide. A "vulnerability score" was used to quantify impact.

Figure 20.2 Salt marshes in Southern California and in many other locations around the globe have been obliterated with reclamation.

modifying our marine world. Water quality has decreased, fisheries are imperiled, recreational areas are at risk, and new health hazards are developing.

Pollution is not the only or most important way we impact the marine environment. Dredging, landfilling, and even the use of explosives modify or destroy habitats, the places where organisms live. The effects of such physical disturbances are direct and immediate, as opposed to indirect effects like those of pollutants released somewhere else. The indirect effects of destruction of habitats, however, can be over much larger areas, as when the nursery grounds of fishes are destroyed.

Most destruction of habitats takes place along the coast close to where people live, as in estuaries and mangrove forests. It results mostly from unplanned or poorly planned coastal development (Fig. 20.2). The problem is more acute in many developing nations, where population growth, poverty, and the absence of effective management are harmfully combined. Growing cities, increased tourism, and development for industrial and recreational uses have radically altered the coastlines of the richer nations as well, as in the case of Singapore and Dubai, where artificial islands have been recently built for residential or industrial purposes.

Human Impact on Coral Reefs

Coral reefs are threatened by human activities all over the tropics. More than a quarter of the world's coral reefs have already been lost or are at high risk, and many reef-building corals face **extinction**. All corals, particularly reef-building corals, are listed as vulnerable to exploitation by the Convention on International Trade in Endangered Species (CITES) (see Table 20.1, p. 639). Though coral reefs

DIVE IN!

What is the problem with dredging? Many ports and harbors are not naturally deep enough to accomodate large ships, so they must be deepened by excavating and removing bottom sediments, a process known as dredging. Dredging can change the alter sea-floor habitats, and remove or bury soft bottom communities such as seagrass meadows. It can also change the way the water flows over a habitat, altering the intensity of tides and currents not only of the dredged area, but of surrouding habitats. Routine dredging is common in navigable waters, meaning that soft-bottom communities disturbed by dredging never get a chance to be re-established.

Figure 20.3 Coral mining in the Maldive Islands, Indian Ocean. Reef-building corals are widely used as construction material in many parts of the tropics, even though the islands rely on their coral reefs to attract tourists, an important sector of the economy. Two small islands between India and Sri Lanka recently sank, in part as a result of coral mining.

support luxuriant life that provides much-needed protein and potentially life-saving drugs, they are subjected to much anthropogenic stress. This includes pollution by excess nutrients from sewage and agricultural and stormwater runoff as well as overfishing. Like tropical rain forests, coral reefs are among the oldest and richest environments on Earth. Ironically, the rapid disappearance of tropical rain forests threatens coral reefs, too. The clearing, or deforestation, of the forests for agriculture, logging, and urban development increases the amount of soil washed out to sea by the plentiful tropical rain. The resuspension of sediments by nearshore dredging also increases sediment loads.

Physical Damage Corals can tolerate moderate levels of sediment, but the higher levels produced by human activities can smother them, and young corals do not settle on surfaces that are covered with sediment. Sediment in the water makes it more murky, reducing the amount of light that reaches the corals, which depend largely on food produced by zooxanthellae. Due to these various effects, sedimentation is a serious threat to many of the world's reefs.

Fishing with explosives, and with poisons like bleach and cyanide, kills corals. Though this is generally banned, such explosives and poisons are used in Southeast Asia and parts of the western Pacific. Other threats are the mining of coral for construction material (Fig. 20.3) and the indiscriminate collection of corals for the aquarium trade and for sale as souvenirs and decorations. Shell collectors can be very destructive to coral reefs by turning over or breaking corals to get their specimens. Damage by anchors, fish traps, reef walking, and scuba diving has also taken its toll.

Another major problem threatening corals is ocean acidification, which is caused by the increase in atmospheric carbon dioxide (CO_2). An increase in the acidity of the water makes the calcium carbonate ($CaCO_3$) skeleton of corals more soluble and it inhibits coral growth.

Stress An indication of stress to coral reefs is the worldwide outbreaks of mass coral bleaching. Shallow parts of more than half of coral reefs around the world showed bleaching in 1998. There is increasing concern that global climate change is causing higher water temperatures and thus more frequent, widespread, and severe bleaching events. Bleaching occurs when corals expel their symbiotic zooxanthellae, causing white patches to form on the colonies (Fig. 20.4). Recovery can take place because corals have different species of zooxanthellae and a new type can replace one that has been expelled. Bleached corals never lose all their symbionts. Even dying colonies have large numbers of zooxanthellae in their tissues, not enough to give color but enough to reestablish normal numbers when conditions improve. Bleached corals, however, do not grow and are vulnerable to disintegration.

Kurt Amsler/ardea.com

An increase in killing diseases is additional evidence of coral stress. Infections typically show as a line of discolored dead tissue that exposes the underlying white skeleton as it advances over the colony. Various infections have been described, and their names ("black-band disease," "white-band disease") refer to the color of the infection. These diseases seem to be caused by bacteria and fungi that take hold on colonies damaged or stressed by excess nutrients.

Trawling

Trawls that are dragged along the bottom for fish and shrimp are a major threat to subtidal habitats. It has been estimated that about 40,000 km² of the bottom are scoured by trawls each day. Trawling scours the sea floor and leaves scars on soft sediments, particularly muddy bottoms. It also causes the resuspension of sediment, which kills suspension feeders. The release of bottom nutrients can induce the growth of bacteria and seaweeds, which deplete oxygen (see "Eutrophication," section 20.2). On hard bottoms trawling breaks off many attached animals, some of which, like sponges and tubeworms, offer shelter to juveniles of fishes and other animals. Trawling also displaces or overturns boulders, harming or killing organisms living on their surface and exposing others to predators. Repeated trawling gives bottom communities little chance of recovery. The total number of species ultimately decreases markedly because the continuous disturbance favors short-living, fast-reproducing species like small worms, while long-living, slow-reproducing animals like sponges, clams, and sea stars tend to disappear. Deep-water trawling and ocean mining, also threaten many vulnerable species that inhabit seamounts in deep water.

Figure 20.4 These colonies of a brain coral (meandrites) in the Caribbean have become almost white after losing their symbiotic algae, or zooxanthellae, a phenomenon known as bleaching.

Section Review

1. **Summarize** Make a table like the one below to summarize the impacts of human activities on coral reefs.

Problem	Human Activity Causing Problem	Impact on Coral Reefs
Increased sediment		
Physical destruction of reef		
Ocean acidification		

2. **Describe** the impact of trawling on marine environments.

20.2 Pollution

Main Idea

Pollution reaching the marine environment results in changes to both abiotic and biotic factors that affect marine life.

Key Questions

1. What is pollution? What are the major types of pollution that affect marine environments?
2. What is biological magnification?

Pollution—visible or invisible and on land, air, or water—is an unwanted but familiar part of our lives. **Pollution** can be described as the introduction by humans of substances, heat energy, or even sound that decreases the quality of the marine environment. Many of these substances, or **pollutants**, are artificial substances that do not occur naturally. Some substances, however, have natural sources, like natural oil seeps and volcanic eruptions, which are not considered to be sources of pollution. By contrast, the liberation by humans of naturally occurring substances, for example, when mining releases metals that are naturally stored in rocks, is considered to be pollution.

The role of humans in decreasing the quality of the marine environment has been enormous. The potentially detrimental effects of pollution can directly or indirectly affect all parts of the ocean, from beaches to great depths. Pollution can also present a hazard to human health when marine organisms are eaten or when we go swimming, diving, or surfing.

Pollution is the introduction or release of substances or energy that decrease the quality of the marine environment. Many pollutants are toxic or harmful to marine life and humans.

There is considerable diversity in the types, distribution, sources, and effects of marine pollutants, but most pollution comes from land-based sources. Major land-based pollution sources include urban development, agriculture, transport, industry, and power generation. The most important sea-based sources are shipping and offshore oil drilling, but these account for less than 20% of the total pollution that enters the ocean. Here we will briefly discuss only the most important types of marine pollutants and some of their effects on marine life.

Eutrophication

Fertilizers in agricultural and stormwater runoff and sewage are major anthropogenic sources of nitrate, phosphate, and other nutrients to the marine environment. Atmospheric input from fossil fuel combustion is another major source of nitrogen, the largest, in fact, to the open ocean. Anthropogenic inputs of nitrogen to the oceans now exceed natural inputs. Humans have come to dominate the global nitrogen cycle, and the situation will intensify because of

the ever-expanding use of fertilizers, sewage from more people living along coasts, fossil-fuel combustion, and other human activities that increase nutrient inputs to the sea.

Though nutrients are needed by marine primary producers, excessive amounts encourage too much algal growth, a phenomenon known as eutrophication. Eutrophication is primarily a problem of coastal waters, particularly in shallow, partially enclosed areas. Eutrophication can damage important habitats like seagrass beds and coral reefs by causing a chronic increase in phytoplankton abundance, reducing the penetration of sunlight to the bottom, and accelerating the growth of seaweeds that overgrow the bottom. It can also trigger phytoplankton blooms, the short-term, explosive increases in the abundance of phytoplankton. Blooms of phytoplankton and particularly of bottom-dwelling cyanobacteria are sometimes toxic.

Organic matter in the form of the remains of the phytoplankton blooms and the feces of the zooplankton and fishes that feed on the plankton fall to the bottom. Decay bacteria then break down the organic matter. Bacteria can completely use up the oxygen on the bottom, causing anoxic conditions. Seasonal **hypoxic** (or "dead") zones, areas where the water lacks oxygen, that result from the effects of agricultural runoff have become common in a number of areas, including the Gulf of Mexico (Fig. 20.5), Chesapeake Bay, and the Baltic Sea. About 150 hypoxic zones have been identified in oceans around the world.

Figure 20.5 Hypoxic zones develop on the bottom of coastal waters due to increased bacterial decomposition. This is caused by nutrient pollution, mostly from agricultural runoff as in the case of the Gulf of Mexico. Surface fresh water from the Mississippi River causes stratification, which prevents the mixing of oxygen-rich surface water with the oxygen-poor bottom water. Many fishes and other animals move out. Those who can't simply die. "Dead" zones are not really devoid of life because many microbes and some invertebrates do thrive.

DIVE IN!

What is the "Great Pacific Garbage Patch"? Currents have carried bits of plastic, Styrofoam, shoes, pieces of fishing nets, and other items to areas near Japan on the western side of the Pacific Ocean and between Hawai'i and the continental U.S. on the opposite side of the Pacific. The name is a bit misleading, as it causes most people to form an image of a mountain of visible trash floating on the surface of the ocean. Instead, most of the pieces of garbage are small, some are even microscopic (microplastics), and they float just below the surface, making the water look cloudy. Marine animals can consume the debris by mistake. This can result in two problems: one, the animals do not get the food and nutrients they need, which can cause them to weaken and die, and two, they are exposed to chemicals, such as bisphenol A (BPA), that plastics contain. These chemicals can accumulate in the tissues of higher-level consumers.

Coastal pollution can also trigger frequent red tides and other phytoplankton blooms. These events seem to be occurring more frequently in many coastal waters. Blooms of phytoplankton and other algae have also become a recurrent problem in China, Japan, and the Baltic and Adriatic seas, being responsible for considerable damage to fisheries and tourism.

Stormwater runoff not only carries fertilizers from lawns but harmful substances we deposit on land. Pesticides, oil, as well as bacteria, viruses, and other parasites from pet waste, get carried into the ocean when it rains.

Sewage

Disposing of ever-increasing amounts of sewage is a major problem in cities around the world. **Domestic sewage** carries waste-water from homes and city buildings. It can also carry stormwater runoff. **Industrial sewage** contains a variety of wastes from factories and the like. Most sewage is discharged into the sea, or into rivers that flow to the sea.

Impacts of Sewage Sewage discharges into coastal waters are a serious health hazard, a threat more serious than once thought. Sewage contains many viruses, bacteria, and other parasites that cause disease. Infectious hepatitis, for example, is carried by viruses found in human feces. Several million worldwide cases of infectious hepatitis annually result from people eating oysters, clams, and other shellfish that concentrate the viruses as they filter the water for food. Swimming in sewage-polluted water can also be hazardous; people can get sick from swallowing contaminated water or develop ear, throat, and eye infections just from contact with the water.

The closing of beaches when sewers overflow after a rainstorm has now become routine in many areas. Contamination by failing septic systems that continually leak sewage materials is also a major source. It has been estimated that in Southern California alone there are 1.5 million annual cases of gastrointestinal illnesses among recreational users as a result of sewage pollution, with a health cost of over $400 million. The economic impact due to lost tourism and the closure of shellfish farms can be considerable.

Sewage can also be a hazard to marine life. Chemicals found in prescription drugs and many health products have been found to adversely affect reproduction in some marine animals. Even hormones from the urine of women taking birth-control pills are known to cause sex changes in male fishes. Sewage that contains industrial waste invariably includes pesticides, heavy metals, and other toxic chemicals.

Sewage Treatment and Sludge The harmful effects of sewage can be reduced by sewage treatment. Many countries require by law some form of treatment before sewage is discharged. The sewage can simply be allowed to sit in a basin for a time, so that much of the solid matter settles out. A more advanced, but more expensive, option is to allow decay bacteria and other organisms to break down the organic matter in the sewage. The addition of chemicals or other steps can further purify the sewage. After treatment, the

sewage is often disinfected by using chlorine to kill bacteria and viruses. Ozone treatment and UV irradiation are other methods of disinfection. Advanced forms of treatment can produce water that is pure enough to be used for irrigation or even drinking.

Sewage treatment reduces the impacts of sewage on the marine environment and human health, but it is not without drawbacks. The cost of sewage treatment increases as more advanced treatment methods are used, and many communities cannot afford it. Chlorine used for disinfection can remain in the discharged wastewater and is toxic to marine life. Sewage treatment also creates **sludge**, the wastes that are taken out of the sewage. This semiliquid material is more concentrated than the original sewage and often contains high levels of heavy metals and other toxic substances. If sludge is disposed of in the ocean, it smothers bottom communities, creating black deserts around outfalls or at best dominance by only a few species (Fig. 20.6).

It is impossible for most detritus feeders to handle the massive amounts of organic matter contained in sludge. The organic matter is instead decomposed by bacteria that thrive under these conditions, using oxygen to such an extent that anoxic conditions develop. The total number of species decreases because many of the natural inhabitants disappear. They are replaced by hardier forms of life, like certain species of polychaete worms. Bottom fishes collected around sludge disposal sites tend to show skin tumors, erosion of fins, and other abnormalities, apparently a result of the high concentration of toxic substances, viruses, and bacteria.

It is possible to implement alternatives to the discharge of sewage into the sea. Some communities take advantage of the ability of marshes to recycle nutrients by using the marshes for natural sewage treatment (Fig. 20.7, p. 628). Ideally we could recycle the large amounts of precious water and nutrients that go to waste. Sludge is sometimes recycled into landfill, construction blocks, and compost. It is also spread on farmlands as fertilizer, burned to generate electricity, and converted into fuel.

Figure 20.6 Treated sewage effluents from the Hyperion treatment plant being released at a depth of 60 m at the end of an 8-km outfall in Santa Monica Bay, Southern California. Large sea anemones (*Metridium*) surround the site. Santa Monica Bay is actually getting cleaner. Treatment at the plant has been improved in order to meet standards set by the Environmental Protection Agency.

Oil

Crude oil, or **petroleum**, is a complex mixture of hydrocarbons, long chains of carbon and hydrogen, and certain other chemicals. Crude oil is refined to yield not only fuels but raw materials for making plastics, synthetic fibers, rubber, fertilizers, and countless other products.

Sources Oil is one of the most widespread pollutants in the ocean. An estimated 682 million tons of oil and derivatives like fuel and lubricating oil enter the world oceans each year as a result of pollution. Oil from natural seepage, not a pollutant, contributes another 584 million tons worldwide. Natural seeps are the most important source of oil in North American waters, however, around 160,000 tons a year, while different sources of pollution contribute more than 96,000 tons annually (Fig. 20.8, p. 629). Almost 85% of

Figure 20.7 This lush artificial marsh is part of a network of marshes where domestic sewage from the town of Arcata in Northern California is naturally purified before it is piped into the ocean. Sewage, which is mostly water, is given preliminary treatment, and chlorine is added to kill harmful bacteria. It is then pumped into the marshes, where mud bacteria break down the organic matter. The released nutrients fertilize the marsh plants. The marshes prevent the pollution of the ocean by sewage and attract many species of birds and other wildlife.

the polluting oil in North America comes from stormwater and river runoff, fuel from small boats and jet skis, and fuel jettisoned by planes. The remaining 15% comes from tanker and pipeline spills and as a result of exploration and extraction from the seabed.

The massive oil spills that result from the sinking or collision of supertankers and accidents from drilling platforms, however, are the most devastating to the marine environment. The 1978 grounding of a supertanker, the *Amoco Cadiz*, poured 230,000 tons of crude oil along the coasts of northwestern France. In 1989 more than 35,000 tons of crude oil were spilled by the *Exxon Valdez* along the coast of southern Alaska. The 2010 blowout of the *Deepwater Horizon* drilling platform released an estimated 672,000 tons of crude oil into the Gulf of Mexico, so far the largest marine spill in the United States and second in the world after the Persian Gulf spill during the Iraqi War in 1991.

Most of the components of oil are insoluble in water and float on the surface. They appear as thin, iridescent slicks on the surface or as black deposits on the shore. Some of the other components sink and accumulate in sediments, especially after the lighter ones evaporate. Some residues remain on the surface and form tar balls, which are common along shipping lanes worldwide. Tar balls can persist for many years in the water and have been observed in remote areas far from shipping traffic. Some even have barnacles living on them! You would expect large areas of the ocean to be covered with the oil that has accumulated over the years. Fortunately, some of its lighter components evaporate. Oil is ultimately broken down, or decomposed, by bacteria and thus is almost completely **biodegradable**. The breakdown rate is usually slow but varies among marine communities.

IN CONTEXT

One metric ton of crude oil = 7.33 barrels (bbl), or 308 U.S. gal.

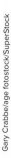

Gary Crabbe/age fotostock/SuperStock

The oil and shipping industries have nevertheless made considerable progress in the protection of the marine environment. Accidents have prompted tighter restrictions, like having double hulls on tankers. Older tankers, however, do not necessarily have double hulls, and many still navigate busy shipping lines.

Effects of Oil on Marine Life Even in small amounts, oil has a variety of effects on marine organisms. Organisms can accumulate oil components, many of which are toxic, from the water, sediments, and their food. There is evidence, for instance, that compounds found in oil interfere with the reproduction, development, growth, and behavior of many organisms. These compounds are released from oil for many decades. Oil also increases susceptibility to diseases in fishes and inhibits the growth of phytoplankton. Crude oils vary markedly in their chemical composition, and therefore in their toxicity and environmental persistence. Refined products like fuel oils tend to be more toxic than crude oil.

Seabirds and marine mammals like the sea otter are particularly susceptible. Many die of exposure when their feathers or hair is coated with oil, which affects their ability to maintain the thin layer of warm air needed for insulation. Birds that rely on flying to catch food are unable to do so and die of starvation. It is difficult to determine the number of birds killed by oil spills because many sink without reaching the coast, where the corpses can be counted. Sea turtles, crabs, and many other marine animals are simply overcome by the oil and die

Sources of Oil Pollution in North America	Estimated Amount (tons)	Percentage
Natural seeps	160,000	60
Land-based pollution and recreational boating	84,000	28
Transportation of oil	9,100	9
Exploration and extraction	3,000	3

Figure 20.8 Annual sources of oil in the marine environment.

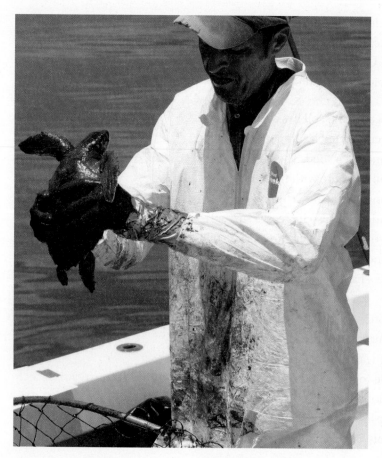

Figure 20.9 A loggerhead sea turtle (*Caretta caretta*) coated with oil from the *Deepwater Horizon* spill in the Gulf of Mexico in 2010.

(Fig. 20.9). The *Exxon Valdez* spill is believed to have killed between 100,000 and 300,000 seabirds and 3,500 to 5,000 sea otters. Estimates are that it will take many decades for the wildlife in the area of the spill to recover fully.

The effect of oil spills on exposed rocky shores is less devastating than it may appear at first sight. Initially there is mortality among many attached inhabitants, but wave action and tides help clean away the oil. Rocky shore communities do recover, though recovery is dependent on factors like the amount of oil, wave action, and temperature. Degradation by bacteria takes place, but usually slowly, especially in cold water. Spills can be degraded more quickly by bacteria if an oil-soluble fertilizer is added to the water or sprayed on rocks and sediment. Experience has shown that recovery begins within months and that an apparently near-normal condition can occur as early as one or two years after the spill. Higher oil concentrations in sediments and isolated pockets, however, have been found to remain for 15 years or longer.

Large spills can have more catastrophic effects when they drift to salt marshes and mangrove forests, as shown by the *Deepwater Horizon* spill. These communities are characteristic of sheltered coasts and estuaries where oil cannot be dispersed by wave action. Oil is absorbed by the fine sediment characteristic of these communities and can persist for more than 30 years in salt marshes. Coral reefs and seagrass beds also are greatly affected by oil spills. Coral colonies show swollen tissues, excessive production of mucus, and areas without tissue. Reproduction and feeding in surviving coral polyps are affected by oil. There is little evidence, however, of dramatic, long-term effects.

Economic Impacts Containing an oil spill is a major headache. Fencing off the spill with floating, fire-resistant booms can prevent the spill from moving into shore. "Skimmers," boats equipped with U-shaped booms, are used to skim off and recover some of the oil. These methods, however, cannot be used in heavy seas. Chemical **dispersants** are added to the spill to break the surface oil into small droplets that can then disperse in the water. Unfortunately, dispersants also are harmful to marine life, as shown by the *Deepwater Horizon* spill. Furthermore, the dispersed oil still remains toxic underwater. The use of powerful streams of hot water to wash oil from beaches after the *Exxon Valdez* spill was also harmful to many forms of life, perhaps as harmful as the oil itself.

The cost of oil spills to local economies can be enormous. Fish and filter-feeding shellfish like oysters and clams that are tainted with oil are unmarketable. The

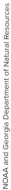

NOAA and Georgia Department of Natural Resources

size of fish catches, particularly of bottom-dwelling species, can decrease because of the initial mortality of adults and juveniles or a drop in the abundance of their food. Furthermore, oil-soaked beaches are disastrous to resort areas dependent on tourism. Claims by those affected plus the cleanup bill ran well past the billion-dollar mark in the *Exxon Valdez* incident. Exxon was ordered to pay a total of $5 billion in fines for damage caused by the *Exxon Valdez*. The fine was halved to $2.5 billion in 2006 and to $507.5 million in 2008. British Petroleum agreed in 2015 to pay $18.7 billion to settle *Deepwater Horizon* claims.

Persistent Toxic Substances

Many important pollutants that reach the sea from land are substances that are said to be **persistent** because they remain in the environment for years, even many decades. Persistent substances are not readily broken down either by decay microorganisms (and for this reason the substances are known as non-biodegradable) or by physical or chemical means in the environment.

Pesticides One group of synthetic chemical pollutants is the chlorinated hydrocarbons. They include many **pesticides**, chemicals used to kill insects and to control weeds, including DDT, dieldrin, and chlordane. Millions of tons of these and many other chlorinated hydrocarbons have been used since the 1940s, when DDT made its debut. These pesticides have been used to protect plant crops from insect pests and to control insects that carry diseases. Pesticides have saved millions of people from disease and starvation, but the unchecked use of chlorinated hydrocarbons has been shown to be harmful to many non-targeted forms of life.

These pesticides have not been used directly in the ocean. They are, however, carried into the ocean by rivers, stormwater runoff, and sewage and are transported long distances, even to the open ocean, in the atmosphere (Fig. 20.10). In the ocean they are absorbed by phytoplankton and particles suspended in the water and thus find their way into other organisms.

Chlorinated hydrocarbon pesticides dissolve in fats and are not excreted, so organisms tend to retain them almost indefinitely. Animals absorb and accumulate the pesticides in the organisms they eat, so their internal concentration of chlorinated hydrocarbons is higher than in their food supply. At each level of the **food chain**, then, the chlorinated hydrocarbons are more concentrated. This phenomenon is known as **biological magnification**. Among marine animals, pesticide concentrations are higher in carnivorous fishes, and even more so in the fish-eating birds and mammals at the top of the trophic pyramid (Fig. 20.10).

The alarming effects of worldwide use of chlorinated hydrocarbons began to be noticed during the 1960s. Fishes caught for human consumption in the United States had to be destroyed because they contained too much pesticide. Pesticides began to accumulate in top carnivores in concentrations that were thousands and even millions of times above those found in seawater.

It was Rachel Carson, an American marine biologist, who in her 1962 book, *Silent Spring,* brought widespread attention of the environmental problems

Figure 20.10 The concentration of chlorinated hydrocarbons increases with the relative position organisms have in the food chain, thus showing biological magnification. In the trophic pyramid that summarizes this generalized food chain, the concentration of pesticides is expressed in ppm, or parts per million.

In the figure: Aerial spraying; Runoff; Wind; Rivers; Phytoplankton; Sediments; Zooplankton; Precipitation; Plankton-feeding fishes; Seals, sea lions, seabirds; Carnivorous fishes. Trophic pyramid: Seals, sea lions, seabirds (10–25 ppm); Carnivorous fishes (1.0–2.0 ppm); Plankton-feeding fishes (0.5 ppm); Zooplankton (0.3 ppm); Phytoplankton (0.1 ppm).

caused by the indiscriminate use of pesticides, and most especially their effects on wildlife and human health. Carson's work was most influential on the environmental movement not only in the United States but throughout the world.

The effects on birds, on land and at sea, were particularly dramatic. The birds were not actually poisoned, but the high concentrations of chlorinated hydrocarbons in their body fat interfered with reproduction, specifically with the deposition of calcium in their eggshells. The eggshells became so thin that they broke during incubation before chicks became fully developed.

The once-abundant brown pelican (Fig. 20.11), for example, became a rare sight in most of the United States due to widespread reproductive failure during the late 1960s and early 1970s. Female pelicans sat on broken eggs, or adults simply did not attempt to nest. The Channel Islands off Southern California, the last remaining nesting colony on the Pacific coast north of Mexico, recorded only one chick in 1970 and seven in 1971. High amounts of DDT and related chemicals were found in the tissues of birds taken from these and other nesting colonies and in other marine animals in the area, from filter feeders like sand, or mole, crabs to top carnivores like sea lions.

By 1972 most uses of DDT and several other chlorinated hydrocarbons were banned in the United States and many other industrialized nations. DDT residues in marine animals and sediments then began to decrease. The brown pelican, once nearly extinct in the United States, has recovered, and its reproduction has returned to normal. DDT and related residues, however, can still be found in bottom-dwelling fishes. One of these residues, DDE, was found to be degraded by bacteria under laboratory conditions. A 2000 international agreement restricting the production and use of certain toxic substances nevertheless allowed continued use of DDT because it provides inexpensive

IN CONTEXT

Food Chain The steps of transfer of energy from producers, the algae and plants, through consumers, the animals.

Trophic Pyramid The relationship of energy, number of individuals, or biomass of the organisms found in a food chain.

- 12.3, The Flow of Energy and Materials

control of insects in developing nations, primarily to control the mosquitoes that transmit malaria.

PCBs and Other Toxic Organic Chemicals Another problematic group of toxic chlorinated hydrocarbons are the PCBs, the polychlorinated biphenyls. PCBs are non-biodegradable and persistent and thus show biological magnification. They were widely used in electrical transformers and capacitors, in the manufacturing of plastics and paints, and in many other products. PCBs proved to be highly toxic, causing cancer and birth defects, and their production and use were gradually regulated or banned by many nations, including the United States in 1979.

This ban, however, came after PCBs had spread throughout the oceans. PCBs and other chlorinated hydrocarbons have been detected in the blubber of whales and other marine mammals. Farmed fish, like salmon, often have concentrations of PCBs and other contaminants that are higher than wild fish. The farmed fish are fed fish meal and contaminants in the food become concentrated in the farmed fish through biological magnification.

PCBs continue to be used in some parts of the world, and electrical equipment containing them is still around us. Such equipment is required to remain sealed and the PCBs carefully disposed of when it wears out. Invariably, though, PCBs are present in large amounts of hazardous wastes that accumulate and have to be disposed of somehow, somewhere. PCBs are widespread in landfills and waste dumps. Like other chlorinated hydrocarbons, decades of use and disposal have also left significant amounts of these persistent pollutants in the marine environment, especially around sewage outfalls and in the sediment of harbors of industrial cities. Dredging to keep shipping lanes open releases persistent pollutants; the dredged material is another source.

Chlorinated dioxins and furans, two other groups of chlorinated hydrocarbons, also enter the marine environment from land. Pulp mills and waste incinerators are important sources of the chemicals, which also occur naturally, for example, as a result of forest fires. Some dioxins are among the most toxic of all chemical pollutants. They are carcinogenic and cause birth defects and damage to the immune system in many vertebrates and humans. Furthermore, they show biological magnification.

PCBs, a group of chemicals used as fire retardants (PBDEs), certain pesticides, and other toxic chemicals evaporate into the atmosphere, where the wind can carry them considerable distances before they condense in the cool upper atmosphere. When they condense, they are carried in rain or snow back to the ground, where the process starts all over again. Because of wind patterns and the fact that more condensation of the chemicals takes place near the poles, where it is cold, the chemicals are concentrated in polar regions. This global system of evaporation and condensation of substances, many of which are toxic, is known as **global distillation**. Global distillation has been implicated in whale kills, and high levels of PCBs, mercury (see below), and other toxic chemicals have been found in seals, polar bears, and even the Inuit and other aboriginal peoples of the Arctic and thousands of miles from where the pollutants were used. As ice melting increases

Figure 20.11 The brown pelican (*Pelecanus occidentalis*) is again a common sight along the coasts of the United States and the Caribbean. Many, however, were affected by the 2010 Gulf of Mexico oil spill.

Ingram Publishing/SuperStock

Effects of Climate Change on Arctic Peoples

The anticipated increases in average temperatures and the reduction in sea ice being observed as a consequence of global climate change will be particularly sever along the coasts of the Arctic Ocean. Inuit and other aboriginal peoples have carved out a precarious existence this region for thousands of years.

Climate change in the Arctic The melting of Arctic sea ice is advancing at a rate faster than originally predicted. The consequences of global warming are especially worrisome to the communities of the Canadian Arctic, a vast region of islands, glaciers, and fjords. The shrinkage of the coastal sea ice is responsible for the loss of habitat to Arctic marine life like the polar bear, harp seal, and walrus, which are important sources of food, shelter, clothing, and tools for the Inuit. This change in marine life will severely disrupt traditional hunting by the Inuit. Also being affected are land animals like the caribou, which are also used for food and many kinds of implements. The thawing of permanently frozen soil is making the Arctic shore crumble, damaging buildings and disrupting transportation. Such is the case in the Beaufort Sea north of Alaska, where ice bluffs along the coast retreated from an annual average of 6.8 m in 1955 to 19 m at the present, forcing some communities to relocate. The loss of coastal sea ice will also increase exposure of coastal communities to storms.

Economic impacts The retreating ice is already revealing deposits of minerals like zinc and iron, and has made exploration for oil and gas easier. The opening of a Northwest Passage shipping route across the Canadian Arctic (see foldout map) and a northern sea route across the Siberian Arctic is now possible without the use of icebreakers during summer. Traffic is rapidly increasing: 34 vessels crossed the Arctic in 2011 whereas 71 crossed in 2013. In 2014 a cargo ship crossed for the first time without the help of icebreakers. It is a positive development for commerce that promises economic development, but one that will threaten traditional ways of life and increase the danger of oil spills. The question of sovereignty of the Northwest Passage and of possible claims for the ocean bottom beyond the continental shelf are also potential sources of tension.

The future of Arctic peoples The vulnerability of the Canadian Inuit when facing the transformation of the Arctic environment has triggered concern about their future. A multidisciplinary research program, ArcticNet, was created to study not only the environmental costs of climate change but also the socio-economic consequences to the Inuit. The program involves research teams from several Canadian universities and government agencies, Inuit organizations and teams from several nations. Epidemiologists are documenting how the health of coastal Arctic inhabitants is impacted by climate change and modernization. A network of eight research stations is being created. Perhaps the most serious problem for the Inuit is the availability of wildlife. Disruption of traditional hunting would lead to reliance on alien foods, affecting both human health and causing cultural changes. Researchers are working to analyze impact that landscape change may have on daily activities and how an alteration in the social benefits of hunting may lead to changes in the way of life.

in Antarctica and the Arctic as a result of global climate change, so is the release of the toxic pollutants that have accumulated in the ice through global distillation.

Even at low concentrations years after being banned, PCBs and other chlorinated hydrocarbons have been implicated in abnormal sexual behavior and reproduction in seabirds, marine mammals, and fishes. These pollutants form chemicals that are similar to sex hormones (also called "environmental estrogen") and seem to disrupt reproduction. Atlantic tomcod (*Microgadus tomcod*) in PCBs-contaminated Hudson River in New York state have nevertheless rapidly evolved an increased resistance to the effects of the pollutant.

Heavy Metals An additional category of chemical contaminants of the world's oceans is metals, particularly those classified as **heavy metals**. Very minute amounts of some metals are needed by most if not all organisms, but an excess can be toxic.

One particularly troublesome heavy metal is mercury, which reaches the ocean through several natural processes: the weathering of rocks, volcanic activity, rivers, and dust particles from the atmosphere. Mercury is also used in the production of chlorine and plastics and other chemical processes, in batteries, fluorescent lamps, drugs, and even tooth fillings. Discharges from industries and cities and the burning of coal, which contains traces of mercury, have increased the concentration of mercury in the marine environment.

Pure, liquid mercury, like that in thermometers, is harmless unless its vapors are inhaled. It is a different story when mercury combines with organic chemicals, as when transformed by some bacteria and other microbes in the water or sediment. These organic compounds, like methyl mercury, are persistent and accumulate in the food chain. Levels of mercury too high for human consumption have been found in large fishes like tuna and swordfish. The older the fish, the higher the mercury content. Methyl mercury also can be found in coastal sediments, particularly around areas where wastes are dumped. Mercury compounds also undergo global distillation. They are very difficult to eliminate and are highly toxic to practically all forms of life. In humans they can cause brain, kidney, and liver damage and birth defects.

The dangers of the presence of mercury in seafood were demonstrated by the appearance in the 1950s and 1960s of a crippling neurological disorder among the inhabitants of a town in southern Japan (Fig. 20.12, p. 636). The victims were poisoned by eating fish and shellfish that contained concentrated mercury discharged at sea by a chemical plant. More recently, high levels of mercury in fish prompted the United States government in 2004 to warn pregnant women and young children to limit their intake of canned tuna, swordfish, and other fishes.

Lead is another widespread heavy metal pollutant. As with mercury, organic lead compounds are persistent and concentrate in the tissues of organisms. Lead is toxic to humans, causing nervous disorders and death. The principal source of lead pollution in the marine environment is the exhaust of vehicles run with leaded fuels. The lead reaches the water by way of rain and wind-blown dust. Lead has also found its way into all sorts of products, like paints and ceramics, that eventually make their way to the ocean.

The removal of lead from gasoline has been responsible for a decrease in lead levels in surface water, especially in the North Atlantic. This represents a success story that demonstrates that environmental problems can be addressed.

Cadmium and copper are other toxic heavy metals that are slowly concentrated in marine life. Mining and refining operations are major sources. Cadmium is also present in the waste from battery manufacturing and in discarded batteries. Together with lead, mercury, and other toxic metals, cadmium is common in discarded computers and other electronic components. These toxic metals often seep into rivers or the ocean from disposal sites. Sources of copper

(a) 1945 1955 1965 1975

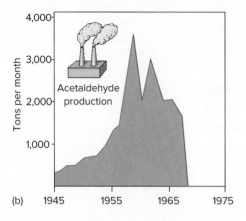

(b) 1945 1955 1965 1975

(c) 1945 1955 1965 1975

Figure 20.12 In the 1950s and 1960s a neurological disorder that often ends in severe brain damage, paralysis, or death appeared among the inhabitants of Minamata, Japan. The number of new cases (a) was directly related to the production of acetaldehyde by a chemical plant (b). A mercury compound was used in the production of acetaldehyde, which is used in making plastics. As a result, an estimated 200 to 600 tons of waste mercury were discharged into Minamata Bay between 1952 and 1968, when it was finally stopped. The disease, now called Minamata disease, was caused by eating seafood contaminated with mercury, here indicated by its concentration in the edible clam *Venus* (c).

include wood treatment and other industrial processes. Unlike lead and mercury, which are transported in the atmosphere, unnaturally high levels of cadmium, copper, and other heavy metals remain near their source.

Radioactive Wastes Pollution by radioactive substances has been a problem in the marine environment since the first atomic bomb explosions in the early 1940s. **Radioactivity** is a property exhibited by certain atoms that emit radiation in the form of energy or particular types of particles. Some radioactive isotopes occur in nature but in minute amounts, and some harmful radiation reaches us from outer space. Exposure to some types of radiation is harmful to all forms of life. In humans it causes leukemia and other cancers, as well as other disorders. Radioactive waste does not need to be ingested to have effects because it can penetrate through living matter. Radioactive material also can continue to emit radiation for thousands of years.

A source of radioactivity is the waste from nuclear power plants. Radioactive waste must be disposed of somewhere. Some was dumped into designated areas of the ocean. Sunken nuclear-powered submarines and ships, fallen satellites, and crashed planes carrying nuclear weapons are other potential sources of radioactive waste. Accidents in nuclear reactors along the coast, as in the case of the Japan earthquake and tsunami in 2011, and radioactive waste in industrial effluents add to the risk. Nevertheless, the use of radioactive material is highly regulated, so it generally poses no major threat to the marine environment.

Solid Waste

A look at the upper reaches of many beaches will reveal an amazing assortment of trash brought in by high water. Most of it is plastic: bottles, bags, foam cups and packaging, nets, and thousands of other items. Solid waste from land still flows to sea at an alarming rate, even if dumping of solid waste and other pollutants is banned by the London Convention of 1972.

Plastic is especially troublesome because it is strong, durable, and non-biodegradable. A huge garbage patch has been formed in the North Pacific gyre thousands of miles from land. At least four other giant patches are known elsewhere. Styrofoam and other plastics eventually break down into tiny particles that are now found in every remote corner of the ocean, finding their way to the guts of many animals that ingest them by mistake. Seabirds are particularly vulnerable. **Microplastics**, which are used in toothpaste and skin scrubs, have been found in the gills and guts of crabs and fishes. Microplastics are found practically everywhere, in sediments as well as in the water column, all impossible to remove. Larger

plastic debris is also a major threat. Sea turtles, seabirds, seals, and others are maimed or killed after getting entangled in plastic fishing line (Fig. 20.13). Many die with their digestive tract clogged with plastic bags and other debris.

Plastic bags, water bottles, 6-pack rings, cups, packaging material, and many other types of plastic debris are discarded everywhere by beach visitors, ocean vessels, pleasure boats, or carried by rivers and wind from land. Only a minute fraction of the huge amount of plastic bags and water bottles used daily are recycled. The best solution is to ban or restrict the use of plastics rather than undertaking the impossible task of removing all of the plastic from the environment. A worldwide effort to such effect is beginning to take hold but it will represent only a small fraction of the billions of plastic bags and water bottles used and discarded every year.

Thermal Pollution

Seawater is often used as a coolant in power plants, oil refineries, and other industries along the coast. The heated water that results from the cooling process is pumped back to sea, causing alterations to the environment known as **thermal pollution**. Local warm pockets can be created in poorly mixed bays. Even though some fishes can be attracted to the site, such temperature increases are known to adversely affect some organisms. Higher temperatures also decrease the ability of water to dissolve oxygen. The effects of thermal pollution on marine life are especially pronounced in the tropics. In contrast to organisms living in temperate and polar regions, corals and some other tropical species normally live just below the highest tolerable temperatures. Reef-building corals are particularly sensitive (see Fig. 20.4).

Saline brines from desalination plants that are used to obtain fresh water from seawater are a major pollutant in the Persian Gulf and other coastal regions. Brines are typically warmer than coastal waters and highly saline as well.

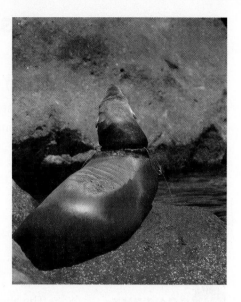

Figure 20.13 It has been estimated that plastic debris kills as many as 2 million seabirds and 100,000 marine mammals every year. Plastic bags kill sea turtles that swallow them, thinking they have caught a jellyfish. Six-pack rings easily ensnare seabirds, which, unable to feed or fly, face a slow death. This California sea lion (*Zalophus californianus*) is being slowly strangled by nylon fishing line.

Section Review

1. **Describe** What are the impacts of sewage and sludge on marine organisms?

2. **Summarize** Make a table like the one below to summarize the sources and impacts of persistent toxic substances.

Persistent Toxic Substance	Sources	Impacts on Marine Environment
Pesticides		
PCBs		
Heavy metals		
Radioactive waste		

3. **Describe** the impacts of plastics on marine animals.

20.3 Threatened and Endangered Species

VOCABULARY

extermination

rare

threatened

endangered

Main Idea
The alteration of marine habitats by humans can lead to marine organisms becoming threatened, endangered, or extinct.

. .

Key Questions
1. What is extinction?
2. What are the types of statuses used to classify species facing extinction?

IN CONTEXT

Natural Selection The production of more offspring by those individuals in a population that are genetically best-adapted to the environment.
• 6.4, The Diversity of Life in the Sea

The alteration or destruction of habitats by humans can have another disastrous effect: driving species to their eventual disappearance, or extinction. Organisms become adapted to changes in the environment as a result of natural selection. If they cannot adapt, they go extinct. Extinction is therefore a natural consequence of the process of evolution. To make a distinction, some biologists refer to human-induced extinction as **extermination**.

How Humans Cause Extinction
Species that face extinction are classified as **rare** when they are not in immediate danger but are at risk, **threatened** when their numbers have become low, and **endangered** when in immediate danger of disappearing.

Marine species can be threatened by overexploitation for food, hides, and other products. Many species of fishes, sea turtles, and other marine animals are killed and discarded as fisheries by-catch (see "Optimal Yields and Overfishing," in 19.1, Fisheries). Habitat destruction, pollution, and the introduction of pests (see "Biological Invasions: The Uninvited Guests") and diseases also place species

Figure 20.14 The Steller's sea cow (*Hydrodamalis gigas*), weighing an estimated 10 tons or more, ate mostly kelp. Demand for its meat, which was considered "as good as the best cuts of beef," led to its extermination at the hands of whalers. The species was slaughtered to extinction soon after its discovery; the last known live individual was taken in 1768.

Fortean/TopFoto/The Image Works

Table 20.1 Endangered, Threatened, and Vulnerable Marine Species

Group of Marine Animals	Red List of Threatened Species, 2014[1] (Endangered and Vulnerable Species Only)	CITES List of Species Threatened with Extinction, 2013[2]	CITES List of Species Vulnerable to Exploitation, 2013[3]
Corals and other cnidarians	234	0	All scleractinian, black, blue, organ-pipe, and hydrozoan corals
Other invertebrates	68	0	9
Marine fishes	439	9	66
Sea turtles	5	7 (all)	0
Marine iguana	1	0	1
Seabirds	170	8	1
Sirenians	4 (all)	4 (all)	1
Seals, sea lions, walrus	8	4	8
Sea otter, marine otter	2	2	0
Polar bear	1	0	0
Whales, marine dolphins	14	32	0

[1]Compiled by the International Union for Conservation of Nature and Natural Resources (IUCN).

[2]Appendix I, Convention on International Trade in Endangered Species of Wild Flora and Fauna.

[3]Appendix II, Convention on International Trade in Endangered Species of Wild Flora and Fauna.

at risk. The Steller's sea cow, a sirenian, is a shocking example of a marine animal that was rapidly exterminated because of unregulated hunting for food (Fig. 20.14).

Species at Risk

The future of many marine species is at stake (Table 20.1). Perhaps the most widely known case is that of the whales, but there are many other examples. Giant clams are taken for food and for their shells in such numbers that they have become rare or even locally extinct in most of the tropical Pacific. Coral colonies that take decades to grow and are essential parts of threatened coral reefs are collected and bleached for sale (Fig. 20.15). Marine snails like cowries

DIVE IN!

How are corals protected by international law? The importation of all species of reef-building coral into the United States is illegal under the Convention on International Trade in Endangered Species (CITES). When traveling in other countries, people may see pieces or whole corals for sale and be tempted to buy them as souvenirs. Coral colonies take years to grow, and were likely collected by being broken off from a reef before ending up on a merchant's table. Bringing pieces of coral back into the United States, even if they are part of jewelry, can lead to fines and the coral being confiscated.

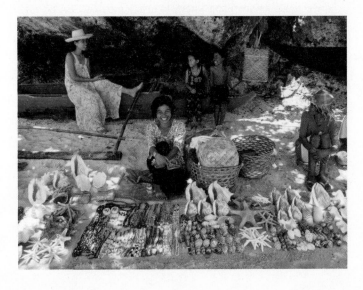

Figure 20.15 The selling of corals and shells to tourists in the Philippines.

Biological Invasions: The Uninvited Guests

The introduction of invasive species in areas where they do not naturally occur can have devastating effects on the natural functioning of an ecosystem. Native species can suffer due to competition with invasive species for space or resources, as well as from parasites brought in by invasives. There has been an increase in the frequency of invasive marine species during the past few decades as a result of growth in shipping, aquaculture, and the aquarium trade. An estimated 84% of coasts worldwide have been invaded by invasive species, mostly as accidental introductions.

Shipping Busy ports are particular hot spots for invasive species. Fouling organisms, such as encrusting algae and barnacles, can traverse an ocean on the underside of a ship or on a movable oil platform. Many other species have been introduced as planktonic larvae in ballast water, which is used to fill large tanks in ships to provide stability while at sea. Upon entering a new port, the ship's ballast water is emptied and the planktonic foreigners are introduced to an ecosystem they previously did not inhabit.

The Asian clam (*Potamocorbula amurensis*) was accidentally introduced in San Francisco Bay, California, after the United States opened trade with China. These clams did not live in the bay in 1985, but by 1990 they literally covered whole sections of the bay's muddy bottom. As many as 10,000 clams per square meter carpet the bottom in some areas. Over 250 marine invasive species have been found in San Francisco Bay, considered the most invaded estuary in the world.

Animals are not the only kinds of organisms considered invasive species. *Codium fragile,* a green alga accidentally introduced from Europe to the Pacific coast, has become a pest by growing in masses on rocks and oysters on the northeastern United States. Another green alga, *Caulerpa taxifolia*, was accidentally introduced in the Mediterranean from the Caribbean in 1984. It grows into branches as long as

FIVE SPINES

FLAT REAR LEG

Up to 3 inches wide across the back of the shell
Color variable, dark, mottled, often green or orange

ALERT!
Have You Seen This Crab?

The green crab, *Carcinus maenas*, is a voracious predator of clams and oysters and a potential threat to these commercial species. Although not naturally occurring in B.C. water, green crab have now migrated here. As of 2010, green crabs have been trapped extensively along the west coast of Vancouver Island from Winter Harbour to Barkley Sound. We need your help to document their range, expansion, and growth rates. If you see what you believe is a green crab please collect it, freeze it, record the precise date and location found, and contact:

Tom Therriault
Pacific Biological Station
Fisheries and Oceans
Nanaimo, B.C.
P: (250) 756-7394
F: (250) 756-7138

Green Crab

Common Shore Crab

3 m, and has spread rapidly, particularly in the western Mediterranean, where it smothers native seagrasses and depletes oxygen from the bottom.

Transplantation The intentional introduction, or transplantation, of species for commercial purposes can also bring in invasive species. The Japanese oyster (*Crassostrea gigas*) was introduced as spat, or young individuals, on the Pacific coast of North America. It was a successful transplantation in the sense that the oyster is now of significant commercial value. Unfortunately, many species living on the spat shells were also introduced, like the oyster drill (*Ceratostoma inornatum*), a gastropod that preys on oysters and other native bivalves.

Aquarium trade The aquarium trade is also responsible for the introduction of invasive species. The accidental release of the beautiful lionfish (*Pterois*), native to the Indian and Pacific Ocean, from an aquarium in Florida is the leading hypothesis for the rapidly expanding population in the United States, Caribbean, and South American coasts. Lionfish are voracious consumers of small reef fish that do not recognize the lionfish as predators, which use an ambush strategy to attack prey. SCUBA divers in some areas catch lionfish by spearfishing and feed them to sharks and large grouper in an effort to teach them to eat lionfish. Lionfish are delicious for human consumption, as well, once the venomous spines are removed. Perhaps fishing can mitigate the spread of this invasive species.

It is difficult to prevent the introduction of uninvited guests. The use of filtered or sterilized ballast water or the exchange of ballast water in mid-ocean far from land has been explored. The transplantation of species from different regions must also be strictly regulated and studied before they are moved. However, the more humans move around the world, the more other species follow them.

Courtesy Department of Fisheries and Oceans, Nanaimo, British Columbia, Canada

and cones, whose shells are eagerly sought by collectors, have similarly disappeared from many areas.

Intense commercial fishing to satisfy the growing markets for shark meat and fins, recreational fishing, and by-catch on longlines threatens many species of sharks. Sharks, like whales, reproduce slowly. It is feared that many shark species could be pushed to the brink of extinction in a decade or two. Manta rays are hunted for their gill, which are used in traditional Chinese medicine as an unproven cure for cancer and other ailments.

All seven species of sea turtles are threatened. Adults as well as eggs have been exploited for food and tortoiseshell (Fig. 20.16). Their nesting sites have been overrun by development, and they drown in fishing nets. Whales, although mostly protected, are still threatened by ship strikes. The populations of polar bears in eastern Alaska and western Canada, threatened by shrinking ice, have decreased by 40% from 2001 to 2010.

Seabirds have not fared well, either. Overfishing has been responsible for a decline in the number of seabirds in many parts of the world as a result of the decline in food. Longline fishing kills thousands of albatrosses a year in Antarctic waters alone. Nineteen of the 21 species of albatrosses are threatened or endangered, some critically so. Plastic debris, when ingested, kills many seabirds and other marine animals.

Other marine mammals that have low reproductive rates also face extermination. Some species of seals, sea lions, and the walrus have been decimated for their skins, meat, blubber, or the ivory of their tusks. The walrus, like polar bears, are threatened by diminishing Arctic ice. Monk seals are endangered; one species, the Caribbean monk seal (*Monachus tropicalis*), is extinct. The number of Steller sea lions (*Eumetopias jubatus*) has sharply declined in the northeastern Pacific. The sea otter (*Enhydra lutris*) had made a comeback, but it is still threatened. Manatees (*Trichechus*) and the dugong (*Dugong dugon*), the smaller cousins of the extinct Steller's sea cow, are both in danger of extinction.

Figure 20.16 Green turtles (*Chelonia mydas*) confiscated from poachers in Bali, Indonesia. Poaching of endangered or threatened marine animals is a problem worldwide, with many of these species being part of an illegal multimillion-dollar trade.

Section Review

1. **Explain** Which actions by humans have affected the following groups of marine animals: giant clams, snails, sharks, sea turtles, seabirds?

2. **Explain** How is the reproductive rate of marine mammals related to their status in regards to extinction?

20.4 Conserving and Enhancing the Marine Environment

DIVE IN!

Where are the world's largest artificial reefs? In 2006, the decommissioned USS *Oriskany* was sunk off the coast of Florida, near Pensacola. At almost 300 m long and 45 m wide, it is nicknamed the Great Carrier Reef and is a popular scuba diving site. In 2014, a team of engineers, scientists, and divers began building what will be, upon completion, the world's largest artificial reef. The reef, called Kan-Kanán, is being built off of the Yucatán Peninsula in Mexico. It will consist of hollow, concrete pyramids that will allow for coral and other organisms to settle out on its surface. The Mexican government hopes that the new reef will reduce damage to the natural reefs in the area, which are impacted by tourism and climate change, as well as reduce erosion along the coastline. Once finished, the reef is expected to be approximately 2 km in length.

Main Idea

Through conservation and habitat restoration efforts, humans are working to reverse the negative impacts of previous actions.

Key Questions

1. What are the aims of conservation of biodiversity?
2. What is a marine protected area (MPA)? Give two examples of MPAs.

Conservation

The protection, or conservation, of coastal areas from the advancing wave of development is of crucial importance. The apparent conflict between economic development and the protection of marine resources occurs everywhere. Development must be **sustainable**, that is, it must meet the needs of today without affecting the ability of future generations to meet their needs. Many governments have passed laws, but there is still a lot to be done. Coastal management aims to promote the wise use of our coasts while ensuring that the benefits of this use can be sustained for future generations. Coastal management requires judicious planning to balance the often conflicting interests of developers, fishers, beach-goers, and nesting seabirds. It also deals with issues as varied as historical preservation, beach access, and tourism.

Conservation efforts include projects and legislation across disciplines and boundaries. Oil drilling, for instance, has been banned along some coasts, as in California. Commercial fisheries are regulated by national governments through exclusive economic zones (EEZ) and by international commissions.

Governments have also established marine protected areas (MPA) for the protection and management of areas of ecological significance. The National Estuarine Reserve Research System of the United States designated 27 reserves in 21 states and Puerto Rico protecting almost 5,370 km² of estuarine habitat. Australia established the world's largest network of marine reserves, a total area of 3.1 million km². The largest single marine protected area in the world is the Natural Park of the Coral Sea in New Caledonia, southwestern Pacific, established in 2014, with a total area of 3.1 million km².

Marine protected areas are richer in both the number and the abundance of species compared with unprotected waters around them. There is evidence that protected areas benefit surrounding waters through either "seeding," in which the offspring of individuals inside the protected area settle elsewhere, or "spillover," in which juveniles and adults move out of the protected area. Less than 1% of the ocean, however, is protected.

Many national and international organizations play an active role in marine conservation through education, sponsoring projects, and lobbying for legislation. Some concentrate on issues like biodiversity, habitat conservation, and sustainable fishing, whereas others aim at particular animal groups, environments, or geographical regions.

Restoration of Habitats

Habitat restoration is another strategy for improving the quality of the environment. Habitat restoration helps recovery from stress by transplanting or restocking key species from healthier areas. The loss of salt marshes and mangrove forests through landfilling or the development can be partially compensated for by creating or improving a similar but unstressed habitat elsewhere. These efforts, of course, assume that the new location meets the physical requirements (tides, salinity, type of substrate) for the development of a healthy biological community.

Successful examples of habitat restoration include the restoration of salt marshes and of native oyster reefs in Chesapeake Bay, eastern United States. Young giant kelp, with their root-like holdfasts tied to concrete blocks or to submerged nylon lines, have been used to help restore kelp forests, particularly in artificial reefs built of large rocks (Fig. 20.17). New techniques in the transplantation of coral and seagrasses promise the reseeding of damaged reefs and seagrasses. Seabirds are even being attracted to old nesting grounds by using painted decoys and solar-powered tape recorders that play bird calls! Habitat restoration is nevertheless a poor alternative to preventing the degradation of marine environments. Restoration rarely if ever restores habitats to their natural condition, and it is usually expensive.

Artificial Reefs

Biodiversity can be enhanced by building **artificial reefs**. The structure provided by the reefs attract fishes, lobsters, and other life. Shellfish and seaweeds can thrive on their surfaces. Everything from concrete blocks, tires, and toilets to scuttled ships, and an old 727 jet have been used to build artificial reefs. They increase commercial yields of fishes, abalone, sea urchins, and seaweeds in Japan, where some artificial reefs have been shown to support a fish biomass that is 20 times greater than that in natural reefs. Offshore oil rigs have been turned into artificial reefs after "reefing," the removal of the sections above the water. Life on and around these oil rigs has similarly been shown to be far richer than in nearby rocky reefs, and can also serve as nursery grounds for fishes.

Unfortunately, some organisms do not settle on painted surfaces, metal, plastic, or rubber, and some artificial reefs disintegrate faster than originally thought. It has also been argued that artificial reefs only concentrate fishes and lobsters in one spot, making them easier to catch and thus contributing to their depletion. The use of long-lasting concrete modules for artificial reefs (Fig. 20.18), however, has been more successful. Round ones, called "reef balls," can be positioned to maximize photosynthesis and allow for an effective flow of plankton. A rough, more natural surface promotes the settlement of coral larvae in the tropics.

IN CONTEXT

Exclusive Economic Zone (EEZ)
A zone 370 km wide along the coast, in which nations have exclusive rights to fishing and other resources.
• 19.1, Fisheries

Figure 20.17 Restoration of marine habitats has included the transplantation of young giant kelp plants (*Macrocystis pyrifera*) into areas where kelp forests have diminished or vanished. Removal of grazing sea urchins by volunteers can also be effective.

Sand on the Run, or What to Do with Our Shrinking Beaches

Sandy beaches constitute one of our most valued coastal resources. Millions use them for recreation, and their importance as a tourist attraction is evident from Miami Beach to Sydney. Unfortunately, beaches have been shrinking and disappearing worldwide. Sand shifts, so disruptions like storms, hurricanes, winds, and currents periodically modify the shore.

Barrier islands The Atlantic and Gulf coasts of the United States are protected by barrier islands—low, sandy islands that run parallel to the coast. Barrier islands are found along the coastlines of wide continental shelves. Waves and wind pushed bottom sediments and formed bars on the flat shelf, which eventually became barrier islands.

Effects of Hurricane Katrina, August 2005 (right), on a barrier island off the Louisiana coast.

Barrier islands protect coastlines by absorbing the stress of hurricanes and currents. In addition, their sand dunes are inhabited by salt-resistant plants and land animals, and seabirds often use them as nesting sites. Barrier islands may contain a variety of habitats. Unfortunately, because bridges and roads have made some barrier islands more accessible untouched barrier islands are becoming rare.

How do we protect our beaches? Development is not the only concern. Barrier islands are by their very nature ever-changing. Wind and longshore currents erode their seaward side by continually shifting sand from one tip to the other. Channels between islands may fill up, and previously damaged islands may be eroded away, as in the case of Hurricane Katrina in 2005.

Not everybody realizes that sandy beaches and barrier islands will not tolerate permanent structures. Seawalls, breakwaters, jetties, and groins may be built as safeguards, but they often make the problem worse. Encouraging the natural development of dunes or planting vegetation that helps stabilize the sand is a better solution. Replenishing sand by bringing it periodically from offshore is one expensive, but temporary, way to save a beach. This is what is being done in highly urbanized Southern California, where rivers that once brought in sand to the shore now end behind dams or have been transformed into concrete-bottomed channels. Efforts to manage beaches and barrier islands, however, may only disrupt the natural coastal processes that sooner or later will triumph over our ingenuity.

Prospects for the Future

It is impossible to predict what lies ahead for our world ocean. Optimists have visions of cities built underwater or self-ruling cities built on floating islands, food from aquaculture, unlimited energy from the water and seabed, and no pollution to speak of. Most, however, are less confident. They point out that our impact on the marine environment will escalate as the world's population expands. Modern technology will lead to new means of exploitation, and for resources never imagined before.

The oceans, the heritage of all humankind, must be safeguarded to save them from overexploitation and pollution. This has been one of the aims of the United Nations Convention on the Law of the Sea (UNCLOS), the 1982 treaty that culminated many years of deliberations by the many nations that were represented. Most nations have ratified the treaty and implemented many of its

AP Images

Figure 20.18 Artificial reefs can be built of various types of prefabricated concrete blocks.

provisions, but the United States, who was one of the original signatories of the convention, has yet to ratify it.

The treaty, for example, led to the establishment of the exclusive economic zones (EEZs). This agreement has had significant consequences for the management of marine resources, particularly fisheries. Nations that ratified the treaty before May 13, 1999, had 10 years in which to claim any extensions of their continental shelf beyond the 200 nautical miles as long as the extensions were not more than 100 nautical miles from the point at which the sea reached a depth of 2,500 m and no more than 350 nautical miles from land. Many have done so, except the United States, which has not ratified UNCLOS. Nations owning tiny islands and reefs barely above high water can now claim large portions of the surrounding seabed, which could hold oil, gas, minerals, and fisheries. Such is the case of the Paracel and Spratly islands and surrounding waters in the South China Sea, now claimed by several nations in the region, and in potential conflicts over the exploitation of resources in the Arctic Ocean.

Other international agreements have attempted to solve other problems. The United Nations Conference on Environment and Development at Rio de Janeiro, known as the Rio Earth Summit of 1992, is one example. It resulted in an agreement to preserve biodiversity on the planet.

From the early beginnings the oceans have influenced humankind by providing food and avenues of commerce and culture. It is clear, however, that the relationship has been reversed. The future of the dynamic, ever-expanding field of marine science is thus facing many challenges on the ocean we share. These challenges include a global food crisis that demands new and more efficient uses of food resources from the ocean, an energy crisis that similarly calls for innovative sources of renewable energy, the decrease in biodiversity and the disappearance of marine habitats brought about by humankind, global climate change and its effects on sea level and ocean acidification, and the effects of overfertilization and other types of pollution.

Section Review

1. **List** examples of successful habitat restoration in marine environments.

2. **Identify** What are the advantages and disadvantages of artificial reefs?

Ten Simple Things You Can Do to Save the Oceans

We all impact the oceans, no matter how far we live from the seashore. You can do a lot to help save the oceans and the rest of the planet. Here are 10 simple things inspired by *50 Ways to Save the Ocean* by D. Helvarg.

1. **Take care of the environment.** Do not disturb the environment in any way if you go to the seashore, go snorkeling, or diving. Respect natural habitats. Return any overturned rocks to their original positions. Leave all forms of life where they are.

2. **Eat sustainably caught seafood.** Ask questions when buying fish or ordering it at a restaurant: Where is it from and how was it caught? Buy fish certified as sustainably caught or farmed. Eat lower on the food chain (shellfish, squid) and less or no swordfish, tuna, shark, and grouper. If you go fishing, target species that are abundant and aren't susceptible to overfishing, even if it's legal. Know the regulations and take only what you really need for food.

3. **Beware of what you buy.** If you travel overseas, don't buy products made from endangered or threatened species like tortoiseshell or ivory. Corals and shells should be left alone and alive, in their natural home. Be sure to tell merchants you object to their sale of shells, corals, sand dollars, and other marine life, which were assumedly collected alive and killed for sale. If you buy fishes, corals, and other marine life for your marine aquarium, be sure to check before buying if they were obtained in an environmentally responsible way.

4. **Save gas.** Saving gas keeps carbon dioxide, which promotes global warming and acidification of seawater, out of the atmosphere. Tailpipe pollutants add to the nitrogen nutrients that cause eutrophication of coastal waters. Use public transportation, carpool, or a bike more often; get a smaller car or a hybrid.

5. **Save electricity.** Most power plants still burn coal to produce electricity, releasing more carbon dioxide (an electric drier produces an average of 3.6 pounds of carbon dioxide a day but drying clothes on a clothesline produces none!), plus more nitrogen compounds as well as mercury. Saving gasoline and electricity also reduces the need for oil and thus the threats of more offshore oil drilling and oil spills.

6. **Dispose of hazardous materials properly.** Reduce your use of toxic chemicals at home, which are in discarded computers and electronics, batteries, cleaners, and many other types of household trash. Dumping them or letting them accumulate in landfills is likely to pollute the oceans. Flushing them down the drain simply adds them to sewage, which may ultimately flush, untreated, into the ocean. Your local recycling center can tell you how to recycle hazardous materials.

7. **Recycle plastics. Use less of them.** Reduce the risk of plastics being washed to sea by recycling them. Use reusable canvas bags for shopping instead of plastic bags, thousands of which get blown or thrown to the sea every day around the world. Don't forget to pick up any plastic you may have left around the beach. Cut open six-pack rings before disposal; you don't know where they may end up. Properly dispose of snarled fishing line; never toss it at sea or on shore.

8. **Keep it clean.** Clean after your pets. Their feces may carry parasites that, if washed to sea, can infect marine mammals as well as humans.

9. **Get involved. Keep informed.** Be aware of and keep up with environmental issues. Read, ask questions, listen carefully. Numerous organizations and agencies are committed to help save our planet. Hear what they have to say and sponsor them. Make your voice heard. Vote for the oceans and the planet.

10. **Endorse a new ocean ethic.** Treat the ocean as a valuable resource, not as a garbage dump. Think globally.

Chapter 20 Review

Multiple Choice

1. Disturbance to natural environments caused by humans is called

 a. anthropogenic impact.

 b. biological magnification.

 c. ecological succession.

 d. greenhouse effect.

2. The process by which increased inputs of nutrients leads to an undesirable increase in algal growth is called

 a. biological magnification.

 b. eutrophication.

 c. extinction.

 d. ocean acidification.

Short Answer

3. Which are the indicators that coral reefs are under stress?

4. What are the effects of oil on marine organisms?

Critical Thinking

5. Wastes from duck farms used to wash into two shallow-water bays on Long Island, New York. The wastes, rich in nutrients such as nitrate and phosphate, polluted the water. What do you suppose was the immediate effect of the pollutants? Can you speculate on the likely effects on the commercially valuable shellfish of the area?

6. Tourism and its effects (for example, pollution from hotels and the impact of boats and tourists on fragile habitats) often clash with conservation efforts. Sometimes, however, tourism can help. The economic impact of banning the hunting of harp seals in eastern Canada has been compensated for, in part, by the influx of tourists who now go to see the seals. Can you think of other examples? What recommendations can you make to minimize the impact of tourism on unspoiled marine environments?

What's happening with hypoxic zones in the Gulf of Mexico? Eutrophication has been linked to seasonal or sometimes permanent hypoxic zones. Hypoxic zones form when large amounts of phytoplankton and the organic matter they produce sink and decay in deeper water, using up all the oxygen. Scientists have been tracking the size of the hypoxic zone in the Gulf of Mexico each summer for several decades.

Data and Observations

The graph shows the size of the hypoxic zone in the Gulf of Mexico every summer from 1990 through 2014.

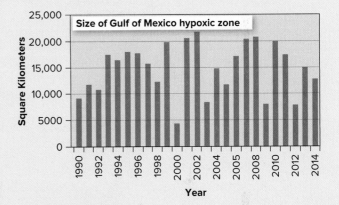

Think Critically

1. Describe the general trend in the size of the hypoxic zone since 1990.

2. What do you think happens to fish and other marine organisms that need oxygen to survive during the summer in the Gulf of Mexico?

3. Worldwide, hypoxic zones have doubled in number every decade since the 1960s. Propose a solution to reducing the size and number of hypoxic zones.

Data obtained from: Nancy N. Rabalais, LUMCON, and R. Eugene Turner, LSU. Funding Sources: NOAA Center for Sponsored Coastal Ocean Research and U.S. EPA. SGulf of Mexico Program.

Appendix A

Units of Measurement

10 meters
= 32.8 feet

Blue whale
(length: 26 m)

1 meter (m)
= 100 centimeters
= 39.37 inches
= 3.28 feet

Leatherback turtle
(length: 2 m)

1 centimeter (cm)
= 10 millimeters
= .01 meter
= .39 inch

Flatworm
(length: 1 cm)

1 millimeter (mm)
= 1,000 microns
= .001 meter
= .039 inch

Copepod
(length: 1.2 mm)

1 micron (μm)
= .001 millimeter
= .000039 inch

Bacterium
(length: 1μm)

1 nanometer (nm)
= .001 μm
= .000000039 inch

Virus
(diameter: 100 nm)

Appendix B

References for the Identification of Marine Organisms in North America

General

Harrison, P., 1985. Seabirds: An Identification Guide. Houghton Mifflin, Boston, MA.

Lieske, E. and R. Myers, 2001. Coral Reef Fishes. Indo-Pacific and Caribbean. Princeton Pocket Guides, Princeton, NJ.

Meinkoth, N. A., 1981. The Audubon Society Field Guide to North American Seashore Creatures. Knopf, New York.

Shirihai, H. and B. Jarrett, 2006. Whales, Dolphins and Other Marine Mammals of the World. Princeton University Press, Princeton, NJ.

Smith, D. L. and K. B. Johnson, 1996. A Guide to Marine Coastal Plankton and Marine Invertebrate Larvae. Kendall-Hunt, Dubuque, IA.

Atlantic Coast of North America (Including the Gulf of Mexico, the Caribbean, and Bermuda)

Gosner, K. L., 1999. A Field Guide to the Atlantic Seashore. Houghton Mifflin, Boston, MA.

Hoese, H. D. and R. H. Moore, 1998. Fishes of the Gulf of Mexico. Texas A&M Press, College Station.

Johnson, W. S. and D. M. Allen, 2012. Zooplankton of the Atlantic and Gulf Coasts: A Guide to their Identification and Ecology. Johns Hopkins University Press, Baltimore, MD.

Kaplan, E. H., 1999. A Field Guide to Coral Reefs: Caribbean and Florida. Houghton Mifflin, Boston, MA.

Kaplan, E. H., 1999. A Field Guide to Southeastern and Caribbean Seashores. Houghton Mifflin, Boston, MA.

Katona, S. K., V. Rough, and D. T. Richardson, 1993. A Field Guide to the Whales, Porpoises, and Seals from Cape Cod to Newfoundland. Smithsonian Institution Press, Washington, DC.

Lippson, A. J. and R. L. Lippson, 1997. Life in the Chesapeake Bay. Johns Hopkins University Press, Baltimore, MD.

Littler, D. S., M. M. Littler, K. E. Bucher, and J. N. Norris, 1989. Marine Plants of the Caribbean. Smithsonian Institution Press, Washington, DC.

Proctor, N. S. and P. J. Lynch, 2005. A Field Guide to North Atlantic Wildlife: Marine Mammals, Seabirds, Fish, and Other Sea Life. Yale University Press, New Haven, CT.

Robins, C. R. and G. C. Ray, 1999. A Field Guide to Atlantic Coast Fishes. Houghton Mifflin, Boston, MA.

Ruppert, E. and R. Fox, 1988. Seashore Animals of the Southeast. University of South Carolina Press, Columbia.

Schneider, C. W. and R. B. Searles, 1991. Seaweeds of the Southeastern United States: Cape Hatteras to Cape Canaveral. Duke University Press, Durham, NC.

Smith, C. L., 1997. National Audubon Society Field Guide to Tropical Marine Fishes of the Caribbean, Gulf of Mexico, Florida, the Bahamas, and Bermuda. Knopf, New York.

Snyderman, M. and C. Wiseman, 1996. Guide to Marine Life: Caribbean Bahamas Florida. Aqua Quest, Locust Valley, NY.

Spalding, M. D., 2004. A Guide to the Coral Reefs of the Caribbean. University of California Press, Berkeley.

Pacific Coast of North America (Including the Gulf of California and Hawai'i)

Allen, G. R. and D. R. Robertson, 1994. Fishes of the Tropical Eastern Pacific. University of Hawai'i Press, Honollu.

Dawson, E. Y. and M. S. Foster, 1982. Seashore Plants of California. University of California Press, Berkeley.

Druhel, L. D., 2001. Pacific Seaweeds. Harbour Publishing, Madeira Park, BC, Canada.

Ebert, D. A., 2003. Sharks, Rays, and Chimaeras of California. University of California Press, Berkeley.

Eschmeyer, W. N., E. S. Herald, and H. E. Hammann, 1999. A Field Guide to Pacific Coast Fishes. Houghton Mifflin, Boston, MA.

Goodson, G., 1988. Fishes of the Pacific Coast. Alaska to Peru, Including the Gulf of California and the Galapagos Islands. Stanford University Press, Stanford, CA.

Gotshall, D. W., 2000. Pacific Coast Inshore Fishes. Sea Challengers, Monterey, CA.

Gotshall, D. W., 2005. Guide to Marine Invertebrates. Alaska to Baja California. Sea Challengers, Monterey, CA.

Hinton, S., 1988. Seashore Life of Southern California. University of California Press, Berkeley.

Hoover, J. P., 2012. Hawai'i's Sea Creatures. A Guide to Hawai'i's Marine Invertebrates. Mutual Publishing, Honolulu, HI.

Kerstitch, A. and H. Bertsch, 2007. Sea of Cortez Marine Invertebrates. Sea Challengers, Monterey, CA.

Kozloff, E. N., 1983. Seashore Life of the Northern Pacific Coast. University of Washington Press, Seattle.

Kozloff, E. N., 1996. Marine Invertebrates of the Pacific Northwest. University of Washington Press, Seattle.

Mondragon, J. and J. Mondragon, 2003. Seaweeds of the Pacific Coast. Common Marine Algae from Alaska to Baja California. Sea Challengers, Monterey, CA.

Orr, R. T., R. C. Helm, and J. Schoehnwald, 1989. Marine Mammals of California. University of California Press, Berkeley.

Ricketts, E. F., J. Calvin, J.W. Hedgpeth, and D. W. Phillips, 1985. Between Pacific Tides (5th ed). Stanford University Press, Stanford, CA.

Glossary

A

abdomen The portion of a decapod body between the cephalothorax and the tail.

abiotic A non-living (physical or chemical) component of the environment.

aboral surface The surface opposite the mouth in cnidarians, comb jellies, and echinoderms.

abyssal hills Topographic features of the deep-ocean floor.

abyssal plain The nearly flat region of the deep-sea floor.

abyssal zone The bottom from a depth of approximately 4,000 m to 6,000 m.

abyssopelagic zone The pelagic environment from a depth of 4,000 m to 6,000 m.

accessory photosynthetic pigment Chemical compound that captures solar energy together with chlorophyll.

active continental margin A continental margin that is colliding with another plate and as a result is geologically active.

active transport The transfer of substances across membranes by a cell against a concentration gradient.

adductor muscle One of two muscles that close a bivalve's shells.

adhesion The sticking together of particles of different substances.

aerobic Requiring oxygen.

aerobic respiration The chemical process involved in the release of energy from organic matter.

aerosols Microscopic particles in the atmosphere.

agar A commercially important phycocolloid extracted from red algae.

ahermatypic corals Corals that do not build reefs and often lack zooxanthellae.

albedo The reflectivity of Earth's surface. The higher the albedo, the more solar energy is reflected back to space.

alga Any of several groups of eukaryotic, autotrophic protists that lack the structural features of plants.

algal ridge A ridge of coralline algae that is found on the outer edge of some coral reefs.

algal turf A dense growth of often filamentous algae.

algin A phycocolloid extracted from brown algae that is widely used in food processing.

alternation of generations A reproductive cycle in which a sexual stage alternates with an **asexual** one, as in the case of a gametophyte alternating with a sporophyte.

ambergris Undigested material that accumulates in sperm whale intestines.

ambulacral groove Each of the radiating channels of echinoderms through which tube feet protrude.

amino acid One of the 20 nitrogen-containing molecules that make up proteins.

amphibians Vertebrates that lay eggs in fresh water: frogs, salamanders, and allies.

amphipods A group of small, laterally compressed crustaceans that includes beach hoppers and others.

ampulla Each of the muscular sacs that extend inside the body opposite the tube feet of echinoderms.

ampulla of Lorenzini One of several sensory structures in the head of sharks that detect weak electric fields.

anadromous Marine fishes that migrate to fresh water to breed. Compare catadromous.

anaerobic in the absence of oxygen

anaerobic respiration The breaking down of organic matter by organisms in the absence of oxygen. Also see aerobic respiration.

anal fin The last ventral fin of fishes.

animals Members of the kingdom *Animalia*, which consists of heterotrophic, eukaryotic, multicellular organisms.

anoxic Lacking oxygen. Compare hypoxic.

antenna A sensory appendage on the head of arthropods.

anterior The defined front end of a bilaterally symmetric animal.

anthozoans Cnidarians whose life cycle consists of a complex polyp and no medusa.

anthropogenic Human-derived.

antinodes The high and low points at each end of a standing wave.

apneustic breathing Breathing pattern shown in many marine mammals that involves deep breaths followed by exhalation.

aquaculture The farming of marine, brackish-water, and freshwater organisms.

archaea Prokaryotic, unicellular microorganisms in the domain *Archaea*.

Aristotle's lantern Jaws and associated muscles used by sea urchins to bite off food.

arrow worms (chaetognaths) Planktonic invertebrates characterized by a streamlined, transparent body.

arthropods Invertebrates that have jointed appendages and a segmented exoskeleton.

artificial reefs Irregular hard structures that provide habitat for marine life.

asexual reproduction The type of reproduction that takes place without the formation of gametes. Compare sexual reproduction.

asthenosphere The layer of upper mantle that lies below the lithosphere.

atoll A coral reef that develops as a ring around a central lagoon.

atom The smallest unit into which an element can be divided and still retain its properties.

ATP (adenosine triphosphate) A molecule that stores energy and releases it to power chemical reactions in organisms.

autotroph An organism that manufactures its own organic matter by using energy from the sun or other sources. Compare heterotroph.

auxospore The resistant stage of diatoms that restores the maximum size characteristic to the species.

B

bachelor groups Groups of male pinnipeds that are not in control of harems.

backbone A series of articulating bones surrounding the nerve cord of vertebrates; a spine or vertebral column.

back reef The inner part of a barrier reef or an atoll.

bacteria Prokaryotic, unicellular microorganisms in the domain *Bacteria.*

bacteriophages Viruses that infect bacteria.

baleen The filtering plates that hang from the upper jaws of baleen whales.

baleen whales The filter-feeding whales.

barbels Whisker-like sensory organs near the mouth of some bottom feeding fishes.

bar-built estuary An estuary that is formed when a barrier island separates a section of the coast where fresh water enters.

barnacles Crustaceans that live attached to surfaces and are typically enclosed by heavy **calcareous** plates.

barophilic Pressure-loving.

barrier island A long and narrow island that is built by waves along the coast.

barrier reef A type of coral reef that develops at some distance from the coast.

basalt The dark-colored rock that forms the sea floor, or oceanic crust.

bathyal zone The bottom between the shelf break and a depth of approximately 4,000 m.

bathypelagic zone The pelagic environment from a depth of 1,000 m to 4,000 m.

bayous Networks of slow flowing tidal creeks found in low-lying deltas such as the Mississippi River.

beard worms (pogonophorans) Tube-dwelling annelids that lack a digestive system.

bell The characteristic rounded body of scyphozoan jellyfishes.

bends A painful condition in divers caused by nitrogen bubbles in the blood as a result of surfacing too quickly.

benthic Organisms that live on the bottom.

biological species concept The definition of species as a population of organisms that have common characteristics and can successfully breed with each other.

Big Bang theory The theory that a cosmic explosion produced clouds of dust and gas from which Earth and the solar system originated.

bilateral symmetry The arrangement of body parts in such a way that there are only two identical halves, with different anterior and posterior ends and dorsal and ventral surfaces.

binomial nomenclature A system of naming species using two names, the first of which refers to the genus.

biodegradable Able to be broken down by bacteria or other organisms.

bioerosion Erosion of hard structures such as reefs by grazing animals.

biofouling Unwanted settlement of organisms on submerged surfaces of boats, pilings, and other structures.

biofuel A fuel obtained from a biological renewable resource.

biogenous sediment Sediment that is made up of the skeletons and shells of marine organisms. Also see calcareous and siliceous ooze.

biological clock A repeated rhythm that is synchronized with time.

biological magnification The increased concentration of non-biodegradable chemicals in the higher levels of the food chain.

bioluminescence The production of light by living organisms.

biomass The total mass of living organisms.

biotechnology Techniques such as genetic engineering which are used to change or enhance biological processes.

biotic A living component of an environment.

bioturbator A member of the infauna that moves sediment while burrowing or feeding.

birds Vertebrates that have feathers and lay eggs with calcified shells on land.

bivalves Molluscs that possess a two-valved shell, filtering gills, and a shovel-like foot.

black corals Colonial anthozoans that secrete a black protein skeleton.

black smoker A chimney-like accumulation of mineral deposits that is found at hydrothermal vents.

blade The leaf-like portion of the thallus of a seaweed.

bleaching The expulsion of zooxanthellae by reef corals in response to stress.

bloom A sudden increase in the abundance of an alga or phytoplankton.

blowhole The nostrils of cetaceans.

blubber A thick layer of fat under the skin of many marine mammals.

bony fishes (*Osteichthyes*) Fishes with a skeleton made mostly of bone; they also have opercula and fin rays.

boring sponges Sponges that bore through calcareous skeletons and shells.

brackish Water of intermediate salinity.

bradychardia The slowing down of heart rate during diving in marine mammals.

breaching Leaping into the air by whales.

brittle stars Echinoderms with five flexible arms that radiate from a conspicuous central disk and tube feet that are used in feeding.

broadcast spawning Spawning by releasing gametes into the air or water.

brown algae Seaweeds with a predominance of yellow and brown pigments.

bryozoans Small, colonial, encrusting invertebrates with delicate skeletons.

budding A type of asexual reproduction by which a separate, new individual is produced from a small outgrowth, or bud.

buffering The shift in the proportion of carbon dioxide and carbonate species which keeps the ocean pH from changing too drastically.

by-catch Non-target catch that is taken while fishing for other species.

byssal threads Strong fibers secreted by mussels for attachment.

C

calcareous Made of calcium carbonate.

calcareous algae Algae with a hard exterior.

calcareous ooze A type of biogenous sediment that is made of the calcium carbonate shells and skeletons of marine organisms.

calcium carbonate ($CaCO_3$) A mineral that is the major component of the shell, skeleton, and other parts of many organisms.

calorie The amount of energy required to raise the temperature of one gram of water one degree.

calcified Structures containing calcium, usually calcium carbonate.

canopy The thick mat of kelp fronds which float at the surface of the water.

capillaries Thin blood vessels that allow for gas exchange.

capillary waves Waves that form at the surface of a body of water, caused by friction between wind and water.

capsid The outer protein coat of viruses.

carapace (1) The shield-like structure that covers the anterior portion of some crustaceans. (2) The shell of sea turtles.

carbohydrate An organic compound that consists of chains or rings of carbon with hydrogen and oxygen attached to them.

carbon (C) An element that is an essential constituent of all organic compounds.

carbon capture The storage of carbon dioxide (CO_2) away from the atmosphere.

carbon cycle The cyclic conversion of carbon between carbon dioxide and various organic compounds.

carbon dating A procedure used to determine the age of recent fossils.

carbon dioxide (CO_2) A colorless gas that is required in the process of photosynthesis.

carbon fixation The conversion of inorganic carbon (CO_2) into energy-rich organic carbon, usually by photosynthesis.

carnivore An animal that eats other animals.

carotenoid One of a group of yellow, orange, and red plant pigments.

carpospores Non-motile, diploid spores produced by some red algae.

carrageenan A phycocolloid extracted from red algae that is widely used in food processing.

carrying capacity The maximum population size that can be sustained by the available resources in a given environment.

cartilage A rigid material that is still lighter and more flexible than bone that makes up the skeletons of sharks, rays, and skates.

cartilaginous fishes (*Chondrichthyes*) Fishes with a skeleton made of cartilage.

catadromous Freshwater fishes that migrate to sea to breed.

catch-effort curve A plot of sustainable yield versus fishing effort.

caudal fin The posterior, or tail, fin of fishes.

cell membrane The layer that separates the inside of a cell from the outside environment.

cellulose A complex carbohydrate that is the main component of fibers and other support structures in plants.

cell wall A rigid barrier found outside the cell membrane of prokaryote cells.

central nervous system The brain and one or more nerve cords.

central rift valley A depression in the mid-ocean ridge.

cephalopods Molluscs that possess a foot modified into arms that surround the head.

cephalothorax The anterior portion of the body of many arthropods, which consists of the head fused with other body segments.

cetaceans Marine mammals with anterior flippers, no posterior limbs, and a dorsal fin.

chemical equilibrium The point in a chemical reaction where amounts of reactants and products is relatively stable and unlikely to change.

chemical precipitation Chemical reactions which convert phosphate into insoluable phosphorus which settle to the sea floor.

chemosynthetic prokaryotes Autotrophic bacteria and archaea that use energy by releasing it from particular chemical compounds.

chitin The main component of the exoskeleton of many animals.

chitons Molluscs that have a shell divided into eight overlapping plates.

chloride cells Cells in the gills of fishes that are involved in the excretion of excess salts.

chlorinated hydrocarbons A group of non-biodegradable, synthetic chemicals. Some are toxic and become pollutants.

chlorofluorocarbons (CFCs) Chemicals, used in sprays, air conditioners, and other products, that affect the ozone layer.

chlorophyll A green photosynthetic pigment.

chloroplast The organelle where photosynthesis takes place.

chordates Animals that display a hollow dorsal nerve cord, gill slits, and a notochord. Includes protochordates and vertebrates.

chromatophore specialized cell in the skin of bony fishes which contain colored pigments.

chromosome The cell structure where DNA is located.

ciguatera A type of food poisoning contracted from eating certain tropical fishes.

ciliary comb One of eight bands of cilia fused at the base that is found in comb jellies.

ciliates A group of protozoans with cilia.

cilium (pl. cilia) A short, hair-like flagellum that is found in large numbers and used for movement and feeding.

clasper A copulatory organ along the inner edge of each pelvic fin in male sharks and other cartilaginous fishes.

cleaning associations A symbiotic association in which a smaller partner regularly removes parasites from fishes.

climax community The final stage in an ecological succession.

cloaca The common opening for the intestine and the excretory and reproductive systems of cartilaginous fishes and other animals.

clone A series of genetically identical cells or individuals that have developed from a single cell or individual.

closed circulatory system A circulatory system where the blood or circulatory fluid is contained in distinct vessels.

clupeoid Small, plankton-feeding fishes which account for the largest catches in the world.

cnidarians Invertebrates with nematocysts and radial symmetry.

cnidocyte A cell that contains a nematocyst.

coastal management The use of coastal resources with the intention of preserving them.

coccolithophorids Unicellular, eukaryotic members of the phytoplankton that have calcareous, button-like structures, or coccoliths.

coelacanths A group of lobed-fin fishes. *Latimeria* was first discovered alive in 1952.

coelom The body cavity found in structurally complex animals.

coelomic fluid The fluid that fills the coelom of echinoderms and other invertebrates.

coevolution The process in which one species evolves in response to another.

cohesion The sticking together of particles of the same substance.

cold seeps Areas where hydrogen sulfide and methane produced by decomposition seep out from the sea floor.

collar cell A flagellated, food-trapping cell of sponges.

comb jellies The invertebrates with a gelatinous body, radial symmetry, and eight rows of ciliary combs.

commensalism The type of symbiosis in which one species obtains shelter, food, or other benefits without affecting the other, or host.

community The populations that live and interact in an area.

competition The interaction that results when a resource is in short supply and one organism uses the resource at the expense of another.

competitive exclusion The elimination of one species by another as a result of competition.

compound eye An eye that consists of numerous light-sensitive units.

conditioning A form of learning in which a behavior is associated with a reward.

constant proportions, rule of A principle that states that the relative amounts of ions in seawater are always the same.

consumer A heterotroph

continental drift The movement of continental masses on the surface of Earth.

continental margin The edge of a continent; the zone between a continent and the deep-sea floor. Also see active and passive continental margins.

continental rise The gently sloping area at the base of the continental slope.

continental shelf The shallow, gently sloping section of the continental margin that extends from the shore to the point where the slope gets steeper.

continental slope The steeper, seaward section of the continental margin.

controlled Variables prevented from affecting an experiment.

convection The movement that results when heat is transferred in a fluid, like the heat-driven motion observed in the Earth's mantle.

convergent evolution Evolution in which two different species evolve similar structures because of similar lifestyles.

copepods Small, mostly planktonic crustaceans.

copulation The sexual union undertaken to transfer gametes.

coral knoll A column of coral within the lagoon of an atoll.

coral reef A massive deposit of calcium carbonate by colonial stony corals and other organisms.

coralline algae Red algae that deposit calcium carbonate in their thallus.

cordgrasses Salt-tolerant grasses that inhabit salt marshes.

core The innermost layer of Earth.

Coriolis effect The tendency of objects moving large distances on Earth's surface to bend to the right in the northern hemisphere and to the left in the southern.

counterillumination The emission of light by midwater animals to match the background light.

countercurrent system of flow An adaptation of bony fishes where the blood in the gills flows in the opposite direction of the water passing over them, increasing the efficiency of gas exchange.

countershading A color pattern that results in a dark back and a light belly; most common in epipelagic fishes.

courtship behavior Behavior related to attracting the opposite sex and mating.

crinoids Echinoderms with a small, cup-shaped body and feathery arms.

crown-of-thorns sea star A predator of reef corals.

crust The outermost layer of Earth.

crustaceans Arthropods that have two pairs of antennae and an exoskeleton hardened by calcium carbonate.

cryptic coloration A color pattern that allows an organism to blend with the surroundings.

cryptophytes Unicellular, eukaryotic members of the phytoplankton that have two flagella and no skeleton.

crystal A solid that consists of a regular pattern of molecules.

crystalline style An enzyme-releasing rod in the stomach of bivalves.

ctenoid Overlapping spined scales of bony fishes.

cubomedusae Cnidarians having a cube-shaped medusa with only four tentacles.

current A horizontal movement of water.

cyanobacteria A group of generally photosynthetic bacteria. Formerly known as blue-green algae.

cycloid Overlapping smooth scales of bony fishes.

cyclonic storms Rotating storms that are generated over warm waters in the tropics.

cyst An inert or resting stage of a dinoflagellate.

cytoplasm The gelatinous interior of a cell.

cytoskeleton A complex framework inside cells made of protein fibers.

D

daughter cell Genetically identical cells that are the product of cell division.

decapods Crustaceans with five pairs of walking legs and a well-developed carapace.

decay bacteria Bacteria that carry out decomposition.

decomposers Organisms that break down dead organic matter into smaller molecules.

deduction Reasoning from general principles to specific conclusions.

deep (bottom) layer The deepest and coldest of the three layers of the ocean.

deep scattering layer (DSL) A sound-reflecting layer made up of many types of organisms that migrate daily from the mesopelagic zone to the epipelagic zone.

deep sea The dark waters below the mesopelagic zone.

deep-sea gigantism The pattern of deep-sea animals to be significantly larger than their shallow water relatives.

deep sound channel The depth at which sound waves travel slowest, allowing low-frequency waves to travel vast distances.

deep-water wave A wave moving through water greater than ½ of its wavelength.

delayed implantation The delay in the attachment of the early embryo to the womb in pinnipeds and some other mammals in order to time birth with favorable conditions.

demersal fish A bottom-dwelling fish.

denitrification The conversion of fixed nitrogen back to nitrogen gas.

density The mass of a given volume of a substance.

deposit feeder An animal that feeds on organic matter that settles on the bottom.

desalination The conversion of seawater into fresh water.

desiccate To dry out.

detritus Particles of dead organic matter.

diatomaceous earth Siliceous material derived from the skeletons of diatoms.

diatomaceous ooze A biogenous sediment that consists mostly of the siliceous frustules of diatoms.

diatoms Unicellular and eukaryotic autotrophs with a siliceous frustule.

diffusion The movement of molecules from an area of high concentration to an area of low concentration.

digestive gland An enzyme-producing gland in several groups of invertebrates where digestion and absorption take place.

dinoflagellates Unicellular, eukaryotic, mostly autotrophic protists with two unequal flagella.

dioxins A group of toxic chlorinated hydrocarbon pollutants.

diploid (2n) cell A cell that contains two similar sets of chromosomes, one from each parent.

directional selection Natural selection that favors extreme expression of a trait.

dispersal The manner in which organisms get from place to place.

dispersants Chemicals added to an oil spill to break the surface oil into small droplets that can then disperse in the water.

disruptive coloration A color pattern that helps break the outline of an organism.

dissociation The breaking up of a salt molecule into ions when placed in water or other solvents.

dissolved organic matter (DOM) Organic matter that is dissolved in water rather than being in particles.

diurnal tide A tidal pattern with a high and a low tide each day.

diversity The total number of species inhabiting a particular environment.

DNA (deoxyribonucleic acid) Nucleic acids that contain the inherited genetic code that specifies the structure and function of an organism.

domestic sewage Wastewater from homes and non-industrial buildings.

domoic acid Toxin produced by some planktonic diatoms.

dorsal The upper or back surface of an animal with bilateral symmetry.

downwelling The sinking of surface water that is denser than underlying water layers.

drag Resistance to movement through water or any other medium.

drift net A fishing net that is allowed to drift for a long time before it is pulled on board.

drowned river valley estuary An estuary that is formed when sea level rose at the end of the last glacial age.

E

echinoderms Invertebrates with five-way radial symmetry and a water vascular system.

echiurans Burrowing invertebrates that have an unsegmented body and a non-retractable proboscis.

echolocation The ability of some animals to sense their surroundings by analyzing the reflection of sound waves, or clicks, they emit.

ecological niche The full range of ecological characteristics of a species, like its feeding habits, specific habitat, and reproductive strategy.

ecological succession The regular replacement of species by others in a given area.

ecology The study of the interactions between organisms and their environment.

ecosystem A community or communities plus the physical environment, interacting in a large, more or less self-contained area.

ectotherm An organism that loses metabolic heat to the environment without it affecting its body temperature.

eggs Female gametes or reproductive cells.

Ekman spiral The spiral change in the movement of water in the water column when the water is pushed by wind. The Ekman layer is that part of the water column affected by wind; Ekman transport is the net water movement 90° from the wind direction.

element A substance that consists of atoms of the same kind and that cannot be decomposed by ordinary chemical means.

El Niño–Southern Oscillation (ENSO) Large-scale changes in atmospheric and ocean current patterns in which, among other things, warm surface water in the Pacific Ocean moves further to the east than normal. El Niño refers specifically to the warming of the surface water in the Eastern Pacific.

embryo The early developmental stage that ultimately becomes an adult individual.

embryo sac The female gametophyte in flowering plants.

emersion Being out of water and exposed to air.

encrusting Describing an organism that grows as a crust over rocks and other hard surfaces.

endangered species A species that is in immediate danger of extermination.

endolithic Cyanobacteria that burrow into rocks or corals.

endophyte A photosynthetic organism that lives within the cells or tissues of algae or plants.

endoplasmic reticulum An extensive system of folded membranes present in most eukaryotic cells.

endoskeleton A skeleton under the external surface of an animal.

endotherm An organism that retains some metabolic heat, which raises its body temperature.

energy The ability to do work.

enzyme A protein that speeds up a specific chemical reaction.

epifauna Animals that live on the surface of the substrate.

epipelagic zone The pelagic environment from the surface to a depth of 100 to 200 m.

epiphyte A photosynthetic organism that lives on algae or plants.

equatorial currents Major ocean currents that move parallel to the Equator.

erythrocyte A specialized type of blood cell that carries hemoglobin in vertebrates.

estuary A semi-enclosed area where fresh water and seawater meet and mix.

eukaryote An organism that consists of one or more eukaryotic cells.

eukaryotic cell A cell that contains a nucleus and other organelles.

euryhaline An organism that can tolerate a wide range of salinities.

eutrophication The process by which increased inputs of nutrients lead to an undesirable increase in algal growth.

evaporation The escape of molecules from the liquid phase into the gaseous phase.

evaporative cooling The lower speed and hence lower temperature of molecules remaining in the liquid phase after evaporation of the fastest molecules.

evisceration The expulsion of internal organs in sea cucumbers.

evolution A change in the genetic makeup of a species, usually as a result of natural selection favoring some individual characteristics over others.

evolutionary adaptation The genetic adaptation of a population to its environment through evolution.

exclusive economic zone (EEZ) A zone 370 km wide along the coast where nations have exclusive rights to any resource.

excurrent siphon The secondary opening on a tunicate which expels water from the gut.

exoskeleton A skeleton that forms the external surface of an animal, as in arthropods.

experiment An artificially created situation that is used to test a hypothesis.

exponential growth Unrestricted population growth where the population grows at a faster rate with each successive generation.

extermination The extinction of a species that is caused by humans.

external fertilization The release of gametes into the water.

extinction The disappearance of a species.

extracellular Occurring outside of a cell.

F

facultative symbionts Organisms that can, but do not need, to live in association with a symbiont.

fall bloom A burst of production caused by the high nutrient levels and light in autumn.

fault A crack in Earth's crust usually formed when pieces of crust move past each other.

fecal pellets The solid waste excreted by zooplankton, which serves as food for detritivores.

femtoplankton The smallest size category of plankton, containing viruses and microbes in the 0.02–0.2 μm size range.

fertilization The union of gametes.

fetch The span of the sea surface over which the wind blows to form wind-driven waves.

filter feeder A suspension feeder that actively filters food particles.

finfish The fisheries term for commercially harvested fishes.

fin ray Each of the bony spines in the fins of bony fishes.

fishing effort A measure of how much effort is put into a catch from a fishery. This includes factors such as how many boats and fishers there are and how long they stay at sea.

fish meal A fish-protein supplement used in animal feeds.

fish protein concentrate A fish protein supplement for human consumption.

fission A type of asexual reproduction that results when one individual splits into two individuals.

fjord An estuary that is formed in a deep valley created by a retreating glacier.

flagellum A long, whip-like organelle that is usually involved in locomotion.

flatworms Invertebrates that are dorsoventrally flattened and have an incomplete digestive tract, true organs, and organ systems.

flood tide Incoming, or rising, tide.

flowering plants Plants that have flowers, seeds, and true leaves, stems, and roots.

fluke The fin-like tail of cetaceans.

flukes A group of parasitic flatworms.

food chain The steps of transfer of energy from primary producers through consumers.

food web All of the interconnecting feeding relationships in a community.

foot The muscular locomotory structure of molluscs.

foraminiferans (forams) Protozoans with a calcareous *test,* and pseudopodia.

fore reef The outer part of a barrier reef or an atoll.

fringing reef A coral reef that develops as a narrow band close to a shore.

fronds The leaf-like blades of kelp.

frustule The siliceous, box-like cell wall of diatoms.

fucoxanthin The yellow to golden brown photosynthetic pigment of brown algae.

fungi Non-photosynthetic, mostly multicellular organisms, many of which are decomposers; members of the kingdom *Fungi.*

G

gamete A haploid reproductive cell that develops into a new individual after its union with another gamete.

gametophyte The haploid, gamete-producing generation in many seaweeds.

ganglia Clusters of nerve cells that serve as "local brains" in gastropods and bivalves.

gas exchange The movement of oxygen and other gases between the atmosphere and the ocean, or between the water or atmosphere and living organisms, in which case it is often called respiratory exchange.

gastropods Molluscs that typically possess a coiled dorsal shell and a ventral creeping foot.

generating forces A disturbance that starts a wave.

genetic engineering The artificial alteration of the genetic information of an individual.

genital slit The genital opening of cetaceans.

genome The complete genetic information of an organism.

genus A group of similar species.

germ tissue Tissue found in sexually reproducing organisms that divides by meiosis.

gestation The length of time between fertilization and birth in mammals.

ghost fishing Trapping of animals by abandoned fishing gear.

gill Thin-walled extensions of the body that are used in gas exchange.

gill arch A supporting structure of fish gills.

gill filament The thin projection of a fish gill where gas exchange takes place.

gill nets A fishing method that uses a drifting net. Fishes swim into the net which catches on their gills, trapping them.

gill raker Each of the projections along the inner surface of fish gills.

glass sponges Deep-water sponges with a skeleton of fused silica spicules.

global distillation Evaporation and condensation of substances on a global scale, leading to the net transport of certain pollutants to the poles.

global warming A long-term increase in the global average temperature. Anthropogenic (human-caused) global warming results from an intensified greenhouse effect due to human emissions of greenhouse gases.

glucose A simple sugar that plays an important role in the metabolism of most organisms.

Golgi apparatus An arrangement of sacs and membranes in many eukaryotic cells that is involved in collecting and transporting molecules.

gonad An organ that contains the germ tissue that produces the gametes.

Gondwana One of two large continents, the southern one, that formed when the supercontinent Pangaea broke up about 180 million years ago.

gorgonians Colonial anthozoans that secrete a skeleton made of protein.

granite The light-colored rock that forms most of the continental crust.

grazer An organism that feeds primarily on plants.

great ocean conveyor A global circulation pattern in which water cycles throughout the ocean basins.

great whales Large whales: the sperm and the baleen whales.

green algae Seaweeds in which chlorophyll is not masked by other pigments.

greenhouse effect The retention of heat in the lower atmosphere due to the presence of carbon dioxide, water vapor, and other greenhouse gases.

greenhouse gases Gases that block transmission of heat back to space and insulate Earth.

gross primary production The total amount of organic carbon manufactured by primary producers.

guano The accumulation of the excrement of seabirds.

Gulf Stream The current that carries water from the tropics up the east coast of the United States before turning east and bringing warm water to Europe.

guyot A flat-topped seamount.

gyre A large, nearly circular system of wind-driven surface currents that center around latitude 30° in both hemispheres.

H

habitat The natural environment where an organism lives.

habitat restoration The recovery of stressed or destroyed habitats.

hadal zone The bottom below 6,000 m.

hadopelagic zone The pelagic environment below 6,000 m.

halophyte A salt-tolerant terrestrial plant.

haploid cell A cell that contains only half the chromosomes found in diploid cells.

harem A large group of females herded by a large male pinniped for the purpose of mating; harems also occur in some fishes.

harmful algal blooms Unusual population explosions, or blooms, of phytoplankton that are detrimental to humans.

heat The total kinetic energy, in calories, of all the particles in a substance or object.

heat capacity The amount of heat that must be added to a substance to raise its temperature by a given amount, which reflects the substance's ability to store heat.

heavy metals A group of toxic metals: mercury, lead, and others.

hemichordates Invertebrates with a dorsal nerve cord and gill slits.

hemoglobin A blood protein that transports oxygen in many animals.

herbivore An animal that eats plants.

heredity The transmission of genetic characteristics from one generation to the next.

hermaphrodite An organism that has both male and female gonads.

hermatypic coral Corals that build coral reefs.

heterotroph An organism that obtains energy from organic matter.

holdfast The root-like portion of the thallus of a seaweed.

holoplankton Organisms that spend their entire lives in the plankton. Compare meroplankton.

homeostasis Regulation of an organism's internal environment to maintain conditions needed for life.

homeotherm An organism that is able to keep its body temperature more or less constant regardless of the temperature of the environment.

homing The ability of an animal to find its home area.

hormone A substance that acts as a chemical messenger in an organism.

horseshoe crabs Arthropods with a large, horseshoe-shaped carapace.

hydrogen An element that is one of the constituents of water, organic matter, and many other chemicals.

hydrogen bond Weak bonds between water molecules that result from an attraction between the **hydrogen** atoms of one molecule and the **oxygen** atoms of adjacent molecules.

hydrogen sulfide (H_2S) The gas that is produced in anoxic sediments.

hydrostatic skeleton A system that uses water pressure against the body wall to maintain body shape and aid in locomotion.

hydrothermal vent A deep-sea hot spring where heated seawater forces its way up through the crust.

hydrozoans Cnidarians whose life cycle typically includes a polyp, which is often colonial, and a medusa.

hyphae Filaments of fungi.

hypothesis A statement that might be true.

hypoxic Sharp reduction of water oxygen.

I

ice age A period of time when significant amounts of ice form on the continents, inducing a fall in sea level.

ichthyology The scientific study of fishes.

immersion Being out of air and submerged.

incurrent siphon The "mouth" of tunicates.

Indo-West Pacific region The tropical Indian and western and central Pacific oceans.

inducible defenses Defense mechanisms that an organism uses only in response to predators.

induction Reasoning from specific observations to a general conclusion.

industrial sewage Wastewater from industries.

infauna Animals that burrow in the substrate.

ink sac A gland found in some cephalopods that secretes a dark fluid used to discourage predators.

inner ear A paired, sound-sensitive organ in vertebrates.

insects Arthropods with three pairs of legs and one pair of antennae.

interglacial period A geological period between ice ages when, as at present, the Earth's climate is relatively warm.

intermediate layer Of the three main ocean layers, the one below the surface, or mixed, layer. It includes the main thermocline.

intermediate waves Waves traveling in water between ½ and 1/20th of its wavelength.

internal fertilization Sperm is directly transferred from males to females through copulation.

internal wave A wave that forms between waters of two different densities.

International Whaling Commission (IWC) An agency that regulates whaling.

interspecific competition Competition among members of different species.

interstitial fauna Animals living between sediment particles.

interstitial water The water contained between sediment particles.

intertidal (littoral) zone The area between the highest and lowest tide.

intracellular digestion Digestion that takes place within cells, usually those lining the gut or digestive tract.

intraspecific competition Competition among members of the same species.

invasive species A species introduced into a new environment by humans.

invertebrates Animals that lack a backbone.

ion An atom or a group of atoms that is electrically charged.

iridophore A chromatophore with light-reflecting crystals which gives fishes their structural colors.

island arc A curved chain of volcanic islands that form along a trench.

isopods Small, dorsoventrally flattened crustaceans.

isotopes Different atomic forms of an element.

J

jawless fishes (*Agnatha*) Fishes that lack jaws and paired fins.

K

kelp Brown algae characterized by their large size and complexity. Some form dense kelp beds or kelp forests.

keystone predator A predatory species whose effects on its community are proportionately much greater than its abundance.

krill Planktonic crustaceans that are an important food of whales and other animals.

L

lactic acid Harmful product of anaerobic respiration in muscles.

lagoon A shallow, sheltered body of water separated from the open sea by coral reefs, sand bars, and/or barrier islands.

lamella Each of the many thin plates that make up the gill filaments of fish gills.

lamp shells Invertebrates that have a lophophore and a shell that consists of two valves.

lancelets Chordates with three basic chordate characteristics but lacking a backbone.

La Niña Large-scale changes in atmospheric and ocean current patterns which include warm surface water in the Pacific Ocean moveing further to the west than normal.

larva Immature stage of an animal that looks different from the adult.

larvaceans Tunicates that retain the body of a tadpole larva throughout life.

latent heat of evaporation The amount of heat energy that is needed to evaporate a substance, that is, to change it from a liquid to a gas.

latent heat of melting The amount of heat energy needed to melt a substance, that is, to change it from a solid to a liquid.

lateral line A system of canals and sensory cells on the sides of fishes that helps them detect vibrations in the water.

Laurasia One of the two large continents, the northern one, that formed when the supercontinent Pangaea broke up about 180 million years ago.

leeches Segmented worms that are specialized predators and parasites.

leeward The sheltered side of a reef or island.

leptocephalous larva The leaf-shaped larva of freshwater eels and other fishes.

lichen The organism that results from the symbiosis of a fungus and an autotroph like a green alga.

light-limited When primary production is reduced because there is not sufficient light for photosynthesis.

limiting resource An essential factor whose short supply limits the growth of a population.

lipid A group of organic molecules that are often used by organisms in the long-term storage of energy, waterproofing, buoyancy, and insulation.

lithogenous sediment Sediment that is derived from the weathering of rocks.

lithosphere The crust and the top part of the mantle that covers Earth's surface.

logistic growth Population growth that is constrained by biotic or abiotic factors such as competition or nutrients.

longlines A fishing method that uses a long line with several hooks that can be towed across the surface or the bottom.

longshore transport The movement of water or sediment down a shore in the direction of wave action.

lophophore A feeding structure that consists of ciliated tentacles.

lorica The test of a ciliate.

loriciferans Tiny invertebrates that live among sediment particles and have a body enclosed by six plates.

lysogenic viruses Viruses that reproduce when their nucleic acid becomes part of the genome of the host cells.

M

macoalgae The formal name for seaweed; multicellular algae.

macrofauna Larger animals of the deep sea benthos.

macroplankton The component of the plankton that consists of large organisms 2 to 20 cm in size.

madreporite A porous plate that connects the water vascular system of echinoderms to the exterior.

magnetic anomalies Magnetic bands in the sea floor that run parallel to the mid-ocean ridge.

male parasitism The permanent attachment of a male to a female in some deep-sea fishes.

mammals Vertebrates that have hair and mammary glands.

mammary glands The milk-secreting glands of mammals.

mangals A forest of mangrove trees.

mangroves Shrubs and trees that live along the seashore in tropical and subtropical regions and tolerate inundation by seawater.

mantle (1) The semiliquid region between the crust and the core of Earth. (2) The outer layer of tissue that secretes the shell of molluscs.

mantle cavity The space lined by the mantle of molluscs.

mariculture The culture of marine and brackish-water organisms.

marine natural products Chemical compounds obtained from marine organisms.

marine snow Large, fluffy detritus particles that are found in the water column.

massive sulfide deposits Mineral deposits on the sea floor that are created when hot water emerges from the crust and hits the cold ocean water.

maxillipeds The food-sorting appendages of some crustaceans.

maximum sustainable yield The maximum catch of a stock that can be harvested year after year without diminishing the stock.

medusa The bell-shaped, free-swimming stage of cnidarians.

megaplankton The component of the plankton that consists of very large organisms over 20 cm in size.

meiofauna Microscopic animals that live on the bottom; often used as a synonym of interstitial fauna.

meiosis Cell division that results in the formation of gametes.

melon A fatty structure on the forehead of some cetaceans that is used to direct sound waves emitted during echolocation.

meroplankton Planktonic organisms that spend only part of their life in the plankton. Compare holoplankton.

mesenterial filament Any of the long, thin tubes attached to the gut of corals and other cnidarians that are involved in digestion and absorption.

mesoderm The middle layer of embryonic tissue that gives rise to muscles, the reproductive system, and other organs..

mesopelagic zone The pelagic environment from a depth of approximately 200 m to 1,000 m.

mesoplankton The component of the plankton that consists of organisms 0.2 to 2 mm in size.

mesosphere The layer of the atmosphere between 55-90 km above Earth's surface.

metabolism All the chemical reactions that take place in an organism.

metamorphosis A marked change in form during embryological development.

methane hydrate A potential source of energy in the form of methane trapped in frozen water.

microbial loop A component of pelagic food webs in which dissolved organic matter is cycled through the picoplankton and nanoplankton back into the main food web.

microfossils The microscopic shells and other remains of marine organisms that make up biogenous sediments.

microherbivores Small invertebrates such as snails, chitons, crustaceans and polychaete worms.

microorganisms Organisms from all three domains that are microscopic in size.

microplankton The component of the plankton that consists of organisms 20 to 200 µm in size.

microplastics Small plastic particles which are commonly found in personal care products and which end up as pollutants in marine environments.

mid-ocean ridge The continuous chain of volcanic submarine mountains that extends around Earth.

midwater Pertaining to the mesopelagic zone.

migration The regular movement of one organism from one place to another.

mitochondria The organelle in which respiration takes place in eukaryotes.

mitosis The cell division in which a cell divides into two daughter cells that are identical to the original cell.

mixed semidiurnal tide A tidal pattern with two successive high tides of different heights each day.

molecule A combination of two or more atoms.

molluscs Invertebrates with a soft, unsegmented body, a muscular foot, and, with some exceptions, a calcareous shell.

molt The exoskeleton that is shed during the molting process.

monsoon Winds in the northern Indian Ocean that blow from the southwest in summer and from the northeast in winter.

mudflat A muddy bottom that is exposed at low tide.

mutualism The type of symbiosis in which both partners benefit from the relationship.

myoglobin A type of muscle protein that stores oxygen in vertebrates.

myomere Each of the bands of muscles along the sides of fishes.

N

nanoplankton The component of the plankton that consists of very small organisms 2 to 20 µm in size.

native species A species that naturally occurs in a given area.

natural selection A mechanism of evolutionary change that results when individuals that are better adapted than others in meeting the challenges of the environment produce more offspring.

nauplius A planktonic larva of many crustaceans.

neap tides The tides with a small tidal range. They occur around the times when the moon is in quarter.

negative estuaries Estuaries where loss of fresh water through evaporation is higher than fresh water input from rivers.

nekton Organisms that swim strongly enough to move against the current.

nematocyst The stinging structure of cnidarians.

nematodes Invertebrates with a cylindrical body, a conspicuous body cavity, and a complete digestive tract.

neritic zone The pelagic environment above the continental shelf.

nerve cord A long, compact bundle of neurons that is part of the central nervous system.

nerve net The network of interconnecting neurons in cnidarians and other invertebrates.

net plankton Plankton that can be caught in a standard plankton net.

net primary production The amount of organic carbon left over to fuel the food web after accounting for producer's energetic needs.

neuromasts Sensory cells in bony fishes that are sensitive to vibrations.

neuron A cell specialized to originate or transmit nerve impulses.

neuston Organisms that live right at the sea surface but remain underwater.

nictitating membrane A thin layer of tissue that can be drawn across the eye in sharks and some other vertebrates.

nitrate (NO_3^{-1}) An important nutrient in the ocean.

nitrogen An element that is one of the constituents of proteins.

nitrogen cycle The cycling of nitrogen among various nitrogen compounds.

nitrogen fixation The conversion of gaseous nitrogen (N_2) into nitrogen compounds that can be utilized by autotrophs.

node The point or points around which a standing wave oscillates.

non-biodegradable Unable to be broken down by bacteria or other organisms.

non-renewable resource A resource that is not naturally replaced.

notochord A flexible rod that lies below the nerve cord of chordates.

nucleic acid Organic molecules that store and transmit genetic information. Also see deoxyribonucleic acid, ribonucleic acid.

nucleus The organelle of eukaryotic cells that contains the chromosomes.

nudibranchs Gastropods that lack a shell and have exposed gills.

nutrient A raw material other than carbon dioxide and water that is needed by an autotroph to produce organic matter.

nutrient regeneration The release of nutrients from organic matter by decomposers.

O

obligate symbionts Organisms that need to live in association with a symbiont to survive.

ocean acidification Decrease of seawater pH as a result of increased carbon dioxide.

ocean thermal energy conversion (OTEC) A process for obtaining energy by exploiting depth differences in temperature.

oceanic zone The pelagic environment beyond the shelf break.

olfactory sacs Structures on both sides of the head of fishes that are sensitive to chemical stimuli.

open circulatory system A system where the blood or circulatory fluid is not contained in vessels and can flow into the body cavity.

operculum (1) The tough lid that closes the shell opening of many gastropods when the body is withdrawn. (2) The flap of bony plates that covers the gills of bony fishes.

organ A group of tissues specialized for one function.

organelle A membrane-bound, specialized structure within a cell.

organic compound A molecule that contains carbon, hydrogen, and usually oxygen.

organization, level of The extent of specialization and organization of the cells of an organism. Organization may be at the cellular, tissue, or organ level.

oscillations Cyclical changes in ocean circulation that affect climate.

osculum A large opening in many sponges.

osmoconformer An organism that allows its internal salt concentration to change with the salinity of the surrounding water.

osmoregulator An organism that controls its internal salt concentration.

osmosis The movement of water across a selectively permeable membrane which allows only certain molecules to pass through.

otoliths Calcareous ear stones that fishes use to detect change in body position.

outwelling The export of detritus and other organic matter from estuaries to other ecosystems.

overfishing Over-exploitation of a fishery, resulting in dwindling catches of smaller fishes, leading to unprofitability.

overturn The sinking of surface water that has become more dense than water below.

oviparous An animal that releases eggs.

ovoviviparous An animal that produces eggs that hatch inside the female immediately before birth.

ovulation The release of an egg from the ovary.

oxygen An element that is one of the constituents of water, organic matter, and many other chemicals.

oxygen minimum zone A layer of water at a depth of approximately 500 m where oxygen is depleted.

oyster reef A dense oyster bed present in some estuaries and other marine environments.

ozone layer Ozone (O_3) in the atmosphere that deflects ultraviolet radiation, which is harmful to life.

P

Pangaea The single large supercontinent that broke up to form today's continents.

Panthalassa The large ocean that surrounded the supercontinent Pangaea and that was the ancestor of the modern Pacific Ocean.

paralytic shellfish poisoning A condition caused when humans eat shellfish that have become contaminated with the toxin present in certain dinoflagellates that cause red tides.

parapodium A usually flat lateral extension present on each of the body segments of polychaetes.

parasitism The type of symbiosis in which one partner, the parasite, derives benefit from the other, the host.

parthenogenesis Development of an egg into an embryo without fertilization by a sperm.

passive continental margin A continental margin that is located at the "trailing edge" of a continent and as a result shows little geological activity.

passive suspension feeder A suspension feeder that uses mucus or cilia to move food particles to the mouth.

patchy The grouping of organisms in clumps or patches.

peanut worms Burrowing invertebrates with an unsegmented body and an anterior end that can be pulled into the body.

pectoral fin Each of the pair of fins just behind the head of fishes.

pedicellaria One of the minute, pincer-like organs of some echinoderms that help keep the body surface clean.

pelagic organisms Organisms that live in the water column, away from the bottom, including the plankton and nekton.

pelvic fin Each of the second pair of ventral fins of fishes.

pen The reduced, thin shell of squids.

pentamerous Radial symmetry based on five parts.

perennial A plant that lives more than two successive years.

pereopods Paired legs in decapods.

permafrost A layer of frozen polar soil.

persistent Pollutants that remain in an environment for many years as they are not readily broken down either by decay bacteria or by physical or chemical means.

pesticides Chemicals intended to kill insects and to control weeds.

petroleum Crude oil.

pharyngeal slits Small openings along the anterior part of the gut; one of the four key characteristics of chordates.

pharynx The anterior portion of the digestive tract of many animals.

pheromone A chemical that organisms use to communicate with other members of their species.

phoronids Tube-dwelling, unsegmented invertebrates that possess a horseshoe-shaped or circular lophophore.

phosphate (PO_4^{-3}) A source of phosphorus important to marine primary producers.

phosphate cycle The cycling of phosphorus among various phosphorus compounds.

phosphorus (P) An element that is essential to life.

photic zone The surface layer where there is enough light for photosynthesis to occur.

photophore An organ that produces bioluminescence.

photosynthesis The chemical process involved in the transformation of solar energy into glucose.

photosynthetic Autotrophs which get energy from light from the Sun.

photosynthetic pigment A molecule like chlorophyll that is responsible for capturing solar energy in photosynthesis.

phycobilins A group of photosynthetic pigments that includes phycocyanin, a bluish pigment in cyanobacteria, and phycoerythrin, a red pigment in red algae.

phycocolloid One of several starch-like chemicals found in some seaweeds.

phylogenetics The study of evolutionary relationships.

phylogeny The evolutionary history of a species or other taxon.

phylum The taxon that represents a main division of a kingdom. The equivalent term division is used in the kingdom Plantae.

physiological adaptation Non-inherited adaptation to the environment in an individual.

phytoplankton The photosynthetic component of plankton consisting primarily of single-celled algae and bacteria.

picoplankton The component of the plankton that consists of extremely small organisms, 0.2 to 2 μm in size.

pinacocyte The flat cells on the outer surface of sponges.

pinnipeds Mammals with paddle-shaped flippers.

placenta A membrane that connects the embryo with the mother's womb to provide nourishment.

placoid scales Small rough scales that make up the skin of cartilaginous fishes.

planktivore An animal that feeds on plankton.

plankton Organisms that drift in the water.

planktonic Organisms that drift in the water column (cannot swim against a current).

plants Members of the kingdom *Plantae*, which consists of photosynthetic, eukaryotic, and multicellular organisms.

planula The ciliated larva of cnidarians.

plate tectonics The process involved in the movement of large plates on Earth's crust.

Pleistocene A geological period, which began about 2.6 million years ago, characterized by a series of ice ages.

pleopods Specialized appendages beneath the body of a decapod, also "swimmerets."

pleuston Organisms that live right at the sea surface, with part of the body projecting into the air.

pneumatocyst A gas-filled bladder in seaweeds.

pneumatophore Upward extension of roots of some mangroves.

pod A school of cetaceans.

poikilotherm An organism whose body temperature varies with that of its surroundings.

polar easterlies Variable winds that blow at high latitudes.

pollen grain The male gametophyte in flowering plants.

pollutants Natural and artificial substances which cause pollution.

pollution The introduction of harmful substances, heat energy, or sound into the environment by humans.

polychaetes Segmented worms that typically have parapodia.

polychlorinated biphenyls (PCBs) A group of chlorinated hydrocarbon pollutants.

polymetallic nodules Lumps of minerals such as manganese, nickel, or copper, found on the deep ocean floor beyond the shelf.

polyp The cylindrical, typically attached stage of cnidarians.

population A group of organisms of the same species and living in the same place.

pore cell The tube-like cell of sponges that forms a pore.

post-anal tail A tail that extends beyond the anus; one of the four key characteristics of chordates.

post-settlement hypothesis A hypothesis that holds that most species have abundant larvae, and the nature of reef fish communities is determined by the competitive ability of the different species.

posterior The defined rear end of a bilaterally symmetric animal.

practical salinity units (PSU) Measurement of seawater salinity based on electrical conductivity, measured in parts per thousand.

precious corals Gorgonians that secrete a red or pink skeleton consisting of fused calcareous spicules.

precipitation Atmospheric water that falls to the ground, such as rain and snow.

predation The act of an animal, or predator, eating another organism, or prey.

pre-settlement hypothesis A hypothesis that holds that the nature of reef fish communities is determined by the availability of different species' larvae before they settle.

pressure The weight exerted over a unit area of surface.

primary producer An autotroph. An organism that carries out primary production.

primary production The conversion of the inorganic carbon, usually carbon dioxide, into organic compounds by autotrophs.

proboscis An extension near the mouth that in some invertebrates helps in the capture or collection of food.

productivity The rate of primary production.

profile A graph showing changes in temperature, salinity, or other parameter with depth.

progressive wind waves Waves that are started by wind, restored by gravity, and move in a particular direction.

prokaryote An organism, like bacteria, with a prokaryotic cell.

prokaryotic cell The simplest type of cell, one that lacks a nucleus and most other kinds of organelles.

protective coloration Coloration that benefits the individual by providing concealment from predators.

protein A large group of complex nitrogen-containing organic molecules that play many crucial roles in organisms.

protists Collection of different groups of relatively simple eukaryotes, including unicellular forms and multicellular algae.

protochordates Chordates that lack a backbone.

protozoans Animal-like protists that are mostly heterotrophic.

pseudopodium A thin or blunt extension of the cytoplasm.

psychrophilic Cold-temperature-loving.

pteropods Pelagic gastropods in which the foot is modified for swimming and the shell is reduced or absent.

purse seines Large nets that surround and trap schools of fishes.

pyloric caecum Each of the slender tubes found in the intestines of many bony fishes.

pyramid of biomass The decrease in biomass that is observed in each higher level of a food chain.

pyramid of energy Decrease in energy observed in each higher level of a food chain.

pyramid of numbers The decrease in the number of individuals that is observed in each higher level of a food chain.

R

radial symmetry The regular arrangement of similar body parts around a central axis.

radioactivity Emission of radiation by unstable atoms in the form of particles and rays.

radiolarian ooze A type of biogenous sediment that consists mostly of the silica skeletons of radiolarians.

radiolarians Protozoans with a silica skeleton and pseudopodia.

radula The ribbon-like band of teeth of molluscs.

rare species A species at risk but not in immediate danger of extinction.

recombination The formation of new genetic combinations, like the one that takes place during fertilization.

recruitment The entry of young individuals into a population or, in fisheries biology, into a fished stock.

rectal gland A gland in the cloaca of cartilaginous fishes that excretes excess salts.

red algae Seaweeds that contain characteristic red pigments.

red clay A fine sediment that is the most common type of lithogenous sediment in the open ocean floor.

red tide An extensive bloom of phytoplankton that discolors the water.

reef crest The shallow outer edge of the reef slope of a coral reef.

reef flat The wide and shallow upper surface of a coral reef.

reef slope The outer, steep margin of a coral reef. Also see fore reef.

refraction The change in the direction of a wave as it moves into shallow water.

regenerated When nutrients are released into the water column as bacteria decompose organic matter.

regeneration The ability of an organism to grow a body part that has been lost.

remote sensing Collecting data through the use of satellites.

renewable resource A resource that is naturally replaced.

replication The copying of a cell's deoxyribonucleic acid (DNA) before the cell divides.

reproductive isolation The inability of separate populations to interbreed.

reproductive strategy The reproductive patterns followed by a particular species.

reptiles Vertebrates with scales on their skin and leathery eggs that are laid on land.

resource partitioning The sharing of resources by specialization.

respiration In this text the term "respiration" by itself refers to aerobic respiration.

respiratory tree The branched extensions of the posterior end of the digestive tract of sea cucumbers that are involved in gas exchange.

restoring force A force that causes the surface of the water to return to an undisturbed state; gravity or surface tension.

rete mirabile A network of blood vessels that in some fishes functions as a heat-exchange system to help keep the internal body temperature higher than that of the water.

retina The light-sensitive part of an eye.

retroviruses Viruses that contain RNA instead of DNA.

rhizomes Underground horizontal stems found in salt marsh plants.

ribbon (nemertean) worms Invertebrates with a complete digestive tract, a true circulatory system, and a long proboscis to capture prey.

ribosomes Organelles where a cell's proteins are made.

rift A crack in Earth's crust formed as pieces of the crust separate.

rip currents An area of rapid seaward-flowing water.

RNA (ribonucleic acid) Nucleic acids that translate the genetic information contained in DNA (deoxyribonucleic acid) into proteins and that have many other functions in the cell.

rockweeds (wracks) Brown algae like *Fucus* that inhabit rocky shores in temperate areas.

rorquals Blue, fin, and the other baleen whales that display long grooves on the underside.

S

saline brines Warm and highly saline waste water from desalination plants.

salinity The total amount of salts dissolved in seawater. It is generally expressed in parts per thousand (‰).

salivary gland Any of the glands in molluscs, vertebrates, and other animals that release digestive enzymes into the mouth.

salmon farming Aquaculture where salmon are raised from young stages in offshore enclosures.

salmon ranching Practice in which cultured juvenile salmon are released into fresh water and allowed to migrate to sea so they can be harvested later when they return as adults.

salps Pelagic tunicates with a transparent, cylindrical body, sometimes forming long colonies.

salt A substance that consists of ions that have opposite electrical charges.

salt gland A gland that secretes excess salts in seabirds and sea turtles.

salt marsh A grassy area that extends along the shores of estuaries and sheltered coasts in temperate and subpolar regions.

salt wedge A layer of denser, saltier seawater that flows along the bottom in estuaries.

sand cays Small islands formed on reef flats by wind and waves.

sand dollars Sea urchins with a flat, round test and short spines that live partly buried in soft sediments.

Sargasso Sea The region of the Atlantic Ocean north of the West Indies that is characterized by floating masses of Sargasso weed, a brown alga.

scavenger An animal that feeds on dead organic matter.

school A well-defined group of fishes or cetaceans of the same species.

scientific method The set of procedures by which scientists learn about the world.

scientific theory A comprehensive explanation of a phenomenon that is backed up by considerable evidence.

sclerites Small sharp calcium carbonate needles found in some soft corals to deter predation.

sclerosponges (coralline sponges) Sponges with a massive calcareous skeleton.

scuba (*self-contained underwater breathing apparatus*) Equipment that allows breathing underwater from tanks of compressed air.

scyphozoans Cnidarians whose life cycle includes a conspicuous medusa and a much reduced or absent polyp.

sea A wave with a sharp peak and a relatively flat wave trough. Seas are found in areas where waves are generated by the wind.

sea anemones Anthozoans that consist of one large polyp.

sea cucumbers Echinoderms with a soft, elongated body that lacks spines.

sea-floor spreading The process by which new sea floor is formed as it moves away from spreading centers in mid-ocean ridges.

sea spiders Arthropods that have a reduced body and four pairs of legs.

sea squirts Tunicates with a typically sac-like, attached body as adults.

sea stars Echinoderms with five or more radiating arms and tube feet that are used in locomotion.

sea urchins Echinoderms with a round or flattened test and movable spines.

seagrasses Grass-like flowering plants like eelgrass that are adapted to live at sea.

seamount A submarine volcano in the abyssal plain.

seaweeds (macrophytes) Large, multicellular algae.

sediment Loose material like sand and mud that settles on the bottom.

sedimentation The settling of particles from the water column.

seeding In mariculture, the release of cultured juvenile individuals to enrich a natural stock.

segmentation A body consisting of similar compartments or segments.

segmented worms Invertebrates that display an elongate body with distinct segmentation and a digestive tract that lies in a coelom.

selectively permeable A membrane that allows some substances to enter and exit the cell but blocks others.

self-regulating population A population with a growth rate that is dependent on its own numbers.

self-shading Reduction in the amount of light available to phytoplankton that live below other phytoplankton.

semidiurnal tide A tidal pattern with two high and two low tides each day.

septa The thin tissue partitions in the polyp of anthozoans.

sequence (of nucleic acids) The order in which nucleotides occur on a nucleic acid molecule, which specifies the information contained in the molecule.

sequential hermaphroditism An animal that changes from male to female (or female to male) over the course of its life.

sessile An organism that lives attached to the bottom or to a surface.

seta (pl. setae) The bristles of polychaetes.

sex hormone A hormone that controls the timing of reproduction and sexual characteristics in vertebrates.

sexual reproduction Reproduction that involves the union of gametes. Compare asexual reproduction.

shallow-water wave A wave moving through water less than 1/20th of its wavelength.

shear boundary The boundary between two plates that move past each other on Earth's surface.

shelf break The section of the continental shelf where the slope abruptly becomes steeper, usually at a depth of 120 to 200 m.

shellfish The fisheries term for molluscs and crustaceans.

silica (SiO₂) A mineral similar to glass that is the major component of the cell wall, shell, or skeleton of many marine organisms.

siliceous Made of silica.

siliceous ooze A type of biogenous sediment that consists mostly of the silica shell and skeletons of marine organisms. Also see diatomaceous ooze and radiolarian ooze.

silicoflagellates Unicellular and eukaryotic members of the phytoplankton that have a star-shaped silica skeleton.

simple sugar A sugar, like glucose, that cannot be broken down into simpler sugar molecules.

simultaneous hermaphroditism Animals that can produce eggs and sperm at the same time.

siphon (funnel) The tube-like extension through which water flows in and out of the mantle cavity in bivalves and cephalopods, and in tunicates.

siphonophores Hydrozoans that exist as drifting colonies.

sirenians (sea cows) Marine mammals with anterior flippers, no rear limbs, and a paddle-shaped tail.

sludge Concentrated wastes extracted from sewage during sewage treatment.

snowblower vents Hydrothermal vents where chemosynthetic bacteria living in the rocks produce filaments of sulfur that coalesce into snowflake like particles.

soft corals Colonial anthozoans with no hard skeleton.

solute Any material dissolved in a solvent.

sonar (sound *n*avigation and *r*anging) A technique or equipment used to locate objects underwater by the detection of echoes.

spawning The release of gametes or eggs into the water.

species Individuals of the same kind that cannot breed with those of other kinds.

species epithet The species name of an organism.

specific heat The amount of energy required to raise a mass of substance one degree Celsius.

sperm Male gametes or reproductive cells.

spermaceti organ The melon of sperm whales. It contains spermaceti, an oil once widely used in candle making.

spermatophore A packet of sperm in cephalopods.

spicule Any of the small calcareous or siliceous bodies embedded among the cells of sponges or in the tissues of other invertebrates.

spiracle One of two openings behind the eyes of cartilaginous fishes.

spiral valve A spiral portion in the intestine of cartilaginous fishes.

sponges Invertebrates that consist of a complex aggregation of cells, including collar cells, and have a skeleton of fibers and/or spicules.

spongin The resistant fibers of sponges.

spore The asexual, sometimes resistant structure produced by some algae. Also see zoospore.

sporophyte The diploid, spore-producing generation in many seaweeds.

spout (blow) The water vapor and seawater that is observed when whales surface and exhale.

spring bloom A seasonal peak in phytoplankton growth brought on by longer days and plenty of nutrients.

spring tides The tides with a large tidal range; they occur around the times of full or new moon.

spur-and-grove formation A characteristic formation on barrier reefs with finger-like projections of coral interspersed with sand channels.

stabilizing selection Natural selection that acts against individuals with extreme expression of a trait.

stable A water column where warmer, less-dense surface water floats on top of denser water below.

standing stock (standing crop) The total amount, or biomass, of an organism at a given time.

standing waves Waves that do not progress in a particular direction, but rather form around a node.

starch A carbohydrate that consists of chains of simple sugars.

statocyst A fluid-filled cavity provided with sensitive hairs and a small free body that is used to orient animals with respect to gravity.

stenohaline Organism that can tolerate a narrow range of salinities.

stipe The stem-like portion of the thallus of seaweeds.

stock The size of a population.

storm surges Waters that flood coastal areas as cyclonic storms make landfall.

stormwater runoff Water that often carries harmful substances such as fertilizers, pesticides, oil, and parasites that gets carried into the oceans after it rains.

stratification The separation of the water column into layers, with the most dense water at the bottom and the least dense water at the surface.

stratopause The transition between the stratosphere and the mesosphere, 50 km above Earth's surface.

stratosphere The layer of the atmosphere between 20–50 km above Earth's surface.

stromatolites Massive calcareous skeletons formed by cyanobacteria.

structural color A color that results when light is reflected by a particular surface.

structural molecule A molecule, like cellulose, that provides support and protection.

subduction The downward movement of a plate into the mantle that occurs in trenches, which are also known as subduction zones.

submarine canyon A narrow, deep depression in the continental shelf formed by the erosion of rivers or glaciers before the shelf was submerged.

subsidence The sinking of a landmass.

subsiding Slow sinking.

substrate The type of bottom or material on or in which an organism lives.

subtidal zone The bottom above the continental shelf.

succulent A fleshy plant that accumulates water.

sulfide One of the minerals that is abundant in the hot water that seeps through hydrothermal vents.

surf A wave that becomes so high and steep as it approaches the shoreline that it breaks.

surface layer The upper layer of water that is mixed by wind, waves, and currents.

surface tension The elastic tendency of liquid.

surface-to-volume ratio (S/V ratio) The amount of surface area relative to the total volume of an organism.

suspension feeder An animal that feeds on particles suspended in the water.

sustainable Development which meets the needs of today without affecting the ability of future generations to meet their needs.

swarming The aggregation of individuals for spawning or other purposes.

sweeper tentacle A type of tentacle in corals that is used to sting neighboring colonies.

swell A wave with a flatter, rounded wave crest and trough. Swells are found away from the area where waves are generated by the wind.

swim bladder The gas-filled sac in the body cavity of bony fishes that is involved in the adjustment of buoyancy.

symbiosis A close relationship between two species, including commensalism, mutualism, and parasitism.

system (organ system) A group of organs specialized for one function.

T

tadpole larva The larva of tunicates.

tapeworms Parasitic flatworms, typically consisting of a chain of repeated units.

taste buds Structures in the mouth and other locations of fishes that are sensitive to chemical stimuli.

taxon (pl. taxa) A group of organisms that share a common ancestry.

tectonic estuary An estuary that results from the sinking of land due to movements of the crust.

tectonic plates Regions of Earth's lithosphere which move together over the asthenosphere. There are approximately 20 total plates.

temperature The average kinetic energy of the particles in a substance. Measured in degrees (C, K, F) usually with a thermometer.

temperature-dependent sex determination In reptiles, where the sex of offspring is determined by the temperature at which they were incubated.

tentacle A flexible, elongated appendage.

territory A home area that is defended by an animal.

test The calcium carbonate "shell" of marine protozoans such as foraminiferans.

Tethys Sea A shallow sea that once separated the Eurasian and African sections of the supercontinent Pangaea. It eventually gave rise to the modern Mediterranean Sea.

tetrapods Vertebrates that have two pairs of legs, a group that includes the cetaceans.

thallus The complete body of a seaweed.

theory A hypothesis that is accepted as "true" for the time being because it has passed test after test and is supported by a large body of evidence.

theory of evolution The unifying theory of biology that posits that all living organisms evolved from a common ancestor.

thermal expansion When a substance expands with increased temperature.

thermal pollution Pollution by heated water.

thermocline A zone in the water column that shows a sudden change in temperature with depth. The main thermocline is the zone where the temperature change marks the transition between the warm surface water and the cold deep water.

thermohaline circulation Ocean circulation driven by differences in water density, due to variations in water temperature and salinity, rather than by the wind or tides.

thermosphere The outer layer of the atmosphere, that extends from 90–500 km above Earth's surface.

threatened species A species that occurs in unnaturally low numbers.

tidal bore A steep wave generated as high tides move up some estuaries and rivers.

tidal current A current that is generated by tides.

tidal energy Energy that can be harnessed as a result of the movement of tides.

tidal range The difference in water level between successive high and low tides.

tide The periodic, rhythmic rise and fall of the sea surface.

tide pool A depression that holds seawater at low tide.

tide table A table that gives the predicted time and height of tides for particular points along a coast.

tintinnids Ciliates that secrete vase-like tests, or loricas.

tissue A group of cells specialized for one function.

trade winds Steady winds that blow from east to west toward the Equator, replacing the hot air that rises at the Equator.

transform fault A large horizontal displacement in the mid-ocean ridge.

transplantation The intentional introduction of a species.

trawls Nets dragged along the bottom to catch fishes such as cod, walleye, or Pollock.

trench A narrow, deep depression in the sea floor.

trochophore A planktonic larva found in polychaetes, some molluscs, and other invertebrates.

trophic cascade When changes in the abundance of an organism result in changes at other trophic levels of the food chain.

trophic level Each of the steps in a food chain.

trophic pyramid The pyramid-like relationship found in the pyramids of biomass, energy, and numbers of organisms in a food chain.

tropopause The division between the troposphere and the stratosphere 12–20 km above Earth's surface.

troposphere The layer of the atmosphere closest to the surface of Earth (up to 12 km high).

tsunami (seismic sea waves) Long, fast waves produced by earthquakes and other seismic disturbances of the sea floor.

tube foot Any of the external muscular extensions of the water vascular system of echinoderms.

tubular eyes Specialized eyes of many midwater animals that allow acute upward or downward vision.

tunic The outer covering of sea squirts.

tunicates Chordates that show the three basic chordate characteristics only in the larva.

turbellarians Mostly free-living flatworms.

turbulence Water motion.

tusk shells Molluscs that have an elongated, tapered shell that is open at both ends.

U

umbo The region around the hinge of a bivalve's shell.

understory The area of a kelp bed below the canopy.

unvegetated Lacking in seaweeds or plants.

upwelling The process by which colder water rich in nutrients rises from a lower to a higher depth. It includes coastal and equatorial upwelling.

urea A toxic waste product of some vertebrates.

urogenital opening The common opening for urine and gametes in bony fishes and other animals.

uterus (womb) The portion of female mammals' reproductive tract in which the embryo develops.

V

valve Any of the two shells of bivalves and lamp shells.

variables Factors which might affect observations.

vegetative reproduction See asexual reproduction.

veliger A planktonic larva in gastropods and bivalves.

ventral The underside or belly surface of an animal with bilateral symmetry.

vertebra (pl. vertebrae) Each of the bones that make up the backbone.

vertebrates Chordates with a backbone.

vertical migration Moving from the surface to deeper water and back on a regular cycle.

vertical zonation The pattern of banding caused by intertidal organisms growing only at a particular height within the intertidal.

vestigial Undeveloped structure that is fully developed in ancestral species.

vestimentiferans Polychaetes that are common at hydrothermal vents.

viruses Non-cellular particles containing nucleic acid but unable to reproduce unless infecting living cells.

viscosity A liquid's resistance to flow. The higher the viscosity, the higher the resistance to flow.

viviparous An animal whose eggs develop inside the female while the embryo derives nutrition from the mother directly from the womb (aplacental viviparity) or by a placenta (placental viviparity) as in most mammals.

W

warming signal Consistent variation in temperature over a long period of time.

warning coloration Coloration that allows organisms to escape from predators by advertising something harmful or distasteful.

water column The vertical column of seawater that extends from the surface to the bottom.

water mass A body of water that can be identified by its temperature and salinity.

water vascular system A network of water-filled canals in echinoderms used in locomotion and food gathering.

wave The undulation that forms as a disturbance moves along the surface of the water. Waves can be described by their *height* (the vertical distance between wave crest and trough), *amplitude* (one-half the wave height), *wavelength* (the horizontal distance between adjacent crests), *period* (the time the wave takes to move past a given point) and *frequency* (inverse of the period).

wave cancellation When the crest of one wave meets the trough of another, effectively canceling out the wave.

wave crest The highest part of a wave.

wave energy A source of energy derived from wind-generated waves and strong ocean currents.

wave reinforcement When the crests or troughs of two waves sync up, causing much larger waves.

wave shock The intensity of the impact of a wave.

wave trough The lowest part of a wave.

weathering The physical and chemical breakdown of rocks.

westerlies The winds that blow from west to east at middle latitudes.

wetlands Areas flooded by tides or with visible standing water like salt marshes, mangrove forests, and freshwater marshes.

white smokers Hydrothermal vents that occur further away from the active rift and are cooler.

wind The movement of air over Earth's surface.

windward The side of a reef or island that faces the prevailing winds.

world ocean A concept that is used to indicate that all oceans on Earth are interconnected.

Y

yolk sac A yolk-containing sac that is attached to the embryo of fishes and other vertebrates.

Z

zonation The presence of organisms within a particular range, as in the vertical zonation observed in the intertidal.

zooid Each member of a colony of bryozoans and other colonial invertebrates.

zooplankton The heterotrophic, animal and protozoan component of plankton. Compare phytoplankton.

zoospore A spore provided with one or more flagella.

zooxanthellae Dinoflagellates that live within the tissues of reef corals and other marine animals.

zygote The diploid cell that results from fertilization, that is, a fertilized egg.begin each column with a new glossary

Index

Note: Page numbers followed by *f* refer to figures and *t* refer to tables.

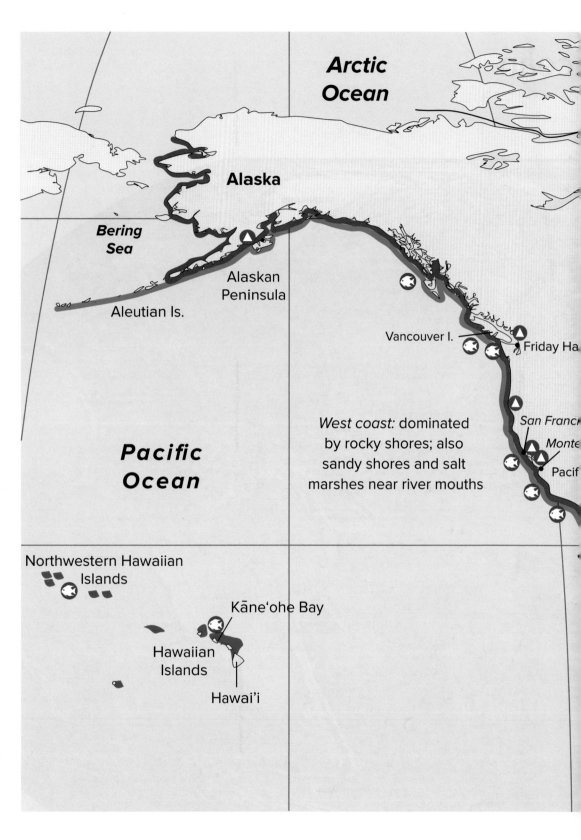

Arctic
Ocean

Alaska

Bering
Sea

Alaskan
Peninsula

Aleutian Is.

Vancouver I.

Friday Ha

West coast: dominated
by rocky shores; also
sandy shores and salt
marshes near river mouths

San Franc

Monte

Pacif

Pacific
Ocean

Northwestern Hawaiian
Islands

Kāne'ohe Bay

Hawaiian
Islands

Hawai'i

Rocky shores ▬▬▬ Kelp forests ▬▬▬ Mangrove

🌐 National Marine Sanctuaries (U.S.), Marine Protected Areas (Canada

🌐 Other protected areas

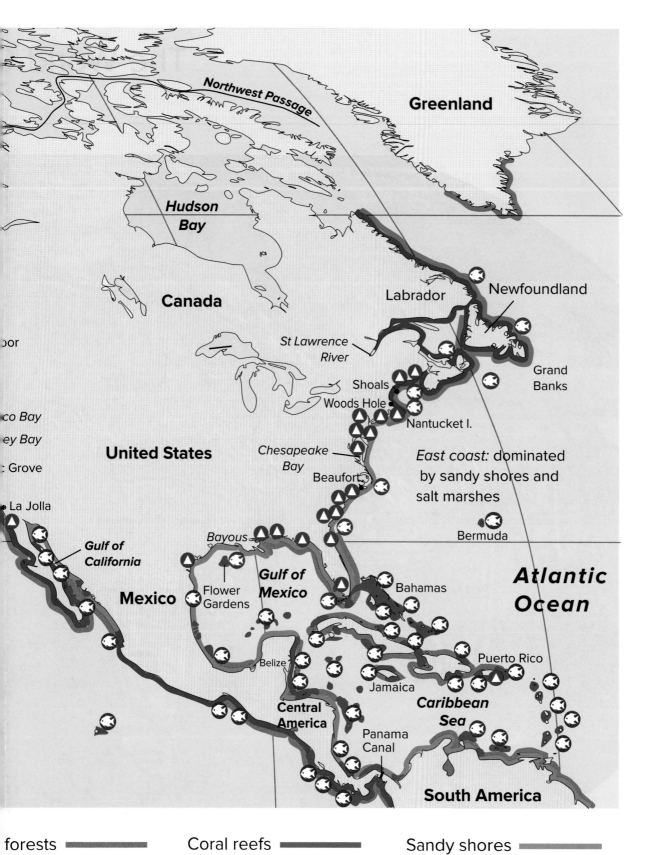

Northwest Passage

Greenland

Hudson Bay

Canada

Labrador

Newfoundland

St Lawrence River

Grand Banks

Shoals

Woods Hole

Nantucket I.

or

co Bay

ey Bay

Grove

United States

Chesapeake Bay

Beaufort

East coast: dominated by sandy shores and salt marshes

La Jolla

Gulf of California

Bermuda

Bayous

Flower Gardens

Gulf of Mexico

Mexico

Bahamas

Atlantic Ocean

Belize

Puerto Rico

Jamaica

Central America

Caribbean Sea

Panama Canal

South America

forests ▬▬▬ Coral reefs ▬▬▬ Sandy shores ▬▬▬

🔺 National Estuarine Research Reserve System (U.S.)